Yuki Nakamura • Yonghua Li-Beisson
Editors

Lipids in Plant and Algae Development

Springer

Editors
Yuki Nakamura
Institute of Plant and Microbial Biology
Academia Sinica
Taipei, Taiwan

Yonghua Li-Beisson
Institut de Biologie Environnementale
et Biotechnologie
UMR 7265 CEA - CNRS - Université Aix
Marseille, CEA Cadarache
Saint-Paul-lez-Durance, France

ISSN 0306-0225
Subcellular Biochemistry
ISBN 978-3-319-25977-2   ISBN 978-3-319-25979-6 (eBook)
DOI 10.1007/978-3-319-25979-6

Library of Congress Control Number: 2016930683

Springer Cham Heidelberg New York Dordrecht London

Printed on acid-free paper

Springer International Publishing AG Switzerland is part of Springer Science+Business Media (www.springer.com)

Preface

This book titled *Lipids in Plant and Algae Development* aims at summarizing recent advances in function of lipids in plant and algal development.

As a primary biomolecule, lipids have structural as well as diverse physiological functions such as essential constituents of biological membranes, sustainable carbon energy storage, and active signal transducer in cellular processes. In the past few decades, plant and algal lipids have gone through an established biochemistry, enzymology, and analytical chemistry, which revealed distinct functions of specific lipid classes based on their physical and biochemical properties. Upon entry into the postgenomic era, gene knockout studies on the lipid-related genes rapidly uncover functional aspects of lipids. Owing to a pile of recent publications that discuss how lipids modulate critical biological processes in plants and algae, a new axis is being developed which classifies global lipidome on a physiological basis.

Our conceptual novelty in this book is to summarize our recent understanding of lipids from the viewpoint of developmental context. Furthermore, we put together the discussion of algal and plant lipids so as to contrast the lipid function and underlying evolutionary context in photosynthetic unicellular and multicellular organisms. The first chapter gives a general overview on lipids, including their structures, metabolism, and analytical tools. The subsequent chapters can be grouped into three major parts: part I, lipids in photosynthesis (Chaps. 2, 3, 4, 5, 6, and 7); part II, lipids in development and signaling (Chaps. 8, 9, 10, 11, 12, 13, 14, 15, 16, and 17); and part III, lipids in industrial applications (Chaps. 18, 19, and 20). The subjects of each chapter cover the fast-moving topics in the field of plant and algal lipids, inviting contribution by the internationally recognized expert groups, so that the book provides refreshing viewpoint in addition to the solid discussion by established authority. Given the current fascination of plants and algae in carbon fixation and their potential as an alternative source for production of energies or novel chemical molecules, this book would also encompass what plants or algae can do in the field of industrial applications.

This book would not have been published without valuable contributions by many of our friends and colleagues. First of all, we are grateful to the authors of the individual chapters who kindly agreed to devote their time and effort to provide this book with the highest degree of expertise. Thanks are also due to the scientists in the relevant research fields who have contributed tremendous original articles and form the intellectual basis of this book. We thank Thijs van Vlijmen, Sara Germans, and other editorial staffs of Springer for their professional support to make the publication of this book possible.

Taipei, Taiwan
Saint-Paul-lez-Durance, France
22 July 2015

Yuki Nakamura
Yonghua Li-Beisson

Contents

1 **Lipids: From Chemical Structures, Biosynthesis, and Analyses to Industrial Applications** 1
Yonghua Li-Beisson, Yuki Nakamura, and John Harwood

Part I Lipids in Photosynthesis

2 **Roles of Lipids in Photosynthesis** 21
Koichi Kobayashi, Kaichiro Endo, and Hajime Wada

3 **DGDG and Glycolipids in Plants and Algae** 51
Barbara Kalisch, Peter Dörmann, and Georg Hölzl

4 **Thylakoid Development and Galactolipid Synthesis in Cyanobacteria** 85
Koichiro Awai

5 **Role of Lipids in Chloroplast Biogenesis** 103
Koichi Kobayashi and Hajime Wada

6 **Role of MGDG and Non-bilayer Lipid Phases in the Structure and Dynamics of Chloroplast Thylakoid Membranes** 127
Győző Garab, Bettina Ughy, and Reimund Goss

7 **Chemical Genetics in Dissecting Membrane Glycerolipid Functions** 159
Florian Chevalier, Laura Cuyàs Carrera, Laurent Nussaume, and Eric Maréchal

Part II Lipids in Development and Signaling

8 **Triacylglycerol Accumulation in Photosynthetic Cells in Plants and Algae** 179
Zhi-Yan Du and Christoph Benning

9 Cellular Organization of Triacylglycerol Biosynthesis in Microalgae 207
Changcheng Xu, Carl Andre, Jilian Fan, and John Shanklin

10 High-Throughput Genetics Strategies for Identifying New Components of Lipid Metabolism in the Green Alga *Chlamydomonas reinhardtii* 223
Xiaobo Li and Martin C. Jonikas

11 Plant Sphingolipid Metabolism and Function 249
Kyle D. Luttgeharm, Athen N. Kimberlin, and Edgar B. Cahoon

12 Plant Surface Lipids and Epidermis Development 287
Camille Delude, Steven Moussu, Jérôme Joubès, Gwyneth Ingram, and Frédéric Domergue

13 Role of Lipid Metabolism in Plant Pollen Exine Development 315
Dabing Zhang, Jianxin Shi, and Xijia Yang

14 Long-Distance Lipid Signaling and its Role in Plant Development and Stress Response 339
Allison M. Barbaglia and Susanne Hoffmann-Benning

15 Acyl-CoA-Binding Proteins (ACBPs) in Plant Development 363
Shiu-Cheung Lung and Mee-Len Chye

16 The Rise and Fall of Jasmonate Biological Activities 405
Thierry Heitz, Ekaterina Smirnova, Emilie Widemann, Yann Aubert, Franck Pinot, and Rozenn Ménard

17 Green Leaf Volatiles in Plant Signaling and Response 427
Kenji Matsui and Takao Koeduka

Part III Lipids in Industrial Application

18 Omics in *Chlamydomonas* for Biofuel Production 447
Hanna R. Aucoin, Joseph Gardner, and Nanette R. Boyle

19 Microalgae as a Source for VLC-PUFA Production 471
Inna Khozin-Goldberg, Stefan Leu, and Sammy Boussiba

20 Understanding Sugar Catabolism in Unicellular Cyanobacteria Toward the Application in Biofuel and Biomaterial Production 511
Takashi Osanai, Hiroko Iijima, and Masami Yokota Hirai

Index 525

Chapter 1
Lipids: From Chemical Structures, Biosynthesis, and Analyses to Industrial Applications

Yonghua Li-Beisson, Yuki Nakamura, and John Harwood

Abstract Lipids are one of the major subcellular components, and play numerous essential functions. As well as their physiological roles, oils stored in biomass are useful commodities for a variety of biotechnological applications including food, chemical feedstocks, and fuel. Due to their agronomic as well as economic and societal importance, lipids have historically been subjected to intensive studies. Major current efforts are to increase the energy density of cell biomass, and/or create designer oils suitable for specific applications. This chapter covers some basic aspects of what one needs to know about lipids: definition, structure, function, metabolism and focus is also given on the development of modern lipid analytical tools and major current engineering approaches for biotechnological applications. This introductory chapter is intended to serve as a primer for all subsequent chapters in this book outlining current development in specific areas of lipids and their metabolism.

Keywords Fatty acids • Lipid biotechnology • Lipid metabolism • Lipid analysis • Algae • Plants

---

Y. Li-Beisson (✉)
Institut de Biologie Environnementale et Biotechnologie, UMR 7265 CEA - CNRS - Université Aix Marseille, CEA Cadarache, Saint-Paul-lez-Durance 13108, France
e-mail: yonghua.li@cea.fr

Y. Nakamura
Institute of Plant and Microbial Biology, Academia Sinica, Taipei 11529, Taiwan
e-mail: nakamura@gate.sinica.edu.tw

J. Harwood
School of Biosciences, Cardiff University, Cardiff CF10 3AX, UK
e-mail: harwood@cardiff.ac.uk

Y. Nakamura, Y. Li-Beisson (eds.), *Lipids in Plant and Algae Development*, Subcellular Biochemistry 86, DOI 10.1007/978-3-319-25979-6_1

Introduction

This article is the first chapter opening our current book on 'Lipids in plant and algae development' including chapters on specific topics contributed by leading experts in their respective fields. We consider it necessary to use the first chapter to introduce what are lipids, their chemical structures and biosynthesis. We also cover topics that are less explored in the other chapters, i.e. definition, classical and new development in lipid analysis technologies, and also summarize the applications of lipids and current biotechnological goals in a variety of domains. Due to limitations of space and scope, focus is given here on knowledge gained specifically from the study of higher plants and algae.

Definition and Importance

Whereas most lipid biochemists have a working understanding of what is meant by the term lipid (derived from the Greek *lipos*- or fat) there is no generally-accepted definition. Often lipids are described as hydrophobic or amphipathic small molecules that are readily soluble in organic solvents such as chloroform, ethers or alcohols. In fact, the solubility definition can be misleading because many lipids are nearly as soluble in water as in organic solvents (Christie 2013).

Although there are a huge number of different classes of lipids, the most abundant in most organisms, including algae and higher plants, are glycerolipids (which are based on the trihydric alcohol glycerol). Other types of lipids which have often been used for the classification of algae are the pigments, principal of which are the carotenoids and the chlorophylls (or their derivatives).

Lipids play four major roles in different organisms (Gurr et al. 2002). First, they naturally form membrane structures where they represent 30–70 % of the total dry weight. Phosphoglycerides and glycosylglycerides are the main such lipids in plants and algae. Sometimes, ether lipids are important for this purpose in algae. Second, where storage lipids are present they are usually triacylglycerols (TAGs) although some algae can accumulate hydrocarbons, such as in *Botryococcus braunii* (Banerjee et al. 2002). A few plants (e.g. jojoba seeds) accumulate wax esters for storage. Third, lipids or their metabolites can act as signaling molecules. For algae this area is still in its infancy but, by comparison with higher plants, is likely to expand rapidly in the next few years. Fourth, in many organisms lipids contribute to the surface coverings. This function has not really been examined in depth for algae but is likely to be more important for macro-species. In plants, the surface coverings (including surface waxes, cutin and suberin) have been well examined (Kunst and Samuels 2009; Pollard et al. 2008; Kolattukudy 2001).

There are a number of rather specialized functions for individual classes of lipids of which photosynthesis is clearly of vital importance in algae and in many higher plant tissues (Mizusawa and Wada 2012).

Structures of Important Lipids in Algae and Higher Plants

Depending on the alga concerned, the quantitative importance of different lipids will vary. However, as mentioned above phosphoglycerides, glycosylglycerides and some ether lipids are important (Guschina and Harwood 2013). In extra-chloroplastic membranes, phosphatidylcholine (PC) and phosphatidylethanolamine (PE) (Fig. 1.1) are prominent while the chloroplast thylakoids typically contain three glycosylglycerides (monogalactosyldiacylglycerol, MGDG; digalactosyldiacylglycerol, DGDG; sulfoquinovosyldiacylglycerol, sulfolipid, SQDG) and phosphatidylglycerol (PG) (Fig. 1.1) (Harwood and Guschina 2009; Harwood and Jones 1989). For higher plants, the same general remarks apply (Table 1.1) although ether lipids are not usually significant.

In non-photosynthetic tissues such as roots the relative contribution of the glycosylglycerides is reduced because plastids are less abundant organelles compared with leaves (Gunstone et al. 2007).

The structure of some important algal ether lipids is shown in Fig. 1.2. Of these, DGTS (diacylglyceryltrimethylhomoserine ether lipid) is found in many green algae while DGTA (diacylglycerylhydroxymethyl trimethyl-β-alanine) and DGCC (diacylglycerylcarboxy(hydroxylmethyl) choline) are significant in some brown algae (Guschina and Harwood 2006).

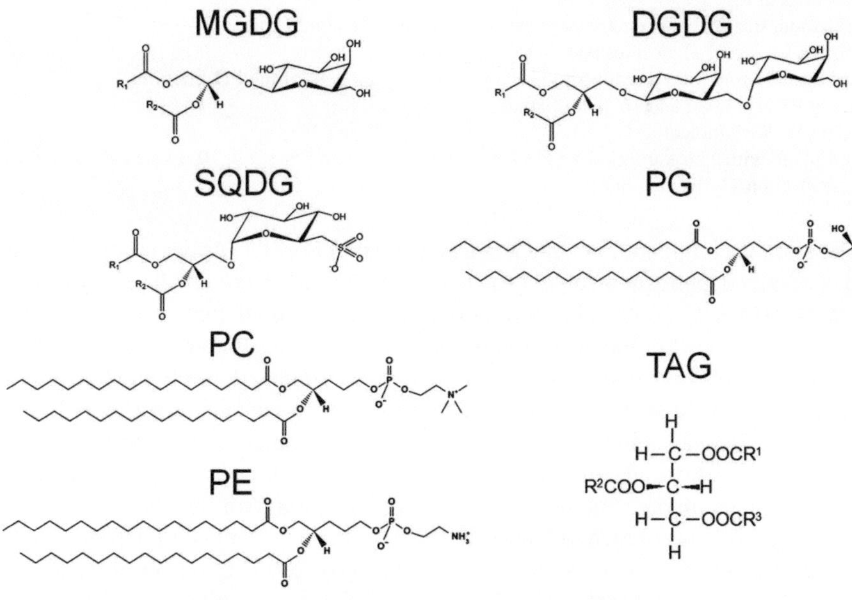

Fig. 1.1 Some examples of major membrane and storage lipid structures. *R* represents any acyl group. Abbreviations: *MGDG* monogalactosyldiacylglycerol, *DGDG* digalactosyldiacylglycerol, *PC* phosphatidylcholine, *PE* phosphatidylethanolamine, *PG* phosphatidylglycerol, *SQDG* sulfoquinovosyldiacylglycerol, *TAG* triacylglycerol

Table 1.1 Lipid composition of some plant tissues

	% total lipids									
	MGDG	DGDG	SQDG	PC	PE	PG	PI	DPG	TAG	Other
Potato tuber	6	16	1	26	13	1	6	1	15	15[a]
Apple fruit	1	5	1	23	11	1	6	1	5	48[b]
Soybean	tr.	tr.	tr.	4	2	tr.	2	tr.	88	4
Maize leaf	42	31	5	6	3	7	1	tr.	tr.	5
Clover leaf	46	28	4	7	5	6	1	tr.	tr.	3

	Fatty acid composition (% total)						
	16:0	16:1[c]	18:0	18:1	18:2	18:3	Other
Rapeseed (LEAR) oil	4	–	2	62	20	10	2
Palm oil	44	tr.	4	39	11	1	1
Palm kernel oil	9	tr.	2	15	2	tr.	72[d]
Castor bean oil	1	–	tr.	3	5	–	91[e]
Maize leaf	8	4	2	7	8	66	5
Pea leaf	12	3	1	2	22	56	4
Spinach leaf	13	3	tr.	7	16	56	5[f]

For more information see Gunstone et al. (2007), Gurr et al. (2002), and Harwood (1980)

Fatty acids are abbreviated with the numbers before and after the colon showing the number of carbon atoms and the number of double bonds, respectively. 18:1 is oleate, 18:2 is linoleate and 18:3 is α-linolenate

Abbreviations: *PI* phosphatidylinositol, *DPG* diphosphatidylglycerol (cardiolipin), *tr.* trace (<0.5), the others as in Fig. 1.1

[a]Cerebroside, sterol glycoside and acyl sterol glycosides are significant

[b]Cerebrosides, sterol glycosides and sterols are significant

[c]In leaf tissue, *trans*-3-hexadecenoate is a significant component of the PG fraction

[d]Contains 48 % laurate and 16 % myristate

[e]Contains 90 % ricinoleate

[f]a '16:3-plant' with a prokaryotic-like fatty acid metabolism and 5 % *cis*-7, 10, 13-hexadecatrienoic acid, almost entirely located in MGDG

The storage compound, i.e. TAG, is also shown in Fig. 1.1. Accumulation of this lipid is usually dependent on algal growth conditions and can be enhanced by (nutrient) stress (Hu et al. 2008). This is important for the potential commercial exploitation of algae such as for the provision of long-chain polyunsaturated fatty acids. For higher plants, TAG is almost the exclusive lipid accumulated and typical oil seeds may have up to 70 % dry weight as lipid. Plants contain a spectacular variety of fatty acids within their accumulated oils although most commercially-important crops have palmitate, oleate and linoleate as major components. However, many plant TAGs have niche markets including as renewable materials for the chemical industry. Two examples are shown in Table 1.1 where palm kernel oil is used in the detergent industry and castor oil as a lubricant (Murphy 2005; Carlsson et al. 2011; Vanhercke et al. 2013). On the other hand, rapeseed and palm oils contain the usual fatty acids (above) as their main components.

All the above compounds are acyl lipids. Therefore, within each lipid class there will be a host of molecular species, each containing different combinations and

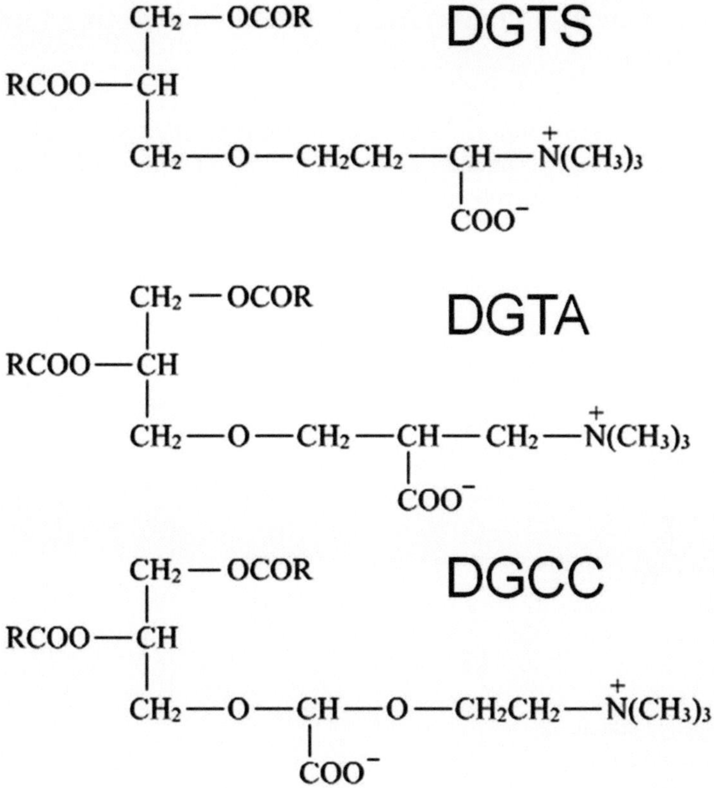

Fig. 1.2 Examples of structures of ether lipids found in algae. *R* represents any acyl group. Abbreviations: *DGTS* 1,2-diacylglyceryl-3-*O*-4′-(*N,N,N*-trimethyl)-homoserine, *DGTA* 1,2-diacylglyceryl-3-*O*-2′-(hydroxymethyl)-(*N,N,N*-trimethyl)-β-alanine,*DGCC*1,2-diacylglyceryl-3-*O*-carboxy-(hydroxymethyl)-choline

positional distributions of fatty acids. Freshwater algae contain similar fatty acids to terrestrial plants although their proportions vary considerably. In general, freshwater algae contain higher proportions of 16C (carbon) and less 18C fatty acids than plants. Unlike marine algae, 20C (or 22C) polyunsaturated fatty acids (PUFA) are usually absent (or only present in trace amounts) in freshwater species. However, in those algae growing in high-salt environments, such as the Dead Sea, arachidonate (ARA) and n-3 eicosapentaenoate (EPA) are often major components (Pohl and Zurheide 1979; Harwood and Jones 1989).

The leaves, shoots and roots of higher plants are dominated by palmitate as the main saturated fatty acid, oleate as the principal monounsaturated and linoleate and α-linolenate as the prominent polyunsaturated acids. The latter is particularly enriched in chloroplast thylakoids and, hence, in leaf tissue (Table 1.1).

Different Algae and Plants Often Contain Contrasting Lipid Compositions

Algae are a very diverse group of organisms so, unsurprisingly, their lipid compositions can vary markedly. The fatty acid compositions of a range of different algae are shown in Table 1.2. As mentioned above, 16C unsaturated fatty acids are often

Table 1.2 Fatty acid composition of different algae

	Fatty acid composition (% total)												
	14:0	16:0	16:1	16:2	16:3	16:4	18:1	18:2	18:3	18:4	20:4	20:5	22:6
Freshwater spp.													
Scenedesmus obliquus	1	35	2	tr.	tr.	15	9	6	30	20	–	–	–
Chorella vulgaris	2	26	8	7	2	–	2	33	20	–	–	–	–
Chlamydomonas reinhardtii	1	21	4	1	4	23	7	6	31	2	–	–	–
Salt-tolerant spp.													
Ankistrodesmus sp.	tr.	13	3	1	1	14	25	2	29	2	–	1	–
Isochrysis sp.	16	15	4	1	1	–	21	5	4	17	tr.	tr.	9
Nannochlorsis atomus	1	20	10	4	15	–	5	10	22	3	2	3	tr.
Marine phytoplankton													
Monochrysis lutheri (Chrysophyceae)	10	13	22	5	7	1	3	1	tr.	2	1	18	7
Olisthodiscus spp. *(Xanthophyceae)*	8	14	10	2	2	1	4	4	6	18	2	19	2
Lauderia borealis (Bacillariophyceae)	7	12	21	3	12	1	2	1	tr.	–	1	3	–
Amphidinium carterale (Dinophyceae)	3	24	1	1	tr.	–	5	1	2	15	–	14	25
Dunaliella salina (Chlorophyceae)	tr.	41	15	tr.	–	–	11	8	19	–	–	–	–
Hemiselmis brunescens (Cryptophyceae)	1	13	3	3	tr.	tr.	2	tr.	9	30	tr.	14	–
Marine macroalgae													
Fucus vesiculosus (Phaeophyceae)	tr.	21	2	tr.	–	tr.	26	10	7	4	15	8	–
Chondrus crispus (Rhodophyceae)	tr.	34	6	tr.	–	–	9	1	1	4	18	22	–
Ulva lactuca (Chlorophyceae)	1	18	2	tr.	1	18	9	2	17	24	1	2	tr.

Fatty acids are abbreviated with the figure before the colon indicating the number of carbon atoms and the figure after the colon showing the number of double bonds. See Harwood and Jones (1989) for sources of information. Although the precise structure of individual fatty acids reported was not always determined the unsaturated components are likely to be mainly 16:1n-7, 16:2n-4, 16:3n-4 or 16:3n-3, 16:4n-3, 18:1n-9, 18:2n-6, 18:3n-3, 18:4n-3, 20:4n-6, 20:5n-3, 22:6n-3

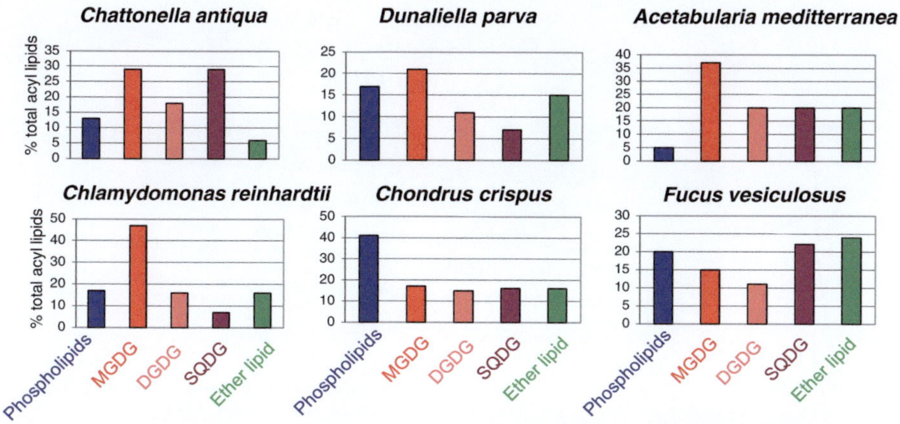

Fig. 1.3 The acyl lipid composition of some algae. Abbreviations: Phospholipids include PC and PE; ether lipids include DGTS and DGTA. Other abbreviations as detailed in Fig. 1.1 legend

significant while 20 or 22C PUFA are prominent in marine or salt-tolerant species. Since algae are often at the bottom of food chains then nutritionally-important PUFA (EPA or DHA i.e. docosahexaenoic acid), often referred to as 'fish oil components', are extremely important and may be more exploited commercially in the future.

Just as the fatty acid composition of algae varies considerably so does their relative acyl lipid content. Some examples are shown in Fig. 1.3. For more information, the reader is referred to Ratledge and Cohen (2010), Guschina and Harwood (2006), Harwood and Scrimgeour (2007), and Harwood and Jones (1989). By comparison, the acyl lipid compositions of higher plant tissues are much more consistent. Thus, in Table 1.1, monocotyledon leaves (e.g. maize) are rather similar to those of dicotyledons (e.g. clover). Moreover, the overall fatty acid compositions of leaves are also consistent even when a particular species (e.g. spinach) may have some subtle differences in metabolism (see Table 1.1). More details of higher plant lipids and their distribution will be found in Harwood and Gunstone (2007) and Murphy (2005). Nonetheless, little is known about the evolutionary advantage or physiological needs whereby particular cell types preferentially synthesize one type of fatty acid rather than another.

Biosynthesis of Major Glycerolipids and Their Cellular Functions

Lipid biosynthesis is a complex hub, so the detailed lipid metabolic map may be somewhat overwhelming. Here, we present a simplified view of lipid biosynthesis (Fig. 1.4). We emphasize the aspects that will be covered by the subsequent chapters

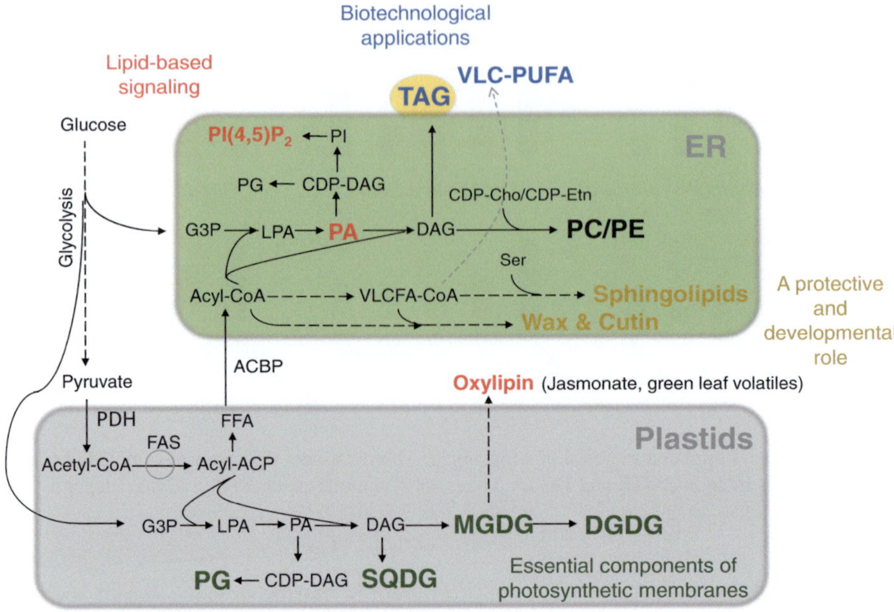

Fig. 1.4 A simplified view of lipid biosynthesis in green photosynthetic cells. Abbreviations: *ACBP* acyl-CoA binding protein, *CDP-DAG* cytidinediphosphate-diacylglycerol, *CoA* coenzyme A, *FAS* fatty acid synthase, *FFA* free (non-esterified) fatty acid, *PA* phosphatidic acid, *PI* phosphatidylinositol, *VLC-PUFA* very long chain (>20C) polyunsaturated fatty acids, other abbreviations as detailed in Fig. 1.1 legend

in this book. For more complete information on current knowledge of acyl-lipid metabolism, readers are referred to Li-Beisson et al. (2013) for the higher plant model *Arabidopsis thaliana*; and to Li-Beisson et al. (2015) and Liu and Benning (2013) for the model green microalga *Chlamydomonas reinhardtii*.

Carbon Source and Two-Carbon by Two-Carbon Fatty Acid Elongation in the Plastid

Most of lipid classes discussed in this book contain both glycerol and fatty acids, thus their biosyntheses share some common pathways. After carbon fixation through photosynthetic reactions, glycolysis provides two major substrates for biosynthesis of glycerolipids, i.e. glycerol 3-phosphate (G3P) and acetyl-Coenzyme A (CoA). Acetyl-CoA is the major source for plastidial fatty acid biosynthesis. Although two reactions have been proposed to supply acetyl-CoA for de novo fatty acid synthesis, now several lines of evidence indicate that pyruvate is the major source (Schwender et al. 2003; Bao et al. 1998; Lin et al. 2003). This reaction is catalyzed by pyruvate dehydrogenase (PDH). The contribution of plastidial acetyl-CoA synthetase (ACS) to supply acetyl-CoA for fatty acid synthesis is considered marginal in several plant tissue types (Bao et al. 1998; Lin et al. 2003). In algae, genes encoding orthologous enzymes to PDH and ACS occur, but their detailed contribution to fatty acid synthesis awaits experimental evidence.

The first committed step is catalyzed by acetyl-CoA carboxylase (ACCase) producing malonyl-CoA. The chain elongates 2C at a time, and these reactions are catalyzed by four well characterized enzymes together called fatty acid synthase (FAS), and the reactions terminate, in most species, when the chain length reaches 16C or 18C. The acyl chain length beyond 16C or 18C can be further elongated by ER-resident fatty acid elongases to produce very-long-chain fatty acid (VLCFA)-CoA (>20C), which are substrates for sphingolipids or surface lipids such as waxes and cutin (Kunst and Samuels 2009).

Glycerolipid Assembly

Acyl-chains produced in the chloroplasts serve either as substrates for chloroplastic glycerolipid biosynthesis or are exported to the ER where they act as an acyl donor for ER-located glycerolipid synthesis pathways. Glycerolipids are assembled by two sequential transfers of acyl chains to G3P resulting in production of phosphatidic acid (PA), which occurs in both plastids and the ER (Kennedy pathway) (Kennedy 1956) (Fig. 1.4). PA is a substrate for both diacylglycerol (DAG) and cytidine diphosphate-DAG (CDP-DAG), which are further converted to different classes of glycerolipids. In addition, PA itself functions as a signaling molecule. In chloroplasts, four major glycerolipid classes, PG, MGDG, DGDG, SQDG, are synthesized. MGDG, DGDG and SQDG are synthesized from DAG, whereas PG is produced from CDP-DAG. These lipid classes are synthesized exclusively in chloroplasts except PG, which can be also produced in the ER and mitochondria. In the ER, major phospholipid classes are synthesized, including PA, PC, PE, PG, PI and PS. PC and PE are produced from DAG using CDP-Cho or CDP-Etn, respectively, while PS is synthesized from PE by a base-exchange reaction (Yamaoka et al. 2011). PI is synthesized from CDP-DAG, which is also a substrate of PG biosynthesis. The major storage lipid TAG is synthesized from DAG by two complementary reactions, acyl-CoA dependent and acyl-CoA independent reactions catalyzed by a diacylglycerol acyltransferase (DGAT) and a phospholipid: diacylglycerol acyltransferase (PDAT), respectively (Zhang et al. 2009).

Major Cellular Functions of Lipids

The galactolipids (MGDG and DGDG) are the most predominant glycerolipid class in photosynthetic tissues of plants as well as algal cells under photoautotrophic growth. MGDG and DGDG are essential in the development and function of thylakoids and entire plastids. PG is the only phospholipid produced in the chloroplasts and it is an essential component in the center of photosystem II.

As well as their role in plastid function, galactolipids are also substrates for production of oxylipins, because polyunsaturated fatty acids released from the galactolipids are preferred substrates for oxidation reactions catalyzed by lipoxygenases (Feussner and Wasternack 2002). Initial biosynthetic steps to jasmonic acid (JA), a representative oxylipin signal, occur in chloroplasts. Moreover, galactolipids are the substrates of various volatile oxylipin derivatives, named green leaf volatiles, in plant signaling, response and communications. PI can be further phosphorylated to produce phosphatidylinositol phosphates (or phosphoinositides). A representative phosphoinositide, phosphatidylinositol 4,5-bisphosphate (PI(4,5)P_2), has functions in lipid signaling (Boss and Im 2012).

Surface lipids represent an essential role of lipids to coat the outermost surfaces of plant tissues to protect from environmental stresses. Therefore, their metabolism is tightly linked to the developmental regulation of plant body shape. Sphingolipids are a lipid class containing a backbone of aliphatic amino alcohols. They are major components in plasma membranes, tonoplast and endomembranes, playing structural and bioactive roles.

Major Tools in Lipid Analyses: From Chemical Composition to Lipid Imaging

As a first step toward attributing functions to lipids, chemical composition analyses of the tissue studied are obviously important. Due to their hydrophobic nature, lipid analyses usually start with their extraction into an organic solvent, which remove many hydrophilic cellular components such as carbohydrate, sugars and proteins. The extracted lipids contained in the solvent are usually dried down with a stream of nitrogen gas. An estimation of total lipid content can be determined by dry weight at this stage; but this measurement takes into account the weight of all solvent-soluble components and, thus, represents a total of lipophilic compounds including fatty acids, glycerolipids, sphingolipids, triterpenoids and chlorophyll or other pigments as well as non-lipid contaminants. To further evaluate the quantity and composition of each class of lipids, chromatographic methods have been developed since the 1950s. Two most popular and affordable ones are those of Thin Layer Chromatography (TLC) and Gas Chromatography (GC).

A recent method that has been developed for plant tissues in 2003 is that of direct infusion electron spray ionization-mass spectrometry (ESI-MS/MS), as well as liquid chromatography-mass spectrometry (ESI-MS/MS). Both methods are referred to as lipidomics, allowing a snapshot of all lipid molecular species present in a sample. Mass spectrometry imaging has also been developed and used in analyzing plant cells, which gives additional information in relation to lipid chemical composition. The major characteristics of each technique are summarized in Fig. 1.5.

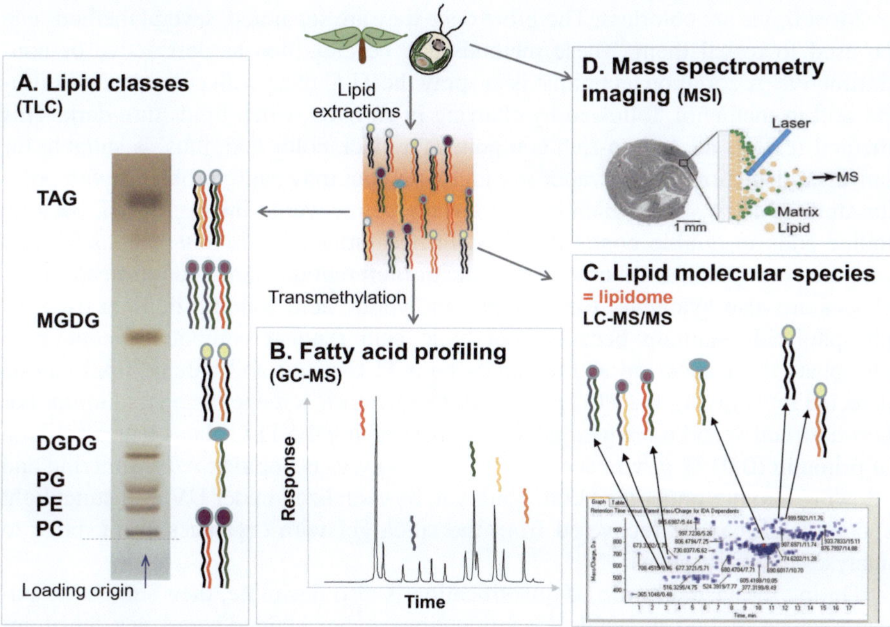

Fig. 1.5 A summary of major lipid analytical tools. Abbreviations: *GC-MS* gas chromatography – mass spectrometry, *LC-MS/MS* liquid chromatography-tandem mass spectrometry. Other abbreviations as detailed in Fig. 1.1. We acknowledge Dr. Fred Beisson in conception of this figure

These methods can be used alone or in combination. Below we describe in more detail some of the above mentioned analytical tools.

Thin Layer Chromatography (TLC)

TLC is probably one of the most highly used lipid analytical tools. It is highly reproducible and low cost, which makes it accessible to all laboratories interested in lipids and was developed over 60 years ago. It is simple in setup requiring a glass tank, a glass plate pre-coated with a thin layer of silica gel (stationary phase), and organic solvents (mobile phase). After the sample (i.e. the lipid extracts) has been applied to one end of the plate, a solvent or solvent mixture is drawn up the plate via capillary action. Because different lipid classes possess different polarity, each lipid class ascends the TLC plate at different rates. Thus separation is achieved. An example of separation of polar lipids is shown in Fig. 1.5a. For separation of some complex lipids, two-dimensional TLC plate can also be applied.

Most lipids are colorless. Therefore once they are separated, several methods can be used to reveal them. These methods can be classified as destructive or non-destructive. A common technique is to spray the TLC plate with concentrated sulfuric acid in methanol, followed by charring in an oven, when lipids turn dark. This method renders all carbon-rich compounds a black color and, thus, is suitable for universal detection. Iodine vapor is widely used but may cause problems with poly-unsaturated fatty acids which cannot always be recovered intact. Indeed, because iodine adds to double bonds it will stain unsaturated lipids more intensely than saturated compounds. Otherwise, specific or preferential staining of particular lipid classes are also available; for example, molybolic acid staining is often used for phospholipids staining because molybolic acid renders compounds containing phosphate ester show up as blue spots on a TLC plate; and betaine lipid can be detected by spraying the TLC plate with Dragendorff's reagent spray solution. For non-destructive and more general detection, spraying the TLC plates with a solution of primulin (0.01 % w/v in acetone-water 60:40 v/v) is popular. After spraying and allowing it to dry under the hood, lipids can be visualized under UV as distinct light spots, which can be recovered from the silica gel with organic solvent, prior to analyses by other means.

During the visualization, if quantification is also desirable, then authentic standards can be applied alongside. Quantification can be obtained based on a densitometry method. More accurate quantification can be further obtained if the lipids are to be recovered, and then analyzed by GC-FID. Other information related to TLC methods can be found in Christie et al. (2007) and the AOCS Lipid Library website (http://lipidlibrary.aocs.org/index.html).

Gas Chromatography (GC)

In plants and algal cells, the majority of lipids contain fatty acids, such as glycerolipids, sphingolipids, wax esters and monomers of cutin and suberins. These fatty acid-containing lipids can be converted to fatty acid methyl esters (FAMEs), which can be analyzed by GC. GC is often coupled to a Flame Ionization Detector (GC-FID) or Mass Spectrometer (GC-MS). Fatty acids are provisionally identified based on their retention time and then the mass split patterns used to compare with authentic standards. Quantification can be achieved with addition of an internal standard, often heptadecanoic acid (17:0 fatty acid) because this odd-chain fatty acid is usually absent in eukaryotic cells.

Conversion of acyl-lipids to their respective FAMEs can be catalyzed either by an acid- or a base- mediated transmethylation reaction. In most cases, both methods are inter-changeable. Having said these, it is worth noting that base catalysts do not esterify free (non-esterified) fatty acids (FFAs) and neither trans-esterify amide-bond fatty acids in sphingolipids; whereas acid catalysts are usually too strong and are not suited for certain fatty acids with sensitive functional groups, such as epoxy, cyclopropane or cylopropene rings (Bao et al. 2002). Although free fatty acids are only very minor cellular components, largely due to their detergent properties, derivation of FFAs themselves can be achieved by using a mixture of 1-ethyl-3-3-dimethylaminopropylcarbodiimide containing methanolic solution (Kaczmarzyk and Fulda 2010). This ability to measure the content of FFA alone (without the bulk of glycerolipids) is desirable in some genetic engineering studies.

FAMEs are then subjected to separation, identification and quantification by GC-FID or GC-MS. An example of elution of FAMEs from cell extracts of *Chlamydomonas reinhardtii* is shown in Fig. 1.5b. For detailed methods on preparation of fatty acid derivatives and GC conditions, please consult the AOCS Lipid Library site (http://lipidlibrary.aocs.org/topics/ester_93/index.htm).

Lipidomics

With the development in modern analytical tools, especially in mass spectrometry, we can now not only determine the total lipid content, lipid classes, or fatty acid profile for a given sample, but also we can determine the type, number and quantity of each lipid molecular species present in a tissue (Fig. 1.5c). The determination of the total number and type of lipid molecular species present in a sample is called 'lipidomics'. In general well over 200 lipid molecular species have been identified in plants and algal cells (Welti et al. 2002, 2007; Nguyen et al. 2013; Liu et al. 2013). Application of modern lipidomic tools to the green lineage was first established in the Kansas Lipidomics Centre (http://www.k-state.edu/lipid/lipidomics/) funded and directed by Professor Ruth Welti since 2003 and has since also been setup and routinely operated in a few other laboratories including the author's laboratory at CEA Cadarache.

Lipidomic analysis is one sub-branch of metabolomics, in that the former deal principally with hydrophobic molecules. Currently, two major types of lipidomics are in use, direct-infusion electron spray-MS/MS (ESI-MS/MS) or separation by liquid chromatography followed by mass spectrometry (LC-MS/MS). Most lipid types can be analyzed by either method. Organelle or sub-cellular lipidomics can be achieved by extracting lipids from isolated subcellular organelles or fractions. Due to the requirement for authentic standards, lipidomics are often used in a comparative context, for example, comparing wild-type to a mutant, or comparing one developmental stage to another, comparing one condition to a stress condition etc. The power of lipidomics in linking genes to function can be seen in numerous publications (Welti et al. 2002; Yoon et al. 2012). As compared to other analytical tools, lipidomics profiling gives an instaneous snapshot of the lipidomic state of a cell, interpretation of which can be combined together with transcriptomic and proteomic dataset thus allowing a systematic understanding of cell physiology. Nonetheless, it has also the disadvantages that the apparatus is still quite expensive, and the handling and processing of data needs specialist knowledge. This means that it is not accessible to most laboratories.

Lipid Imaging: From Lipophilic Stains to Mass Spectrometry Imaging

The above mentioned methods have largely focused on the chemical analyses of lipid compositions. As with other subcellular components, lipids can also be visualized. Classical techniques of lipid imaging include first staining with a lipophilic dye such as Nile red followed by observation under a microscope; or can also be visualized by electron microscopic techniques. Examples of these analytical tools in advancing our understanding of gene functions are abundant, and some examples can be found in Li-Beisson et al. (2013), Cagnon et al. (2013), and Xie et al. (2014).

Recently a coordinated analysis of the lipid composition with their sub-cellular localizations at a high spatial resolution have been developed, and one such approach is called mass spectrometry imaging (MSI) (Fig. 1.5d). MSI integrates in situ visualization with chemical based lipidomics. Three major types of MSI are now available, which differ in their ionization sources, i.e. secondary ion MS (SIMS), desorption electrospray ionization (DESI), laser desorption/ionization (MALDI)-MS. For detailed description of these platforms, readers are referred to Horn et al. (2012), Horn and Chapman (2014) and other references there-in. These technologies have been applied to a number of plant tissues.

Biotechnological Applications of Lipids Derived from Algae and Plants

Over several hundred fatty acid structures have been identified in nature, many of which are present only in particular species. For example, ricinoleic acid is found almost exclusively in castor bean. Due to their large structural diversity, lipid applications are also diverse including sectors like food, chemicals, bioenergy, and nutraceuticals and the cosmetics industry. Among all lipid structures presented, oils are the most exploitable ones for two major reasons: oils are one of the most abundant lipids stored in cells; and industrial processes of oil extraction and processing are affordable and represent a mature technology as compared to extraction and recovery of membrane lipids (i.e. galactolipids and phospholipids).

The most abundant fatty acid species present in most plant and algae cells are less than ten. Fatty acids with secondary modifications (for example oxygenation) or unusual chain length were often called unusual fatty acids. These 'unusual' fatty acids can be of particular industrial value. However, they are often synthesized naturally only in some non-agronomic plant species, or are present in a less-easily accessible form such as polymers coating plant surfaces (Li and Beisson 2009).

The current major biotechnological objectives in the area of plant/algal lipid research include: (i), increasing the productivity of current oilseed crops to meet the increasing demand of the world population; (ii), introducing novel genes thus allowing synthesis and production of unusual fatty acids in the oil fraction of existing

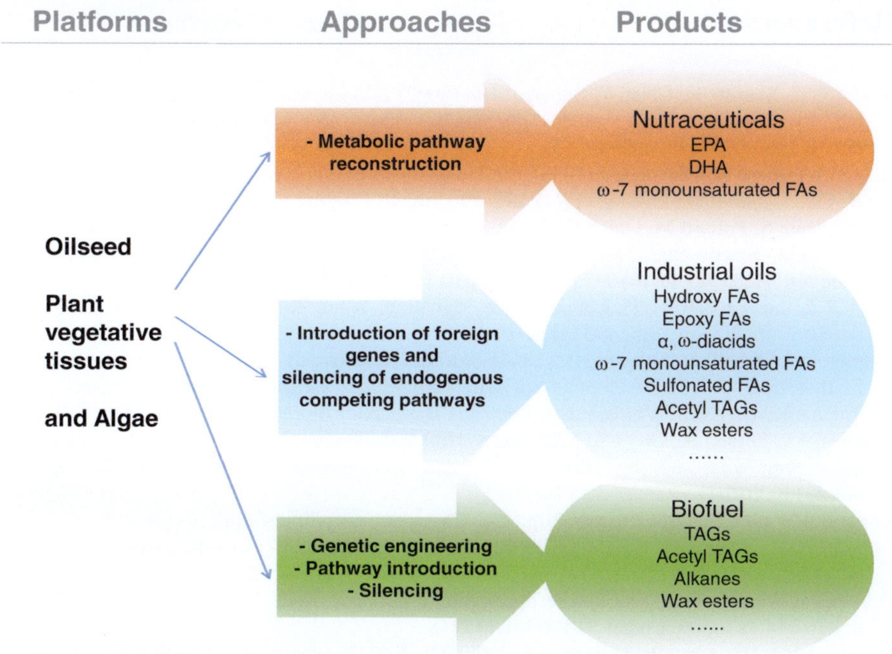

Fig. 1.6 Some examples of biotechnological applications of lipids and the approaches used toward these purposes

crop species; (iii), increasing oil/energy content of plant vegetative tissues or algal cells, thus establishing new platforms for oil production and having the advantage of not competing with oil crops for food usage; and (iv), pathway reconstruction for production of very long chain polyunsaturated fatty acids for the nutraceutical industry (Fig. 1.6). Abundant successful proof-of-the-concept examples are available in the literature (Napier and Graham 2010; Jaworski and Cahoon 2003; Lu et al. 2011). Now one major challenge for the next step is the commercialization of these findings, thus requiring field trials, techno-economic and life cycle analyses.

Conclusion

In this chapter, we have highlighted a few key points related to lipids: structures, biosynthesis, analysis and applications, which we consider important to appreciate the following chapters in this book. Part of this book describes the latest developments in the functions of lipids in photosynthesis, and part of this book deals with the roles of lipids in plant and algal development and signaling; emerging platforms for industrial applications of lipids are also covered.

References

Banerjee A, Sharma R, Chisti Y, Banerjee UC (2002) *Botryococcus braunii*: a renewable source of hydrocarbons and other chemicals. Crit Rev Biotechnol 22(3):245–279. doi:10.1080/07388550290789513

Bao XM, Pollard M, Ohlrogge J (1998) The biosynthesis of erucic acid in developing embryos of *Brassica rapa*. Plant Physiol 118(1):183–190. doi:10.1104/pp. 118.1.183

Bao X, Katz S, Pollard M, Ohlrogge J (2002) Carbocyclic fatty acids in plants: biochemical and molecular genetic characterization of cyclopropane fatty acid synthesis of *Sterculia foetida*. Proc Natl Acad Sci 99(10):7172–7177. doi:10.1073/pnas.092152999

Boss WF, Im YJ (2012) Phosphoinositide signaling. Annu Rev Plant Biol 63(1):409–429. doi:10.1146/annurev-arplant-042110-103840

Cagnon C, Mirabella B, Nguyen HM, Beyly-Adriano A, Bouvet S, Cuine S, Beisson F, Peltier G, Li-Beisson Y (2013) Development of a forward genetic screen to isolate oil mutants in the green microalga *Chlamydomonas reinhardtii*. Biotechnol Biofuels 6(1):178

Carlsson AS, Yilmaz JL, Green AG, Stymne S, Hofvander P (2011) Replacing fossil oil with fresh oil – with what and for what? Eur J Lipid Sci Technol 113(7):812–831. doi:10.1002/ejlt.201100032

Christie WW (2013) A lipid primer. http://lipidlibrary.aocs.org/lipids.html

Christie WW, Dijkstra AJ, Knothe G (2007) Analysis. In: The lipid handbook with CD-ROM, 3rd edn. CRC Press, Boca Raton, Florida, USA, pp 415–469. doi:10.1201/9781420009675.ch6

Feussner I, Wasternack C (2002) The lipoxygenase pathway. Annu Rev Plant Biol 53:275–297. doi:10.1146/annurev.arplant.53.100301.135248

Gunstone FD, Harwood JL, Dijkstra AJ (2007) The lipid handbook with CD-ROM. CRC Press, Boca Raton

Gurr MI, Harwood JL, Frayn KN (2002) Lipid biochemistry. Wiley, Malden

Guschina IA, Harwood JL (2006) Lipids and lipid metabolism in eukaryotic algae. Prog Lipid Res 45(2):160–186. doi:10.1016/j.plipres.2006.01.001

Guschina IA, Harwood JL (2013) Algal lipids and their metabolism. Algae Biofuels Energy 17:201–214

Harwood JL (1980) 1 – plant acyl lipids: structure, distribution, and analysis. In: Stumpf PK (ed) Lipids: structure and function. Academic, New York, pp 1–55. doi:http://dx.doi.org/10.1016/B978-0-12-675404-9.50007-2

Harwood JL, Gunstone FD (2007) Occurrence and characterisation of oils and fats. In: The lipid handbook with CD-ROM, 3rd edn. CRC Press, Boca Raton, pp 37–141. doi: 10.1201/9781420009675.ch2

Harwood JL, Guschina IA (2009) The versatility of algae and their lipid metabolism. Biochimie 91(6):679–684. doi:10.1016/j.biochi.2008.11.004

Harwood JL, Jones AL (1989) Lipid metabolism in algae. In: Callow JA (ed) Advances in botanical research, vol 16. Academic, London, pp 1–53. doi:http://dx.doi.org/10.1016/S0065-2296(08)60238-4

Harwood JL, Scrimgeour CM (2007) Fatty acid and lipid structure. In: The lipid handbook with CD-ROM, 3rd edn. CRC Press, Boca Raton, pp 1–36. doi:10.1201/9781420009675.ch1

Horn PJ, Chapman KD (2014) Lipidomics in situ: insights into plant lipid metabolism from high resolution spatial maps of metabolites. Prog Lipid Res 54(0):32–52. doi:http://dx.doi.org/10.1016/j.plipres.2014.01.003

Horn PJ, Korte AR, Neogi PB, Love E, Fuchs J, Strupat K, Borisjuk L, Shulaev V, Lee YJ, Chapman KD (2012) Spatial mapping of lipids at cellular resolution in embryos of cotton. Plant Cell 24(2):622–636. doi:10.1105/tpc.111.094581

Hu Q, Sommerfeld M, Jarvis E, Ghirardi M, Posewitz M, Seibert M, Darzins A (2008) Microalgal triacylglycerols as feedstocks for biofuel production: perspectives and advances. Plant J 54(4):621–639

Jaworski J, Cahoon EB (2003) Industrial oils from transgenic plants. Curr Opin Plant Biol 6(2):178–184

Kaczmarzyk D, Fulda M (2010) Fatty acid activation in cyanobacteria mediated by acyl-acyl carrier protein synthetase enables fatty acid recycling. Plant Physiol 152(3):1598–1610. doi:10.1104/pp. 109.148007

Kennedy EP (1956) The biological synthesis of phospholipids. Can J Biochem Physiol 34(2):334–348

Kolattukudy PE (2001) Polyesters in higher plants. In: Babel W, Steinbuschel A (eds) Advances in biochemical engineering/biotechnology, vol 71, Biopolyesters. Springer, Berlin, pp 1–49

Kunst L, Samuels L (2009) Plant cuticles shine: advances in wax biosynthesis and export. Curr Opin Plant Biol 12(6):721–727. doi:10.1016/j.pbi.2009.09.009

Li YH, Beisson F (2009) The biosynthesis of cutin and suberin as an alternative source of enzymes for the production of bio-based chemicals and materials. Biochimie 91(6):685–691. doi:10.1016/j.biochi.2009.03.016

Li-Beisson Y, Shorrosh B, Beisson F, Andersson MX, Arondel V, Bates PD, Baud S, Bird D, DeBono A, Durrett TP, Franke RB, Graham IA, Katayama K, Kelly AA, Larson T, Markham JE, Miquel M, Molina I, Nishida I, Rowland O, Samuels L, Schmid KM, Wada H, Welti R, Xu C, Zallot R, Ohlrogge J (2013) Acyl-lipid metabolism. Arabidopsis Book 11:e0161. doi:10.1199/tab.0161

Li-Beisson Y, Beisson F, Riekhof W (2015) Metabolism of acyl-lipids in *Chlamydomonas reinhardtii*. Plant J 82(3):504–522. doi:10.1111/tpj.12787

Lin M, Behal R, Oliver DJ (2003) Disruption of plE2, the gene for the E2 subunit of the plastid pyruvate dehydrogenase complex, in Arabidopsis causes an early embryo lethal phenotype. Plant Mol Biol 52(4):865–872. doi:10.1023/a:1025076805902

Liu B, Benning C (2013) Lipid metabolism in microalgae distinguishes itself. Curr Opin Biotechnol 24(2):300–309. doi:http://dx.doi.org/10.1016/j.copbio.2012.08.008

Liu B, Vieler A, Li C, Daniel Jones A, Benning C (2013) Triacylglycerol profiling of microalgae *Chlamydomonas reinhardtii* and *Nannochloropsis oceanica*. Bioresour Technol 146(0):310–316. doi:http://dx.doi.org/10.1016/j.biortech.2013.07.088

Lu CF, Napier JA, Clemente TE, Cahoon EB (2011) New frontiers in oilseed biotechnology: meeting the global demand for vegetable oils for food, feed, biofuel, and industrial applications. Curr Opin Biotechnol 22(2):252–259. doi:10.1016/j.copbio.2010.11.006

Mizusawa N, Wada H (2012) The role of lipids in photosystem II. Biochim Biophys Acta-Bioenerg 1817(1):194–208. doi:10.1016/j.bbabio.2011.04.008

Murphy DJ (2005) Plant lipids: biology, utilisation, and manipulation, Biological Sciences Series. Wiley, Oxford

Napier JA, Graham IA (2010) Tailoring plant lipid composition: designer oilseeds come of age. Curr Opin Plant Biol 13(3):330–337. doi:10.1016/j.pbi.2010.01.008

Nguyen HM, Cuiné S, Beyly-Adriano A, Légeret B, Billon E, Auroy P, Beisson F, Peltier G, Li-Beisson Y (2013) The green microalga *Chlamydomonas reinhardtii* has a single ω-3 fatty acid desaturase that localizes to the chloroplast and impacts both plastidic and extraplastidic membrane lipids. Plant Physiol 163(2):914–928. doi:10.1104/pp. 113.223941

Pohl P, Zurheide F (1979) Fatty acids and lipids of marine algae and the control of their biosynthesis by environmental factors. In: Hoppe HA, Tore L, Yukio T (eds) Marine algae in pharmaceutical science. W. de Gruyter, Berlin

Pollard M, Beisson F, Li YH, Ohlrogge JB (2008) Building lipid barriers: biosynthesis of cutin and suberin. Trends Plant Sci 13(5):236–246. doi:10.1016/j.tplants.2008.03.003

Ratledge C, Cohen Z (2010) Single cell oils: microbial and algal oils. AOCS Press, Urbana

Schwender J, Ohlrogge JB, Shachar-Hill Y (2003) A flux model of glycolysis and the oxidative pentosephosphate pathway in developing *Brassica napus* embryos. J Biol Chem 278(32):29442–29453. doi:10.1074/jbc.M303432200

Vanhercke T, Tahchy AE, Shrestha P, Zhou X-R, Singh SP, Petrie JR (2013) Synergistic effect of WRI1 and DGAT1 coexpression on triacylglycerol biosynthesis in plants. FEBS Lett 4:364. doi:10.1016/j.febslet.2012.12.018

Welti R, Li WQ, Li MY, Sang YM, Biesiada H, Zhou HE, Rajashekar CB, Williams TD, Wang XM (2002) Profiling membrane lipids in plant stress responses – role of phospholipase D alpha in freezing-induced lipid changes in Arabidopsis. J Biol Chem 277(35):31994–32002. doi:10.1074/jbc.M205375200

Welti R, Shah J, Li WQ, Li MY, Chen JP, Burke JJ, Fauconnier ML, Chapman K, Chye ML, Wang XM (2007) Plant lipidomics: discerning biological function by profiling plant complex lipids using mass spectrometry. Front Biosci 12:2494–2506. doi:10.2741/2250

Xie B, Stessman D, Hart JH, Dong H, Wang Y, Wright DA, Nikolau BJ, Spalding MH, Halverson LJ (2014) High-throughput fluorescence-activated cell sorting for lipid hyperaccumulating *Chlamydomonas reinhardtii* mutants. Plant Biotechnol J 12:872–882

Yamaoka Y, Yu Y, Mizoi J, Fujiki Y, Saito K, Nishijima M, Lee Y, Nishida I (2011) Phosphatidylserine synthase1 is required for microspore development in *Arabidopsis thaliana*. Plant J Cell Mol Biol 67(4):648–661. doi:10.1111/j.1365-313X.2011.04624.x

Yoon K, Han D, Li Y, Sommerfeld M, Hu Q (2012) Phospholipid:diacylglycerol acyltransferase is a multifunctional enzyme involved in membrane lipid turnover and degradation while synthesizing triacylglycerol in the unicellular green microalga *Chlamydomonas reinhardtii*. Plant Cell 24(9):3708–3724. doi:10.1105/tpc.112.100701

Zhang M, Fan J, Taylor DC, Ohlrogge JB (2009) DGAT1 and PDAT1 acyltransferases have overlapping functions in Arabidopsis triacylglycerol biosynthesis and are essential for normal pollen and seed development. Plant Cell 21(12):3885–3901. doi:10.1105/tpc.109.071795

Part I
Lipids in Photosynthesis

Chapter 2
Roles of Lipids in Photosynthesis

Koichi Kobayashi, Kaichiro Endo, and Hajime Wada

Abstract Thylakoid membranes in cyanobacterial cells and chloroplasts of algae and higher plants are the sites of oxygenic photosynthesis. The lipid composition of the thylakoid membrane is unique and highly conserved among oxygenic photosynthetic organisms. Major lipids in thylakoid membranes are glycolipids, monogalactosyldiacylglycerol, digalactosyldiacylglycerol and sulfoquinovosyldiacylglycerol, and the phospholipid, phosphatidylglycerol. The identification of almost all genes involved in the biosynthesis of each lipid class over the past decade has allowed the generation and isolation of mutants of various photosynthetic organisms incapable of synthesizing specific lipids. Numerous studies using such mutants have revealed that these lipids play important roles not only in the formation of the lipid bilayers of thylakoid membranes but also in the folding and assembly of the protein subunits in photosynthetic complexes. In addition to the studies with the mutants, recent X-ray crystallography studies of photosynthetic complexes in thylakoid membranes have also provided critical information on the association of lipids with photosynthetic complexes and their activities. In this chapter, we summarize our current understanding about the structural and functional involvement of thylakoid lipids in oxygenic photosynthesis.

Keywords Digalactosyldiacylglycerol • Monogalactosyldiacylglycerol • Phosphatidylglycerol • Photosynthesis • Sulfoquinovosyldiacylglycerol • Thylakoid membrane

Abbreviations

BN-PAGE	Blue native polyacrylamide gel electrophoresis
Chl	Chlorophyll
Cyt	Cytochrome
DCMU	3-(3, 4-dichlorophenyl)-1,1-dimethylurea

K. Kobayashi • K. Endo • H. Wada (✉)
Department of Life Sciences, Graduate School of Arts and Sciences,
The University of Tokyo, Komaba 3-8-1, Meguro-ku, Tokyo 153-8902, Japan
e-mail: hwada@bio.c.u-tokyo.ac.jp

Y. Nakamura, Y. Li-Beisson (eds.), *Lipids in Plant and Algae Development*,
Subcellular Biochemistry 86, DOI 10.1007/978-3-319-25979-6_2

DGDG	Digalactosyldiacylglycerol
GlcADG	Glucuronosyldiacylglycerol
LHC	Light-harvesting chlorophyll-protein complex
MGDG	Monogalactosyldiacylglycerol
MGlcDG	Monoglucosyldiacylglycerol
PG	Phosphatidylglycerol
PGP	Phosphatidylglycerophosphate
Pi	Phosphate
PQ	Plastoquinone
PSI	Photosystem I
PSII	Photosystem II
SQDG	Sulfoquinovosyldiacylglycerol
UDP	Uridine 5′-diphosphate

Introduction

The thylakoid membrane is the site of photochemical and electron transport reactions of oxygenic photosynthesis in cyanobacteria, algae, and higher plants. The photosynthetic electron transport chain consists of photosystem (PS) II, the cytochrome (Cyt) b_6/f and PSI complexes, with the mobile electron carriers plastoquinone (PQ) and plastocyanin. PSII performs light-induced electron transfer reactions by splitting water molecules into protons and molecular oxygen to reduce PQ molecules. Electrons excited in PSII are stepwise transferred to the PQ pool, Cyt b_6/f complex, plastocyanin and PSI, where another charge separation creates a strong reductant capable of reducing $NADP^+$ (Rochaix 2011). The core photosynthetic protein–cofactor complexes are embedded with ATP synthase in a membrane lipid bilayer, which avoids free diffusion of ions and allows for generating an electrochemical potential difference across the membrane for ATP synthesis. The lipid composition of the thylakoid membrane is unique and highly conserved among oxygenic photosynthetic organisms; uncharged galactolipids, monogalactosyldiacylglycerol (MGDG) and digalactosyldiacylglycerol (DGDG) account for 60–80 mol% of total thylakoid lipids, with anionic lipids, sulfoquinovosyldiacylglycerol (SQDG) and phosphatidylglycerol (PG) constituting the remainder (Fig. 2.1). As revealed by X-ray crystallography studies, photosynthetic complexes include these lipid molecules as structural and functional components. Thus, thylakoid glycerolipids are essential for forming the lipid bilayer and for activity of photosynthetic complexes, as described in detail later.

Each lipid in a different class has a specific role in thylakoid membrane biogenesis and photosynthesis depending on the nature of its head group. MGDG has a cone-like shape with a small galactose head group and flexible poly-unsaturated fatty acid tails, which allow for the formation of an inverted hexagonal II structure and a non-bilayer lipid phase. In contrast to MGDG, DGDG has a more cylindrical

Fig. 2.1 Composition of major lipids in thylakoid membranes from the cyanobacterium *Synechocystis* sp. PCC 6803 and the higher plant *Spinacia oleracea*. The original data of lipid analysis of thylakoid membranes are from Sakurai et al. (2006)

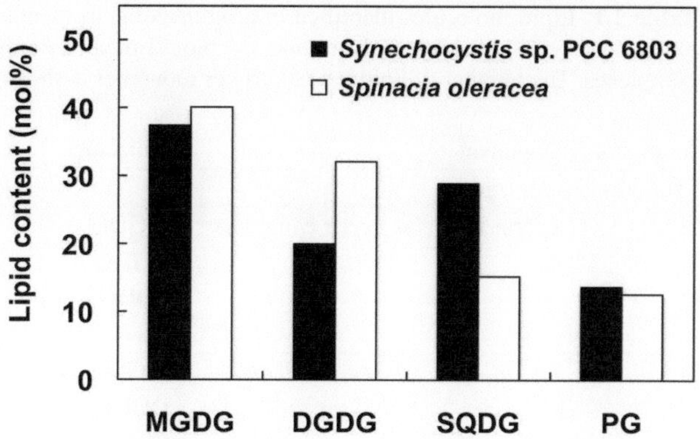

shape, with two galactoses in the polar head, and forms bilayer lamellar phases in mixture with water (Shipley et al. 1973). These neutral galactolipids constitute the bulk of the thylakoid lipid bilayer, and the MGDG/DGDG ratio may be important for the structure and stability of the thylakoid membrane (Murphy 1982; Demé et al. 2014).

SQDG and PG are classified as acidic lipids, with their negative charge at neutral pH. PG is the only major phospholipid in thylakoid membranes (Dorne et al. 1990; Wada and Murata 1998) and has indispensable roles in photosynthesis. SQDG too has an important function in photosynthesis, although the requirement for this lipid differs among species. PG and SQDG are at least partially functionally redundant, which may be related to maintenance of an anionic charge on the surface of the thylakoid membrane (Apostolova et al. 2008).

Almost all genes involved in the biosynthesis of each lipid class have been identified, and many mutants have been made in various photosynthetic organisms over the past decade, which has allowed for elucidating the specific roles of lipids in oxygenic photosynthesis in addition to the common function as building blocks of the membrane lipid bilayer. In parallel with molecular genetic studies, biochemical and biophysical analyses as well as X-ray crystallography studies have provided critical information on the association of lipids with photosynthetic complexes and their activities. In this chapter, we highlight the current understanding about the structural and functional involvement of thylakoid membrane lipids in oxygenic photosynthesis.

Composition and Localization of Lipids in Photosynthetic Protein–Cofactor Complexes

X-ray crystallography studies revealed lipid molecules in PSI (Jordan et al. 2001), PSII (Loll et al. 2005; Guskov et al. 2009; Umena et al. 2011), Cyt b_6/f (Stroebel et al. 2003; Kurisu et al. 2003), and light-harvesting chlorophyll-protein complex

Table 2.1 Lipid molecules identified in protein–cofactor complexes from thylakoid membranes by X-ray crystallography analysis and by biochemical analysis of lipids extracted from the complexes. The number of lipid molecules per monomer is shown for each protein complex

Protein complex	Organism	No. of lipid molecules		References
X-ray crystallography				
PSII	*Thermosynechococcus elongatus*	MGDG	11	Guskov et al. (2009)
		DGDG	7	
		SQDG	5	
		PG	2	
PSII	*Thermosynechococcus vulcanus*	MGDG	6	Umena et al. (2011)
		DGDG	5	
		SQDG	4	
		PG	5	
PSI	*T. elongatus*	MGDG	1	Jordan et al. (2001)
		PG	3	
LHCII	*Spinacia oleracea*	DGDG	1	Liu et al. (2004)
		PG	1	
LHCII	*Pisum sativum*	DGDG	1	Standfuss et al. (2005)
		PG	1	
Cyt *b6/f*	*Mastigocladus laminosus*	PC	2	Kurisu et al. (2003)
Cyt *b6/f*	*Chlamydomonas reinhardtii*	MGDG(?)	2	Stroebel et al. (2003)
		SQDG	1	
Lipid analysis				
PSI	*Synechocystis* sp. PCC 6803	MGDG	2	Kubota et al. (2010)
		DGDG	1	
		SQDG	1	
		PG	2	
PSII	*Synechocystis* sp. PCC 6803	MGDG	6	Sakurai et al. (2006)
		DGDG	3	
		SQDG	5	
		PG	6	
	T. vulcanus	MGDG	8	Sakurai et al. (2006)
		DGDG	6	
		SQDG	6	
		PG	8	

(LHC) II (Liu et al. 2004; Standfuss et al. 2005). The presence of thylakoid lipids in photosynthetic complexes was confirmed by lipid analyses of these complexes purified from various organisms (Nussberger et al. 1993; Sakurai et al. 2006; Kubota et al. 2010; Hasan et al. 2011). Table 2.1 summarizes the lipid molecules that have been identified in photosynthetic complexes.

PSI

X-ray crystallography analysis of PSI from *Thermosynechococcus elongatus* at 2.5-Å resolution demonstrated one molecule of MGDG and three molecules of PG present in the PSI complex, together with approximately 15 protein subunits and many cofactors such as chlorophylls (Chls) and carotenoids (Jordan et al. 2001). In the PSI crystal structure, lipids are positioned at the stromal side of the membrane plane forming hydrogen bonds with PsaA/PsaB. One of the 3 PG molecules (lipid I) and the MGDG molecule (lipid II) are located symmetrically relative to one another in the vicinity of the core of PSI. These two lipids may be specifically important for the formation of the photochemical reaction center of the PSI complex. One of the other two PG molecules, lipid III, is close to the monomer–monomer interface and associates with the antenna Chl *a* in its phosphodiester group, whereas the other molecule, lipid IV, is in the contact site with PsaX. Because severe PG deprivation in a *Synechocystis* sp. PCC 6803 mutant caused dissociation of PSI trimers and concomitant accumulation of the PSI monomer (Domonkos et al. 2004), the PG (lipid III) located at the monomer–monomer interface may participate in the PSI trimerization.

The content of lipid molecules in the PSI structure was determined in *Synechocystis* sp. PCC 6803 (Kubota et al. 2010). The PSI complex was purified from transgenic strains expressing an His-tagged PsaF or PsaJ subunit by Ni-affinity column chromatography. Lipid analysis of the purified PSI trimer identified 6 lipid molecules (2 MGDGs, 1 DGDG, 1 SQDG, and 2 PGs) per monomer (Table 2.1). Although the number of lipids in *Synechocystis* PSI was close to that in the crystal structure of *T. elongatus* PSI (Jordan et al. 2001), the lipid composition differs between these two cyanobacteria. The lipid composition of PSI may differ among species, and DGDG and SQDG molecules found in *Synechocystis* PSI may be replaced by PG in *T. elongatus*.

PSII

PSII comprises approximately 20 protein subunits in addition to many cofactors, such as pigments, metals, and lipids (Murata et al. 1984; Nanba and Satoh 1987; Boekema et al. 1995; Hankamer et al. 2001). Crystallography analysis of the PSII dimer complex from *T. elongatus* at 2.9-Å resolution identified 25 lipid molecules (11 MGDGs, 7 DGDGs, 5 SQDGs, and 2 PGs) per monomer in the structure (Guskov et al. 2009). Meanwhile, in the crystal structure of the *Thermosynechococcus vulcanus* PSII dimer resolved at 1.9 Å, 20 lipid molecules (6 MGDGs, 5 DGDGs, 4 SQDGs, and 5 PGs) per monomer were assigned, with three unknown diglycerides (Table 2.1) (Umena et al. 2011). Gas chromatography revealed that the lipid composition of PSII dimers from *Synechocystis* sp. PCC 6803 (Sakurai et al. 2006) was similar to that in the crystal structure of *T. vulcanus* PSII. These data indicate enriched PG molecules in the PSII dimer as compared with the relatively lower abundance of PG in the thylakoid membrane.

Of the 20 lipid molecules found in the crystal structure of *T. vulcanus* PSII (Umena et al. 2011), 3 (1 MGDG and 2 SQDGs) were located at the monomer–monomer interface, 4 (2 MGDGs, 1 DGDG, and 1 SQDG) were at the periphery of PSII, and 13 (3 MGDGs, 4 DGDGs, 1 SQDG, and 5 PGs) were buried in the vicinity of the D1/D2 heterodimer. The head groups of all anionic lipids in PSII faced the cytoplasmic side, whereas those of all galactolipids except for one MGDG molecule faced the lumenal side. Of the 13 inner lipids surrounding the reaction center, 3 PGs (PG664, PG694, and PG702) were located around the Q_A binding site, whereas 2 MGDGs (MGDG729 and MGDG751), 3 DGDGs (DGDG657, DGDG660, and DGDG661), 1 SQDG (SQDG659) and 1 PG (PG714) were present at the interface of D1 and CP43. PG molecules in the vicinity of the Q_A binding site may participate in electron transport activity (described in section "Roles of PG in photosynthesis"), whereas other lipids at the D1–CP43 interface may be important for the binding of these proteins. The lipid-rich environment in PSII would provide a structural flexibility around the reaction center. PSII damaged during the photochemical reaction is repaired via a multi-step process that includes replacing degraded D1 protein with newly synthesized protein (Komenda et al. 2012; Järvi et al. 2015). Lipids abundantly present in PSII may facilitate efficient turnover of the D1 protein.

Cyt b$_6$/f

The crystal structure of the dimeric Cyt *b$_6$/f* complex, which mediates electron transfer between PSII and PSI by generating an electrochemical proton gradient across the thylakoid membrane, has been determined for *C. reinhardtii* (Stroebel et al. 2003) and the cyanobacteria *Mastigocladus laminosus* (Kurisu et al. 2003) and *Nostoc* sp. PCC 7120 (Baniulis et al. 2009). From these data, eight lipid-binding sites per monomer were defined by the presence of two natural lipids (SQDG and eicosane) and detergents or synthetic lipids used for purification and crystallization (Hasan et al. 2011). In the *C. reinhardtii* Cyt *b$_6$/f* crystal structure, two neutral lipids at the lumenal side were modeled as MGDGs. Mass spectroscopy analysis of Cyt *b$_6$/f* purified from spinach detected PG and DGDG, which were not identified in the crystal structures, as well as SQDG and MGDG, so PG and DGDG may also occupy some lipid binding sites in the complex (Hasan et al. 2011). Lipids in the lumenal side, including an MGDG, may function as anchors that link peripheral transmembrane helixes to the core and stabilize the complex. On the stromal side, SQDG may function as a cross-linker between transmembrane helixes of the Cyt *f* and Rieske [2Fe-2S] iron-sulfur proteins. The SQDG molecule has a mobile head group and disordered fatty acid tails in the crystal and forms an inter-monomer cavity with two lipid-binding sites, which may function as a pathway for quinone/quinol diffusion (Hasan et al. 2011). Because mutations of amino acid residues that interact with the SQDG molecule resulted in disassembly of the Rieske iron-sulfur protein

(De Vitry et al. 2004) and impaired regulation of Cyt *f* synthesis (Choquet et al. 2003), SQDG may have an important role in the formation and organization of the Cyt b_6/f complex.

LHCII

In plants and green algae, LHCs function in collecting and transferring solar energy to the reaction centers and avoiding photodamage from excess light energy. The crystal structure of LHCII, the antenna complex associated with PSII, was resolved in an icosahedral proteoliposome assembly from spinach at 2.7 Å (Liu et al. 2004) and in thin hexagonal plates from pea at 2.5 Å (Standfuss et al. 2005). In the crystal structure of spinach LHCII, 2 DGDG molecules are located at the contact surfaces between adjacent trimers and mediate their interactions in the icosahedral structure. Meanwhile, in pea LHCII, 3 DGDG molecules occupy a hydrophobic cavity at the threefold axis on the lumenal side. An *Arabidopsis thaliana* DGDG-deficient mutant (*dgd1*) showed a decreased level of the LHCII trimer with an increased level of the monomer (Hölzl et al. 2006). Thus, DGDG molecules associated with LHCII may play a role in its trimerization. In both LHCII crystal structures from spinach and pea (Liu et al. 2004; Standfuss et al. 2005), one PG molecule was buried at the monomer–monomer interface with its phosphodiester group coordinating a Chl *a*. Of the two fatty acid chains of the PG molecule, the *trans*-Δ^3-hexadecenoic acid (t16:1), which is exclusively found at the *sn*-2 position of chloroplast PG, penetrates the deep binding pocket of the trimer. The amino acid sequence WYXXXR of LHCII, which contributes to the binding of the PG molecule (Pan et al. 2013), is essential for trimer formation (Hobe et al. 1995), and removal of the N-terminal domain including the WYXXXR motif resulted in loss of PG and complete dissociation of the trimer into monomers (Nussberger et al. 1993). Moreover, phospholipase A_2 hydrolysis of the LHCII-binding PGs disrupted their trimeric form (Nussberger et al. 1993; Kim et al. 2007). Mutant analysis in *C. reinhardtii* supports these data; trimeric LHCII level was decreased in a PG-deficient mutant but was partially recovered by supplementing t16:1-binding PG to the mutant (Dubertret et al. 1994). This evidence demonstrates an essential role of PG in LHCII trimerization.

Roles of MGDG in Photosynthesis

Plants

MGDG in plants is synthesized by MGDG synthase, which transfers galactose from uridine 5′-diphosphate (UDP)-galactose to diacylglycerol (Shimojima et al. 1997). MGDG is not only the main constituent of thylakoid membranes; it is also a substrate for DGDG synthesis. Of the three MGDG synthase genes (*MGD1*, *MGD2* and *MGD3*) in *A. thaliana* (Miège et al. 1999; Awai et al. 2001), *MGD2* and *MGD3* have only a marginal role in thylakoid membrane biogenesis but conditionally contribute to DGDG accumulation under phosphate (Pi)-deficient conditions, presumably to economize Pi in membranes (Kobayashi et al. 2009). MGD1 is the major isoform in photosynthetic tissues (Awai et al. 2001; Kobayashi et al. 2004) and contributes to the bulk of galactolipid synthesis in chloroplasts (Jarvis et al. 2000; Kobayashi et al. 2007).

Identification of an *Arabidopsis mgd1-1* mutant, which carries a T-DNA insertion in the *MGD1* promoter region, allowed for detailed analysis of the role of MGDG in photosynthesis (Jarvis et al. 2000). The *mgd1-1* mutant showed 75 % reduced *MGD1* expression and MGDG synthase activity as compared with the wild type, which resulted in 42 % reduced MGDG content. The reduced *MGD1* expression caused a 50 % deficiency in total Chl levels per plant and impaired development of thylakoid membranes but did not affect etioplast development in dark-grown seedlings. In the *mgd1-1* mutant, the maximal PSII efficiency (F_V/F_M) was almost the same as in the wild type, so the 42 % reduced MGDG content had no effect on the intrinsic PSII activity (Aronsson et al. 2008). Meanwhile, the partial MGDG deficiency in *mgd1-1* increased the conductivity of thylakoid membranes at high light intensity, which led to a reduced proton motive force and less acidity in the thylakoid lumen. The *mgd1-1* mutant contained less zeaxanthin and more violaxanthin than the wild type, presumably because of decreased pH-dependent conversion from violaxanthin to zeaxanthin by the violaxanthin de-epoxidase (Aronsson et al. 2008). Consequently, the mutant was more susceptible to photoinhibition of PSII with the impaired capacity for thermal dissipation of excess light. In vitro analysis revealed that MGDG and phosphatidylethanolamine, which have small head groups and form the hexagonal lipid structure, solubilize the xanthophylls diadinoxanthin and violaxanthin more efficiently than the bilayer-forming lipids DGDG and phosphatidylcholine (Goss et al. 2005; Yamamoto 2006). Moreover, MGDG mediates binding of violaxanthin de-epoxidase to the thylakoid membrane (Jahns et al. 2009) and strongly enhances the de-epoxidation of LHCII-associated violaxanthin (Schaller et al. 2010). These data indicate an essential role of MGDG in the photoprotective mechanism of photosynthesis by promoting the xanthophyll cycle.

In contrast to the moderate defects with the leaky *mgd1-1* mutation, a knockout *MGD1* mutation (*mgd1-2*) caused severe loss of both MGDG and DGDG, for absence of thylakoid membranes and photosynthetic components under nutrient-sufficient conditions (Kobayashi et al. 2007). However, under P-deficient conditions, the alternative MGD2/MGD3 pathway was activated and induced biosynthesis of thylakoid-like membranes with a large accumulation of DGDGs in *mgd1-2* (Kobayashi et al. 2013). Despite increased levels of photosynthetic membrane proteins in Pi-deficient *mgd1-2* seedlings, photosynthetic electron transport was still fully dysfunctional. The formation of LHC–PS complexes was severely perturbed, with disordered LHC aggregates, which suggests that MGDG is crucial for the proper assembly of these complexes (Kobayashi et al. 2013). This observation is consistent with the addition of MGDG in vitro modifying the disorganized structures of the lipid-depleted LHCII and inducing formation of the ordered macro-arrays of LHCII (Simidjiev et al. 2000; Schaller et al. 2011). Furthermore, MGDG strongly induced dimerization of the monomeric PSII in vitro (Kansy et al. 2014), which indicates the importance of MGDG in organization of LHC–PS complexes. The involvement of MGDG in the interaction between LHC and core complexes was confirmed in inducible *MGD1*-knockdown mutants (*amiR-MGD1*), which express an artificial microRNA targeting *MGD1* under the control of a dexamethasone-inducible promoter (Fujii et al. 2014). Chl fluorescence analysis at 77 K suggested that MGDG deficiency weakens interactions between PS core complexes and LHC. Consistent with the mutant analysis, in vitro assembly analysis revealed that MGDG intensifies the physical interactions between LHCII and PSII core complexes and increases their energy coupling (Zhou et al. 2009).

Although the partial MGDG loss in the *mgd1-1* mutant had no impact on intrinsic PSII activity (Aronsson et al. 2008), stronger reduced MGDG content by *amiR-MGD1* decreased PSII activity in *Arabidopsis* cotyledons (Fujii et al. 2014). Because the reduced PSII activity was correlated with the severity of *MGD1* suppression, MGDG biosynthesis would be closely associated with the PSII activity. Induction analysis of Chl fluorescence revealed that the strong MGDG deficiency by *amiR-MGD1* decreased the electron-accepting capacity of Q_A and impaired electron transfer from excited Chls to downstream within the PSII, which may reduce the efficiency of light utilization in PSII (Fujii et al. 2014). In addition to the defective intrinsic PSII activity, actual PSII efficiency under light was decreased by the *MGD1* suppression via impaired electron transfer downstream of PSII. The tobacco *MGD1* mutant M18, in which MGDG abundance was decreased by 53 % as compared with the wild type, showed blocked intersystem electron transport with decreased Cyt b_6/f complex levels (Wu et al. 2013). MGDG is structurally involved in the Cyt b_6/f complex (Hasan et al. 2011) and specifically interacts with the complex (Georgiev et al. 2012), so it may be required for the structure or activity of the Cyt b_6/f complex. In *amiR-MGD1* plants, excess light energy caused by the small electron transport capacity was mainly dissipated in non-regulated forms of heat and fluorescence, which reflects a limitation of photoprotective capacity with the MGDG deficiency (Fujii et al. 2014). The reduced photoprotective capacity may reduce photostability of the PSII apparatus and cause cumulative photodamage under long-term exposure to high light, as was observed in the tobacco M18 mutant (Wu et al. 2013).

Cyanobacteria

Plants synthesize MGDG from UDP-galactose and diacylglycerol in one step, whereas cyanobacteria first synthesize monoglucosyldiacylglycerol (MGlcDG) using UDP-glucose and then epimerize it to MGDG (Awai et al. 2006, 2014). Identification of the epimerase (MgdE) that converts MGlcDG to MGDG in *Synechocystis* sp. PCC 6803 allowed for generation of the *mgdE* mutant, which accumulates MGlcDG instead of MGDG and DGDG (Awai et al. 2014).

Photosynthetic activity of the *mgdE* mutant was comparable to that of the wild type at saturating light intensity but significantly lower at low light conditions because of greater nonphotochemical energy dissipation than in the wild type. Pulse amplitude modulation fluorescence analysis revealed that the *mgdE* mutation partially decreased F_V/F_M and the coefficient of photochemical quenching (qP), which suggests that the electron transport within and downstream of PSII were both partially inhibited in the *mgdE* mutant (Awai et al. 2014). MGlcDG can partially complement loss of galactolipids in *Synechocystis* sp. PCC 6803, and thus galactose head groups are not a prerequisite for oxygenic photosynthesis.

Leng et al. (2008) examined the effects of lipase treatment on the structure and function of a PSII dimer isolated from *T. vulcanus*. Lipase treatment degraded MGDG from ten to five molecules per PSII monomer without changing the content of other lipid classes. Despite the degradation of half of the total MGDG in PSII, oxygen-evolving activity was reduced only by 16 % as compared with the untreated control, so MGDG molecules digested by lipase have no crucial roles in PSII activity. The remaining half of the MGDG molecules resistant to lipase treatment may be deeply buried inside PSII and involved in maintaining the structure and function of PSII.

Roles of DGDG in Photosynthesis

Plants

DGDG biosynthesis in plants is predominantly catalyzed by DGDG synthase, which transfers a second galactose to MGDG from UDP-galactose (Kelly and Dörmann 2002; Kelly et al. 2003). Of two homologous genes (*DGD1* and *DGD2*) encoding DGDG synthase in *A. thaliana*, *DGD1* plays a major role in DGDG biosynthesis, as illustrated by more than 90 % reduced DGDG content in the *dgd1* mutant, which has a point mutation in the *DGD1* gene (Dörmann et al. 1995). The substantial decrease in DGDG content in *dgd1* causes a dwarf phenotype, pale green leaves, altered chloroplast structure with a highly curved thylakoid membrane, and decreased quantum yield of photosynthesis. Although a T-DNA insertional mutant of *DGD2* (*dgd2*) grew normally without significantly reduced DGDG content under optimal conditions, the double *dgd1dgd2* mutant contained only negligible amounts of DGDG and showed more severe growth retardation than *dgd1*, with a reduced ability for photoautotrophic growth (Kelly et al. 2003). Interestingly, transgenic expression of a bacterial glucosyltransferase and consequent accumulation of glucosylgalactosyldiacylglycerol (GlcGalDG) almost completely rescued the growth retardation and disordered chloroplast ultrastructure in *dgd1* and *dgd1dgd2* (Hölzl et al. 2006, 2009). These data suggest that the second galactose residue of DGDG has no essential function unique to its chemical property in the structural organization of the thylakoid membrane and efficient plant growth.

Meanwhile, photosynthetic dysfunction with DGDG deficiency was not fully complemented by the GlcGalDG accumulation, as described later.

The levels of LHCII relative to PSII core proteins (D1 and α-subunit of Cyt b_{559}) and inner antenna complexes (Lhcb4 and Lhcb5) were higher in the *dgd1* mutant than the wild type (Härtel et al. 1997). Consistent with the increased LHCII levels, levels of carotenoids involved in xanthophyll cycle were increased in the mutant (Steffen et al. 2005). In *dgd1*, the PSII/PSI ratio was reduced to 60 %, although oxygen-evolving activity was almost the same as in the wild type (Härtel et al. 1997). Fluorometric analysis of *dgd1* revealed that the properties of the oxygen-evolving complex were modified by the decreased DGDG content, with the acceptor side of PSII only marginally affected (Reifarth et al. 1997). The data suggest that DGDG does not play a key role in establishing the structure of the Q_B binding site but is required for the functional and structural integrity of the oxygen-evolving complex. Further analysis with a laser flash fluorometer demonstrated that the DGDG deficiency increased the probability of the dissipative recombination reaction between $P680^+$ and Q_A^-, presumably because of slower electron donation to PSII from water (Steffen et al. 2005). A comparison of *dgd1* and *dgd1dgd2* mutants revealed a nonlinear relationship between DGDG content and functional effects on PSII (Steffen et al. 2005). The results indicate that the total DGDG pool is highly heterogeneous in its functional relevance for PSII; only a small fraction is closely associated with PSII, whereas the bulk of DGDG has no particular role in PSII.

Complementation analysis of DGDG-deficient mutants with GlcGalDG biosynthesis demonstrated specific roles of DGDG in photosynthesis. In the *dgd1* mutant, the level of LHCII trimers, which would be functionally relevant structures in vivo, was decreased along with reduced photosynthetic efficiency (Hölzl et al. 2006). The reduced photosynthetic efficiency of PSII in *dgd1* and *dgd1dgd2* mutants was partially attributed to increased nonphotochemical quenching by photoinhibition. In particular, *dgd1dgd2* was highly susceptible to high light stress (Hölzl et al. 2009). These defects in photosynthetic components were only partially restored by the accumulation of GlcGalDG, so the second galactose residue of DGDG has a specific function in maintaining efficient photosynthetic machinery (Hölzl et al. 2006, 2009).

DGDG plays an important role in the structure and function of PSI. In vitro protein analysis demonstrated that on treatment with chaotropic salts, the PSI subunits PsaD and PsaE on the stromal side are less stable in *dgd1* than the wild type (Guo et al. 2005). Blue native polyacrylamide gel electrophoresis (BN-PAGE) analysis showed accumulation of a PSI supercomplex lacking LHCI and PsaD in *dgd1* thylakoids. Therefore, DGDG is required for stability of the PSI complex. Ivanov et al. (2006) reported substantially lower levels of PSI core proteins (PsaA and PsaB) along with PsaC, PsaL and PsaH subunits in *dgd1* than the wild type, with no change in levels of PSII-related proteins (PsbA, Lhcb1, Lhcb2, Lhcb3 and Lhcb5). The *dgd1* mutation caused a PSI acceptor-side limitation, which was accompanied by restricted intersystem electron transport, increased reduction rate of the PQ pool, increased capacity for cyclic electron flow around PSI, and decreased capacity for state transitions. These modifications were associated with increased susceptibility of PSI to photoinhibition (Ivanov et al. 2006).

Severe DGDG deficiency in the *dgd1* mutant increased thermal instability of PSII affecting Q_A re-oxidation by Q_B (Reifarth et al. 1997). At high temperature, disassembly of the LHCII-PSII – containing macrodomain, degradation of PSI complexes, excitation energy quenching, and impaired energization of thylakoid membranes with increased permeability were significantly greater in *dgd1* than the wild type (Krumova et al. 2010). Thus, DGDG plays a crucial role in thermostability of photosynthetic complexes and the thylakoid membrane, which is consistent with increased DGDG level and DGDG/MGDG ratio being associated with the ability of plants to acquire thermotolerance (Chen et al. 2006).

Cyanobacteria

In cyanobacteria, the *dgdA* gene encodes a DGDG synthase, which presumably transfers a second galactose to MGDG but not MGlcDG (Awai et al. 2007; Sakurai et al. 2007b). Although a *dgdA* mutant of *Synechocystis* sp. PCC 6803 failed to accumulate DGDG, its growth was not greatly affected under standard conditions, so DGDG is not essential for normal growth of this cyanobacterium (Awai et al. 2007; Sakurai et al. 2007b). However, in the mutant, in vitro PSII activity at the donor side was strongly inhibited, along with a slight decrease in net photosynthetic activity from H_2O to CO_2 (Sakurai et al. 2007b), which is consistent with the important role of DGDG in the oxygen-evolving complex in *A. thaliana*. In the *dgdA* mutant, the extrinsic PSII proteins PsbU, PsbV, and PsbO, which stabilize the oxygen-evolving complex, were dissociated in PSII purified from the mutant. Comparison of *dgdA* with mutants for *psbO*, *psbV*, or *psbU* suggested that PsbU but not psbO and psbV is dissociated from PSII in vivo on DGDG deficiency (Sakurai et al. 2007b). Thus, DGDG may mediate PsbU binding to PSII, which subsequently stabilizes the association of PsbO and PsbV with PSII and increases the integrity of the oxygen-evolving complex.

Cell growth of *dgdA* was impaired under high temperature (38 °C) (Mizusawa et al. 2009a), high light and $CaCl_2$-limited conditions (Mizusawa et al. 2009b). As observed in mutants lacking the extrinsic PSII proteins (Clarke and Eaton-Rye 1999; Kimura et al. 2002; Roose et al. 2007), sensitivity of oxygen-evolving activity to heat and hydroxylamine treatment was increased in the *dgdA* mutant (Sakurai et al. 2007a; Mizusawa et al. 2009a). In these mutants, deficiency of extrinsic proteins could lead to an unprotected manganese cluster, so oxygen-evolving activity may be easily inactivated by heat (Nishiyama et al. 1994, 1997; Shen et al. 1995) or hydroxylamine treatment (Ghanotakis et al. 1984). Furthermore, the *dgdA* mutant was susceptible to photoinhibition especially at elevated growth temperature (Mizusawa et al. 2009a, b). Although both photodamage and repair processes of photosynthesis were affected in *dgdA*, the repair process was more severely impaired. In a recent photoinhibition model, photodamage was suggested to proceed in two steps (Tyystjärvi 2008; Murata et al. 2012). First, the manganese cluster of the oxygen-evolving complex is disrupted with release of manganese ions by light. Then, the positive charge is overaccumulated on $P680^+$ because of the lack of electron donors, which causes irreversible damage to the D1 protein at the PSII reaction center. In *dgdA*, manganese ions may be easily released from the PsbU-dissociated oxygen-evolving complex particularly at high temperature, which leads to irreversible photodamage to P680. This finding is consistent with a report that the photodamage proceeds with high efficiency in the PsbU-lacking mutant (Inoue-Kashino et al. 2005). Although only the photodamage process was affected in the PsbU-lacking mutant, the repair process was also affected in *dgdA*. Therefore, DGDG may be required for multiple processes of the photoprotective machinery other than the binding of PsbU protein to the oxygen-evolving complex.

The repair process of inactivated PSII includes monomerization of dimers, dissociation of the core antenna CP43 protein and extrinsic proteins from the reaction center, replacement of damaged D1 with newly synthesized D1 precursor, re-association of CP43 and extrinsic proteins, D1 protein maturation, assembly of the manganese cluster, and PSII dimerization (Nixon et al. 2010; Komenda et al. 2012). Biochemical characterization of monomeric PSII core complexes isolated from *T. elongatus* revealed that Psb27 was associated with a monomeric PSII intermediate that lacks PsbO, PsbU, and PsbV proteins and oxygen-evolving activity (Nowaczyk et al. 2006). Another extrinsic protein, Psb28, is mainly associated with the monomeric PSII lacking CP43 (Sakata et al. 2013). Interestingly, the *Synechocystis dgdA* mutant showed marked accumulation of an inactive PSII monomer that binds Psb27 and Psb28 (Sakurai et al. 2007b). Moreover, BN-PAGE analysis revealed a notable accumulation of the monomeric PSII intermediate lacking CP43 in *dgdA*, presumably because of inhibited association of CP43 with the monomeric PSII intermediate during the repair process (Sakata et al. 2013). Thus, DGDG may mediate reassembly of the active PSII dimer from monomer intermediates during the repair process, which is crucial for the recovery of PSII from photoinhibition.

Crystallography analysis of the dimeric PSII complex revealed that the polar head group of all DGDG molecules faced the lumenal side, with 3 DGDG molecules located between D1 and CP43 subunits (Umena et al. 2011). None of the DGDG molecules directly interact with the extrinsic proteins or the manganese cluster. In the *dgdA* mutant, complete loss of DGDG (~3.5 molecules per monomer) from PSII complexes was compensated by an equivalent increase in MGDG content, with PG and SQDG contents unchanged, which suggests that all DGDG molecules integrated in the wild-type PSII may be replaced by MGDG in *dgdA* (Sakurai et al. 2007b). In *dgdA*, the absence of one galactose moiety from each DGDG molecule may inhibit interactions among the PSII proteins, especially between CP43 and D1, and cause a conformational change on the lumenal side of PSII.

Roles of PG in Photosynthesis

Plants

Roles of PG in plant photosynthesis were initially studied in phospholipase-treated thylakoid membranes in which PG was specifically degraded. Jordan et al. (1983) showed that phospholipase A_2 treatment eliminated approximately 70 % of the original PG from thylakoid membranes isolated from *Pisum sativum* (pea), which led to dysfunctional electron transport in PSII without any significant effect on PSI activity. Similarly, Droppa et al. (1995) reported that degradation of approximately half of the original PG in pea thylakoid membranes by phospholipase C abolished electron transport in PSII. Moreover, the PSII dimer prepared from spinach was dissociated to monomers by phospholipase A_2 treatment, whereas monomers were reversibly converted to dimers by the addition of PG (Kruse et al. 2000). In addition, LHCII trimers in pea thylakoid membranes were dissociated into monomers by phospholipase A_2 treatment (Nussberger et al. 1993). These data were confirmed in thylakoid membranes from *A. thaliana*; phospholipase A_2 treatment inhibited the PSII electron transport at both donor and acceptor sides and disassembled the PSII–LHCII complexes into PSII and LHCII monomers (Kim et al. 2007). These data suggest a crucial role of PG in electron transport and structural organization of the plant PSII complex.

In addition to the biochemical approach, molecular genetics approaches have revealed roles of PG in plant photosynthesis. In the *C. reinhardtii* mutants *mf1* and *mf2*, complete lack of t16:1-binding PG, which resulted in a 60–70 % reduction in overall PG content, abolished PSII activity with decreasing trimeric LHCII levels and grana stacking of thylakoid membranes (Dubertret et al. 1994; El Maanni et al. 1998). These data suggest a requirement of t16:1-binding PG for LHCII trimerization. However, the t16:1-binding species itself may not be essential for the role of PG in LHCII trimerization; an *Arabidopsis* mutant that lacks a desaturase required for t16:1 synthesis showed no significant impairment of trimeric LHCII formation or energy transfer from LHC antenna to reaction centers (McCourt et al. 1985). Because PG is required for the LHCII trimerization (see section "LHCII"), the substantial decrease in overall PG content in *Chlamydomonas mf1* and *mf2* mutants may impair trimeric LHCII formation.

In *A. thaliana*, phosphatidylglycerophosphate (PGP) synthase1 (PGP1) is essential for PG biosynthesis in chloroplasts (Hagio et al. 2002; Babiychuk et al. 2003). The PG content in leaves of the *pgp1-1* mutant, with a point mutation in *PGP1*, was decreased by 30 % from the wild-type level, with an 80 % reduction in plastidic PGP synthase activity (Xu et al. 2002). The mutant had pale green leaves and a slightly reduced capacity for photosynthesis. Hagio et al. (2002) and Babiychuk et al. (2003) isolated null mutants for *PGP1*, in which the level of PG was decreased by ~80 % as compared with the wild type. These mutants had pale yellow-green leaves with severely undeveloped chloroplasts and required sucrose supplementation for growth. In-depth analysis of one of the null *pgp1* mutants (*pgp1-2*) revealed that PG deficiency in chloroplasts strongly perturbed the formation of PS–LHC complexes and photosynthetic electron transport (Kobayashi et al. 2015). Interestingly, Pi limitation activated glycolipid biosynthesis and increased the proportion of glycolipids (MGDG, DGDG and SQDG) with a further decrease in PG content in *pgp1-2*. Although the thylakoid membrane was developed along with accumulation of Chls and photosynthetic proteins in the Pi-limited mutant leaves, PSII photochemical reaction was completely abolished (Kobayashi et al. 2015). The data suggest that glycolipids cannot compensate the function of PG in photosynthesis, although they can form thylakoid membrane networks without PG.

Cyanobacteria

Among the genes for the three enzymes specifically involved in the biosynthesis of PG in cyanobacteria, the *cdsA* gene for CDP-DG synthase (Sato et al. 2000) and the *pgsA* gene for PGP synthase (Hagio et al. 2000) have been identified in *Synechocystis* sp. PCC 6803. The gene for PGP phosphatase, which catalyzes the last step of PG biosynthesis, has not yet been identified. Identification of these genes has allowed targeted mutagenesis to generate mutants of *Synechocystis* sp. PCC 6803 that cannot synthesize PG. The *cdsA* (Sato et al. 2000) and *pgsA* (Hagio et al. 2000) mutants were both unable to synthesize PG and required an exogenous supply of PG for growth under photoautotrophic conditions. Other phospholipids, phosphatidylcholine, phosphatidylethanolamine, phosphatidylserine, phosphatidic acid and cardiolipin, which except for phosphatidic acid are not present in cyanobacterial cells, failed to support the growth of the mutant cells (Hagio et al. 2000). Therefore, PG can be taken up by *Synechocystis* sp. PCC 6803 cells from the growth medium and plays an indispensable function for their growth.

With the mutants of *Synechocystis* sp. PCC 6803, the role of PG in photosynthesis was comprehensively revealed (Sato et al. 2000, 2004; Hagio et al. 2000; Gombos et al. 2002; Sakurai et al. 2003, 2007a; Domonkos et al. 2004). The content of PG in the mutant cells decreased after transfer of cells from PG-replete to -depleted media. The photosynthetic activity of the *pgsA* and *cdsA* mutants decreased markedly with a concomitant decrease of PG content in thylakoid membranes after short-term PG deprivation (3–7 days), whereas the decreased activity was fully recovered to the original level after the re-addition of PG to the growth medium. The impaired photosynthetic activity by the short-term PG deprivation was attributed to the decrease of PSII but not PSI activity, which suggests that PG plays an important role in PSII (Sato et al. 2000; Hagio et al. 2000). A decrease in PSI activity was observed only after a longer period of PG deprivation (>2 weeks) and further longer deprivation of PG induced cell death (Domonkos et al. 2004).

PG plays a crucial role in both acceptor- and donor-side activities of PSII. Measurements of fluorescent yield of Chl indicated that electron transport from Q_A to Q_B was inhibited in *pgsA* mutant cells after PG deprivation, which resulted in the accumulation of the reduced form of Q_A (Q_A^-) (Gombos et al. 2002).

Thermoluminescence measurements confirmed the accumulation of Q_A^- with PG deprivation in the mutant. These results are consistent with the observation that Q_B-mediated electron transport was inhibited on digestion of PG by treating the PSII complex with phospholipase A_2 (Sakurai et al. 2006; Leng et al. 2008). The effect of PG deprivation in the donor side of PSII was revealed in the purified PSII complex; PSII from PG-depleted *pgsA* mutant cells sustained only ~50 % of the oxygen-evolving activity as compared with the wild-type PSII (Sakurai et al. 2007a). Extrinsic proteins PsbO, PsbV, and PsbU, which are essential for the stabilization of the manganese cluster and maintain oxygen-evolving activity, were dissociated from the PSII core purified from the PG-depleted *pgsA* mutant, whereas the released PsbO re-bound to PSII when PG was added back to the PG-depleted mutant cells, even in the presence of lincomycin, which inhibits de novo protein synthesis (Sakurai et al. 2007a). As observed in Δ*psbO*, Δ*psbV*, and Δ*psbU* mutant cells, PG-depleted *pgsA* mutant cells showed decreased oxygen-evolving activity with heat treatment (Sakurai et al. 2007a). This property of the *pgsA* mutant resembles that of the *dgdA* mutant, which is unable to synthesize DGDG (Sakurai et al. 2007b). However, *psgA* but not *dgdA* cells were susceptible to dark inactivation of oxygen-evolving activity in addition to heat treatment. Because the inactivation of oxygen-evolving activity under darkness indicates the dissociation of PsbO and/or PsbV in vivo (Burnap et al. 1996; Shen et al. 1998), PSII of the PG-depleted mutant cells may be unable to functionally bind extrinsic proteins, possibly PsbO and/or PsbV, even in vivo. By contrast, in *dgdA*, only PsbU protein appears to be dissociated from PSII in vivo (Sakurai et al. 2007b), which leads to sensitivity to heat treatment but not dark inactivation of oxygen-evolving activity.

In PG-depleted *pgsA* mutant cells, the dimerization of the PSII core monomers was impaired, which implies an involvement of PG in dimerization of PSII (Sakurai et al. 2003). Dimeric formation of PSII was also severely impaired in the *ΔpsbO* mutant and partially inhibited in the *ΔpsbV* and *ΔpsbU* mutants (Sakurai et al. 2007a). Therefore, dissociation of extrinsic proteins from PG-depleted PSII may lead to monomerization of the PSII complex, although a direct involvement of PG in the PSII dimerization cannot be excluded.

Lipid analysis revealed that the PSII dimer from PG-depleted *pgsA* mutant cells contained 3 PG molecules per monomer, whereas that from wild-type cells had ~6 PG molecules (Sakurai et al. 2006, 2007a). Thus, the 3 PG molecules lost from the mutant PSII complex may play an important role in electron transport from Q_A to Q_B and assembly of extrinsic proteins in the oxygen-evolving complex.

The structure of the PSII dimer complex from *Thermosynechococcus elongatus* was analyzed by X-ray crystallography at 2.9-Å resolution (Guskov et al. 2009): 25 lipid molecules (11 MGDGs, 7 DGDGs, 5 SQDGs, 2 PGs) per monomer were bound to the PSII dimer (Table 2.1). Recently, a more precise PSII dimer crystal structure, determined in *T. vulcanus* at 1.9-Å resolution, contained 20 lipid molecules (6 MGDGs, 5 DGDGs, 4 SQDGs, 5 PGs) per monomer (Umena et al. 2011) (Table 2.1, Fig. 2.2b). The lipid composition in the crystal structure at 1.9-Å resolution was similar to that obtained by biochemical analysis of PSII complex purified from the same cyanobacterial strain (Sakurai et al. 2006). In the crystal structure,

3 PG molecules (PG664, PG694, PG702) are located near the Q_A binding site, in the vicinity of PsbT, PsbL and PsbM, and surrounded by α-helixes of D2 protein at the monomer–monomer interface (Fig. 2.2c). The other PG molecules, PG772 and PG714, are located near the Q_B binding site and the region between CP43 and the D1/D2 heterodimer, respectively (Fig. 2.2d). The head groups of all PG molecules face the cytoplasmic side. Therefore, loss of the former 3 PG molecules are expected to cause dysfunction of Q_A, whereas that of the latter 2 PG molecules would induce impaired function of Q_B.

Itoh et al. (2012) speculated that the inhibition of electron transfer from Q_A to Q_B observed in the *pgsA* mutant is caused by dissociation of PG772 because this PG molecule is close to the Q_B binding site. They assumed that PG772 is a low-affinity PG and is dissociated easily from PSII; thereby, a loss of the PG772 at the Q_B-binding site inhibits electron transfer from Q_A to Q_B and makes Q_B function sensitive to *p*-benzoquinone. However, we recently performed site-directed mutagenesis of amino-acid residues of PsbE, Thr5 and Ser11, interacting with PG772, and found that the mutations of PsbE do not affect the electron transfer from Q_A to Q_B (Kaichiro Endo, Naoki Mizusawa, Jian-Ren Shen, Hsiu-An Chu, Koichi Kobayashi and Hajime Wada, unpublished data). We also mutagenized amino-acid residues of D1, Ser232 and Asp234, interacting with PG664 and PG694, and found that the mutations of D1 affected the redox potential of Q_B but not Q_A, and approximately one PG molecule, PG664 or PG694, was lost from the PSII complex of the mutants (Endo et al. 2015). These PG molecules are located near the Q_A binding site, and therefore loss of any of the PG molecules would likely affect the function of Q_A. Moreover, as described above, these PG molecules are located near PsbT, PsbL, and PsbM and are surrounded by α-helixes of D2 protein at the monomer–monomer interface. Therefore, dissociation of these PG molecules is expected to cause monomerization of PSII complexes along with release of the small subunit proteins. However, the site-directed mutations mainly affected the function of Q_B but not Q_A and dimerization of PSII. Therefore, even loss of PG close to the Q_A binding site affects the structure around the Q_B binding site and the function of Q_B. Modifications of the S232 and N234 residues are assumed to induce a small change in the structure of the Q_A site, located on the D2 protein, and a larger effect on the Q_B site, located farther away on the same D1 protein, but with a more flexible structure.

From the results obtained with the *pgsA* mutant of *Synechocystis* sp. PCC 6803, we suggest that some PG molecules are involved in the binding of extrinsic proteins in PSII. However, in the crystal structure of PSII at 1.9-Å resolution (Umena et al. 2011), all PG molecules face the cytoplasmic side, with no direct interaction of PG molecules with extrinsic proteins. Thus, loss of PG molecules likely induces conformational changes in the lumenal sides of proteins such as D1 and CP43 that interact with the extrinsic proteins and cause the dissociation of extrinsic proteins. Nevertheless, some PG molecules within the crystal structure that directly interact with extrinsic proteins may remain to be identified.

In addition to PSII, PSI was affected after a longer period of PG deprivation (>2 weeks) in the *pgsA* mutant (Domonkos et al. 2004); decreased PSI activity and accumulation of PSI monomer were observed. Jordan et al. (2001) analyzed the

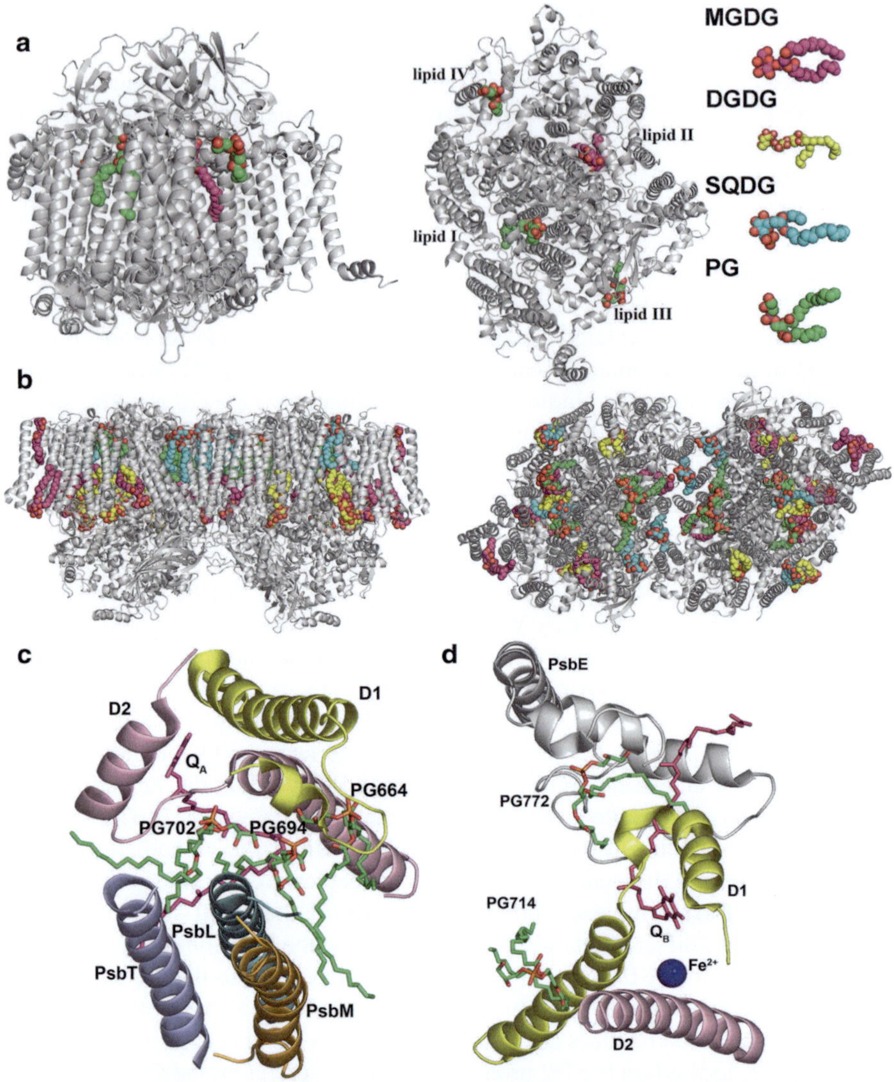

Fig. 2.2 Lipids in photosystem I (PSI) and PSII complexes. (**a**) Location of lipid molecules in PSI trimer complex from *Thermosynechococcus elongatus* (Jordan et al. 2001). Only the monomer structure in the PSI trimer is shown in the figures. *Left figure*, side view; *right figure*, top view. Monogalactosyldiacylglycerol (MGDG) (*purple spheres*), digalactosyldiacylglycerol (DGDG) (*yellow spheres*), sulfoquinovosyldiacylglycerol (SQDG) (*aqua spheres*), phosphatidylglycerol (PG) (*green spheres*). The polar head groups of all lipid molecules (four lipid molecules; one MGDG and three PGs) face the cytoplasmic side. Oxygen atoms in each lipid molecule are represented by *orange spheres*. (**b**) Location of lipid molecules in the PSII dimer complex from *Thermosynechococcus vulcanus* (Umena et al. 2011). Lipid molecules (20 lipid molecules; 6 MGDGs, 5 DGDGs, 4 SQDGs, 5 PGs) are as shown in (**a**). *Left figure*, side view; *right figure*, top view. The polar head groups of all SQDG and PG molecules face the cytoplasmic side, whereas those of all MGDG and DGDG molecules except for one MGDG molecule face the lumenal side. (**c**) Location and structure of 3 PG molecules, PG664, PG694, and PG702, located near the Q_A binding site in PSII. In the structure of PG molecules, oxygen, phosphorus, and carbon atoms are shown in *red*, *purple*, and *green*, respectively. (**d**) Location and structure of 2 PG molecules, PG772 and PG714, located near the Q_B binding site in PSII. PG molecules are as shown in (**c**)

structure of the PSI trimer complex from *T. elongatus* by X-ray crystallography at 2.5-Å resolution and identified one molecule of MGDG and three molecules of PG per monomer of the complex (Fig. 2.2a). One (lipid I) of the 3 PG molecules and the MGDG molecule (lipid II) were located symmetrically relative to one another in the complex, which suggests that these two lipid molecules might be specifically important in the formation of the photochemical reaction center of the PSI complex. A second PG molecule (lipid III) was located near the monomer–monomer interface in the trimeric structure of the PSI complex. This PG molecule might participate in the trimerization of the complex. A third PG molecule (lipid IV) was found between the PsaB and PsaX subunits, so it might contribute to the binding of PsaX to the PSI complex. The presence and location of PG molecules in the crystal structure of PSI are consistent with the observation that PSI activity decreased and PSI monomer accumulated after a long period of PG deprivation in the *pgsA* mutant of *Synechocystis* sp. PCC 6803 (Domonkos et al. 2004).

Roles of SQDG in Photosynthesis

Plants

Roles of SQDG in oxygenic photosynthesis were comprehensively analyzed in a *Chlamydomonas* SQDG-deficient mutant, *hf-2*, isolated from a population of UV-induced mutants showing high-Chl fluorescence yields (Sato et al. 1995b). The *hf-2* cells contained extremely curled thylakoid membranes (Sato et al. 1995b) with decreased PSII activity, 30–40 % of the wild-type level (Sato et al. 1995a). The lowered PSII activity in the *hf-2* cells was recovered by SQDG supplementation in growth media (Minoda et al. 2002), which indicates a requirement of SQDG for the PSII function in *C. reinhardtii*. In the *hf-2* mutant, the sensitivity to 3-(3, 4-dichlorophenyl)-1,1-dimethylurea (DCMU), which inhibits electron transport from Q_A to Q_B, was increased as compared with the wild type, with no effect on sensitivity toward atrazine, another inhibitor of the Q_A to Q_B electron transport (Minoda et al. 2002; Sato et al. 2003). Therefore, a limited region around the Q_B binding site, where DCMU but not atrazine can interact, may be modified by SQDG deficiency. In-depth analyses revealed that the reduced PSII activity in the mutant was derived from decreased efficiency of electron donation from water to Y_Z^+ (i.e., redox-active Tyr_{161} of the D1 protein) rather than from retarded electron transport from Q_A to Q_B (Minoda et al. 2003). Because the PSII activity in the *hf-2* mutant was also sensitive to treatment with hydroxylamine (Minoda et al. 2003) or high temperature (Sato et al. 2003), binding of the manganese cluster and extrinsic proteins to PSII may be unstabilized by SQDG deficiency. In contrast to the strong impairment of PSII activity, PSI function was not affected by the *hf-2* mutation, so SQDG has no crucial roles in PSI.

In *A. thaliana*, SQDG deficiency has no notable effect on photosynthesis under optimal growth conditions. Yu et al. (2002) isolated a T-DNA insertional mutant of the *SQD2* gene (*sqd2*), which is responsible for the final step of SQDG biosynthesis, namely the transfer of sulfoquinovose from UDP-sulfoquinovose to diacylglycerol. Although the *sqd2* mutant completely lacked SQDG, it showed no obvious growth defects under nutrient-sufficient conditions. Chl fluorescence measurements revealed that the F_V/F_M value in *sqd2* was similar to that of the wild type, with the actual quantum efficiency of PSII under light only slightly decreased. However, SQDG deficiency strongly affected photosynthesis and growth when chloroplast PG biosynthesis was partially reduced by the *pgp1-1* mutation at the same time (Yu and Benning 2003). Chl content was further decreased in the *sqd2pgp1-1* double mutant as compared with the pale-green *pgp1-1* single mutant. The double mutant showed a strong decrease in actual PSII quantum efficiency with a decreased fraction of oxidized Q_A. Moreover, sensitivity to DCMU treatment was enhanced in the *sqd2pgp1-1* double mutant as compared with the wild type and each mutant alone. These data suggest that SQDG is not essential under normal conditions but is required for photosynthetic electron transport with limited availability of PG. This finding is consistent with increased SQDG content under Pi-limited conditions in parallel with decreased PG content, so the proportion of total anionic lipids remains constant (Essigmann et al. 1998). Unlike in the wild type, in the *sqd2* mutant, the proportion of PG was not decreased by Pi deficiency (Yu and Benning 2003), which demonstrates the existence of a regulatory mechanism that maintains constant levels of total anionic lipids in chloroplasts. Therefore, the main role of SQDG in higher plants may be substitution for PG under Pi-limited conditions to maintain the proper balance of anionic charge in the thylakoid membrane.

The *sqd2* mutant showed more severe growth retardation than the wild type under Pi-limited conditions (Yu et al. 2002; Okazaki et al. 2013). However, this finding may not be attributed to SQDG deficiency, because knockout mutants for SQD1, which catalyze UDP-sulfoquinovose synthesis essential for SQDG biosynthesis, did not show such growth defects under Pi-limited conditions despite complete lack of SQDG (Okazaki et al. 2013). Levels of another anionic glycolipid, glucuronosyldiacylglycerol (GlcADG), were increased, as were SQDG levels in Pi-limited wild-type and *sqd1* seedlings. However, *sqd2* could not accumulate GlcADG under Pi-limited conditions, presumably because SQD2 is required for GlcADG biosynthesis as a glucuronic acid transferase (Okazaki et al. 2013). Loss of GlcADG along with SQDG in *sqd2* may cause strong growth retardation under Pi-limited conditions, although an involvement of GlcADG in photosynthesis remains unknown.

Cyanobacteria

The proportion and importance of SQDG differ considerably among cyanobacterial species. SQDG accounts for up to 60 % of the total lipids in *Prochlorococcus* strains but is absent in *Gloeobacter violaceus* (Van Mooy et al. 2006). SQDG-deficient mutants were generated in *Synechocystis* sp. PCC 6803 (Aoki et al. 2004) and *Synechococcus* sp. PCC 7942 (Güler et al. 1996; Aoki et al. 2004) by disrupting *sqdB* genes encoding a UDP-sulfoquinovose synthase. The *Synechocystis sqdB* mutant, designated SD1, lost the ability to produce SQDG and required supplementation with SQDG for photoautotrophic growth. SQDG deprivation in SD1 cells strongly decreased PSII activity and increased sensitivity to DCMU and atrazine, so SQDG is crucial for the photosynthetic electron transport activity in *Synechocystis* (Aoki et al. 2004). By contrast, two independent studies revealed that SQDG deficiency did not severely affect PSII activity and growth of *Synechococcus* sp. PCC 7942 (Güler et al. 1996; Aoki et al. 2004). SQDG is also used for photosynthetic membranes in an anoxygenic photosynthetic bacterium, *Rhodobacter sphaeroides*. The *Rhodobacter* SQDG-deficient mutant with *sqdB* disruption was not dysfunctional in the photosynthetic electron transport chain, although its growth was inhibited under Pi-limited conditions (Benning et al. 1993). Thus, the requirement of SQDG for photosynthesis is species-dependent among photosynthetic organisms.

Conclusions and Future Perspectives

Recent advances in the identification of genes involved in the biosynthesis of thylakoid lipids in cyanobacteria and higher plants have provided important information on the biosynthesis of thylakoid lipids and regulation and have allowed for reverse genetic approaches by targeted mutagenesis in cyanobacteria and T-DNA insertional mutagenesis in higher plants. The obtained gene-manipulated cyanobacteria and higher plants have been powerful tools to study the functions of thylakoid lipids in photosynthesis. Extensive studies with the mutants have revealed that depletion or reduction of thylakoid lipids, MGDG, DGDG, SQDG, and PG, cause defects in photosynthesis, and that each of these lipids plays important functions in photosynthesis. PG is required for structural integrity of the Q_B binding site in PSII and development of thylakoid membranes in chloroplasts. DGDG and PG are involved in the binding of extrinsic proteins, which stabilize the manganese cluster in PSII. Galactolipids, MGDG and DGDG as well as PG are required for development of thylakoid membranes in chloroplasts. Galactolipids are not essential for photosynthesis and can be substituted with a glucolipid MGlcDG in cyanobacteria. SQDG plays an important role in substituting for PG under Pi-limited conditions to maintain the proper balance of anionic charge in thylakoid membranes. However, the mutants used for studies of the roles of thylakoid lipids in photosynthesis lack or contain a reduced amount of one of four lipid classes, and modification of lipid composition occurs in thylakoid membranes and in other membranes, such as the cytoplasmic membrane in cyanobacterial cells and envelope membranes of chloroplasts in higher plants. Such changes in lipids in genetically manipulated cyanobacterial cells and higher plants affect photosynthesis and also many other processes, inducing a variety of phenotypic alterations. Although changes in phenotype can suggest important roles for individual lipids in photosynthesis, they do not necessarily reveal the specific function of each lipid because changes in the phenotype can be induced indirectly and directly.

Recent X-ray crystallography analyses of PSI, Cyt b_6/f, PSII, and LHCII complexes have provided a great deal of information regarding the structural and functional roles of lipids in photosynthetic complexes. Lipids are essential as the major components of the thylakoid matrix for embedding the complexes and as integral components of the complexes. Although lipid molecules have been identified in the crystal structures of the complexes, each complex likely binds larger numbers of lipid molecules than those identified in the crystal structures. In the crystal structure of the PSII dimer, about ten lipid molecules per monomer remain to be assigned even at 1.9-Å resolution. X-ray crystallography analysis at higher resolution may reveal additional lipid molecules in the photosynthetic complexes. The lipids in the complexes might play important roles in the folding of protein subunits and in the assembly and stabilization of the complexes. Future studies will need to clarify the roles of individual lipid molecules bound to the photosynthetic complexes by site-directed mutagenesis directed toward substitution of the specific amino acid residues interacting with each lipid molecule. Structural studies of the photosynthetic complexes from various photosynthetic organisms are needed. By comparing the conserved and non-conserved lipid binding sites among different photosynthetic organisms, we can understand the universality and diversity of roles of individual lipid molecules in the photosynthetic complexes.

Methods for the analysis of lipids have greatly improved in recent years, and detailed analysis of lipids can now involve very small amounts of materials. Improvements in analytical methods have allowed us to monitor even a tiny change in the lipid composition of thylakoid membranes. However, analytical methods that reveal the distribution and topology of lipid molecules in thylakoid membranes remain rudimentary, and molecular probes that allow for monitoring the location of individual lipid molecules in thylakoid membranes are urgently needed. Further analyses are necessary to fully understand the contribution of lipids to the structures, functions, and assembly of the photosynthetic complexes and development of thylakoid membranes.

Acknowledgments The work performed in the authors' laboratory was supported by Grants-in-Aid for Scientific Research and JST CREST from the Ministry of Education, Culture, Sports, Science, and Technology of Japan.

References

Aoki M, Sato N, Meguro A, Tsuzuki M (2004) Differing involvement of sulfoquinovosyl diacylglycerol in photosystem II in two species of unicellular cyanobacteria. Eur J Biochem 271:685–693. doi:10.1111/j.1432-1033.2003.03970.x

Apostolova EL, Domonkos I, Dobrikova AG et al (2008) Effect of phosphatidylglycerol depletion on the surface electric properties and the fluorescence emission of thylakoid membranes. J Photochem Photobiol B Biol 91:51–57. doi:10.1016/j.jphotobiol.2008.02.002

Aronsson H, Schöttler MA, Kelly AA et al (2008) Monogalactosyldiacylglycerol deficiency in Arabidopsis affects pigment composition in the prolamellar body and impairs thylakoid membrane energization and photoprotection in leaves. Plant Physiol 148:580–592. doi:10.1104/pp.108.123372

Awai K, Maréchal E, Block MA et al (2001) Two types of MGDG synthase genes, found widely in both 16:3 and 18:3 plants, differentially mediate galactolipid syntheses in photosynthetic and nonphotosynthetic tissues in *Arabidopsis thaliana*. Proc Natl Acad Sci U S A 98:10960–10965. doi:10.1073/pnas.181331498

Awai K, Kakimoto T, Awai C et al (2006) Comparative genomic analysis revealed a gene for monoglucosyldiacylglycerol synthase, an enzyme for photosynthetic membrane lipid synthesis. Plant Physiol 141:1120–1127. doi:10.1104/pp. 106.082859.1120

Awai K, Watanabe H, Benning C, Nishida I (2007) Digalactosyldiacylglycerol is required for better photosynthetic growth of *Synechocystis* sp. PCC6803 under phosphate limitation. Plant Cell Physiol 48:1517–1523. doi:10.1093/pcp/pcm134

Awai K, Ohta H, Sato N (2014) Oxygenic photosynthesis without galactolipids. Proc Natl Acad Sci U S A 111:13571–13575. doi:10.1073/pnas.1403708111

Babiychuk E, Müller F, Eubel H et al (2003) Arabidopsis phosphatidylglycerophosphate synthase 1 is essential for chloroplast differentiation, but is dispensable for mitochondrial function. Plant J 33:899–909. doi:10.1046/j.1365-313X.2003.01680.x

Baniulis D, Yamashita E, Whitelegge JP et al (2009) Structure-function, stability, and chemical modification of the cyanobacterial cytochrome b_6f complex from *Nostoc* sp. PCC 7120. J Biol Chem 284:9861–9869. doi:10.1074/jbc.M809196200

Benning C, Beatty JT, Prince RC, Somerville CR (1993) The sulfolipid sulfoquinovosyldiacylglycerol is not required for photosynthetic electron transport in *Rhodobacter sphaeroides* but enhances growth under phosphate limitation. Proc Natl Acad Sci U S A 90:1561–1565

Boekema EJ, Hankamer B, Bald D et al (1995) Supramolecular structure of the photosystem II complex from green plants and cyanobacteria. Proc Natl Acad Sci U S A 92:175–179

Burnap RL, Qian M, Pierce C (1996) The manganese-stabilizing protein of photosystem II modifies the *in vivo* deactivation and photoactivation kinetics of the H_2O oxidation complex in *Synechocystis* sp. PCC6803. Biochemistry 35:874–882. doi:10.1021/bi951964j

Chen J, Burke JJ, Xin Z et al (2006) Characterization of the *Arabidopsis* thermosensitive mutant *atts02* reveals an important role for galactolipids in thermotolerance. Plant Cell Environ 29:1437–1448. doi:10.1111/j.1365-3040.2006.01527.x

Choquet Y, Zito F, Wostrikoff K, Wollman F-A (2003) Cytochrome *f* translation in Chlamydomonas chloroplast is autoregulated by its carboxyl-terminal domain. Plant Cell 15:1443–1454. doi:10.1105/tpc.011692

Clarke SM, Eaton-Rye JJ (1999) Mutation of Phe-363 in the photosystem II protein CP47 impairs photoautotrophic growth, alters the chloride requirement, and prevents photosynthesis in the absence of either PSII-O or PSII-V in *Synechocystis* sp. PCC 6803. Biochemistry 38:2707–2715. doi:10.1021/bi981981j

De Vitry C, Ouyang Y, Finazzi G et al (2004) The chloroplast rieske iron-sulfur protein. At the crossroad of electron transport and signal transduction. J Biol Chem 279:44621–44627. doi:10.1074/jbc.M406955200

Demé B, Cataye C, Block MA et al (2014) Contribution of galactoglycerolipids to the 3-dimensional architecture of thylakoids. FASEB J 28:3373–3383. doi:10.1096/fj.13-247395

Domonkos I, Malec P, Sallai A et al (2004) Phosphatidylglycerol is essential for oligomerization of photosystem I reaction center. Plant Physiol 134:1471–1478. doi:10.1104/pp.103.037754

Dörmann P, Hoffmann-Benning S, Balbo I, Benning C (1995) Isolation and characterization of an Arabidopsis mutant deficient in the thylakoid lipid digalactosyl diacylglycerol. Plant Cell 7:1801–1810. doi:10.1105/tpc.7.11.1801

Dorne A, Joyard J, Douce R (1990) Do thylakoids really contain phosphatidylcholine? Proc Natl Acad Sci U S A 87:71–74

Droppa M, Horváth G, Hideg É, Farkas T (1995) The role of phospholipids in regulating photosynthetic electron transport activities: treatment of thylakoids with phospholipase C. Photosynth Res 46:287–293. doi:10.1007/BF00020442

Dubertret G, Mirshahi A, Mirshahi M et al (1994) Evidence from in vivo manipulations of lipid composition in mutants that the Δ3-*trans*-hexadecenoic acid-containing phosphatidylglycerol is involved in the biogenesis of the light-harvesting chlorophyll *a/b*-protein complex of *Chlamydomonas reinhardtii*. Eur J Biochem 226:473–482. doi:10.1111/j.1432-1033.1994.tb20072.x

El Maanni A, Dubertret G, Delrieu M-J et al (1998) Mutants of Chlamydomonas reinhardtii affected in phosphatidylglycerol metabolism and thylakoid biogenesis. Plant Physiol Biochem 36:609–619. doi:10.1016/S0981-9428(98)80009-0

Endo K, Mizusawa N, Shen J-R et al. (2015) Site-directed mutagenesis of amino acid residues of D1 protein interacting with phosphatidylglycerol affects the function of plastoquinone Q in photosystem II. Photosynth Res 126:385–397. doi:10.1007/s11120-015-0150-9

Essigmann B, Güler S, Narang RA et al (1998) Phosphate availability affects the thylakoid lipid composition and the expression of *SQD1*, a gene required for sulfolipid biosynthesis in *Arabidopsis thaliana*. Proc Natl Acad Sci U S A 95:1950–1955

Fujii S, Kobayashi K, Nakamura Y, Wada H (2014) Inducible knockdown of MONOGALACTOSYLDIACYLGLYCEROL SYNTHASE1 reveals roles of galactolipids in organelle differentiation in Arabidopsis cotyledons. Plant Physiol 166:1436–1449. doi:10.1104/pp. 114.250050

Georgiev GA, Ivanova S, Jordanova A et al (2012) Interaction of monogalactosyldiacylglycerol with cytochrome b_6f complex in surface films. Biochem Biophys Res Commun 419:648–651. doi:10.1016/j.bbrc.2012.02.067

Ghanotakis DF, Topper JN, Youcum CF (1984) Structural organization of the oxidizing side of photosystem II. Exogenous reductants reduce and destroy the Mn-complex in photosystems II membranes depleted of the 17 and 23 kDa polypeptides. Biochim Biophys Acta 767:524–531. doi:10.1016/0005-2728(84)90051-3

Gombos Z, Várkonyi Z, Hagio M et al (2002) Phosphatidylglycerol requirement for the function of electron acceptor plastoquinone Q_B in the photosystem II reaction center. Biochemistry 41:3796–3802. doi:10.1021/bi011884h

Goss R, Lohr M, Latowski D et al (2005) Role of hexagonal structure-forming lipids in diadinoxanthin and violaxanthin solubilization and de-epoxidation. Biochemistry 44:4028–4036. doi:10.1021/bi047464k

Güler S, Seeliger A, Härtel H et al (1996) A null mutant of *Synechococcus* sp. PCC7942 deficient in the sulfolipid sulfoquinovosyl diacylglycerol. J Biol Chem 271:7501–7507

Guo J, Zhang Z, Bi Y et al (2005) Decreased stability of photosystem I in *dgd1* mutant of *Arabidopsis thaliana*. FEBS Lett 579:3619–3624. doi:10.1016/j.febslet.2005.05.049

Guskov A, Kern J, Gabdulkhakov A et al (2009) Cyanobacterial photosystem II at 2.9-Å resolution and the role of quinones, lipids, channels and chloride. Nat Struct Mol Biol 16:334–342. doi:10.1038/nsmb.1559

Hagio M, Gombos Z, Várkonyi Z et al (2000) Direct evidence for requirement of phosphatidylglycerol in photosystem II of photosynthesis. Plant Physiol 124:795–804. doi:10.1104/pp.124.2.795

Hagio M, Sakurai I, Sato S et al (2002) Phosphatidylglycerol is essential for the development of thylakoid membranes in *Arabidopsis thaliana*. Plant Cell Physiol 43:1456–1464. doi:10.1093/pcp/pcf185

Hankamer B, Morris E, Nield J et al (2001) Subunit positioning and transmembrane helix organisation in the core dimer of photosystem II. FEBS Lett 504:142–151. doi:10.1016/S0014-5793(01)02766-1

Härtel H, Lokstein H, Dörmann P et al (1997) Changes in the composition of the photosynthetic apparatus in the galactolipid-deficient *dgd1* mutant of *Arabidopsis thaliana*. Plant Physiol 115:1175–1184. doi:10.1104/pp.115. 3.1175

Hasan SS, Yamashita E, Ryan CM et al (2011) Conservation of lipid functions in cytochrome *bc* complexes. J Mol Biol 414:145–162. doi:10.1016/j.jmb.2011.09.023

Hobe S, Förster R, Klingler J, Paulsen H (1995) N-proximal sequence motif in light-harvesting chlorophyll *a/b*-binding protein is essential for the trimerization of light-harvesting chlorophyll *a/b* complex. Biochemistry 34:10224–10228

Hölzl G, Witt S, Kelly AA et al (2006) Functional differences between galactolipids and glucolipids revealed in photosynthesis of higher plants. Proc Natl Acad Sci U S A 103:7512–7517. doi:10.1073/pnas.0600525103

Hölzl G, Witt S, Gaude N et al (2009) The role of diglycosyl lipids in photosynthesis and membrane lipid homeostasis in Arabidopsis. Plant Physiol 150:1147–1159. doi:10.1104/pp.109.139758

Inoue-Kashino N, Kashino Y, Satoh K et al (2005) PsbU provides a stable architecture for the oxygen-evolving system in cyanobacterial photosystem II. Biochemistry 44:12214–12228. doi:10.1021/bi047539k

Itoh S, Kozuki T, Nishida K et al (2012) Two functional sites of phosphatidylglycerol for regulation of reaction of plastoquinone Q_B in photosystem II. Biochim Biophys Acta 1817:287–297. doi:10.1016/j.bbabio.2011.10.002

Ivanov AG, Hendrickson L, Krol M et al (2006) Digalactosyl-diacylglycerol deficiency impairs the capacity for photosynthetic intersystem electron transport and state transitions in *Arabidopsis thaliana* due to photosystem I acceptor-side limitations. Plant Cell Physiol 47:1146–1157. doi:10.1093/pcp/pcj089

Jahns P, Latowski D, Strzalka K (2009) Mechanism and regulation of the violaxanthin cycle: the role of antenna proteins and membrane lipids. Biochim Biophys Acta 1787:3–14. doi:10.1016/j.bbabio.2008.09.013

Järvi S, Suorsa M, Aro E-M (2015) Photosystem II repair in plant chloroplasts – regulation, assisting proteins and shared components with photosystem II biogenesis. Biochim Biophys Acta 1847(9):900–909. doi:10.1016/j.bbabio.2015.01.006

Jarvis P, Dörmann P, Peto CA et al (2000) Galactolipid deficiency and abnormal chloroplast development in the *Arabidopsis MGD synthase 1* mutant. Proc Natl Acad Sci U S A 97:8175–8179. doi:10.1073/pnas.100132197

Jordan BR, Chow W, Baker AJ (1983) The role of phospholipids in the molecular organisation of pea chloroplast membranes. Effect of phospholipid depletion on photosynthetic activities. Biochim Biophys Acta 725:77–86. doi:10.1016/0005-2728(83)90226-8

Jordan P, Fromme P, Witt HT et al (2001) Three-dimensional structure of cyanobacterial photosystem I at 2.5 Å resolution. Nature 411:909–917. doi:10.1038/35082000

Kansy M, Wilhelm C, Goss R (2014) Influence of thylakoid membrane lipids on the structure and function of the plant photosystem II core complex. Planta 240:781–796. doi:10.1007/s00425-014-2130-2

Kelly AA, Dörmann P (2002) *DGD2*, an *Arabidopsis* gene encoding a UDP-galactose-dependent digalactosyldiacylglycerol synthase is expressed during growth under phosphate-limiting conditions. J Biol Chem 277:1166–1173. doi:10.1074/jbc.M110066200

Kelly AA, Froehlich JE, Dörmann P (2003) Disruption of the two digalactosyldiacylglycerol synthase genes *DGD1* and *DGD2* in Arabidopsis reveals the existence of an additional enzyme of galactolipid synthesis. Plant Cell 15:2694–2706. doi:10.1105/tpc.016675

Kim E-H, Razeghifard R, Anderson JM, Chow WS (2007) Multiple sites of retardation of electron transfer in photosystem II after hydrolysis of phosphatidylglycerol. Photosynth Res 93:149–158. doi:10.1007/s11120-006-9126-0

Kimura A, Eaton-Rye JJ, Morita EH et al (2002) Protection of the oxygen-evolving machinery by the extrinsic proteins of photosystem II is essential for development of cellular thermotolerance in *Synechocystis* sp. PCC 6803. Plant Cell Physiol 43:932–938

Kobayashi K, Awai K, Takamiya K, Ohta H (2004) Arabidopsis type B monogalactosyldiacylglycerol synthase genes are expressed during pollen tube growth and induced by phosphate starvation. Plant Physiol 134:640–648. doi:10.1104/pp. 103.032656

Kobayashi K, Kondo M, Fukuda H et al (2007) Galactolipid synthesis in chloroplast inner envelope is essential for proper thylakoid biogenesis, photosynthesis, and embryogenesis. Proc Natl Acad Sci U S A 104:17216–17221. doi:10.1073/pnas.0704680104

Kobayashi K, Awai K, Nakamura M et al (2009) Type-B monogalactosyldiacylglycerol synthases are involved in phosphate starvation-induced lipid remodeling, and are crucial for low-phosphate adaptation. Plant J 57:322–331. doi:10.1111/j.1365-313X.2008.03692.x

Kobayashi K, Narise T, Sonoike K et al (2013) Role of galactolipid biosynthesis in coordinated development of photosynthetic complexes and thylakoid membranes during chloroplast biogenesis in Arabidopsis. Plant J 73:250–261. doi:10.1111/tpj.12028

Kobayashi K, Fujii S, Sato M et al (2015) Specific role of phosphatidylglycerol and functional overlaps with other thylakoid lipids in Arabidopsis chloroplast biogenesis. Plant Cell Rep 34(4):631–642. doi:10.1007/s00299-014-1719-z

Komenda J, Sobotka R, Nixon PJ (2012) Assembling and maintaining the photosystem II complex in chloroplasts and cyanobacteria. Curr Opin Plant Biol 15:245–251. doi:10.1016/j.pbi.2012.01.017

Krumova SB, Laptenok SP, Kovács L et al (2010) Digalactosyl-diacylglycerol-deficiency lowers the thermal stability of thylakoid membranes. Photosynth Res 105:229–242. doi:10.1007/s11120-010-9581-5

Kruse O, Hankamer B, Konczak C et al (2000) Phosphatidylglycerol is involved in the dimerization of photosystem II. J Biol Chem 275:6509–6514. doi:10.1074/jbc.275.9.6509

Kubota H, Sakurai I, Katayama K et al (2010) Purification and characterization of photosystem I complex from *Synechocystis* sp. PCC 6803 by expressing histidine-tagged subunits. Biochim Biophys Acta 1797:98–105. doi:10.1016/j.bbabio.2009.09.001

Kurisu G, Zhang H, Smith JL, Cramer WA (2003) Structure of the cytochrome b_6f complex of oxygenic photosynthesis: tuning the cavity. Science 302:1009–1014. doi:10.1126/science.1090165

Leng J, Sakurai I, Wada H, Shen J-R (2008) Effects of phospholipase and lipase treatments on photosystem II core dimer from a thermophilic cyanobacterium. Photosynth Res 98:469–478. doi:10.1007/s11120-008-9335-9

Liu Z, Yan H, Wang K et al (2004) Crystal structure of spinach major light-harvesting complex at 2.72 Å resolution. Nature 428:287–292. doi:10.1038/nature02373

Loll B, Kern J, Saenger W et al (2005) Towards complete cofactor arrangement in the 3.0 Å resolution structure of photosystem II. Nature 438:1040–1044. doi:10.1038/nature04224

McCourt P, Browse J, Watson J et al (1985) Analysis of photosynthetic antenna function in a mutant of *Arabidopsis thaliana* (L.) lacking *trans*-hexadecenoic acid. Plant Physiol 78:853–858

Miège C, Maréchal E, Shimojima M et al (1999) Biochemical and topological properties of type A MGDG synthase, a spinach chloroplast envelope enzyme catalyzing the synthesis of both prokaryotic and eukaryotic MGDG. Eur J Biochem 265:990–1001. doi:10.1046/j.1432-1327.1999.00801.x

Minoda A, Sato N, Nozaki H et al (2002) Role of sulfoquinovosyl diacylglycerol for the maintenance of photosystem II in *Chlamydomonas reinhardtii*. Eur J Biochem 269:2353–2358. doi:10.1046/j.1432-1033.2002.02896.x

Minoda A, Sonoike K, Okada K et al (2003) Decrease in the efficiency of the electron donation to tyrosine Z of photosystem II in an SQDG-deficient mutant of *Chlamydomonas*. FEBS Lett 553:109–112. doi:10.1016/S0014-5793(03)00981-5

Mizusawa N, Sakata S, Sakurai I et al (2009a) Involvement of digalactosyldiacylglycerol in cellular thermotolerance in *Synechocystis* sp. PCC 6803. Arch Microbiol 191:595–601. doi:10.1007/s00203-009-0486-7

Mizusawa N, Sakurai I, Sato N, Wada H (2009b) Lack of digalactosyldiacylglycerol increases the sensitivity of *Synechocystis* sp. PCC 6803 to high light stress. FEBS Lett 583:718–722. doi:10.1016/j.febslet.2009.01.021

Murata N, Miyao M, Omata T et al (1984) Stoichiometry of components in the photosynthetic oxygen evolution system of Photosystem II particles prepared with Triton X-100 from spinach chloroplasts. Biochim Biophys Acta 765:363–369. doi:10.1016/0005-2728(84)90177-4

Murata N, Allakhverdiev SI, Nishiyama Y (2012) The mechanism of photoinhibition *in vivo*: re-evaluation of the roles of catalase, α-tocopherol, non-photochemical quenching, and electron transport. Biochim Biophys Acta 1817:1127–1133. doi:10.1016/j.bbabio.2012.02.020

Murphy DJ (1982) The importance of non-planar bilayer regions in photosynthetic membranes and their stabilisation by galactolipids. FEBS Lett 150:19–26. doi:10.1016/0014-5793(82)81297-0

Nanba O, Satoh K (1987) Isolation of a photosystem II reaction center consisting of D-1 and D-2 polypeptides and cytochrome *b-559*. Proc Natl Acad Sci U S A 84:109–112. doi:10.1073/pnas.84.1.109

Nishiyama Y, Hayashi H, Watanabe T, Murata N (1994) Photosynthetic oxygen evolution is stabilized by cytochrome c_{550} against heat inactivation in *Synechococcus* sp. PCC 7002. Plant Physiol 105:1313–1319. doi:10.1104/pp. 105.4.1313

Nishiyama Y, Los DA, Hayashi H, Murata N (1997) Thermal protection of the oxygen-evolving machinery by PsbU, an extrinsic protein of photosystem II, in *Synechococcus* species PCC 7002. Plant Physiol 115:1473–1480. doi:10.1104/pp. 115.4.1473

Nixon PJ, Michoux F, Yu J et al (2010) Recent advances in understanding the assembly and repair of photosystem II. Ann Bot 106:1–16. doi:10.1093/aob/mcq059

Nowaczyk MM, Hebeler R, Schlodder E et al (2006) Psb27, a cyanobacterial lipoprotein, is involved in the repair cycle of photosystem II. Plant Cell 18:3121–3131. doi:10.1105/tpc.106.042671

Nussberger S, Dörr K, Wang DN, Kühlbrandt W (1993) Lipid-protein interactions in crystals of plant light-harvesting complex. J Mol Biol 234:347–356. doi:10.1006/jmbi.1993.1591

Okazaki Y, Otsuki H, Narisawa T et al (2013) A new class of plant lipid is essential for protection against phosphorus depletion. Nat Commun 4:1510. doi:10.1038/ncomms2512

Pan X, Liu Z, Li M, Chang W (2013) Architecture and function of plant light-harvesting complexes II. Curr Opin Struct Biol 23:515–525. doi:10.1016/j.sbi.2013.04.004

Reifarth F, Christen G, Seeliger AG et al (1997) Modification of the water oxidizing complex in leaves of the *dgd1* mutant of *Arabidopsis thaliana* deficient in the galactolipid digalactosyldiacylglycerol. Biochemistry 36:11769–11776. doi:10.1021/bi9709654

Rochaix J-D (2011) Regulation of photosynthetic electron transport. Biochim Biophys Acta 1807:375–383. doi:10.1016/j.bbabio.2010.11.010

Roose JL, Kashino Y, Pakrasi HB (2007) The PsbQ protein defines cyanobacterial photosystem II complexes with highest activity and stability. Proc Natl Acad Sci U S A 104:2548–2553. doi:10.1073/pnas.0609337104

Sakata S, Mizusawa N, Kubota-Kawai H et al (2013) Psb28 is involved in recovery of photosystem II at high temperature in *Synechocystis* sp. PCC 6803. Biochim Biophys Acta 1827:50–59. doi:10.1016/j.bbabio.2012.10.004

Sakurai I, Hagio M, Gombos Z et al (2003) Requirement of phosphatidylglycerol for maintenance of photosynthetic machinery. Plant Physiol 133:1376–1384. doi:10.1104/pp. 103.026955

Sakurai I, Shen J-R, Leng J et al (2006) Lipids in oxygen-evolving photosystem II complexes of cyanobacteria and higher plants. J Biochem 140:201–209. doi:10.1093/jb/mvj141

Sakurai I, Mizusawa N, Ohashi S et al (2007a) Effects of the lack of phosphatidylglycerol on the donor side of photosystem II. Plant Physiol 144:1336–1346. doi:10.1104/pp. 107.098731

Sakurai I, Mizusawa N, Wada H, Sato N (2007b) Digalactosyldiacylglycerol is required for stabilization of the oxygen-evolving complex in photosystem II. Plant Physiol 145:1361–1370. doi:10.1104/pp. 107.106781

Sato N, Sonoike K, Tsuzuki M, Kawaguchi A (1995a) Impaired photosystem II in a mutant of *Chlamydomonas reinhardtii* defective in sulfoquinovosyl diacylglycerol. Eur J Biochem 234:16–23. doi:10.1111/j.1432-1033.1995.016_c.x

Sato N, Tsuzuki M, Matsuda Y et al (1995b) Isolation and characterization of mutants affected in lipid metabolism of *Chlamydomonas reinhardtii*. Eur J Biochem 230:987–993. doi:10.1111/j.1432-1033.1995.0987g.x

Sato N, Hagio M, Wada H, Tsuzuki M (2000) Requirement of phosphatidylglycerol for photosynthetic function in thylakoid membranes. Proc Natl Acad Sci U S A 97:10655–10660

Sato N, Aoki M, Maru Y et al (2003) Involvement of sulfoquinovosyl diacylglycerol in the structural integrity and heat-tolerance of photosystem II. Planta 217:245–251. doi:10.1007/s00425-003-0992-9

Sato N, Suda K, Tsuzuki M (2004) Responsibility of phosphatidylglycerol for biogenesis of the PSI complex. Biochim Biophys Acta 1658:235–243. doi:10.1016/j.bbabio.2004.06.008

Schaller S, Latowski D, Jemioła-Rzemińska M et al (2010) The main thylakoid membrane lipid monogalactosyldiacylglycerol (MGDG) promotes the de-epoxidation of violaxanthin associated with the light-harvesting complex of photosystem II (LHCII). Biochim Biophys Acta 1797:414–424. doi:10.1016/j.bbabio.2009.12.011

Schaller S, Latowski D, Jemioła-Rzemińska M et al (2011) Regulation of LHCII aggregation by different thylakoid membrane lipids. Biochim Biophys Acta 1807:326–335. doi:10.1016/j.bbabio.2010.12.017

Shen J-R, Vermaas W, Inoue Y (1995) The role of cytochrome c-550 as studied through reverse genetics and mutant characterization in Synechocystis sp. PCC 6803. J Biol Chem 270:6901–6907. doi:10.1074/jbc.270.12.6901

Shen J-R, Qian M, Inoue Y, Burnap RL (1998) Functional characterization of *Synechocystis* sp. PCC 6803 Δ*psb*U and Δ*psb*V mutants reveals important roles of cytochrome c-550 in cyanobacterial oxygen evolution. Biochemistry 37:1551–1558. doi:10.1021/bi971676i

Shimojima M, Ohta H, Iwamatsu A et al (1997) Cloning of the gene for monogalactosyldiacylglycerol synthase and its evolutionary origin. Proc Natl Acad Sci 94:333–337

Shipley GG, Green JP, Nichols BW (1973) The phase behavior of monogalactosyl, digalactosyl, and sulphoquinovosyl diglycerides. Biochim Biophys Acta 311:531–544. doi:10.1016/0005-2736(73)90128-4

Simidjiev I, Stoylova S, Amenitsch H et al (2000) Self-assembly of large, ordered lamellae from non-bilayer lipids and integral membrane proteins *in vitro*. Proc Natl Acad Sci U S A 97:1473–1476

Standfuss J, Terwisscha van Scheltinga AC, Lamborghini M, Kühlbrandt W (2005) Mechanisms of photoprotection and nonphotochemical quenching in pea light-harvesting complex at 2.5 Å resolution. EMBO J 24:919–928. doi:10.1038/sj.emboj.7600585

Steffen R, Kelly AA, Huyer J et al (2005) Investigations on the reaction pattern of photosystem II in leaves from *Arabidopsis thaliana* wild type plants and mutants with genetically modified lipid content. Biochemistry 44:3134–3142. doi:10.1021/bi048465f

Stroebel D, Choquet Y, Popot J-L, Picot D (2003) An atypical haem in the cytochrome b_6f complex. Nature 426:413–418. doi:10.1038/nature02155

Tyystjärvi E (2008) Photoinhibition of photosystem II and photodamage of the oxygen evolving manganese cluster. Coord Chem Rev 252:361–376. doi:10.1016/j.ccr.2007.08.021

Umena Y, Kawakami K, Shen J-R, Kamiya N (2011) Crystal structure of oxygen-evolving photosystem II at a resolution of 1.9 Å. Nature 473:55–60. doi:10.1038/nature09913

Van Mooy BAS, Rocap G, Fredricks HF et al (2006) Sulfolipids dramatically decrease phosphorus demand by picocyanobacteria in oligotrophic marine environments. Proc Natl Acad Sci U S A 103:8607–8612. doi:10.1073/pnas.0600540103

Wada H, Murata N (1998) Membrane lipids in cyanobacteria. In: Siegenthaler P, Murata N (eds) Lipids in photosynthesis: structure, function and genetics, vol 6, Advances in photosynthesis and respiration. Springer, Heidelberg, pp 65–81. doi:10.1007/0-306-48087-5_4

Wu W, Ping W, Wu H et al (2013) Monogalactosyldiacylglycerol deficiency in tobacco inhibits the cytochrome b_6f-mediated intersystem electron transport process and affects the photostability of the photosystem II apparatus. Biochim Biophys Acta 1827:709–722. doi:10.1016/j.bbabio.2013.02.013

Xu C, Härtel H, Wada H et al (2002) The *pgp1* mutant locus of Arabidopsis encodes a phosphatidylglycerolphosphate synthase with impaired activity. Plant Physiol 129:594–604. doi:10.1104/pp.002725

Yamamoto HY (2006) Functional roles of the major chloroplast lipids in the violaxanthin cycle. Planta 224:719–724. doi:10.1007/s00425-006-0257-5

Yu B, Benning C (2003) Anionic lipids are required for chloroplast structure and function in *Arabidopsis*. Plant J 36:762–770. doi:10.1046/j.1365-313X.2003.01918.x

Yu B, Xu C, Benning C (2002) *Arabidopsis* disrupted in *SQD2* encoding sulfolipid synthase is impaired in phosphate-limited growth. Proc Natl Acad Sci U S A 99:5732–5737. doi:10.1073/pnas.082696499

Zhou F, Liu S, Hu Z et al (2009) Effect of monogalactosyldiacylglycerol on the interaction between photosystem II core complex and its antenna complexes in liposomes of thylakoid lipids. Photosynth Res 99:185–193. doi:10.1007/s11120-008-9388-9

Chapter 3
DGDG and Glycolipids in Plants and Algae

Barbara Kalisch, Peter Dörmann, and Georg Hölzl

Abstract Photosynthetic organelles in plants and algae are characterized by the high abundance of glycolipids, including the galactolipids mono- and digalactosyldiacylglycerol (MGDG, DGDG) and the sulfolipid sulfoquinovosyldiacylglycerol (SQDG). Glycolipids are crucial to maintain an optimal efficiency of photosynthesis. During phosphate limitation, the amounts of DGDG and SQDG increase in the plastids of plants, and DGDG is exported to extraplastidial membranes to replace phospholipids. Algae often use betaine lipids as surrogate for phospholipids. Glucuronosyldiacylglycerol (GlcADG) is a further glycolipid that accumulates under phosphate deprived conditions. In contrast to plants, a number of eukaryotic algae contain very long chain polyunsaturated fatty acids of 20 or more carbon atoms in their glycolipids. The pathways and genes for galactolipid and sulfolipid synthesis are largely conserved between plants, Chlorophyta, Rhodophyta and algae with complex plastids derived from secondary or tertiary endosymbiosis. However, the relative contribution of the endoplasmic reticulum- and plastid-derived lipid pathways for glycolipid synthesis varies between plants and algae. The genes for glycolipid synthesis encode precursor proteins imported into the photosynthetic organelles. While most eukaryotic algae contain the plant-like galactolipid (MGD1, DGD1) and sulfolipid (SQD1, SQD2) synthases, the red alga *Cyanidioschyzon* harbors a cyanobacterium-type DGDG synthase (DgdA), and the amoeba *Paulinella*, derived from a more recent endosymbiosis event, contains cyanobacterium-type enzymes for MGDG and DGDG synthesis (MgdA, MgdE, DgdA).

Keywords Chlorophyta • Chloroplast • Endosymbiosis • Eukaryotic • Galactolipid • Phosphate deprivation • Prokaryotic • Rhodophyta

---

B. Kalisch • P. Dörmann (✉) • G. Hölzl
Institute of Molecular Physiology and Biotechnology of Plants (IMBIO), University of Bonn, Karlrobert-Kreiten-Straße 13, 53115 Bonn, Germany
e-mail: bkalisch@uni-bonn.de; doermann@uni-bonn.de; georghoelzl@uni-bonn.de

Y. Nakamura, Y. Li-Beisson (eds.), *Lipids in Plant and Algae Development*, Subcellular Biochemistry 86, DOI 10.1007/978-3-319-25979-6_3

Introduction

Photosynthetic membranes of plants, eukaryotic algae and cyanobacteria contain large amounts of the two galactolipids MGDG and DGDG. A further important glycolipid is the negatively charged SQDG. Together with the phospholipid phosphatidylglycerol (PG), they build the matrix for the photosynthetic apparatus in the thylakoids and are integral components in the photosystems and in light-harvesting complex II (Siegenthaler 1998; Jones 2007; Douce and Joyard 1980). In contrast to the highly conserved lipid composition of photosynthetic membranes, the composition of non-photosynthetic, extraplastidial membranes differs between plants and algae. Besides, extraplastidial membranes mainly contain phospholipids, sphingolipids and sterol lipids. Interestingly, lipid backbones derived from extraplastidial lipids oftentimes serve as precursors for the synthesis of plastidial glycolipids.

Extraplastidial membranes of many nonseed plants and algae contain betaine lipids which are phosphate-free glycerolipids structurally related to phosphatidylcholine (PC), e.g. diacylglyceryl-trimethylhomoserine (DGTS), diacylglyceryl-hydroxymethyltrimethyl-β-alanine (DGTA) and diacylglyceryl-carboxyhydroxymethylcholine (DGCC). Table 3.1 presents an overview on the lipid composition of representative species of plants and algae. DGTS is found in many green algae, ferns and mosses, and in different algal species with complex red plastids, which also contain DGTA or DGCC (Eichenberger and Gribi 1997; Kato et al. 1996; Araki et al. 1991a; Sato 1992). Algae with high levels of betaine lipids often have low

Table 3.1 Glycerolipid composition (mol%) of plants and algae

	A. thaliana[a]	*C. reinhardtii*[b]	*Euglena gracilis*[c]	*Gracilaria verrucosa*[d]	*Dictyota dichotoma*[e]	*Pavlova lutheri*[f]
MGDG	48.9	38.2	37.0	23.6	32.5	34.0
DGDG	14.5	14.8	17.3	26.1	15.1	26.5
SQDG	1.6	5.0	3.0	16.7	24.3	14.6
GlcADG	–	–	–	–	–	2.1
DGTS[g]	–	25.8	–	–	–	–
DGTA	–	–	–	–	19.7	7.7
DGCC	–	–	–	–	–	14.0
PG	8.6	7.7	6.7	2.8	2.6	1.0
PC	19.7	–	24.4	28.9	–	–
PE	6.7	6.1	11.6	1.8	–	–
PI	–	2.3	–	–	0.7	–
PS	–	–	–	–	1.2	–

[a]Kelly et al. (2003)
[b]Riekhof et al. (2003)
[c]Kaoua and Laval-Martin (1995)
[d]Araki et al. (1990)
[e]Araki et al. (1991)
[f]Eichenberger and Gribi (1997)
[g]DGTS and acylated SQDG were not separated in this study
– not measured or not detected

amounts or lack PC, suggesting a reciprocal relationship of DGTS/DGTA/DGCC and PC. Betaine lipids are of extraplastidial origin and therefore resemble phospholipids in their fatty acid distribution (Künzler et al. 1997; Sato 1992).

Fatty acids are synthesized in the stroma of plant and algal plastids by a fatty acid synthase of type II (FASII) (Ohlrogge and Browse 1995). The products are mainly 16:0 and 18:0 bound to acyl carrier protein (ACP). Δ9-acyl-ACP desaturase (FAB2, stearoyl-ACP desaturase, SAD) introduces a double bond into 18:0-ACP to form 18:1-ACP (Shanklin and Somerville 1991). The newly synthesized fatty acids are employed for the production of plastidial lipids (prokaryotic pathway), or are exported to the endoplasmatic reticulum (ER) to produce extraplastidial lipids (eukaryotic pathway). Lipids originating from the prokaryotic or eukaryotic pathways can be distinguished by a C16 or C18 fatty acid linked to the *sn*2 position of glycerol, respectively. ER-derived lipids can be re-imported into the plastid to serve as precursors for galactolipids or SQDG with eukaryotic structure. Plants which employ the two pathways for the production of chloroplast lipids are called "16:3-plants", because they contain galactolipids enriched in 16:3 fatty acids, while "18:3"-plants use mainly the eukaryotic pathway for galactolipid synthesis and therefore lack 16:3 fatty acids (Ohlrogge and Browse 1995; Browse et al. 1986). Arabidopsis and spinach (16:3 plants) derive diacylglycerol (DAG) for glycolipid biosynthesis from the prokaryotic and the eukaryotic pathway, while pea or corn (18:3 plants) produce galactolipids only from the eukaryotic pathway. The plastidial pathway was lost independently during evolution in several Angiosperm families (Mongrand et al. 1998).

In algae, further pathways (designated as ω3/ω6 pathways) exist to synthesize very long chain polyunsaturated fatty acids (VLCPUFAs), particularly 20:4 (arachidonic acid), 20:5 (eicosapentaenoic acid) and 22:6 (docosahexaenoic acid) (Goss and Wilhelm 2009). The formation of VLCPUFAs requires the involvement of elongases and desaturases from the ER. However, information about the transport form and import mechanism of these fatty acids into the plastid is scarce. Some algae with complex plastids might also use a dual system of FASII in combination with a cytosolic FASI for fatty acid production (Petroutsos et al. 2014). Therefore, the origin of fatty acids and precursors for glycolipid synthesis is more complicated in these algae as compared to plants. Furthermore, the definition of prokaryotic and eukaryotic structures as outlined above for plants might not be applicable to algae with complex plastids.

Cyanobacteria, which according to the endosymbiont theory are related to the progenitors of chloroplasts, use a different pathway for galactolipid biosynthesis. First, monoglucosyldiacylglycerol (MGlcDG) is formed by action of MgdA. MGlcDG is epimerized to MGDG by MgdE and galactosylated by a cyanobacterial-type DGDG synthase (DgdA) to DGDG (Awai et al. 2006a, 2007, 2014) (Chapter 4).

Glycolipids in Plants

Glycolipids including MGDG, DGDG and SQDG are prevalent in the plant kingdom, representing around 70–85 % of membrane lipids in chloroplasts (Block et al. 1983). As characteristic chloroplast lipids, lower amounts of MGDG, DGDG and

SQDG are also found in non-photosynthetic plastids such as chromoplasts and amyloplasts (Kleinig and Liedvogel 1978; Fishwick and Wright 1980). In general, glycolipids are highly enriched with polyunsaturated fatty acids (PUFAs), in particular hexadecatrienoic acid (16:3) and α-linolenic acid (18:3). Angiosperms contain a characteristic set of acyl groups in each glycolipid (Table 3.1). In Arabidopsis, MGDG contains prokaryotic and eukaryotic molecular species (70 mol% C16, 30 mol% C18 at *sn*2; Table 3.2) (Browse et al. 1986). In contrast, DGDG is primarily derived from the eukaryotic pathway (88 mol% C18, 12 mol% C16 at *sn*2) (Browse et al. 1986). The mechanism leading to the differences in fatty acid composition of the two galactolipids is not understood. SQDG is composed of more saturated fatty acids in comparison to galactolipids, and comprises eukaryotic and prokaryotic species in varying amounts depending on the plant species (Heinz 1977). The main SQDG molecular species in spinach are 18:3/16:0- (~2/3) and 16:0/18:3-SQDG (~1/3) (Murata and Hoshi 1984; Okazaki et al. 2009).

In contrast to Angiosperms, galactolipids from mosses like *Pleurozium schreberi* and *Ceratodon purpureus* (Bryophyta) are highly enriched with VLCPUFAs (C20 – C24), comparable to algae (Aro and Karunen 1979). Other mosses like *Sphagnum fimbriatum* contain only traces of VLCPUFAs bound to glycolipids (Koskimies-Soininen and Nyberg 1991). Ferns like *Polypodium vulgare* or *Pteridium aquilinum* harbor glycolipids with high amounts of 16:0, 16:3 and 18:3 comparable to Angiosperms, whereas VLCPUFAs are only present in negligible amounts (Jamieson and Reid 1975). Galactolipids of conifers like *Pinus sylvestris* and *Picea abies* contain the unusual C18-fatty acids $18{:}3^{\Delta 5,9,12}$ (pinolenic acid) and $18{:}4^{\Delta 5,9,12,15}$ (coniferonic acid), in addition to $18{:}3^{\Delta 9,12,15}$ (α-linolenic acid) (Jamieson and Reid 1972).

Enzymes, Localization and Regulation of Glycolipid Biosynthesis in Plants

In plants, MGDG is synthesized by transfer of galactose onto DAG by MGDG synthases, and DGDG by galactosylation of MGDG by DGDG synthases, with UDP-galactose (UDP-Gal) as sugar donor (Miège et al. 1999; Kelly et al. 2003; Neufeld and Hall 1964). The sugar in MGDG is linked in β-anomeric configuration, while the outermost galactose in DGDG is α-anomeric (Fig. 3.1).

Arabidopsis contains three MGDG synthase isoforms that can be organized into two types (A and B) (Awai et al. 2001). Type A MGDG synthase MGD1 catalyzes the bulk of MGDG synthesis and is highly expressed in photosynthetic tissues and throughout leaf development (Awai et al. 2001; Kobayashi et al. 2004). MGDG production by MGD1 is essential for embryogenesis and for thylakoid assembly and therefore for photosynthesis (Kobayashi et al. 2007; Jarvis et al. 2000; Fujii et al. 2014). The precursor protein sequence contains a cleavable transit peptide guiding MDG1 to the inner chloroplast envelope (Maréchal et al. 2000). Heterologously

Table 3.2 Fatty acid profiles of glycolipids from plants and algae (mol%)

Fatty acid[a]	Higher plants[b]	Mosses[c]	Green algae[d]	Red algae[e]	*Cyanidio-schyzon*[f]	Diatoms[g]	Brown algae[h]	*Pavlova*[i]
MGDG								
14:0	–	0.8	–	0.4	–	1.9	5.0	3.1
16:0	2.1	6.6	1.1	10.5	24.4	4.1	5.5	8.3
16:1	1.4	0.7	0.6	1.4	–	22.1	4.0	8.7
16:2	0.5	–	–	–	–	1.9	–	–
16:3	30.9	2.9	1.0	–	–	25.4	–	–
16:4	–	–	43.7	–	–	3.6	–	–
18:0	0.7	1.3	0.2	0.9	1.9	–	0.5	–
18:1	0.1	2.7	2.2	9.6	1.6	1.8	9.6	–
18:2	2.3	9.7	2.1	0.6	72.1	0.8	11.1	2.5
18:3	61.7	23.6	24.4	–	–	0.7	16.7	7.5
18:4	–	–	24.3	–	–	–	20.3	26.1
20:4	–	34.9	–	36.8	–	2.7	9.9	tr
20:5	–	12.5	tr	36.8	–	33.5	15.9	43.8
22:5	–	–	–	–	–	–	–	–
22:6	–	–	–	–	–	–	–	–
Others	–	1.3	–	1.6	–	1.6	–	–
DGDG								
14:0	–	0.8	–	0.6	0.1	6.1	9.0	4.6
16:0	16.4	8.4	24.8	31.5	28.8	24.4	20.0	19.0
16:1	2.1	0.6	3.4	0.9	–	29.9	20.5	10.5
16:2	0.9	–	–	0.1	–	4.6	–	–

(continued)

Table 3.2 (continued)

Fatty acid[a]	Higher plants[b]	Mosses[c]	Green algae[d]	Red algae[e]	*Cyanidio-schyzon*[f]	Diatoms[g]	Brown algae[h]	*Pavlova*[i]
16:3	1.8	tr	3.6	–	–	1.6	–	–
16:4	–	–	2.6	–	–	–	–	–
18:0	2.3	1.3	0.9	4.1	1.2	2.7	3.3	–
18:1	0.1	4.5	6.6	25.4	2.2	4.3	14.1	–
18:2	4.1	7.2	14.2	1.0	67.7	2.4	10.9	tr
18:3	71.7	40.0	13.8	–	–	3.2	7.0	3.5
18:4	–	–	23.8	–	–	–	3.2	13.3
20:4	–	19.6	–	19.7	–	0.8	2.1	tr
20:5	–	14.6	–	13.2	–	17.3	3.4	48.6
22:5	–	–	–	–	–	–	–	–
22:6	–	–	–	–	–	–	–	0.7
Others	–	2.1	–	2.3	–	3.0	–	–
SQDG								
14:0	–	3.0	–	4.2	0.4	33.9	3.6	34.7
16:0	48.8	30.3	58.7	32.3	71.3	23.3	45.2	44.8
16:1	7.8	tr	1.5	3.4	–	30.7	4.3	9.7
16:2	4.1	–	–	–	–	2.0	–	–
16:3	–	tr	–	–	–	1.2	–	tr
16:4	–	–	–	–	–	–	–	–
18:0	4.9	6.0	1.6	8.7	7.0	–	3.6	tr
18:1	8.3	7.2	15.4	12.5	1.4	3.0	21.9	5.8
18:2	6.1	8.6	2.1	1.2	19.5	0.6	7.7	tr
18:3	20.1	24.3	18.6	–	–	0.2	3.9	tr

18:4	–	–	1.6	–	–	–	0.9	0.6
20:4	–	9.6	–	20.1	–	–	3.0	tr
20:5	–	3.4	–	13.0	–	5.2	1.4	2.6
22:5	–	–	–	–	–	–	–	0.5
22:6	–	–	–	–	–	–	–	tr
Others	–	1.9	–	2.0	0.5	–	–	1.3

GlcADG/*Pavlova*[i]

Fatty acid	14:0	16:0	16:1	18:0	18:1	18:2	18:3	18:4	20:4	20:5	22:5	22:6	Others
mol%	1.1	4.9	tr	3.3	19.2	1.9	tr	0.8	0.5	3.5	44.4	18.9	1.3

[a]The different fatty acids may comprise different isomers, which are not separately listed. *tr* present in trace amounts, –, not measured or not detected
[b]*Arabidopsis thaliana* (Kelly et al. 2003; Härtel et al. 1998)
[c]*Pleurozium schreberi* (Gellerman et al. 1975)
[d]*Ulva fenestrata* (Sanina et al. 2004)
[e]*Anfeltia tobuchiensis* (Sanina et al. 2004)
[f]*Cyanidioschyzon merolae* (Sakurai et al. 2007)
[g]*Phaeotactylum tricornutum* (Arao et al. 1987)
[h]*Laminaria japonica* (Sanina et al. 2004)
[i]*Pavlova lutheri* (Eichenberger and Gribi 1997)

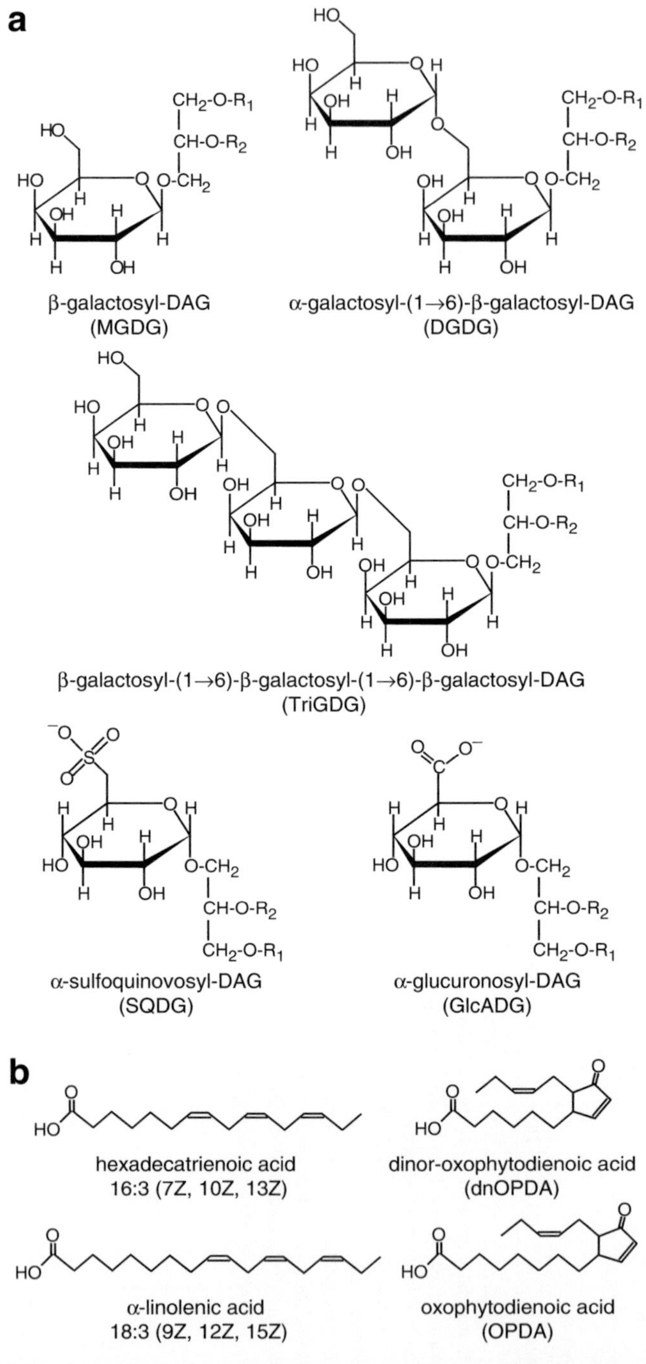

Fig. 3.1 Structures of glycolipids and oxylipins in plants and algae. (**a**) The major glycolipids are MGDG, DGDG and SQDG. Minor glycolipids present only under certain conditions are GlcADG and oligogalactolipids, such as TriGDG. (**b**) Two representatives of the oxylipins, dn-OPDA and OPDA, derived from hexadecatrienoic and α-linolenic acid, respectively, accumulate after wounding in plants. R1, R2, fatty acid residues bound to *sn*1 and *sn*2 of glycerol

expressed MGD1 does not show preference for prokaryotic or eukaryotic DAG, and its activity depends on the anionic lipids PA and PG (Dubots et al. 2010; Awai et al. 2001). Type B MGDG synthases (MGD2 and MGD3) harbor shorter N-terminal transit peptides and localize to the outer chloroplast envelope. The sequences of MGD2 and MGD3 are highly similar, and presumably MGD2 emerged from gene duplication of MGD3 (Awai et al. 2001). MGD2/MGD3 expression is elevated in whole seedlings during germination, whereas in adult plants, expression is restricted to the leaf tip and non-photosynthetic tissues like inflorescences (particularly stigmata and pollen grains) and roots (Awai et al. 2001; Kobayashi et al. 2004). During phosphate starvation, the expression of type B MGDG synthases is increased to provide MGDG as precursor for DGDG synthesis (see below) (Awai et al. 2001; Kobayashi et al. 2004, 2009). Recombinant MGD2 and MGD3 proteins show preference for eukaryotic DAG consistent with the increase of eukaryotic galactolipids under phosphate deprivation (Awai et al. 2001).

In Arabidopsis, DGDG is synthesized by two DGDG synthases, DGD1 and its paralog DGD2. The two proteins differ in their N-terminal portions. DGD1 as the main activity includes an N-terminal domain of 338 amino acids comprising the targeting information for the outer envelope (Dörmann et al. 1999; Froehlich et al. 2001). In photosynthetic tissues, DGD1 is responsible for DGDG production for thylakoid biogenesis (Dörmann et al. 1995; Härtel et al. 1997; Ivanov et al. 2006). DGD1 also plays an important role during phosphate starvation (see below) (Kelly et al. 2003). The second DGDG synthase, DGD2, is synthesized as a precursor including a short cleavable targeting sequence guiding the protein to the outer envelope (Kelly et al. 2003). Under standard growth conditions, DGD2 expression is low, while under phosphate deprivation, the amount of DGD2 transcripts is increased (Kelly and Dörmann 2002). MGD2/MGD3 and DGD2 establish an additional galactolipid biosynthetic pathway active under phosphate limited conditions to provide DGDG as a substitute for extraplastidial phospholipids (Härtel et al. 2000) (see below). Besides DGD1 and DGD2, a further activity exists in plants, which produces oligogalactolipids under freezing conditions, the galactolipid: galactolipid galactosyltransferase (GGGT or sensitive to freezing2, SFR2). SFR2 localizes to the outer envelope (Fourrier et al. 2008; Moellering et al. 2010) where it transfers galactose from MGDG onto galactolipids producing DGDG and oligogalactolipids with β-anomeric bonds. The second product of this reaction is DAG which is acylated to triaycylglycerol (TAG). The formation of oligogalactolipids and TAGs is thought to prevent membrane aggregation under freezing conditions (Moellering and Benning 2011).

SQDG synthesis is catalyzed by UDP-glucose pyrophsophorylase (UGP3), UDP-sulfoquinovose synthase (SQD1) and sulfoquinovose transferase (SQD1) (Shimojima 2011). UGP3 converts glucose-1-phosphate and UTP into UDP-Glucose (UDP-Glc) in the stroma (Okazaki et al. 2009). UDP-Glc and sulfite are employed for UDP-sulfoquinovose (UDP-SQ) synthesis by the stroma-localized SQD1 (Essigmann et al. 1998; Sanda et al. 2001). The third step of SQDG production is the transfer of sulfoquinovose to DAG by SQD2 at the inner envelope (Yu et al. 2002). SQD2 is also involved in the synthesis of GlcADG, a glycolipid with glucuronic acid in the head group (Fig. 3.1) (Okazaki et al. 2013b). Expression of SQD1, SQD2 and UGP3 is strongly increased when phosphate is limited, leading to an increased synthesis of SQDG and GlcADG, which is not detectable under normal growth conditions (Fig. 3.2) (see below).

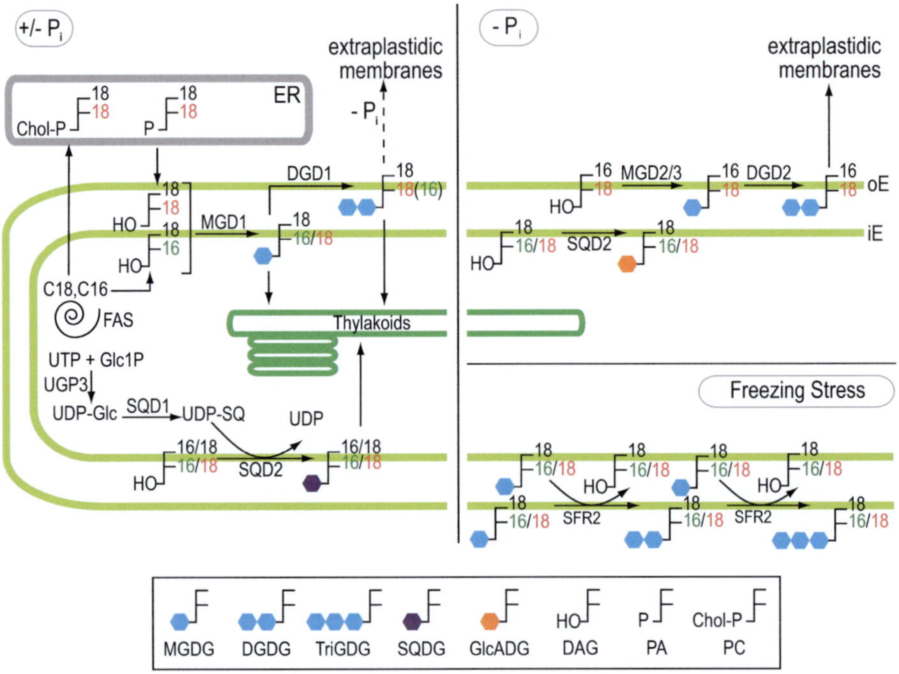

Fig. 3.2 Biosynthesis of glycolipids in Arabidopsis. The bulk of galactolipids present under phosphate replete conditions (+P) is synthesized by MGD1 and DGD1 in the inner and outer envelope, respectively. DAG, the substrate for MGDG synthesis, is formed in the plastid (prokaryotic pathway, C16 at *sn*2, *green*) or derived from a phospholipid precursor formed in the ER (eukaryotic pathway, C18 at *sn*2, *red*). While MGDG is composed of prokaryotic and eukaryotic species, DGDG synthesized by DGD1 mainly consists of eukaryotic species. The pathway for SQDG biosynthesis comprises the production of UDP-Glc by UGP3 in the chloroplast, conversion of UDP-Glc to UDP-SQ by SQD1, and the glycosylation of DAG by SQD2 with UDP-SQ. Phosphate deprivation (-P) leads to the increased synthesis of DGDG after induced expression of DGD1, DGD2 and MGD2, MGD3, and of SQDG after induction of SQD2 and SQD1 expression. The additional DGDG formed accumulates in the thylakoids and is exported to replace extraplastidial phospholipids. A fraction of PG in the plastid is replaced by SQDG. GlcADG is also formed under phosphate deprivation by SQD2. Oligogalactolipids with all sugars in β-anomeric configuration occur only under stress conditions like freezing. Oligogalactolipids are synthesized by SFR2 via glycosylation of galactolipids with MGDG as galactosyl donor

Characteristics and Function of Plant Glycolipids

The three glycolipids MGDG, DGDG, and SQDG in chloroplasts of plants exhibit different biophysical and biochemical characteristics. MGDG with its small head group is self-organized in a reversed hexagonal phase, and thus is bilayer destabilizing, whereas DGDG and the anionic SQDG form lamellar phases (bilayer-forming) (Shipley et al. 1973; Webb and Green 1991). Lipid mixtures containing MGDG, DGDG, PG and SQDG, mimicking the thylakoid lipid composition, self-organize as regular stacks of bilayers that can switch between hexagonal and lamellar phases upon hydration in vitro (Demé et al. 2014). In particular, the DGDG content is critical for membrane stacking due to the formation of hydrogen bonds between adjacent head groups. DGDG confers thermotolerance to plants due to its bilayer-stabilizing properties as demonstrated by the failure of DGDG-deficient *dgd1* mutant plants to adapt to high growth temperatures (Chen et al. 2006). In wild type plants, the MDGD to DGDG ratio is decreased upon heat stress to stabilize plastidial membranes. The opposite effect was observed after decreasing the growth temperature to 5 °C (Hendrickson et al. 2006). At this temperature, *dgd1* plants show a partially complemented phenotype, with an increase of chlorophyll content and photosynthetic efficiency.

Glycolipids play an important role in chloroplast development and morphology, as revealed by studies on Arabidopsis mutants. Decreased galactolipid synthesis capacity affects chloroplast ultrastructure, with less and disordered thylakoid stacks as well as reduced chlorophyll content and photosynthetic efficiency (Jarvis et al. 2000; Kobayashi et al. 2007; Dörmann et al. 1995). Due to their role in photosynthetic membranes, galactolipid biosynthesis is indispensable for embryogenesis and germination (Kobayashi et al. 2007; Fujii et al. 2014) (Chap. 4). During flower development and pollen tube elongation, an increase in the amounts of galactolipids, especially DGDG, was observed (Nakamura et al. 2009) (Chap. 12). The *mgd1* mutant deficient in MGDG synthesis contains only trace amounts of galactolipids and is impaired in photosynthetic ability and photoautotrophic growth (Jarvis et al. 2000; Kobayashi et al. 2007). The depletion of DGDG in the *dgd1* or *dgd1dgd2* mutants leads to strong morphological changes with retarded growth and abnormal thylakoid morphology (Dörmann et al. 1995; Kelly et al. 2003). Interestingly, defects in growth and chloroplast morphology of *dgd1* can be complemented with the bacterial lipid glucosylgalactosyldiacylglycerol (GlcGalDG) after expression of a GlcGalDG synthase from *Chloroflexus aurantiacus* in the *dgd1* or *dgd1dgd2* mutants (Hölzl et al. 2006, 2009). Therefore, GlcGalDG can fulfill the bulk membrane functions of DGDG, but it cannot fully complement the impaired photosynthetic efficiency of PSII, indicating the presence of specific interactions of the sugar headgroup with photosynthetic protein complexes.

The *sqd2* mutant of Arabidopsis inhibited in SQDG synthesis shows normal growth under phosphate replete conditions, presumably because it still contains PG as anionic lipid. However, *sqd2* grows poorly under phosphate deprivation when additional amounts of SQDG are required to replace phospholipids (Yu et al. 2002).

The *pgp1* mutant carrying a leaky mutation in the PGP1 (phosphatidylglycerol-phosphate synthase) gene involved in PG synthesis contains reduced amounts of PG. Growth and photosynthesis of the *pgp1* mutant are affected to a minor extent (Xu et al. 2002). However, *sqd2 pgp1* double mutants are strongly impaired in their photoautotrophic growth and photosynthetic capacity (Yu and Benning 2003). SQDG and PG can be mutually replaced in the single mutants, but not in the double mutant, and therefore, the presence of at least one anionic lipid in the chloroplast is crucial for thylakoid membrane integrity.

Phosphate is an essential macronutrient for plants. Around one third of the organic phosphate is bound to phospholipids (Poirier et al. 1991), thus representing as a potential phosphate reservoir. During phosphate limited growth, plants reorganize their membranes by partially replacing the phospholipids with glycolipids to save phosphate for other metabolic processes. Next to the activity of the main pathway involving MGD1 and DGD1, plants activate an additional pathway involving MGD2/MGD3 and DGD2 for the synthesis of galactolipids (Fig. 3.1) (Awai et al. 2001; Kobayashi et al. 2004; Härtel et al. 2000; Kelly and Dörmann 2002). The MGDG content is not altered in phosphate starved plants, as the additional MGDG produced is employed for the production of DGDG (Kelly et al. 2003). In thylakoids, DGDG increases during phosphate deprivation due to the induced expression of DGD1. The additional DGDG synthesized by the MGD2/MGD3/DGD2 pathway is transported to extraplastidial membranes including the plasma membrane, the tonoplast or mitochondrial membranes, for the replacement of phospholipids (Fig. 3.1) (Härtel et al. 2000; Andersson et al. 2005; Jouhet et al. 2004). In the plasma membrane, this replacement is restricted to the inner leaflet, while in the apoplastic leaflet phospholipids are substituted with sterols presumably to maintain a high degree of lipid order (Tjellström et al. 2010). DGDG also occurs in specialized extraplastidial membranes, e.g. the peribacteroid membrane surrounding the rhizobial bacteria in nodules as shown for *Lotus japonicus* and *Glycine max* to save phosphate for nodule formation (Gaude et al. 2004). Furthermore, PG is partially replaced with SQDG during phosphate limitation (Essigmann et al. 1998; Yu et al. 2002; Siegenthaler 1998). As mentioned above, SQD2 is also involved in the biosynthesis of GlcADG under phosphate deprivation (Okazaki et al. 2013b, 2015).

Different studies suggest the involvement of glycolipids in the response to various abiotic stresses. For example, an elevated expression of MGD1 was observed in Arabidopsis shoots after wounding. Expression of MGDG synthases is upregulated during high salinity treatment as well as submergence and drought in Arabidopsis and rice (Qi et al. 2004; Klecker et al. 2014). The increase in MGD1 transcript abundance might provide the means for increased galactolipid synthesis for membrane repair, or as oxylipin defense precursors (Kobayashi et al. 2009). Oxylipins such as oxophytodienoic acid (OPDA) or dinor-OPDA (dnOPDA) are synthesized from 18:3 to 16:3 by the consecutive action of a 13-lipoxygenase, allene oxide-synthase and a cyclase. The fatty acids remain esterified to the galactolipids during conversion into oxylipins, or are released by lipases prior to the lipoxygenation (Schaller and Stintzi 2009; Nilsson et al. 2012; Bonaventure 2014). In particular, the oxylipin-galactolipids or "arabidopsides", galactolipids with one or two oxylipins esterified to glycerol were detected in Arabidopsis after wounding (Böttcher and Weiler 2007). Free OPDA and dnOPDA serve as precursors for the biosynthesis of other oxylipins including the phytohormone jasmonate (Schaller and Stintzi 2009).

Another glycolipid modification observed as a response to stress is the acylation of the C6 carbon of the galactose of MGDG and DGDG in various plant species. The function of 6-acyl-MGDG or 6-acyl-DGDG is unknown (Vu et al. 2014; Heinz 1967a). Synthesis of 6-acyl-MGDG is mediated by acyl transfer from DGDG to MGDG (Heinz 1967b).

Lipid Trafficking in Plants

The integration of chloroplast lipid synthesis into host cell metabolism requires the transfer of numerous lipid precursors and metabolites between the ER and the envelope membranes, and between the envelopes and the thylakoids. In plants, fatty acid de novo synthesis localizes to the stroma of plastids. C16 and C18 fatty acids entering the eukaryotic pathway have to pass the inner and outer envelope and are transported to the ER (Hurlock et al. 2014). Fatty acid export from the chloroplast is believed to be mediated by FAX1 in the inner envelope (Li et al. 2015). Fatty acids are converted into acyl-CoA esters by long-chain acyl-CoA synthases (Joyard and Stumpf 1981; Roughan and Slack 1977) and imported into the ER presumably via ABCA9, an ATP-binding cassette (ABC) transporter (Kim et al. 2013). At the ER, acyl groups are incorporated into phospholipids and desaturated.

The lipid backbones of a proportion of the eukaryotic phospholipids are re-imported to the chloroplast to provide precursors for glycolipid synthesis. An ABC transporter consisting of three subunits, TGD1, TGD2 and TGD3, in the inner envelope is involved in the transfer of lipid moieties from the ER to the plastid. TGD1 encodes a permease-like protein, TGD2 carries the substrate-binding site, and TGD3 the ATPase (Xu et al. 2003; Awai et al. 2006b; Lu et al. 2007). TGD4 which localizes to the outer envelope is required for the import of eukaryotic lipid precursors to the chloroplast (Xu et al. 2008; Wang et al. 2012). TGD2 and TGD4 bind to PA, but it is not clear whether PA is also transported by these proteins (Awai et al. 2006b; Wang et al. 2012). Because Arabidopsis mutants of *TGD* genes accumulate TriGDG, they were named *tgd1* to *tgd4*. Alternative lipid moieties potentially imported into the chloroplast are PC or lyso-PC (Mongrand et al. 2000; Hurlock et al. 2014; Jessen et al. 2015). Lipid transport is presumably facilitated at contact sites between the ER and the outer chloroplast envelope (Andersson et al. 2007).

It is assumed that the transfer of glycolipids from the envelopes, the site of their biosynthesis, to the thylakoids could be mediated by vesicles. The vesicle-inducing protein in plastids 1 (VIPP1) is required for the formation of cold-induced vesicles at the inner envelope and for envelope integrity, but the in vivo function of this protein remains unclear (Kroll et al. 2001; Zhang et al. 2012). Alternatively, the formation of protrusions derived from the inner envelope into the stroma might be involved in thylakoid membrane formation (Kobayashi et al. 2007). During phosphate deprivation, DGDG is exported from the plastid envelopes to extraplastidial membranes. The mechanism of this transport is unknown, but it was shown that lipid export is not specific, because GlcGalDG produced in Arabidopsis *dgd1* plants expressing the *Chloroflexus* GlcGalDG synthase, is also exported to extraplastidial membranes (Hölzl et al. 2009).

Glycolipids in Eukaryotic Algae

Eukaryotic algae comprise a highly diverse group of photosynthetic organisms. Their evolution started with the uptake of a free-living ancient cyanobacterium by a heterotrophic eukaryotic host cell. The transition of this symbiotic cyanobacterium into the plastidial ancestor according to the endosymbiont theory occurred about 1.2 billion years ago (Parfrey et al. 2011). Primary endosymbiosis can be traced back to a common ancestor (Rodríguez-Ezpeleta et al. 2005) leading to three lineages of Glaucophyta, Chlorophyta (green algae and land plants) and Rhodophyta (red algae). The best evidence for the single origin of primary plastids is the similarity of the protein import machinery, the TOC and TIC systems (translocon of the inner and outer chloroplast membrane) (Kalanon and McFadden 2008). Primary plastids are located in the cytosol of the cell and surrounded by two membranes, both derived from the cyanobacterial endosymbiont. The transition into an organelle involved the transfer of most of the endosymbiont genes to the host nucleus. The protein products of many of these genes carry transit peptides at their N-terminus and are targeted back to the organelle. Indeed, most of the enzymes of galactolipid and sulfolipid synthesis in eukaryotic algae are predicted to carry such N-terminal, plastid-targeting sequences.

Subsequent to primary endosymbiosis, primary eukaryotic algae were incorporated by other eukaryotic host cells in secondary endosymbiosis (Keeling 2013). The secondary endosymbiont was reduced to a photosynthetic organelle, the "complex plastid". This process included the loss and the transfer of many endosymbiont genes to the host nucleus. Complex plastids are surrounded by three or four membranes. The two innermost membranes represent the original plastidial membranes of the primary endosymbiont. A further (second outermost) membrane originates from the plasma membrane of the algal endosymbiont, which is lost in some complex plastids, while the outermost membrane is derived from the phagosomal or ER membrane of the secondary host. The secondary plastids are located within the endomembrane system. Consequently, secondary algae of the red and green lineages established a different plastidial targeting and import system as primary plastids. Nuclear-encoded plastidial proteins are composed of a bipartite leader peptide, targeting the protein first to the endomembrane system and further to the plastid (Keeling 2013). Indeed, MGDG and DGDG synthases of complex plastids often harbor a bipartite targeting sequence implicating their transport to the innermost envelope membrane (Petroutsos et al. 2014). The absence of targeting sequences might localize the enzymes to the outermost membranes of the complex plastid. Secondary endosymbiosis with green and red algae occurred at least three times during evolution. Secondary endosymbiosis involving Glaucophyta is not known (Gould et al. 2008).

Genes involved in plastidial glycolipid metabolism of plants (MGD1, DGD1, SQD1, SQD2) often show high similarity with the orthologs in algae. For example, MGDG synthases from eukaryotic algae and higher plants are monophyletic and can be traced back to a single ancestor (Petroutsos et al. 2014). Furthermore, most eukaryotic algae employ plant-type DGDG synthases for galactolipid synthesis (Riekhof et al. 2003; Sato et al. 1995). Despite of the availability of numerous genome sequences, gene expression data for algae are scarce. Furthermore, the complexity of secondary plastids complicates membrane separation and thus analysis of lipid composition and localization of enzymes (Petroutsos et al. 2014).

Glycolipids in Glaucophyta

The Glaucophyta are a small group of microalgae containing a relict of a peptidoglycan layer between the two plastid envelopes. Their photosynthetic chlorophyll a-containing organelles are termed cyanelles (Keeling 2010; Gould et al. 2008). Glaucophyta contain MGDG, DGDG and SQDG (Leblond et al. 2010). No sequence data are available for the enzymes involved in MGDG synthesis (Botté et al. 2005), and no eukaryotic DGDG synthase orthologs were detected. Contradictory reports exist on the presence or absence of DgdA orthologs, e.g. in *Cyanophora paradoxa* (Botté et al. 2011; Awai et al. 2007). Therefore, the question of the involvement of plant-like or cyanobacterium-type galactolipid synthases in Glaucophyta remains open. Because of the conservation of the SQDG synthesis pathway in plants, eukaryotic algae and cyanobacteria, SQDG biosynthesis in Glaucophyta might be similar to plants and *Chlamydomonas* (see below) with the involvement of orthologs of SQD1 and SQD2.

Glycolipids in Chlorophyta

The photosynthetic organelles in Chlorophyta ("chloroplasts") contain chlorophyll a and b (Keeling 2010). The lipid and fatty acid composition of Chlorophyta is highly similar to plants with large amounts of polyunsaturated C16 and C18 fatty acids (Table 3.2). The green alga *Chlamydomonas* contains the unusual fatty acids $16{:}4^{\Delta 4,7,10,13}$, $18{:}3^{\Delta 5,9,12}$ (pinolenic acid) and $18{:}4^{\Delta 5,9,12,15}$ (coniferonic acid), which are absent from Arabidopsis (Sato et al. 1995). The Δ5 double bonds are introduced by an ω13 desaturase with $18{:}2^{\Delta 9,12}$ or $18{:}3^{\Delta 9,12,15}$ as substrates, while $16{:}2^{\Delta 7,10}$ and $16{:}3^{\Delta 7,10,13}$ are not accepted by this enzyme (Kajikawa et al. 2006). $16{:}4^{\Delta 4,7,10,13}$ is exclusively found in galactolipids, especially in MGDG. The Δ4 double bond is introduced by a Δ4 desaturase specific for $16{:}3^{\Delta 7,10,13}$ bound to MGDG (Zäuner et al. 2012). This Δ4 desaturase presumably has a regulatory role in MGDG synthesis, because over-expression or knock down in *Chlamydomonas* affects the content of MGDG. Another C18 fatty acid with 4 double bonds, often found in Chlorophyta (e.g. *Ulva* sp.), is stearidonic acid ($18{:}4^{\Delta 6,9,12,15}$) (Kumari et al. 2013). VLCPUFAs, like 20:4, 20:5, are in general absent or only found in traces in fresh water green algae, while these two fatty acids are often present in low proportions in marine green algae, such as *Ulva fenestrata* where they are mainly found in DGTS, but not in glycolipids (Thompson 1996). In contrast, considerable amounts of 20:4 are found in membrane lipids and in TAGs of the oleaginous alga *Parietochloris incisa* (Bigogno et al. 2002).

Green algae can be divided into two groups, with the first group producing glycolipids structurally similar to glycolipids in 16:3 plants involving the plastidial and ER pathway, while the second group forms only prokaryotic lipids, exclusively derived from the plastidial pathway. *Chlorella kessleri* (Sato et al. 2003b) belongs to the first group, with a lipid biosynthesis pathway comparable to 16:3 plants. The precursor for eukaryotic galactolipids in *Chlorella* is derived from PC. Most of the green algae (e.g. *Chlamydomonas*, *Dunaliella*) belong to the second group, using only the prokaryotic pathway for glycolipid biosynthesis (Fig. 3.3a). In contrast to plants, prokaryotic lipid backbones can also be found in extraplastidial lipids like DGTS (14 %) and PE (11 %) in *Chlamydomonas*. This curiosity may be explained by a less pronounced fatty acid specificity of the microsomal lysophosphatidic acid acyltransferase.

The characteristic MGDG species in most green algae are 18:3/16:4- and to a less amount 18:3/16:3-MGDG (Cho and Thompson 1987a; Giroud et al. 1988; Sanina et al. 2004), which are produced by sequential desaturation of 18:1/16:0-MGDG as shown for *Chlamydomonas* or *Dunaliella*. DGDG is more saturated, and

Fig. 3.3 Pathways for galactolipid biosynthesis in green and red algae. (**a**) *Dunaliella salina* (green alga) employs only the prokaryotic pathway for glycolipid synthesis (Cho and Thompson 1987b). 18:1/16:0-MGDG is desaturated to 18:3/16:4-MGDG. DGDG synthesis follows two routes. The first rout starts with galactosylation of 18:1/16:0-MGDG, with further desaturation leading to 18:3/16:0-DGDG. The second route starts with 18:1/16:1-MGDG or higher unsaturated MGDG leading to the formation of the respective DGDG molecular species. Further desaturation of DGDG results in production of 18:3/16:3- or 18:3/16:4-DGDG. (**b**) MGDG in *Porphyridium cruentum* (red alga) consists of eukaryotic and prokaryotic molecules, while DGDG has a prokaryotic structure with 20:5/16:0 as the main species (Cohen et al. 1997). Eukaryotic MGDG contains C20 fatty acids at *sn*1 and *sn*2, derived from PC. Prokaryotic MGDG is synthesized via a coupled pathway, with 16:0 derived from the chloroplast, and C20 imported from the ER. 20:4 fatty acids are synthesized by desaturation and elongation of C18 fatty acids at the ER. (**c**) *Cyanidioschyzon merolae* (red alga) lacks a plastidial Δ9-acyl-ACP desaturase. Therefore fatty acid desaturation occurs only at the ER (Sakurai et al. 2007). The precursors for galactolipid synthesis are derived from 18:2/18:2-PC or from a coupled pathway with 18:2 originating from 16:0/18:2-PC and 16:0 from the plastid leading to eukaryotic (18:2/18:2) and prokaryotic (18:2/16:0) galactolipids, respectively. For simplification, the two envelope membranes are presented as one line (*red* or *green*). *Grey line*, ER; MGDS and DGDS, plant-type MGDG and DGDG synthases, respectively; DgdA, cyanobacterial-type DGDG synthase; Des, Desaturase

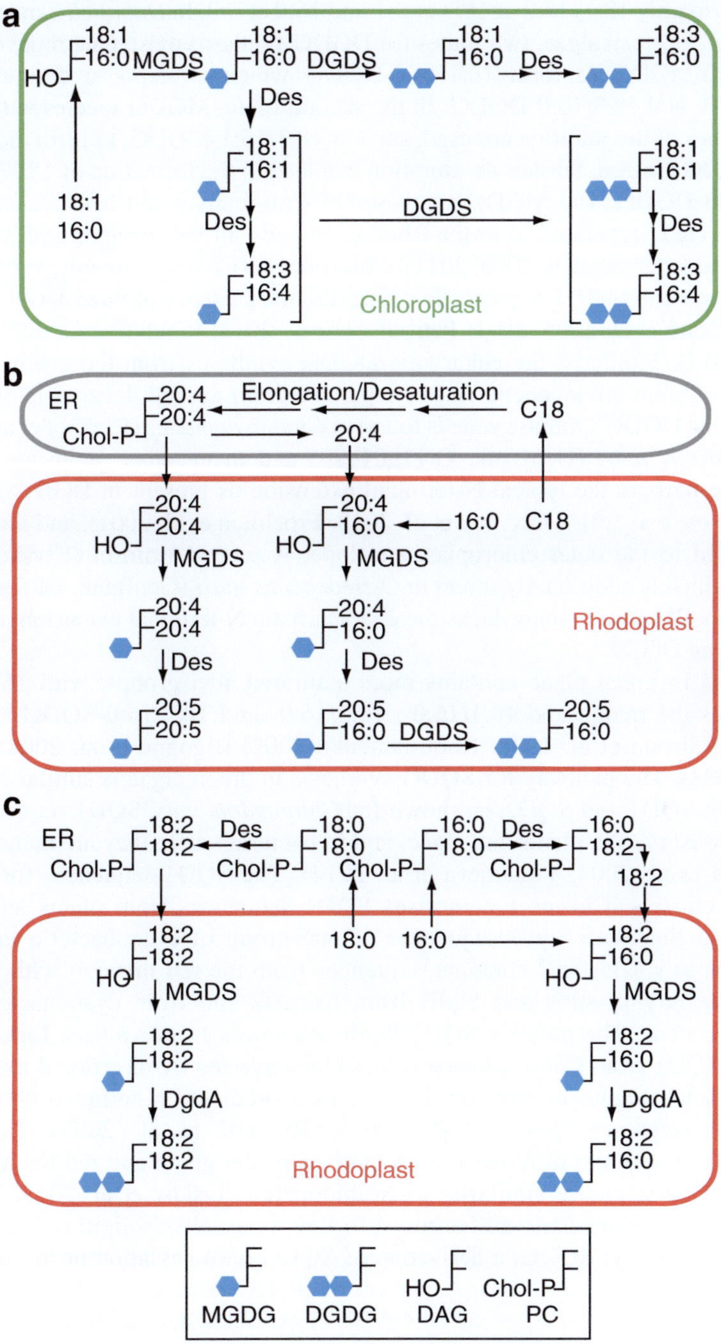
a
18:1
16:0
HO
MGDS
18:1
16:0
DGDS
18:1
16:0
Des
18:3
16:0
Des
18:1
16:1
18:1
16:0
Des
DGDS
18:1
16:1
Des
18:3
16:4
Chloroplast
18:3
16:4
b
ER
20:4
20:4
Chol-P
Elongation/Desaturation
C18
20:4
20:4
20:4
HO
MGDS
20:4
16:0
HO
16:0
C18
MGDS
20:4
20:4
20:4
16:0
Rhodoplast
Des
Des
20:5
20:5
20:5
16:0
DGDS
20:5
16:0
c
ER
18:2
18:2
Chol-P
Des
18:0
18:0
Chol-P
16:0
18:0
Chol-P
Des
16:0
18:2
Chol-P
18:2
18:2
18:2
HO
18:0
16:0
18:2
16:0
HO
MGDS
MGDS
18:2
18:2
18:2
16:0
DgdA
DgdA
18:2
18:2
Rhodoplast
18:2
16:0
MGDG
DGDG
HO
DAG
Chol-P
PC

contains mainly 18:1, 18:2 or 18:3 at *sn*1 and 16:0 at *sn*2. In *Dunaliella* and possibly also in other green algae, two routes for DGDG synthesis exist. One route employs 18:1/16:0-MGDG to form 18:1/16:0-DGDG which is stepwise desaturated to 18:2/16:0- and 18:3/16:0-DGDG. In the second route, MGDG species with different degrees of desaturation are used, such as 18:1/16:1-MGDG, to form the respective DGDG species. Further desaturation can lead to the formation of 18:3/16:3- or 18:3/16:4-DGDG. The MGDG synthases of Chlorophyta and land plants form a different clade separated from the Rhodophyta and all the lineages with red complex plastids (Botté et al. 2005, 2011). Chlorophyta (*Chlamydomonas*, *Volvox*) usually contain one MGD1 sequence in contrast to the presence of three *MGD* genes in plants (Botté et al. 2005, 2011; Petroutsos et al. 2014; Riekhof et al. 2005; Abida et al. 2015). Similarly, the eukaryotic DGDG synthases from the green lineages form a different phylogenetic cluster separated from the red lineages (Botté et al. 2011). One DGDG synthase gene is found in *Chlamydomonas* (Riekhof et al. 2005), *Ostreococcus tauri* (GenBank: Ot11g01480) and many other green algae. This sequence harbors the typical N-terminal extension as present in DGD1 of plants (Petroutsos et al. 2014; Ulvskov et al. 2013; Froehlich et al. 2001), and thus might be located to the outer chloroplast envelope. A second isoform (Ulvskov et al. 2013), which is additionally found in *Ostreococcus tauri* (GenBank: Ot07g02970), but not in *Chlamydomonas*, lacks the characteristic N-terminal extension, and thus resembling DGD2.

SQDG in green algae contains more saturated acyl groups, with 16:0/16:0-SQDG as the major, and 18:1/16:0-, 18:2/16:0- and 18:3/16:0-SQDG as minor species (Giroud et al. 1988; Khotimchenko 2002; Bigogno et al. 2002; Sanina et al. 2004). The pathway for SQDG synthesis in green algae is similar to plants involving SQD1 and SQD2, as shown for *Chlamydomonas*. SQD1 sequences are highly conserved in plants and algae, and in bacteria where they are named SqdB (Riekhof et al. 2003; Villanueva et al. 2014). The SQD1 sequences form three distinct clusters. Cluster 1 comprises SQD1 sequences from plants and green algae, together with SqdB sequences of one group of cyanobacteria including *Synechocystis*. Cluster 2 contains sequences from the red lineages with primary and secondary plastids and SqdB from bacteria and from cyanobacteria like *Synechococcus*. The putative SQD1/SqdB sequences from Archaea form a third cluster. SQD1 from *Chlamydomonas* was characterized by insertional mutagenesis and heterologous expression. It contains a predicted N-terminal chloroplast targeting sequence (Sato et al. 2003a; Riekhof et al. 2003). SQD2 in *Chlamydomonas* and different representatives of the green and red lineages was identified by sequence similarity to Arabidopsis (Ulvskov et al. 2013; Riekhof et al. 2005). *Chlamydomonas* harbors a further, unusual sulfolipid with unknown function, 2-*O*-acyl-SQDG, a derivative of SQDG with acylation of the sulfoquinovose (Riekhof et al. 2003). Acylation of SQDG occurs only with C18/16:0-SQDG, employing $18{:}3^{\Delta 5,9,12}$ or $18{:}4^{\Delta 5,9,12,15}$ fatty acids derived from the ER, but the acyltransferase is unknown (Riekhof et al. 2005).

Glycolipids in Rhodophyta

Rhodoplasts, the photosynthetic organelles in the red lineage, contain phycoerythrin and chlorophyll a as photosynthetic pigments (Keeling 2010). The main fatty acids in red algae are 16:0, 18:1 (Δ9 and Δ11), 20:4 and 20:5 (Table 3.2) (Li et al. 2002; Khozin-Goldberg et al. 2000; Sanina et al. 2004; Khotimchenko 2002). The presence of high amounts of C20 fatty acids sets Rhodophyta apart from Chlorophyta and higher plants. These differences are also reflected in the structure and biosynthesis of galactolipids and sulfolipids. Similar to Arabidopsis, Rhodophyta employ two pathways for galactolipid biosynthesis with eukaryotic (C20/C20) or prokaryotic (C20/C16) structure (Fig. 3.3b). In plants, the fatty acids at *sn*1 of prokaryotic lipids are synthesized exclusively in the chloroplast. In red algae, however, the C20 fatty acids at *sn*1 of prokaryotic-type lipids originate from the extraplastidial compartment. Therefore, the galactolipids with prokaryotic structure in *Porphyridium cruentum* and other Rhodophyta are structurally, but not biosynthetically related to prokaryotic lipids of plants. The main molecular MGDG species in red algae such as *Porphyridium cruentum* or *Porphyra yezoensis* are 20:5/20:5- and 20:5/16:0-MGDG (Khozin et al. 1997; Araki et al. 1987). The precursor for the synthesis of eukaryotic (20:4/20:4) lipids is derived from PC, and after import into the rhodoplast, converted into 20:4/20:4-MGDG. Additionally, single fatty acids (20:4) can be imported and inserted into the *sn*1 position to synthesize prokaryotic-type 20:4/16:0-MGDG (Khozin et al. 1997). Besides PC, TAG is a further source of C20 fatty acids and it serves as a reservoir of C20 for the rapid production of eukaryotic MGDG involving an unknown mechanism (Khozin-Goldberg et al. 2000). Interestingly, DGDG in different Rhodophyta is mainly prokaryotic (20:5/16:0), in contrast to plants where the major fraction of DGDG is eukaryotic. The major SQDG species in red algae is also 20:5/16:0 (Araki et al. 1987).

MGDG synthases and DGDG synthases of Rhodophyta and all the lineages with red complex plastids form a clade separated from the Chlorophyta and plants (Botté et al. 2005, 2011). Rhodophyta contain one MGD1 sequence and most of them one or two plant-like DGDG synthases. DGD2-like isoforms are found in many red algae, such as *Chondrus crispus* (Petroutsos et al. 2014). Sulfolipid genes (*SQD1*, *SQD2*) in Rhodophyta are highly related to the plant orthologs.

In contrast to most Rhodophyta, a small group of primitive red algae of the order Cyanidiales (*Cyanidioschyzon merolae*, *Cyanidium caldarium*) are characterized by a very simple fatty acid composition (Allen et al. 1970; Sato and Moriyama 2007) with 16:0 and 18:2 predominating. They further contain 18:0 and 18:1, but lack the characteristic C20 fatty acids found in most Rhodophyta. The membrane lipid composition of Cyanidiales is similar to plants with the same set of glycolipids and phospholipids. The galactolipids comprise a mixture of prokaryotic (mainly 18:2/16:0) and eukaryotic (mainly 18:2/18:2) species, while SQDG is only prokaryotic and more saturated (16:0/16:0, 18:2/16:0). As shown for *Cyanidioschyzon merolae*, Cyanidiales lack the Δ9-acyl-ACP desaturase (SAD) producing 18:1-ACP as in plants. Desaturation of 18:0 and 18:1 occurs only at the

ER. Therefore, the production of prokaryotic-type MGDG (18:2/16:0) employs a coupled pathway with 18:2 derived from the ER and 16:0 produced in the plastid (Fig. 3.3c). This pathway might operate in a similar manner as in other Rhodophyta, employing 20:5 fatty acids for the synthesis of prokaryotic MGDG (see above). It is not known whether all red algae lack the plastidial Δ9-acyl-ACP desaturase (Sato and Moriyama 2007 #7831) because candidate genes for Δ9 acyl-ACP desaturases were identified in complex plastids of the diatoms (Stramenopiles) (Abida et al. 2015). Members of the Cyanidiales like *Cyanidioschyzon merolae* lack the plant-like DGDG synthases, but contain an ortholog of the cyanobacterial *dgdA* gene in their plastid genome (Petroutsos et al. 2014; Sato and Moriyama 2007; Botté et al. 2011; Awai et al. 2007). Homology searches in *Cyanidioschyzon* led to the identification of genes coding for a plant-like MGDG synthase and a putative cyanobacterial-type MGlcDG synthase (Sato and Moriyama 2007). *Cyanidioschyzon* synthesizes MGDG via the typical MGDG synthase pathway of plants with UDP-Gal as substrate (Awai et al. 2007). The function of the putative MGlcDG synthase is unknown, because MGlcDG could not be detected (Sato and Moriyama 2007; Petroutsos et al. 2014). The localization of the DgdA ortholog and the MGDG synthase in *Cyanidioschyzon* is unknown (Botté et al. 2011). It has been suggested that cyanobacterium-derived DgdA was present in the ancestors of eukaryotic algae, but was lost during evolution and replaced with the eukaryotic DGDG synthases (Botté et al. 2011). Consistent with this scenario, DgdA sequences are absent from multicellular red algae and green lineages including plants (Awai et al. 2007; Petroutsos et al. 2014). SQD1 and SQD2 from *Cyanidioschyzon merolae* are highly related to the plant orthologs.

Algae Derived from Secondary Endosymbiosis with Chlorophyta

Two independent endosymbiotic events with green algae led to the evolution of Chlorarachniophyta and Euglenophyta. The plastidial genomes of these two lineages are not closer related as compared to the green algae, and the hosts are also phylogenetically distant. Further differences concern the organization and structure of their plastids. While Euglenophyta are characterized by three plastidial membranes, Chlorarachniophyta plastids are surrounded by four membranes. They further contain a relict nucleus called nucleomorph, originating from the green algal endosymbiont, which is absent from Euglenophyta. *Euglena* contains MGDG, DGDG and SQDG similar to plants and green algae, but with higher amounts of VLCPUFAs (Regnault et al. 1995; Kaoua and Laval-Martin 1995).

Algae Derived from Secondary Endosymbiosis with Rhodophyta

The initial step of secondary red plastid evolution was monophyletic (Keeling 2013; Zimorski et al. 2014). The further radiation of the complex red plastids is not fully resolved, because it is likely that further downstream hosts are involved and therefore possible additional endosymbiotic events have to be considered (Keeling 2013; Zimorski et al. 2014). Secondary red plastids are found in Cryptophyta, Haptophyta, Heterokonta (or Stramenopiles) and Dinoflagellata. The plastids in the first three groups are surrounded by four, in Dinoflagellata by three membranes. In addition, the Cryptophyta contain a nucleomorph. All four groups are characterized by the presence of chlorophyll a and c.

Glycolipids in Haptophyta

The haptophyte *Pavlova lutheri* is characterized by its conspicuous lipid composition (Eichenberger and Gribi 1997). Besides of galactolipids and SQDG, it contains a further glycolipid, GlcADG, as described for Arabidopsis and a few algae and more recently for other plants (Eichenberger and Gribi 1994; Okazaki et al. 2013a, 2015). GlcADG from *Pavlova* has a eukaryotic structure, is localized outside of the plastid and synthesized by an unknown enzyme, while the plant counterpart is produced in the chloroplast by SQD2. Moreover, *Pavlova* contains DGTA and DGCC, while the only phospholipid PG is present in very low amounts. DGCC is the precursor for redistribution of fatty acids for plastidial glycolipids. It has been suggested that C18 and C20 fatty acids, but not intact DAG entities, are individually transported to the plastid to form eukaryotic-type MGDG by an unknown mechanism. A major fraction of MGDG and DGDG are of eukaryotic structure (20:5/18:4). SQDG contains mainly of 16:0 and 14:0 fatty acids. Further fatty acids in *Pavlova* are 22:5 (docosapentaenoic acid) and 22:6 (docosahexaenoic acid), which are especially enriched in GlcADG, but hardly detectable in MGDG, DGDG, and SQDG.

Glycolipids in Heterokonts (Strameopiles)

The main fatty acids in brown algae (Phaeophyta, Stramenopiles) are 14:0, 16:0, different C18 fatty acids (including stearidonic acid, $18{:}4^{\Delta 6,9,12,15}$), 20:4 and 20:5 (Li et al. 2002; Khotimchenko 2002; Sanina et al. 2004; Dembitsky et al. 1990; Kumari et al. 2013). The high amounts of C18 fatty acids typical for Phaeophyta separate them from the Rhodophyta. As in other algae, MGDG is the most unsaturated and SQDG the most saturated glycolipid (Hofmann and Eichenberger 1997). $18{:}4^{\Delta 6,9,12,15}$ is almost completely restricted to *sn*2 of MGDG and DGDG, and to a minor extent to SQDG which instead contains more $18{:}3^{\Delta 9,12,15}$ (Hofmann and Eichenberger 1997). Galactolipids further contain C16 fatty acids and are enriched with 20:5. 16:0 is dominant in SQDG and 14:0 is found at *sn*1 of galactolipids, SQDG and DGTA (Araki et al. 1991a; Khotimchenko 2002). The desaturation of 18:1 and 20:4

fatty acids occurs on MGDG, which is mainly of eukaryotic structure. The main species in *Dictyopteris membranacea* are 18:3/18:4-, 18:4/18:4- and 20:5/18:4-MGDG (Hofmann and Eichenberger 1997). In contrast, DGDG comprises both eukaryotic and prokaryotic (16:0 at *sn*2) species. The prokaryotic species with more saturated fatty acids predominate in SQDG (Hofmann and Eichenberger 1997). Many brown algae contain DGTA. If present, then they often lack PC or both PC and PE (Kato et al. 1996; Araki et al. 1991a). DGTA possibly plays a key role in the formation of eukaryotic-type glycolipids, similar to PC in plants (Ohlrogge and Browse 1995; Roughan and Slack 1982; Hofmann and Eichenberger 1998).

Diatoms (Bacillariophyta, Stramenopiles) contain the same glycolipids and phospholipids as plants. However, compared to brown algae, most diatoms lack betaine lipids, while DGTA was recently detected in *Phaeodactylum tricornutum* (Abida et al. 2015). The majority of diatom species investigated synthesize glycolipids with a prokaryotic structure. The characteristic fatty acids of diatoms are 14:0, 16:0, 16:1, 16:2, $16{:}3^{\Delta 6,9,12}$, $16{:}4^{\Delta 6,9,12,15}$ and 20:5 (Arao et al. 1987; Alonso et al. 2000; Abida et al. 2015), whereas C18 fatty acids are in general low. The desaturation degree and the amount of C16 fatty acids vary in the different diatom species (Yongmanitchai and Ward 1993; d'Ippolito et al. 2004; Xu et al. 2010; Liang et al. 2014). 16:3 and 16:4 are only present in glycolipids, while 20:5 is found in all lipids, but in glycolipids restricted to *sn*1. The main MGDG species in many diatoms (*Phaeodactylum tricornutum*, *Stephanodiscus* sp., *Skeletonema costatum*, *Fistulifera solaris*) are 20:5/C16- and C16/C16-MGDG. In general, DGDG is more saturated than MGDG, and SQDG is characterized by high amounts of 14:0. Similar to *Chlamydomonas*, an acylated form of SQDG was found in *Phaeodactylum tricornutum* with 20:5 linked to the sugar (Abida et al. 2015). Some diatoms like *Cyclotella meneghiniana* contain additionally eukaryotic-type galactolipids such as 18:4/18:4-MGDG and 18:3/18:4-DGDG (Vieler et al. 2007). Interestingly, the non-photosynthetic diatom *Nitzschia alba* also contains high levels of galactolipids in contrast to the Apicomplexa, which also harbor non-photosynthetic plastids (see below) (Anderson et al. 1978). PC plays a central role in the synthesis of galactolipids in diatoms as shown for *Phaeodactylum* or *Fistulifera solaris* (Fig. 3.4) (Arao et al. 1994; Liang et al. 2014). 20:5/16:1-PC provides the DAG moiety for the synthesis of 20:5/16:1-MGDG, which can be further desaturated to 20:5/16:3-MGDG. C16/C16-MGDG is suggested to be synthesized in the plastid. MGDG molecules carrying C16 at *sn*1 (such as 16:1/16:1-MGDG in *Phaeodactylum*) are very unusual, because in analogy to Arabidopsis, it is believed that the presence of *sn*1 bound C16 fatty acids is characteristic for extraplastidial lipids (16:0/C18). Interestingly, in diatoms C16 fatty acids also occur at *sn*2 of extraplastidial lipids such as 20:5/16:1- or 16:0/16:0-PC leading to a prokaryotic-type structure of extraplastidial phospholipids. These unusual lipid structures in diatoms raise the question about the nature of lipid precursors and their transport. It is suggested that 20:5/C16-PC or 20:5/C16-DAG are possible precursors or 20:5-acyl donors for galactolipid synthesis, and they are formed by acyl-editing or head group exchange (Liang et al. 2014). Diatoms harbor one or several candidate genes for Δ9-acyl-CoA desaturases at the ER and for plastidial Δ9-acyl-ACP desaturases (DES) (Muto et al.

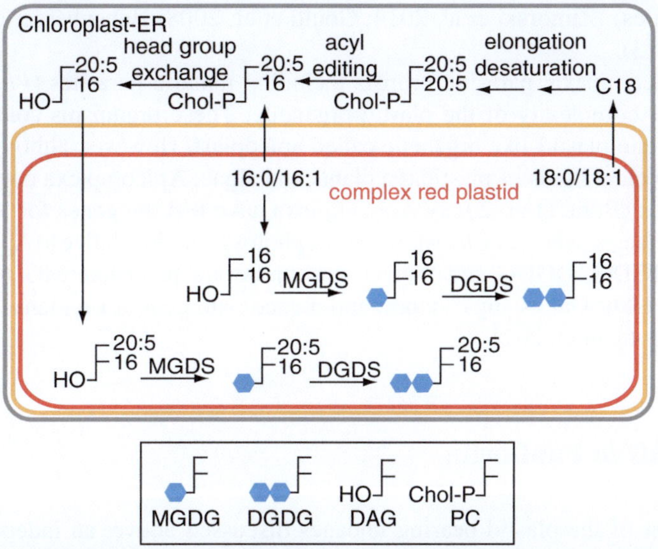

Fig. 3.4 Two routes of galactolipid biosynthesis in complex red plastids of diatoms (*Fistulifera solaris*, *Phaeodactylum tricornutum*). The first route employs C16 fatty acids synthesized in the plastid to form 16/16-MGDG and 16/16-DGDG with different desaturation degrees. The second route involves 20:5 imported from the ER, which may be transported as single fatty acid or in lipid bound form to synthesize 20:5/16-MGDG and 20:5/16-DGDG. 20:5 is produced in the ER with C18 fatty acids as starting molecules. Prokaryotic-type PC (20:5/16) is formed by acyl editing (Liang et al. 2014). For simplification, the two innermost membranes of the complex plastid are presented as one *red line*; the *yellow line* represents the second outermost and the *grey line* the outermost or chloroplast-ER membrane. MGDS and DGDS, plant-type MGDG and DGDG synthases, respectively

2013). The plastidial Δ9-acyl-ACP desaturases in *Phaeodactylum* presumably act on 16:0-ACP rather than on 18:0-ACP (Abida et al. 2015). The galactolipid and SQDG synthesis in diatoms is closely related to that of Rhodophyta. Interestingly, *Phaeodactylum tricornutum* contains up to three paralogs for MGDG synthases and for DGDG synthases, 2 isoforms of SQD2 (Abida et al. 2015) and one isoform of SQD1 (see above).

Glycolipids in Alveolata (Dinoflagellata, Apicomplexa)

Plastid bearing eukaryotes of the Dinoflagellata and Apicomplexa together with further groups form the superphylum of Alveolata (Janouškovec et al. 2010). Some Dinoflagellata went a step further in evolution and replaced their secondary complex plastid with an endosymbiont of secondary red origin leading to tertiary endosymbiosis, or with a green alga to form a serial secondary endosymbiosis. The serial secondary endosymbiosis occurred only once with a Chlorophyta as endosymbiont. The tertiary plastids were acquired from Haptophyta, Cryptophyta or diatoms

(Stamenopiles) (Zimorski et al. 2014; Gould et al. 2008; Hehenberger et al. 2014; Keeling 2013).

The Apicomplexa, parasitic protists including malaria parasites (*Plasmodium*), reduced the complexity of the plastid organelle. These organisms contain a non-photosynthetic plastid-like organelle called apicoplast. However, although galactolipids are typical lipids of plastids in plants and algae, Apicomplexa do not contain galactolipids (Botté et al. 2013). Apicomplexa have lost the genes for MGDG and DGDG synthesis, whereas *Chromera velia* a photosynthetic relative to Apicomplexa, contains MGDG, DGDG and SQDG, and harbors a plant-like MGDG synthase employing UDP-Gal as sugar donor, and a candidate gene for a plant-like DGDG synthase (Botté et al. 2011).

Glycolipids in Paulinella

Independent of the plastid bearing lineages discussed above, an independent primary endosymbiosis event occurred more recently. The unicellular eukaryote *Paulinella chromatophora* acquired its photosynthetic organelle about 60 million years ago (Nowack et al. 2008). *Paulinella* is a fresh-water amoeba of the phylum Cercozoa (in the Rhizaria supergroup), contains one or two blue-green organelles called chromatophores and is only distantly related to other eukaryotic algae (Nakayama and Archibald 2012). The chromatophores resemble free-living cyanobacteria especially *Synechococcus*. However, the chromatophore is not just an endosymbiotic cyanobacterium, but a true photosynthetic organelle. It already lost 75 % of its coding genes, and at least 30 endosymbiont-derived genes are expressed in the host nucleus (Bodył et al. 2012). Nuclear encoded proteins are imported into the chromatophore via vesicle transport from the ER/Golgi, in contrast to the TOC/TIC import system of plastids. Classical plastid transit peptides are not involved, and the nature of the targeting system is far from clear (Mackiewicz et al. 2012; Nakayama and Archibald 2012). Therefore, the current developmental stage of the chromatophore can be regarded as an early step in organelle evolution. *Paulinella* contains the typical galactolipids as found in algae and plants, but it might employ a pathway for MGDG and DGDG biosynthesis as found in present-day cyanobacteria. *Paulinella* contains genes coding for a putative cyanobacterium-type MGlcDG synthase (MgdA), an epimerase converting MGlcDG into MGDG (MgdE), and DgdA. Orthologs of plant-type MGDG synthases have not been detected in Rhizaria (Petroutsos et al. 2014; Awai et al. 2014).

Lipid Response of Algae to Phosphate Deprivation

Phosphate supply is often a growth limiting factor not only for plants, but also for algae. The replacement of phospholipids with non-phospholipids is a general mechanism of plants, algae and some bacteria to reduce the phosphate requirement in phosphate deficient environments. Marine ecosystems are often characterized by low amounts of phosphate, with the phytoplankton as major consumer. Phytoplankton communities are dominated by cyanobacteria, and under certain conditions also by small eukaryotic algae. Phytoplankton can reduce its physiological phosphate demand by 50 % in response to phosphate deprivation (Van Mooy et al. 2009). The most important surrogates for phospholipids in phytoplankton are SQDG and betaine lipids. PG is replaced with SQDG in cyanobacteria and to a minor extent in eukaryotic algae, which mainly use betaine lipids as structural counterparts for PC. Galactolipids do not seem to be involved in the response to phosphate deprivation in many marine algae that contain betaine lipids, because betaine lipids act as the main surrogates for phospholipids in extraplastidial membranes, while SQDG replaces PG in the plastids (Abida et al. 2015). In contrast, DGTS is not involved in the response to phosphate deprivation in *Chlamydomonas*, which might be explained by the fact that this fresh water alga a priori has no PC. Therefore, SQDG plays the most important role during phosphate deprivation in *Chlamydomonas* (Riekhof et al. 2003; Sato et al. 2000). As in plants, PG and SQDG in *Chlamydomonas* contents are in a reciprocal relationship to maintain a certain level of anionic lipids. DGDG participates only moderately in the response to phosphate deprivation, but it can quantitatively compensate for the loss of SQDG in the *Chlamydomonas* Δ*sqd1* mutant. A stronger response to phosphate deprivation can be observed in the diatom *Phaeodactylum tricornutm* with a complete reduction of all phospholipids, accompanied by the accumulation of large amounts of DGTA and SQDG and minor amounts of DGDG. Therefore, PG can be fully replaced with SQDG during phosphate limitation in *Phaeodactylum*. In contrast, the reduction of PG in the Arabidopsis *pgp1* mutant affects growth and photosynthesis (Xu et al. 2002). DGDG, DGTS and SQDG play major roles as surrogates for phospholipids in the freshwater Eustigmatophyta *Monodus subterraneus* (Stramenopiles) (Khozin-Goldberg and Cohen 2006). Interestingly, two prokaryotic-type DGDG species are involved in the lipid response to phosphate deprivation, with 20:5/16:1-DGDG being strongly increased and 16:1/16:0-DGDG strongly reduced. In contrast to plants, where eukaryotic 16:0/C18-DGDG is exported to extraplastidial membranes under phosphate deprivation (Härtel et al. 2000), no information is available about the localization or possible export of 20:5/16:1-DGDG in *Mondus*, as well as for other DGDG molecular species in algae.

References

Abida H, Dolch LJ, Meï C, Villanova V, Conte M, Block MA, Finazzi G, Bastien O, Tirichine L, Bowler C, Rébeillé F, Petroutsos D, Jouhet J, Maréchal E (2015) Membrane glycerolipid remodeling triggered by nitrogen and phosphorus starvation in *Phaeodactylum tricornutum*. Plant Physiol 167:118–136

Allen CF, Good P, Holton RW (1970) Lipid composition of *Cyanidium*. Plant Physiol 46:748–751

Alonso DL, Belarbi EH, Fernández-Sevilla JM, Rodriguez-Ruiz J, Molina Grima E (2000) Acyl lipid composition variation related to culture age and nitrogen concentration in continuous culture of the microalga *Phaeodactylum tricornutum*. Phytochemistry 54:461–471

Anderson R, Livermore BP, Kates M, Volcani BE (1978) The lipid composition of the non-photosynthetic diatom *Nitzschia alba*. Biochim Biophys Acta 528:77–88

Andersson MX, Larsson KE, Tjellström H, Liljenberg C, Sandelius AS (2005) Phosphate-limited oat. The plasma membrane and the tonoplast as major targets for phospholipid-to-glycolipid replacement and stimulation of phospholipases in the plasma membrane. J Biol Chem 280:27578–27586

Andersson MX, Goksor M, Sandelius AS (2007) Optical manipulation reveals strong attracting forces at membrane contact sites between endoplasmic reticulum and chloroplasts. J Biol Chem 282:1170–1174

Araki S, Sakurai T, Kawaguchi A, Murata N (1987) Positional distribution of fatty acids in glycerolipids of the marine red alga, *Phorphyra yezoensis*. Plant Cell Physiol 28:761–766

Araki S, Sakurai T, Oohusa T, Kayama M, Nisizawa K (1990) Content of arachidonic and eicosapentaenoic acids in polar lipids from *Gracilaria* (Gracilariales, Rhodophyta). Hydrobiologia 204(205):513–519

Araki S, Eichenberger W, Sakurai T, Sato N (1991) Distribution of diacylglycerylhydroxymethylt rimethyl-β-alanine (DGTA) and phosphatidylcholine in brown algae. Plant Cell Physiol 32:623–628

Arao T, Kawaguchi A, Yamada M (1987) Positional distribution of fatty acids in lipids of the marine diatom *Phaeodactylum tricornutum*. Phytochemistry 26:2573–2576

Arao T, Sakaki T, Yamada M (1994) Biosynthesis of polyunsaturated lipids in the diatom *Phaeodactylum tricornutum*. Phytochemistry 36:629–635

Aro E-M, Karunen P (1979) Effect of changed environmental conditions on glycolipids of the mosses *Pleurozium Schreberi* and *Ceratodon purpureus*. Physiol Plant 45:201–206

Awai K, Maréchal E, Block MA, Brun D, Masuda T, Shimada H, Takamiya K, Ohta H, Joyard J (2001) Two types of MGDG synthase genes, found widely in both 16:3 and 18:3 plants, differentially mediate galactolipid syntheses in photosynthetic and nonphotosynthetic tissues in *Arabidopsis thaliana*. Proc Natl Acad Sci U S A 98:10960–10965

Awai K, Kakimoto T, Awai C, Kaneko T, Nakamura Y, Takamiya K, Wada H, Ohta H (2006a) Comparative genomic analysis revealed a gene for monoglucosyldiacylglycerol synthase, an enzyme for photosynthetic membrane lipid synthesis in cyanobacteria. Plant Physiol 141:1120–1127

Awai K, Xu C, Tamot B, Benning C (2006b) A phosphatidic acid-binding protein of the chloroplast inner envelope membrane involved in lipid trafficking. Proc Natl Acad Sci U S A 103:10817–10822

Awai K, Watanabe H, Benning C, Nishida I (2007) Digalactosyldiacylglycerol is required for better photosynthetic growth of *Synechocystis* sp. PCC6803 under phosphate limitation. Plant Cell Physiol 48:1517–1523

Awai K, Ohta H, Sato N (2014) Oxygenic photosynthesis without galactolipids. Proc Natl Acad Sci U S A 111:13571–13675

Bigogno C, Khozin-Goldberg I, Boussiba S, Vonshak A, Cohen Z (2002) Lipid and fatty acid composition of the green oleaginous alga *Parietochloris incisa*, the richest plant source of arachidonic acid. Phytochemistry 60:497–503

Block MA, Dorne AJ, Joyard J, Douce R (1983) Preparation and characterization of membrane fractions enriched in outer and inner envelope membranes from spinach chloroplasts. II. Biochemical characterization. J Biol Chem 258:13281–13286

Bodył A, Mackiewicz P, Gagat P (2012) Organelle evolution: *Paulinella* breaks a paradigm. Curr Biol 22:R304–R306

Bonaventure G (2014) Lipases and the biosynthesis of free oxylipins in plants. Plant Signal Behav 9:e28429

Böttcher C, Weiler E (2007) cyclo-Oxylipin-galactolipids in plants: occurrence and dynamics. Planta 226:629–637

Botté C, Jeanneau C, Snajdrova L, Bastien O, Imberty A, Breton C, Maréchal E (2005) Molecular modeling and site-directed mutagenesis of plant chloroplast monogalactosyldiacylglycerol synthase reveal critical residues for activity. J Biol Chem 280:34691–34701

Botté CY, Yamaryo-Botté Y, Janouskovec J, Rupasinghe T, Keeling PJ, Crellin P, Coppel RL, Maréchal E, McConville MJ, McFadden GI (2011) Identification of plant-like galactolipids in *Chromera velia*, a photosynthetic relative of malaria parasites. J Biol Chem 286:29893–29903

Botté CY, Yamaryo-Botté Y, Rupasinghe TWT, Mullin KA, MacRae JI, Spurck TP, Kalanon M, Shears MJ, Coppel RL, Crellin PK, Maréchal E, McConville MJ, McFadden GI (2013) Atypical lipid composition in the purified relict plastid (apicoplast) of malaria parasites. Proc Natl Acad Sci U S A 110:7506–7511

Browse J, Warwick N, Somerville CR, Slack CR (1986) Fluxes through the prokaryotic and eukaryotic pathways of lipid synthesis in the '16:3' plant *Arabidopsis thaliana*. Biochem J 235:25–31

Chen J, Burke JJ, Xin Z, Xu C, Velten J (2006) Characterization of the *Arabidopsis* thermosensitive mutant *atts02* reveals an important role for galactolipids in thermotolerance. Plant Cell Environ 29:1437–1448

Cho SH, Thompson GA Jr (1987) On the metabolic relationships between monogalactosyldiacylglycerol and digalactosyldiacylglycerol molecular species in *Dunaliella salina*. J Biol Chem 262:7586–7593

Cohen Z, Shiran D, Khozin I, Heimer YM (1997) Fatty acid unsaturation in the red alga *Porphyridium cruentum*. Is the methylene interrupted nature of polyunsaturated fatty acids an intrinsic property of the desaturases? Biochim Biophys Acta 1344:59–64

d'Ippolito G, Tucci S, Cutignano A, Romano G, Cimino G, Miralto A, Fontana A (2004) The role of complex lipids in the synthesis of bioactive aldehydes of the marine diatom *Skeletonema costatum*. Biochim Biophys Acta 1686:100–107

Dembitsky VM, Rozentsvet OA, Pechenkina EE (1990) Glycolipids, phospholipids and fatty acids of brown algae species. Phytochemistry 29:3417–3421

Demé B, Cataye C, Block MA, Maréchal E, Jouhet J (2014) Contribution of galactoglycerolipids to the 3-dimensional architecture of thylakoids. FASEB J 28:3373–3383

Dörmann P, Hoffmann-Benning S, Balbo I, Benning C (1995) Isolation and characterization of an *Arabidopsis* mutant deficient in the thylakoid lipid digalactosyl diacylglycerol. Plant Cell 7:1801–1810

Dörmann P, Balbo I, Benning C (1999) *Arabidopsis* galactolipid biosynthesis and lipid trafficking mediated by DGD1. Science 284:2181–2184

Douce R, Joyard J (1980) Plant galactolipids. In: Stumpf PK (ed) The biochemistry of plants. Lipids: structure and function. Academic, New York, pp 321–362

Dubots E, Audry M, Yamaryo Y, Bastien O, Ohta H, Breton C, Maréchal E, Block MA (2010) Activation of the chloroplast monogalactosyldiacylglycerol synthase MGD1 by phosphatidic acid and phosphatidylglycerol. J Biol Chem 285:6003–6011

Eichenberger W, Gribi C (1994) Diacylglyceryl-α-D-glucuronide from *Ochromonas danica* (Chrysophyceae). J Plant Physiol 144:272–276

Eichenberger W, Gribi C (1997) Lipids of *Pavlova lutheri*: cellular site and metabolic role of DGCC. Phytochemistry 45:1561–1567

Essigmann B, Güler S, Narang RA, Linke D, Benning C (1998) Phosphate availability affects the thylakoid lipid composition and the expression of *SQD1*, a gene required for sulfolipid biosynthesis in *Arabidopsis thaliana*. Proc Natl Acad Sci U S A 95:1950–1955

Fishwick MJ, Wright AJ (1980) Isolation and characterization of amyloplast envelope membranes from *Solanum tuberosum*. Phytochemistry 19:55–59

Fourrier N, Bédard J, Lopez-Juez E, Barbrook A, Bowyer J, Jarvis P, Warren G, Thorlby G (2008) A role for *SENSITIVE TO FREEZING2* in protecting chloroplasts against freeze-induced damage in Arabidopsis. Plant J 55:734–745

Froehlich JE, Benning C, Dörmann P (2001) The digalactosyldiacylglycerol (DGDG) synthase DGD1 is inserted into the outer envelope membrane of chloroplasts in a manner independent of the general import pathway and does not depend on direct interaction with monogalactosyldiacylglycerol synthase for DGDG biosynthesis. J Biol Chem 276:31806–31812

Fujii S, Kobayashi K, Nakamura Y, Wada H (2014) Inducible knockdown of *MONOGALACTOSYLDIACYLGLYCEROL SYNTHASE1* reveals roles of galactolipids in organelle differentiation in Arabidopsis Cotyledons. Plant Physiol 166:1436–1449

Gaude N, Tippmann H, Flemetakis E, Katinakis P, Udvardi M, Dörmann P (2004) The galactolipid digalactosyldiacylglycerol accumulates in the peribacteroid membrane of nitrogen-fixing nodules of soybean and *Lotus*. J Biol Chem 279:34624–34630

Gellerman J, Anderson W, Schlenk H (1975) Synthesis and analysis of phytyl and phytenoyl wax esters. Lipids 10:656–661

Giroud C, Gerber A, Eichenberger W (1988) Lipids of *Chlamydomonas reinhardtii*. Analysis of molecular species and intracellular site(s) of biosynthesis. Plant Cell Physiol 29:587–595

Goss R, Wilhelm C (2009) Lipids in algae, lichens and mosses. In: Wada H, Murata N (eds) Lipids in photosynthesis, vol 30, Advances in photosynthesis and respiration. Springer, Dordrecht, pp 117–135

Gould SB, Waller RF, McFadden GI (2008) Plastid evolution. Annu Rev Plant Biol 59:491–517

Härtel H, Lokstein H, Dörmann P, Grimm B, Benning C (1997) Changes in the composition of the photosynthetic apparatus in the galactolipid-deficient *dgd1* mutant of *Arabidopsis thaliana*. Plant Physiol 115:1175–1184

Härtel H, Essigmann B, Lokstein H, Hoffmann-Benning S, Peters-Kottig M, Benning C (1998) The phospholipid-deficient pho1 mutant of *Arabidopsis thaliana* is affected in the organization, but not in the light acclimation, of the thylakoid membrane. Biochim Biophys Acta 1415:205–218

Härtel H, Dörmann P, Benning C (2000) DGD1-independent biosynthesis of extraplastidic galactolipids after phosphate deprivation in *Arabidopsis*. Proc Natl Acad Sci U S A 97:10649–10654

Hehenberger E, Imanian B, Burki F, Keeling PJ (2014) Evidence for the retention of two evolutionary distinct plastids in dinoflagellates with diatom endosymbionts. Genome Biol Evol 6:2321–2334

Heinz E (1967a) Acylgalaktosyldiglycerid aus Blatthomogenaten. Biochim Biophys Acta 144:321–332

Heinz E (1967b) Über die enzymatische Bildung von Acylgalaktosyldiglycerid. Biochim Biophys Acta 144:333–343

Heinz E (1977) Enzymatic reactions in galactolipid biosynthesis. In: Tevini M, Lichtenthaler HK (eds) Lipids and lipid polymers in higher plants. Springer, Berlin, pp 102–120

Hendrickson L, Vlčková A, Selstam E, Huner N, Öquist G, Hurry V (2006) Cold acclimation of the Arabidopsis *dgd1* mutant results in recovery from photosystem I-limited photosynthesis. FEBS Lett 580:4959–4968

Hofmann M, Eichenberger W (1997) Lipid and fatty acid composition of the marine brown alga *Dictyopteris membranacea*. Plant Cell Physiol 38:1046–1052

Hofmann M, Eichenberger W (1998) Radiolabelling studies on the lipid metabolism in the marine brown alga *Dictyopteris membranacea*. Plant Cell Physiol 39:508–515

Hölzl G, Witt S, Kelly AA, Zähringer U, Warnecke D, Dörmann P, Heinz E (2006) Functional differences between galactolipids and glucolipids revealed in photosynthesis of higher plants. Proc Natl Acad Sci U S A 103:7512–7517

Hölzl G, Witt S, Gaude N, Melzer M, Schöttler MA, Dörmann P (2009) The role of diglycosyl lipids in photosynthesis and membrane lipid homeostasis in Arabidopsis. Plant Physiol 150:1147–1159

Hurlock AK, Roston RL, Wang K, Benning C (2014) Lipid trafficking in plant cells. Traffic 15:915–932

Ivanov AG, Hendrickson L, Krol M, Selstam E, Öquist G, Hurry V, Huner NP (2006) Digalactosyldiacylglycerol deficiency impairs the capacity for photosynthetic intersystem electron transport and state transitions in *Arabidopsis thaliana* due to photosystem I acceptor-side limitations. Plant Cell Physiol 47:1146–1157

Jamieson GR, Reid EH (1972) The leaf lipids of some conifer species. Phytochemistry 11:269–275

Jamieson GR, Reid EH (1975) The fatty acid composition of fern lipids. Phytochemistry 14:2229–2232

Janouškovec J, Horák A, Oborník M, Lukeš J, Keeling PJ (2010) A common red algal origin of the apicomplexan, dinoflagellate, and heterokont plastids. Proc Natl Acad Sci U S A 107:10949–10954

Jarvis P, Dörmann P, Peto CA, Lutes J, Benning C, Chory J (2000) Galactolipid deficiency and abnormal chloroplast development in the *Arabidopsis* MGD synthase 1 mutant. Proc Natl Acad Sci U S A 97:8175–8179

Jessen D, Roth C, Wiermer M, Fulda M (2015) Two activities of long-chain acyl-coenzyme A synthetase are involved in lipid trafficking between the endoplasmic reticulum and the plastid in Arabidopsis. Plant Physiol 167:351–366

Jones MR (2007) Lipids in photosynthetic reaction centres: structural roles and functional holes. Prog Lipid Res 46:56–87

Jouhet J, Maréchal E, Baldan B, Bligny R, Joyard J, Block MA (2004) Phosphate deprivation induces transfer of DGDG galactolipid from chloroplast to mitochondria. J Cell Biol 167:863–874

Joyard J, Stumpf PK (1981) Synthesis of long-chain acyl-CoA in chloroplast envelope membranes. Plant Physiol 67:250–256

Kajikawa M, Yamato KT, Kohzu Y, Shoji S, Matsui K, Tanaka Y, Sakai Y, Fukuzawa H (2006) A front-end desaturase from *Chlamydomonas reinhardtii* produces pinolenic and coniferonic acids by ω13 desaturation in methylotrophic yeast and tobacco. Plant Cell Physiol 47:64–73

Kalanon M, McFadden GI (2008) The chloroplast protein translocation complexes of *Chlamydomonas reinhardtii*: a bioinformatic comparison of Toc and Tic components in plants, green algae and red algae. Genetics 179:95–112

Kaoua ME, Laval-Martin DL (1995) Evolution of PS IIα and PS IIβ centers during the greening of *Euglena gracilis* Z: correlations with changes in lipid content. Photosynth Res 43:155–163

Kato M, Sakai M, Adachi K, Ikemoto H, Sano H (1996) Distribution of betaine lipids in marine algae. Phytochemistry 42:1341–1345

Keeling PJ (2010) The endosymbiotic origin, diversification and fate of plastids. Philos Trans R Soc Lond B Biol Sci 365:729–748

Keeling PJ (2013) The number, speed, and impact of plastid endosymbioses in eukaryotic evolution. Annu Rev Plant Biol 64:583–607

Kelly AA, Dörmann P (2002) DGD2, an *Arabidopsis* gene encoding a UDP-galactose-dependent digalactosyldiacylglycerol synthase is expressed during growth under phosphate-limiting conditions. J Biol Chem 277:1166–1173

Kelly AA, Froehlich JE, Dörmann P (2003) Disruption of the two digalactosyldiacylglycerol synthase genes DGD1 and DGD2 in *Arabidopsis* reveals the existence of an additional enzyme of galactolipid synthesis. Plant Cell 15:2694–2706

Khotimchenko SV (2002) Distribution of glyceroglycolipids in marine algae and grasses. Chem Nat Compd 38:223–229
Khozin I, Adlerstein D, Bigongo C, Heimer YM, Cohen Z (1997) Elucidation of the biosynthesis of eicosapentaenoic acid in the microalga *Porphyridium cruentum* (II. Studies with radiolabeled precursors). Plant Physiol 114:223–230
Khozin-Goldberg I, Cohen Z (2006) The effect of phosphate starvation on the lipid and fatty acid composition of the fresh water eustigmatophyte *Monodus subterraneus*. Phytochemistry 67:696–701
Khozin-Goldberg I, Yu HZ, Adlerstein D, Didi-Cohen S, Heimer YM, Cohen Z (2000) Triacylglycerols of the red microalga *Porphyridium cruentum* can contribute to the biosynthesis of eukaryotic galactolipids. Lipids 35:881–889
Kim S, Yamaoka Y, Ono H, Kim H, Shim D, Maeshima M, Martinoia E, Cahoon EB, Nishida I, Lee Y (2013) AtABCA9 transporter supplies fatty acids for lipid synthesis to the endoplasmic reticulum. Proc Natl Acad Sci U S A 110:773–778
Klecker M, Gasch P, Peisker H, Dörmann P, Schlicke H, Grimm B, Mustroph A (2014) A shoot-specific hypoxic response of Arabidopsis sheds light on the role of the phosphate-responsive transcription factor PHOSPHATE STARVATION RESPONSE1. Plant Physiol 165:774–790
Kleinig H, Liedvogel B (1978) Fatty acid synthesis by isolated chromoplasts from the daffodil [^{14}C]acetate incorporation and distribution of labelled acids. Eur J Biochem 83:499–505
Kobayashi K, Awai K, Takamiya K, Ohta H (2004) *Arabidopsis* type B monogalactosyldiacylglycerol synthase genes are expressed during pollen tube growth and induced by phosphate starvation. Plant Physiol 134:640–648
Kobayashi K, Kondo M, Fukuda H, Nishimura M, Ohta H (2007) Galactolipid synthesis in chloroplast inner envelope is essential for proper thylakoid biogenesis, photosynthesis, and embryogenesis. Proc Natl Acad Sci U S A 104:17216–17221
Kobayashi K, Awai K, Nakamura M, Nagatani A, Masuda T, Ohta H (2009) Type-B monogalactosyldiacylglycerol synthases are involved in phosphate starvation-induced lipid remodeling, and are crucial for low-phosphate adaptation. Plant J 57:322–331
Koskimies-Soininen K, Nyberg H (1991) Effects of temperature and light on the glycolipids of *Sphagnum fimbriatum*. Phytochemistry 30:2529–2536
Kroll D, Meierhoff K, Bechtold N, Kinoshita M, Westphal S, Vothknecht UC, Soll J, Westhoff P (2001) VIPP1, a nuclear gene of *Arabidopsis thaliana* essential for thylakoid membrane formation. Proc Natl Acad Sci U S A 98:4238–4242
Kumari P, Bijo AJ, Mantri VA, Reddy CR, Jha B (2013) Fatty acid profiling of tropical marine macroalgae: an analysis from chemotaxonomic and nutritional perspectives. Phytochemistry 86:44–56
Künzler K, Eichenberger W, Radunz A (1997) Intracellular localization of two betaine lipids by cell fractionation and immunomicroscopy. Z Naturforsch C 52:487–495
Leblond JD, Timofte HI, Roche SA, Porter NM (2010) Mono- and digalactosyldiacylglycerol composition of glaucocystophytes (Glaucophyta): a modern interpretation using positive-ion electrospray ionization/mass spectrometry/mass spectrometry. Phycol Res 58:222–229
Li X, Fan X, Han L, Lou Q (2002) Fatty acids of some algae from the Bohai Sea. Phytochemistry 59:157–161
Li N, Gügel IL, Giavalisco P, Zeisler V, Schreiber L, Soll J, Philippar K (2015) FAX1, a novel membrane protein mediating plastid fatty acid export. PLoS Biol 13:e1002053
Liang Y, Maeda Y, Yoshino T, Matsumoto M, Tanaka T (2014) Profiling of polar lipids in marine oleaginous diatom *Fistulifera solaris* JPCC DA0580: prediction of the potential mechanism for eicosapentaenoic acid-incorporation into triacylglycerol. Mar Drugs 12:3218–3230
Lu B, Xu C, Awai K, Benning C (2007) TGD3, an ATPase protein of Arabidopsis, functions in ER-to-plastid lipid trafficking. FASEB J 21:A236
Mackiewicz P, Bodył A, Gagat P (2012) Possible import routes of proteins into the cyanobacterial endosymbionts/plastids of *Paulinella* chromatophora. Theory Biosci 131:1–18

Maréchal E, Awai K, Block M, Brun D, Masuda T, Shimada H, Takamiya K, Ohta H, Joyard J (2000) The multigenic family of monogalactosyl diacylglycerol synthases. Biochem Soc Trans 28:732–738

Miège C, Maréchal E, Shimojima M, Awai K, Block MA, Ohta H, Takamiya K, Douce R, Joyard J (1999) Biochemical and topological properties of type A MGDG synthase, a spinach chloroplast envelope enzyme catalyzing the synthesis of both prokaryotic and eukaryotic MGDG. Eur J Biochem 265:990–1001

Moellering ER, Benning C (2011) Galactoglycerolipid metabolism under stress: a time for remodeling. Trends Plant Sci 16:98–107

Moellering ER, Muthan B, Benning C (2010) Freezing tolerance in plants requires lipid remodeling at the outer chloroplast membrane. Science 330:226–228

Mongrand S, Bessoule J-J, Cabantous F, Cassagne C (1998) The C16:3/C18:3 fatty acid balance in photosynthetic tissues from 468 plant species. Phytochemistry 49:1049–1064

Mongrand S, Cassagne C, Bessoule JJ (2000) Import of lyso-phosphatidylcholine into chloroplasts likely at the origin of eukaryotic plastidial lipids. Plant Physiol 122:845–852

Murata N, Hoshi H (1984) Sulfoquinovosyl diacylglycerols in chilling-sensitive and chilling-resistant plants. Plant Cell Physiol 25:1241–1245

Muto M, Kubota C, Tanaka M, Satoh A, Matsumoto M, Yoshino T, Tanaka T (2013) Identification and functional analysis of Delta-9 desaturase, a key enzyme in PUFA synthesis, isolated from the oleaginous diatom Fistulifera. PLoS One 8:e73507

Nakamura Y, Kobayashi K, Ohta H (2009) Activation of galactolipid biosynthesis in development of pistils and pollen tubes. Plant Physiol Biochem 47:535–539

Nakayama T, Archibald JM (2012) Evolving a photosynthetic organelle. BMC Biol 10:35

Neufeld EF, Hall CW (1964) Formation of galactolipids by chloroplasts. Biochem Biophys Res Commun 14:503–508

Nilsson AK, Fahlberg P, Ellerström M, Andersson MX (2012) Oxo-phytodienoic acid (OPDA) is formed on fatty acids esterified to galactolipids after tissue disruption in *Arabidopsis thaliana*. FEBS Lett 586:2483–2487

Nowack EC, Melkonian M, Glöckner G (2008) Chromatophore genome sequence of *Paulinella* sheds light on acquisition of photosynthesis by eukaryotes. Curr Biol 18:410–408

Ohlrogge J, Browse J (1995) Lipid biosynthesis. Plant Cell 7:957–970

Okazaki Y, Shimojima M, Sawada Y, Toyooka K, Narisawa T, Mochida K, Tanaka H, Matsuda F, Hirai A, Hirai MY, Ohta H, Saito K (2009) A chloroplastic UDP-glucose pyrophosphorylase from *Arabidopsis* is the committed enzyme for the first step of sulfolipid biosynthesis. Plant Cell 21:892–909

Okazaki Y, Otsuki H, Narisawa T, Kobayashi M, Sawai S, Kamide Y, Kusano M, Aoki T, Hirai MY, Saito K (2013) A new class of plant lipid is essential for protection against phosphorus depletion. Nat Commun 4:1510

Okazaki Y, Nishizawa T, Takano K, Ohnishi M, Mimura T, Saito K (2015) Induced accumulation of glucuronosyldiacylglycerol in tomato and soybean under phosphorus deprivation. Physiol Plant 155(1):33–42

Parfrey LW, Lahr DJ, Knoll AH, Katz LA (2011) Estimating the timing of early eukaryotic diversification with multigene molecular clocks. Proc Natl Acad Sci U S A 108:13624–13629

Petroutsos D, Amiar S, Abida H, Dolch LJ, Bastien O, Rébeillé F, Jouhet J, Falconet D, Block MA, McFadden GI, Bowler C, Botté C, Maréchal E (2014) Evolution of galactoglycerolipid biosynthetic pathways – from cyanobacteria to primary plastids and from primary to secondary plastids. Prog Lipid Res 54:68–85

Poirier Y, Thoma S, Somerville C, Schiefelbein J (1991) Mutant of *Arabidopsis* deficient in xylem loading of phosphate. Plant Physiol 97:1087–1093

Qi Y, Yamauchi Y, Ling J, Kawano N, Li D, Tanaka K (2004) Cloning of a putative monogalactosyldiacylglycerol synthase gene from rice (*Oryza sativa* L.) plants and its expression in response to submergence and other stresses. Planta 219:450–458

Regnault A, Chervin D, Chammai A, Piton F, Calvayrac R, Mazliak P (1995) Lipid composition of *Euglena gracilis* in relation to carbon-nitrogen balance. Phytochemistry 40:725–733

Riekhof WR, Ruckle ME, Lydic TA, Sears BB, Benning C (2003) The sulfolipids 2′-O-acyl-sulfoquinovosyldiacylglycerol and sulfoquinovosyldiacylglycerol are absent from a *Chlamydomonas reinhardtii* mutant deleted in *SQD1*. Plant Physiol 133:864–874

Riekhof WR, Sears BB, Benning C (2005) Annotation of genes involved in glycerolipid biosynthesis in *Chlamydomonas reinhardtii*: discovery of the betaine lipid synthase BTA1Cr. Eukaryot Cell 4:242–252

Rodríguez-Ezpeleta N, Brinkmann H, Burey SC, Roure B, Burger G, Löffelhardt W, Bohnert HJ, Philippe H, Lang BF (2005) Monophyly of primary photosynthetic eukaryotes: green plants, red algae, and glaucophytes. Curr Biol 15:1325–1330

Roughan PG, Slack CR (1977) Long-chain acyl-coenzyme a synthetase activity of spinach chloroplasts is concentrated in the envelope. Biochem J 162:457–459

Roughan PG, Slack CR (1982) Cellular organization of glycerolipid metabolism. Annu Rev Plant Physiol Plant Mol Biol 33:97–132

Sakurai I, Mizusawa N, Wada H, Sato N (2007) Digalactosyldiacylglycerol is required for stabilization of the oxygen-evolving complex in photosystem II. Plant Physiol 145:1361–1370

Sanda S, Leustek T, Theisen MJ, Garavito RM, Benning C (2001) Recombinant *Arabidopsis* SQD1 converts UDP-glucose and sulfite to the sulfolipid head group precursor UDP-sulfoquinovose *in vitro*. J Biol Chem 276:3941–3946

Sanina NM, Goncharova SN, Kostetsky EY (2004) Fatty acid composition of individual polar lipid classes from marine macrophytes. Phytochemistry 65:721–730

Sato N (1992) Betaine lipids. Bot Mag Tokyo 105:185–197

Sato N, Moriyama T (2007) Genomic and biochemical analysis of lipid biosynthesis in the unicellular rhodophyte *Cyanidioschyzon merolae*: lack of a plastidic desaturation pathway results in the coupled pathway of galactolipid synthesis. Eukaryot Cell 6:1006–1017

Sato N, Tsuzuki M, Matsuda Y, Ehara T, Osafune T, Kawaguchi A (1995) Isolation and characterization of mutants affected in lipid metabolism of *Chlamydomonas reinhardtii*. Eur J Biochem 230:987–993

Sato N, Hagio M, Wada H, Tsuzuki M (2000) Environmental effects on acidic lipids of thylakoid membranes. Biochem Soc Trans 28:912–914

Sato N, Sugimoto K, Meguro A, Tsuzuki M (2003a) Identification of a gene for UDP-sulfoquinovose synthase of a green alga, *Chlamydomonas reinhardtii*, and its phylogeny. DNA Res 10:229–237

Sato N, Tsuzuki M, Kawaguchi A (2003b) Glycerolipid synthesis in *Chlorella kessleri* 11h. I. Existence of a eukaryotic pathway. Biochim Biophys Acta 1633:27–34

Schaller A, Stintzi A (2009) Enzymes in jasmonate biosynthesis – structure, function, regulation. Phytochemistry 70:1532–1538

Shanklin J, Somerville C (1991) Stearoyl-acyl-carrier-protein desaturase from higher plants is structurally unrelated to the animal and fungal homologs. Proc Natl Acad Sci U S A 88:2510–2514

Shimojima M (2011) Biosynthesis and functions of the plant sulfolipid. Prog Lipid Res 50:234–239

Shipley GG, Green JP, Nichols BW (1973) The phase behavior of monogalactosyl, digalactosyl, and sulphoquinovosyl diglycerides. Biochim Biophys Acta 311:531–544

Siegenthaler P (1998) Molecular organization of acyl lipids in photosynthetic membranes of higher plants. In: Siegenthaler P, Murata N (eds) Lipids in photosynthesis: structure, function and genetics, vol 6, Advances in Photosynthesis. Kluwer Academic Publishers, Dordrecht, pp 119–144

Thompson GA Jr (1996) Lipids and membrane function in green algae. Biochim Biophys Acta 1302:17–45

Tjellström H, Hellgren LI, Wieslander A, Sandelius AS (2010) Lipid asymmetry in plant plasma membranes: phosphate deficiency-induced phospholipid replacement is restricted to the cytosolic leaflet. FASEB J 24:1128–1138

Ulvskov P, Paiva DS, Domozych D, Harholt J (2013) Classification, naming and evolutionary history of glycosyltransferases from sequenced green and red algal genomes. PLoS One 8:e76511
Van Mooy BA, Fredricks HF, Pedler BE, Dyhrman ST, Karl DM, Koblížek M, Lomas MW, Mincer TJ, Moore LR, Moutin T, Rappé MS, Webb EA (2009) Phytoplankton in the ocean use non-phosphorus lipids in response to phosphorus scarcity. Nature 458:69–72
Vieler A, Wilhelm C, Goss R, Süss R, Schiller J (2007) The lipid composition of the unicellular green alga *Chlamydomonas reinhardtii* and the diatom *Cyclotella meneghiniana* investigated by MALDI-TOF MS and TLC. Chem Phys Lipids 150:143–155
Villanueva L, Bale N, Hopmans EC, Schouten S, Damsté JS (2014) Diversity and distribution of a key sulpholipid biosynthetic gene in marine microbial assemblages. Environ Microbiol 16:774–787
Vu HS, Roth MR, Tamura P, Samarakoon T, Shiva S, Honey S, Lowe K, Schmelz EA, Williams TD, Welti R (2014) Head-group acylation of monogalactosyldiacylglycerol is a common stress response, and the acyl-galactose acyl composition varies with the plant species and applied stress. Physiol Plant 150:517–528
Wang Z, Xu C, Benning C (2012) TGD4 involved in endoplasmic reticulum-to-chloroplast lipid trafficking is a phosphatidic acid binding protein. Plant J 70:614–623
Webb MS, Green BR (1991) Biochemical and biophysical properties of thylakoid acyl lipids. Biochim Biophys Acta 1060:133–158
Xu C, Härtel H, Wada H, Hagio M, Yu B, Eakin C, Benning C (2002) The *pgp1* mutant locus of Arabidopsis encodes a phosphatidylglycerolphosphate synthase with impaired activity. Plant Physiol 129:594–604
Xu C, Fan J, Riekhof W, Froehlich JE, Benning C (2003) A permease-like protein involved in ER to thylakoid lipid transfer in *Arabidopsis*. EMBO J 22:2370–2379
Xu C, Fan J, Cornish A, Benning C (2008) Lipid trafficking between the endoplasmic reticulum and the plastid in Arabidopsis requires the extraplastidic TGD4 protein. Plant Cell 20:2190–2204
Xu J, Chen D, Yan X, Chen J, Zhou C (2010) Global characterization of the photosynthetic glycerolipids from a marine diatom *Stephanodiscus* sp. by ultra performance liquid chromatography coupled with electrospray ionization-quadrupole-time of flight mass spectrometry. Anal Chim Acta 663:60–68
Yongmanitchai W, Ward OP (1993) Positional distribution of fatty acids, and molecular species of polar lipids, in the diatom *Phaeodactylum tricornutum*. J Gen Microbiol 139:465–472
Yu B, Benning C (2003) Anionic lipids are required for chloroplast structure and function in *Arabidopsis*. Plant J 36:762–770
Yu B, Xu C, Benning C (2002) *Arabidopsis* disrupted in *SQD2* encoding sulfolipid synthase is impaired in phosphate-limited growth. Proc Natl Acad Sci U S A 99:5732–5737
Zäuner S, Jochum W, Bigorowski T, Benning C (2012) A cytochrome b_5-containing plastid-located fatty acid desaturase from *Chlamydomonas reinhardtii*. Eukaryot Cell 11:856–863
Zhang L, Kato Y, Otters S, Vothknecht UC, Sakamoto W (2012) Essential role of VIPP1 in chloroplast envelope maintenance in *Arabidopsis*. Plant Cell 24:3695–3707
Zimorski V, Ku C, Martin WF, Gould SB (2014) Endosymbiotic theory for organelle origins. Curr Opin Microbiol 22C:38–48

Chapter 4
Thylakoid Development and Galactolipid Synthesis in Cyanobacteria

Koichiro Awai

Abstract Cyanobacteria carry out oxygenic photosynthesis and share many features with chloroplasts, including thylakoid membranes, which are mainly composed of membrane lipids and protein complexes that mediate photosynthetic electron transport. Although the functions of the various thylakoid protein complexes have been well characterized, the details underlying the biogenesis of thylakoid membranes remain unclear. Galactolipids are the major constituents of the thylakoid membrane system, and all the genes involved in galactolipid biosynthesis were recently identified. In this chapter, I summarize recent advances in our understanding of the factors involved in thylakoid development, including regulatory proteins and enzymes that mediate lipid biosynthesis.

Keywords Thylakoid membrane • Galactolipid • Glycosyltransferase • Epimerase • Photosynthetic protein complex • VIPP1

Introduction

Cyanobacteria are the only prokaryotes known to conduct oxygenic photosynthesis, which is very similar to that carried out in chloroplasts of photosynthetic eukaryotic cells such as land plants and algae. Because cyanobacteria and chloroplasts share many aspects, e.g., oxygenic photosynthesis, genome structure and lipid composition, it is generally accepted that cyanobacteria and chloroplasts share a common ancestor. Oxygenic photosynthesis produces oxygen via the electron transport chain, which comprises three large membrane protein complexes, namely photosystem II, cytochrome b_6f and photosystem I. These complexes are also conserved in

K. Awai (✉)
Department of Biological Science, Faculty of Science, Shizuoka University, 836 Ohya, Suruga-ku, Shizuoka 422-8529, Japan

Research Institute of Electronics, Shizuoka University, 3-5-1 Johoku, Naka-ku, Hamamatsu 432-8011, Japan

JST, PRESTO, 4-1-8 Honcho, Kawaguchi, Saitama 332-0012, Japan
e-mail: awai.koichiro@shizuoka.ac.jp

Y. Nakamura, Y. Li-Beisson (eds.), *Lipids in Plant and Algae Development*, Subcellular Biochemistry 86, DOI 10.1007/978-3-319-25979-6_4

both cyanobacteria and chloroplasts and are embedded in thylakoid membranes. The word "thylakoid" comes from the Greek word θύλακος (thylakos), which means "a bag, sack or pocket". Indeed, thylakoid membranes surround the thylakoid lumen where oxygen is produced during photosynthesis. Thylakoid membranes are composed of proteins (~60 % by weight) and lipids (~40 %) (Chapman et al. 1983). Thus, it is obvious that proteins and lipids are important for thylakoid membrane biogenesis.

Despite the accumulated research on the biochemical, chemical and physical properties of the photosynthetic apparatus, the mechanisms underlying thylakoid membrane biogenesis have not been examined in depth, probably because of difficulties inherent in the analysis of *de novo* synthesis of a membrane system. In plants, if thylakoid membranes do not develop, there is no photosynthetic activity and plants cannot survive. Indeed, a mutant of the major thylakoid lipid synthase is seedling lethal (Kobayashi et al. 2007). If plants are kept in the dark before germination, plastids will differentiate into etioplasts and develop prolamellar bodies. This inter-organelle structure is formed by crystallization of protochlorophyllide, NADPH-protochlorophyllide oxidoreductase, and lipids (mainly galactolipids), and thylakoid differentiation occurs upon illumination (Selstam 1998). Thus, this non-natural paradigm of thylakoid development cannot be considered to reflect *de novo* biogenesis. As for cyanobacteria, they essentially cannot survive in the dark. Under appropriate light intensity, cyanobacteria divide into two daughter cells that share thylakoid membranes, and thus it is difficult to observe *de novo* biogenesis of thylakoids. Even with these difficulties, however, recent advances in omics techniques have fostered the identification of key components involved in thylakoid membrane development.

This chapter summarizes the structure of thylakoid membranes as well as the proteins and lipids involved in thylakoid biogenesis. For further details on individual topics, please refer to the following reviews: thylakoid membrane organization/structure (Liberton and Pakrasi 2008); lipid biosynthesis/regulation in cyanobacteria (Sato and Wada 2010); evolution of galactolipid synthesis in photosynthetic organisms (Petroutsos et al. 2014); biogenesis of thylakoid membranes (Rast et al. 2015); and function(s) of VIPP1 (Zhang and Sakamoto 2015).

Structure of Thylakoid Membranes in Oxygenic Photosynthetic Organisms

Thylakoid membrane structure is relatively conserved in chloroplasts of land plants. In the photosynthetic mesophyll cells, thylakoid membranes have a multi-layer structure with highly stacked areas called grana (Fig. 4.1a). The detailed structure of grana is still a matter of debate, but its basic structure comprises stacks of 300- to 600-nm membrane discs that presumably are connected to each other (Mustardy and Garab 2003; Mullineaux 2005). Another component of thylakoids is the lamella,

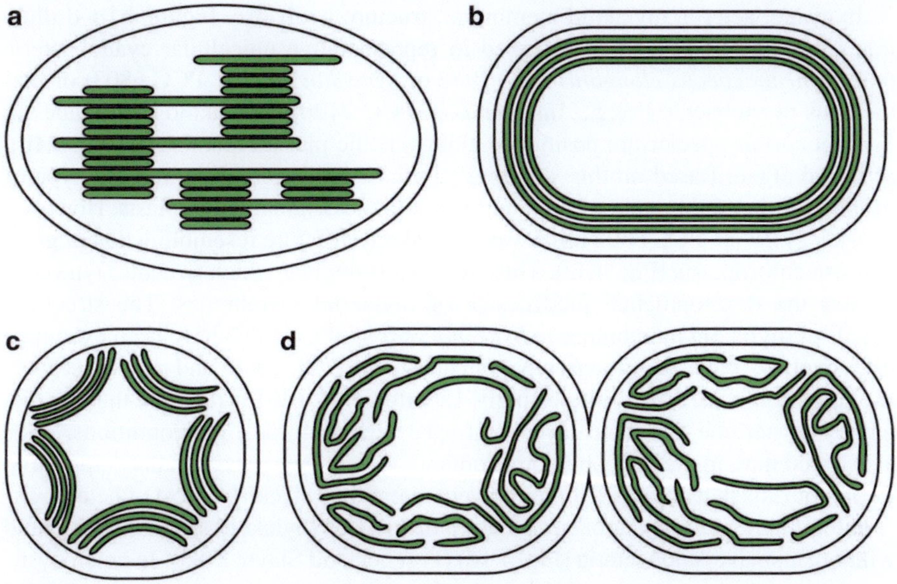

Fig. 4.1 Schematic representation of typical thylakoid structure in plants, algae and cyanobacteria. (**a**) Thylakoid membranes of plant chloroplasts with highly stacked grana structure. (**b**) Thylakoid membranes of unicellular cyanobacteria (e.g., *Synechococcus elongatus* PCC 7942) and red algal chloroplasts. (**c**) Thylakoid membranes of unicellular cyanobacteria (e.g., *Synechocystis* sp. PCC 6803) and glaucophyte chloroplasts. (**d**) Thylakoid membranes of filamentous cyanobacteria (e.g., *Anabaena* sp. PCC 7120)

which connects adjacent stacks of grana. Interestingly, lamellar structure can be reconstituted by mixing thylakoid lipids with thylakoid-specific integral membrane proteins (Simidjiev et al. 2000). A similar grana structure is also seen in chloroplasts of green algae such as *Chlamydomonas reinhardtii*. Deme et al. (2014) proposed that grana stacking is aided by the hydroxyl groups of galactolipids, and this may explain why galactolipids are the predominant lipid type in thylakoid membranes (see below). In red algae, glaucophytes and cyanobacteria, however, grana are not observed in thylakoid membranes. For example, in the primitive red alga *Cyanidioschyzon merolae*, thylakoid membranes lie along the inner-envelope membranes of chloroplasts and are not stacked (Fig. 4.1b). Chloroplasts of the glaucophyte *Cyanophora paradoxa* have thylakoid membranes with a "wavy" structure (Fig. 4.1c; Miyagishima et al. 2014) but have no grana. These algae utilize soluble phycobilisomes for the antenna protein complex to carry out photosynthesis. Each phycobilisome comprises a large complex of ~600 proteins and pigments (e.g., phycocyanobilin and phycoerythrobilin). Because this protein complex attaches to the stroma site of photosynthetic protein complexes (mainly photosystem II), it is widely believed that phycobilisomes physically prevent the stacking of thylakoid membranes in these algae.

In cyanobacteria, thylakoid membrane structure is diverse. Figure 4.1b–d illustrates the various structures observed in representative unicellular cyanobacteria (e.g., *Synechococcus elongatus* PCC 7942 or *Synechocystis* sp. PCC 6803) or filamentous cyanobacteria (e.g., *Anabaena* sp. PCC 7120). Thylakoid membranes in *Synechococcus* species are positioned alongside the plasma membrane (Fig. 4.1b), as in red algae. Based on this structural similarity, it is plausible that this type of cyanobacteria might retain a common feature to the original chloroplasts. However, *Synechocystis* sp. PCC 6803 has a wavy thylakoid structure resembling that of glaucophyte chloroplasts (Fig. 4.1c). Thus, it remains unclear which structure type constitutes the developmental predecessor of thylakoid membranes. The structural details of thylakoid membranes in *Synechocystis* sp. PCC 6803 have been examined intensively by electron tomography (van de Meene et al. 2006) and neutron scattering (Liberton et al. 2013). Interestingly, Liberton et al. (2013) reported that the distance between the thylakoid layers can vary depending on light conditions. They also found that, in the phycobilisome mutants, this distance was smaller, supporting the hypothesis that phycobilisomes prevent membrane stacking. Most of heterocyst-forming filamentous cyanobacteria have a disordered thylakoid structure compared with other such cyanobacteria (Fig. 4.1d) (Peschek and Sleytr 1983). In heterocysts, which specialize in nitrogen fixation, the thylakoid membranes are more contorted (Lang and Fay 1971). The molecular basis for the diversity of thylakoid structure in cyanobacteria has not been explored, and thus how these structures mature remains an open question.

Proteins Involved in the Biogenesis of Thylakoid Membranes

Besides the major constituent protein complexes of thylakoid membranes, such as photosystem I/II, cytochrome b_6f and ATP synthase, Vesicle-Inducing Protein in Plastid 1 (VIPP1) is the most widely acknowledged protein reported to be involved in thylakoid development from cyanobacteria to plants. VIPP1 was first identified as a protein localized to both the inner envelope and thylakoid membranes of plastids (Li et al. 1994). Later, the gene encoding VIPP1 was identified in both *Arabidopsis thaliana* and *Synechocystis* sp. PCC 6803 (Kroll et al. 2001; Westphal et al. 2001). A mutant of Arabidopsis VIPP1, which can survive only on synthetic medium containing sucrose, has a T-DNA insertion in the promoter region of the gene and expresses only ~20 % of the VIPP1 of wild-type plants (Kroll et al. 2001). Cyanobacteria have a multi-copy genome, and Vipp1 null mutants cannot be segregated even after 4 years on selective medium (Fuhrmann et al. 2009). These results indicate that the *VIPP1* gene is essential in both plants and cyanobacteria. These mutants have facilitated analysis of the physiological functions of VIPP1. In response to cold stress, plastids accumulate vesicles in the stroma (Carde et al. 1982). Arabidopsis *vipp1* mutants have lesser vesicle formation than wild type under the same cold stress. These data indicated that VIPP1 may be involved in vesicle formation in plastids (Kroll et al. 2001). However, VIPP1 is now known to have promiscuous function. Zhang et al. (2012) proposed that VIPP1 is involved in the maintenance of chloroplast envelopes. They found that, in *vipp1* knockout and knockdown mutants of Arabidopsis, chloroplasts have a balloon-like structure and thylakoids are located in a limited area (Zhang and Sakamoto 2013). They also showed that VIPP1 moves along the surface of the envelope depending on whether there is osmotic stress, and they concluded that VIPP1 makes lattice-like macrocomplexes that facilitate the resealing of osmotically damaged envelope membranes of chloroplasts. The amino-terminal α-helix of recombinant VIPP1, when produced in *Escherichia coli*, binds tightly to *E. coli* lipids (Otters et al. 2013). In the green alga *Chlamydomonas*, VIPP1 is purportedly involved in the biogenesis and/or assembly of photosystem core complexes by delivering specific structural lipids required for core complex assembly (Nordhues et al. 2012). It is not known which lipid species attaches to VIPP1, but phosphatidylglycerol is a bona fide candidate because it is the only membrane lipid that exists in both thylakoid and *E. coli* membranes. Zhang and Sakamoto (2015) suggested that VIPP1 can also bind galactolipids for the maintenance of envelope membranes. Future work, such as carrying out lipid binding assays, is required to determine which lipid species binds VIPP1.

Recently, a *vipp1* knockout mutant was isolated from the cyanobacterium *Synechococcus* sp. PCC 7002 using a sophisticated strategy (Zhang et al. 2014). They first found that merodiploid mutants of *vipp1* had a reduced level of photosystem I. Based on this result, the gene *psaAB*, which encodes core proteins of photosystem I, were knocked out and then the mutants were grown in medium supplemented with glycerol as the only carbon and energy source. Using this strain, *vipp1* was disrupted and the *psaAB* genes were rescued to obtain a *vipp1* null mutant. This mutant grew photoheterotrophically with glycerol under dim light and had vestigial, yet clear, thylakoid membranes. This result indicates that Vipp1 is not essential for the biogenesis of thylakoid membranes at least in *Synechococcus* sp. PCC 7002. A similar result was obtained using an inducible promoter system in *Synechocystis* sp. PCC 6803. Under conditions in which VIPP1 content was low, the thylakoid membranes were still well developed but photosynthetic activity was much less than that of wild type (Gao and Xu 2009), indicating that Vipp1 is important for photosynthesis rather than thylakoid development. It remains unclear whether VIPP1 has a central role(s) in thylakoid development via vesicle trafficking, but VIPP1 seems to be involved in the assembly of photosystem complexes and/or membrane maintenance, and failure of these functions results in retarded thylakoid development. According to these results, Vothknecht et al. (2012) proposed that the protein be renamed as <u>V</u>ery <u>I</u>mportant <u>P</u>rotein in <u>P</u>lastids 1.

A few other proteins have also been found to be involved in thylakoid development. The Arabidopsis CURVATURE THYLAKOID 1 (CURT1) helps to ensure proper grana architecture (Armbruster et al. 2013). CURT1 is conserved from plants to cyanobacteria but not in the particular cyanobacterium *Gloeobacter violaceus* PCC 7421. *Gloeobacter* species lack thylakoid structure (Rippka et al. 1974), suggesting that CURT1 functions in thylakoid development and that this particular function is evolutionarily conserved (Armbruster et al. 2013). A knockout mutant of a CURT1 homolog (*slr0483*) in *Synechocystis* sp. PCC 6803 could not be segregated as a null mutant, indicating that CURT1 is essential in cyanobacteria. Because cyanobacteria do not have grana structure, the exact function of CURT1 in cyanobacteria is unknown.

Stengel et al. (2012) reported that Processing-Associated Tetratricopeptide Repeat protein (PratA) in *Synechocystis* sp. PCC 6803 is indirectly involved in thylakoid development. They proposed that PratA has a role in delivering Mn^{2+} to photosystem II and localizes to the thylakoid center—the site of thylakoid biogenesis (Kunkel 1982). The structure of the thylakoid center is altered in a PratA knockout mutant (Klinkert et al. 2004), suggesting that PratA not only is involved in photosystem II assembly but also has a role in maintaining the integrity of the thylakoid center. However, a PratA homolog also exists in *Gloeobacter* species, and thus this protein may have a function other than Mn^{2+} delivery in these thylakoid-less cyanobacteria.

Lipids in Thylakoid Membranes of Cyanobacteria

As in plant chloroplasts, cyanobacterial thylakoid membranes contain four main membrane lipids, namely monogalactosyldiacylglycerol (MGDG), digalactosyldiacylglycerol (DGDG), sulfoquinovosyldiacylglycerol (SQDG) and phosphatidylglycerol (Table 4.1, Fig. 4.2) (Stanier and Cohen-Bazire 1977). One exception is *Gloeobacter* species, which do not have SQDG (Selstam and Campbell 1996) and lack thylakoids (Rippka et al. 1974). The *Gloeobacter* lineage is predicted to have diverged from other cyanobacteria before the initial chloroplast endosymbiosis (Nelissen et al. 1995; Shih et al. 2013). Recently, Rexroth et al. (2011) found that cytoplasmic membranes of *Gloeobacter violaceus* contain domains enriched with photosynthetic and respiratory complexes, which might have the same evolutionary origin(s) as thylakoid membranes.

Table 4.1 Lipid composition of representative cyanobacteria

	GlcDG	MGDG	DGDG	SQDG	PG	TGDG	Ref.
Synechocystis sp. PCC 6803	0.6	46.2	17.9	25.0	10.3		Sato (1994)
Synechocystis sp. PCC 6803 (+5 mM glucose)	12.4	39.0	15.6	23.8	9.3		Sato (1994)
Synechococcus elongatus PCC 7942		57	13	9	21		Murata et al. (1992)
Anabaena sp. PCC 7120		63.8	14.9	8.2	10.2	2.9	Nakaya and Awai unpublished

GlcDG monoglucosyldiacylglycerol (βGlcDG), *MGDG* monogalactosyldiacylglycerol (βMGDG), *DGDG* digalactosyldiacylglycerol (αβDGDG), *SQDG* sulfoquinovosyldiacylglycerol, *PG* phosphatidylglycerol, *TGDG* trigalactosyldiacylglycerol (cyanobacterial ααβTGDG)

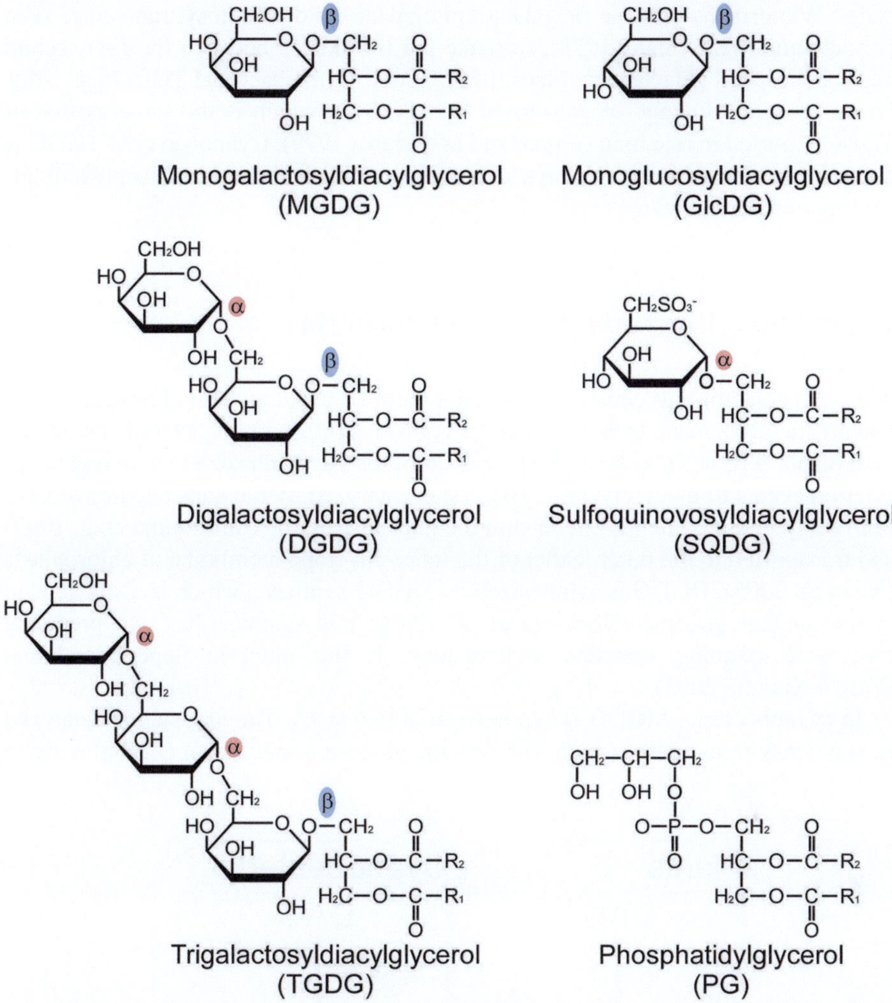

Fig. 4.2 Lipids of thylakoid membranes in cyanobacteria. Five glycolipids and one phospholipid (phosphatidylglycerol) are shown, and the various α and β linkages are indicated

In addition to the four major membrane lipids, cyanobacteria also contain another glycolipid, monoglucosyldiacylglycerol (GlcDG), which is an intermediate of MGDG synthesis (Feige et al. 1980; Sato and Murata 1982b). This glycolipid does not accumulate substantially in cyanobacteria, e.g., generally <2 % of total cellular lipids (Petroutsos et al. 2014). In some filamentous cyanobacteria, trigalactosyldiacylglycerol (TGDG) is a minor lipid (Table 4.1) (Nakaya and Awai unpublished data). TGDG is also found in land plants, but the anomeric configuration depends on the plant species. In Arabidopsis, TGDG is involved in freezing tolerance (Moellering et al. 2010). This lipid is synthesized by a β-galactosidase, originally called Wintermans enzyme or galactolipid:galactolipid galactosyltransferase (van Besouw and Wintermans 1978), to make β-1,6 linkages between the first-second and second-third galactose residues (βββTGDG). Cyanobacterial TGDG has a different anomeric linkage of galactoses (ααβTGDG), which is the same as that of TGDG reported in rice bran (Fujino and Miyazawa 1979). Cyanobacterial TGDG is not induced by cold stress (Nakaya and Awai, unpublished data), and its physiological function is unknown.

Galactolipid Biosynthesis in Cyanobacteria

Although galactolipids constitute the major membrane component of oxygenic photosynthetic organisms, their biosynthetic pathways differ among cyanobacteria and land plants (Fig. 4.3). In land plants, galactolipids are synthesized via two galactosylation events of diacylglycerol. The first is catalyzed by the galactosyltransferase MGDG synthase, which is encoded in the nuclear genome (Shimojima et al. 1997) and transported to the outer leaflet of the inner-envelope membrane of chloroplasts (Xu et al. 2005). DGDG is synthesized by DGDG synthase, which is also encoded in the nuclear genome (Dörmann et al. 1999); this synthase has two predicted membrane-spanning domains and resides in the outer-envelope membrane (Froehlich et al. 2001).

In cyanobacteria, MGDG is synthesized in two steps. The first step is catalyzed by a glucosyltransferase, which transfers the glucose moiety from UDP-glucose to

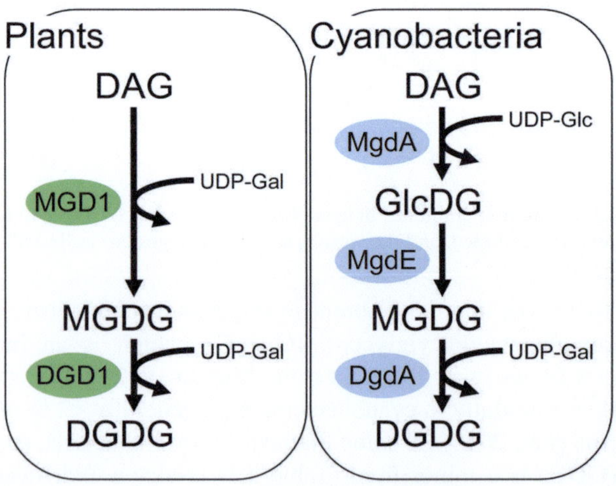

Fig. 4.3 Biosynthetic pathway for galactolipids in plant chloroplasts and cyanobacteria. *DAG* diacylglycerol, *DGD1* plant-type DGDG synthase, *DgdA* cyanobacterial DGDG synthase, *MGD1* plant-type MGDG synthase, *MgdA* cyanobacterial GlcDG synthase, *MgdE* cyanobacterial GlcDG epimerase, *UDP-Gal* UDP-galactose, *UDP-Glc* UDP-glucose

diacylglycerol to yield GlcDG. In the second step, an epimerase isomerizes the glucose moiety of GlcDG to galactose, yielding MGDG. This pathway was first proposed by Feige et al. (1980), who identified GlcDG as a cyanobacteria-specific lipid, and this was confirmed by Sato and Murata (1982b) via pulse-chase labeling. Awai et al. (2006) later confirmed this pathway at the molecular level. The gene encoding GlcDG synthase was identified by comparative genomic analysis using genomic sequences of the representative unicellular cyanobacterium *Synechocystis* sp. PCC 6803 (Kaneko et al. 1996) and filamentous cyanobacterium *Anabaena* sp. PCC 7120 (Kaneko et al. 2001). Only 40 % of genes in both genomes are orthologous, e.g., genes involved in basic reactions such as photosynthesis, nucleic acid metabolism, and fatty acid synthesis (Kaneko et al. 2001). It was hypothesized that, because the cyanobacterial galactolipid synthetic pathway is conserved only among cyanobacteria, the genes corresponding to the enzymes should be conserved in cyanobacterial genomes. To identify the gene for GlcDG synthase, genes encoding enzymes containing a sugar transferase motif were selected in *Synechocystis* sp. PCC 6803. Those genes were then analyzed with regard to their conservation in *Anabaena* sp. PCC 7120, and four conserved genes were flagged. One of them was shown to encode a GlcDG synthase (MgdA) based on an *in vitro* enzyme assay with subsequent NMR-based structure analysis of the accumulated glycolipid in *E. coli* cell membranes (Awai et al. 2006).

Sato and Murata (1982b) proposed that DGDG biosynthesis occurs by the same mechanism in cyanobacteria and plants, i.e., via galactosylation of MGDG. However, the enzyme for the reaction was predicted to differ between cyanobacteria and plants because no homolog for the plant-type DGDG synthase has been found in the cyanobacterial genomes that have been sequenced. The gene for DGDG synthase (*dgdA*) was later identified in *Synechocystis* sp. PCC 6803 (Awai et al. 2007) using the same method used to identify *mgdA*. Sakurai et al. (2007) also identified *dgdA* from *Synechocystis* sp. PCC 6803 using comparative genomics. Knockout mutants of *dgdA* have no detectable DGDG, indicating that this galactolipid is not essential for thylakoid membranes—at least in *Synechocystis* sp. PCC 6803. DGDG was seen in the crystal structure of the photosystem II complex, i.e., five molecules of DGDG per complex (Umena et al. 2011). Although these DGDG molecules are not located in the periphery of the oxygen-evolving complex, the *dgdA* mutants of *Synechocystis* sp. PCC 6803 have a defect in stabilization of the oxygen-evolving complex of photosystem II (Sakurai et al. 2007); this leads to increased sensitivity to light and high temperature (Mizusawa et al. 2009a, b). These results suggest that some DGDGs are not tightly bound to photosystem II but rather are loosely associated with the oxygen-evolving complex.

The last part of the cyanobacterial galactolipid synthetic pathway to be identified was the epimerase. The epimerase-encoding gene was recently identified by Awai et al. (2014), who used cluster analysis because the method used to identify *mgdA* and *dgdA* proved inefficient. The key aspect in this cluster analysis was the complete genome sequence obtained from the chromatophore of the photosynthetic cercozoan *Paulinella chromatophora* (Nowack et al. 2008). The chromatophore is a chloroplast-like organelle, phylogenetically close to the *Prochlorococcus* species of cyanobacteria. In the genome of the chromatophore, all the genes reported to be involved in cyanobacterial lipid biosynthesis, including *mgdA* and *dgdA*, are conserved. Because the epimerase reaction should couple with glucosylation of diacylglycerol to synthesize MGDG, the analysis identified candidate genes with a similar phylogenetic cluster to MgdA with a redox reaction motif. One of the candidates, *mgdE*, was co-expressed with *mgdA* in *E. coli*, and the epimerase activity was thus confirmed.

MgdE belongs to the short-chain dehydrogenase/reductase family (Kramm et al. 2012) and is predicted to have two domains—a fatty acid hydroxylase domain at the amino-terminus and a Rossmann fold domain at the carboxy-terminus. In the hydroxylase domain, all the typical histidine residues involved in the hydroxylase reaction are conserved (Shanklin et al. 1994). However, hydroxylated fatty acids were not detected in either the wild-type *Synechocystis* sp. PCC 6803 or the *E. coli* that expressed *mgdE*. Thus, the exact function of this domain remains unknown. The Rossmann fold domain is involved in redox reactions that use the cofactor NAD. Because epimerases catalyze a redox reaction, it is not surprising that MgdE contains this motif.

Substrate Specificities and Localization of Galactolipid Synthases

The substrate specificity of MgdE was first analyzed by expressing glycolipid synthases of other bacteria. Hölzl et al. (2005) introduced a glycosyltransferase from the bacteria *Deinococcus radiodurans* or *Thermotoga maritima* (α-GlcT) into *S. elongatus* PCC 7942. These enzymes are distant homologs of a glycosyltransferase of the mycobacterium *Acholeplasma laidlawii*, and they synthesize GlcDG—but this GlcDG has an α linkage between glucose and glycerol (αGlcDG), which is different from GlcDG of cyanobacteria (βGlcDG) (Berg et al. 2001). Interestingly, cyanobacteria expressing α-GlcT accumulated both αGlcDG and αMGDG, indicating that MgdE also can epimerize αGlcDG. This was confirmed by co-expression of MgdE with α-GlcT from *A. laidlawii* in *E. coli* (Inoue and Awai, unpublished data). These results indicate that MgdE alone is sufficient to epimerize αGlcDG, i.e., no other cyanobacterial factors are required.

MgdA has restricted substrate specificity. MgdA transfers the glucose of UDP-Glc and makes a β-linkage at the *sn*-3 position of glycerol. Using cyanobacterial membrane fractions and recombinant MgdA, two groups confirmed that UDP-Gal is not the substrate for MGDG synthesis (Sato and Murata 1982a; Awai et al. 2006). MgdA is activated by high temperature (Shimojima et al. 2009), whereas MgdE is inactivated by high temperature (Awai et al. 2014), leading to the accumulation of GlcDG at high temperature *in vivo* (Balogi et al. 2005). GlcDG is also reported to accumulate upon addition of 5 mM glucose to the culture medium (Table 4.1) (Sato

1994). However, the mechanism underlying this accumulation, e.g., activation of MgdA and/or inactivation of MgdE, is unknown.

Because no *in vitro* enzyme assay for purified or recombinant cyanobacterial DgdA is yet available, the exact substrates for DGDG synthesis in cyanobacteria remain unknown. However, using the membrane fraction of the cyanobacterium *Anabaena variabilis*, Sato and Murata (1982a) proposed that DGDG is synthesized from MGDG and UDP-galactose, as seen in plants (Kelly and Dormann 2002). As for the sugar acceptor, DgdA is probably specific for MGDG. In the *mgdE* mutant, MGDG cannot be detected yet GlcDG accumulates. However, neither DGDG nor galactosyl-glucosyl-diacylglycerol accumulates in this mutant, indicating that DgdA does not utilize GlcDG as a substrate (Awai et al. 2014). Hölzl et al. (2005) expressed α-GlcT from *D. radiodurans* or *T. maritima* in *S. elongatus* PCC 7942 as described above. These transformants accumulated αGlcDG and αMGDG but not α-galactosyl-α-glucosyl-diacylglycerol or α-galactosyl-α-galactosyl-diacylglycerol, again indicating that MGDG is the specific substrate for DgdA.

Localization of GlcDG synthase activity in cyanobacteria was first analyzed in *S. elongatus* PCC 6301 (formally called *Anacystis nidulans*), which is a very close relative of *S. elongatus* PCC 7942 (Omata and Murata 1986). Separation of thylakoid membranes, cytoplasmic membranes, and cell walls containing the outer membranes was done by sucrose gradient centrifugation; subsequent assessment of enzyme activity revealed that the glucosyltransferase exists both in the thylakoid and cytoplasmic membranes. Later, Selao et al. (2014) carried out a similar analysis in *Synechocystis* sp. PCC 6803 and also found that the glucosyltransferase activity and MgdA localize to both the thylakoid and cytoplasmic membranes separated by sucrose gradient centrifugation coupled with two-phase partitioning. Interestingly, DgdA was detected only in the cytoplasmic membrane, indicating that the site of DGDG synthesis does not completely overlap with that of GlcDG synthesis.

Structure of Thylakoid Membranes in Mutants of Genes for Galactolipid Biosynthesis and Components of Photosystem Complexes

Thylakoid structure in the *mgdE* mutant was analyzed by transmission electron microscopy (Awai et al. 2014). The mutant had normal thylakoid structure but did not have detectable amounts of MGDG and DGDG, indicating that these two galactolipids are not necessary for thylakoid membrane development. This result also showed that the glucolipid GlcDG can substitute for MGDG and DGDG during thylakoid development. The knockout of *mgdA* was found to be lethal, and thus it was not possible to assess the effect of *mgdA* deletion on thylakoid structure (Awai et al. 2006; Shimojima et al. 2009).

The ultrastructure of thylakoid membranes in DGDG-deficient mutants has not been determined in cyanobacteria. As described above, however, the *mgdE* mutant produced neither MGDG nor DGDG yet had a thylakoid structure similar to that of wild type. Thus, a *dgdA* mutant probably would not have a noticeably altered thylakoid structure. In plants, DGDG deficiency reportedly causes the dwarf phenotype and altered thylakoid structure (Dörmann et al. 1995). In the *dgd1* mutant, which has a stop codon in the glycosyltransferase motif in the principal DGDG synthase DGD1 (Dörmann et al. 1999), the thylakoid membranes are highly curved and located closer to the inner-envelope membranes. The number of grana per chloroplast and the number of thylakoids per granum were increased, with concomitant increases in the length of both the lamella and grana. As such, it will be interesting to assess the morphological consequences of altering the expression and/or localization of CURT1 in the *dgd1* mutant.

Thylakoid structure in a knockout mutant of photosystem II was also investigated in *Synechocystis* sp. PCC 6803 (Nilsson et al. 1992). Three *psbA* genes encoding the D1 protein of photosystem II were disrupted toward the goal of isolating a D1 null mutant. The mutant had a thylakoid structure similar to that of wild type. Because the mutant retained other subunits of photosystem II, i.e., CP43, cytochrome b_{559}, and the 33- and 22-kDa proteins, these proteins might be integral for maintaining the structure of thylakoid membranes.

On the other hand, a mutant of photosystem I had an altered thylakoid structure compared with wild type (van de Meene et al. 2012). The mutant lacking the *psaAB* genes (Shen et al. 1993) was grown under dim light and analyzed with electron microscopy; the thylakoid membranes had a tubular structure. Furthermore, in a photosystem I/II-less mutant (Ermakova-Gerdes et al. 1995), the tubular structures were seen but much less so compared with the photosystem I mutant (van de Meene et al. 2012). As described above, Vipp1 seems to be involved in the assembly of photosystem I, and thus it is likely that photosystem I, but not photosystem II, is important for thylakoid development.

De Novo Biogenesis of Thylakoid Membranes

The mechanism underlying thylakoid membrane development remains uncertain. It will be very informative if one could temporarily halt thylakoid biogenesis in living cells and then later allow biogenesis to proceed so as to facilitate observation of *de novo* development. Cyanobacteria are sensitive to glucose concentration, and most cyanobacteria cannot grow heterotrophically. Anderson and McIntosh (1991) developed a method to grow *Synechocystis* sp. PCC 6803 in the dark using 5 min of full-light illumination per day and 5 mM glucose in the medium. Using this method, Barthel et al. (2013) analyzed thylakoid membrane development; they grew cells in the dark and exposed them to light. Time-course electron microscopy revealed that the dark-grown cells had rudimentary thylakoids, indicating that thylakoid membranes do not arise via *de novo* biogenesis but are rebuilt using the existing structure as a site of nucleation. Interestingly, both photosystems I and II were present in the rudimentary thylakoid and/or cytoplasmic membranes of dark-grown cells, but photosystem I was active and photosystem II was inactive. Subsequently, proteomics and lipidomics analyses of these dark-grown cells were conducted and the results compared with those obtained for light-grown cells (Plohnke et al. 2015). In accordance with the results above, the levels of proteins involved in photosystem II—including PratA—decreased, whereas those of photosystem I increased or were unchanged. On the other hand, there was no significant difference in lipid composition between the dark- and light-grown cells except for a slight increase in SQDG in the dark-grown cells. These lipid results were based on relative content and not absolute values, and thus it would be of interest to determine how lipid content increases along with thylakoid membrane development.

Many key components, including the genes involved in galactolipid synthesis, have been identified, and thus comprehensive analyses of the development mechanism underlying thylakoid membranes are now possible.

Acknowledgments The author highly appreciate Dr. Naoki Sato for reading the manuscript and constructive comments. This work was supported in part by the Precursory Research for Embryonic Science and Technology program of the Japanese Science and Technology Agency.

References

Anderson SL, McIntosh L (1991) Light-activated heterotrophic growth of the cyanobacterium Synechocystis sp. strain PCC 6803: a blue-light-requiring process. J Bacteriol 173(9):2761–2767

Armbruster U, Labs M, Pribil M, Viola S, Xu W, Scharfenberg M, Hertle AP, Rojahn U, Jensen PE, Rappaport F, Joliot P, Dormann P, Wanner G, Leister D (2013) Arabidopsis CURVATURE THYLAKOID1 proteins modify thylakoid architecture by inducing membrane curvature. Plant Cell 25(7):2661–2678. doi:10.1105/tpc.113.113118

Awai K, Kakimoto T, Awai C, Kaneko T, Nakamura Y, Takamiya K, Wada H, Ohta H (2006) Comparative genomic analysis revealed a gene for monoglucosyldiacylglycerol synthase, an enzyme for photosynthetic membrane lipid synthesis in cyanobacteria. Plant Physiol 141(3):1120–1127. doi:10.1104/pp. 106.082859

Awai K, Watanabe H, Benning C, Nishida I (2007) Digalactosyldiacylglycerol is required for better photosynthetic growth of Synechocystis sp. PCC6803 under phosphate limitation. Plant Cell Physiol 48(11):1517–1523. doi:10.1093/pcp/pcm134

Awai K, Ohta H, Sato N (2014) Oxygenic photosynthesis without galactolipids. Proc Natl Acad Sci U S A 111(37):13571–13575. doi:10.1073/pnas.1403708111

Balogi Z, Torok Z, Balogh G, Josvay K, Shigapova N, Vierling E, Vigh L, Horvath I (2005) "Heat shock lipid" in cyanobacteria during heat/light-acclimation. Arch Biochem Biophys 436(2):346–354. doi:10.1016/j.abb.2005.02.018

Barthel S, Bernat G, Seidel T, Rupprecht E, Kahmann U, Schneider D (2013) Thylakoid membrane maturation and PSII activation are linked in greening Synechocystis sp. PCC 6803 cells. Plant Physiol 163(2):1037–1046. doi:10.1104/pp. 113.224428

Berg S, Edman M, Li L, Wikstrom M, Wieslander A (2001) Sequence properties of the 1,2-diacylglycerol 3-glucosyltransferase from Acholeplasma laidlawii membranes. Recognition of a large group of lipid glycosyltransferases in eubacteria and archaea. J Biol Chem 276(25):22056–22063. doi:10.1074/jbc.M102576200

Carde JP, Joyard J, Douce R (1982) Electron-microscopic studies of envelope membranes from spinach plastids. Biol Cell 44(3):315
Chapman DJ, De-Felice J, Barber J (1983) Growth temperature effects on thylakoid membrane lipid and protein content of pea chloroplasts. Plant Physiol 72(1):225–228
Deme B, Cataye C, Block MA, Marechal E, Jouhet J (2014) Contribution of galactoglycerolipids to the 3-dimensional architecture of thylakoids. FASEB J 28(8):3373–3383. doi:10.1096/fj.13-247395
Dörmann P, Hoffmann-Benning S, Balbo I, Benning C (1995) Isolation and characterization of an Arabidopsis mutant deficient in the thylakoid lipid digalactosyl diacylglycerol. Plant Cell 7(11):1801–1810. doi:10.1105/tpc.7.11.1801
Dörmann P, Balbo I, Benning C (1999) Arabidopsis galactolipid biosynthesis and lipid trafficking mediated by DGD1. Science 284(5423):2181–2184
Ermakova-Gerdes S, Shestakov S, Vermaas WFJ (1995) Development of photosystem I-less strain of Synechocystis sp. PCC 6803 for analysis of mutations in the photosystem II proteins D2 and CP43. Photosynthesis: from biology to biosphere, vol 1. Kluwer, Dordrecht
Feige G, Heinz E, Wrage K, Cochems N, Ponzelar E (1980) Biogenesis and function of plant lipids. Elsevier/North Holland Biomedical Press, Amsterdam
Froehlich JE, Benning C, Dormann P (2001) The digalactosyldiacylglycerol (DGDG) synthase DGD1 is inserted into the outer envelope membrane of chloroplasts in a manner independent of the general import pathway and does not depend on direct interaction with monogalactosyldiacylglycerol synthase for DGDG biosynthesis. J Biol Chem 276(34):31806–31812. doi:10.1074/jbc.M104652200
Fuhrmann E, Gathmann S, Rupprecht E, Golecki J, Schneider D (2009) Thylakoid membrane reduction affects the photosystem stoichiometry in the cyanobacterium Synechocystis sp. PCC 6803. Plant Physiol 149(2):735–744. doi:10.1104/pp. 108.132373
Fujino Y, Miyazawa T (1979) Chemical structures of mono-, di-, tri- and tetraglycosyl glycerides in rice bran. Biochim Biophys Acta 572(3):442–451. doi:10.1016/0005-2760(79)90151-6
Gao H, Xu X (2009) Depletion of Vipp1 in Synechocystis sp. PCC 6803 affects photosynthetic activity before the loss of thylakoid membranes. FEMS Microbiol Lett 292(1):63–70. doi:10.1111/j.1574-6968.2008.01470.x
Hölzl G, Zahringer U, Warnecke D, Heinz E (2005) Glycoengineering of cyanobacterial thylakoid membranes for future studies on the role of glycolipids in photosynthesis. Plant Cell Physiol 46(11):1766–1778. doi:10.1093/pcp/pci189
Kaneko T, Sato S, Kotani H, Tanaka A, Asamizu E, Nakamura Y, Miyajima N, Hirosawa M, Sugiura M, Sasamoto S, Kimura T, Hosouchi T, Matsuno A, Muraki A, Nakazaki N, Naruo K, Okumura S, Shimpo S, Takeuchi C, Wada T, Watanabe A, Yamada M, Yasuda M, Tabata S (1996) Sequence analysis of the genome of the unicellular cyanobacterium Synechocystis sp. strain PCC6803. II. Sequence determination of the entire genome and assignment of potential protein-coding regions. DNA Res 3(3):109–136
Kaneko T, Nakamura Y, Wolk CP, Kuritz T, Sasamoto S, Watanabe A, Iriguchi M, Ishikawa A, Kawashima K, Kimura T, Kishida Y, Kohara M, Matsumoto M, Matsuno A, Muraki A, Nakazaki N, Shimpo S, Sugimoto M, Takazawa M, Yamada M, Yasuda M, Tabata S (2001) Complete genomic sequence of the filamentous nitrogen-fixing cyanobacterium Anabaena sp. strain PCC 7120. DNA Res 8(5):205–213, 227–253
Kelly AA, Dormann P (2002) DGD2, an arabidopsis gene encoding a UDP-galactose-dependent digalactosyldiacylglycerol synthase is expressed during growth under phosphate-limiting conditions. J Biol Chem 277(2):1166–1173. doi:10.1074/jbc.M110066200
Klinkert B, Ossenbuhl F, Sikorski M, Berry S, Eichacker L, Nickelsen J (2004) PratA, a periplasmic tetratricopeptide repeat protein involved in biogenesis of photosystem II in Synechocystis sp. PCC 6803. J Biol Chem 279(43):44639–44644. doi:10.1074/jbc.M405393200
Kobayashi K, Kondo M, Fukuda H, Nishimura M, Ohta H (2007) Galactolipid synthesis in chloroplast inner envelope is essential for proper thylakoid biogenesis, photosynthesis, and embryogenesis. Proc Natl Acad Sci U S A 104(43):17216–17221. doi:10.1073/pnas.0704680104

Kramm A, Kisiela M, Schulz R, Maser E (2012) Short-chain dehydrogenases/reductases in cyanobacteria. FEBS J 279(6):1030–1043. doi:10.1111/j.1742-4658.2012.08494.x
Kroll D, Meierhoff K, Bechtold N, Kinoshita M, Westphal S, Vothknecht UC, Soll J, Westhoff P (2001) VIPP1, a nuclear gene of Arabidopsis thaliana essential for thylakoid membrane formation. Proc Natl Acad Sci U S A 98(7):4238–4242. doi:10.1073/pnas.061500998
Kunkel DD (1982) Thylakoid centers: structures associated with the cyanobacterial photosynthetic membrane system. Arch Microbiol 133(2):97–99. doi:10.1007/BF00413518
Lang NJ, Fay P (1971) The heterocysts of blue-green algae. II. Details of ultrastructure. Proc R Soc Lond B 178(1051):193–203
Li HM, Kaneko Y, Keegstra K (1994) Molecular cloning of a chloroplastic protein associated with both the envelope and thylakoid membranes. Plant Mol Biol 25(4):619–632
Liberton M, Pakrasi HB (2008) Membrane systems in cyanobacteria. The cyanobacteria. Molecular biology, genomics and evolution. Caister Academic Press, Norfolk
Liberton M, Page LE, O'Dell WB, O'Neill H, Mamontov E, Urban VS, Pakrasi HB (2013) Organization and flexibility of cyanobacterial thylakoid membranes examined by neutron scattering. J Biol Chem 288(5):3632–3640. doi:10.1074/jbc.M112.416933
Miyagishima SY, Kabeya Y, Sugita C, Sugita M, Fujiwara T (2014) DipM is required for peptidoglycan hydrolysis during chloroplast division. BMC Plant Biol 14:57. doi:10.1186/1471-2229-14-57
Mizusawa N, Sakata S, Sakurai I, Sato N, Wada H (2009a) Involvement of digalactosyldiacylglycerol in cellular thermotolerance in Synechocystis sp. PCC 6803. Arch Microbiol 191(7):595–601. doi:10.1007/s00203-009-0486-7
Mizusawa N, Sakurai I, Sato N, Wada H (2009b) Lack of digalactosyldiacylglycerol increases the sensitivity of Synechocystis sp. PCC 6803 to high light stress. FEBS Lett 583(4):718–722. doi:10.1016/j.febslet.2009.01.021
Moellering ER, Muthan B, Benning C (2010) Freezing tolerance in plants requires lipid remodeling at the outer chloroplast membrane. Science 330(6001):226–228. doi:10.1126/science.1191803
Mullineaux CW (2005) Function and evolution of grana. Trends Plant Sci 10(11):521–525. doi:10.1016/j.tplants.2005.09.001
Murata N, Wada H, Gombos Z (1992) Modes of fatty-acid desaturation in cyanobacteria. Plant Cell Physiol 33(7):933–941
Mustardy L, Garab G (2003) Granum revisited. A three-dimensional model – where things fall into place. Trends Plant Sci 8(3):117–122. doi:10.1016/S1360-1385(03)00015-3
Nelissen B, Van de Peer Y, Wilmotte A, De Wachter R (1995) An early origin of plastids within the cyanobacterial divergence is suggested by evolutionary trees based on complete 16S rRNA sequences. Mol Biol Evol 12(6):1166–1173
Nilsson F, Simpson DJ, Jansson C, Andersson B (1992) Ultrastructural and biochemical characterization of a Synechocystis 6803 mutant with inactivated psbA genes. Arch Biochem Biophys 295(2):340–347
Nordhues A, Schottler MA, Unger AK, Geimer S, Schonfelder S, Schmollinger S, Rutgers M, Finazzi G, Soppa B, Sommer F, Muhlhaus T, Roach T, Krieger-Liszkay A, Lokstein H, Crespo JL, Schroda M (2012) Evidence for a role of VIPP1 in the structural organization of the photosynthetic apparatus in Chlamydomonas. Plant Cell 24(2):637–659. doi:10.1105/tpc.111.092692
Nowack EC, Melkonian M, Glockner G (2008) Chromatophore genome sequence of Paulinella sheds light on acquisition of photosynthesis by eukaryotes. Curr Biol 18(6):410–418. doi:10.1016/j.cub.2008.02.051
Omata T, Murata N (1986) Glucolipid synthesis activities in cytoplasmic and thylakoid membranes from the cyanobacterium Anacystis nidulans. Plant Cell Physiol 27(3):485–490
Otters S, Braun P, Hubner J, Wanner G, Vothknecht UC, Chigri F (2013) The first alpha-helical domain of the vesicle-inducing protein in plastids 1 promotes oligomerization and lipid binding. Planta 237(2):529–540. doi:10.1007/s00425-012-1772-1
Peschek GA, Sleytr UB (1983) Thylakoid morphology of the cyanobacteria Anabaena variabilis and Nostoc MAC grown under light and dark conditions. J Ultrastruct Res 82(2):233–239

Petroutsos D, Amiar S, Abida H, Dolch LJ, Bastien O, Rebeille F, Jouhet J, Falconet D, Block MA, McFadden GI, Bowler C, Botte C, Marechal E (2014) Evolution of galactoglycerolipid biosynthetic pathways – from cyanobacteria to primary plastids and from primary to secondary plastids. Prog Lipid Res 54:68–85. doi:10.1016/j.plipres.2014.02.001

Plohnke N, Seidel T, Kahmann U, Rogner M, Schneider D, Rexroth S (2015) The proteome and lipidome of Synechocystis sp. PCC 6803 cells grown under light-activated heterotrophic conditions. Mol Cell Proteomics 14(3):572–584. doi:10.1074/mcp.M114.042382

Rast A, Heinz S, Nickelsen J (2015) Biogenesis of thylakoid membranes. Biochim Biophys Acta. doi:10.1016/j.bbabio.2015.01.007

Rexroth S, Mullineaux CW, Ellinger D, Sendtko E, Rogner M, Koenig F (2011) The plasma membrane of the cyanobacterium Gloeobacter violaceus contains segregated bioenergetic domains. Plant Cell 23(6):2379–2390. doi:10.1105/tpc.111.085779

Rippka R, Waterbury J, Cohen-Bazire G (1974) A cyanobacterium which lacks thylakoids. Arch Microbiol 100(1):419–436. doi:10.1007/BF00446333

Sakurai I, Mizusawa N, Wada H, Sato N (2007) Digalactosyldiacylglycerol is required for stabilization of the oxygen-evolving complex in photosystem II. Plant Physiol 145(4):1361–1370. doi:10.1104/pp. 107.106781

Sato N (1994) Effect of exogenous glucose on the accumulation of monoglucosyl diacylglycerol in the cyanobacterium Synechocystis PCC 6803. Plant Physiol Biochem 32(1):121–126

Sato N, Murata N (1982a) Lipid biosynthesis in the blue-green algae (Cyanobacterium), Anabaena variabilis III. UDP glucose: diacylglycerol glucosyltransferase activity in vitro. Plant Cell Physiol 23(6):1115–1120

Sato N, Murata N (1982b) Lipid biosynthesis in the blue-green alga, Anabaena variabilis: I. Lipid classes. Biochim Biophys Acta 710(3):271–278. doi:10.1016/0005-2760(82)90109-6

Sato N, Wada H (2010) Lipid biosynthesis and its regulation in cyanobacteria. In: Wada H, Murata N (eds) Lipids in photosynthesis, vol 30, Advances in photosynthesis and respiration. Springer, Dordrecht, pp 157–177. doi:10.1007/978-90-481-2863-1_8

Selao TT, Zhang L, Arioz C, Wieslander A, Norling B (2014) Subcellular localization of monoglucosyldiacylglycerol synthase in Synechocystis sp. PCC6803 and its unique regulation by lipid environment. PLoS One 9(2):e88153. doi:10.1371/journal.pone.0088153

Selstam E (1998) Development of thylakoid membranes with respect to lipids. In: Paul-André S, Norio M (eds) Lipids in photosynthesis: structure, function and genetics, vol 6, Advances in photosynthesis and respiration. Springer, Dordrecht, pp 209–224. doi:10.1007/0-306-48087-5_11

Selstam E, Campbell D (1996) Membrane lipid composition of the unusual cyanobacterium Gloeobacter violaceus sp. PCC 7421, which lacks sulfoquinovosyl diacylglycerol. Arch Microbiol 166(2):132–135. doi:10.1007/s002030050367

Shanklin J, Whittle E, Fox BG (1994) Eight histidine residues are catalytically essential in a membrane-associated iron enzyme, stearoyl-CoA desaturase, and are conserved in alkane hydroxylase and xylene monooxygenase. Biochemistry 33(43):12787–12794

Shen G, Boussiba S, Vermaas WF (1993) Synechocystis sp PCC 6803 strains lacking photosystem I and phycobilisome function. Plant Cell 5(12):1853–1863

Shih PM, Wu D, Latifi A, Axen SD, Fewer DP, Talla E, Calteau A, Cai F, Tandeau de Marsac N, Rippka R, Herdman M, Sivonen K, Coursin T, Laurent T, Goodwin L, Nolan M, Davenport KW, Han CS, Rubin EM, Eisen JA, Woyke T, Gugger M, Kerfeld CA (2013) Improving the coverage of the cyanobacterial phylum using diversity-driven genome sequencing. Proc Natl Acad Sci U S A 110(3):1053–1058. doi:10.1073/pnas.1217107110

Shimojima M, Ohta H, Iwamatsu A, Masuda T, Shioi Y, Takamiya K (1997) Cloning of the gene for monogalactosyldiacylglycerol synthase and its evolutionary origin. Proc Natl Acad Sci U S A 94(1):333–337

Shimojima M, Tsuchiya M, Ohta H (2009) Temperature-dependent hyper-activation of monoglucosyldiacylglycerol synthase is post-translationally regulated in Synechocystis sp. PCC 6803. FEBS Lett 583(14):2372–2376. doi:10.1016/j.febslet.2009.06.033

Simidjiev I, Stoylova S, Amenitsch H, Javorfi T, Mustardy L, Laggner P, Holzenburg A, Garab G (2000) Self-assembly of large, ordered lamellae from non-bilayer lipids and integral membrane proteins in vitro. Proc Natl Acad Sci U S A 97(4):1473–1476

Stanier RY, Cohen-Bazire G (1977) Phototrophic prokaryotes: the cyanobacteria. Annu Rev Microbiol 31:225–274. doi:10.1146/annurev.mi.31.100177.001301

Stengel A, Gugel IL, Hilger D, Rengstl B, Jung H, Nickelsen J (2012) Initial steps of photosystem II de novo assembly and preloading with manganese take place in biogenesis centers in Synechocystis. Plant Cell 24(2):660–675. doi:10.1105/tpc.111.093914

Umena Y, Kawakami K, Shen JR, Kamiya N (2011) Crystal structure of oxygen-evolving photosystem II at a resolution of 1.9 A. Nature 473(7345):55–60. doi:10.1038/nature09913

van Besouw A, Wintermans JFGM (1978) Galactolipid formation in chloroplast envelopes: I. Evidence for two mechanisms in galactosylation. Biochim Biophys Acta 529(1):44–53. doi:10.1016/0005-2760(78)90102-9

van de Meene AM, Hohmann-Marriott MF, Vermaas WF, Roberson RW (2006) The three-dimensional structure of the cyanobacterium Synechocystis sp. PCC 6803. Arch Microbiol 184(5):259–270. doi:10.1007/s00203-005-0027-y

van de Meene AML, Sharp WP, McDaniel JH, Friedrich H, Vermaas WFJ, Roberson RW (2012) Gross morphological changes in thylakoid membrane structure are associated with photosystem I deletion in Synechocystis sp. PCC 6803. Biochim Biophys Acta 1818(5):1427–1434. doi:10.1016/j.bbamem.2012.01.019

Vothknecht UC, Otters S, Hennig R, Schneider D (2012) Vipp1: a very important protein in plastids?! J Exp Bot 63(4):1699–1712. doi:10.1093/jxb/err357

Westphal S, Heins L, Soll J, Vothknecht UC (2001) Vipp1 deletion mutant of Synechocystis: a connection between bacterial phage shock and thylakoid biogenesis? Proc Natl Acad Sci U S A 98(7):4243–4248. doi:10.1073/pnas.061501198

Xu C, Fan J, Froehlich JE, Awai K, Benning C (2005) Mutation of the TGD1 chloroplast envelope protein affects phosphatidate metabolism in Arabidopsis. Plant Cell 17(11):3094–3110. doi:10.1105/tpc.105.035592

Zhang L, Sakamoto W (2013) Possible function of VIPP1 in thylakoids: protection but not formation? Plant Signal Behav 8(2):e22860. doi:10.4161/psb.22860

Zhang L, Sakamoto W (2015) Possible function of VIPP1 in maintaining chloroplast membranes. Biochim Biophys Acta. doi:10.1016/j.bbabio.2015.02.013

Zhang L, Kato Y, Otters S, Vothknecht UC, Sakamoto W (2012) Essential role of VIPP1 in chloroplast envelope maintenance in Arabidopsis. Plant Cell 24(9):3695–3707. doi:10.1105/tpc.112.103606

Zhang S, Shen G, Li Z, Golbeck JH, Bryant DA (2014) Vipp1 is essential for the biogenesis of Photosystem I but not thylakoid membranes in Synechococcus sp. PCC 7002. J Biol Chem 289(23):15904–15914. doi:10.1074/jbc.M114.555631

Chapter 5
Role of Lipids in Chloroplast Biogenesis

Koichi Kobayashi and Hajime Wada

Abstract Chloroplasts are plant organelles that develop the thylakoid membrane inside to perform oxygenic photosynthesis. The biogenesis of the thylakoid membrane requires coordinated synthesis and assembly of proteins, pigments and many photosynthetic cofactors with membrane glycerolipids. The lipid bilayer of the thylakoid membrane mainly consists of four lipid classes; monogalactosyldiacylglycerol (MGDG), digalactosyldiacylglycerol (DGDG), sulfoquinovosyldiacylglycerol (SQDG) and phosphatidylglycerol (PG), each of which has specific roles in biogenesis and maintenance of thylakoid membranes and photosynthesis. Galactolipids MGDG and DGDG constitute the bulk of membrane lipids in chloroplasts and are essential as major structural components of the thylakoid membrane. In addition to galactolipids, the thylakoid membrane requires a certain level of anionic lipids SQDG and PG for its function. Although SQDG and PG substitute for each other to maintain the amount of total anionic lipids in chloroplasts, PG has specific roles in photosynthesis that cannot be compensated by SQDG and galactolipids.

In this chapter, we summarize roles of lipids in chloroplast functions with their biosynthetic pathways, which have been mainly established in *Arabidopsis*, and discuss an involvement of lipid biosynthesis in coordinated development of photosynthetic machinery during chloroplast biogenesis.

Keywords Arabidopsis • Digalactosyldiacylglycerol • Monogalactosyldiacylglycerol • Phosphatidylglycerol • Photosynthesis • Sulfoquinovosyldiacylglycerol • Thylakoid membrane

---

K. Kobayashi (✉) • H. Wada
Department of Life Sciences, Graduate School of Arts and Sciences, The University of Tokyo, Komaba 3-8-1, Meguro-ku, Tokyo 153-8902, Japan
e-mail: kkobayashi@bio.c.u-tokyo.ac.jp

Y. Nakamura, Y. Li-Beisson (eds.), *Lipids in Plant and Algae Development*, Subcellular Biochemistry 86, DOI 10.1007/978-3-319-25979-6_5

Introduction

Plastids are semi-autonomous plant organelles that originated from a photosynthetic cyanobacterium engulfed by the eukaryotic ancestor of algae and land plants (McFadden 2014). Chloroplasts, which have the main role in photosynthesis, are the major form of plastids in most plants and algae. However, plant plastids have diversified from their original function as photosynthetic organelles to fulfill a variety of other important roles depending on the host cell type. Plastid family members include undifferentiated proplastids in meristems, carotenoid-accumulating chromoplasts in flowers and fruits, and starch-rich amyloplasts in storage tissues and root tips (Lopez-Juez and Pyke 2005). In addition to specific roles related to their unique characteristics, plastids have fundamental functions in support of plant growth and life cycle that involve the synthesis of lipids, amino acids, purine and pyrimidine bases, various hormones and secondary metabolites as well as the assimilation of nitrogen and sulfur (Neuhaus and Emes 2000).

All plant plastids are bounded by a double-membrane system with an inner and outer envelope. In addition, chloroplasts have an extensively developed internal membrane system, the thylakoid membrane, where the photosynthetic electron transport reaction occurs. Compared to typical cellular membranes of animals, fungi, or nonphotosynthetic bacteria, the chloroplast thylakoid membrane is unique in lipid composition, with nonphosphorous mono- and digalactosyldiacylglycerol (MGDG and DGDG, respectively) as major constituents (Fig. 5.1). MGDG and DGDG account for approximately 50 % and 25 % of total thylakoid membrane lipids, respectively (Fig. 5.2a) (Dorne et al. 1990). The remainder comprises the anionic glycerolipids, sulfoquinovosyldiacylglycerol (SQDG) and phosphatidylglycerol (PG) (Fig. 5.1), with a marginal proportion of phosphatidylinositol. These lipids are also the main constituents of envelope membranes of chloroplasts and non-photosynthetic plastids, with galactolipids accounting for ~60 % of total envelope lipids (Fig. 5.2a) (Block et al. 1983; Alban et al. 1988); galactolipids and

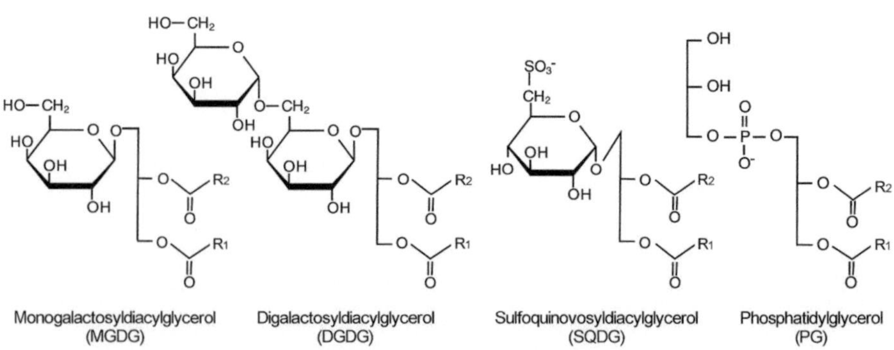

Fig. 5.1 Structure of glycerolipids in thylakoid membranes. R_1 and R_2 denote hydrocarbon chains of fatty acids

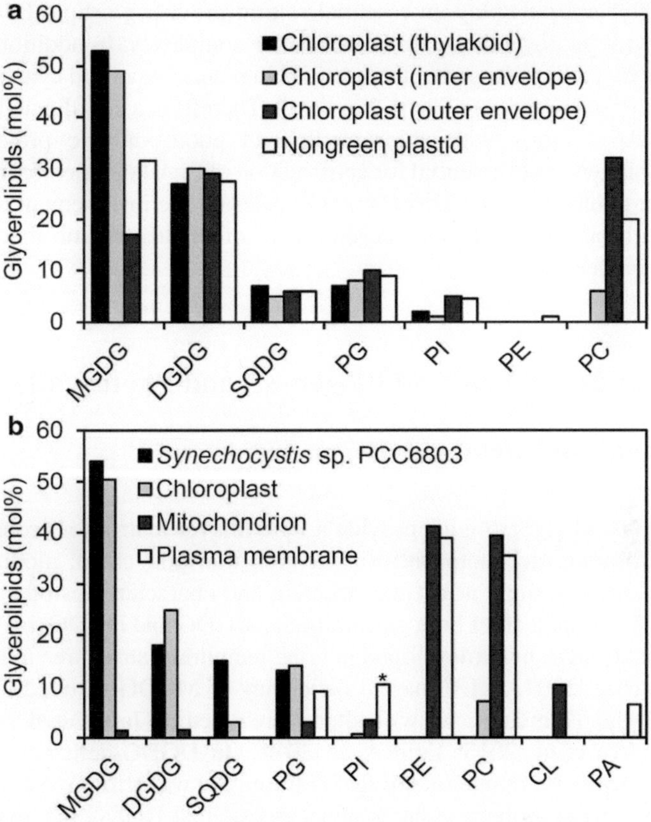

Fig. 5.2 Composition of polar glycerolipids. (**a**) Glycerolipid composition in the thylakoid membrane (Dorne et al. 1990) and in inner and outer envelope membranes (Block et al. 1983) from spinach chloroplasts and total envelope membranes from cauliflower bud plastids (Alban et al. 1988). (**b**) Comparison of glycerolipid composition in chloroplasts (Yu and Benning 2003), mitochondria (Jouhet et al. 2004), and plasma membranes (Uemura et al. 1995) from *Arabidopsis* and that in the total membranes from the cyanobacterium *Synechocystis* sp. PCC 6803 (Wada and Murata 1989). *MGDG* monogalactosyldiacylglycerol, *DGDG* digalactosyldiacylglycerol, *SQDG* sulfoquinovosyldiacylglycerol, *PG* phosphatidylglycerol, *PI* phosphatidylinositol, *PE* phosphatidylethanolamine, *PC* phosphatidylcholine, *CL* cardiolipin, *PA* phosphatidic acid. * PI + phosphatidylserine

SQDG are minor constituents or are absent in extraplastidic membranes of plant cells (Fig. 5.2b). The membrane lipid composition of cyanobacteria, which share a common ancestor with plant plastids, is similar to that of plant thylakoids (Fig. 5.2b) (Omata and Murata 1983; Wada and Murata 1998). The similarity in lipid composition between chloroplasts and cyanobacteria may reflect the same origin of their unique lipids from the common ancestor, although enzymes and pathways for galactolipid biosynthesis differ between plants and cyanobacteria (see Chap. 4).

The highly conserved lipid composition among chloroplasts and cyanobacteria suggests a special requirement of these lipids for oxygenic photosynthesis.

Thylakoid glycerolipids play an essential role in providing a lipid bilayer matrix embedded with photosynthesis protein-cofactor complexes. In addition, thylakoid lipids are involved in photosynthetic reactions in plants, algae and cyanobacteria, as described later in brief and in Chap. 2 in detail. Therefore, a coordinated biosynthesis of thylakoid lipids with chlorophylls and photosynthetic proteins during chloroplast biogenesis is essential for construction of thylakoid membrane networks and efficient photosynthesis. This chapter focuses on the involvement of thylakoid lipid biosynthesis in coordinated development of photosynthetic machinery and chloroplast biogenesis.

Role of Lipids in Thylakoid Biogenesis and Photosynthesis

Role of Lipids in Thylakoid Biogenesis

Because thylakoid glycerolipids provide a lipid bilayer matrix for protein-cofactor complexes forming the photosynthetic electron transport chain, modifications of lipid composition strongly affect the structure and characteristics of the thylakoid membrane. A crucial lack of both galactolipids, MGDG and DGDG, in *Arabidopsis* caused no or severely underdeveloped internal membrane structures in leaf plastids (Kobayashi et al. 2007, 2013); partial deficiency of MGDG resulted in decreased amount of thylakoid membranes with altered architecture (Jarvis et al. 2000; Myers et al. 2011; Wu et al. 2013; Fujii et al. 2014). In DGDG-deficient *Arabidopsis* mutants, the thylakoid membrane of leaf chloroplasts was strongly bent, with large thylakoid-free areas in the stroma, which suggests that DGDG is required for the proper formation and maintenance of the thylakoid membrane in plant chloroplasts (Dörmann et al. 1995; Hölzl et al. 2009). *In vitro* analysis revealed that DGDG contributes to the membrane stacking via formation of hydrogen bonds between polar heads of adjacent bilayers (Demé et al. 2014). The distorted thylakoid structure in the *Arabidopsis* DGDG-deficient mutant was complemented by transgenic expression of a bacterial glucosyltransferase and consequent accumulation of glucosylgalactosyldiacylglycerol (Hölzl et al. 2009); thus, disaccharide head groups are important for the flat thylakoid architecture. MGDG has a cone-like shape with a small head group, which allows it to form an inverted hexagonal phase, whereas DGDG has a more cylindrical shape and forms a bilayer lamellar phase in mixtures with water (Shipley et al. 1973). The ratio of bilayer-forming DGDG to non-bilayer-forming MGDG affects the properties of chloroplast membranes by altering the lipid bilayer from hexagonal II to lamellar phases (Demé et al. 2014) and thus is tightly regulated in response to various stresses (Moellering and Benning 2011).

Acidic lipids with negative charge in their head groups (PG and SQDG) also affect the organization of thylakoid membranes. PG has a crucial role in proper thylakoid biogenesis; deficiency of chloroplast PG in *Arabidopsis* resulted in the formation of vesicles, enlarged vacuolated structures, or underdeveloped membrane fragments instead of thylakoid membrane networks (Hagio et al. 2002; Babiychuk et al. 2003; Haselier et al. 2010; Kobayashi et al. 2015). Meanwhile, deficiency of SQDG caused strong curling of thylakoid membranes with little effect on grana stacking in *Chlamydomonas* (Sato et al. 1995b). Although lack of SQDG biosynthesis in *Arabidopsis* had no effect on the thylakoid architecture, loss of SQDG together with a small reduction in PG content decreased grana stacking in leaf chloroplasts (Yu and Benning 2003).

Role of Lipids in Photosynthesis

In addition to providing a lipid bilayer matrix, thylakoid lipids are required for the structure and function of photosynthetic complexes. X-ray crystallography studies of cyanobacterial photosystem I (PSI) and PSII revealed that both complexes involve thylakoid lipids in their crystal structures, with MGDG and PG near reaction centers in both cases (Jordan et al. 2001; Guskov et al. 2009; Umena et al. 2011). In-depth analyses of PG-deficient mutants in a cyanobacterium, *Synechocystis* sp. PCC 6803, demonstrated that PG is essential for the PSII electron transport from the primary plastoquinone Q_A to the secondary Q_B, stabilization of the oxygen-evolving complex via extrinsic proteins, and trimerization and activity of the PSI complex (Sato et al. 2000; Hagio et al. 2000; Gombos et al. 2002; Sakurai et al. 2003, 2007a). PG is required for chloroplast development and photosynthesis, as revealed by mutant analyses in *Arabidopsis* (Xu et al. 2002; Hagio et al. 2002; Babiychuk et al. 2003; Haselier et al. 2010; Kobayashi et al. 2015) and *Chlamydomonas* (Dubertret et al. 1994; Maanni et al. 1998). In addition, another anionic thylakoid lipid, SQDG, is crucial for activity of PSII at both donor and acceptor sides in *Chlamydomonas* (Sato et al. 1995a; Minoda et al. 2002, 2003) and *Synechocystis* sp. PCC 6803 (Aoki et al. 2004) but does not affect PSI activity. SQDG was not important for growth and photosynthesis in the cyanobacterium *Synechococcus* sp. PCC 7942 (Güler et al. 1996; Aoki et al. 2004). Although SQDG is not required for photosynthesis and chloroplast biogenesis in *Arabidopsis* under nutrient-sufficient conditions (Yu et al. 2002), this lipid would be required to maintain the amount of total anionic lipids in chloroplasts when PG content is decreased under phosphate (P)-deficient conditions (Essigmann et al. 1998; Yu and Benning 2003). Thus, the requirement of SQDG for photosynthesis differs among organisms.

Analyses of a DGDG-deficient mutant (*dgdA*) of *Synechocystis* sp. PCC6803 revealed that DGDG is required for PSII activity by binding extrinsic proteins, which stabilize the oxygen-evolving complex (Sakurai et al. 2007b). The *dgdA* mutation increased the sensitivity to photoinhibition under high temperature and high light, with the repair cycle of photoinhibition more severely affected than the photodamage process (Mizusawa et al. 2009a, b). In *Arabidopsis*, DGDG is important for the structure and function of PSII (Härtel et al. 1997; Hölzl et al. 2006, 2009), primarily on the donor side, which catalyzes the oxidative water cleavage

(Reifarth et al. 1997; Steffen et al. 2005), and for the function and stability of the PSI complex (Guo et al. 2005; Ivanov et al. 2006). As in *Synechocystis* sp. PCC 6803, in *Arabidopsis*, DGDG plays a role in decreasing photoinhibition under high light (Hölzl et al. 2009). DGDG is also required for trimerization of the light-harvesting complex II (LHCII) in *Arabidopsis* (Hölzl et al. 2006), which may be related to DGDG molecules in the LHCII found in spinach and pea (Liu et al. 2004; Standfuss et al. 2005). Because accumulation of glucosylgalactosyldiacylglycerol in DGDG-deficient mutants could not fully restore the PSII functionality and trimerization of the LHCII complex (Hölzl et al. 2006, 2009), the galactose moiety of DGDG may be important for interactions with photosynthetic protein complexes.

A ~40 % decrease in MGDG level in an *Arabidopsis* mutant (*mgd1-1*) had no effect on intrinsic PSII activity but impaired thylakoid membrane energization and photoprotection by increased conductivity of the thylakoid membrane (Aronsson et al. 2008). A similar loss of MGDG in tobacco decreased levels of the cytochrome b_6f complex and blocked intersystem electron transport (Wu et al. 2013). Further MGDG loss (~80 % from wild-type levels) with partial DGDG reduction strongly impaired the PSII photochemical reaction and disrupted energy coupling between reaction centers and antenna complexes (Fujii et al. 2014). Moreover, almost depleted MGDG levels (~5 % of wild-type levels) caused severely disordered PS complexes and complete deficiency of photosynthetic electron transport (Kobayashi et al. 2013), which is consistent with *in vitro* findings of MGDG inducing ordered oligomerization of LHCII (Schaller et al. 2011) and PSII dimerization (Kansy et al. 2014) and enhancing the energy coupling between LHCII and PSII core complexes (Zhou et al. 2009). These data demonstrate a critical role for MGDG in photosynthetic activities in plants. However, identification of a mutant for the epimerase converting monoglucosyldiacylglycerol (MGlcDG) to MGDG in *Synechocystis* sp. PCC 6803 revealed that the accumulation of MGlcDG instead of MGDG and DGDG was not fully but was largely sufficient for oxygenic photosynthesis or development of the thylakoid membrane structure, which may revise the long-standing belief that galactolipids are among the prerequisite components for oxygenic photosynthesis (Awai et al. 2014). However, considering that galactolipid biosynthetic pathways are largely different between plants and cyanobacteria – plants produce MGDG from diacylglycerol in one step (Fig. 5.3), whereas cyanobacteria synthesize it in two steps via MGlcDG (Chap. 4) – whether MGlcDG can substitute MGDG in plant chloroplasts remains elusive.

Biosynthesis of Thylakoid Membrane Lipids in *Arabidopsis*

Figure 5.3 shows the biosynthetic pathway of thylakoid membrane lipids in *Arabidopsis*. In plants, MGDG is synthesized by MGDG synthase, which transfers galactose from UDP-galactose to diacylglycerol in plastid envelopes (Shimojima et al. 1997). Of three MGDG synthase homologs (MGD1, MGD2, and MGD3) identified in *Arabidopsis* (Miège et al. 1999; Awai et al. 2001), inner

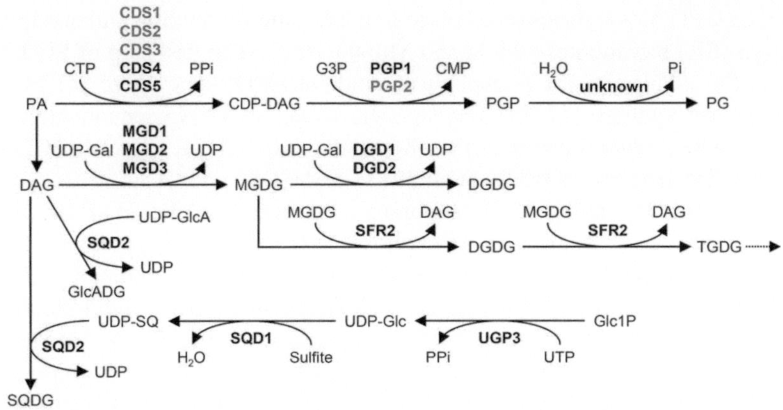

Fig. 5.3 Biosynthetic pathway of membrane glycerolipids from phosphatidic acid (PA) in the *Arabidopsis* plastid. *PA* phosphatidic acid, *DAG* diacylglycerol, *PG* phosphatidylglycerol, *PGP* phosphatidylglycerol phosphate, *MGDG* monogalactosyldiacylglycerol, *DGDG* digalactosyldiacylglycerol, *TGDG* trigalactosyldiacylglycerol, *SQDG* sulfoquinovosyldiacylglycerol, *GlcADG* glucuronosyldiacylglycerol, *CMP* cytidine 5′-monophosphate, *CDP* cytidine 5′-diphosphate, *CTP* cytidine 5′-triphosphate, *UDP* uridine 5′-diphosphate, *G3P* glycerol 3-phosphate, *Gal* galactose, *Glc* glucose, *Glc1P* glucose 1-phosphate, *PPi* pyrophosphate, *SQ* sulfoquinovose, *GlcA* glucuronic acid. Enzymes involved in each step are in **bold**. All enzymes except for those in *gray*, which are localized to the endoplasmic reticulum, are targeted to the plastid

envelope-localized MGD1 has a major role in galactolipid synthesis in chloroplasts (Jarvis et al. 2000; Kobayashi et al. 2007), whereas MGD2 and MGD3, which are localized to the outer envelope membrane (Awai et al. 2001), constitute an alternative pathway that conditionally functions under P-limited conditions (Kobayashi et al. 2004, 2009a). In addition to being a building block for constituting chloroplast membranes, MGDG is used for DGDG biosynthesis as a substrate. DGDG biosynthesis is predominantly catalyzed by DGDG synthase, which transfers a second galactose to MGDG from UDP-galactose in the plastid outer-envelope membrane (Froehlich et al. 2001; Kelly and Dörmann 2002; Kelly et al. 2003). Similar to MGD, two paralogous enzymes (DGD1 and DGD2) have been identified for DGDG synthase in *Arabidopsis* and other angiosperms. DGD1 catalyzes the bulk of DGDG synthesis in chloroplasts in combination with MGD1, whereas DGD2 functions in the alternative pathway with MGD2 and MGD3 (Benning and Ohta 2005; Kobayashi et al. 2009b). Yet another enzyme, SENSITIVE TO FREEZING 2 (SFR2), which is enzymatically different from DGDG synthase, functions to form DGDG and oligogalactolipids. SFR2, located on the plastid envelope membrane, transfers the galactose head group from MGDG to a second MGDG or to an oligogalactolipid, which results in increased number of galactose moieties in the polar head (Moellering et al. 2010). SFR2 does not contribute significantly to DGDG synthesis during normal growth but is required for freezing tolerance in cold-acclimated *Arabidopsis*, presumably via stabilizing the chloroplast membrane both by increasing the hydration of the membrane and by adjusting the ratio of bilayer-forming to non-bilayer-forming lipids (Moellering et al. 2010).

In plants, PG biosynthesis takes place in plastids, mitochondria and endoplasmic reticulum (ER) membranes (Wada and Murata 2007). The first step of PG biosynthesis is the conversion of phosphatidic acid into CDP-diacylglycerol by CDP-diacylglycerol synthase (CDS). Five genes, *CDS1* to *CDS5*, encode functional CDSs in the *Arabidopsis* genome (Haselier et al. 2010). CDS1, CDS2 and CDS3 are targeted to ER membranes (Zhou et al. 2013), and CDS4 and CDS5 are localized to plastids and function in thylakoid membrane biogenesis (Haselier et al. 2010). Then CDP-diacylglycerol is converted to PG phosphate (PGP) with glycerol 3-P by PGP synthase. The *Arabidopsis* genome encodes two functional homologs, PGP1 and PGP2, for PGP synthase. PGP1 is dual-localized in plastids and mitochondria (Babiychuk et al. 2003), whereas PGP2 is targeted to the ER (Tanoue et al. 2014). PGP1 is responsible for thylakoid membrane biogenesis and its disruption greatly reduces total PG content in leaves, with a severe defect in thylakoid membrane development (Hagio et al. 2002; Babiychuk et al. 2003; Kobayashi et al. 2015). PGP2 has a minor role in PG biosynthesis, and the *pgp2* single mutant shows no obvious growth defects. However, a *pgp1pgp2* double mutant could not synthesize PG and therefore showed an embryonic lethal phenotype (Tanoue et al. 2014). Despite strongly impaired chloroplast development in *PGP1*-knockout mutants, loss of PGP1 did not affect mitochondria features (Hagio et al. 2002; Babiychuk et al. 2003), which indicates that ER-localized PGP2 can complement the PGP1 function in mitochondria. The last step for PG biosynthesis is dephosphorylation of PGP by PGP phosphatase. Genes responsible for this step in plants remain unidentified.

For SQDG synthesis in plants, UDP-glucose is synthesized from glucose 1-P and UTP by UDP-glucose pyrophosphorylase 3 in the plastid (Okazaki et al. 2009). Then UDP-sulfoquinovose (SQ) synthase (SQD1) converts UDP-glucose to UDP-SQ with sulfite. The final step is transfer of SQ from UDP-SQ to diacylglycerol, which is catalyzed by SQDG synthase (SQD2). Both SQD1 and SQD2 are encoded by single loci in the *Arabidopsis* genome (Essigmann et al. 1998; Yu et al. 2002; Okazaki et al. 2009). Mutant analyses of *Arabidopsis* revealed that SQDG is not required for chloroplast biogenesis during normal growth conditions but is important under P-deficient conditions (Yu et al. 2002; Yu and Benning 2003). The expression of both SQD1 and SQD2 is greatly upregulated during P deficiency (Essigmann et al. 1998; Yu et al. 2002). Therefore, the amount of SQDG increases in response to P deficiency in parallel with decreased PG content, which may help maintain proper balance of anionic charge in the thylakoid membrane. SQD2 is also involved in the formation of another acidic glycolipid, glucuronosyldiacylglycerol (GlcADG). GlcADG level is increased in response to P deficiency and may play a role in protecting plants against P limitation stress, but its function and subcellular localization remain unclear (Okazaki et al. 2013).

Regulation of Galactolipid Biosynthesis in Coordination with Formation of Photosynthetic Machinery

The development of chloroplasts and photosynthetic machinery is strictly regulated in response to various developmental and environmental cues (Jarvis and López-Juez 2013). Thylakoid membrane biogenesis is one of the most remarkable changes during chloroplast development and requires a substantial increase in content of galactolipids. Thus, the biosynthesis of galactolipids could be regulated in coordination with chlorophyll and photosynthetic protein synthesis and formation of photosynthetic machinery. In cucumber seedlings, the expression of the *MGD1* homolog (*csMGD1*) and MGDG synthesis activity were induced in response to light, which led to greatly increased galactolipid content (Yamaryo et al. 2003). Although the light response of *csMGD1* was weakened in detached cotyledons, galactolipid synthesis was still increased by illumination. The data suggest that light is not sufficient for full upregulation of *csMGD1* and requires other factors presumably transported from the hypocotyl or roots but can activate galactolipid biosynthesis in a post-transcriptional manner. By contrast, exogenous cytokinin treatment to detached cotyledons in the dark induced the *csMGD1* expression with a small increase in galactolipid content (Yamaryo et al. 2003). Thus, cytokinin, which may be transported from the hypocotyl or roots to cotyledons, may play a pivotal role in the upregulation of *csMGD1* but is not sufficient for full accumulation of galactolipids. Taken together, the marked accumulation of galactolipids for chloroplast development during photomorphogenesis may require transcriptional activation of *csMGD1* by cytokinin signaling and post-transcriptional activation of MGDG synthesis by light in an additive manner (Yamaryo et al. 2003).

In vitro analysis revealed that the enzymatic activity of csMGD1 is increased by chloroplast thioredoxins in a redox-dependent manner (Yamaryo et al. 2006). Redox regulation of MGD enzymes was also reported in sesame (Shimojima et al. 2013). Thioredoxin is a class of small oxidoreductase enzymes containing a redox-reactive cysteine pair within the active site that facilitate the reduction of target proteins by cysteine thiol-disulfide exchange. Various important biological processes, which include transcription and translation of nuclear and plastid genes, carbon metabolism including the Calvin cycle, synthesis of chlorophylls and fatty acids, and assembly of photosynthetic protein-cofactor complexes, are regulated in a light-dependent manner by the thioredoxin redox system during chloroplast development (Geigenberger et al. 2005; Serrato et al. 2013). In parallel with these processes, the thioredoxin system may regulate galactolipid synthesis in response to light (Yamaryo et al. 2006). Fatty acid synthesis is also activated in response to light via redox regulation of acetyl-CoA carboxylase, which catalyzes the first committed step of *de novo* fatty acid synthesis (Sasaki et al. 1997). Thus, galactolipid synthesis and *de novo* fatty acid synthesis may be linked with photosynthetic electron transport via the thioredoxin system. In addition to redox regulation, activation of plant MGDG synthases may occur by anionic phospholipids, phosphatidic acid and PG (Covés et al. 1988; Ohta et al. 1995; Dubots et al. 2010; Shimojima et al. 2013).

Complementation of the *Arabidopsis MGD1* knockout mutant (*mgd1-2*) with MGDG synthase from the green sulfur bacterium *Chlorobaculum tepidum*, which could be insensitive to redox and acidic lipid regulation, resulted in disorganized thylakoid formation in some chloroplasts (Masuda et al. 2011), so the enzymatic regulation of MGDG synthesis may be important for the proper development of the thylakoid membrane in *Arabidopsis*. Although the effects of acidic lipids on galactolipid accumulation in response to light remain unknown, the post-translational regulation of MGDG synthesis may play a role in coordinating galactolipid synthesis with the formation of the photosynthetic machinery during thylakoid biogenesis. In addition to regulation at the enzymatic activity level, localization and affinity of the MGD1 protein to the plastid envelope may be affected by local lipid composition, because MGD1 is attracted to lipid monolayers enriched with MGDG but excluded from those enriched with DGDG (Sarkis et al. 2014).

In addition to post-translational regulation, transcriptional regulation may be involved in coordinating galactolipid biosynthesis with chloroplast development. As seen in cucumber (Yamaryo et al. 2003), cytokinin signaling plays a role in upregulation of galactolipid biosynthesis genes in *Arabidopsis* (Awai et al. 2001; Kobayashi et al. 2014). The expression of *MGD1* and *DGD1* was induced by light, but the induction was suppressed by loss of cytokinin receptors, so cytokinin signaling may be required for light-induced upregulation of these galactolipid genes, as observed in other photosynthesis-associated nuclear genes (Kobayashi et al. 2014). A basic Leu zipper transcription factor, HY5, a central regulator of photomorphogenesis functioning downstream of photoreceptors (Bae and Choi 2008), is essential for the upregulation of *MGD1* and *DGD1* in response to light (Kobayashi et al. 2014). *DGD1* but not *MGD1* has been identified as a putative direct target of HY5 along with many other photosynthesis-associated nuclear genes (Lee et al. 2007). Other transcription factors functioning in chloroplast development downstream of light and cytokinin signaling also affect the expression of galactolipid genes. Particularly, GOLDEN2-LIKE (GLK) transcription factors, which directly upregulate genes involved in chlorophyll biosynthesis and light harvesting (Waters et al. 2009), strongly affect the transcript level of *DGD1* (Kobayashi et al. 2014). Although regulatory mechanisms may differ somewhat between *MGD1* and *DGD1*, greening-promoting signaling such as light and cytokinin may be important for the coordinated formation of the thylakoid membrane system at the transcriptional level.

Transcriptional regulation may be also involved in fine-tuning galactolipid biosynthesis to the varying needs of the thylakoid membrane in response to developmental and environmental conditions surrounding chloroplasts. In fact, the expression of *MGD1* and *DGD1* was partially suppressed in mutants deficient in chlorophyll biosynthesis (Kobayashi et al. 2014), so the expression of these galactolipid genes could be coordinated with chlorophyll biosynthesis. Many photosynthesis-associated nuclear genes are downregulated when the photosynthetic machinery is exposed to disrupted stress and impaired chloroplast biogenesis (Woodson and Chory 2008; Larkin and Ruckle 2008). Thus, the functional state of chloroplasts may be monitored by the nucleus via signals retrogradely transmitted from chloroplasts to optimize the expression of nuclear genes involved in chloroplast functions. Various *Arabidopsis* mutants that fail to downregulate photosynthesis-associated nuclear genes in response to chloroplast dysfunction have been identified and named *genomes-uncoupled* (*gun*) mutants, which supports the existence of the plastid-to-nucleus retrograde signaling, simply called "plastid signaling" (Woodson and Chory 2008; Larkin and Ruckle 2008). As observed for many photosynthesis-associated genes, the expression of *MGD1* and *DGD1* in *Arabidopsis* was suppressed in response to chloroplast dysfunction on treatment with the herbicide norflurazon, which inhibits carotenoid biosynthesis. However, the norflurazon-induced suppression of *MGD1* and *DGD1* was attenuated in the mutant for GUN1, a central regulator of plastid signaling (Kobayashi et al. 2014). Therefore, *MGD1* and *DGD1* expression is under plastid signal regulation, with GUN1 playing a pivotal role. GUN1 is a plastidic pentatricopeptide repeat protein that integrates multiple plastid signals to downregulate photosynthesis-associated nuclear genes in response to chloroplast defects (Koussevitzky et al. 2007). Thus, galactolipid biosynthesis may be coordinated with the formation of the photosynthetic machinery at the transcriptional level according to the chloroplast functionality via GUN1-mediated plastid signaling. Consistent with these data, the expression of *MGD1* and *DGD1* was downregulated in a PG-deficient mutant that failed to develop functional chloroplasts (Kobayashi et al. 2015). Furthermore, other genes involved in galactolipid and SQDG biosynthesis were downregulated along with *MGD1* and *DGD1*. Thus, chloroplast functionality or lipid homeostasis in chloroplasts may globally affect the transcription of genes involved in biosynthesis of thylakoid membrane glycolipids.

Role of Thylakoid Lipid Synthesis in Regulating Chloroplast Biogenesis

Photosynthesis with highly photoreactive chlorophylls allows plants to convert light energy from the sun into chemical energy to grow photoautotrophically, but with simultaneous threat of photooxidative damage. To develop functional chloroplasts while preventing photooxidative damage from intermediate complexes or unbalanced electron transfer systems, plants strictly regulate the construction of the photosynthetic machinery by orchestrating myriad components with the formation of the thylakoid lipid bilayer (Walters 2005; Takahashi and Badger 2011). The photoprotection machinery includes the light-dependent xanthophyll cycle, which switches LHC systems between a light-harvesting state under low light and a thermal dissipating state under high light via reversible conversion between violaxanthin and zeaxanthin. MGDG plays a crucial role in the xanthophyll cycle via solubilization of violaxanthin and mediating the binding of violaxanthin de-epoxidase to the thylakoid membrane (Jahns et al. 2009). The hexagonal lipid structure, formed by MGDG in the thylakoid membrane, is essential for de-epoxidation of xanthophylls and therefore involved in photoprotective nonphotochemical quenching of photosynthesis. In thylakoid membranes, an MGDG phase surrounding the LHCII strongly enhances the de-epoxidation of LHCII-associated violaxanthin (Schaller et al. 2010). These findings agree with the report of inhibited conversion of violaxanthin to zeaxanthin in the *mgd1-1* mutant, for increased PSII photoinhibition (Aronsson et al. 2008).

MGDG biosynthesis is indispensable for thylakoid membrane biogenesis: a knockout mutation of *MGD1* (*mgd1-2*) by a T-DNA insertion in the first exon resulted in loss of thylakoid membranes, chlorophylls, and membrane photosynthetic proteins in leaf plastids, along with a substantial reduction in content of both galactolipids (Kobayashi et al. 2007). In the *mgd1-2* plastids, invaginated inner-envelope membranes, which are scarcely observed in mature wild-type chloroplasts, were observed instead of thylakoid membrane structures (Kobayashi et al. 2007, 2013). Similar membrane structures from the inner envelope were observed in plastids treated with galvestine-1, a specific inhibitor of MGDG synthase (Botté et al. 2011). Because the invagination of the inner envelope membrane is sometimes observed at the very early stage of chloroplast differentiation (Vothknecht and Westhoff 2001), the impaired MGDG synthesis may inhibit an early stage of thylakoid biogenesis initiated from invagination of inner-envelope membranes. In parallel with the inhibition of thylakoid biogenesis in the *mgd1-2* mutant, plastid nucleoids appeared aggregated in the stroma (Kobayashi et al. 2013). Because the morphology and distribution of plastid nucleoids are associated with development of the thylakoid membrane (Powikrowska et al. 2014), loss of the thylakoid membrane may cause nucleoid aggregation in *mgd1-2* plastids. Moreover, in the *mgd1-2* mutant, photosynthesis-associated genes were strongly downregulated both in the nucleus and plastids (Kobayashi et al. 2013), which suggests that galactolipid biosynthesis and subsequent thylakoid biogenesis are involved in regulating photosynthesis-associated gene expression in both the nucleus and plastids.

Although MGD2 and MGD3 in the outer-envelope membrane could not compensate for loss of MGD1 under nutrient-sufficient growth conditions, they were strongly induced by P deficiency and produced DGDG in the *mgd1-2* mutant in collaboration with DGDG synthase (Kobayashi et al. 2013). Activation of the alternative galactolipid pathway in response to P deficiency resulted in thylakoid-like internal membrane biogenesis in *mgd1-2* plastids (Kobayashi et al.. 2013; Fig. 5.4). Moreover, the partial formation of the internal membranes in Pi-deficient *mgd1-2* plastids was accompanied by a change in nucleoid morphologic features and upregulation of photosynthesis-associated genes encoded in the nucleus and plastids. Although the internal-membrane biogenesis under P deficiency induced chlorophyll accumulation and partial formation of PS complexes in *mgd1-2*, photosynthetic electron transport was still dysfunctional in the mutant (Kobayashi et al. 2013). Thus, upregulation of photosynthesis-associated genes in P-deficient *mgd1-2* could be independent of photosynthetic activities.

The involvement of thylakoid lipids in regulating chloroplast biogenesis was also evidenced by analysis of a *PGP1*-knockout mutant (*pgp1-2*) (Kobayashi et al. 2015). The *pgp1-2* mutant has a pale yellow-green phenotype with severely underdeveloped internal membranes in leaf plastids under nutrient-sufficient conditions

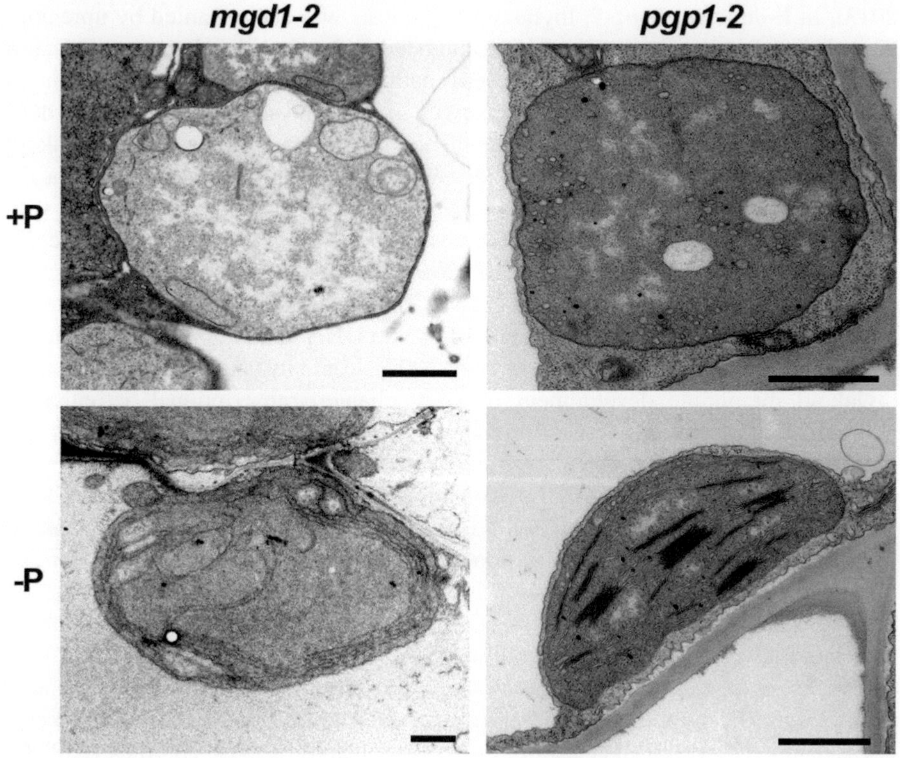

Fig. 5.4 Ultrastructure of leaf plastids in *mgd1-2* and *pgp1-2* under phosphate-sufficient (+P) or –deficient (–P) conditions. Bars = 1.0 μm

(Fig. 5.4). Similar to the *mgd1-2* mutant, in the *pgp1-2* mutant, the expression of photosynthesis-associated genes encoded in the plastid and nucleus was strongly suppressed. Moreover, plastid nucleoids in the *pgp1-2* mutant were larger than those in wild-type *Arabidopsis*, although to a lesser extent than in the *mgd1-2* mutant (Kobayashi et al. 2015). These results suggest that lipid biosynthesis is essential for nucleoid modification during thylakoid membrane biogenesis and the expression of photosynthesis-associated genes in the nucleus and plastids.

Under P-deficient conditions, *MGD2*, *MGD3*, *DGD1*, *DGD2*, *SQD1* and *SQD2*, all inducible by P deficiency, were upregulated in the *pgp1-2* mutant. In addition to these genes, the expression of *MGD1*, which is unresponsive to P deficiency in wild-type *Arabidopsis* (Awai et al. 2001; Kobayashi et al. 2004), was increased with P deficiency (Kobayashi et al. 2015). Consequently, the proportion of all glycolipids was substantially increased in the P-deficient *pgp1-2* mutant in parallel with the decrease in content of phospholipids including PG (Kobayashi et al. 2015). These drastic changes in the glycolipid metabolism led to the development of thylakoid membranes and changes in nucleoid morphologic features in plastids of the *pgp1-2* leaves (Kobayashi et al. 2015; Fig. 5.4). As observed in *mgd1-2* (Kobayashi et al.

2013), in P-deficient *pgp1-2*, thylakoid biogenesis was accompanied by upregulation of photosynthesis-associated genes encoded in the plastid and nucleus. Because mutants for the chlorophyll biosynthetic pathway did not show upregulation of photosynthesis-associated genes or changes in nucleoid morphology with P deficiency (Kobayashi et al. 2013), P deficiency would specifically affect thylakoid lipid mutants in terms of plastid biogenesis by activating glycolipid biosynthesis. These data suggest that formation of the thylakoid membrane is tightly linked with nucleoid morphology and distribution, consistent with findings that several nucleoid-associated proteins are associated with the thylakoid membrane and may tether nucleoids there (Powikrowska et al. 2014). Considering that plastid transcription is regulated by the structural organization of plastid DNA in addition to the activity of RNA polymerase complexes (Sekine et al. 2002), lipid biosynthesis and subsequent thylakoid membrane development possibly change transcriptional activities in nucleoids. Furthermore, transcription and translation activities in plastids are linked to the expression of photosynthesis-associated genes in the nucleus via the nucleoid-associated protein GUN1 (Koussevitzky et al. 2007; Woodson and Chory 2008). Thus, lipid-mediated thylakoid biogenesis may affect the expression of photosynthesis-associated nuclear genes at the onset of chloroplast biogenesis through a plastid signaling pathway, which could be independent of photosynthetic activity because both *mgd1-2* and *pgp1-2* showed no photosynthetic electron transport activities under P-deficient conditions (Kobayashi et al. 2013, 2015).

Interestingly, reciprocal regulation of plastid protein translation and fatty acid synthesis has been reported in *Chlamydomonas*. The E2 dihydrolipoyl transacetylase subunit (DLA2) of the chloroplast pyruvate dehydrogenase complex (cpPDC) binds *psbA* mRNA, encoding the PSII D1, and targets it to a translation zone in an acetate- and light-dependent manner (Bohne et al. 2013). The interaction of *psbA* mRNA and DLA2 induces D1 translation but reduces PDC activity. PDC catalyzes the first step of fatty acid synthesis by converting pyruvate into acetyl-CoA, which suggests a reciprocal regulation of D1 translation and fatty acid synthesis. This regulation may link lipid synthesis and photosynthetic protein synthesis during thylakoid biogenesis.

Involvement of Thylakoid Membrane Biogenesis in Cell and Tissue Development in Leaves

Role of Galactolipid Biosynthesis in Coordinated Development of Chloroplasts and Other Organelles during Cotyledon Development

During the transition from heterotrophic to photoautotrophic growth, a tight metabolic coordination occurs between chloroplasts and other organelles such as peroxisomes and mitochondria in leaf cells. In oilseed plants such as *Arabidopsis*, which store triacylglycerols in seed cells, peroxisomes convert fatty acids bound to triacylglycerols to succinate via β-oxidation and the glyoxylate cycle during initial growth stages after germination (Graham 2008). Mitochondrial metabolism and gluconeogenesis in cytosol are required for progression of the glyoxylate cycle to provide sugars, carbon and energy sources for heterotrophic growth. At the onset of activation of photosynthesis in cotyledon cells, mitochondria and peroxisomes transform their function in coordination with chloroplast biogenesis to operate photorespiratory metabolism, which recycles 2-phosphoglycolate, the product of oxygenation reaction instead of carboxylation by Rubisco in chloroplasts (Peterhansel et al. 2010). Thus, close cooperation among these organelles with strict regulation is essential for plant growth.

Arabidopsis transgenic lines that suppress *MGD1* expression by an *MGD1*-specific artificial microRNA (*amiR-MGD1*) under the control of a dexamethasone (DEX)-inducible promoter can reveal the impact of galactolipid biosynthesis on the initial stage of chloroplast biogenesis (Fujii et al. 2014). The induction of *amiR-MGD1* by DEX treatment suppressed *MGD1* expression in cotyledons to 25 % of the untreated control level, for substantial reduction in MGDG synthesis rate and galactolipid contents. As observed in the *mgd1-2* mutant (Kobayashi et al. 2013), in *amiR-MGD1* lines, decreased galactolipid contents and impaired thylakoid membrane biogenesis led to downregulated photosynthesis-associated genes encoded in plastids and the nucleus during cotyledon development. Furthermore, nuclear genes involved in photorespiratory metabolism in peroxisomes and mitochondria were downregulated concomitantly with photosynthesis-associated genes by *MGD1* suppression (Fujii et al. 2014). The data are consistent with photorespiration being associated with photosynthetic activity and thus peroxisomal and mitochondrial functions changing in concert with the chloroplast functionality (Peterhansel et al. 2010). By contrast, the expression of genes involved in the glyoxylate cycle, which is rapidly downregulated after cotyledon development in the wild type (Graham 2008), remained at high levels in *amiR-MGD1* lines grown with DEX (Fujii et al. 2014). The expression of glyoxylate cycle genes was not simply associated with triacylglycerol levels in cotyledons but rather appeared to be linked to lipid metabolism in plastids. The data suggest that galactolipid biosynthesis plays an important role in the coordinated transcriptional regulation of chloroplasts and other organelles during cotyledon development.

Role of Thylakoid Lipid Biosynthesis in Leaf Development

Mutations that impair chloroplast differentiation are repeatedly observed to perturb leaf cell differentiation. In fact, impaired thylakoid lipid biosynthesis strongly affects mesophyll cell organization and leaf development. In the *mgd1-2* mutant, embryogenesis was stunted at the globular stage, thus resulting in loss of cotyledons after seedling germination (Kobayashi et al. 2007; Kobayashi and Ohta 2008). This mutant could develop true leaves but with large intercellular spaces and reduced numbers of mesophyll cells. The inducible knockdown of *MGD1* by *amiR-MGD1* further revealed the effect of galactolipid biosynthesis in cotyledon development (Fujii et al. 2014). Postgerminative growth of both epidermal and mesophyll cells in cotyledons was affected by the *MGD1* suppression, as illustrated in their irregular shapes with indented cell outlines. Moreover, the palisade and spongy cell alignment in cotyledon mesophylls was largely disorganized in *amiR-MGD1* lines treated with DEX. Galactolipid biosynthesis and consequent chloroplast development may play a role in regulating cell development during embryogenesis, true leaf biogenesis, and even postgerminative development of cotyledons.

In addition to galactolipids, anionic thylakoid lipids are essential for leaf development. In the *pgp1-2* mutant deficient in plastid PG biosynthesis, leaf mesophyll cells were observed only around vascular structures with large intercellular spaces under nutrient-sufficient conditions (Hagio et al. 2002; Kobayashi et al. 2015). Epidermal cell layers of the *pgp1-2* leaves were also distorted and abnormally crooked. However, under P-deficient conditions, intercellular spaces were decreased and the distortion of epidermal cell layers was moderated in *pgp1-2* leaves (Kobayashi et al. 2015). Because glycolipid biosynthesis in the *pgp1-2* mutant was strongly activated in response to P deficiency without functional photosynthesis, glycolipid accumulation in leaf plastids would lead to leaf cell development independent of photosynthetic activity. Impaired mesophyll cell development was observed even in the leaky mutant carrying a point mutation in the *PGP1* gene (*pgp1-1*). In the *pgp1-1* mutant, a 30 % reduction in the PG content with a 80 % reduction in plastidic PGP activity caused decreased number of mesophyll cells with intercellular spaces in leaves, although thylakoid membrane biogenesis inside leaf chloroplasts was not markedly affected (Xu et al. 2002; Yu and Benning 2003). The double mutation of *pgp1-1* and *sqd2* further decreased number of mesophyll cells and increased intercellular spaces in leaves (Yu and Benning 2003). These results suggest a special requirement of anionic lipids for differentiation, development and alignment of mesophyll cells. An involvement of plastid signaling in coordinating the development of leaf cells with chloroplast functionalities has been proposed. When chloroplast biogenesis was impaired in young true leaves under norflurazon treatment, the transition of leaf cells from the proliferation to expansion stage was inhibited, for decreased cell size (Andriankaja et al. 2012). Likewise, impaired chloroplast biogenesis by the plastid translation inhibitor lincomycin inhibited cotyledon opening and expansion (Ruckle and Larkin 2009). However, disruption of plastid signaling by the *gun1* mutation attenuated the lincomycin-inhibited cotyledon development. These data suggest that GUN1-mediated plastid signaling inhibits leaf cell development in response to chloroplast dysfunction. Although the association of thylakoid lipids and nucleoid-localized GUN1 protein is unknown, thylakoid lipid biosynthesis may be involved in plastid signaling, which coordinates leaf cell development and chloroplast functionalities.

Conclusions and Perspectives

Concerted synthesis and assembly of membrane lipids with proteins, pigments and cofactors are required for the formation of efficient photosynthetic systems and chloroplast biogenesis. Thylakoid lipid biosynthesis can trigger the expression of photosynthesis-associated genes in both the nucleus and plastids. In addition, fatty acid synthesis may be linked to plastid protein synthesis. Thus, formation of thylakoid membrane precursors may function as a determinant process for chloroplast biogenesis and leaf development. Meanwhile, biosynthesis of thylakoid lipids is coordinated with biosynthesis of photosynthetic protein-cofactor complexes at transcriptional and post-translational levels in response to endogenous and environmental cues. The reciprocal regulation coordinating biosynthesis of lipids, proteins and pigments would guarantee efficient formation of photosynthetic systems but avoid the unbalanced accumulation of thylakoid constituents and intermediates that are potential sources of photooxidative stresses.

Thylakoid lipid biosynthesis in regulating chloroplast biogenesis and leaf development has been revealed, but the detailed mechanisms remain elusive. There are few reports about transcription factors and related regulators involved in thylakoid lipid biosynthesis as compared with the increasing knowledge about regulators for chlorophyll and photosynthetic protein synthesis. The molecular mechanism of how the functional state of lipid biosynthesis is linked to photosynthesis-associated gene expression in plastids and nucleus is also unknown. In addition, the physiological importance of post-translational regulation of galactolipid biosynthesis needs to be addressed. Elucidation of the lipid trafficking mechanism involved in thylakoid membrane biogenesis (e.g., via envelope invagination, vesicle transport) is required to understand the molecular processes of the assembly of entire sets of photosynthetic complexes during chloroplast biogenesis.

Acknowledgements This work was supported by Grants-in-Aid for Scientific Research on Priority Areas (No. 24770055) in Japan.

References

Alban C, Joyard J, Douce R (1988) Preparation and characterization of envelope membranes from nongreen plastids. Plant Physiol 88:709–717. doi:10.1104/pp.88.3.709

Andriankaja M, Dhondt S, De Bodt S et al (2012) Exit from proliferation during leaf development in *Arabidopsis thaliana*: a not-so-gradual process. Dev Cell 22:64–78. doi:10.1016/j.devcel.2011.11.011

Aoki M, Sato N, Meguro A, Tsuzuki M (2004) Differing involvement of sulfoquinovosyl diacylglycerol in photosystem II in two species of unicellular cyanobacteria. Eur J Biochem 271:685–693. doi:10.1111/j.1432-1033.2003.03970.x

Aronsson H, Schöttler MA, Kelly AA et al (2008) Monogalactosyldiacylglycerol deficiency in Arabidopsis affects pigment composition in the prolamellar body and impairs thylakoid membrane energization and photoprotection in leaves. Plant Physiol 148:580–592. doi:10.1104/pp.108.123372

Awai K, Maréchal E, Block MA et al (2001) Two types of MGDG synthase genes, found widely in both 16:3 and 18:3 plants, differentially mediate galactolipid syntheses in photosynthetic and nonphotosynthetic tissues in *Arabidopsis thaliana*. Proc Natl Acad Sci U S A 98:10960–10965. doi:10.1073/pnas.181331498

Awai K, Ohta H, Sato N (2014) Oxygenic photosynthesis without galactolipids. Proc Natl Acad Sci U S A 111:13571–13575. doi:10.1073/pnas.1403708111

Babiychuk E, Müller F, Eubel H et al (2003) Arabidopsis phosphatidylglycerophosphate synthase 1 is essential for chloroplast differentiation, but is dispensable for mitochondrial function. Plant J 33:899–909. doi:10.1046/j.1365-313X.2003.01680.x

Bae G, Choi G (2008) Decoding of light signals by plant phytochromes and their interacting proteins. Annu Rev Plant Biol 59:281–311. doi:10.1146/annurev.arplant.59.032607.092859

Benning C, Ohta H (2005) Three enzyme systems for galactoglycerolipid biosynthesis are coordinately regulated in plants. J Biol Chem 280:2397–2400. doi:10.1074/jbc.R400032200

Block MA, Dorne A-J, Joyard J, Douce R (1983) Preparation and characterization of membrane fractions enriched in outer and inner envelope membranes from spinach chloroplasts. II. Biochemical characterization. J Biol Chem 258:13281–13286

Bohne A-V, Schwarz C, Schottkowski M et al (2013) Reciprocal regulation of protein synthesis and carbon metabolism for thylakoid membrane biogenesis. PLoS Biol 11:e1001482. doi:10.1371/journal.pbio.1001482

Botté CY, Deligny M, Roccia A et al (2011) Chemical inhibitors of monogalactosyldiacylglycerol synthases in *Arabidopsis thaliana*. Nat Chem Biol 7:834–842. doi:10.1038/nchembio.658

Covés J, Joyard J, Douce R (1988) Lipid requirement and kinetic studies of solubilized UDP-galactose:diacylglycerol galactosyltransferase activity from spinach chloroplast envelope membranes. Proc Natl Acad Sci U S A 85:4966–4970. doi:10.1073/pnas.85.14.4966

Demé B, Cataye C, Block MA et al (2014) Contribution of galactoglycerolipids to the 3-dimensional architecture of thylakoids. FASEB J 28:3373–3383. doi:10.1096/fj.13-247395

Dörmann P, Hoffmann-Benning S, Balbo I, Benning C (1995) Isolation and characterization of an Arabidopsis mutant deficient in the thylakoid lipid digalactosyl diacylglycerol. Plant Cell 7:1801–1810. doi:10.1105/tpc.7.11.1801

Dorne A, Joyard J, Douce R (1990) Do thylakoids really contain phosphatidylcholine ? Proc Natl Acad Sci U S A 87:71–74

Dubertret G, Mirshahi A, Mirshahi M et al (1994) Evidence from in vivo manipulations of lipid composition in mutants that the Δ^3-*trans*-hexadecenoic acid-containing phosphatidylglycerol is involved in the biogenesis of the light-harvesting chlorophyll a/b-protein complex of *Chlamydomonas reinhardtii*. Eur J Biochem 226:473–482. doi:10.1111/j.1432-1033.1994.tb20072.x

Dubots E, Audry M, Yamaryo Y et al (2010) Activation of the chloroplast monogalactosyldiacylglycerol synthase MGD1 by phosphatidic acid and phosphatidylglycerol. J Biol Chem 285:6003–6011. doi:10.1074/jbc.M109.071928

El Maanni A, Dubertret G, Delrieu M-J et al (1998) Mutants of *Chlamydomonas reinhardtii* affected in phosphatidylglycerol metabolism and thylakoid biogenesis. Plant Physiol Biochem 36:609–619. doi:10.1016/S0981-9428(98)80009-0

Essigmann B, Güler S, Narang RA et al (1998) Phosphate availability affects the thylakoid lipid composition and the expression of *SQD1*, a gene required for sulfolipid biosynthesis in *Arabidopsis thaliana*. Proc Natl Acad Sci U S A 95:1950–1955

Froehlich JE, Benning C, Dörmann P (2001) The digalactosyldiacylglycerol (DGDG) synthase DGD1 is inserted into the outer envelope membrane of chloroplasts in a manner independent of the general import pathway and does not depend on direct interaction with monogalactosyldiacylglycerol synthase for DGDG biosynthesis. J Biol Chem 276:31806–31812. doi:10.1074/jbc.M104652200

Fujii S, Kobayashi K, Nakamura Y, Wada H (2014) Inducible knockdown of MONOGALACTOSYLDIACYLGLYCEROL SYNTHASE1 reveals roles of galactolipids in organelle differentiation in Arabidopsis cotyledons. Plant Physiol 166:1436–1449. doi:10.1104/pp.114.250050

Geigenberger P, Kolbe A, Tiessen A (2005) Redox regulation of carbon storage and partitioning in response to light and sugars. J Exp Bot 56:1469–1479. doi:10.1093/jxb/eri178

Gombos Z, Várkonyi Z, Hagio M et al (2002) Phosphatidylglycerol requirement for the function of electron acceptor plastoquinone Q_B in the photosystem II reaction center. Biochemistry 41:3796–3802. doi:10.1021/bi011884h

Graham IA (2008) Seed storage oil mobilization. Annu Rev Plant Biol 59:115–142. doi:10.1146/annurev.arplant.59.032607.092938

Güler S, Seeliger A, Härtel H et al (1996) A null mutant of *Synechococcus* sp. PCC7942 deficient in the sulfolipid sulfoquinovosyl diacylglycerol. J Biol Chem 271:7501–7507

Guo J, Zhang Z, Bi Y et al (2005) Decreased stability of photosystem I in *dgd1* mutant of *Arabidopsis thaliana*. FEBS Lett 579:3619–3624. doi:10.1016/j.febslet.2005.05.049

Guskov A, Kern J, Gabdulkhakov A et al (2009) Cyanobacterial photosystem II at 2.9-Å resolution and the role of quinones, lipids, channels and chloride. Nat Struct Mol Biol 16:334–342. doi:10.1038/nsmb.1559

Hagio M, Gombos Z, Várkonyi Z et al (2000) Direct evidence for requirement of phosphatidylglycerol in photosystem II of photosynthesis. Plant Physiol 124:795–804. doi:10.1104/pp.124.2.795

Hagio M, Sakurai I, Sato S et al (2002) Phosphatidylglycerol is essential for the development of thylakoid membranes in *Arabidopsis thaliana*. Plant Cell Physiol 43:1456–1464. doi:10.1093/pcp/pcf185

Härtel H, Lokstein H, Dörmann P et al (1997) Changes in the composition of the photosynthetic apparatus in the galactolipid-deficient *dgd1* mutant of *Arabidopsis thaliana*. Plant Physiol 115:1175–1184. doi:10.1104/pp.115.3.1175

Haselier A, Akbari H, Weth A et al (2010) Two closely related genes of Arabidopsis encode plastidial cytidinediphosphate diacylglycerol synthases essential for photoautotrophic growth. Plant Physiol 153:1372–1384. doi:10.1104/pp.110.156422

Hölzl G, Witt S, Kelly AA et al (2006) Functional differences between galactolipids and glucolipids revealed in photosynthesis of higher plants. Proc Natl Acad Sci U S A 103:7512–7517. doi:10.1073/pnas.0600525103

Hölzl G, Witt S, Gaude N et al (2009) The role of diglycosyl lipids in photosynthesis and membrane lipid homeostasis in Arabidopsis. Plant Physiol 150:1147–1159. doi:10.1104/pp.109.139758

Ivanov AG, Hendrickson L, Krol M et al (2006) Digalactosyl-diacylglycerol deficiency impairs the capacity for photosynthetic intersystem electron transport and state transitions in *Arabidopsis thaliana* due to photosystem I acceptor-side limitations. Plant Cell Physiol 47:1146–1157. doi:10.1093/pcp/pcj089

Jahns P, Latowski D, Strzalka K (2009) Mechanism and regulation of the violaxanthin cycle: the role of antenna proteins and membrane lipids. Biochim Biophys Acta 1787:3–14. doi:10.1016/j.bbabio.2008.09.013

Jarvis P, López-Juez E (2013) Biogenesis and homeostasis of chloroplasts and other plastids. Nat Rev Mol Cell Biol 14:787–802. doi:10.1038/nrm3702

Jarvis P, Dörmann P, Peto CA et al (2000) Galactolipid deficiency and abnormal chloroplast development in the *Arabidopsis MGD synthase 1* mutant. Proc Natl Acad Sci U S A 97:8175–8179. doi:10.1073/pnas.100132197

Jordan P, Fromme P, Witt HT et al (2001) Three-dimensional structure of cyanobacterial photosystem I at 2.5 Å resolution. Nature 411:909–917. doi:10.1038/35082000

Jouhet J, Maréchal E, Baldan B et al (2004) Phosphate deprivation induces transfer of DGDG galactolipid from chloroplast to mitochondria. J Cell Biol 167:863–874. doi:10.1083/jcb.200407022

Kansy M, Wilhelm C, Goss R (2014) Influence of thylakoid membrane lipids on the structure and function of the plant photosystem II core complex. Planta 240:781–796. doi:10.1007/s00425-014-2130-2

Kelly AA, Dörmann P (2002) DGD2, an *Arabidopsis* gene encoding a UDP-galactose-dependent digalactosyldiacylglycerol synthase is expressed during growth under phosphate-limiting conditions. J Biol Chem 277:1166–1173. doi:10.1074/jbc.M110066200

Kelly AA, Froehlich JE, Dörmann P (2003) Disruption of the two digalactosyldiacylglycerol synthase genes *DGD1* and *DGD2* in Arabidopsis reveals the existence of an additional enzyme of galactolipid synthesis. Plant Cell 15:2694–2706. doi:10.1105/tpc.016675

Kobayashi K, Ohta H (2008) Possible requirement of galactolipids for embryogenesis. In: Allen J, Gantt E, Golbeck J, Osmond B (eds) Photosynthesis. Energy from the sun. Springer, Heidelberg, pp 783–786. doi:10.1007/978-1-4020-6709-9

Kobayashi K, Awai K, Takamiya K, Ohta H (2004) Arabidopsis type B monogalactosyldiacylglycerol synthase genes are expressed during pollen tube growth and induced by phosphate starvation. Plant Physiol 134:640–648. doi:10.1104/pp.103.032656

Kobayashi K, Kondo M, Fukuda H et al (2007) Galactolipid synthesis in chloroplast inner envelope is essential for proper thylakoid biogenesis, photosynthesis, and embryogenesis. Proc Natl Acad Sci U S A 104:17216–17221. doi:10.1073/pnas.0704680104

Kobayashi K, Awai K, Nakamura M et al (2009a) Type-B monogalactosyldiacylglycerol synthases are involved in phosphate starvation-induced lipid remodeling, and are crucial for low-phosphate adaptation. Plant J 57:322–331. doi:10.1111/j.1365-313X.2008.03692.x

Kobayashi K, Nakamura Y, Ohta H (2009b) Type A and type B monogalactosyldiacylglycerol synthases are spatially and functionally separated in the plastids of higher plants. Plant Physiol Biochem 47:518–525. doi:10.1016/j.plaphy.2008.12.012

Kobayashi K, Narise T, Sonoike K et al (2013) Role of galactolipid biosynthesis in coordinated development of photosynthetic complexes and thylakoid membranes during chloroplast biogenesis in Arabidopsis. Plant J 73:250–261. doi:10.1111/tpj.12028

Kobayashi K, Fujii S, Sasaki D et al (2015) Transcriptional regulation of thylakoid galactolipid biosynthesis coordinated with chlorophyll biosynthesis during the development of chloroplasts in *Arabidopsis*. Front Plant Sci 5:272. doi:10.3389/fpls.2014.00272

Kobayashi K, Fujii S, Sato M et al (2015) Specific role of phosphatidylglycerol and functional overlaps with other thylakoid lipids in Arabidopsis chloroplast biogenesis. Plant Cell Rep 34(4):631–642. doi:10.1007/s00299-014-1719-z

Koussevitzky S, Nott A, Mockler TC et al (2007) Signals from chloroplasts converge to regulate nuclear gene expression. Science 316:715–719. doi:10.1126/science

Larkin RM, Ruckle ME (2008) Integration of light and plastid signals. Curr Opin Plant Biol 11:593–599. doi:10.1016/j.pbi.2008.10.004

Lee J, He K, Stolc V et al (2007) Analysis of transcription factor HY5 genomic binding sites revealed its hierarchical role in light regulation of development. Plant Cell 19:731–749. doi:10.1105/tpc.106.047688

Liu Z, Yan H, Wang K et al (2004) Crystal structure of spinach major lightharvesting complex at 2.72Å resolution. Nature 428:287–292. doi:10.1038/nature02373

Lopez-Juez E, Pyke KA (2005) Plastids unleashed: their development and their integration in plant development. Int J Dev Biol 49:557–577. doi:10.1387/ijdb.051997el

Masuda S, Harada J, Yokono M et al (2011) A monogalactosyldiacylglycerol synthase found in the green sulfur bacterium *Chlorobaculum tepidum* reveals important roles for galactolipids in photosynthesis. Plant Cell 23:2644–2658. doi:10.1105/tpc.111.085357

McFadden GI (2014) Origin and evolution of plastids and photosynthesis in eukaryotes. Cold Spring Harb Perspect Biol 6:a016105. doi:10.1101/cshperspect.a016105

Miège C, Maréchal E, Shimojima M et al (1999) Biochemical and topological properties of type A MGDG synthase, a spinach chloroplast envelope enzyme catalyzing the synthesis of both prokaryotic and eukaryotic MGDG. Eur J Biochem 265:990–1001. doi:10.1046/j.1432-1327.1999.00801.x

Minoda A, Sato N, Nozaki H et al (2002) Role of sulfoquinovosyl diacylglycerol for the maintenance of photosystem II in *Chlamydomonas reinhardtii*. Eur J Biochem 269:2353–2358. doi:10.1046/j.1432-1033.2002.02896.x

Minoda A, Sonoike K, Okada K et al (2003) Decrease in the efficiency of the electron donation to tyrosine Z of photosystem II in an SQDG-deficient mutant of *Chlamydomonas*. FEBS Lett 553:109–112. doi:10.1016/S0014-5793(03)00981-5

Mizusawa N, Sakata S, Sakurai I et al (2009a) Involvement of digalactosyldiacylglycerol in cellular thermotolerance in *Synechocystis* sp. PCC 6803. Arch Microbiol 191:595–601. doi:10.1007/s00203-009-0486-7

Mizusawa N, Sakurai I, Sato N, Wada H (2009b) Lack of digalactosyldiacylglycerol increases the sensitivity of *Synechocystis* sp. PCC 6803 to high light stress. FEBS Lett 583:718–722. doi:10.1016/j.febslet.2009.01.021

Moellering ER, Benning C (2011) Galactoglycerolipid metabolism under stress: a time for remodeling. Trends Plant Sci 16:98–107. doi:10.1016/j.tplants.2010.11.004

Moellering ER, Muthan B, Benning C (2010) Freezing tolerance in plants requires lipid remodeling at the outer chloroplast membrane. Science 330:226–228. doi:10.1126/science.1191803

Myers AM, James MG, Lin Q et al (2011) Maize *opaque5* encodes monogalactosyldiacylglycerol synthase and specifically affects galactolipids necessary for amyloplast and chloroplast function. Plant Cell 23:2331–2347. doi:10.1105/tpc.111.087205

Neuhaus HE, Emes MJ (2000) Nonphotosynthetic metabolism in plastids. Annu Rev Plant Physiol Plant Mol Biol 51:111–140. doi:10.1146/annurev.arplant.51.1.111

Ohta H, Shimojima M, Arai T et al (1995) UDP-galactose: diacylglycerol galactosyltransferase in cucumber seedlings: purification of the enzyme and the activation by phosphatidic acid. In: Kader JC, Mazliak P (eds) Plant lipid metabolism. Springer, Heidelberg, pp 152–155. doi:10.1007/978-94-015-8394-7

Okazaki Y, Shimojima M, Sawada Y et al (2009) A chloroplastic UDP-glucose pyrophosphorylase from *Arabidopsis* is the committed enzyme for the first step of sulfolipid biosynthesis. Plant Cell 21:892–909. doi:10.1105/tpc.108.063925

Okazaki Y, Otsuki H, Narisawa T et al (2013) A new class of plant lipid is essential for protection against phosphorus depletion. Nat Commun 4:1510. doi:10.1038/ncomms2512

Omata T, Murata N (1983) Isolation and characterization of the cytoplasmic membranes from the blue-green alga (Cyanobacterium) *Anacystis nidulans*. Plant Cell Physiol 24:1101–1112

Peterhansel C, Horst I, Niessen M et al (2010) Photorespiration. Arabidopsis Book 12:e0130. doi:10.1199/tab.0130

Powikrowska M, Oetke S, Jensen PE, Krupinska K (2014) Dynamic composition, shaping and organization of plastid nucleoids. Front Plant Sci 5:424. doi:10.3389/fpls.2014.00424

Reifarth F, Christen G, Seeliger AG et al (1997) Modification of the water oxidizing complex in leaves of the *dgd1* mutant of *Arabidopsis thaliana* deficient in the galactolipid digalactosyldiacylglycerol. Biochemistry 36:11769–11776. doi:10.1021/bi9709654

Ruckle ME, Larkin RM (2009) Plastid signals that affect photomorphogenesis in *Arabidopsis thaliana* are dependent on GENOMES UNCOUPLED 1 and cryptochrome 1. New Phytol 182:367–379. doi:10.1111/j.1469-8137.2008.02729.x

Sakurai I, Hagio M, Gombos Z et al (2003) Requirement of phosphatidylglycerol for maintenance of photosynthetic machinery. Plant Physiol 133:1376–1384. doi:10.1104/pp.103.026955

Sakurai I, Mizusawa N, Ohashi S et al (2007a) Effects of the lack of phosphatidylglycerol on the donor side of photosystem II. Plant Physiol 144:1336–1346. doi:10.1104/pp.107.098731

Sakurai I, Mizusawa N, Wada H, Sato N (2007b) Digalactosyldiacylglycerol is required for stabilization of the oxygen-evolving complex in photosystem II. Plant Physiol 145:1361–1370. doi:10.1104/pp.107.106781

Sarkis J, Rocha J, Maniti O et al (2014) The influence of lipids on MGD1 membrane binding highlights novel mechanisms for galactolipid biosynthesis regulation in chloroplasts. FASEB J 28:3114–3123. doi:10.1096/fj.14-250415

Sasaki Y, Kozaki A, Hatano M (1997) Link between light and fatty acid synthesis: thioredoxin-linked reductive activation of plastidic acetyl-CoA carboxylase. Proc Natl Acad Sci U S A 94:11096–11101. doi:10.1073/pnas.94.20.11096

Sato N, Sonoike K, Tsuzuk M, Kawaguchi A (1995a) Impaired photosystem II in a mutant of *Chlamydomonas reinhardtii* defective in sulfoquinovosyl diacylglycerol. Eur J Biochem 234:16–23. doi:10.1111/j.1432-1033.1995.016_c.x

Sato N, Tsuzuki M, Matsuda Y et al (1995b) Isolation and characterization of mutants affected in lipid metabolism of *Chlamydomonas reinhardtii*. Eur J Biochem 230:987–993. doi:10.1111/j.1432-1033.1995.0987g.x

Sato N, Hagio M, Wada H, Tsuzuki M (2000) Requirement of phosphatidylglycerol for photosynthetic function in thylakoid membranes. Proc Natl Acad Sci U S A 97:10655–10660

Schaller S, Latowski D, Jemioła-Rzemińska M et al (2010) The main thylakoid membrane lipid monogalactosyldiacylglycerol (MGDG) promotes the de-epoxidation of violaxanthin associated with the light-harvesting complex of photosystem II (LHCII). Biochim Biophys Acta 1797:414–424. doi:10.1016/j.bbabio.2009.12.011

Schaller S, Latowski D, Jemioła-Rzemińska M et al (2011) Regulation of LHCII aggregation by different thylakoid membrane lipids. Biochim Biophys Acta 1807:326–335. doi:10.1016/j.bbabio.2010.12.017

Sekine K, Hase T, Sato N (2002) Reversible DNA compaction by sulfite reductase regulates transcriptional activity of chloroplast nucleoids. J Biol Chem 277:24399–24404. doi:10.1074/jbc.M201714200

Serrato AJ, Fernández-Trijueque J, Barajas-López J-D et al (2013) Plastid thioredoxins: a "one-for-all" redox-signaling system in plants. Front Plant Sci 4:463. doi:10.3389/fpls.2013.00463

Shimojima M, Ohta H, Iwamatsu A et al (1997) Cloning of the gene for monogalactosyldiacylglycerol synthase and its evolutionary origin. Proc Natl Acad Sci 94:333–337

Shimojima M, Watanabe T, Madoka Y et al (2013) Differential regulation of two types of monogalactosyldiacylglycerol synthase in membrane lipid remodeling under phosphate-limited conditions in sesame plants. Front Plant Sci 4:469. doi:10.3389/fpls.2013.00469

Shipley GG, Green JP, Nichols BW (1973) The phase behavior of monogalactosyl, digalactosyl, and sulphoquinovosyl diglycerides. Biochim Biophys Acta 311:531–544. doi:10.1016/0005-2736(73)90128-4

Standfuss J, Terwisscha van Scheltinga AC, Lamborghini M, Kühlbrandt W (2005) Mechanisms of photoprotection and nonphotochemical quenching in pea light-harvesting complex at 2.5Å resolution. EMBO J 24:919–928. doi:10.1038/sj.emboj.7600585

Steffen R, Kelly AA, Huyer J et al (2005) Investigations on the reaction pattern of photosystem II in leaves from *Arabidopsis thaliana* wild type plants and mutants with genetically modified lipid content. Biochemistry 44:3134–3142. doi:10.1021/bi048465f

Takahashi S, Badger MR (2011) Photoprotection in plants: a new light on photosystem II damage. Trends Plant Sci 16:53–60. doi:10.1016/j.tplants.2010.10.001

Tanoue R, Kobayashi M, Katayama K et al (2014) Phosphatidylglycerol biosynthesis is required for the development of embryos and normal membrane structures of chloroplasts and mitochondria in Arabidopsis. FEBS Lett 588:1680–1685. doi:10.1016/j.febslet.2014.03.010

Uemura M, Joseph RA, Steponkus PL (1995) Cold acclimation of Arabidopsis thaliana (effect on plasma membrane lipid composition and freeze-induced lesions). Plant Physiol 109:15–30. doi:10.1104/pp.109.1.15

Umena Y, Kawakami K, Shen J-R, Kamiya N (2011) Crystal structure of oxygen-evolving photosystem II at a resolution of 1.9 Å. Nature 473:55–60. doi:10.1038/nature09913

Vothknecht UC, Westhoff P (2001) Biogenesis and origin of thylakoid membranes. Biochim Biophys Acta 1541:91–101. doi:10.1016/S0167-4889(01)00153-7

Wada H, Murata N (1989) *Synechocystis* PCC6803 mutants defective in desaturation of fatty acids. Plant Cell Physiol 30:971–978

Wada H, Murata N (1998) Membrane lipids in cyanobacteria. In: Siegenthaler P, Murata N (eds) Lipids in photosynthesis: structure, function and genetics, vol 6, Advances in photosynthesis and respiration. Springer, Heidelberg, pp 65–81. doi:10.1007/0-306-48087-5_4

Wada H, Murata N (2007) The essential role of phosphatidylglycerol in photosynthesis. Photosynth Res 92:205–215. doi:10.1007/s11120-007-9203-z

Walters RG (2005) Towards an understanding of photosynthetic acclimation. J Exp Bot 56:435–447. doi:10.1093/jxb/eri060

Waters MT, Wang P, Korkaric M et al (2009) GLK transcription factors coordinate expression of the photosynthetic apparatus in Arabidopsis. Plant Cell 21:1109–1128. doi:10.1105/tpc.108.065250

Woodson JD, Chory J (2008) Coordination of gene expression between organellar and nuclear genomes. Nat Rev Genet 9:383–395. doi:10.1038/nrg2348

Wu W, Ping W, Wu H et al (2013) Monogalactosyldiacylglycerol deficiency in tobacco inhibits the cytochrome b_6f-mediated intersystem electron transport process and affects the photostability of the photosystem II apparatus. Biochim Biophys Acta 1827:709–722. doi:10.1016/j.bbabio.2013.02.013

Xu C, Härtel H, Wada H et al (2002) The *pgp1* mutant locus of Arabidopsis encodes a phosphatidylglycerolphosphate synthase with impaired activity. Plant Physiol 129:594–604. doi:10.1104/pp.002725

Yamaryo Y, Kanai D, Awai K et al (2003) Light and cytokinin play a co-operative role in MGDG synthesis in greening cucumber cotyledons. Plant Cell Physiol 44:844–855. doi:10.1093/pcp/pcg110

Yamaryo Y, Motohashi K, Takamiya K et al (2006) In vitro reconstitution of monogalactosyldiacylglycerol (MGDG) synthase regulation by thioredoxin. FEBS Lett 580:4086–4090. doi:10.1016/j.febslet.2006.06.050

Yu B, Benning C (2003) Anionic lipids are required for chloroplast structure and function in *Arabidopsis*. Plant J 36:762–770. doi:10.1046/j.1365-313X.2003.01918.x

Yu B, Xu C, Benning C (2002) *Arabidopsis* disrupted in *SQD2* encoding sulfolipid synthase is impaired in phosphate-limited growth. Proc Natl Acad Sci U S A 99:5732–5737. doi:10.1073/pnas.082696499

Zhou F, Liu S, Hu Z et al (2009) Effect of monogalactosyldiacylglycerol on the interaction between photosystem II core complex and its antenna complexes in liposomes of thylakoid lipids. Photosynth Res 99:185–193. doi:10.1007/s11120-008-9388-9

Zhou Y, Peisker H, Weth A et al (2013) Extraplastidial cytidinediphosphate diacylglycerol synthase activity is required for vegetative development in Arabidopsis thaliana. Plant J 75:867–879. doi:10.1111/tpj.12248

Chapter 6
Role of MGDG and Non-bilayer Lipid Phases in the Structure and Dynamics of Chloroplast Thylakoid Membranes

Győző Garab, Bettina Ughy, and Reimund Goss

Abstract In this chapter we focus our attention on the enigmatic structural and functional roles of the major, non-bilayer lipid monogalactosyl-diacylglycerol (MGDG) in the thylakoid membrane. We give an overview on the state of the art on the role of MGDG and non-bilayer lipid phases in the xanthophyll cycles in different organisms. We also discuss data on the roles of MGDG and other lipid molecules found in crystal structures of different photosynthetic protein complexes and in lipid-protein assemblies, as well as in the self-assembly of the multilamellar membrane system. Comparison and critical evaluation of different membrane models – that take into account and capitalize on the special properties of non-bilayer lipids and/or non-bilayer lipid phases, and thus to smaller or larger extents deviate from the 'standard' Singer-Nicolson model – will conclude this review. With this chapter the authors hope to further stimulate the discussion about, what we think, is perhaps the most exciting question of membrane biophysics: the why and wherefore of non-bilayer lipids and lipid phases in, or in association with, bilayer biological membranes.

Keywords Bilayer phase • Inverted hexagonal phase • Light-harvesting complex • Membrane models • Merocyanine-540 • Non-lamellar lipid phases • ^{31}P-NMR • Photosystem • Xanthophyll cycle • Zeaxanthin

G. Garab (✉) • B. Ughy
Biological Research Center, Hungarian Academy of Sciences, Szeged, Hungary
e-mail: gyozo@brc.hu; ughy.bettina@brc.mta.hu

R. Goss
Institute of Biology, Department of Plant Physiology, University of Leipzig, Leipzig, Germany
e-mail: rgoss@rz.uni-leipzig.de

Y. Nakamura, Y. Li-Beisson (eds.), *Lipids in Plant and Algae Development*, Subcellular Biochemistry 86, DOI 10.1007/978-3-319-25979-6_6

Introduction

In oxygenic photosynthetic organisms the light reactions of photosynthesis occur in the thylakoid membrane, flattened lipid vesicles, which are densely packed with pigment-protein complexes and other constituents of the photosynthetic machinery. They contain the two photosystems: PSII (photosystem II or water-plastoquinone oxidoreductase) – responsible for the production of molecular oxygen, and PSI (photosystem I or plastocyanin:ferredoxin oxidoreductase) – responsible for the reduction of NADP to $NADPH_2$, which is then used as reducing power to assimilate carbon dioxide. The two photochemical reaction centers are fed by excitation energy via their associated core antenna complexes and the main, peripheral light-harvesting antenna complexes, LHCII and LHCI, for PSII and PSI, respectively. The membranes also embed the cytochrome *b6f* complex and some further constituents of the electron transport system. The operation of the vectorial electron transport, linked to proton transport, generate the energized state of the thylakoid membrane: an electrochemical potential gradient, which consists of a transmembrane ΔpH and an electric potential gradient ($\Delta\Psi$) – which are utilized for the synthesis of the energy carrier molecule ATP with the aid of ATP-synthase.

The build-up of the energized state of thylakoid membranes, and thus the ATP synthesis, requires the separation of the two aqueous phases, the lumenal and stromal sides, and requires, in particular, the impermeability of membranes to most water-soluble molecules and to ions. This is warranted by the organization of thylakoid membranes as bilayers (Williams 1998), a requirement in all energy-converting membranes.

Given this strong restriction of the bilayer state of the functional membrane, for the first glance (and perhaps also for second and third glances), it is difficult to understand that the major lipid species in thylakoid membranes is the non-bilayer lipid monogalatosyl-diacylglycerol (MGDG), which constitutes about half of the total lipid content of thylakoid membranes. This is because, non-lamella-forming or non-bilayer lipids are not capable to self-assemble into bilayers in aqueous media under physiologically relevant conditions; instead, they are assembled into different non-lamellar or non-bilayer lipid phases (see section "Phase behaviour of thylakoid membrane lipids"). Only about the other half of the thylakoid membrane lipids, digalactolsyl-diacylglycerol (DGDG), sulfoquinosyl-diacylglycerol (SQDG) and phosphatidylglycerol (PG) are bilayer lipids (see section "Thylakoid Lipids").

The problem is not limited to thylakoid membranes. All energy converting membranes are (and evidently must be, for their functioning) organized as bilayers. On the other hand, their dominant lipid species are non-bilayer lipids. This is most remarkable because the chemical composition of the major constituent lipids and proteins are quite different: mainly galactolipids and pigment-protein complexes in the thylakoid membranes; while e.g. the inner mitochondrial membranes are composed mainly of phosopholipids, phosphatidylcholine (PC) and phosphatidylethanolamine (PE), and cardiolipin (CL), and respiratory proteins. Only in small part they both contain essentially identical or similar protein complexes, the ATP-synthase and the cytochrome *b6f/bc* complex. Further, albeit at lower concentrations, apparently all biological membranes contain non-bilayer lipid species (Epand 1998). Hence, while probably all non-bilayer lipids participate in essential lipid-protein interactions, the general answer to their roles in lipid bilayers of biological membranes must be related to their non-lamella-forming property rather than in their chemical composition. Also, while the significance of specific lipid-protein interactions, e.g. between MGDG and the key protein complexes of the thylakoid membrane should not be ignored (cf. MGDG bound to protein complexes.), explanation must be offered for the behaviour of bulk lipids, about 60 % of the total thylakoid lipids, which are found in a fluid-like phase at room temperature (Páli et al. 2003 and references therein), i.e. which constitute the bulk bilayer membrane structure.

Because of the presence of non-bilayer lipids in the bulk phase and the absence of lateral heterogeneity on mesoscopic scale (Van Eerden et al. 2015), the bulk phase lipid mixture acquires a strong non-bilayer propensity (see also Garab et al. 2000). Several membrane models have been proposed in the past years to answer this enigmatic question, i.e. on the roles of non-bilayer lipids in the bilayer membranes, and on the possible physiological roles of non-bilayer lipid phases in, or in association with, the thylakoid membranes, and biomembranes in general. In the present chapter, while we will also outline recent experimental data on the role of MGDG and non-bilayer lipid phases in the xanthophyll cycle, as well as on bound MGDG and other lipid molecules found in different crystal structures, our attention will be focused on the above basic question. For many details, on the composition of thylakoid membranes, the phase behaviour of different isolated lipids and lipid mixtures, the physical mechanisms and physiological significances of phase transitions etc., the readers are referred to excellent reviews in the literature both on thylakoid membranes and on broader subjects, on non-bilayer lipids and lipid phases (Epand 1998; Sackman 1995; Seddon and Templer 1995; Selstam 1998; Williams 1998; Van den Brink-van der Laan et al. 2004; Vigh et al. 2005; Quinn 2012; Koynova and Tenchov 2013; Jouhet 2013; Boudière et al. 2014).

Thylakoid Lipids

Thylakoid membranes contain glycolipids, MGDG, DGDG and SQDG, and only one phospholipid, PG. The neutral galactolipids, DGDG and mainly MGDG, are the major lipid components and only smaller amounts of negatively charged lipids SQDG and PG contribute to the thylakoid membranes. This membrane composition is different from other types of eukaryotic and bacterial membranes, which are built up mainly from phospholipids and may also contain sphingolipids and sterols. Thus, the lipid composition of thylakoid membranes is quite unique; nevertheless, as pointed out in the Introduction, with their high non-bilayer lipid content they resemble the energy-converting mitochondrial and retinal membranes. As most biological membranes, thylakoids also consist of lipid bilayers and membrane-intrinsic proteins and protein complexes and some other compounds.

The galactolipid MGDG is the main thylakoid membrane lipid of higher plants and green algae, which accounts for about 40–55 % of the total membrane lipid content (Murata and Siegenthaler 1998; Goss and Wilhelm 2009; Lepetit et al. 2012). The second most abundant lipid is the other galactolipid, DGDG, which is normally found in a concentration of around 25–30 %. The negatively charged lipids, SQDG and PG, contribute with about 10–15 % to the total lipid content of the thylakoid membrane. This typical lipid composition results in a ratio of neutral to negatively charged lipids of around 4 (Goss et al. 2009; Lepetit et al. 2012). This composition also means that thylakoids, to a significant degree, are built up from the non-bilayer lipid MGDG. This lipid has been proposed to be 'forced' into the bilayer by proteins – as shown by the self-assembly of LHCII:MGDG membranes (Simidjiev et al. 2000). This, i.e. the formation of a bilayer membrane, composed of thylakoid lipid mixture and photosynthetic proteins, however, also depends on the nature of proteins. As it has been thoroughly documented, essentially the same thylakoid lipids, and the proteins that are involved in the phototransformation of protochlorophylls assemble into cubic, rather than lamellar phases (Selstam 1998).

In the absence of proteins, the thylakoid lipid mixture, under physiologically relevant conditions is not capable of forming lamellar phases; instead, it assumes different non-bilayer structures (Williams 1998). The non-bilayer propensity of MGDG-containing lipid mixtures strongly depends on the proportion of MGDG in the lipid mixture. For instance, in the binary mixture of MGDG and DMPC (dimyristoylphosphatidylcholine) it was found that lipid bilayers are formed only when the mole fraction of MGDG is ≤ 0.6 (Castro et al. 2007). The fatty acid composition of MGDG also strongly influences the non-bilayer propensity of lipid mixtures; in particular, polyunsaturated lipids facilitate the transition from bilayer to non-bilayer lipid structures (Van Eerden et al. 2015 and references therein).

The negatively charged PG is the only phospholipid of thylakoid membranes; although represented in relatively small amount, it plays a crucial role in the structure formation and function of photosynthetic complexes (Sato et al. 2000; Wada and Murata 2007; Domonkos et al. 2008; Sozer et al. 2011). It appears that specific PG species are needed for the proper performance of the cyanobacterial cellular processes since the PG deficient *Synechocystis* cells retailored the supplied artificial dioleoyl PG molecules, and created more natural molecular species (Laczko-Dobos et al. 2010).

SQDG, the other negatively charged lipid of the thylakoids in some cases could substitute for PG – apparently replacing its negative charge (Yu and Benning 2003; Benning et al. 1993; Güler et al. 1996). SQDG could contribute to the activity of PSII by associating with PSII and LHCII complexes (Sato et al. 1995; Minoda et al. 2002). It also had an impact on the assembly and the function of photosynthetic complexes (Kansy et al. 2014; Mizusawa and Wada 2012).

The galactolipids, DGDG and MGDG, are the dominant thylakoid lipids and they appear to be very important for functional photosynthesis. DGDG plays an important role in the stability and activity of PSII and PSI and also in the stabilization of LHCII trimers (Dörmann and Hölzl, Chap. 3). Changes in the ratio of MGDG per DGDG affected the membrane organization and the protein folding and insertion, which exerted an impact on the photosynthetic performance (Dörmann and Benning 2002; Kobayashi et al. 2013). Proteoliposome experiments have revealed that MGDG enhances the assembly and energetic coupling of LHCII to the core of PSII (Zhou et al. 2009). As suggested by the experiments of Wang et al. (2014), using MGDG synthase mutants, the two galactolipids play an important structural role in salt-stress response of vascular plants. MGDG deficiency also affected the photoprotection capability of plants, evidently via hindering the functioning of the xanthophyll-cycle (Aronsson et al. 2008) (cf. "Roles of MGDG in xanthophyll cycles"). Changes in the relative amounts of non-bilayer lipids are usually observed in plants in response to variations of the environmental conditions (Harwood 1998). Such alterations are of special physiological importance in photosynthetic organisms, which encounter a wide range of temperatures and other environmental stresses, such as drought and high light.

The lipid composition of the thylakoid membranes of diatoms is different from that of higher plants and green algae, and MGDG may not be the main membrane lipid (Vieler et al. 2007; Goss and Wilhelm 2009; Goss et al. 2009; Lepetit et al. 2012). Diatom thylakoid membranes are strongly enriched in the negatively charged lipids, SQDG and PG, and high light illumination during the cultivation of the algal cells further increased the amount of the negatively charged lipids (Lepetit et al. 2012). Thus, the most abundant lipid in diatom thylakoids is the anionic lipid SQDG, which is found in equal or even higher concentrations than MGDG. The high amount of SQDG and PG in the diatom thylakoid membranes results in a ratio of neutral to negatively charged lipids of around 1 (Goss et al. 2009; Lepetit et al. 2012), whereas this ratio has a value between 3 and 4 in higher plants and green algae due to the high MGDG concentration. It has been proposed that the high amount of SQDG and PG in the diatom thylakoids has significant consequences for the structure of the diatom thylakoid membrane and for the function of the diatom xanthophyll cycle (see section "Roles of MGDG in xanthophyll cycles"). A recent model for the distribution of the membrane lipids and photosynthetic pigment-protein complexes (Lepetit et al. 2012) predicts an enrichment of MGDG in the inner membranes of the stacks of three thylakoid membranes that are typical for the diatom chloroplasts (for further details see section "Localization and operation of xanthophyll cycles in the native thylakoid membrane").

Phase Behaviour of Thylakoid Membrane Lipids

The large amounts of the non-bilayer lipid MGDG in the thylakoid membranes lends its lipid mixture a high propensity to form non-lamellar (non-bilayer) phases. This propensity, as it has been thoroughly documented in the literature (for reviews see e.g. Epand 1998; Williams 1998) arises from the conical shape of the non-bilayer lipid molecules, as opposed to the cylindrical shapes of the lamella-forming lipids. For thylakoid membrane lipids, this is illustrated in Fig. 6.1, which shows the

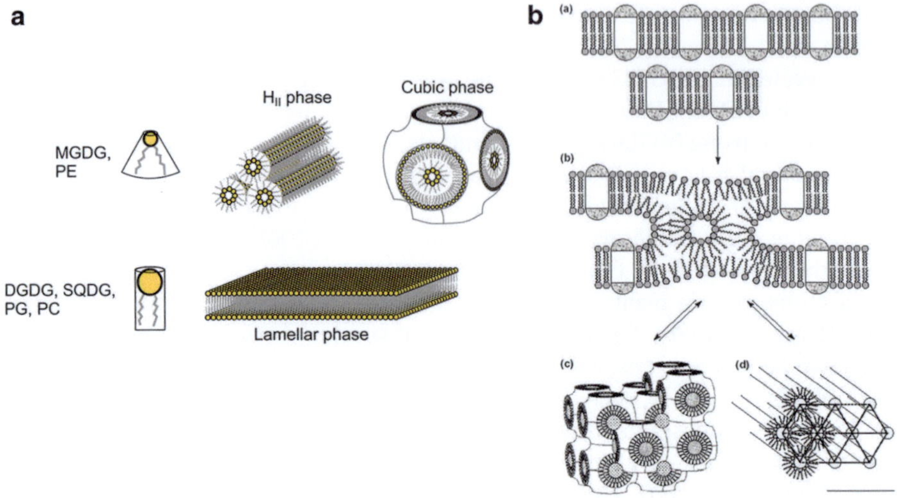

Fig. 6.1 (**a**) Structure formation of thylakoid lipid molecules. Small polar-headgroup containing lipid molecules possess a conical shape and thus form inverted hexagonal (H_{II}) phases or cubic (bicontinuous) structures. Lipids that have similar headgroup and tail size look like cylinders and form lamellar phases, with no curvatures. (**b**) Simplified model of thylakoid membrane (*a*) containing a bilayer and membrane-embedded proteins, which via (*b*) local, transient non-bilayer structures extrudes lipids which, in turn, assemble into (*c*) cubic or (*d*) H_{II} phases (From Garab et al. (2000) with permission)

preferred, inverted hexagonal (H_{II}) phase of MGDG in aqueous media, while DGDG, PG and SQDG spontaneously form lamellar phases.

Depending on their composition and the physico-chemical and external environmental conditions (e.g. ionic strength, pH, hydration state, and ambient temperature and drought stress, respectively), thylakoid membrane lipids *in vitro* and *in vivo* can form different lamellar and non-lamellar phases. For the phase behavior and transitions of different thylakoid lipids the reader is referred to the reviews by Williams (1998) and Selstam (1998); and for membrane lipids and phase transitions, in general, to the reviews of Sackman (1995), Seddon and Templer (1995), and Koynova and Tenchov (2013). These latter authors display the structure of six different lamellar phases, five micellar structures and six non-lamellar liquid-crystalline phases with various topologies, including the inverted hexagonal (H_{II}) phase and some cubic phases of different space group symmetry, and show how these different phases are capable to transform into each other – a key feature of lipid membranes.

As concerns the detection of these different phases the most straightforward and most commonly used techniques are X-ray diffraction (or SAXS/WAXS small-angle/wide-angle X-ray scattering) (Tyler et al. 2015; Pan et al. 2015) and neutron diffraction (also measured as SANS, small-angle neutron scattering) (Deme et al. 2014). They detect the periodic structures and measure all lattice constants. Hence, these techniques can be used primarily on lipid aggregates, rather than on biological membranes, where diffractions may be smeared or the lattice-structures perturbed. Freeze fracture electron microscopy, which is often used in combination with SAXS (e.g. Csiszár et al. 2003), can more readily be used to detect e.g. H_{II} phases even in thylakoid membrane preparations (Kirchhoff et al. 2007 and references therein). This, i.e. the detectability on biomembranes, is an important issue because, as already pointed out above, the phase behaviour of thylakoid membranes (and of biological membranes, in general) is largely influenced by their protein contents. The best example is the cubic phase of the prolamellar body of etiolated plastids, as opposed to the lamellar phase of thylakoids in mature chloroplasts – despite the virtually identical lipid composition in the two lipid phases (Selstam 1998).

Another powerful and commonly used technique, which can be applied both for model and native systems, is ^{31}P-NMR (Fig. 6.2) (Gruner 1985; Castro et al. 2007). The signature of ^{31}P-NMR is based on the sensitivity of this measurement on the degree of diffusional freedom of the phospholipid molecule in the lipid phase. There is a 'weakness' of this technique, namely, that it can be used for phospholipids only – and PG is only a minor lipid in thylakoid membranes (see section "Thylakoid

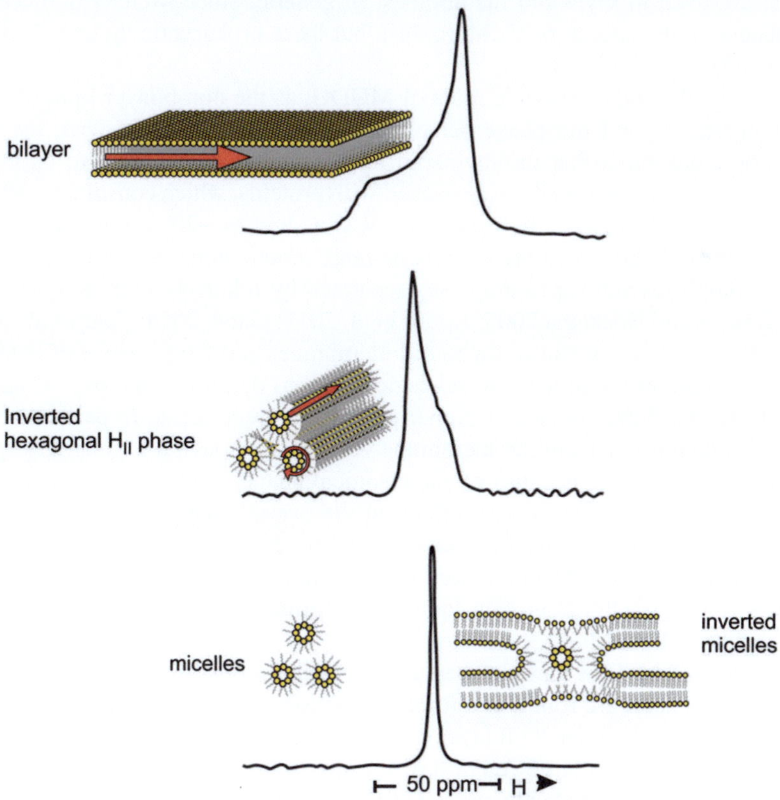

Fig. 6.2 ^{31}P-NMR signatures of different lipid phases (Based on NMR data of Cullis and de Kruijff (1979) with permission)

Lipids"). Nevertheless, as shown first by Haranczyk et al. (1995) it can be used for thylakoid membrane preparations. As demonstrated recently on model membranes composed of glycolipids and phospholipids, it readily detects the co-existence of lipid phases, a feature of special interest, which will be discussed in section "Membrane models using the properties of non-bilayer lipids" in conjunction with ^{31}P-NMR data of Krumova et al. (2008a) obtained on intact thylakoid membranes.

Non-bilayer Lipids and Lipid Phases: Facts and Hypotheses

The most widely established role of MGDG in the thylakoid membranes is in the operation of the xanthophyll cycle (XC). This will be discussed in depth in section "Roles of MGDG in xanthophyll cycles" – focusing on molecular mechanisms and some open questions. Although the role of MGDG and the involvement of a non-bilayer lipid phase have been firmly established in this case, it evidently does not answer the why and wherefore of non-bilayer lipids in energy converting membranes; not even in thylakoid membranes, in general, since MGDG is present in cyanobacteria, the ancestors of chloroplast, but these prokaryotic thylakoids do not operate a XC.

With regard to other possible roles of MGDG, as the non-bilayer lipid of thylakoid membranes, and the phase behaviour of thylakoid lipids *in vivo*, the main boundary condition is that the thylakoid membranes (again, similar to all energy converting membranes) are densely packed with proteins, which constitute ~80 % of the dry mass and likewise the lipid content is as low as ~20 % (Williams 1998). Accordingly, thylakoid membranes contain large protein domains, ordered arrays of protein complexes and supercomplexes separated by relatively narrow lipid 'channels' (Dekker and Boekema 2005; Kouril et al. 2013; Garab 2014; Tietz et al. 2015).

The high protein content of thylakoid membranes and the presence of extended ordered protein arrays in the stacked region (grana) determine the overall appearance of thylakoid membranes of higher plants and green algae. In particular, they self-assemble into stacks of flat membrane vesicles. Although it is generally agreed that non-bilayer lipids, because of their conical shape, tend to induce curvatures (Gruner 1985), this does not take effect on the overall shape of thylakoids – evidently the protein arrays span the membrane sheet (cf. Garab 2014). Further, recent data show that the extreme curvatures at the margins depend on the CURVATURE THYLAKOID1 (CURT1) protein family (Armbruster et al. 2013), rather than on the non-bilayer lipid MGDG.

As indicated by EPR measurements, somewhat more than half of the total thylakoid lipid content are found in a fluid-like phase, i.e. not bound to proteins and/or not found in their inner shell layer (cf. Páli et al. 2003; Lee 2003 and references therein). Conversely, a large fraction of the lipids are bound tightly or loosely to proteins. Indeed, as revealed by crystal structures, the membrane-intrinsic proteins bind a number of lipids – while it must also be realized that often only a fraction of them is resolved in the X-ray structure. These lipid-protein interactions are of interest from the point of view of protein functions – examples will be shown in section "MGDG bound to protein complexes". More often, however, bound lipids do not appear to have a well discernible, specific role related to the structure and function of proteins – the majority of cases in section "MGDG bound to protein complexes". In fact, as pointed out by Lee (2003), because the transmembrane domains of membrane-embedded proteins contain protrusions, fatty acyl chains of lipids fill the grooves. By this means, the permeability barrier of membranes can be maintained, which is of paramount importance with regard to the energization of thylakoid membranes.

Another fraction of the bound lipids are found in the immediate molecular environment of protein complexes but are dynamically exchanging with the bulk lipids. These lipids surrounding the membrane-intrinsic proteins, non-bilayer lipids like MGDG, in particular, might exert strong effects on the structure and function of proteins. Different models have been proposed which are 'using' the special features of MGDG and non-bilayer lipids in the bilayer membranes and complex membrane systems (see section "Membrane models using the properties of non-bilayer lipids").

The loose association of non-bilayer lipids with proteins might play important roles. Zick et al. (2014) have demonstrated that the membrane fusions of yeast vacuoles, which had been known to be mediated by SNARE proteins, require the presence of three types of non-bilayer phospholipid molecules. This finding is perfectly in line with earlier data and models suggesting the crucial role of non-bilayer lipids in membrane fusion (Seddon and Templer 1995; Lohner 1996; Epand 1998), which is proposed to occur via the formation of local and temporal non-bilayer structures (Pomorski et al. 2014 and references therein).

As follows from previous arguments, the thylakoid membranes also contain areas occupied by 'free' lipids – which, in turn, can be directly exposed to water, sugars etc., factors that influence their phase behavior. In section "Membrane models using the properties of non-bilayer lipids", a membrane model will be outlined, which is based on the experimentally proven co-existence of bilayer and non-bilayer lipid phases and their 'communication'.

MGDG Bound to Protein Complexes

Importance of MGDG in LHCII Structures

The crystal structure of LHCII shows the existence of one PG and one DGDG molecule per LHCII monomer (Liu et al. 2004). The main thylakoid membrane lipid MGDG is not observed as intrinsic LHCII lipid in the crystal structures. However, LHCII preparations obtained by solubilisation of the thylakoid membrane, followed by protein separation, contain significant amounts of this galactolipid (Schaller et al. 2010). The close association between MGDG and LHCII also appears to be important in the higher order, multilamellar, onion-like structures adopted by LHCII, somewhat similar to those in the grana (Simidjiev et al. 1998). Addition of DGDG to the isolated LHCII induced the formation of stacked unilamellar vesicles, while in the presence of externally added PG and SQDG lamellar sheets were observed (Simidjiev et al. 1998), suggesting specific roles of each of the lipid species (cf. also Zhou et al. 2009). The structural flexibility of lamellar aggregates of LHCII enriched with MGDG exhibited the most prominent structural flexibility, which manifested in intense light-induced reversible changes in their macrodomain-associated CD signals (Simidjiev et al. 1998). MGDG and DGDG have also been shown to play crucial roles in the structurally flexible multibilayer LHCII-containing macro-assemblies (Janik et al. 2013).

Recent measurements have investigated the influence of purified lipids on the isolated, lipid-depleted and highly aggregated LHCII (Schaller et al. 2011). Addition of MGDG (and DGDG) induced a modification of the disorganized structures of the lipid-depleted LHCII and supported the aggregated state of the complex. In contrast, the negatively charged lipids SQDG and PG exerted a strong disaggregating effect on the isolated LHCII. LHCII disaggregation was suppressed under a high proton concentration and in the presence of cations (Mg^{2+}). This suggested that the negative charge of the anionic lipids in conjunction with negatively charged domains of the LHCII proteins was responsible for the disaggregation. Additional measurements by photon correlation spectroscopy, which were used to gain information about the size of the LHCII aggregates, showed that in the presence of MGDG (and DGDG) an increased number of LHCII aggregates with large particle sizes in the μm-range were formed. This observation is in line with results that reported the establishment of LHCII macrodomains upon the interaction with MGDG (Simidjiev et al. 2000, see above). The incubation with the negatively charged lipids, on the other hand, led to much smaller LHCII particles (around 40 nm in the case of PG) with a homogeneous distribution (Schaller et al. 2011).

Association of MGDG with the Light-Harvesting Complexes of Diatoms

The peripheral antenna system of the diatom PSII core complex is composed of the so-called fucoxanthin chlorophyll protein (FCP) complexes (for recent review, see Büchel 2015). Depending on the detergent used for the preparation of isolated FCP complexes, i.e. dodecylmaltoside or Triton X-100, these antenna systems can be isolated with a shield of MGDG molecules (Lepetit et al. 2010, see also Dörmann and Hölzl Chap. 2) or in a form that contains high amounts of surrounding SQDG and PG molecules (Schaller et al. 2014). For the latter FCP preparation from the diatom *P. tricornutum* it was shown that it was less sensitive to Mg^{2+} and low pH than LHCII, which was characterized by lower amounts of SQDG and a higher concentration of MGDG. High $MgCl_2$ concentrations and pH-values below pH 6 induced significant changes in the spectroscopic features of LHCII, and were accompanied by a strong aggregation, which was also visible as a pellet after centrifugation on a sucrose cushion. The FCP enriched in SQDG and PG responded with less pronounced changes in the absorption and fluorescence spectra to low pH and Mg^{2+} incubation, and did not show a visible pellet after incubation with either low pH or high Mg^{2+} concentrations. Only the combined action of Mg^{2+} and pH 5 led to FCP aggregates of a size that could be pelleted by centrifugation. The decreased sensitivity of the SQDG- and PG-containing FCP to aggregation at high Mg^{2+} and low pH is in line with other data that show that neutral galactolipids, and especially MGDG, are able to stabilize or induce the formation of large, ordered aggregates of light-harvesting complexes (Simidjiev et al. 1998, 2000). Another study on the lipid dependence of FCP aggregation employed artificial lipid membranes, i.e. liposomes, to incorporate isolated FCP complexes of the diatom *C. meneghiniana* (Gundermann and Büchel 2012). The lipid composition of the liposomes was comparable to the lipid composition of the diatom thylakoid membrane, which was shown to contain significantly higher concentrations of the negatively charged lipids SQDG and PG and a reduced amount of the neutral galactolipids MGDG and DGDG (Goss et al. 2009; Lepetit et al. 2012, see Wada Chap. 2). In this study (Gundermann and Büchel 2012) it was shown that high lipid concentrations in general reduced the ability of FCP complexes to form aggregates. Reduction of the lipid per protein ratio led to an aggregation of FCP complexes and a quenching of Chl *a* fluorescence. Additionally, FCP complexes at low lipid concentrations were sensitive to low pH.

Roles of MGDG in the Structure and Function of PSII

As revealed by the crystal structures of the PSII core complex of the cyanobacterium *T. elongatus* (Guskov et al. 2009; Kern and Guskov 2011) a PSII monomer binds 11 molecules of MGDG. In addition, seven DGDG, five SQDG and two PG are found within the PSII core complex. In contrast to the negatively charged lipids SQDG and PG, whose headgroups are only oriented towards the lumenal side, the headgroups of the MGDG molecules are oriented towards both the cytoplasmic and lumenal side of the thylakoid membrane (Guskov et al. 2009). Umena and coworkers identified six MGDG, five DGDG, four SQDG and five PG molecules in the crystal structure of oxygen-evolving PSII at a resolution of 1.9 Å of *T. vulcanus* (Umena et al. 2011). The MGDG and DGDG molecules, except one MGDG molecule are located on the lumenal side, while the headgroups of SQDG and PG molecules were located at the cytoplasmic surface of the thylakoid membrane. Together with SQDG, MGDG shows an enrichment at the monomer-monomer interface of the PSII dimer, where 14 lipids are found (Guskov et al. 2009; Kern and Guskov 2011). Six of these 14 lipids are arranged in such a way that they form contact with protein subunits from both monomers, six other lipids mediate interactions between a protein subunit of one monomer and a lipid of the other monomer. The lipids mediating the dimerization of PSII core complexes show an enrichment of MGDG and 8 of the 14 lipids located at the interface are MGDG molecules. The enrichment of MGDG at the monomer-monomer interface points to an important role of MGDG in PSII dimerization. This was recently supported by experiments where isolated monomeric PSII core complexes were incubated with a surplus of purified MGDG

(Kansy et al. 2014). Size exclusion chromatography showed that after the incubation with MGDG the complete fraction of PSII monomers had been transformed into dimeric PSII core complexes. The ability of MGDG to induce dimerization of PSII monomers was in sharp contrast to the effect of the negatively charged lipids SQDG and PG. Incubation with the anionic lipids led to further destabilization of the monomeric PSII by the dissociation of the inner antenna protein CP43 and part of the D1 protein of the reaction center. A second important cluster of lipids within the PSII structure is found at the PSII acceptor side and is composed of eight lipids (Loll et al. 2007; Guskov et al. 2009; Kern and Guskov 2011). These lipids are located close to the Q_B binding site and form a lipid bilayer within the PSII core complex. The cluster of lipids is thought to act as a transfer pathway, which targets Q_B molecules to their PSII binding site at the D1 protein. Alternative mechanisms for the entry and exit of Q_B molecules have been proposed (Kern and Guskov 2011). In addition, the lipids also serve as a storage pool for additional Q_B (plastoquinone) molecules. Interestingly, the area around Q_A, which is tightly bound to the D2 protein, is dominated by MGDG whereas the Q_B region is enriched in the negatively charged lipids SQDG and PG (Kern and Guskov 2011). It has been proposed that different lipids within the vicinity of the two quinone sites modulate the properties of Q_A and Q_B in different ways without the need to change the protein environment. With respect to lipid-protein interactions at the donor side of PSII it seems that MGDG does not play a crucial role. Three DGDG molecules are located in the area of OEC (Guskov et al. 2009) and it has been shown that depletion of DGDG has a negative influence on the functionality of water splitting (Reifarth et al. 1997; Steffen et al. 2005). MGDG is most probably also not involved in the interaction between the inner antenna proteins CP43 and CP47 with the PSII reaction center. In this case, PG seems to play a major role in this interaction via two PG molecules located at the interface between CP43 and the D1/D2 heterodimer (Guskov et al. 2009; Kern and Guskov 2011). Depletion of PG in mutant strains of *Synechocystis* sp. PCC6803 led to the dissociation of CP43 from the PSII core complex (Laczko-Dobos et al. 2008).

With regard to the influence of lipids on the functionality of PSII a recent study (Kansy et al. 2014) has shown that neutral galactolipids have a tendency to stabilize the PSII electron transport. MGDG and DGDG molecules are involved in the maintenance of the flexibility and assembly of PSII even under stress conditions (Loll et al. 2005; Mizusawa and Wada 2012; Wang et al. (2014). It should be also noted that lipase treatment of isolated PSII leads to the degradation of only half of the bound MGDG molecules (Leng et al. 2008), which might indicate that the other half are more strongly bound to the protein subunits and therefore protected from the lipase. Conversely, about half of the MGDG molecules are found in the bulk lipid phase. The lipase treatment did not induce significant changes in the function of PSII, although, the strongly bound MGDG molecules may play more important role in the structure and function of PSII, as it was suggested by Mizusawa and Wada (2012). This finding is in good agreement with reports showing that phase separation of lipids with non-bilayer propensity from the thylakoid membrane, upon the addition of cosolutes, does not inhibit PSII but increases its thermal stability (Williams et al. 1992).

MGDG Bound to Cytochrome *b6f*

Kurisu and co-workers identified two phosphatidylcholine molecules in the cytochrome *b6f* monomers of *Mastogocladus laminosus* (Kurisu et al. 2003), while Stroebel and co-workers found one SQDG and two other lipid, most probably MGDG, molecules at the lumenal side in the crystalized cytochrome *b6f* of *Chlamydomonas reinhardtii* (Stroebel et al. 2003). Investigations using MGDG Langmuir monolayer and isolated cytochrome *b6f* indicated specific interaction between MGDG and cytochrome *b6f* that could be attributed to hydrogen bonding between the galactose headgroup of MGDG and the protein (Georgiev et al. 2012). Characterization of transgenic M18 tobacco plants containing reduced MGDG content by 53 % indicated that MGDG deficiency reduces the amount of cytochrome *b6f* in the mutant plants, which may be due to unstable integration of the complex into the membrane (Wu et al. 2013). The lowered MGDG content also could be the reason of the inefficient intersystem electron transport in the mutant plants. *In vitro* experiments also indicated that MGDG efficiently stimulates the electron transfer activity of cytochrome *b6f* (Yan et al. 2000).

Roles of MGDG in the Structure and Function of PSI

One MGDG molecule and three PG molecules were identified in the PSI complex of *T. elongatus* by X-ray crystallography at 2.5 Å resolution (Jordan et al. 2001). The MGDG molecule and two of the PG molecules are bound to PsaA and PsaB (Jordan et al. 2001; Domonkos et al. 2008). One of the PG molecule and the MGDG located symmetrically to each other in the complex may have an important role in the photochemical reaction center of PSI (Mizusawa and Wada 2012). Investigation of the *dgd1* mutant of *Arabidopsis thaliana* indicated that MGDG could have a role in the stability of the PSI complex (Guo et al. 2005; see also Krumova et al. 2010). Recently in the architecture of the PSI-LHCI supercomplex six PG, three MGDG and one DGDG molecules were assigned (Qin et al. 2015). Two of the MGDG molecules were identified as cofactors in the four Lhca subunits and the third MGDG was found in PsaG.

Roles of MGDG in Xanthophyll Cycles

Xanthophyll Cycles of Higher Plants and Algae

The xanthophyll cycles (XC) of higher plants and algae are important photoprotective mechanisms, which help to prevent an overexcitation of the photosynthetic apparatus during illumination with high light intensities (for recent reviews see Goss and Jakob 2010; Lavaud and Goss 2014; Goss and Lepetit 2015). The operation of the XCs is associated with the thylakoid membrane and strongly depends on the presence of MGDG (see sections “Role of MGDG in the solubilisation of xanthophyll cycle pigments”, “Importance of the MGDG-dependent non-bilayer phases for the activity of xanthophyll cycles”, and “Localization and operation of xanthophyll cycles in the native thylakoid membrane”). The conversion of special xanthophyll molecules in the XCs plays a key role in switching LHCII from a light-harvesting state into a state where the excessive excitation energy is dissipated as heat (reviewed by Horton and Ruban 2005; Jahns and Holzwarth 2012; Niyogi and Truong 2013; Domonkos et al. 2013). This is made possible by a structural change, i.e. an aggregation of the antenna complexes. The enhanced heat dissipation from the aggregated LHCII is visible as a strong quenching of the chlorophyll (Chl) *a* fluorescence (Krause and Jahns 2004), termed non-photochemical quenching of Chl *a* fluorescence (NPQ) (Demmig-Adams et al. Eds. 2014).

The main XCs that are presently known are the violaxanthin, antheraxanthin, zeaxanthin (VAZ) cycle of higher plants, green and brown algae, and the diadinoxathin, diatoxanthin (DD/Dt) cycle of diatoms, haptophytes and dinophytes (for recent review see Goss and Jakob 2010; Lavaud and Goss 2014; Goss and Lepetit 2015). Both cycles consist of a forward reaction that is taking place during illumination with high light intensities and a back reaction that is observed during low light or dark periods that follow the high light phase. In the VAZ cycle V, which contains two epoxy-groups, is de-epoxidized in two steps to the epoxy-free Z with A as intermediate (Yamamoto et al. 1962; Hager 1967a). The forward reaction of the DD/Dt cycle comprises only one de-epoxidation step from DD to Dt since DD contains only one epoxy-group (Hager and Stransky 1970; Stransky and Hager 1970). The back reaction of the cycle reintroduces the epoxy group into Z or Dt and converts these molecules back to V or DD, respectively (Hager 1967b; Hager and Stransky 1970; Stransky and Hager 1970). The forward reaction is triggered by the enzymes V or DD de-epoxidase (VDE, DDE). These enzymes are water-soluble and located in the thylakoid lumen in their inactive form (Hager and Holocher 1994; Arnoux et al. 2009; Saga et al. 2010). After their activation, which is induced by the decrease of the lumenal pH during photosynthetic electron transport and possibly includes a dimerization (Arnoux et al. 2009; Saga et al. 2010), the enzymes bind to the thylakoid membrane (see Chap. 2.2.4) and convert the respective epoxy-xanthophylls (Hager and Holocher 1994; Schaller et al. 2010). Both enzymes have a pH-optimum at around pH 5 (VDE, pH 5.2; DDE, pH 5.5) and use ascorbate as cosubstrate, i.e. as electron donor for the reduction of the epoxy-group (Hager 1969; Pfündel et al.

1994; Jakob et al. 2001). The back reaction of the cycles is catalysed by the Z or Dt epoxidase (ZEP, DEP). These enzymes are most probably peripheral membrane proteins located at the stromal side of the thylakoid membrane (Schaller et al. 2012). Recent results have indicated that DEP might contain an additional membrane spanning domain (Coesel et al. 2008). Both enzymes have a broad pH-optimum in the neutral to slightly basic pH-range and use O_2 and NADPH+H^+ as cosubstrates to reintroduce the epoxy-group (Hager 1975; Siefermann and Yamamoto 1975; Büch et al. 1995). The de-epoxidases as well as the epoxidases belong to the lipocalin family of proteins (Hieber et al. 2000; Grzyb et al. 2006), a protein family that mainly binds small hydrophobic substrates. The catalytic site is composed of eight β-strands which form a hydrophobic barrel-like structure. The binding of the hydrophobic xanthophylls takes place in the central cavity of the barrel and is realized by a penetration of the catalytic site into MGDG-enriched regions of the thylakoid membrane where the XC pigments are located (see sections "Role of MGDG in the solubilisation of xanthophyll cycle pigments", "Importance of the MGDG-dependent non-bilayer phases for the activity of xanthophyll cycles", and "Localization and operation of xanthophyll cycles in the native thylakoid membrane").

Role of MGDG in the Solubilisation of Xanthophyll Cycle Pigments

In the thylakoid membrane a part of the XC pigments is bound to the antenna proteins of PSII and PSI whereas another part is localized in the lipid phase of the membrane. In higher plants and green algae the majority of the VAZ cycle pigments are located in LHCII and the minor Chl *a*/*b* binding proteins of the PSII antenna, CP29, CP26 and CP24 (Bassi et al. 1993; Ruban et al. 1994; Goss et al. 1997). In diatoms the DD/Dt cycle pigments are bound to the fucoxanthin chlorophyll proteins (FCP) that form the peripheral antenna complex for both PSII and PSI (Büchel 2003; Lavaud et al. 2003; Lepetit et al. 2007). Data of Lepetit et al. (2008) have indicated that special FCP proteins which are associated with PSI, i.e. form a PSI specific antenna, are enriched in DD/Dt cycle pigments. In addition to the XC pigments bound to the respective antenna proteins a part of the hydrophobic xanthophylls are freely located in the lipid phase of the membrane. The existence of free Z in the thylakoid membrane of higher plants was proposed to account for the antioxidative function of the de-epoxidized XC pigments against the harmful action of reactive oxygen species (ROS) (Havaux and Niyogi 1999). More recent results obtained with isolated LHCII of higher plants have shown that part of the total pool of VAZ cycle pigments is bound to the LHCII apoproteins whereas another part of the VAZ pool is localized in a lipid shield purified with the trimeric LHCII (Schaller et al. 2010). This lipid shield consisted mainly of MGDG molecules and it was proposed that in the native thylakoid membrane an MGDG phase is surrounding the antenna complexes and serves to solubilise the VAZ cycle pigments. In diatoms a comparable separation of DD/Dt cycle pigments between protein bound and lipid solubilised pigments has been detected (Lepetit et al. 2010). Especially in high-light cultivated diatom cells a large pool of DD/Dt cycle pigments was found. Since the additional high-light synthesized DD/Dt pigments were unable to participate in the process of NPQ (Schumann et al. 2007) it was hypothesized that these pigments are not bound to the FCP complexes. Later, comparing isolated FCP complexes from high-light and low-light cultures, it could be shown that the additional DD/Dt synthesized during high-light cultivation has the same spectroscopic features as DD/Dt dissolved in purified MGDG (Lepetit et al. 2010). This observation together with the finding that the isolated FCP complexes were enriched in MGDG led to the conclusion that especially in high-light cultivated diatoms a significant part of the total XC pigment pool is located in an MGDG phase surrounding the antenna complexes. This observation is in line with the finding that, similar to Z in higher plants, the lipid-dissolved Dt has an antioxidative function (Lepetit et al. 2010).

The localization of XC pigments in MGDG phases of the thylakoid membrane is in line with *in vitro* experiments with isolated XC pigments, isolated de-epoxidases and different membrane lipids. The first studies were conducted by Yamamoto and Higashi (1978), who found that MGDG is able to enhance the de-epoxidation of V to Z, and concluded that one of the functions of MGDG is the solubilisation of the XC pigments. More recent experiments have shown that non-bilayer lipids which form inverted hexagonal structures (H_{II} phases) in aqueous media, i.e. MGDG and PE, are much better suited to solubilise the VAZ and DD/Dt cycle pigments than lipids which form bilayer structures, i.e. DGDG, SQDG or PC (Goss et al. 2005). In addition, it was demonstrated that the XC pigments are preferentially located in MGDG or PE phases in liposome systems composed of non-bilayer and bilayer lipids that served as a model for the native membrane (Goss et al. 2007). This, together with the data on the isolated LHCII and FCP complexes indicate that a large part of the free XC pigments is localized within MGDG phases in the native thylakoid membrane of higher plants and diatoms. Furthermore, the experiments (Goss et al. 2005, 2007) showed that solubilisation of V or DD is essential for the operation of the VAZ and DD/Dt cycles, respectively. Without solubilisation of the xanthophylls in MGDG the pigments form aggregates in an aqueous medium which cannot be converted by the enzymes VDE and DDE, respectively.

Importance of the MGDG-Dependent Non-bilayer Phases for the Activity of Xanthophyll Cycles

The higher solubilisation capacity of the non-bilayer lipids is one of the two main functions of these lipids with respect to xanthophyll cycling (see section "Role of MGDG in the solubilisation of xanthophyll cycle pigments"). The second main function is the provision of a non-bilayer phase, possibly inverted hexagonal phase, itself which has first been observed by Latowski et al. (2002, 2004). Later, it has been demonstrated that bilayer lipids are able to completely solubilise the XC pigments V and DD, albeit at a significantly higher lipid concentrations (Goss et al. 2005). However, despite the complete solubilisation of the pigments no or only a very minor conversion of V to Z and DD to Dt have been observed in lipid bilayer systems (Goss et al. 2005). The situation is comparable in liposome system consisting of bilayer and non-bilayer lipids, where despite the complete pigment solubilisation a certain percentage of non-bilayer lipids are needed before xanthophyll de-epoxidation can be observed (Goss et al. 2007). These results have implied that not only the efficient solubilisation of the xanthophyll but also the three-dimensional lipid structure is important for the activity of the XC enzymes VDE and DDE. It has been proposed that only a non-bilayer phase allows the penetration of the enzyme into the hydrophobic membrane interior where the XC pigments are located (Goss et al. 2007). The penetration has to be deep enough to allow the catalytic site of the enzyme, i.e. the central cavity of the barrel-like structure (see section "Xanthophyll cycles of higher plants and algae") to come into contact with its substrate. Furthermore, it has been suggested that the penetration of VDE or DDE into the lipid phase is made possible by a reduced surface tension in the non-bilayer phase due to the reduced size of the headgroup of MGDG (Goss et al. 2007, for a review concerning lipids and the lateral pressure profile see Van den Brink-van der Laan et al. 2004). In contrast, bilayer lipids like DGDG are characterized by a comparable size of the headgroup area and the area which is occupied by the fatty acid chains; leading to a rather tightly closed surface of the lipid bilayer (Van den Brink-van der Laan et al. 2004).

At present it is not clear to what extent non-bilayer phases exist in the native thylakoid membrane of higher plants and diatoms and where these phases may be located. It has been proposed that non-bilayer phases may be established within the membrane (Jahns et al. 2009) or that these phases are excluded from the membrane but remain in tight local and functional association with the membrane bilayer (Garab et al. 2000; Goss et al. 2007, see also section "Localization and operation of xanthophyll cycles in the native thylakoid membrane"). Data on the interaction between LHCII and MGDG (Simidjiev et al. 1998, 2000), which show that this interaction forces MGDG into the membrane bilayer, suggest a co-existence and interaction of bilayer and non-bilayer phases in the native membrane (for further arguments, see section "Membrane models using the properties of non-bilayer lipids").

Localization and Operation of Xanthophyll Cycles in the Native Thylakoid Membrane

Recent models of NPQ predict a dissociation and aggregation of LHCII and FCP complexes from the PSII core complex in higher plants and diatoms, respectively (Miloslavina et al. 2009; Holzwarth et al. 2009; Lavaud and Goss 2014; Goss and Lepetit 2015). This dissociation and aggregation leads to the formation of the so-called quenching site Q1 during high-light illumination of plants or algal cells. Since both LHCII and FCPs have been shown to be surrounded by an MGDG shield that incorporates the VAZ or DD/Dt cycle pigments (Schaller et al. 2010; Lepetit et al. 2010, see sections "Role of MGDG in the solubilisation of xanthophyll cycle pigments", and "Importance of the MGDG-dependent non-bilayer phases for the activity of xanthophyll cycles") it has been proposed that the aggregation of antenna complexes leads to a dissociation of MGDG from the complexes (Goss et al. 2007). The dissociation of a higher amount of MGDG molecules is thought to result in an increased non-bilayer phase, possibly an inverted hexagonal phase, in the thylakoids. With respect to the operation of the XC (Fig. 6.3) it has been proposed that after the activation of the VDE and DDE the enzymes bind to the non-bilayer lipid phase, which contains the free XC pigments (Goss et al. 2007; Schaller et al. 2010; Goss and Lepetit 2015). These pigments can exist as free pigments in the lipid phase of the membrane or can detach from the LHCII/FCP during illumination and diffuse into the non-bilayer phase. The de-epoxidases can then penetrate into the hydrophobic core of the non-bilayer phase, where they gain access to V or DD. The conversion results in the generation of Z in higher plants and Dt in diatoms. It is reasonable to believe that the establishment of the non-bilayer phase takes place in the vicinity of the light-harvesting complexes because the de-epoxidized XC pigments have to rebind to the antenna proteins to participate in the process of NPQ (Horton and Ruban 2005; Lavaud and Goss 2014; Goss and Lepetit 2015). This reasoning is supported by the finding that the de-epoxidized DD/Dt cycle pigment Dt exhibits a decreased solubility in MGDG compared to the epoxidized DD thus facilitating the rebinding of Dt to the antenna apoproteins (Goss et al. 2007).

With respect to the possible localization of the non-bilayer phase in diatom thylakoids it has to be taken into account that the diatom membrane is not differentiated into grana and stroma thylakoids as in higher plants (for a review see Garab and Mustárdy 1999) but shows regular stacks of three membranes (Gibbs 1962, 1970). Based on the finding that MGDG forms a lipid shield around the FCP complexes, which incorporates the free DD/Dt cycle pigments (Lepetit et al. 2010), and the observation that the negatively charged lipid SQDG, which is present in high concentration in the diatom membranes, inhibits the de-epoxidation of DD (Goss et al. 2009) a model for the lipid and protein arrangement in the thylakoids has been presented (Lepetit et al. 2012). This model predicts that in the inner membranes of the stacks of three PSII with its peripheral antenna complexes is located. Since the FCP complexes are surrounded by an MGDG phase, MGDG may also be enriched in the inner membranes. The enrichment of MGDG in the inner membranes, together with an aggregation of FCP complexes, may lead to the formation of non-bilayer structures in the inner membranes, which then represent an attraction site for the enzyme DDE and support the efficient conversion of DD to Dt. Finally, it has been demonstrated that the efficiency of the VAZ cycle is hindered in mutants deficient in MGDG, which displayed reduced XC activity and contained less Z after a high-light illumination compared to the wild type (Aronsson et al. 2008). The lower Z concentration resulted in a decreased capacity for NPQ and thus a higher photoinhibition of PSII. The authors ascribed the reduced Z concentration to a higher conductivity of the thylakoid membrane for protons, resulting in a decreased activation of the VDE. However, taking into account the results described in sections above, it is likely that MGDG deficiency had an impact on the formation and extent of non-bilayer phases in the thylakoids, thus affecting both the solubilisation of V and the activity of VDE.

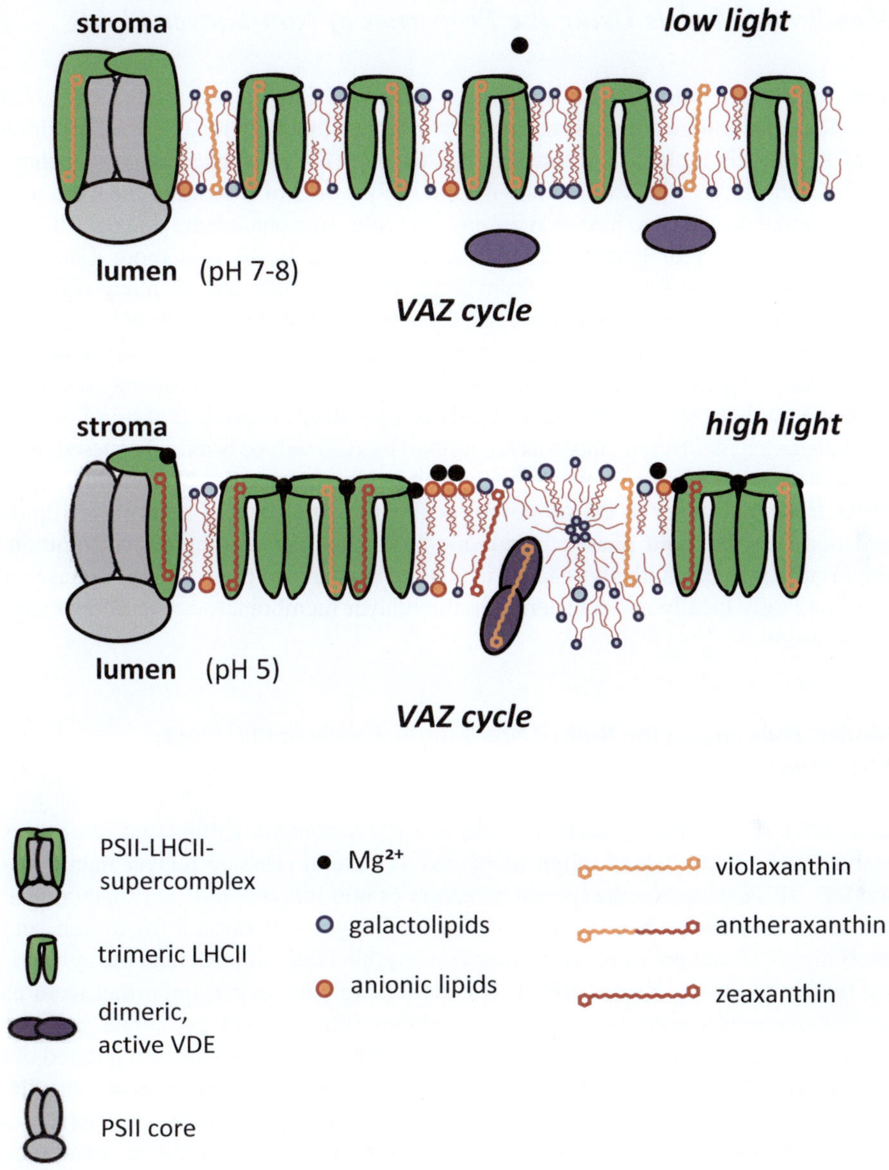

Fig. 6.3 Model for the MGDG dependence of xanthophyll de-epoxidation in thylakoid membranes. High-light illumination of thylakoid membranes of higher plants that contain the VAZ cycle leads to a disconnection and aggregation of LHCII. The MGDG molecules surrounding the LHCII dissociate and form non-bilayer phases. The VAZ cycle pigment V is disconnected from its binding site at the LHCII apoproteins and diffuses into the non-bilayer phase. The non-bilayer phase acts as an attraction site for the enzyme VDE, which forms a dimer upon activation by the pH-drop of the thylakoid lumen. The non-bilayer phase with its reduced surface tension allows the penetration of the enzyme and its catalytic site into the hydrophobic area of the lipid phase, where it gains access to the hydrophobic V which is preferentially solubilised in MGDG. After the conversion of V to Z, Z rebinds to the LHCII, possibly due to a reduced solubility in the non-bilayer phase compared to V. Z bound to the LHCII increases NPQ and thus the protection against an excess of excitation energy under high-light illumination

Membrane Models Using the Properties of Non-bilayer Lipids

The 'standard', fluid mosaic membrane model of Singer and Nicolson (1972) describes the biological membranes as a two-dimensional liquid, a bilayer, in which lipid and protein molecules diffuse easily. Although the existence and phase behaviour of non-bilayer lipids were known from early works of Luzzati (1968), the proposed model provides no hint on their possible roles (the phase behaviour of MGDG was described a year later by Shipley et al. (1973)). In the past more than four decades the standard model has been refined many times and by many research groups in order to accommodate special features, such as the asymmetry of the two membrane leaflets, protein-lipid interactions, lipid rafts, membrane curvature etc. (https://en.wikipedia.org/wiki/Fluid_mosaic_model). In the following sections, focusing on the roles of non-bilayer lipids and possible occurrence of non-lamellar lipid phases, two different approaches, refined models will be briefly discussed, noting that they appear to us as mutually non-exclusive models. Nonetheless, they differ from each other in the proposed localization and roles of non-bilayer lipids and non-lamellar lipid phases in relation to the bilayer membrane. A common, important feature in both approaches is that the appearance of non-bilayer phases is restricted only locally and transiently in the bilayer membranes, e.g. to allow membrane fusion.

Models Focusing on the Role of Non-bilayer Lipids in the Bilayer Membrane

In a series of systematic works Ben de Kruijff, Antoinette Killian and coworkers elaborated a model, which might be named as "lateral pressure bilayer membrane model". This is based on the special geometry of non-bilayer lipids, i.e. their conical shapes, which thus, when incorporated in the bilayer will cause a frustration, and exert higher lateral pressure on the membrane-embedded proteins than the cylindrical bilayer lipids (de Kruijff 1997). The model explains several important features of biomembranes and properties of membrane-intrinsic and peripheral proteins (reviewed by van den Brink-van der Laan et al. 2004). In particular, it is pointed out that in order to understand the properties of a protein, the lateral pressure profile, which largely depends on the presence of non-bilayer lipids, must be understood. Their experimental data also suggest that non-bilayer lipids facilitate membrane binding of peripheral membrane proteins via changes in the lateral pressure profiles (for further details see the above cited review). In thylakoid membranes, the regulation of the MGDG/DGDG ratio might be correlated with the lateral pressure – which evidently affects the structural flexibility of membranes (see section "Importance of MGDG in LHCII structures"); this might be of high physiological significance in light adaptation and stress tolerance mechanisms.

The flexible surface model (FSM) proposed by Brown (2012) also focuses on lipid-protein interactions in the membrane. It challenges the standard model, to a large extent via assigned structural roles of non-bilayer lipid molecules in the bilayer membrane and roles of curvature forces. The FSM model also offers explanation on a number of earlier experimental results on membranes enriched in non-bilayer lipids. It emphasizes that the "polymorphism of membrane lipids is connected with their spontaneous curvature (or molecular packing)" and points out that both curvature matching and hydrophobic thickness matching must be taken into account. This concept would probably help the construction of theoretical models on lipid:LHCII assemblies and eventually of the bilayer thylakoid membranes.

A Model Proposing the Co-existence and Close Association of Bilayer and Non-bilayer Lipid Phases

The model presented here can be considered as a refined model proposed earlier by Garab et al. (2000). It focuses on the properties of bulk lipids, i.e. which are found in fluid-like phase and constitute at least half of the total lipid content of thylakoid membranes; this bulk lipid mixture, on mesoscopic scale, displays no lateral heterogeneity (see Introduction). The model described below on thylakoid membranes is thought to be applicable, *mutatis mutandis*, for other energy-converting membranes and might be adapted to other biological membranes with substantial amounts of non-bilayer lipids and thus lending a significant degree of non-bilayer propensity to their lipids. The model is based on several premises, which will be outlined below.

Although it might look trivial, we (here and below, GG and his coworkers) proposed that the most important property of non-bilayer lipids, what we should take into account in the first place, is that, if the lipids are in surplus they segregate from the bilayer (Garab et al. 2000). This offered an explanation on the apparently constant and high protein to lipid ratio in the thylakoids and in all energy-converting membranes. This self-regulatory (physical) mechanism prevents the 'dilution' of membranes, which otherwise could easily occur in the absence of strict synchronization of lipid and protein syntheses. The dilution of membranes could easily exert devastating effects on the protein macro-assemblies and the entire, highly organized membrane system (Garab 2014).

As already mentioned (section "Thylakoid Lipids"), we have shown, by using purified MGDG and LHCII that non-bilayer lipids can be forced into the bilayer, apparently in a frustrated state (Simidjiev et al. 1998, 2000). These experimental findings suggested to us that non-bilayer lipids can be trafficking between the non-bilayer phase and the bilayer (Garab et al. 2000). In other terms we assumed that non-bilayer lipids, via this putative dynamic exchange mechanism, contribute to the structural flexibility of thylakoid membranes. It must be stressed, however, that with this assumption we hypothesized that the two phases, the bilayer phase and a non-bilayer phase, co-exist in the thylakoids; further, the non-bilayer lipids (lipid mixtures with high non-bilayer propensity) are hypothesized to enter the membrane not only occasionally (as examples have shown), or become excluded from the membrane not only under special conditions (known in the literature), but dynamically.

To test these hypotheses, first, experimental evidence had to be found that the bilayer and non-bilayer lipid phases co-exist in a way which allows their interaction. Although EM data showed the presence of non-bilayer lipid phases in mature chloroplasts, they, in lipid droplets, were remotely located from the thylakoid membranes, allowing no interaction between the two structures (see Williams 1998, Garab et al. 2000 and references therein). Thus, we turned to the technique of ^{31}P-NMR and performed experiments on isolated intact thylakoid membranes (Krumova et al. 2008a). Earlier, using Tris-washed thylakoids, which, as we also showed, severely affects the organization of thylakoid membranes, Haranczyk et al. (1995) reported on the appearance of H_{II} phases in the sample. Our experiments provided clear evidence that in intact thylakoid membranes at low (but still physiological) temperatures the lamellar phase co-existed with a non-bilayer phase (Krumova et al. 2008a) (Fig. 6.4a). Surprisingly, however, the NMR signature of the lamellar phase disappeared already around 20 °C, while we had convincing experimental evidences that the bilayer structure was maintained (Fig. 6.4b). These experimental findings could only be attributed to the interconnection of the two lipid phases – the bilayer membrane and a non-bilayer lipid phase thought to be located primarily in the lumenal aqueous phase. (The lipids that were expelled from the membranes on the stromal side could easily detach in the absence of the native stroma liquid, which was replaced by a buffered medium.) Figure 6.4c shows the irreversible heat-induced formation of the isotropic and non-bilayer phases at 0 and ~4 ppm, respectively. Additional experiments, using the fluorescent lipid probe Merocyanine-540 corroborated the NMR data by showing complex, temperature-dependent phase behaviour of lipids, which could not be accounted for by the sole, bilayer phase (Krumova et al. 2008b). These experiments also revealed that at elevated temperature

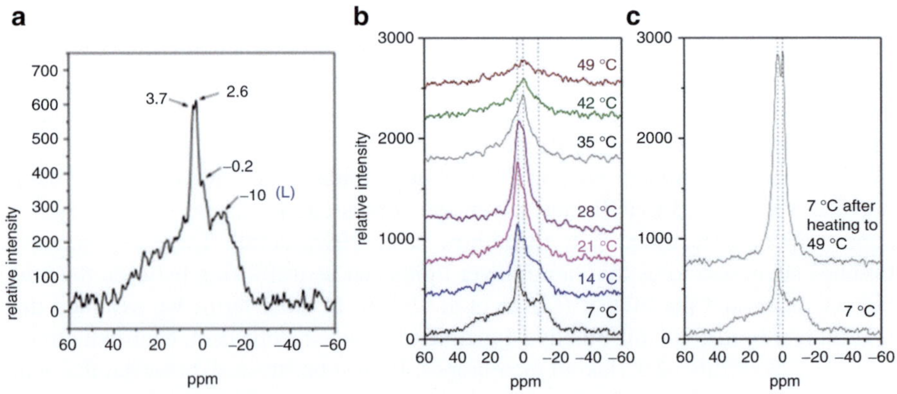

Fig. 6.4 ^{31}P-NMR spectrum of thylakoid membranes at 7 °C. (**a**) The positions (in ppm) of the different resonances are denoted by *arrows* and are assigned to the lamellar (L) (bilayer) phase, isotropic structures and a non-bilayer lipid structure. (**b**) ^{31}P-NMR spectra of thylakoid membranes at different temperatures, as indicated, and, for comparison, (**c**) the spectra at 7 °C obtained before and after heating of the sample to 49 °C. The *dashed lines* indicate the positions of the different resonances (Based on figures from Krumova et al. (2008a) with permission)

the lipids tend to be found in a lipid phase that could be assigned to a non-bilayer phase. Further, using a DGDG-deficient Arabidopsis mutant, Krumova et al. (2010) have also shown that the substantially increased relative abundance of MGDG brings about significant alterations in the overall organization of the thylakoid membranes and decreased thermal stabilities and an enhanced lipid extrusion from the bilayer membrane of the mutant thylakoids.

These data are explained within the frameworks of a simplified model shown in Fig. 6.5 illustrating that the thylakoid lumen contains a non-bilayer lipid phase, that is associated with the bilayer phase and the two phases form a continuum. Because of the narrow space the extruded lipids in the aqueous phase will not be able to assume a periodic structure, such as a 'true' H_{II} but will nevertheless be distinguishable from the bilayer phase. The model also points to the proposed roles of VDE and other lipocalin proteins in the lumen, which are capable of binding lipid molecules (see section "Roles of MGDG in xanthophyll cycles"). These proteins might be capable of mediating between the two phases. Recent molecular dynamics simulations on two adjacent bilayers that contain plant lipid mixtures have "demonstrated that the thylakoid membrane is close to the formation of an inverted hexagonal phase". Further, the simulations revealed that these thylakoid model lipid membranes have a strong, hydration-dependent propensity to the formation of stalk, the main intermediate between bilayer and hexagonal phases (Van Eerden et al. 2015). These findings are fully consistent with our model, which nevertheless requires further rigorous experimental and molecular dynamics tests.

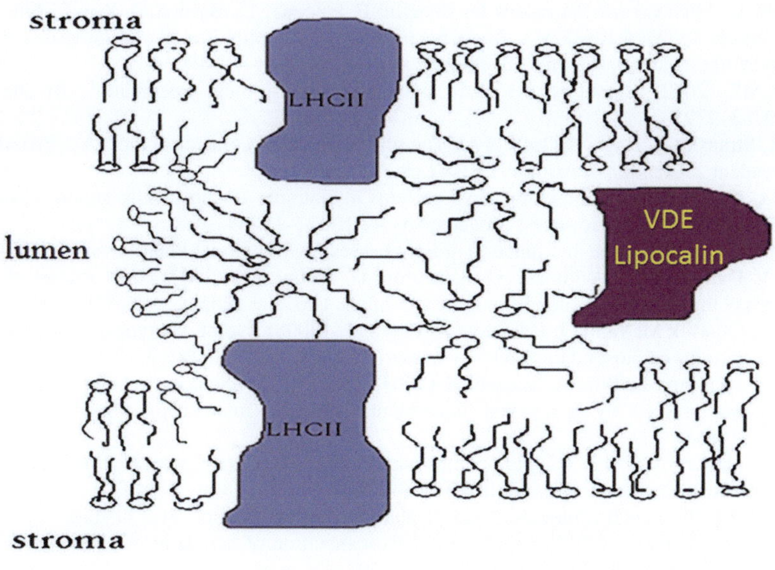

Fig. 6.5 Simplified thylakoid membrane model illustrating that the bilayer and non-bilayer lipid phases are closely associated with each other and interact both with membrane proteins (LHCII) and lumenal proteins (VDE/lipocalin) (The model is based on the experimental findings of Krumova et al. 2008a, b. Drawing, courtesy of Dr. Sashka B. Krumova)

Acknowledgements The authors are indebted to Dr. Sashka B. Krumova for fruitful discussions and for the drawing of the simplified membrane model outline in section "A model proposing the co-existence and close association of bilayer and non-bilayer lipid phases". Financial supports from OTKA (K112688) and TÁMOP (422D-15/1/KONV-2015-0024) to GG and from the DFG (Go 818/7-1) to RG are also acknowledged. Susann Schaller-Laudel is acknowledged for the composition of Fig. 6.3. The authors thank Profs. Lászlo Vígh and Kazimierz Strzalka for inspiring discussions.

References

Armbruster U, Labs M, Pribil M, Viola S, Xu W, Scharfenberg M, Hertle AP, Rojahn U, Jensen PE, Rappaport F, Joliot P, Dörmann P, Wanner G, Leister D (2013) *Arabidopsis* CURVATURE THYLAKOID1 proteins modify thylakoid architecture by inducing membrane curvature. Plant Cell 25:2661–2678

Arnoux P, Morosinotto T, Saga G, Bassi R, Pignol D (2009) A structural basis for the pH-dependent xanthophyll cycle in *Arabidopsis thaliana*. Plant Cell 21:2036–2044

Aronsson H, Schöttler MA, Kelly AA, Sundqvist C, Dörmann P, Karim S, Jarvis P (2008) Monogalactosyldiacylglycerol deficiency in Arabidopsis affects pigment composition in the prolamellar body and impairs thylakoid membrane energization and photoprotection in leaves. Plant Physiol 148:580–592

Bassi R, Pineau B, Dainese P, Marquardt J (1993) Carotenoid-binding proteins of photosystem II. Eur J Biochem 212:297–303

Benning C, Beatty JT, Prince RC, Somerville CR (1993) The sulfolipid sulfoquinovosyldiacylglycerol is not required for photosynthetic electron transport in *Rhodobacter sphaeroides* but enhances growth under phosphate limitation. Proc Natl Acad Sci U S A 90:1561–1565

Boudière L, Michaud M, Petroutsos D, Rébeillé F, Falconet D, Bastien O, Roy S, Finazzi G, Rolland N, Jouhet J, Block MA, Maréchal E (2014) Glycerolipids in photosynthesis: composition, synthesis and trafficking. Biochim Biophys Acta 1837:470–480

Brown MF (2012) Curvature forces in membrane lipid-protein interactions. Biochemistry 51:9782–9795

Büch K, Stransky H, Hager A (1995) FAD is a further essential cofactor of the NAD(P)H and O_2-dependent zeaxanthin-epoxidase. FEBS Lett 376:45–48

Büchel C (2003) Fucoxanthin-chlorophyll proteins in diatoms: 18 and 19 kDa subunits assemble into different oligomeric states. Biochemistry 42:13027–13034

Büchel C (2015) Evolution and function of light harvesting proteins. J Plant Physiol 172:62–75

Castro V, Dvinskikh SV, Widmalm G, Sandström D, Maliniak A (2007) NMR studies of membranes composed of glycolipids and phospholipids. Biochim Biophys Acta 1768:2432–2437

Coesel S, Obornik M, Varela J, Falciatore A, Bowler C (2008) Evolutionary origins and functions of the carotenoid biosynthetic pathway in marine diatoms. PLoS ONE 3, e2896

Csiszár A, Klumpp E, Bóta A, Szegedi K (2003) Effect of 2,4-dichlorophenol on DPCC/water liposomes studied by X-ray and freez-facture electron microscopy. Chem Phys Lipids 126:155–166

Cullis PR, de Kruijff B (1979) Lipid polymorphism and the functional roles of lipids in biological membranes. Biochim Biophys Acta 559:399–420

de Kruijff B (1997) Biomembranes. Lipids beyond the bilayer. Nature 386:129–130

Dekker JP, Boekema EJ (2005) Supramolecular organization of thylakoid membrane proteins in green plants. Biochim Biophys Acta 1706:12–39

Demé B, Cataye C, Block MA, Maréchal E, Jouhet J (2014) Contribution of galactoglycerolipids to the 3-dimensional architecture of thylakoids. FASEB J 28:3373–3383

Demmig-Adams B, Adams WW III, Garab G, Govindjee (eds) (2014) Non-photochemical quenching and thermal energy dissipation in plants, algae and cyanobacteria, Advances in Photosynthesis and Respiration. Springer, Dordrecht

Domonkos I, Laczko-Dobos H, Gombos Z (2008) Lipid-assisted protein-protein interactions that support photosynthetic and other cellular activities. Prog Lipid Res 47:422–435

Domonkos I, Kis M, Gombos Z, Ughy B (2013) Carotenoids, versatile components of oxygenic photosynthesis. Prog Lipid Res 52:539–561

Dörmann P, Benning C (2002) Galactolipids rule in seed plants. Trends Plant Sci 7:112–118

Epand RM (1998) Lipid polymorphism and protein–lipid interactions. Biochim Biophys Acta 1376:353–368

Garab G (2014) Hierarchical organization and structural flexibility of thylakoid membranes. Biochim Biophys Acta 1837:481–494

Garab G, Mustardy L (1999) Role of LHCII-containing macrodomains in the structure, function and dynamics of grana. Aust J Plant Physiol 27:649–658

Garab G, Lohner K, Laggner P, Farkas T (2000) Self-regulation of the lipid content of membranes by non-bilayer lipids: a hypothesis. Trends Plant Sci 5:489–494

Georgiev GA, Ivanov SI, Jordanova A, Tsanova A, Getov V, Dimitrov M, Lalchev Z (2012) Interaction of monogalactosyldiacylglycerol with cytochrome b_6f complex in surface films. Biochem Biophys Res Commun 419:648–651

Gibbs SP (1962) The ultrastructure of the pyrenoids of algae, exclusive of the green algae. J Ultra Mol Struct R 7:247–261

Gibbs SP (1970) The comparative ultrastructure of the algal chloroplast. Ann NY Acad Sci 175:454–473

Goss R, Jakob T (2010) Regulation and function of xanthophyll cycle-dependent photoprotection in algae. Photosynth Res 106:103–122

Goss R, Lepetit B (2015) Biodiversity of NPQ. J Plant Physiol 172:13–32

Goss R, Wilhelm C (2009) Lipids in algae, lichens and mosses. In: Wada H, Murata N, Govindjee (eds) Lipids in photosynthesis: essential and regulatory functions. Springer, Dordrecht, pp 117–137

Goss R, Richter M, Wild A (1997) Pigment composition of PS II pigment protein complexes purified by anion exchange chromatography. Identification of xanthophyll cycle pigment binding proteins. J Plant Physiol 151:115–119

Goss R, Lohr M, Latowski D, Grzyb J, Vieler A, Wilhelm C, Strzalka K (2005) Role of hexagonal structure-forming lipids in diadinoxanthin and violaxanthin solubilization and de-epoxidation. Biochemistry 44:4028–4036

Goss R, Latowski D, Grzyb J, Vieler A, Lohr M, Wilhelm C, Strzalka K (2007) Lipid dependence of diadinoxanthin solubilization and de-epoxidation in artificial membrane systems resembling the lipid composition of the natural thylakoid membrane. Biochim Biophys Acta 1768:67–75

Goss R, Nerlich J, Lepetit B, Schaller S, Vieler A, Wilhelm C (2009) The lipid dependence of diadinoxanthin de-epoxidation presents new evidence for a macrodomain organization of the diatom thylakoid membrane. J Plant Physiol 166:1839–1854

Gruner SM (1985) Intrinsic curvature hypothesis for biomembrane lipid composition: a role for nonbilayer lipids. Proc Natl Acad Sci U S A 82:3665–3669

Grzyb J, Latowski D, Strzalka K (2006) Lipocalins-a family portrait. J Plant Physiol 163:895–915

Güler S, Seeliger A, Härtel H, Renger G, Benning C (1996) A null mutant of *Synechococcus* sp. PCC7942 deficient in the sulfolipid sulfoquinovosyl diacylglycerol. J Biol Chem 271:7501–7507

Gundermann K, Büchel C (2012) Factors determining the fluorescence yield of fucoxanthin-chlorophyll complexes (FCP) involved in non-photochemical quenching in diatoms. Biochim Biophys Acta 1817:1044–1052

Guo J, Zhang Z, Bi Y, Yang W, Xu Y, Zhang L (2005) Decreased stability of photosystem I in dgd1 mutant of Arabidopsis thaliana. FEBS Lett 579:3619–3624

Guskov A, Kern J, Gabdulkhakov A, Broser M, Zouni A, Saenger W (2009) Cyanobacterial photosystem II at 2.9-angstrom resolution and the role of quinones, lipids, channels and chloride. Nat Struct Mol Biol 16:334–342

Hager A (1967a) Untersuchungen über die lichtinduzierten, reversiblen Xanthophyllumwandlungen an *Chlorella* und *Spinacia oleracea*. Planta 74:148–172

Hager A (1967b) Untersuchungen über die Rückreaktionen im Xanthophyll-Cyclus bei *Chlorella*, *Spinacia* und *Taxus*. Planta 76:138–148

Hager A (1969) Lichtbedingte pH-Erniedrigung in einem Chloroplasten-Kompartiment als Ursache der enzymatischen Violaxanthin-Zeaxanthin-Umwandlung: Beziehungen zur Photophosphorylierung. Planta 89:224–243

Hager A (1975) Die reversiblen, lichtabhängigen Xanthophyllumwandlungen im Chloroplasten. Ber Deutsch Bot Ges 88:27–44

Hager A, Holocher K (1994) Localization of the xanthophyll-cycle enzyme violaxanthin de-epoxidase within the thylakoid lumen and abolition of its mobility by a (light-dependent) pH decrease. Planta 192:581–589

Hager A, Stransky H (1970) Das Carotinoidmuster und die Verbreitung des lichtinduzierten Xanthophyllcyclus in verschiedenen Algenklassen. Archiv Mikrobiol 73:77–89

Harańczyk H, Strzałka K, Dietrich W, Blicharski JS (1995) 31P-NMR observation of the temperature and glycerol induced non-lamellar phase formation in wheat thylakoid membranes. J Biol Phys 21:125–139

Harwood JL (1998) Involvement of chloroplast lipids in the reaction of plants submitted to stress. In: Siegenthaler P-A, Murata N (eds) Advances in photosynthesis. Lipids in photosynthesis. Kluwer Academic Publishers, Dordrecht, pp 287–302

Havaux M, Niyogi KK (1999) The violaxanthin cycle protects plants from photooxidative damage by more than one mechanism. Proc Natl Acad Sci U S A 96:8762–8767

Hieber AD, Bugos RC, Yamamoto HY (2000) Plant lipocalins: violaxanthin de-epoxidase and zeaxanthin epoxidase. Biochim Biophys Acta 1482:84–91

Holzwarth AR, Miloslavina Y, Nilkens M, Jahns P (2009) Identification of two quenching sites active in the regulation of photosynthetic light-harvesting studied by time-resolved fluorescence. Chem Phys Lett 483:262–267

Horton P, Ruban AV (2005) Molecular design of the photosystem II light-harvesting antenna: photosynthesis and photoprotection. J Exp Bot 56:365–373

Jahns P, Holzwarth AR (2012) The role of the xanthophyll cycle and of lutein in photoprotection of photosystem II. Biochim Biophys Acta 1817:182–193

Jahns P, Latowski D, Strzalka K (2009) Mechanism and regulation of the violaxanthin cycle: the role of antenna proteins and membrane lipids. Biochim Biophys Acta 1787:3–14

Jakob T, Goss R, Wilhelm C (2001) Unusual pH-dependence of diadinoxanthin de-epoxidase activation causes chlororespiratory induced accumulation of diatoxanthin in the diatom *Phaeodactylum tricornutum*. J Plant Physiol 158:383–390

Janik E, Bednarska J, Zubik M, Puzio M, Luchowski R, Grudzinski W, Mazur R, Garstka M, Maksymiec W, Kulik A, Dietler G, Gruszecki WI (2013) Molecular architecture of plant thylakoids under physiological and light stress conditions: a study of lipid-light-harvesting complex II model membranes. Plant Cell 25:2155–2170

Jordan P, Fromme P, Witt HT, Klukas O, Saenger W, Krauss N (2001) Three-dimensional structure of cyanobacterial photosystem I at 2.5 A resolution. Nature 411:909–917

Jouhet J (2013) Importance of the hexagonal lipid phase in biological membrane organization. Front Plant Sci 4:494. doi:10.3389/fpls.2013.00494. eCollection 2013

Kansy M, Wilhelm C, Goss R (2014) Influence of thylakoid membrane lipids on the structure and function of the plant photosystem II core complex. Planta 240:781–796

Kern J, Guskov A (2011) Lipids in photosystem II: multifunctional cofactors. J Photochem Photobiol B Biology 104:19–34

Kirchhoff H, Haase W, Wegner S, Danielsson R, Ackermann R, Albertsson PA (2007) Low-light-induced formation of semicrystalline photosystem II arrays in higher plant chloroplasts. Biochemistry 46:11169–11176

Kobayashi K, Narise T, Sonoike K, Hashimoto H, Sato N, Kondo M, Nishimura M, Sato M, Toyooka K, Sugimoto K, Wada H, Masuda T, Ohta H (2013) Role of galactolipid biosynthesis incoordinated development of photosynthetic complexes and thylakoid membranes during chloroplast biogenesis in Arabidopsis. Plant J 73:250–261

Kouril R, Wientjes E, Bultema JB, Croce R, Boekema EJ (2013) High-light vs. low-light: effect of light acclimation on photosystem II composition and organization in Arabidopsis thaliana. Biochim Biophys Acta 1827:411–419

Koynova R, Tenchov B (2013) Recent patents on nonlamellar liquid crystalline lipid phases in drug delivery. Recent Pat Drug Deliv Formul 7:165–173

Krause GH, Jahns P (2004) Non-photochemical energy dissipation determined by chlorophyll fluorescence quenching. Characterization and function. In: Papageorgiou GC, Govindjee (eds) Chlorophyll fluorescence: a signature of photosynthesis. Springer, Dordrecht, pp 463–495

Krumova SB, Dijkema C, de Waard P, Van As H, Garab G, van Amerongen H (2008a) Phase behaviour of phosphatidylglycerol in spinach thylakoid membranes as revealed by 31P-NMR. Biochim Biophys Acta 1778:997–1003

Krumova SB, Koehorst RB, Bóta A, Páli T, van Hoek A, Garab G, van Amerongen H (2008b) Temperature dependence of the lipid packing in thylakoid membranes studied by time- and spectrally resolved fluorescence of Merocyanine 540. Biochim Biophys Acta 1778:2823–2833

Krumova SB, Laptenok SP, Kovács L, Tóth T, van Hoek A, Garab G, van Amerongen H (2010) Digalactosyl-diacylglycerol-deficiency lowers the thermal stability of thylakoid membranes. Photosynth Res 105:229–242

Kurisu G, Zhang H, Smith JL, Cramer WA (2003) Structure of the cytochrome b6f complex of oxygenic photosynthesis: tuning the cavity. Science 302:1009–1014

Laczko-Dobos H, Ughy B, Toth SZ, Komenda J, Zsiros O, Domonkos I, Parducz A, Bogos B, Komura M, Itoh S, Gombos Z (2008) Role of phosphatidylglycerol in the function and assembly of Photosystem II reaction center, studied in a *cdsA*-inactivated PAL mutant strain of *Synechocystis sp.* PCC6803 that lacks phycobilisomes. Biochim Biophys Acta 1777:1184–1194

Laczko-Dobos H, Frycák P, Ughy B, Domonkos I, Wada H, Prokai L, Gombos Z (2010) Remodeling of phosphatidylglycerol in Synechocystis PCC6803. Biochim Biophys Acta 1801:163–170

Latowski D, Kruk J, Burda K, Skrzynecka-Jaskier M, Kostecka-Gugala A, Strzalka K (2002) Kinetics of violaxanthin de-epoxidation by violaxanthin de-epoxidase, a xanthophyll cycle enzyme, is regulated by membrane fluidity in model lipid bilayers. Eur J Biochem 269:4656–4665

Latowski D, Akerlund H-E, Strzalka K (2004) Violaxanthin de-epoxidase, the xanthophyll cycle enzyme, requires lipid inverted hexagonal structures for its activity. Biochemistry 43:4417–4420

Lavaud J, Goss R (2014) The peculiar features of non-photochemical fluorescence quenching in diatoms and brown algae. In: Demmig-Adams B, Adams WW III, Garab G, Govindjee (eds) Non-photochemical quenching and thermal energy dissipation in plants, algae and cyanobacteria, Advances in Photosynthesis and Respiration. Springer, Dordrecht, pp 421–443

Lavaud J, Rousseau B, Etienne A (2003) Enrichment of the light-harvesting complex in diadinoxanthin and implications for the nonphotochemical fluorescence quenching in diatoms. Biochemistry 42:5802–5808

Lee AG (2003) Lipid–protein interactions in biological membranes: a structural perspective. Biochim Biophys Acta 1612:1–40

Leng J, Sakurai I, Wada H, Shen JR (2008) Effect of phospholipase and lipase treatments on photosystem II core dimer from a thermophilic cyanobacterium. Photosynth Res 98:469–478

Lepetit B, Volke D, Szabo M, Hoffmann R, Garab G, Wilhelm C, Goss R (2007) Spectroscopic and molecular characterization of the oligomeric antenna of the diatom *Phaeodactylum tricornutum*. Biochemistry 46:9813–9822

Lepetit B, Volke D, Szabo M, Hoffmann R, Garab G, Wilhelm C, Goss R (2008) The oligomeric antenna of the diatom *P. tricornutum* – localization of diadinoxanthin cycle pigments. In: Allen JF, Gantt E, Golbeck JH, Osmond B (eds) Photosynthesis. Energy from the sun. Springer, Dordrecht, pp 277–280

Lepetit B, Volke D, Gilbert M, Wilhelm C, Goss R (2010) Evidence for the existence of one antenna-associated, lipid-dissolved, and two protein-bound pools of diadinoxanthin cycle pigments in diatoms. Plant Physiol 154:1905–1920

Lepetit B, Goss R, Jakob T, Wilhelm C (2012) Molecular dynamics of the diatom thylakoid membrane under different light conditions. Photosynth Res 111:245–257

Liu ZF, Yan HC, Wang KB, Kuang KY, Zhang JP, Gui LL, An XM, Chang WR (2004) Crystal structure of spinach major light-harvesting complex at 2.72 angstrom resolution. Nature 428:287–292

Lohner K (1996) Is the high propensity of ethanolamine plasmalogens to form non-lamellar lipid structures manifested in the properties of biomembranes? Chem Phys Lipids 81:167–184

Loll B, Kern J, Saenger W, Zouni A, Biesiadka J (2005) Toward complete cofactor arrangement in the 3.0 A resolution structure of photosystem II. Nature 438:1040–1044

Loll B, Kern J, Saenger W, Zouni A, Biesiadka J (2007) Lipids in photosystem II: interactions with protein and cofactors. Biochim Biophys Acta 1767:509–519

Luzzati V (1968) X-ray diffraction studies of lipid–water systems. In: Chapman D (ed) Biological membranes. Academic, New York, pp 71–123

Miloslavina Y, Grouneva I, Lambrev PH, Lepetit B, Goss R, Wilhelm C, Holzwarth AR (2009) Ultrafast fluorescence study on the location and mechanism of non-photochemical quenching in diatoms. Biochim Biophys Acta 1787:1189–1197

Minoda A, Sato N, Nozaki H, Okada K, Takahashi H, Sonoike K, Tsuzuki M (2002) Role of sulfoquinovosyl diacylglycerol for the maintenance of photosystem II in *Chlamydomonas reinhardtii*. Eur J Biochem 269:2353–2358

Mizusawa N, Wada H (2012) The role of lipids in photosystem II. Biochim Biophys Acta 1817:194–208

Murata N, Siegenthaler PA (1998) Lipids in photosynthesis: an overview. In: Siegenthaler PA, Murata N (eds) Lipids in photosynthesis: structure, function and genetics. Kluwer Academic Publishers, Dordrecht, pp 1–20

Niyogi KK, Truong TB (2013) Evolution of flexible non-photochemical quenching mechanisms that regulate light harvesting in oxygenic photosynthesis. Curr Opin Plant Biol 16:307–314

Páli T, Garab G, Horvath LI, Kóta Z (2003) Functional significance of the lipid-protein interface in photosynthetic membranes. Cell Mol Life Sci 60:1591–1606

Pan J, Cheng X, Sharp M, Ho CS, Khadka N, Katsaras J (2015) Structural and mechanical properties of cardiolipin lipid bilayers determined using neutron spin echo, small angle neutron and X-ray scattering, and molecular dynamics simulations. Soft Matter 11:130–138

Pfündel EE, Renganathan M, Gilmore AM, Yamamoto HY, Dilley RA (1994) Intrathylakoid pH in isolated Pea chloroplasts as probed by violaxanthin deepoxidation. Plant Physiol 106:1647–1658

Pomorski TG, Nylander T, Cárdenas M (2014) Model cell membranes: discerning lipid and protein contributions in shaping the cell. Adv Colloid Interface Sci 205:207–220

Qin X, Suga M, Kuang T, Shen JR (2015) Photosynthesis. Structural basis for energy transfer pathways in the plant PSI-LHCI supercomplex. Science 348:989–995

Quinn PJ (2012) Lipid-lipid interactions in bilayer membranes: married couples and casual liaisons. Prog Lipid Res 51:179–198

Reifarth F, Christen G, Seeliger AG, Dormann P, Benning C, Renger G (1997) Modification of the water oxidizing complex in leaves of the *dgd1* mutant of *Arabidopsis thaliana* deficient in the galactolipid digalactosyldiacylglycerol. Biochemistry 36:11769–11776

Ruban AV, Young AJ, Horton P (1994) The effects of illumination on the xanthophyll composition of the photosystem II light-harvesting complexes of spinach thylakoid membranes. Plant Physiol 104:227–234

Sackman E (1995) Biological membranes architecture and function. In: Lipowsky R, Sackmann E (eds) Structure and dynamics of membranes. Elsevier, Amsterdam, pp 1–65

Saga G, Giorgetti A, Fufezan C, Giacometti GM, Bassi R, Morosinotto T (2010) Mutation analysis of violaxanthin de-epoxidase indentifies substrate-binding sites and residues involved in catalysis. J Biol Chem 285:23763–23770

Sato N, Sonoike K, Tsuzuki M, Kwaguchi A (1995) Impaired photosystem II in a mutant of *Chlamydomonas reinhardtii* defective in sulfoquinovosyl diacylglycerol. Eur J Biochem 234:16–23

Sato N, Hagio M, Wada H, Tsuzuki M (2000) Requirement of phosphatidylglycerol for photosynthetic function in thylakoid membranes. Proc Natl Acad Sci U S A 97:10655–10660

Schaller S, Latowski D, Jemiola-Rzeminska M, Wilhelm C, Strzalka K, Goss R (2010) The main thylakoid membrane lipid monogalactosyldiacylglycerol (MGDG) promotes the de-epoxidation of violaxanthin associated with the light-harvesting complex of photosystem II (LHCII). Biochim Biophys Acta 1797:414–424

Schaller S, Latowski D, Jemioła-Rzemioska M, Dawood A, Wilhelm C, Strzałka K, Goss R (2011) Regulation of LHCII aggregation by different thylakoid membrane lipids. Biochim Biophys Acta 1807:326–335

Schaller S, Latowski D, Jemiola-Rzeminska M, Quaas T, Wilhelm C, Strzalka K, Goss R (2012) The investigation of violaxanthin de-epoxidation in the primitive green alga *Mantoniella squamata* (Prasinophyceae) indicates mechanistic differences in xanthophyll conversion to higher plants. Phycologia 51:359–370

Schaller S, Richter K, Wilhelm C, Goss R (2014) Influence of pH, Mg^{2+}, and lipid composition on the aggregation state of the diatom FCP in comparison to the LHCII of vascular plants. Photosynth Res 119:305–317

Schumann A, Goss R, Jakob T, Wilhelm C (2007) Investigation of the quenching efficiency of diatoxanthin in cells of *Phaeodactylum tricornutum* (Bacillariophyceae) with different pool sizes of xanthophyll cycle pigments. Phycologia 46:113–117

Seddon JM, Templer RH (1995) Polymorphism of lipid–water system. In: Lipowsky R, Sackman E (eds) Structure and dynamics of membranes. Elsevier, Amsterdam, pp 97–161

Selstam E (1998) Development of thylakoid membranes with respect to lipids. In: Siegenthaler PA, Murata N (eds) Lipids in photosynthesis: structure, function and genetics. Kluwer Academic Publishers, Dordrecht, pp 209–224

Shipley GG, Green JP, Nichols BW (1973) The phase behavior of monogalactosyl, digalactosyl, and sulphoquinovosyl diglycerides. Biochim Biophys Acta 311:531–544

Siefermann D, Yamamoto H (1975) NADPH and oxygen-dependent epoxidation of zeaxanthin in isolated chloroplasts. Biochem Biopharm Res Co 62:456–461

Simidjiev I, Barzda V, Mustárdy L, Garab G (1998) Role of thylakoid lipids in the structural flexibility of lamellar aggregates of the isolated light-harvesting chlorophyll a/b complex of photosystem II. Biochemistry 37:4169–4173

Simidjiev I, Stoylova S, Amenitsch H, Javorfi T, Mustardy L, Laggner P, Holzenburg A, Garab G (2000) Self-assembly of large, ordered lamellae from non-bilayer lipids and integral membrane proteins *in vitro*. Proc Natl Acad Sci U S A 97:1473–1476

Singer SJ, Nicolson GL (1972) The fluid mosaic model of the structure of cell membranes. Science 175:720–731

Sozer O, Kis M, Gombos Z, Ughy B (2011) Proteins, glycerolipids and carotenoids in the functional photosystem II architecture. Front Biosci (Landmark Ed) 16:619–643

Steffen R, Kelly AA, Huyer J, Dormann P, Renger G (2005) Investigations on the reaction pattern of photosystem II in leaves from *Arabidopsis thaliana* wild type plants and mutants with genetically modified lipid content. Biochemistry 44:3134–3142

Stransky H, Hager A (1970) Das Carotinoidmuster und die Verbreitung des lichtinduzierten Xanthophyllzyklus in verschiedenen Algenklassen. II: Xanthophyceae. Arch Microbiol 71:164–190

Stroebel D, Choquet Y, Popot JL, Picot D (2003) An atypical haem in the cytochrome b(6)f complex. Nature 426:413–418

Tietz S, Puthiyaveetil S, Enlow HM, Yarbrough R, Wood M, Semchonok DA, Lowry T, Li Z, Jahns P, Boekema EJ, Lenhert S, Niyogi KK, Kirchhoff H (2015) Functional implications of photosystem II crystal formation in photosynthetic membranes. J Biol Chem 290:14091–14106

Tyler AI, Law RV, Seddon JM (2015) X-ray diffraction of lipid model membranes. Methods Mol Biol 1232:199–225

Umena Y, Kawakami K, Shen JR, Kamaiya N (2011) Crystal structure of oxygen-evolving photosystem II at a resolution of 1.9 Å. Nature 473:55–60

van den Brink-van der Laan E, Killian JA, de Kruijff B (2004) Non-bilayer lipids affect peripheral and integral membrane proteins via changes in the lateral pressure profile. Biochim Biophys Acta 1666:275–288

Van Eerden FJ, de Jong DH, de Vries AH, Wassenaar TA, Marrink SJ (2015) Characterization of thylakoid lipid membranes from cyanobacteria and higher plants by molecular dynamics simulations. Biochim Biophys Acta 1848:1319–1330

Vieler A, Süß R, Wilhelm C, Schiller J (2007) The lipid composition of two different algae (*Chlamydomonas reinhardtii*, Chlorophyceae and *Cyclotella meneghiniana*, Bacillariophyceae) investigated by MALDI-TOF MS and TLC. Chem Phys Lipids 150:143–155

Vigh L, Escribá PV, Sonnleitner A, Sonnleitner M, Piotto S, Maresca B, Horváth I, Harwood JL (2005) The significance of lipid composition for membrane activity: new concepts and ways of assessing function. Prog Lipid Res 44:303–344

Wada H, Murata N (2007) The essential role of phosphatidylglycerol in photosynthesis. Photosynth Res 92:205–215

Wang S, Uddin MI, Tanaka K, Yin L, Shi Z, Qi Y, Mano J, Matsui K, Shimomura N, Sakaki T, Deng X, Zhang S (2014) Maintenance of chloroplast structure and function by overexpression of the rice MONOGALACTOSYLDIACYLGLYCEROL SYNTHASE gene leads to enhanced salt tolerance in tobacco. Plant Physiol 165:1144–1155

Williams WP (1998) The physical properties of thylakoid membrane lipids and their relation to photosynthesis. In: Siegenthaler PA, Murata N (eds) Advances in photosynthesis. Lipids in photosynthesis. Kluwer Academic Publishers, Dordrecht, pp 103–118

Williams WP, Brain APR, Dominy PJ (1992) Induction of non-bilayer lipid phase separation in chloroplast thylakoid membranes by compatible co-solutes and its relation to the thermal stability of Photosystem II. Biochim Biophys Acta 1099:137–144

Wu W, Ping W, Wu H, Li M, Gu D, Xu Y (2013) Monogalactosyldiacylglycerol deficiency in tobacco inhibits the cytochrome b6f-mediated intersystem electron transport process and affects the photostability oft he photosystem II apparatus. Biochim Biophys Acta 1827:709–722

Yamamoto HY, Higashi RM (1978) Violaxanthin de-epoxidase-lipid-composition and substrate-specificity. Arch of Biochem Biophys 190:514–522

Yamamoto H, Nakayama T, Chichester C (1962) Studies on the light and dark interconversions of leaf xanthophylls. Arch Biochem Biophys 97:168–173

Yan JS, Mao DZ, Chen H, Kuang TY, Li LB (2000) Effects of membrane lipids on the electron transfer activity of cytochrome b6f complex from spinach. Acta Bot Sin 42:1267–1270

Yu B, Benning C (2003) Anionic lipids are required for chloroplast structure and function in *Arabidopsis*. Plant J 36:762–770

Zhou F, Liu S, Hu Z, Kuang T, Paulsen H, Yang C (2009) Effect of monogalactosyldiacylglycerol on the interaction between photosystem II core complex and its antenna complexes in liposomes of thylakoid lipids. Photosynth Res 99:185–193

Zick M, Stroupe C, Orr A, Douville D, Wickner WT (2014) Membranes linked by trans-SNARE complexes require lipids prone to non-bilayer structure for progression to fusion. eLife. doi:10.7554/eLife.01879

Chapter 7
Chemical Genetics in Dissecting Membrane Glycerolipid Functions

Florian Chevalier, Laura Cuyàs Carrera, Laurent Nussaume, and Eric Maréchal

Abstract Chemical genetics has emerged as a powerful approach to dissect biological processes, based on the utilization of small molecules disturbing the function of specific target proteins. By analogy with classical genetics, 'reverse chemical genetics' refers to the utilization of drugs acting on a known target, enabling its functional characterization at the levels of the cells, tissues and organisms. Likewise, 'direct chemical genetics' refers to the utilization of a drug of unknown mode of action, but triggering a phenotype of interest. In that case, one has to identify the target(s) possibly blocked (or possibly activated) by the small molecule. This chapter illustrates both approaches, like the analysis of the elongation of fatty acids, the biosynthesis of galactoglycerolipids or the catabolism of phosphoglycerolipids by reverse chemical genetics or the study of the membrane glycerolipid remodeling triggered upon phosphate starvation, by direct chemical genetics.

Keywords Cerulenin • Diacylglycerol • Galactolipids • Galvestine • Halopemide • Haloxyfop • Phosphate starvation • Phosphatidic acid • Phostatin • Thiolactomycin • Triclosan

Introduction

'Chemical genetics' is a recent discipline, making benefit from small molecules (chemicals) that interfere with biological processes (targets) to perform functional studies. The search for active drugs is probably one of the oldest human scientific

---

F. Chevalier • E. Maréchal (✉)
UMR 5168 CNRS-CEA-INRA-Université Grenoble Alpes, Laboratoire de Physiologie Cellulaire et Végétale, iRTSV, CEA-Grenoble, Grenoble, France
e-mail: eric.marechal@cea.fr

L.C. Carrera • L. Nussaume
UMR 7265 CNRS-CEA-Université Aix-Marseille, Biologie Végétale et Microbiologie Environnementale, iBEB, CEA Cadarache, Saint-Paul-lez-Durance, France

Y. Nakamura, Y. Li-Beisson (eds.), *Lipids in Plant and Algae Development*,
Subcellular Biochemistry 86, DOI 10.1007/978-3-319-25979-6_7

activities, with strategies ranging from traditional pharmacopeia to rationalized medicinal chemistry. A ‘small molecule’ can be considered as drug-like if respecting some general criteria, called the Lipinski rule of five (Lipinski et al. 2001; Ganesan 2008; Walters 2012), i.e. more than 5 hydrogen-bond donors, more than 10 hydrogen-bond acceptors, a molecular weight greater than 500 and a calculated Log *P* greater than 5 (*P* corresponding to the water-octanol partition coefficient, i.e. a measure of molecule’s hydrophobicity). The search for ‘drug targets’ has been made possible only after extensive biochemical and functional analyses of biological macromolecules, and particularly proteins. The drugability of a target is a much more complex issue, since the protein molecular surface should allow the docking of drugs, the protein should be located so as to be available to an applied drug and eventually, the interference with the protein function should lead to a validated impact on the phenotype. Blocking or activating the function of a protein with a small molecule is close to an impairment or activation of the expression of its encoding gene: the term ‘chemical genetics’ (Haggarty et al. 2000; Mayer et al. 1999; Marechal 2008) has been proposed in the early 2000s, although pharmacological studies had been performed for decades.

Like for conventional genetics, ‘reverse chemical genetics’ refers to the utilization of drugs acting on a known target, enabling its functional characterization at the levels of the cells, tissues and organisms. Likewise, ‘direct chemical genetics’ refers to the utilization of a drug of unknown mode of action, but triggering a phenotype of interest. In that case, one has to identify the target possibly blocked (or possibly activated) by the small molecule.

What are the advantages over genetics? The first one is the possibility to use all small molecules previously developed in chemistry laboratories or made available in public or private collections of compounds (chemolibraries). Very large collections of small molecules have been developed to act on different model organisms, for a large number of agrochemical or biomedical applications, including herbicides killing weeds or drugs acting on human diseases. They can be tested rapidly on any biological model without any prior genomic knowledge or molecular tools. A second advantage is the versatility of application of the drug and the possibility to provide variable doses, which is particularly useful to study proteins playing vital roles. When a gene knock out (KO) is lethal, it is still possible to use alternative genetic approaches, like inducible gene inactivation systems or RNA interference in contexts that do not silence completely a gene expression. These approaches can be difficult to establish, to reproduce and can ultimately be unsuccessful. Small molecules allow a control of the activity of the gene product at non-lethal concentrations and at variable doses, allowing a refined characterization of the phenotypic response correlated to the level of inhibition of the protein target. It is possible to inhibit a protein target in a single cell or a small part of a tissue or an organ and analyze the impact at the level of the full organism. It is also possible to apply a small molecule at various development stages. A third advantage is to perform time-course experiments to study rapid phenomenon blocked or activated at the protein level at the time of application of the compound. This can be combined with dose variations. A fourth advantage is the possibility to target complete gene families, using a compound docking to a conserved active site. Eventually, a small molecule is a very powerful tool to apply a selective pressure, trigger a phenotype and perform genetic screens based on this phenotype. It is therefore possible to exploit small molecules in many ways so as to advance knowledge and identify novel genes involved in a given biological process.

What are the drawbacks? They are numerous. In the most favorable situation, researchers simply order a commercial compound known to act on a target of interest with high levels of specificity. Some proteins like tubulin subunits are the targets of multiple drugs and cytoskeleton analyses can really benefit from a very rich toolbox of small molecules (e.g. Lafanechere 2008). For other proteins, no active drugs is known and a screening has to be performed. Following the screening process, some 'hit' compounds can be identified and their chemical scaffold has to be optimized by chemists. In all cases, it is important to control that the treatment with the selected small molecule has no impact on other biological processes. Drugs are often promiscuous, and side effects can be observed on so called 'off targets' (e.g. Botte et al. 2012). Small molecules can be also difficult to dissolve in water and need 'vehicles' like dimethyl-sulfoxide (DMSO) or polysorbate/Tween to go across biological barriers. DMSO and polysorbate/Tween can trigger their own side effects. Regardless of these difficulties, a small molecule remains a powerful and flexible tool, one can utilize in different contexts and on different organisms. Chemical genetics allows therefore some functional characterizations, which cannot be undergone by other means.

'Genetics' and 'chemical genetics' can be used in combination, especially when analyzing the effect of a compound on various mutants. When considering a biological process, like glycerolipid metabolism, as a 'system', one can use chemicals as a mean to disturb this system, trigger a phenotype, induce the up-regulation or down-regulation of some genes, etc., and consequently advance knowledge. In this chapter, we shall illustrate both reverse and direct chemical genetic strategies. We introduce some of the chemical tools currently available to study glycerolipid metabolism and highlight some of the progresses that could be performed using small molecules and detail some important perspectives.

Reverse Chemical Genetics

Screening Strategies

When a drug is not available commercially, the development of an active small molecule or a series of small molecules requires a screening of a chemolibrary. Private companies have collected very large chemolibraries, ranging from hundreds of thousands to millions of small molecules (for reviews, Marechal 2008; Saidani et al. 2009; Maréchal et al. 2011; Awale and Reymond 2014; Irwin and Shoichet 2005; Monika et al. 2013). The ZINC database is an electronic repository for the chemical structures of commercially available compounds (>13 million molecules) (Awale and Reymond 2014; Irwin and Shoichet 2005; Monika et al. 2013). It is possible to screen these collections on a specific protein target using an automated system that (i) collects molecules usually stored in 96-well to 384-well microplates, (ii) transfers these molecules with a multi-pipetting device to microplates containing the target and (iii) performs a miniature assay so as to detect the 'hit' molecules acting on the protein function. The screen is designed to operate with a high throughput. The readout is a measure of the activity of the protein in case of an enzyme, a protein-protein interaction assay, etc. The miniature assay is usually designed to allow a simple detection of the enzymatic activity, protein-protein interaction, etc., using colorimetric, fluorescence or chemi-luminescence detections (for review, Maréchal et al. 2011). Alternatively, if the 3D-structure of the protein is known, a virtual screen can be performed using specific docking algorithms (Awale and Reymond 2014; Irwin and Shoichet 2005; Monika et al. 2013).

In academia, access to very large collections of compounds is only possible if the target has an added value for a critical human disease or an objective of industrial interest. Most often, automated screens can be performed using well selected collections of a few hundreds to a few thousands of molecules. Virtual screening of millions of molecules of the ZINC database is possible using the power of supercomputers or computer grids (distributed computers or clusters).

In this section, it is not possible to detail all known molecules acting on plant glycerolipid metabolism, which can be used for reverse chemical genetics. We illustrate therefore different possible strategies, ranging from the use of an existing molecular toolbox made of commercially available drugs acting on a metabolic pathway, i.e. fatty acid biosynthesis in the stroma of chloroplast, to the complete development of a drug, like galvestine-1 inhibiting the synthesis of galactolipids in the plastid envelope membrane.

Inhibition of Fatty Acid Biosynthesis

The structure of membrane glycerolipids is obtained by the assembly of three building blocks: a three-carbon glycerol backbone originating from glycerol-3-phosphate (G3P); fatty acids (FAs) esterified at positions *sn*-1 and *sn*-2 of glycerol and a polar head at position *sn*-3. Storage glycerolipids, or triacylglycerol, correspond to the addition of a third FA at position *sn*-3. The fatty acid synthase of type II (FAS II, Fig. 7.2) has been demonstrated to be an efficient drug target to develop herbicide, in particular because mammals synthesize their own FAs using another system, a cytosolic FAS I. FAS II is a multicatalytic system composed of several proteins, which are targetable by multiple drugs. The biosynthesis of FAs is an iterative process that requires acetyl-CoA as an initiator and malonyl-CoA as a 2C-donor for elongation of the acyl chain. Upstream the FAS I or II systems, malonyl-CoA is generated by the activity of an acetyl-CoA carboxylase (ACCase). FAS II is found in all bacteria and in the plastid of plants, algae and plastid-containing protists

(Goodman and McFadden 2008; Ben Mamoun et al. 2010; Botte et al. 2012). FA synthesis requires the attachment of a malonyl group to acyl carrier protein (ACP), a reaction catalyzed by a malonyl-CoA: ACP malonyltransférase (MAT) also known as FabD. The sequential addition of two carbons begins at this point. Initially, a keto-acyl-ACP synthase (KS; FabH) activity catalyzes the transfer of 2C from malonyl-ACP to acetyl forming an acetoacetyl-ACP (or β-keto(oxo)acyl-ACP for longer substrates). The ketone group is reduced to alcohol by a NADPH-dependent β-keto acyl-ACP reduction (KR; FabB/F). The enoyl-ACP hydrase (DH; FabZ) then catalyzes the formation of a *cis*-2,3-enoyl, which is reduced by enoyl-ACP reductase (ER; FabI). Seven cycles are required for the synthesis of palmitate (16:0) and eight for stearate (18:0). Produced acyl chains then serve for acyl-lipid syntheses.

Multiple herbicides are known to act at the level of FAS II, including the 'fop' herbicides, thiolactomycin, cerulenin and triclosan (Fig. 7.1). They allow the inhibi-

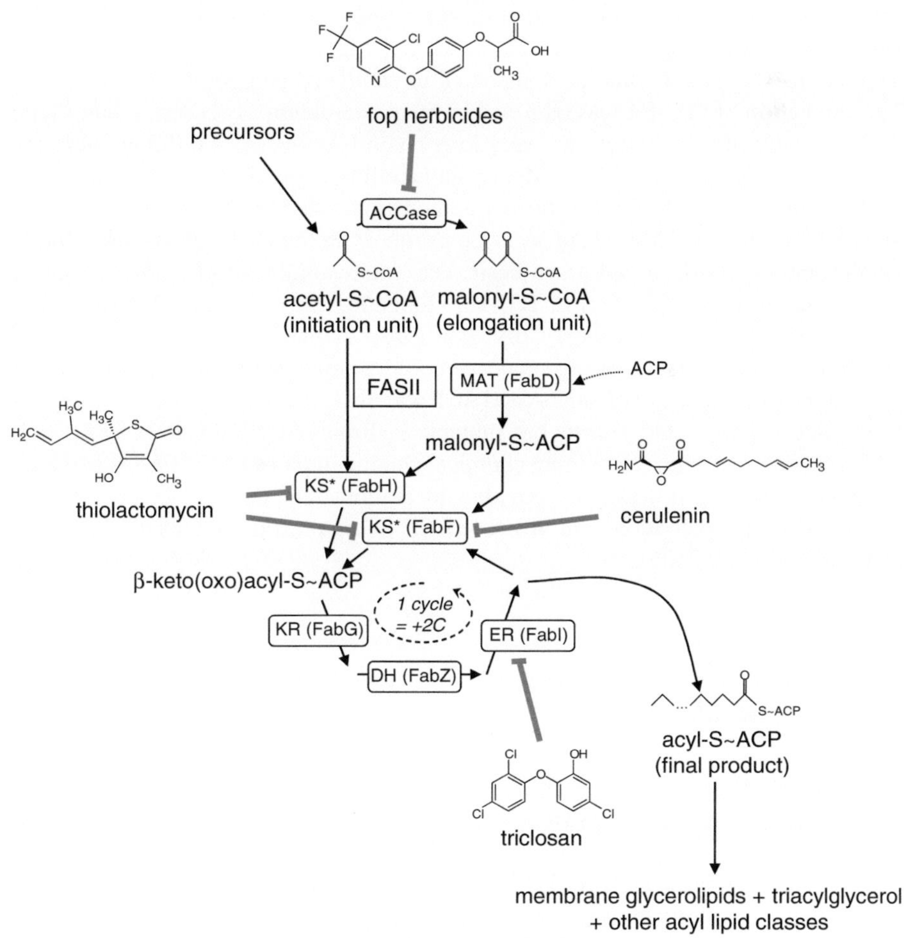

Fig. 7.1 Small molecules acting at the level of fatty acid synthesis

tion of different enzymatic activities of the FAS II system and constitute therefore a complete molecular toolbox for chemical genetics.

The fop herbicides, like clodinafop or haloxyfop, are known to inhibit ACCases, but with a high specificity for the cytosolic ACCase isoform (Xiang et al. 2009). Thiolactomycin is an antibiotic interfering with two activities of FASII: i.e. an inhibition of the ACP acetyltransferase, by competing with ACP binding and being non-competitive with acetyl-CoA; and inhibition of the keto-acyl-ACP synthase activities, competitive with respect to malonyl-ACP and non-competitive with respect to acetyl-ACP (Nishida et al. 1986). Triclosan is a biocide inhibiting FabI (Xiang et al. 2009). The cerulenin scaffold initially unspecific for FASI/FASII, acts at the level of the FabS enzyme (Heath and Rock 2004) (Fig. 7.1).

Numerous molecular tools are therefore available for studying the role of FAS II, without the need of genetic mutations that are made difficult due to the importance of the FAS II system. As an illustration, FAS II inhibitors have proven useful to comprehend how plant cells could elongate exogenously provided FA (Koo et al. 2005). When *Arabidopsis* leaves were incubated with radiolabelled FAs with chain lengths lower than 16C, the label appeared in elongated and desaturated products. The elongation of 12C-FA was inhibited by 50 μm cerulenin. By contrast, haloxyfop inhibited only the elongation into very long chain FAs, whereas it did not alter the synthesis of 16C- and 18C-FAs. Metabolic labelling experiments and biochemical analyses of isolated chloroplast further demonstrated that the elongation of exogenous FAs to 16C- and 18C- FAs occurred primarily in the chloroplasts, most likely via the enzymes of *de novo* FA synthesis. Very interestingly, an *Arabidopsis* mutant displaying an 80 % reduction in 12C-FA elongation into 16C- and 18C-FAs suggested that a long-chain acyl-CoA synthetase homologue, with a predicted N-terminal chloroplast targeting sequence could be involved in this process. Direct acyl-ACP-forming activity from FA and ACP was observed in extracts of *Arabidopsis* leaves and isolated chloroplasts but mutants had markedly reduced *in vitro* acyl-ACP synthesis activity. Combining chemical genetic impairment of FAS II and conventional genetics, it was thus possible to demonstrate that plants possessed a mechanism for direct activation of FA to ACP in the plastid via an acyl-ACP synthetase (Koo et al. 2005).

Inhibition of Enzymes Using or Generating Diacylglycerol and Phosphatidic Acid

Following FA biosynthesis, the first step of the *de novo* biosynthesis of glycerolipids is achieved by the action of two acyl-transferase activities, generating lyso-phosphatidic (lyso-PA) and then phosphatidic (PA) acids. On the one hand, lyso-PA and PA can be synthesized in the ER by the stepwise action of an acyl-CoA:glycerol-3-phosphate acyltransferase (or GPAT) and an acyl-CoA:1-acylglycerol-3-phosphate acyltransferase (or AGPAT). On the other hand, PA neosynthesis can occur in plastids by the action of an acyl-ACP:glycerol-3-phosphate acyltransferase (or ATS1) and an acyl-ACP:1-acylglycerol-3-phosphate acyltransferase (or ATS2)

(Dubots et al. 2012). The following important precursor for glycerolipid synthesis is diacylglycerol (DAG), which is neosynthesized from PA. In this case, DAG is produced by the hydrolysis of PA by phosphatidate phosphatases (PAP). DAG can be phosphorylated into PA by diacylglycerol kinases (DGK). PA and DAG can also be generated from existing phospholipids via the action of phospholipases D (PLDs) and phospholipases C (PLCs), respectively. DAG and cytidine diphosphate diacylglycerol (CDP-DAG), which directly derives from PA, serve as precursors for all membrane glycerolipids (Saidani et al. 2014).

The biosynthesis of glycerolipids is therefore constituted of three major 'metabolic blocks': (1) the neosynthesis/recycling of FAs, (2) a PA ↔ DAG hub, and (3) the synthesis of complex glycerolipids from PA/CDP-DAG and DAG. Two subcellular compartments are involved: one is the ER, synthesizing most phospholipids that are then sorted to compartments of the endomembrane system via vesicular trafficking; the other one is the plastid, synthesizing non-phosphated glyco-glycerolipids, in particular monogalactosyldiacylglycerol (MGDG), and digalactosyldiacylglycerol (DGDG). Given the importance of glycerolipids for membrane biogenesis and dynamics, detailed in other chapters of this book, the metabolic routes for their synthesis have appeared as a target for the search of herbicides that could then serve for chemical genetic studies.

Numerous inhibitors of MGDG synthases have been described, but none was active or selective enough to be used for a chemical genetic strategy. Chemical impairment of galactolipid synthesis has been made possible after the discovery of a specific blocking agent, following a 10-year drug-development program. Firstly, a miniature and robust assay was developed to allow a robotic high throughput screening (Nishiyama et al. 2003). Inhibitors of *Arabidopsis* MGD1 (monogalactolsyldiacylglycerol synthase 1) were then obtained from a screen of a 23,360 compounds and the active structure was optimized based on the synthesis and analysis of 250 analogues. The molecule with optimal activity both *in vitro* and *in vivo* was called galvestine-1 (Botte et al. 2011) (Fig. 7.2).

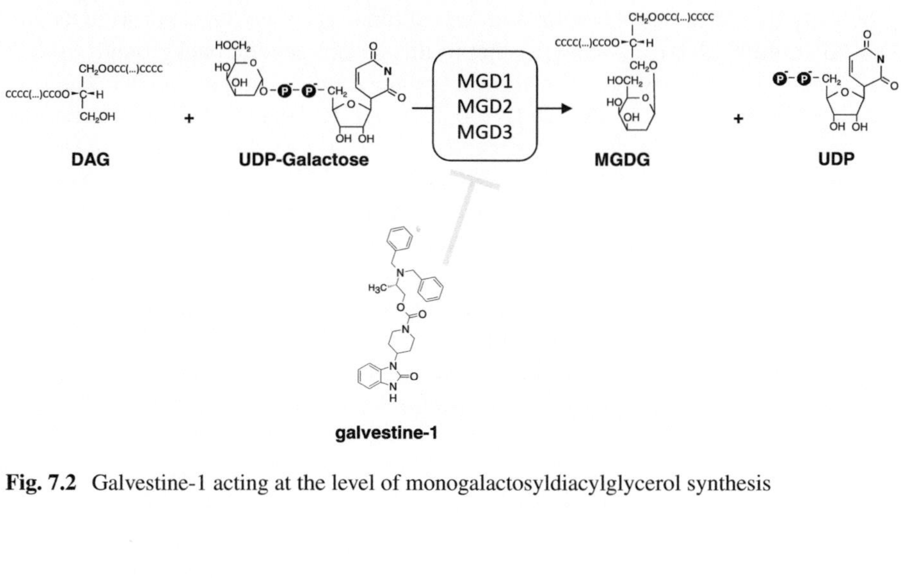

Fig. 7.2 Galvestine-1 acting at the level of monogalactosyldiacylglycerol synthesis

At molecular level, galvestine-1 acts by competing with DAG for the binding on MGD1 (Botte et al. 2011). The inhibitory constant of gavestine-1 (Ki) relative to DAG is 11 times lower than the Km measured for the DAG substrate. Conformational models of DAG and galvestine-1 have been compared and they showed a superposition of structural portions of both molecules, including H-bond acceptors, at the levels of the glycerol backbone of DAG and of the (1,2-dihydro-2-oxobenzo[δ] imidazol-3-yl)piperidine-1-carboxylate scaffold of galvestine-1. Galvestine-1 is thus one of the very few competitive inhibitors of DAG, besides phorbol esters, having a clearly distinct scaffold compared to the acyl-glycerol structure.

Galvestine-1 was shown to inhibit the activity of all the isoforms of MGDs (MGD1, MGD2 and MGD3 multigenic family) and is thus a molecular tool enabling the impairment of all MGDG synthases, which could not be attained by genetic knock out (as discussed below). The addition of increasing amounts of galvestine-1 to the growing medium of *Arabidopsis* led to a dose-dependent control of MGDG level *in planta* (Botte et al. 2011), with a severe chlorosis at highest doses. At low doses, chlorosis was more pronounced around vascular tissues, consistently with an uptake by the roots and a circulation in the xylem and mesophyll. *In vivo* inhibition of MGDG synthesis was measured by metabolic labeling.

Conventional genetics allowed important advances in the comprehension of the complicate processes governing MGDG and DGDG syntheses and relative proportions. In *Arabidopsis*, MGDG mostly contains 16:3 and 18:3 acyls, indicating that the neosynthesized MGDG (C18:1/C16:0) is very rapidly desaturated by FAD5 and then by FAD6 and FAD7/8, producing polyunsaturated MGDG (C18:3/C16:3). By contrast, DGDG mostly contains 16:0 and 18:3 fatty acids, with C16:0 at position *sn*-1 (C16:0/C18:3), indicating that a bifurcation in the utilization of MGDG occurs between DGD1/DGD2 and FAD5. In the knocked down *mgd1-1* mutant (Jarvis et al. 2000), the MGDG content decreases, whereas that of DGDG is stable, indicating that MGDG is not limiting for the production of DGDG. A channeling of C16:0-rich MGDG toward the production of DGDG, via MGD2 or MGD3, might thus occur. In the knocked out *mgd1-2* mutant (Kobayashi et al. 2007), seedling could grow only on sugar-rich medium, and were albino. Very low levels of MGDG and DGDG could be detected, showing that in this severe genetic background, MGD2 and MGD3 were not sufficient to allow a sustained production of DGDG. The acyl profile of MGDG was enriched in C16:0, C18:0, C18:1 and C18:2, indicating that this lipid has been produced using precursors imported from the cytosol. The DGDG acyl profile was identical to that of MGDG supporting, again, a channeling via MGD2 or MGD3 and the DGD1 or DGD2 enzymes. The lack of any C16:3 MGDG indicated that a channeling via MGD1 and FAD5 is very likely to occur in the wild type background. The *mgd2*, *mgd3* and *mgd2*x*mgd3* mutants (Kobayashi et al. 2009) had no striking phenotype A change in the lipid profile could be assessed for *mgd3-1* upon phosphate starvation, showing a role for this gene in response to phosphate variations. Besides this condition, MGD1 apparently complemented the loss of *MGD2* and *MGD3*. A possible function for *MGD2* and *MGD3* was highlighted by the expression pattern during the elongation of the pollen tube (Kobayashi et al. 2004), whereas the expression of *MGD1* in this cell type remained very low. This correlates with an increase of the synthesis of MGDG and DGDG in the growing pollen tube (Nakamura et al. 2009), but the *mgd2*x*mgd3* double mutant (Kobayashi et al. 2009) was not male sterile. Given the impossibility to study a triple mutant, no conclusive role for MGDG synthesis in the pollen tube could be deduced.

Chemical genetics could then highlight some specific properties of MGDG synthases that could not be demonstrated before. In particular, the treatment of *Arabidopsis* with galvestine-1 allowed the analysis of a condition that could not be tested by genetic approaches, i.e. a targeting of all members of the MGD gene family in the same background (Botte et al. 2011). The acyl profile of MGDG was enriched in C16:0, C18:0, C18:1 and C18:2 in a galvestine-1 dose-dependent manner and this evolution was mirrored in the PC acyl profile. Thus, the targeting of all MGDG synthases highlights the stimulation of the 'eukaryotic' route to produce galactolipids, to the expense of the 'prokaryotic' one and a dose-dependent impairment of the MGDG desaturation initiated by FAD5. The phenotype of treated plants was mostly consistent with an impairment of *MGD1*.

Very interestingly, the role of MGDG in specific cell types, such as pollen tube, could eventually be explored in more details. The *in vitro* growth of *Arabidopsis* pollen tubes was partly inhibited by galvestine-1 in a dose-dependent manner (Botte et al. 2011), showing that half of the growth could be attributed to galactolipids. Accumulation of DGDG in the plasma membrane of pollen tube could be immunodetected (Botte et al. 2011) and suggested that in the pollen tube, plastids contribute to the very high demand for lipids requested by the plasma membrane expansion in one of the fastest cellular elongating systems.

The effect of galvestine-1 treatment on the transcriptome could also be studies (Botte et al. 2011). The expression of 525 genes was downregulated in a dose- and time-dependent manner, including genes coding for thylakoid/chloroplast proteins, consistently with the strong effect of galvestine-1 on chloroplast biogenesis. These downregulated genes include *LOX2*, and *AOS* coding for enzymes converting PUFAs hydrolyzed from galactolipids into oxylipin precursors. A relation between the level of 16:3- and 18:3-rich MGDG and the activation of systemic acquired resistance (SAR), a mechanism induced by a broad spectrum of pathogens, had been shown in the *Arabidopsis sfd1* mutant (Lorenc-Kukula et al. 2012). In the list of galvestine-1 downregulated genes, 22 were reported to be upregulated following Pi starvation, including *SQD2*, cording for the enzyme synthesizing sulfoquinovosyldiacylglycerol, *PHO1;H1* and *PHO81* coding for Pi transporters and *SRG3* encoding a glycerophosphoryldiester phosphodiesterase. Among the 182 genes upregulated in a dose- and time-dependent manner, 69 were mostly overexpressed after a short exposure, 90 had an expression increasing with duration and concentration and 23 were strongly overexpressed in all conditions. An important series of genes have been highlighted by the analysis of the transcriptomic response to galvestine-1 and are now analyzed by conventional genetics.

As mentioned in the introduction, the specificity of a drug is critical. Specificity of galvestine-1 on MGD enzymes has been checked by the innocuous effect on glycerolipid metabolism of cell types that do not produce galactolipids, like bacteria systems, and on the absence of other visible impact on plant cells besides an initial decrease in MGDG (Botte et al. 2011). Nevertheless, a survey of the screening projects aiming at discovering drugs acting on glycerolipid metabolism has shown a striking structural similarity between [(*S*)-2-(dibenzylamino)propyl] 4-(2-oxo-3H-benzimidazol-1-yl)piperidine-1-carboxylate (or galvestine-1; PubChem ID CID 25192811) with drugs acting on phospholipases D (Saidani et al. 2014). Indeed, N-[2-[4-(5-chloro-2-oxo-3H-benzimidazol-1-yl)piperidin-1-yl]ethyl]-4-fluorobenzamide (or halopemide; PubChem ID CID 65490) was shown to selectively inhibit humans PLD1 and PLD2 (Monovich et al. 2007; Scott et al. 2009) (Fig. 7.3). Inhibition studies of truncated PLD1 supports that halopemide analogs act by direct binding to the catalytic site, in which phosphatidylcholine (PC) is hydrolyzed into PA. An exploration of the chemical space of halopemide analogs has further shown that compounds could be equally efficient on PLD1 and PLD2 or more specific to one or the other. Thus compounds with a 1-phenyl-1,3,8-triazaspiro[4,5]decan-4-one scaffold, like VU0285655 (PubChem ID CID 44138050), had a higher selectivity for PLD2 (Scott et al. 2009; Selvy et al. 2011).

Chemical tuning of the piperidinyl-benzimidazolidinone scaffold can therefore determine the target specificity of the obtained small molecules. Mammal PLD1 and 2 are similar to *Arabidopsis* PLDζ1 and PLDζ2, and the developed inhibitors

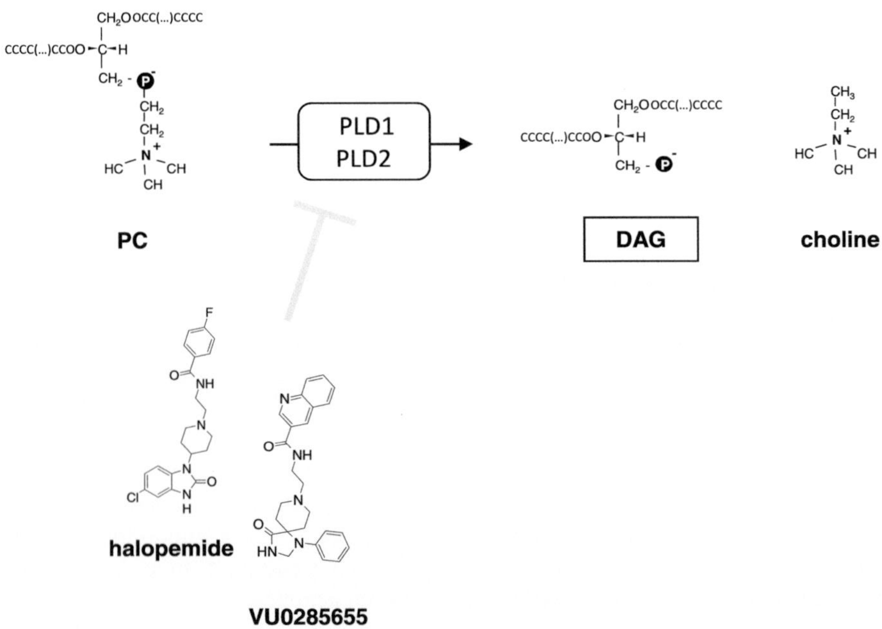

Fig. 7.3 Small molecules acting at the level of phospholipases D

could be used for a chemical genetic analyses of these enzymes in plants. One can attempt to develop specific analogs based on this structure.

The reverse chemical genetic approach can be extremely useful and complementary to other approaches. The search for drugs triggering a phenotype at the level of glycerolipid metabolism can also be achieved, following a direct chemical genetic strategy.

Direct Chemical Genetics

Screening Strategies

Direct chemical genetic strategy essentially refers to the use of drugs of unknown mode of action to screen whole organisms or cells for candidates presenting a phenotype of interest. Once this phenotype of interest is confirmed, next objective becomes the identification of the target, which is usually pursued through biochemical approaches (Sadhukhan et al. 2012).

Several features must be taken into consideration before starting a chemical genetic screen. One of the main limitations of this strategy is the little amount of drug provided in a chemolibrary. So it is essential to optimize conditions in order to carry out the experiment using the minimum amount of drug. As miniaturization is often here a necessity, plants with reduced development or cell culture were favored for these approaches. Roots architecture in plants has evolved for the efficient uptake of mineral and nutrients providing an excellent route for chemical genetic studies (Blackwell and Zhao 2003). The small size combined with the well-established genomics and proteomics tools makes *Arabidopsis* plants ideal for this type of experiments (Raikhel and Pirrung 2005). It is then essential to screen for phenotypes that can be observed during early days of development (germination disorders, alteration in early seedling growth or early responding genes). This strategy has allowed the finding of several valuable compounds like the selective ABA (abscisic acid) agonist pyrabactin that mimics ABA effect (Zhao et al. 2007), morlin which interacts with cellulose synthase causing cytoskeletal alterations by disorders in micro-fibril orientation (DeBolt et al. 2007) or phosphatin, a drug alleviating phosphate starvation response (Arnaud et al. 2014), detailed in the next section. Morlin was identified in a screen of 20,000 compounds using 96-well format filled with Murashige and Skoog (MS) agar media completed with drugs at a final concentration of 20–40 μM. Approximately 10 sterilized seeds were sown per well, and after 3 days of continuous fluorescent light incubation, seedlings were analyzed with a microscope to identify molecules triggering hypocotyl or root swelling (DeBolt et al. 2007). Pyrabactin was identified in a screen for germination inhibitors. Sterilized seeds were suspended in 0.1 % agar and distributed in 96-well plates containing the drugs (10,000 in total, DIVERSet library) at 25 μM in MS agar media. Seeds were scored after 4 days of growth. Pyrabactin caused the etiolation of the seedlings (Zhao et al. 2007).

Other attractive systems for chemical genomics include tobacco BY-2 suspension cell lines or tobacco pollen tubes. These single cells are suitable for cell morphological and subcellular organization observation. Some compounds identified using this approach include cobtorin (Yoneda et al. 2007) or endosidin 1 (Drakakaki et al. 2009). Endosidin 1 blocks the endocytosis of several plasma membrane auxin transporters providing a unique tool to dissect recycling pathways (Robert et al. 2008). Endosidin 1 was isolated in a screen for compounds that perturb tobacco cell growth, which depends essentially on endomembrane trafficking. In this screen, pollen from 30 wild-type tobacco flowers was distributed in 384-well assay plates. Each well contained a compound from a 2,016 chemical library (Microsource Spectrum) that had already been distributed using a pipetting robot. Plates were incubated at room temperature with shaking for a minimum of 3 h to permit germination. Pollen tube morphology was observed by using transmitted confocal imaging (Robert et al. 2008). Cobtorin would also cause swollen phenotype, by the inhibition of the parallel alignment of the cortical microtubules with the cellulose micro-fibrils. In this screen, suspension of BY-2 cells expressing a GFP-α-tubulin fusion protein were cultured for 2 days in 96-well plates in a medium containing 25 μM of each compound (4,000 in total, two different libraries were used, LATCA and Spectrum). Tubulin organization of each cell was analyzed after these 2 days by confocal microscopy (Yoneda et al. 2007).

Search for Molecules Interfering with the Response of Plants to Phosphate Supply, with an Effect on the Lipid Remodeling Phenotype

Phosphate (Pi) is a vital macro-nutriment limiting plants growth in a vast majority of soils. Several physiological adaptive processes allow Pi recovery and utilization. One of these processes consists in the remodelling of membrane phospholids, which are replaced by sulfolipids or galactolipids (for recent reviews, Nakamura 2013; Boudière et al. 2012, 2014; Petroutsos et al. 2014). Such phenomenon was first reported in non-photosynthetic bacterium *Pseudomonas diminuta* (Minnikin et al. 1974) and then in the photosynthetic purple bacterium *Rhodobacter sphaeroides* (Benning et al. 1993). If multiple levels of regulations have been identified, the transcriptional control of this process remains a key step involving the gene family of the PHR1 master regulator (Rubio et al. 2001; Misson et al. 2005; Thibaud et al. 2010). A chemical genetics strategy was considered to identify drugs altering the transcriptional control of Pi starvation response. To that purpose, an *Arabidopsis* gene trap line containing a GUS transcriptional fusion with the high-affinity Pi transporter PHT1;4 (Misson et al. 2004) was used. This reporter gene is strongly induced in primary roots, offering the opportunity to perform the selection of very young plantlets. Four to five seeds germinated in 96-well plates have been used to screen the Library of Active Compounds on Arabidopsis collection of 3,600

Fig. 7.4 Structure of PTN1, or phostatin 1 (Pi-starvation inhibitor 1)

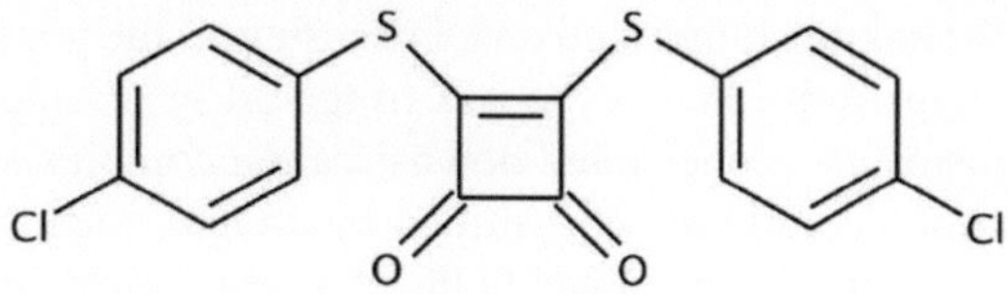

biologically active compounds (Arnaud et al. 2014). Two classes of molecules were identified. The first series, named phostatin analogues (Pi-starvation inhibitor) alleviates phosphate starvation responses (Arnaud et al. 2014), whereas the second series called phostin analogues (Pi inhibitor) trigger phosphate starvation traits despite the presence of Pi in the medium (C. Bonnot et al., manuscript in preparation). The analysis of the activity of phostatin 1 (PTN1) and of commercially available structural analogues has allowed the identification of a common structure composed of a benzene aromatic ring with sulfur and chlorine residues in para positions (Fig. 7.4).

PTN addition reduced various traits of Pi-starvation, such as the accumulation of starch and anthocyanins and most importantly the phospholipid/glycolipid conversion. A transcriptomic analysis revealed the broad impact of PTN addition on the expression of a majority of genes regulated by low Pi availability. The phospholipid/glycolipid ratio varied from 46/54 in Pi-rich conditions to 19/81 in Pi-deficient conditions. The addition of PTN to a Pi-starved medium increased the proportion of phospholipids (the ratio observed reached 25/75). The analysis of the lipid composition revealed that modifications affected mostly DGDG, PI, PE and PC. It is interesting to notice that the extent of these modifications is significant but still lags behind the transcriptomic changes observed for many enzymes of these pathway (such as SQD2, MGD3, SQD1, MGD2 and PLDζ). These drugs offer therefore interesting tools to reveal and/or analyze additional levels of regulation (Arnaud et al. 2014). The search for PTN target(s) should now be undergone and could lead to the unravelling of still unknown processes regulating the complete metabolic scheme of glycerolipids, governing in particular the balance between various lipid classes.

Conclusion

This chapter illustrates the strength of chemical genetics applied to the study of glycerolipid metabolism in plants, with small molecules used as flexible tools for a wide range of purposes, from functional *in vivo* dissection of some lipid intermediates (FAs, glycerolipid classes like MGDG, PC, etc.) to the search of novel genes involved in the regulation of the complete lipid metabolic scheme. Two main approaches can be considered, one from target to drug and the other one from drug to target. The possibility to hit biological processes that cannot be analyzed by classical genetics (lethal genes, multigenic families) can be a clear advantage to search for unknown mechanisms. For instance, the possibility to uncouple the lipid remodeling from other responses triggered by Pi starvation shows that specific mechanisms are orchestrated downstream a complex system sensing environmental cues, which could not be identified by another mean. Small molecules are then tools for research opening novel fields of investigation. In the case of reverse chemical genetics, the analysis of the transcriptome response to variable doses of compounds can point genes/proteins that were not previously positioned in the glycerolipid metabolic scheme or genes/proteins that indicate a connection between the lipid metabolic scheme and other biological processes. The screen of *Arabidopsis* mutants resisting to the developed drugs is an additional approach that can allow the discovery of novel actors. In the case of direct chemical genetics, the search for the target(s), using chemical derivatives grafted onto affinity matrices or via mutant screening strategies is also a way to advance knowledge significantly. The pace of chemical genetic studies is slow, and the risk of failure is not null, but combined with other approaches, the benefits can really lead to important breakthroughs.

Acknowledgements All authors are supported by a grant from Agence Nationale de la Recherche (ANR Reglisse). EM is also supported by the ANR DiaDomOil grant, the OCEANOMICS program from the French Ministry of Research and the Institut Carnot LISA (Lipides pour la santé et l'industrie).

References

Arnaud C, Clement M, Thibaud MC, Javot H, Chiarenza S, Delannoy E, Revol J, Soreau P, Balzergue S, Block MA, Marechal E, Desnos T, Nussaume L (2014) Identification of phosphatin, a drug alleviating phosphate starvation responses in Arabidopsis. Plant Physiol 166(3):1479–1491. doi:10.1104/pp. 114.248112

Awale M, Reymond JL (2014) A multi-fingerprint browser for the ZINC database. Nucleic Acids Res 42(Web Server issue):W234–W239. doi:10.1093/nar/gku379

Ben Mamoun C, Prigge ST, Vial H (2010) Targeting the lipid metabolic pathways for the treatment of malaria. Drug Dev Res 71(1):44–55. doi:10.1002/ddr.20347

Benning C, Beatty JT, Prince RC, Somerville CR (1993) The sulfolipid sulfoquinovosyldiacylglycerol is not required for photosynthetic electron transport in Rhodobacter sphaeroides but enhances growth under phosphate limitation. Proc Natl Acad Sci U S A 90(4):1561–1565

Blackwell HE, Zhao Y (2003) Chemical genetic approaches to plant biology. Plant Physiol 133(2):448–455. doi:10.1104/pp. 103.031138

Botte CY, Deligny M, Roccia A, Bonneau AL, Saidani N, Hardre H, Aci S, Yamaryo-Botte Y, Jouhet J, Dubots E, Loizeau K, Bastien O, Brehelin L, Joyard J, Cintrat JC, Falconet D, Block MA, Rousseau B, Lopez R, Marechal E (2011) Chemical inhibitors of monogalactosyldiacylglycerol synthases in Arabidopsis thaliana. Nat Chem Biol 7(11):834–842. doi:nchembio.658 [pii]

Botte CY, Dubar F, McFadden GI, Marechal E, Biot C (2012) Plasmodium falciparum apicoplast drugs: targets or off-targets? Chem Rev 112:1269–1283. doi:10.1021/cr200258w

Boudière L, Botte C, Saidani N, Lajoie M, Marion J, Bréhélin L, Yamaryo-Botté Y, Satiat-Jeunemaître B, Breton C, Girard-Egrot A, Bastien O, Jouhet J, Falconet D, Block MA, Maréchal E (2012) Galvestine-1, a novel chemical probe for the study of the glycerolipid homeostasis system in plant cells. Mol BioSyst 287:22367–22376

Boudiere L, Michaud M, Petroutsos D, Rebeille F, Falconet D, Bastien O, Roy S, Finazzi G, Rolland N, Jouhet J, Block MA, Marechal E (2014) Glycerolipids in photosynthesis: composition, synthesis and trafficking. Biochim Biophys Acta 1837(4):470–480. doi:10.1016/j.bbabio.2013.09.007

DeBolt S, Gutierrez R, Ehrhardt DW, Melo CV, Ross L, Cutler SR, Somerville C, Bonetta D (2007) Morlin, an inhibitor of cortical microtubule dynamics and cellulose synthase movement. Proc Natl Acad Sci U S A 104(14):5854–5859. doi:10.1073/pnas.0700789104

Drakakaki G, Robert S, Raikhel NV, Hicks GR (2009) Chemical dissection of endosomal pathways. Plant Signal Behav 4(1):57–62

Dubots E, Botté C, Boudière L, Yamaryo-Botté Y, Jouhet J, Maréchal E, Block MA (2012) Role of phosphatidic acid in plant galactolipid synthesis. Biochimie 94(1):86–93

Ganesan A (2008) The impact of natural products upon modern drug discovery. Curr Opin Chem Biol 12(3):306–317. doi:10.1016/j.cbpa.2008.03.016

Goodman CD, McFadden GI (2008) Fatty acid synthesis in protozoan parasites: unusual pathways and novel drug targets. Curr Pharm Des 14(9):901–916

Haggarty SJ, Mayer TU, Miyamoto DT, Fathi R, King RW, Mitchison TJ, Schreiber SL (2000) Dissecting cellular processes using small molecules: identification of colchicine-like, taxol-like and other small molecules that perturb mitosis. Chem Biol 7(4):275–286

Heath RJ, Rock CO (2004) Fatty acid biosynthesis as a target for novel antibacterials. Curr Opin Investig Drugs 5(2):146–153

Irwin JJ, Shoichet BK (2005) ZINC-a free database of commercially available compounds for virtual screening. J Chem Inf Model 45(1):177–182. doi:10.1021/ci049714+

Jarvis P, Dormann P, Peto CA, Lutes J, Benning C, Chory J (2000) Galactolipid deficiency and abnormal chloroplast development in the Arabidopsis MGD synthase 1 mutant. Proc Natl Acad Sci U S A 97(14):8175–8179

Kobayashi K, Awai K, Takamiya K, Ohta H (2004) Arabidopsis type B monogalactosyldiacylglycerol synthase genes are expressed during pollen tube growth and induced by phosphate starvation. Plant Physiol 134(2):640–648

Kobayashi K, Kondo M, Fukuda H, Nishimura M, Ohta H (2007) Galactolipid synthesis in chloroplast inner envelope is essential for proper thylakoid biogenesis, photosynthesis, and embryogenesis. Proc Natl Acad Sci U S A 104(43):17216–17221

Kobayashi K, Awai K, Nakamura M, Nagatani A, Masuda T, Ohta H (2009) Type-B monogalactosyldiacylglycerol synthases are involved in phosphate starvation-induced lipid remodeling, and are crucial for low-phosphate adaptation. Plant J 57(2):322–331

Koo AJ, Fulda M, Browse J, Ohlrogge JB (2005) Identification of a plastid acyl-acyl carrier protein synthetase in Arabidopsis and its role in the activation and elongation of exogenous fatty acids. Plant J 44(4):620–632. doi:10.1111/j.1365-313X.2005.02553.x

Lafanechere L (2008) Chemogenomics and cancer chemotherapy: cell-based assays to screen for small molecules that impair microtubule dynamics. Comb Chem High Throughput Screen 11(8):617–623

Lipinski CA, Lombardo F, Dominy BW, Feeney PJ (2001) Experimental and computational approaches to estimate solubility and permeability in drug discovery and development settings. Adv Drug Deliv Rev 46(1–3):3–26

Lorenc-Kukula K, Chaturvedi R, Roth M, Welti R, Shah J (2012) Biochemical and molecular-genetic characterization of SFD1's involvement in lipid metabolism and defense signaling. Front Plant Sci 3:26. doi:10.3389/fpls.2012.00026

Marechal E (2008) Chemogenomics: a discipline at the crossroad of high throughput technologies, biomarker research, combinatorial chemistry, genomics, cheminformatics, bioinformatics and artificial intelligence. Comb Chem High Throughput Screen 11(8):582–582

Maréchal E, Roy S, Lafanechère L (2011) Chemogenomics and chemical genetics: a user's introduction for biologists, chemists and informaticians. Springer, Berlin

Mayer TU, Kapoor TM, Haggarty SJ, King RW, Schreiber SL, Mitchison TJ (1999) Small molecule inhibitor of mitotic spindle bipolarity identified in a phenotype-based screen. Science 286(5441):971–974

Minnikin DE, Abdolrahimzadeh H, Baddiley J (1974) Replacement of acidic phosphates by acidic glycolipids in Pseudomonas diminuta. Nature 249(454):268–269

Misson J, Thibaud MC, Bechtold N, Raghothama K, Nussaume L (2004) Transcriptional regulation and functional properties of Arabidopsis Pht1;4, a high affinity transporter contributing greatly to phosphate uptake in phosphate deprived plants. Plant Mol Biol 55(5):727–741. doi:10.1007/s11103-004-1965-5

Misson J, Raghothama KG, Jain A, Jouhet J, Block MA, Bligny R, Ortet P, Creff A, Somerville S, Rolland N, Doumas P, Nacry P, Herrerra-Estrella L, Nussaume L, Thibaud MC (2005) A genome-wide transcriptional analysis using Arabidopsis thaliana Affymetrix gene chips determined plant responses to phosphate deprivation. Proc Natl Acad Sci U S A 102(33):11934–11939

Monika, Kour J, Singh K (2013) Virtual screening using the ligand ZINC database for novel lipoxygenase-3 inhibitors. Bioinformation 9(11):583–587. doi:10.6026/97320630009583

Monovich L, Mugrage B, Quadros E, Toscano K, Tommasi R, LaVoie S, Liu E, Du Z, LaSala D, Boyar W, Steed P (2007) Optimization of halopemide for phospholipase D2 inhibition. Bioorg Med Chem Lett 17(8):2310–2311. doi:10.1016/j.bmcl.2007.01.059

Nakamura Y (2013) Phosphate starvation and membrane lipid remodeling in seed plants. Prog Lipid Res 52(1):43–50. doi:10.1016/j.plipres.2012.07.002

Nakamura Y, Kobayashi K, Ohta H (2009) Activation of galactolipid biosynthesis in development of pistils and pollen tubes. Plant Physiol Biochem 47(6):535–539

Nishida I, Kawaguchi A, Yamada M (1986) Effect of thiolactomycin on the individual enzymes of the fatty acid synthase system in Escherichia coli. J Biochem 99(5):1447–1454

Nishiyama Y, Hardre-Lienard H, Miras S, Miege C, Block MA, Revah F, Joyard J, Marechal E (2003) Refolding from denatured inclusion bodies, purification to homogeneity and simplified assay of MGDG synthases from land plants. Prot Expr Purif 31(1):79–87

Petroutsos D, Amiar S, Abida H, Dolch LJ, Bastien O, Rebeille F, Jouhet J, Falconet D, Block MA, McFadden GI, Bowler C, Botte C, Marechal E (2014) Evolution of galactoglycerolipid biosynthetic pathways – from cyanobacteria to primary plastids and from primary to secondary plastids. Prog Lipid Res 54:68–85. doi:10.1016/j.plipres.2014.02.001

Raikhel N, Pirrung M (2005) Adding precision tools to the plant biologists' toolbox with chemical genomics. Plant Physiol 138(2):563–564. doi:10.1104/pp. 104.900155

Robert S, Chary SN, Drakakaki G, Li S, Yang Z, Raikhel NV, Hicks GR (2008) Endosidin1 defines a compartment involved in endocytosis of the brassinosteroid receptor BRI1 and the auxin transporters PIN2 and AUX1. Proc Natl Acad Sci U S A 105(24):8464–8469. doi:10.1073/pnas.0711650105

Rubio V, Linhares F, Solano R, Martin AC, Iglesias J, Leyva A, Paz-Ares J (2001) A conserved MYB transcription factor involved in phosphate starvation signaling both in vascular plants and in unicellular algae. Genes Dev 15(16):2122–2133. doi:10.1101/gad.204401

Sadhukhan A, Sahoo L, Panda SK (2012) Chemical genomics in plant biology. Indian J Biochem Biophys 49(3):143–154

Saidani N, Grando D, Valadie H, Bastien O, Marechal E (2009) Potential and limits of in silico target discovery – case study of the search for new antimalarial chemotherapeutic targets. Infect Genet Evol 9(3):359–367

Saidani N, Botte C, Deligny M, Bonneau AL, Reader J, Lasselin R, Merer G, Niepceron A, Brossier F, Cintrat JC, Rousseau B, Birkholtz L, Cesbron-Delauw MF, Dubremetz JF, Mercier C, Vial H, Lopez R, Marechal E (2014) Discovery of compounds blocking the proliferation of Toxoplasma gondii and Plasmodium falciparum in a chemical space based on piperidinylbenzimidazolone analogs. Antimicrob Agents Chemother 58(5):2586–2597

Scott SA, Selvy PE, Buck JR, Cho HP, Criswell TL, Thomas AL, Armstrong MD, Arteaga CL, Lindsley CW, Brown HA (2009) Design of isoform-selective phospholipase D inhibitors that modulate cancer cell invasiveness. Nat Chem Biol 5(2):108–117. doi:10.1038/nchembio.140

Selvy PE, Lavieri RR, Lindsley CW, Brown HA (2011) Phospholipase D: enzymology, functionality, and chemical modulation. Chem Rev 111(10):6064–6119. doi:10.1021/cr200296t

Thibaud MC, Arrighi JF, Bayle V, Chiarenza S, Creff A, Bustos R, Paz-Ares J, Poirier Y, Nussaume L (2010) Dissection of local and systemic transcriptional responses to phosphate starvation in Arabidopsis. Plant J 64(5):775–789. doi:10.1111/j.1365-313X.2010.04375.x

Walters WP (2012) Going further than Lipinski's rule in drug design. Exp Opin Drug Discov 7(2):99–107. doi:10.1517/17460441.2012.648612

Xiang S, Callaghan MM, Watson KG, Tong L (2009) A different mechanism for the inhibition of the carboxyltransferase domain of acetyl-coenzyme A carboxylase by tepraloxydim. Proc Natl Acad Sci U S A 106(49):20723–20727. doi:10.1073/pnas.0908431106

Yoneda A, Higaki T, Kutsuna N, Kondo Y, Osada H, Hasezawa S, Matsui M (2007) Chemical genetic screening identifies a novel inhibitor of parallel alignment of cortical microtubules and cellulose microfibrils. Plant Cell Physiol 48(10):1393–1403. doi:10.1093/pcp/pcm120

Zhao Y, Chow TF, Puckrin RS, Alfred SE, Korir AK, Larive CK, Cutler SR (2007) Chemical genetic interrogation of natural variation uncovers a molecule that is glycoactivated. Nat Chem Biol 3(11):716–721. doi:10.1038/nchembio.2007.32

Part II
Lipids in Development and Signaling

Chapter 8
Triacylglycerol Accumulation in Photosynthetic Cells in Plants and Algae

Zhi-Yan Du and Christoph Benning

Abstract Plant and algal oils are some of the most energy-dense renewable compounds provided by nature. Triacylglycerols (TAGs) are the major constituent of plant oils, which can be converted into fatty acid methyl esters commonly known as biodiesel. As one of the most efficient producers of TAGs, photosynthetic microalgae have attracted substantial interest for renewable fuel production. Currently, the big challenge of microalgae based TAGs for biofuels is their high cost compared to fossil fuels. A conundrum is that microalgae accumulate large amounts of TAGs only during stress conditions such as nutrient deprivation and temperature stress, which inevitably will inhibit growth. Thus, a better understanding of why and how microalgae induce TAG biosynthesis under stress conditions would allow the development of engineered microalgae with increased TAG production during conditions optimal for growth. Land plants also synthesize TAGs during stresses and we will compare new findings on environmental stress-induced TAG accumulation in plants and microalgae especially in the well-characterized model alga *Chlamydomonas reinhardtii* and a biotechnologically relevant genus Nannochloropsis.

Keywords Nutrient deprivation • Photosynthesis • Lipid droplet • Lipid remodeling • Lipid metabolism

Introduction

During the past decade, basic research on lipid metabolism in microalgae and plants has greatly benefitted from funding available for the exploration of sustainable, domestic production of liquid fuels. As fossil carbon-derived fuels are diminishing and will eventually be depleted by the increasing demand of modern societies, the search for sustainable sources of energy has become more urgent. As a result of the burning of fossil fuels, carbon dioxide (CO_2) is released to the atmosphere, which is

Z.-Y. Du • C. Benning (✉)
Department of Biochemistry and Molecular Biology, Michigan State University, East Lansing, MI 48824, USA
e-mail: benning@cns.msu.edu

Y. Nakamura, Y. Li-Beisson (eds.), *Lipids in Plant and Algae Development*, Subcellular Biochemistry 86, DOI 10.1007/978-3-319-25979-6_8

a greenhouse gas and a likely contributor to global warming as its atmospheric concentration has steadily increased since the onset of the industrial revolution (Cheah et al. 2014; Martinez-Boti et al. 2015). Therefore, alternative energy sources should not only be reliable and renewable, but also not further contribute to the increase of atmospheric CO_2. Biofuel products from photosynthetic organisms can potentially meet this challenge. By converting sunlight into chemical energy, photosynthetic organisms such as plants and algae produce biomass and storage compounds, i.e. carbohydrates and TAGs, which can be converted to liquid transportation fuels equivalent to fossil fuels. A big advantage of biofuel is that photosynthetic organisms consume CO_2 from the atmosphere or directly from anthropogenic sources along with the conversion of solar energy, resulting in greenhouse gas reduction as they displace fossil fuels (Barber 2009; Merchant et al. 2012; Ohlrogge and Chapman 2011). Currently, two major forms of biofuels, ethanol and biodiesel, are available in the market and substitute a small portion of global fossil fuels consumed annually (Ohlrogge and Chapman 2011).

The focus here is on biodiesel, one of the commonly used biofuels which is currently primarily produced from edible plant vegetable oils obtained from agricultural crops such as soybean and oil palm (Durrett et al. 2008). Compared with bioethanol, biodiesel has several advantages. First, plant oils as fuel feedstocks have a higher energy density than carbohydrates; biodiesel has a 25 % higher energy content per volume than ethanol (Durrett et al. 2008). Secondly, the net positive energy balance for biodiesel production is much higher than that for ethanol. Although the exact numbers are still debated, by one estimation, biodiesel from soy oil yields 93 % more energy over the total energy input into its production, whereas ethanol from corn starch yields only 25 % more, including extra energy credits from co-products, e.g. for animal feed (Hill et al. 2006). Thirdly, greenhouse gas emissions are reduced 41 % by the production and combustion of biodiesel compared to the fossil fuels it replaced, whereas the number for ethanol is only 12 % (Hill et al. 2006). Biodiesel also generates less air and chemical pollutants derived from pesticides and fertilizers per net energy gain than ethanol (Hill et al. 2006). Fourthly, with respect to transportation fuel, biodiesel can be used directly for diesel engines which are 30 % more efficient than gasoline engines, whereas ethanol has to be blended with conventional petrol/gasoline before it is used. In addition, ethanol needs to be stored separately before use because it can lead to corrosion of pipelines, and it is not practical for certain applications, such as jet fuel or heavy vehicles (Dismukes et al. 2008; Durrett et al. 2008). All these benefits have motivated research into the biosynthesis of TAGs and its regulation in algae and plants, and substantial effort has been spent on the engineering of TAG quality.

Fatty Acid and Triacylglycerol Production in Non-seed Tissues of Plants

Plant oils, primarily TAGs, are abundantly stored in cytosolic lipid droplets of oil seeds, and have been traditionally a common source for edible vegetable oil and biodiesel feedstocks. For example, in 2005, 1.5 % of the soybean harvest in the USA produced 256 million liters of biodiesel, providing 0.09 % of the total USA diesel consumption (US Department of Energy 2007). During the same year, biodiesel contributed ~1.6 % of the EU diesel usage (Commission of the European Communities 2007) and ~0.21 % of that in the USA (US Department of Energy 2007). By 2030, worldwide demand for edible vegetable oils is expected to double due to an increasing population (Bruinsma 2003; Chapman and Ohlrogge 2012). One approach to avoid a direct competition with food supplies has been the introduction of dedicated oleaginous biofuel crops producing non-edible oils such as *Sapindus mukorossi* (soap-nut tree) and *Jatropha curcas* (physic-nut tree) which also can tolerate marginal agricultural land less suitable for the common agricultural crops (Abdulla et al. 2011; Chhetri et al. 2008).

Genetic Engineering of Oil Accumulation in Non-seed Plant Tissues

Another new strategy for non-food oil feedstocks is to produce oil in non-seed plant tissues such as leaves and stems in high biomass crops (Chapman et al. 2013; Durrett et al. 2008). Plant oils can be easily extracted from vegetative tissues, and the residual lignocellulosic biomass can be converted to biofuel feedstocks by deconstruction followed by fermentation or can be burned directly to produce bioelectricity for electric vehicles (Ohlrogge et al. 2009; Vanhercke et al. 2014). If harvestable vegetative tissues accumulate 10 % TAGs on a dry weight basis, the energy yield from the crop would be increased by at least 30 % (Ohlrogge and Chapman 2011).

Most plant leaves already contain ~5 % fatty acids by dry weight in the form of polar membrane lipids that are not easy to use (Yang and Ohlrogge 2009). Neutral lipids such as TAGs can in principle be synthesized in most plant cells, although they primarily accumulate in plant seeds, and only minor amounts are found in leaves, stems, and roots (Yang and Ohlrogge 2009). However, there are intriguing exceptions. Examples include copious amounts of oil in the fruit mesocarp of olive (*Olea europaea*), avocado (*Persea americana*), and oil palm (*Elaeis guineensis*) (Ross et al. 1993; Tranbarger et al. 2011), in tubers of nutsedge (*Cyperus esculentus*) (Stoller and Weber 1975; Zhang et al. 1996), and in stem tissues of Mongolian oilwood (*Tetraena mongolica*) (Wang et al. 2007), suggesting that vegetative-tissue based oil could be a realistic approach for oil production.

In fact, numerous attempts towards the engineering TAG accumulation in vegetative tissues have been carried out over the past decade with various strategies that are generally trying to optimize the flux of carbon into TAG by overexpressing seed transcription factors, increasing TAG/fatty acid synthesis or blocking TAG turnover. For example, transcription factors that normally control plant oil biosynthesis in developing embryos as summarized in recent reviews (Bates et al. 2013; Baud and Lepiniec 2010; Santos-Mendoza et al. 2008) have been explored to produce oil in non-seed tissues by their ectopic production, including *WRINKLED1* (*WRI1*) (Cernac and Benning 2004; Sanjaya et al. 2011), *ABSCISIC ACID INSENSITIVE4* (*ABI4*) (Yang et al. 2011), *LEAFY COTYLEDON1* (*LEC1*) or *LEC2* (Mu et al. 2008; Santos-Mendoza et al. 2005; Stone et al. 2008). In developing embryos of plants, a fraction of TAG is synthesized through acylation of diacylglycerol (DAG) by diacylglycerol acyltransferase (DGAT) (Cases et al. 1998; Lardizabal et al. 2001), and ectopic expression of DGAT or monoacylglycerol acyltransferase (MGAT) increases TAG content in leaves (Andrianov et al. 2010; Bouvier-Nave et al. 2000; Petrie et al. 2012; Sanjaya et al. 2013). In potato tubers, overexpression of the Arabidopsis acetyl-CoA carboxylase leads to increases in TAG and fatty acid synthesis (Klaus et al. 2004). By reducing TAG turnover in a *cgi-58* knockout mutant, mature Arabidopsis leaves show increases in TAG accumulation, while seed storage, germination and plant growth are not affected (James et al. 2010). TAG accumulation has also been observed in leaves and roots in an Arabidopsis *trigalactosyldiacylglycerol1* (*tgd1*) mutant which is disrupted in lipid transfer between the endoplasmic reticulum (ER) and the chloroplast (Xu et al. 2005). Another intriguing Arabidopsis mutant designated *pickle* has embryo-like roots that produce seed storage compounds including oil (Ogas et al. 1997, 1999).

These studies have achieved the goal of engineering oil in non-seed tissues. However, the levels of oil production were initially still far below the 10 % dry weight benchmark most likely because of their single-gene strategies. More recent studies have attempted to further increase TAG content in non-seed tissues by manipulating the expression of multiple genes from different pathways. Various gene combinations have been tried and examined in transgenic plants such as Arabidopsis and tobacco (*Nicotiana tabacum*). Examples include the overexpression of *LEC2* in the fatty acid breakdown mutant *COMATOSE* (*cts2*) (Slocombe et al. 2009), the overexpression of *WRI1* in an Arabidopsis RNAi line designated *AGP*RNAi that reduces the expression of *APS1*, a gene encoding the ADP-glucose pyrophosphorylase (Sanjaya et al. 2011), the coexpression of DGAT1 and oil-body protein oleosin that increases both TAG and leaf biomass (Winichayakul et al. 2013), the disruption of a TAG lipase gene *SUGAR-DEPENDENT1* (*SDP1*) or *PEROXISOMAL TRANSPORTER1* (*PXA1*) in the *tgd1* mutant (Fan et al. 2014), and the coexpression of *WRI1* and *DGAT1* in tobacco (Vanhercke et al. 2013). Compared to these two-gene attempts, three-gene combinations were even more successful in enhancing TAG accumulation. In Arabidopsis, the coexpression of *OLEOSIN1* and the gene encoding phospholipid:diacylglycerol acyltransferase 1 (*PDAT1*) in the *tgd1* mutant boosted leaf TAG content to ~9 % of the dry weight (Fan et al. 2013). In contrast, the coexpression of *WRI1* and *DGAT1* in the lipase mutant sugar-dependent1 (*sdp1*) resulted in TAG accumulation in leaves, stems and roots ranging from 5 to 8 % of dry weight but along with a ~20 % reduction in leaf biomass (Kelly et al. 2013). By the provision of 3 % (w/v) sucrose, TAG content in the roots of these transgenic plants could be further increased up to 17 % of dry weight (Kelly et al. 2013). Another report showed over 10 % TAG by dry weight in mature tobacco leaves and ~15 % TAG in stems and roots by the coexpression of *WRI1*, *DGAT1* and *OLEOSIN*, three genes involved in different aspects of TAG synthesis (Vanhercke et al. 2014).

Environmental Stresses Induce Oil Accumulation in Plant Vegetative Tissues

Besides manipulation of gene expression, various environmental stresses, such as ozone fumigation, freezing and desiccation, appear to stimulate plant oil accumulation in vegetative tissues (Fig. 8.1) (El-Hafid et al. 1989; Moellering et al. 2010; Navari-Izzo and Rascio 1999; Sakaki et al. 1985). For example, after approximately 6 h fumigation with ozone (0.5 μL/L), TAG accumulation in spinach leaves reached a maximum, whereas three polar lipids, phosphatidylcholine (PtdCho), mono- and digalactosyldiacylglycerol (MGDG and DGDG) decreased strongly during the same period (Sakaki et al. 1985, 1990c). Further analysis such as fatty acid profiling and pulse-chase acetate labeling has revealed that these TAGs are derived from diacylglycerol (DAG) and free fatty acids (FFA) from MGDG due to the activity of a galactolipid:galactolipid galactosyltransferase (GGGT) (Sakaki et al. 1990b, c). Subsequent *in vitro* enzyme activity assays of spinach GGGT suggested its activation by FFA and divalent cations such as Mg^{2+}, Mn^{2+} and Ca^{2+} (Sakaki et al. 1990a). However, the significance of the TAG accumulation in relation to GGGT activation in spinach leaves upon ozone treatment is still not clear.

Stress-induced TAG accumulation has also been observed in Arabidopsis during freezing (Moellering and Benning 2011; Moellering et al. 2010). Studies on Arabidopsis *SENSITIVE TO FREEZING 2* (*SFR2*) (Moellering et al. 2010), a gene encoding a protein annotated as a glycosyl hydrolase family 1 protein at the outer chloroplast envelope membrane (Fourrier et al. 2008; Thorlby et al. 2004), suggested that SFR2 likely is the GGGT in plants and participates in the protection of chloroplast under freezing stress due to its activity causing the formation of oligogalactolipids and DAG by processive transfer of galactosyl moieties from MGDG onto an galactolipid acceptor (Moellering et al. 2010). A more recent study confirmed that SFR2 acts solely as a glycosyltransferase rather than a glycosyl hydrolase and provided insights into its reaction mechanism through structure function studies assisted by a structural homology model (Roston et al. 2014). Under freezing stress, Arabidopsis SFR2 converts nonbilayer-MGDG to bilayer-forming membrane lipids such as tri- and tetra-galactosyldiacylglycerol (TGDG and TeGDG, respectively), as well as DAG that is further acylated to TAG (Moellering et al. 2010). In case of ozone fumigation of spinach leaves, the acyl groups are most

Fig. 8.1 Triacylglycerol accumulation occurs in plant vegetative tissues under environmental stresses. The plant model shown in the figure represents plants described in section "Environmental stresses induce oil accumulation in plant vegetative tissues", e.g. *Arabidopsis thaliana*, *Spinacia oleracea* L. (spinach) and *Gossypium hirsutum* (cotton) plants. The *white* snow flake and the *yellow* leaf indicate freezing condition and leaf senescence, respectively. *TAG* triacylglycerol, *-N* nitrogen deprived, *salt* high salt

likely derived from MGDG hydrolysis (Moellering et al. 2010; Sakaki et al. 1990b, c). The remodeling of membrane lipids by SFR2 upon freezing stress, which leads to severe dehydration of the cell as ice forms first in the apoplast, reduces the tendency of the formation of inter-bilayer hexagonal II phase, increases stabilization of the envelope membranes and, therefore, enhances freezing tolerance (Moellering and Benning 2011; Moellering et al. 2010; Roston et al. 2014). Furthermore, TAG accumulation in Arabidopsis leaves can contribute to the removal of excess membrane lipids like MGDG from the envelope membranes as the organelle shrinks in response to dehydration by the combined activity of TAG-biosynthetic enzymes and SFR2 during freezing treatment or general osmotic stress (Moellering and Benning 2011; Moellering et al. 2010).

Under water-deficit stress conditions, a decrease in MGDG and increase in TAG levels has been observed in a variety of plants such as desiccation-tolerant plants *Craterostigma plantagineum* (blue carpet), *Lindernia brevidens* and *Ramonda serbica* (Serbian-phoenix flower), desiccation-sensitive plants Arabidopsis and *Lindernia subracemosa*, as well as the crops *Gossypium hirsutum* (cotton), *Triticum aestivum* (wheat) and *Zea mays* (maize), indicating an important role of lipid remodeling in plant adaptation to desiccation stress (El-Hafid et al. 1989; Gasulla et al. 2013; Navari-Izzo and Rascio 1999). For instance, a recent study has shown that TAGs in *C. plantagineum* increase from 0.146 to 3.11 nmol mg^{-1} of dry cell weight in desiccated leaves and decrease again following rehydration (Gasulla et al. 2013). Analysis of molecular species has revealed that these TAGs are synthesized from DAG derived either directly from MGDG hydrolysis or due to the activity of SFR2 that converts multiple MGDGs into TGDG/TeGDG and DAG. Simultaneously, a fraction of MGDG is converted into DGDG by UDP-Gal-dependent DGDG synthases DGD1/DGD2 (Gasulla et al. 2013). This conversion of MGDG to DGDG/TGDG/TeGDG and TAG is believed to contribute to the stabilization of membranes during desiccation stress. In addition, TAG accumulation can be induced (or TAG degradation delayed) in Arabidopsis seedlings by the treatment with abscisic acid (Yang et al. 2011), an important plant stress hormone also involved in responses to freezing and desiccation.

TAG accumulation is also observed following senescence/dark treatment in Arabidopsis leaves (Slocombe et al. 2009), which is interpreted as the sequestration of FFA derived from galactolipids (Kaup et al. 2002). During leaf senescence or environmental stresses, disintegration of thylakoid membranes, as well as degradation of chlorophyll and galactolipids, results in the accumulation of toxic intermediates such as free phytol and FFA. In Arabidopsis, *PHYTYL ESTER SYNTHASE1 (PES1)* and *PES2* encode two acyltransferases (of the esterase/lipase/thioesterase family) performing both phytyl ester synthesis and diacylglycerol acyltransferase activities with broad substrate specificities (Lippold et al. 2012). During developmental senescence or nitrogen deprivation induced senescence, TAG and phytyl esters accumulate in the wild type, whereas the *pes1 pes2* double mutant contains ~30 % less TAG than the wild type but much higher free phytol. This observation indicates that PES1 and PES2 help avoid the accumulation of toxic products of thylakoid membrane degradation during leaf senescence and nitrogen deprivation, by the removal of free phytol and FFA in the form of phytyl esters and eventually sequestration of FFA into TAG (Lippold et al. 2012). TAG accumulation has also been reported by another group using nitrogen-deprived Arabidopsis seedlings (Yang et al. 2011). These findings are particularly intriguing because nitrogen deficiency is currently widely used in laboratory and aquaculture settings to induce TAG accumulation in microalgae, another attractive source of sustainable feedstocks for bioenergy.

Microalgae Accumulate Triacylglycerol upon Different Stresses

Microalgae are eukaryotic photosynthetic microorganisms that are masters at using sunlight, CO_2 and water to produce biomass (Georgianna and Mayfield 2012; Hu et al. 2008). Under optimal growth conditions, microalgae are very efficient in the utilization of solar energy for the production biochemical compounds such as starch, cellulose and other carbohydrates which can readily be used as feedstocks for bioethanol production due to the absence, or low content of lignin from algal biomass compared to biomass derived from land plants (John et al. 2011; Jones and Mayfield 2012). When placed under stress conditions, many species of microalgae produce large amounts of neutral lipids typically in the form of TAGs as storage products for carbon and energy, and they are therefore referred to as oleaginous microalgae (Chisti 2007; Hu et al. 2008). As this typically happens in photosynthetically active cells, TAG accumulation in microalgae is conceptually more similar to stress-induced TAG accumulation in vegetative tissues of plants than the related process in developing plant seeds. Therefore, gaining an understanding of TAG accumulation in microalgae may also provide important insights for the engineering of TAG accumulation in vegetative tissues of plants.

Algal TAGs usually have acyl chains with 16 or 18 carbons esterified to the glycerol (Fig. 8.2) (Liu et al. 2013). These fatty acyl chains are chemically similar to diesel fuel components that typically have 10–15 carbons per molecule, and alga-derived biodiesel is directly compatible with diesel engines (ASTM 2002; Durrett et al. 2008). Thus, oleaginous microalgae can theoretically be used for producing both biodiesel derived from their oil and bioethanol/bioelectricity derived from the biomass left after oil extraction maximizing the overall energy yield (Jones and Mayfield 2012; Ohlrogge et al. 2009).

Moreover, using microalgae as feedstocks for biofuel products has potential advantages as compared to current biofuel crops such as corn and oil palm. Firstly, microalgae accumulate a substantial amount of TAG, commonly 20–50 % of dry weight (Chisti 2007; Hu et al. 2008). Secondly, microalgae grow fast (usually 1–3 doublings per day) and they are more efficient than terrestrial plants in the conversion of solar energy to biomass (Chisti 2013; Stephenson et al. 2011). Thirdly, in principle microalgae require less land area for the same yield of biofuel compared with terrestrial crops and their cultivation can utilize marginal lands such as deserts and saline lake beds that cannot be used for conventional agriculture (Georgianna and Mayfield 2012; Stephenson et al. 2011). Microalgae can be incubated in enclosed photobioreactors throughout the year independent of seasons (Georgianna and Mayfield 2012; John et al. 2011), and can use nutrient-rich wastewater to meet the relatively high demand for water (Venkata Mohan et al. 2015). Marine microalgae cultivation can utilize abundant sea water, but requires a costal location of the production site (Stephenson et al. 2011). Thus, cultivation of microalgae has the potential to minimize or avoid the competition with food crops for arable land and other crop-based agricultural production schemes for biofuels. Fourthly, microalgae produce a variety of valuable co-products or by-products such as biopolymers, pro-

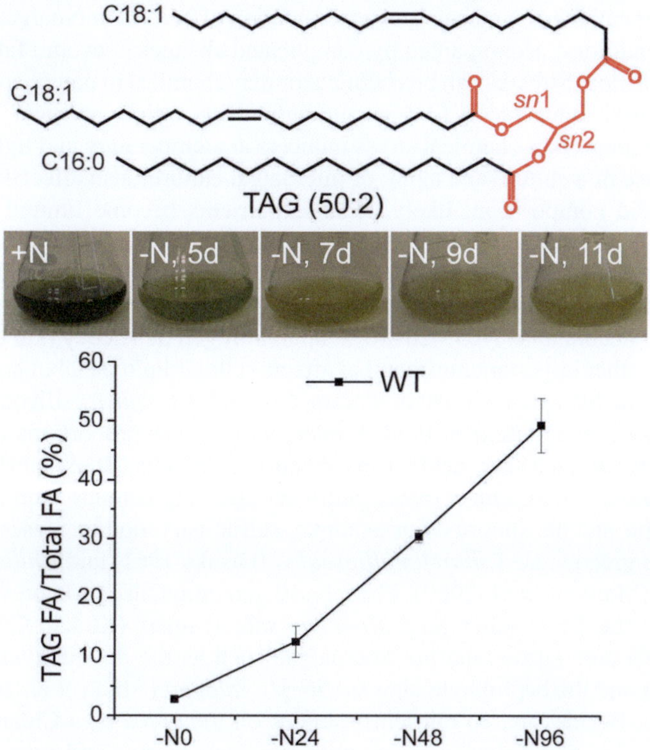

Fig. 8.2 Triacylglycerol is accumulated in nutrient-limited cultures of *Chlamydomonas reinhardtii*. Triacylglycerol (TAG) consists of three acyl chains (in *black*) that are esterified to a glycerol (in *red*). Cells of wild-type Chlamydomonas strain dw 15-1 were incubated in Tris-acetate-phosphate medium supplemented with or without 10 mM NH_4Cl. The ratio of fatty acids (FA) from TAG over total-lipid fatty acids was calculated in Chlamydomonas samples following indicated nitrogen deprivation. *+N* nitrogen replete, *−N* nitrogen deprived. Averages of three independent measurements are given. *Error bars* indicate SD

teins, animal feed and fertilizer (Brennan and Owende 2010; Hu et al. 2008). All these advantages promise a great potential for microalgae-based production of fuels and other products, a promise that, however, still needs to be realized in cost effective ways. But the promise of algae as sustainable feed stocks certainly has globally "fueled" research to explore mechanisms of TAG accumulation in microalgae and to develop algae-based production schemes (Liu and Benning 2013).

Factors Affecting Triacylglycerol Accumulation in Microalgae

Oleaginous microalgae produce only small amounts of TAG during optimal growth or under favorable environmental conditions, under which polar membrane lipids (5–20 % of dry weight) are generally the major lipid compounds (Hu et al. 2008).

As already mentioned, synthesis and accumulation of TAG in microalgae is induced by stress conditions, accompanied by complicated changes in overall fatty acid and lipid composition. Stresses can be chemical or non-chemical in nature and the major chemical-based inducers of TAG accumulation are various nutrient limitations, whereas the major non-chemical stress inducers are temperature and light intensity. Growth phase of a culture and aging of microalgal cultures also affect TAG content and fatty acid composition, likely because nutrients become limited and toxins accumulate as the cultures enter stationary phase.

Nitrogen deficiency is the most frequently studied condition inducing TAG accumulation in different algae (Hu et al. 2008). Microalgae including green algae and diatoms accumulate TAG ~20–50 % upon nitrogen deficiency (Hu et al. 2008). Silicon is another important nutrient that affects cellular lipid metabolism especially in diatoms. In the brackish-water diatom *Cyclotella cryptica*, silicon deficiency induces TAG accumulation in lipid droplets with higher proportions of saturated and mono-unsaturated fatty acids over silicon-replete cells (Roessler 1988; Traller and Hildebrand 2013). Other macro nutrients affecting cellular lipid metabolism include sulfur and phosphorus. For example, sulfur starvation increases the neutral lipids in the green algae *Chlorella ellipsoidea* (Otsuka 1961) and *Chlamydomonas reinhardtii* (Matthew et al. 2009). Phosphorus starvation also promotes TAG accumulation in the fresh water alga *Monodus subterraneus* (Khozin-Goldberg and Cohen 2006) and various marine microalgae such as the diatom *Phaeodactylum tricornutum* and the haptophyte alga *Isochrysis galbana* (Abida et al. 2015; Reitan et al. 1994). Besides macro nutrients, studies on the green alga Chlamydomonas have shown that deficiency in micro nutrients such as zinc and iron also induce TAG accumulation (Kropat et al. 2011; Urzica et al. 2013). In addition, drugs such as Brefeldin A (an ER-stress inducer) have been found to rapidly stimulate TAG accumulation in Chlamydomonas and the freshwater alga *Chlorella vulgaris* (Kim et al. 2013).

Many studies have reported that non-chemical based stresses lead to the formation of TAG in microalgae including unfavorable temperature, light intensity, high salinity and dehydration. For example, increased temperature results in the elevation of lipid content in the freshwater phytoflagellate *Ochromonas danica* (Aaronson 1973), the marine alga *Nannochloropsis salina* (Boussiba et al. 1987) and Chlamydomonas (Hemme et al. 2014). High light intensity increases neutral storage lipid content, mainly TAGs, accompanied by a decrease in total polar lipids (Brown et al. 1996; Khotimchenko and Yakovleva 2005; Napolitano 1994). Furthermore, TAG accumulates in Chlamydomonas during hypoxia in darkness or during extended (24 h) darkness alone (Hemschemeier et al. 2013). Synthesis of TAG can also be induced by high salinity as seen in the green algae *Dunaliella salina* (Takagi et al. 2006) and Chlamydomonas (Siaut et al. 2011), or by dehydration during illumination in the green alga *Chlorella kessleri* (Shiratake et al. 2013).

When reaching the stationary phase, some microalgae such as the green alga *Parietochloris incisa* and the marine dinoflagellate *Gymnodinium* sp. have shown an increase in TAG content (Bigogno et al. 2002; Mansour et al. 2003), which is probably a consequence of nutrient depletion or accumulation of toxic metabolic products.

Culture aging or senescence during prolonged stationary phase also affects lipid metabolism in microalgae. For example, the total lipid content increases along with culture aging in the green alga *Chlorococcum macrostigma* (Collins and Kalnins 1969), and diatoms such as *Thalassiosira fluviatillis* (Conover 1975) and *Coscinodiscus eccentricus* (Pugh 1971). These findings indicate possible roles of TAG accumulation in response to various stimuli, and a better understanding of the mechanism of TAG induction during stress conditions will provide strategies for the engineering of TAG accumulation and TAG quality under conditions favorable for biomass accumulation of microalgae.

Chlamydomonas as a Reference Microalgae to Answer Questions about TAG Accumulation

The unicellular green alga Chlamydomonas traditionally has been used as a model for studies of photosynthesis (Rochaix 1995) or flagella biogenesis (Harris 2001), and has recently been widely adopted as a reference organism for algal TAG metabolism research. Reasons are its simple life cycle and well-developed genetic tools and techniques (Liu and Benning 2013; Merchant et al. 2012). As summarized above, Chlamydomonas cells accumulate TAGs upon various unfavorable conditions such as nutrient deficiency and non-chemical stresses (Liu and Benning 2013; Merchant et al. 2012). Nitrogen deprivation is frequently used to induce TAG accumulation in Chlamydomonas (Fig. 8.2), and will lead to a cessation of cell division and eventually to a cellular state known as cellular quiescence (Tsai et al. 2014).

To understand the mechanism of TAG accumulation in Chlamydomonas, several studies have been carried out to identify key genes by reverse and forward genetic approaches (Liu and Benning 2013; Merchant et al. 2012). For instance, forward genetic screening using insertional mutagenesis (Khozin-Goldberg and Cohen 2011; Li et al. 2012; Merchant et al. 2012; Zhang et al. 2014), deep transcriptome analysis by RNA sequencing (Blaby et al. 2013; Hemschemeier et al. 2013; Juergens et al. 2015; Miller et al. 2010; Park et al. 2015; Schmollinger et al. 2014; Tsai et al. 2014), and proteomics by mass spectrometry (Hemme et al. 2014; Schmollinger et al. 2014; Wang et al. 2009) using wild-type Chlamydomonas or mutant strains have identified genes involved in TAG accumulation and its regulation. Several lipid droplet-focused proteomic studies have reported the presence of a major lipid droplet protein MLDP, which is considered a functional equivalent of plant oleosins (James et al. 2011; Moellering and Benning 2010; Nguyen et al. 2011). Reverse genetic screening based on the analysis of orthologs of characterized genes from yeast, animals and plants have provided candidates for genes involved in TAG biosynthesis (Khozin-Goldberg and Cohen 2011; Merchant et al. 2012; Riekhof and Benning 2009). Previous studies on other eukaryotes such as yeast and Arabidopsis identified two key enzyme families in TAG synthesis, DGATs and PDATs, which are also found in Chlamydomonas (Merchant et al. 2012). Currently, six genes encoding DGATs from two families, type one DGAT and type two DGTT, are known in Chlamydomonas including *DGAT1* and *DGTT1*-to-*DGTT5* (Boyle et al. 2012; La Russa et al. 2012; Miller et al. 2010). In contrast, only one *PDAT* has been described for Chlamydomonas, which participates in membrane lipid turnover and TAG synthesis (Boyle et al. 2012; Yoon et al. 2012).

A recent mutant screening for Chlamydomonas low TAG mutants led to the discovery of a galactoglycerolipid lipase designated PGD1 (PLASTID GALACTOLIPID DEGRADATION 1), and the respective *pgd1* null-mutant provides a tool to experimentally test the role of TAG accumulation following nitrogen deprivation (Fig. 8.3) (Li et al. 2012). A classic hypothesis is that *de novo* TAG synthesis serves as an electron sink to sequester excess electrons from the photosynthetic electron transport chain, thereby counteracting its possible overreduction which can lead to the formation of harmful reactive oxygen species (ROS) at photosystem I through the Mehler reaction (Hu et al. 2008). The study of the *pdg1* mutant supports the hypothesis that

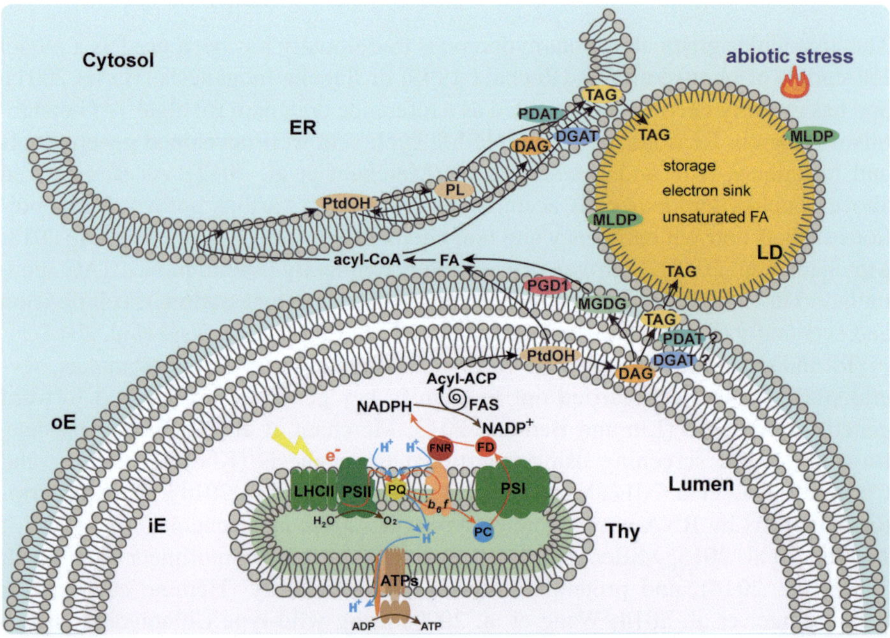

Fig. 8.3 Proposed roles of triacylglycerol accumulation in Chlamydomonas in response to abiotic stresses. *ER* endoplasmic reticulum, *PtdOH* phosphatidic acid, *PL* phospholipid, *DAG* diacylglycerol, *TAG* triacylglycerol, *DGAT* diacylglycerol acyltransferase, *PDAT* phospholipid:diacylglycerol acyltransferase, *LD* lipid droplet, *MLDP* Chlamydomonas major lipid droplet protein, *acyl-CoA* acyl-coenzyme A, *FA* fatty acid, *oE* outer envelope, *iE* inner envelope, *PGD1* PLASTID GALACTOLIPID DEGRADATION 1, *MGDG* monogalactosyldiacylglycerol, *acyl-ACP* acyl-acyl carrier protein, *FAS* fatty acid synthase complex, e^- electron, *FD* ferredoxin, *FNR* ferredoxin:NADP$^+$ reductase, *PQ* plastoquinone and plastoquinol, b_6f, cytochrome b_6f complex, *PC* plastocyanin; H$^+$, proton, *LHCII* light-harvesting complex II, *PSI and II* photosystems I and II, *ATPs* ATP synthase, *ADP/ATP* adenosine di-/triphosphate, *Thy* thylakoid. Different *color arrows* indicate different fluxes: *black*, fatty acid and lipid pathway; *orange*, linear electron flow; *blue*, proton; *brown*, others

TAG accumulation is essential for Chlamydomonas cells to survive under nitrogen starvation (Li et al. 2012). Results of activity assay, *in vivo* pulse-chase and lipid analysis have shown that Chlamydomonas PGD1 hydrolyzes newly incorporated acyl groups at the *sn-1* position of MGDG that can be converted to acyl-CoA for *de novo* TAG synthesis in the cytosol. During nitrogen deprivation, the *pgd1* mutant produces only about 50 % of TAG compared to the wild type, and becomes chlorotic after ~7 days of treatment. Because of the large TAG decrease in the *pgd1* mutant, presumably reduction pressure of the electron transport chain is increased and molecular oxygen at photosystem I (PSI) could serve under these conditions as an alternative electron acceptor to form ROS that will cause damage to thylakoid membranes and chloroplasts and eventually lead to cell death. Indeed, formation of ROS indicated by thiobarbituric acid reactive substances (TBARS) (Baroli et al. 2003) has been observed in the *pgd1* mutant by day 7 of nitrogen deprivation concomitant with chlorosis (Li et al. 2012). Moreover, blocking of electron transfer at the acceptor side of PSII by 3-(3,4-dichlorophenyl)-1,1-dimethylurea (DCMU) can reverse the chlorosis of the *pgd1* mutant, but not TAG accumulation, supporting a possible role of TAG as a sink for electrons (Li et al. 2012). However, it should be noted that comparative transcriptome analysis has shown that following nitrogen deprivation, protein biosynthesis in general and the expression of genes encoding many photosynthesis components is down-regulated (Blaby et al. 2013; Juergens et al. 2015; Miller et al. 2010; Park et al. 2015; Schmollinger et al. 2014). Therefore other explanations for the role of TAG accumulation, such as the transient and safe storage of acyl chains for latter resynthesis of membranes, when conditions improve, need to be considered. Furthermore, these experiments were conducted in acetate medium and hence under photoheterotrophic conditions that may promote TAG formation from acetate. In fact, nitrogen deprivation also leads to a redirection of carbon metabolism such that acetate in the medium is no longer converted to cell building blocks by the glyoxylate cycle and gluconeogenesis, but channeled directly into fatty acid biosynthesis (Miller et al. 2010). Recently, it has been suggested that TAG accumulates to particularly high levels when carbon supply exceeds the capacity of starch synthesis in Chlamydomonas, which is usually the case for microalgae under stress conditions such as nutrient deficiency (Fan et al. 2012). Hence, Chlamydomonas starchless mutants are capable to produce more TAG compared to the wild type (Li et al. 2010; Wang et al. 2009; Work et al. 2010). In some microalgae, TAG synthesis is coordinated with the synthesis of carotenoids such as β-carotene, lutein and astaxanthin that can be sequestered in cytosolic carotenoid-rich lipid droplets. Their peripheral distribution in cells has been proposed to serve as a sun screen to prevent excessive photons from striking the chloroplast and photosynthetic membrane under stress conditions (Rabbani et al. 1998; Zhekisheva et al. 2002). An increase in TAG synthesis has also been observed in the marine alga *Desmodesmus* sp. during high light growth following nitrogen deprivation, which is believed to prevent at least in part photo-oxidative damage under these stress conditions (Gorelova et al. 2015).

Even though Chlamydomonas PGD1 is a galactolipid lipase, it is proposed to not contribute to degradation of membranes but aide in the *de novo* synthesis of TAGs from newly formed fatty acids (Li et al. 2012). However, loss of PGD1 only leads to 50 % reduction in TAG content and other mechanisms to supply precursors of TAG biosynthesis have to be considered. Global transcript analysis indicates that during nitrogen deprivation, genes encoding putative lipases are among those displaying the strongest variations in transcript abundance in Chlamydomonas, which could be a sign for degradation of structural membrane lipids, such as mature MGDG containing 18:3 and 16:4 acyl chains. In fact, a small fraction of TAG contains 16:4 acyl chains, otherwise only found in MDGD of the thylakoid membrane (Liu et al. 2013). These findings are further supported by microscopic observation and lipid analyses indicating that nitrogen deprivation-induced TAG accumulation in lipid droplets occurs concomitantly with the breakdown of thylakoid membranes (Boyle et al. 2012; Iwai et al. 2014). Thus lipid droplet formation is at least in part accompanied by the conversion of polar membrane lipids such as mature MGDG and phosphatidylglycerol (PG) to TAG (Yoon et al. 2012).

Aside from the mechanisms discussed thus far, TAG synthesis involving a PDAT, designated PDAT1, was observed in Chlamydomonas (Fig. 8.3) (Boyle et al. 2012; Yoon et al. 2012). Based on mutants obtained by insertional mutagenesis or artificial microRNA silencing, Chlamydomonas PDAT1 was estimated to contribute up to ~25 % of the total TAG accumulating following nitrogen deprivation due to the turnover of chloroplast membrane lipids, particularly MGDG, sulfoquinovosyldiacylglycerol (SQDG) and PG (Boyle et al. 2012; Yoon et al. 2012). Besides nitrogen deprivation, limitation of iron and zinc in Chlamydomonas cells can also lead to chloroplast/chlorophyll degradation and lipid remodeling (Kropat et al. 2011; Urzica et al. 2013). Under nitrogen deprivation, similar lipid remodeling events that occur in Chlamydomonas have been observed for other algal species such as the model diatom *P. tricornutum* (Abida et al. 2015; Yang et al. 2013), the freshwater alga *M. subterraneus* (Khozin-Goldberg and Cohen 2006), the marine algae *Nannochloropsis gaditana* (Simionato et al. 2013) and *Nannochloropsis oceanica* IMET1 (Jia et al. 2015), and even land plants such as Arabidopsis (Lippold et al. 2012). Thus, TAG accumulation from precursors derived from lipids of the photosynthetic membrane could serve in part as a mechanism to sequester acyl groups for later use, when membranes have to be resynthesized.

Other than nutrient deprivation, stresses such as heat and dark anoxia also trigger TAG accumulation in Chlamydomonas cells (Hemme et al. 2014; Hemschemeier et al. 2013). These stresses lead to the conversion of membrane lipids to TAG which is similar to the observations under nutrient deprivation. However, heat- and dark anoxia-induced TAGs tend to accumulate unsaturated fatty acids, particularly polyunsaturated ones such as linolenic acid (C18:3), compared with TAG produced during nutrient deprivation (Hemme et al. 2014; Hemschemeier et al. 2013). Factors such as temperature and light intensity strongly affect the fatty acid composition in microalgae. For example, it has been observed in many microalgae that increasing temperature leads to more saturated fatty acids whereas decreasing temperature promotes unsaturation of fatty acids (Lynch and Thompson 1982; Renaud et al. 2002; Sato and Murata 1980). High light intensity can also increase the saturation of fatty acids in *Nannochloropsis* sp. cells (Fabregas et al. 2004). It has been reported that

TAG serves as a reservoir of polyunsaturated fatty acids for the rapid formation of membrane lipids upon changes in environmental conditions (e.g. sudden decreases in temperature) in the red alga *Porphyridium cruentum* (Cohen et al. 2000). Thus, in Chlamydomonas polyunsaturated fatty acids derived from the degradation of membrane lipids are likely stored in TAG during non-nutrient deprivation-induced environmental stresses (e.g. heat and dark anoxia) for future utilization. Interestingly, ER stress by Brefeldin A can induce a similar membrane lipid turnover and increase in unsaturation of TAG in Chlamydomonas and *Chlroella vulgaris* (Kim et al. 2013).

Several attempts have been carried out to engineer TAG production in Chlamydomonas (La Russa et al. 2012). For example, three type-two *DGATs*, *DGTT1* to *DGTT3* (also referred to as *DGAT2a*, *b* and *c*) have been independently overexpressed in Chlamydomonas cells but did not increase the intracellular TAG accumulation or significantly alter the composition of the fatty acids compared to the wild type during regular growth condition or under nitrogen or sulfur deprivation (La Russa et al. 2012). In contrast, another research group overexpressed *DGTT4* using a sulphoquinovosyldiacylglycerol 2 (SQD2) promoter that is upregulated during phosphorus starvation, resulting in strongly increased TAG accumulation over the wild type. However, the respective TAG production (~15 mg L^{-1}) was still far below the levels needed for commercial use (Iwai et al. 2014).

A recent study has identified a gene encoding a *SQUAMOSA* promoter-binding-protein-domain containing protein designated *NRR1* (Boyle et al. 2012). It is considered to be an important regulator of TAG synthesis following nitrogen starvation because the *nrr1* mutant produces only ~50 % TAG over the wild type (Boyle et al. 2012). Considering that the single-gene strategies in plant engineering were not very successful whereas triple-gene coexpression of *OLEOSIN*, *WRI1* and *DGAT1* (genes encoding lipid droplet protein, transcription factor and type-one DGAT, respectively) substantially increased TAG production in tobacco vegetative tissues (Vanhercke et al. 2014), multiple-gene coexpression in starchless mutants could be employed to enhance TAG production in Chlamydomonas using its endogenous genes such as *MLDP*, *NRR1* and DGATs in the future. While Chlamydomonas is usually not considered an oleaginous alga for biodiesel production, but a reference organism for the research on lipid metabolism in microalgae, the findings on Chlamydomonas could direct studies on more biotechnologically relevant species such as Nannochloropsis.

Nannochloropsis, an Emerging Model to Study Lipid Metabolism

The oleaginous microalga Nannochloropsis sp., belonging to a genus of unicellular photosynthetic microalgae of the heterokonts, accumulates TAG as the major carbon and energy storage compound under regular or stress conditions (Liu et al. 2013; Meng et al. 2015; Simionato et al. 2013). Nannochloropsis does not accumulate starch (Vieler et al. 2012b; Wang et al. 2014), which is very similar to the starch-less Chlamydomonas mutants that accumulate substantially more TAG than the starch-containing wild type under nitrogen deprivation (Li et al. 2010; Wang et al. 2009; Work et al. 2010). Thus, TAG may serve as an essential sink for photosynthate and as a primary storage compound during stresses in Nannochloropsis and Chlamydomonas. However, unlike Chlamydomonas, Nannochloropsis already synthesizes substantial amounts of TAG (~10 % of dry weight) under nutrient-rich conditions (Taleb et al. 2015). Even though Nannochloropsis produces up to 20 % of dry weight of carbohydrates such as mono- and polysaccharides under nutrient deprivation (Jia et al. 2015; Vieler et al. 2012b; Wang et al. 2014), TAG is the major reserve compound under those conditions (Jia et al. 2015; Simionato et al. 2013; Taleb et al. 2015; Vieler et al. 2012b), making Nannochloropsis an interesting model genus for lipid metabolism research. In terms of lipid metabolism, another big difference between the green alga Chlamydomonas and the oleaginous alga Nannochloropsis is that Chlamydomonas only contains the betaine lipid diacylglyceryl-N,N,N-trimethyl-homoserine (DGTS) that is believed to be a substitution for PtdCho, a common structural lipid in plants and alga with a similar structure as DGTS (Klug and Benning 2001), whereas Nannochloropsis possesses both DGTS and PtdCho (Jia et al. 2015). In addition, Nannochloropsis has been considered one of the more suitable algal species for feedstocks for biofuel production due to its rapid growth and high oil content (Taleb et al. 2015). It also produces large amounts of high-value polyunsaturated fatty acids such as eicosapentaenoic acid (EPA) that can be used for nutritional supplements (Vieler et al. 2012b). Previous studies have focused on six species of Nannochloropsis, and five of them are marine algae that are widely distributed in the marine ecosystems especially in coastal regions (Andersen et al. 1998; Vieler et al. 2012b). In recent years, the genomes of several Nannochloropsis species have been sequenced, e.g. *N. oceanica* CCMP1779 (Vieler et al. 2012b), *N. gaditana* (Radakovits et al. 2012) and *N. oceanica* IMET1 (Wang et al. 2014), and genetic tools and techniques such as nuclear transformation have been developed to facilitate investigations on gene functions and engineering of metabolic pathways (Li et al. 2014a; Radakovits et al. 2012; Vieler et al. 2012b). For example, *N. oceanica* CCMP1779, a small marine alga of ~3 μm in diameter (Fig. 8.4a), has a relative small genome (28.7 Mb) with ~12,000 genes (Vieler et al. 2012b). It can accumulate considerable amounts of TAG in lipid droplets following nitrogen deprivation, and remetabolizes TAG for growth during nitrogen resupply (Fig. 8.4b, c). RNA sequencing analyses of nitrogen-replete or -deprived cells have identified 19 putative genes that are probably directly involved in TAG synthesis, including 13 *DGAT* and 2 *PDAT* genes (Table 8.1) (Vieler et al. 2012b). Most of the genes are up-regulated in response to nitrogen deprivation with the exception of a phosphatidate phosphatase (*PAP*) and *DGAT1* (Table 8.1). Another proteomic study of the lipid droplet of CCMP1779 cells has identified a predominant lipid droplet surface protein, designated as LDSP. Expression of the respective cDNA in the embryo of an Arabidopsis oleosin mutant, *oleo1*, could partially rescue the function of Arabidopsis OLEOSIN1 (Vieler et al. 2012a).

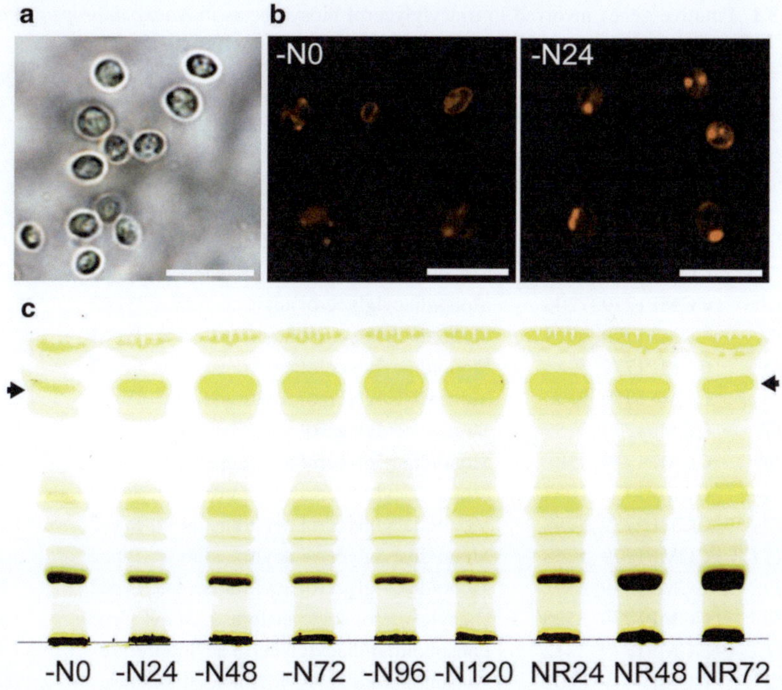

Fig. 8.4 Accumulation of triacylglycerol in *Nannochloropsis oceanica* CCMP1779 following nitrogen deprivation. (**a**) Wild-type *N. oceanica* grown in nitrogen-replete medium. Bar = 10 μm. (**b**) Confocal microscopy images of Nile red-stained wild-type *N. oceanica* cells grown in nitrogen-replete (–N0) or -deprived (–N24) media. Nile red fluorescence (*orange*) indicates lipid droplets (TAG, triacylglycerol). Bars = 10 μm. (**c**) Thin-layer chromatogram of lipid extracts stained for neutral lipids from nitrogen-replete (–N0), -deprived (–N24 to –N120) and -resupplied (NR24–NR72) cultures of wild-type *N. oceanica* at times (hours) indicated. Stained TAG is marked by *black arrows*

Besides *N. oceanica* CCMP1779, genome sequencing and comparative analysis of transcriptomes and lipidomes of other Nannochloropsis species such as *N. gaditana* (Corteggiani Carpinelli et al. 2014) and *N. oceanica* IMET1 (Li et al. 2014b; Wang et al. 2014) have shown putative TAG synthesis pathways and genes involved. Transcriptomic analyses of *N. oceanica* IMET1 cells under nitrogen-replete and -deprived conditions have revealed that many TAG synthesis genes of the Kennedy pathway (acyl-CoA dependent), especially genes encoding seven putative DGATs, are up-regulated upon nitrogen deprivation (Li et al. 2014b). Simultaneously, many genes involved in carbohydrate and protein degradation, as well as genes supplying carbon precursors and energy for *de novo* fatty acid biosynthesis, are increased in their expression and eventually contribute to TAG accumulation. Furthermore, lipidomic analyses using the same Nannochloropsis strain have shown recycling of fatty acids from membrane glycerolipids for TAG biosynthesis following nitrogen deprivation (Li et al. 2014b). Thus, similar lipid remodeling events as shown for

Table 8.1 Putative genes involved in triacylglycerol biosynthesis in *Nannochloropsis oceanica* CCMP1779

Gene	ID	Function	–N/+N
GPAT1	CCMP1779_4533	Glycerol-3-phosphate acyltransferase	1.3
LPAT1	CCMP1779_2512	1-sn-acyl-glycerol-3-phosphate acyltransferase	2.8
LIPIN	CCMP1779_161	Lipin like/ Phosphatidate phosphatase	1.7
PAP	CCMP1779_4742	Phosphatidate phosphatase	1.0
DGAT1	CCMP1779_4340	Diacylglycerol acyltransferase, DGAT Type2	0.6
DGAT2	CCMP1779_3705	Mono/diacylglycerol acyltransferase, Type2	1.2
DGAT3	CCMP1779_7206	Mono/diacylglycerol acyltransferase, Type2	1.7
DGAT4	CCMP1779_9929	Mono/diacylglycerol acyltransferase, Type2	2.8
DGAT5	CCMP1779_3915	Mono/diacylglycerol acyltransferase, Type2	1.4
DGAT6	CCMP1779_9590	Mono/diacylglycerol acyltransferase, Type2	2.5
DGAT7	CCMP1779_3159	Mono/diacylglycerol acyltransferase, Type2	1.5
DGAT8	CCMP1779_358	Mono/diacylglycerol acyltransferase, Type2	1.5
DGAT9	CCMP1779_10272	Mono/diacylglycerol acyltransferase, Type2	2.1
DGAT10	CCMP1779_3159	Mono/diacylglycerol acyltransferase, Type2	1.5
DGAT11	CCMP1779_5368	Mono/diacylglycerol acyltransferase, Type2	3.2
DGAT12	CCMP1779_3592	Mono/diacylglycerol acyltransferase, Type2	3.8
DGAT13	CCMP1779_3520	Diacylglycerol acyltransferase, DGAT Type1	1.8
PDAT1	CCMP1779_2212	Phospholipid/diacylglycerol acyltranferase	1.3
PDAT2	CCMP1779_8602	Phospholipid/diacylglycerol acyltranferase	1.7

–*N* N-deprived for 30 h, +*N* N-replete

Chlamydomonas have been found in Nannochloropsis, indicating that the findings in Chlamydomonas are invaluable for further research on Nannochloropsis (Jia et al. 2015; Martin et al. 2014; Simionato et al. 2013). Taken together, Nannochloropsis information and tool development is rapidly establishing this alga as a potential new reference organism for lipid metabolism research in a biotechnologically relevant microalga.

Conclusions and Perspectives

Plants and algae are highly efficient photosynthetic organisms providing sustainable and clean feedstocks for the production of liquid transportation fuels. Since photosynthetic cells of plants and algae tend to accumulate oils under stress conditions, understanding the mechanism of lipid biosynthesis and metabolism under these condition may provide novel avenues towards the genetic engineering and breeding of stress-tolerant crops and algae. Furthermore, more complete mechanistic insights into stress-induced TAG accumulation will enable the engineering of plants and algae with higher oil content in vegetative cells, but without growth inhibition or yield penalty. Current findings demonstrate similar pathways of TAG biosynthesis in response to stresses for plants and algae, indicating that the discoveries made during plant or algal studies, especially the knowledge gained for reference organisms such as Arabidopsis and Chlamydomonas, may be widely applicable. By broadly screening of naturally occurring species, high oil algae such as Nannochloropsis were selected as candidates for industrial production. The recent genome sequencing of several Nannochloropsis species, as well as establishment of transcriptome, proteome, lipidome, transformation and cultivation, have provided resources and tools for the engineering of oil content and algal biomass quality to eventually overcome the barriers for the commercialization of algal oil. Likewise, findings about stress-induced accumulation of TAGs in algae have the potential to inspire new strategies for the engineering of oil content in vegetative tissues of plants. Taking a multipronged approach learning from algal and plant system is expected to create synergy towards efforts to meet the challenge of green-sustainable biofuel production in the future.

Acknowledgments Work on lipid and TAG metabolism in the Benning lab has been supported by grants from the Center for Advanced Biofuels Systems (CABS), an Energy Frontier Research Center funded by the U.S. Department of Energy, Office of Basic Energy Sciences under award number DE-SC0001295, the U.S. Department of Energy, Office of Basic Energy Sciences Grant DE-FG02-98ER20305, the US National Science Foundation, MCB 0741395 and MCB-1515169, the US Department of Energy-Great Lakes Bioenergy Research Center Cooperative Agreement DE-FC02-07ER64494, the US Air Force Office of Scientific Research, FA9550-11-1-0264, and Michigan State University AgBioResearch.

References

Aaronson S (1973) Effect of incubation temperature on the macromolecular and lipid content of the phytoflagellate *Ochromonas danica*. J Phycol 9:111–113

Abdulla R, Chan ES, Ravindra P (2011) Biodiesel production from *Jatropha curcas*: a critical review. Crit Rev Biotechnol 31:53–64

Abida H, Dolch LJ, Mei C, Villanova V, Conte M, Block MA, Finazzi G, Bastien O, Tirichine L, Bowler C, Rebeille F, Petroutsos D, Jouhet J, Marechal E (2015) Membrane glycerolipid remodeling triggered by nitrogen and phosphorus starvation in *Phaeodactylum tricornutum*. Plant Physiol 167:118–136

Andersen RA, Brett RW, Potter D, Sexton JP (1998) Phylogeny of the Eustigmatophyceae based upon 18S rDNA, with emphasis on Nannochloropsis. Protist 149:61–74

Andrianov V, Borisjuk N, Pogrebnyak N, Brinker A, Dixon J, Spitsin S, Flynn J, Matyszczuk P, Andryszak K, Laurelli M, Golovkin M, Koprowski H (2010) Tobacco as a production platform for biofuel: overexpression of Arabidopsis *DGAT* and *LEC2* genes increases accumulation and shifts the composition of lipids in green biomass. Plant Biotechnol J 8:277–287

ASTM (2002) Standard specification for diesel fuel oils. ASTM D975-02. American Society for Testing and Materials, West Conshohocken

Barber J (2009) Photosynthetic energy conversion: natural and artificial. Chem Soc Rev 38:185–196

Baroli I, Do AD, Yamane T, Niyogi KK (2003) Zeaxanthin accumulation in the absence of a functional xanthophyll cycle protects *Chlamydomonas reinhardtii* from photooxidative stress. Plant Cell 15:992–1008

Bates PD, Stymne S, Ohlrogge J (2013) Biochemical pathways in seed oil synthesis. Curr Opin Plant Biol 16:358–364

Baud S, Lepiniec L (2010) Physiological and developmental regulation of seed oil production. Prog Lipid Res 49:235–249

Bigogno C, Khozin-Goldberg I, Boussiba S, Vonshak A, Cohen Z (2002) Lipid and fatty acid composition of the green oleaginous alga *Parietochloris incisa*, the richest plant source of arachidonic acid. Phytochemistry 60:497–503

Blaby IK, Glaesener AG, Mettler T, Fitz-Gibbon ST, Gallaher SD, Liu B, Boyle NR, Kropat J, Stitt M, Johnson S, Benning C, Pellegrini M, Casero D, Merchant SS (2013) Systems-level analysis of nitrogen starvation-induced modifications of carbon metabolism in a *Chlamydomonas reinhardtii* starchless mutant. Plant Cell 25:4305–4323

Boussiba S, Vonshak A, Cohen Z, Avissar Y, Richmond A (1987) Lipid and biomass production by the halotolerant microalga *Nanochloropsis salina*. Biomass 12:37–47

Bouvier-Nave P, Benveniste P, Oelkers P, Sturley SL, Schaller H (2000) Expression in yeast and tobacco of plant cDNAs encoding acyl CoA:diacylglycerol acyltransferase. Eur J Biochem 267:85–96

Boyle NR, Page MD, Liu B, Blaby IK, Casero D, Kropat J, Cokus SJ, Hong-Hermesdorf A, Shaw J, Karpowicz SJ, Gallaher SD, Johnson S, Benning C, Pellegrini M, Grossman A, Merchant SS (2012) Three acyltransferases and nitrogen-responsive regulator are implicated in nitrogen starvation-induced triacylglycerol accumulation in *Chlamydomonas*. J Biol Chem 287:15811–15825

Brennan L, Owende P (2010) Biofuels from microalgae-a review of technologies for production, processing, and extractions of biofuels and co-products. Renew Sust Energ Rev 14:557–577

Brown MR, Dunstan GA, Norwood SJ, Miller KA (1996) Effects of harvest stage and light on the biochemical composition of the diatom *Thalassiosira pseudonana*. J Phycol 32:64–73

Bruinsma J (ed) (2003) World agriculture: towards 2015/2030. Earthscan, London

Cases S, Smith SJ, Zheng YW, Myers HM, Lear SR, Sande E, Novak S, Collins C, Welch CB, Lusis AJ, Erickson SK, Farese RV Jr (1998) Identification of a gene encoding an acyl CoA:diacylglycerol acyltransferase, a key enzyme in triacylglycerol synthesis. Proc Natl Acad Sci U S A 95:13018–13023

Cernac A, Benning C (2004) *WRINKLED1* encodes an AP2/EREB domain protein involved in the control of storage compound biosynthesis in Arabidopsis. Plant J 40:575–585

Chapman KD, Ohlrogge JB (2012) Compartmentation of triacylglycerol accumulation in plants. J Biol Chem 287:2288–2294

Chapman KD, Dyer JM, Mullen RT (2013) Commentary: why don't plant leaves get fat? Plant Sci 207:128–134

Cheah WY, Show PL, Chang JS, Ling TC, Juan JC (2014) Biosequestration of atmospheric CO_2 and flue gas-containing CO_2 by microalgae. Bioresour Technol. doi:10.1016/j.biortech.2014.1011.1026

Chhetri AB, Tango MS, Budge SM, Watts KC, Islam MR (2008) Non-edible plant oils as new sources for biodiesel production. Int J Mol Sci 9:169–180

Chisti Y (2007) Biodiesel from microalgae. Biotechnol Adv 25:294–306

Chisti Y (2013) Constraints to commercialization of algal fuels. J Biotechnol 167:201–214

Cohen Z, Khozin-Goldberg I, Adlerstein D, Bigogno C (2000) The role of triacylglycerol as a reservoir of polyunsaturated fatty acids for the rapid production of chloroplastic lipids in certain microalgae. Biochem Soc Trans 28:740–743

Collins RP, Kalnins K (1969) The fatty acids of *Cryptomonas ovata* var. palustris. Phyton 26:47–50

Commission of the European Communities (2007) Communication from the commission to the council and the European Parliament: biofuels progress report. Commission of the European Communities, Brussels

Conover SAM (1975) Partitioning of nitrogen and carbon in cultures of the marine diatom *Thalassiosira fluviatillis* supplied with nitrate, ammonium, or urea. Mar Biol 32:231–246

Corteggiani Carpinelli E, Telatin A, Vitulo N, Forcato C, D'Angelo M, Schiavon R, Vezzi A, Giacometti GM, Morosinotto T, Valle G (2014) Chromosome scale genome assembly and transcriptome profiling of *Nannochloropsis gaditana* in nitrogen depletion. Mol Plant 7:323–335

Dismukes GC, Carrieri D, Bennette N, Ananyev GM, Posewitz MC (2008) Aquatic phototrophs: efficient alternatives to land-based crops for biofuels. Curr Opin Biotechnol 19:235–240

Durrett TP, Benning C, Ohlrogge J (2008) Plant triacylglycerols as feedstocks for the production of biofuels. Plant J 54:593–607

El-Hafid L, Pham AT, Zuily-fodil Y, da Silva JV (1989) Enzymatic breakdown of polar lipids in cotton leaves under water stress: I. Degradation of monogalactosyl-diacylglycerol. Plant Physiol Biochem 27:495–502

Fabregas J, Maseda A, Dominquez A, Otero A (2004) The cell composition of *Nannochloropsis* sp. changes under different irradiances in semicontinuous culture. World J Microbiol Biotechnol 20:31–35

Fan J, Yan C, Andre C, Shanklin J, Schwender J, Xu C (2012) Oil accumulation is controlled by carbon precursor supply for fatty acid synthesis in *Chlamydomonas reinhardtii*. Plant Cell Physiol 53:1380–1390

Fan J, Yan C, Zhang X, Xu C (2013) Dual role for phospholipid:diacylglycerol acyltransferase: enhancing fatty acid synthesis and diverting fatty acids from membrane lipids to triacylglycerol in *Arabidopsis* leaves. Plant Cell 25:3506–3518

Fan J, Yan C, Roston R, Shanklin J, Xu C (2014) *Arabidopsis* lipins, PDAT1 acyltransferase, and SDP1 triacylglycerol lipase synergistically direct fatty acids toward beta-oxidation, thereby maintaining membrane lipid homeostasis. Plant Cell 26:4119–4134

Fourrier N, Bedard J, Lopez-Juez E, Barbrook A, Bowyer J, Jarvis P, Warren G, Thorlby G (2008) A role for *SENSITIVE TO FREEZING2* in protecting chloroplasts against freeze-induced damage in Arabidopsis. Plant J 55:734–745

Gasulla F, Vom Dorp K, Dombrink I, Zahringer U, Gisch N, Dormann P, Bartels D (2013) The role of lipid metabolism in the acquisition of desiccation tolerance in *Craterostigma plantagineum*: a comparative approach. Plant J 75:726–741

Georgianna DR, Mayfield SP (2012) Exploiting diversity and synthetic biology for the production of algal biofuels. Nature 488:329–335

Gorelova O, Baulina O, Solovchenko A, Selyakh I, Chivkunova O, Semenova L, Scherbakov P, Burakova O, Lobakova E (2015) Coordinated rearrangements of assimilatory and storage cell compartments in a nitrogen-starving symbiotic chlorophyte cultivated under high light. Arch Microbiol 197:181–195

Harris EH (2001) *Chlamydomonas* as a model organism. Annu Rev Plant Physiol Plant Mol Biol 52:363–406

Hemme D, Veyel D, Muhlhaus T, Sommer F, Juppner J, Unger AK, Sandmann M, Fehrle I, Schonfelder S, Steup M, Geimer S, Kopka J, Giavalisco P, Schroda M (2014) Systems-wide analysis of acclimation responses to long-term heat stress and recovery in the photosynthetic model organism *Chlamydomonas reinhardtii*. Plant Cell 26:4270–4297

Hemschemeier A, Casero D, Liu B, Benning C, Pellegrini M, Happe T, Merchant SS (2013) Copper response regulator1-dependent and -independent responses of the *Chlamydomonas reinhardtii* transcriptome to dark anoxia. Plant Cell 25:3186–3211

Hill J, Nelson E, Tilman D, Polasky S, Tiffany D (2006) Environmental, economic, and energetic costs and benefits of biodiesel and ethanol biofuels. Proc Natl Acad Sci U S A 103:11206–11210

Hu Q, Sommerfeld M, Jarvis E, Ghirardi M, Posewitz M, Seibert M, Darzins A (2008) Microalgal triacylglycerols as feedstocks for biofuel production: perspectives and advances. Plant J 54:621–639

Iwai M, Ikeda K, Shimojima M, Ohta H (2014) Enhancement of extraplastidic oil synthesis in *Chlamydomonas reinhardtii* using a type-2 diacylglycerol acyltransferase with a phosphorus starvation-inducible promoter. Plant Biotechnol J 12:808–819

James CN, Horn PJ, Case CR, Gidda SK, Zhang D, Mullen RT, Dyer JM, Anderson RG, Chapman KD (2010) Disruption of the *Arabidopsis* CGI-58 homologue produces Chanarin-Dorfman-like lipid droplet accumulation in plants. Proc Natl Acad Sci U S A 107:17833–17838

James GO, Hocart CH, Hillier W, Chen H, Kordbacheh F, Price GD, Djordjevic MA (2011) Fatty acid profiling of *Chlamydomonas reinhardtii* under nitrogen deprivation. Bioresour Technol 102:3343–3351

Jia J, Han DX, Gerken HG, Li YT, Sommerfeld M, Hu Q, Xu J (2015) Molecular mechanisms for photosynthetic carbon partitioning into storage neutral lipids in *Nannochloropsis oceanica* under nitrogen-depletion conditions. Algal Res 7:66–77

John RP, Anisha GS, Nampoothiri KM, Pandey A (2011) Micro and macroalgal biomass: a renewable source for bioethanol. Bioresour Technol 102:186–193

Jones CS, Mayfield SP (2012) Algae biofuels: versatility for the future of bioenergy. Curr Opin Biotechnol 23:346–351

Juergens MT, Deshpande RR, Lucker BF, Park JJ, Wang H, Gargouri M, Holguin FO, Disbrow B, Schaub T, Skepper JN, Kramer DM, Gang DR, Hicks LM, Shachar-Hill Y (2015) The regulation of photosynthetic structure and function during nitrogen deprivation in *Chlamydomonas reinhardtii*. Plant Physiol 167:558–573

Kaup MT, Froese CD, Thompson JE (2002) A role for diacylglycerol acyltransferase during leaf senescence. Plant Physiol 129:1616–1626

Kelly AA, van Erp H, Quettier AL, Shaw E, Menard G, Kurup S, Eastmond PJ (2013) The SUGAR-DEPENDENT1 lipase limits triacylglycerol accumulation in vegetative tissues of Arabidopsis. Plant Physiol 162:1282–1289

Khotimchenko SV, Yakovleva IM (2005) Lipid composition of the red alga *Tichocarpus crinitus* exposed to different levels of photon irradiance. Phytochemistry 66:73–79

Khozin-Goldberg I, Cohen Z (2006) The effect of phosphate starvation on the lipid and fatty acid composition of the fresh water eustigmatophyte *Monodus subterraneus*. Phytochemistry 67:696–701

Khozin-Goldberg I, Cohen Z (2011) Unraveling algal lipid metabolism: recent advances in gene identification. Biochimie 93:91–100

Kim S, Kim H, Ko D, Yamaoka Y, Otsuru M, Kawai-Yamada M, Ishikawa T, Oh HM, Nishida I, Li-Beisson Y, Lee Y (2013) Rapid induction of lipid droplets in *Chlamydomonas reinhardtii* and *Chlorella vulgaris* by Brefeldin A. PLoS One 8, e81978

Klaus D, Ohlrogge JB, Neuhaus HE, Dormann P (2004) Increased fatty acid production in potato by engineering of acetyl-CoA carboxylase. Planta 219:389–396

Klug RM, Benning C (2001) Two enzymes of diacylglyceryl-O-4′-(N, N, N,-trimethyl)-homoserine biosynthesis are encoded by btaA and btaB in the purple bacterium Rhodobacter sphaeroides. Proc Natl Acad Sci U S A 98:5910–5915

Kropat J, Hong-Hermesdorf A, Casero D, Ent P, Castruita M, Pellegrini M, Merchant SS, Malasarn D (2011) A revised mineral nutrient supplement increases biomass and growth rate in *Chlamydomonas reinhardtii*. Plant J 66:770–780

La Russa M, Bogen C, Uhmeyer A, Doebbe A, Filippone E, Kruse O, Mussgnug JH (2012) Functional analysis of three type-2 DGAT homologue genes for triacylglycerol production in the green microalga *Chlamydomonas reinhardtii*. J Biotechnol 162:13–20

Lardizabal KD, Mai JT, Wagner NW, Wyrick A, Voelker T, Hawkins DJ (2001) DGAT2 is a new diacylglycerol acyltransferase gene family: purification, cloning, and expression in insect cells of two polypeptides from *Mortierella ramanniana* with diacylglycerol acyltransferase activity. J Biol Chem 276:38862–38869

Li Y, Han D, Hu G, Dauvillee D, Sommerfeld M, Ball S, Hu Q (2010) *Chlamydomonas* starchless mutant defective in ADP-glucose pyrophosphorylase hyper-accumulates triacylglycerol. Metab Eng 12:387–391

Li X, Moellering ER, Liu B, Johnny C, Fedewa M, Sears BB, Kuo MH, Benning C (2012) A galactoglycerolipid lipase is required for triacylglycerol accumulation and survival following nitrogen deprivation in *Chlamydomonas reinhardtii*. Plant Cell 24:4670–4686

Li F, Gao D, Hu H (2014a) High-efficiency nuclear transformation of the oleaginous marine Nannochloropsis species using PCR product. Biosci Biotechnol Biochem 78:812–817

Li J, Han D, Wang D, Ning K, Jia J, Wei L, Jing X, Huang S, Chen J, Li Y, Hu Q, Xu J (2014b) Choreography of transcriptomes and lipidomes of *Nannochloropsis* reveals the mechanisms of oil synthesis in microalgae. Plant Cell 26:1645–1665

Lippold F, vom Dorp K, Abraham M, Holzl G, Wewer V, Yilmaz JL, Lager I, Montandon C, Besagni C, Kessler F, Stymne S, Dormann P (2012) Fatty acid phytyl ester synthesis in chloroplasts of *Arabidopsis*. Plant Cell 24:2001–2014

Liu B, Benning C (2013) Lipid metabolism in microalgae distinguishes itself. Curr Opin Biotechnol 24:300–309

Liu B, Vieler A, Li C, Jones AD, Benning C (2013) Triacylglycerol profiling of microalgae *Chlamydomonas reinhardtii* and *Nannochloropsis oceanica*. Bioresour Technol 146:310–316

Lynch DV, Thompson GA (1982) Low temperature-induced alterations in the chloroplast and microsomal membranes of *Dunaliella salina*. Plant Physiol 69:1369–1375

Mansour MP, Volkman JK, Blackburn SI (2003) The effect of growth phase on the lipid class, fatty acid and sterol composition in the marine dinoflagellate, *Gymnodinium* sp. in batch culture. Phytochemistry 63:145–153

Martin GJ, Hill DR, Olmstead IL, Bergamin A, Shears MJ, Dias DA, Kentish SE, Scales PJ, Botte CY, Callahan DL (2014) Lipid profile remodeling in response to nitrogen deprivation in the microalgae *Chlorella* sp. (Trebouxiophyceae) and *Nannochloropsis* sp. (Eustigmatophyceae). PLoS One 9, e103389

Martinez-Boti MA, Marino G, Foster GL, Ziveri P, Henehan MJ, Rae JW, Mortyn PG, Vance D (2015) Boron isotope evidence for oceanic carbon dioxide leakage during the last deglaciation. Nature 518:219–222

Matthew T, Zhou W, Rupprecht J, Lim L, Thomas-Hall SR, Doebbe A, Kruse O, Hankamer B, Marx UC, Smith SM, Schenk PM (2009) The metabolome of *Chlamydomonas reinhardtii* following induction of anaerobic H_2 production by sulfur depletion. J Biol Chem 284:23415–23425

Meng Y, Jiang J, Wang H, Cao X, Xue S, Yang Q, Wang W (2015) The characteristics of TAG and EPA accumulation in *Nannochloropsis oceanica* IMET1 under different nitrogen supply regimes. Bioresour Technol 179:483–489

Merchant SS, Kropat J, Liu B, Shaw J, Warakanont J (2012) TAG, you're it! Chlamydomonas as a reference organism for understanding algal triacylglycerol accumulation. Curr Opin Biotechnol 23:352–363

Miller R, Wu G, Deshpande RR, Vieler A, Gartner K, Li X, Moellering ER, Zauner S, Cornish AJ, Liu B, Bullard B, Sears BB, Kuo MH, Hegg EL, Shachar-Hill Y, Shiu SH, Benning C (2010) Changes in transcript abundance in *Chlamydomonas reinhardtii* following nitrogen deprivation predict diversion of metabolism. Plant Physiol 154:1737–1752

Moellering ER, Benning C (2010) RNA interference silencing of a major lipid droplet protein affects lipid droplet size in *Chlamydomonas reinhardtii*. Eukaryot Cell 9:97–106

Moellering ER, Benning C (2011) Galactoglycerolipid metabolism under stress: a time for remodeling. Trends Plant Sci 16:98–107

Moellering ER, Muthan B, Benning C (2010) Freezing tolerance in plants requires lipid remodeling at the outer chloroplast membrane. Science 330:226–228

Mu J, Tan H, Zheng Q, Fu F, Liang Y, Zhang J, Yang X, Wang T, Chong K, Wang XJ, Zuo J (2008) *LEAFY COTYLEDON1* is a key regulator of fatty acid biosynthesis in Arabidopsis. Plant Physiol 148:1042–1054

Napolitano GE (1994) The relationship of lipids with light and chlorophyll measurement in freshwater algae and periphyton. J Phycol 30:943–950

Navari-Izzo F, Rascio N (1999) Plant response to water-deficit conditions. In: Pessarakli M (ed) Handbook of plant and crop stress. Dekker, New York, pp 217–231

Nguyen HM, Baudet M, Cuine S, Adriano JM, Barthe D, Billon E, Bruley C, Beisson F, Peltier G, Ferro M, Li-Beisson Y (2011) Proteomic profiling of oil bodies isolated from the unicellular green microalga *Chlamydomonas reinhardtii*: with focus on proteins involved in lipid metabolism. Proteomics 11:4266–4273

Ogas J, Cheng JC, Sung ZR, Somerville C (1997) Cellular differentiation regulated by gibberellin in the *Arabidopsis thaliana pickle* mutant. Science 277:91–94

Ogas J, Kaufmann S, Henderson J, Somerville C (1999) PICKLE is a CHD3 chromatin-remodeling factor that regulates the transition from embryonic to vegetative development in *Arabidopsis*. Proc Natl Acad Sci U S A 96:13839–13844

Ohlrogge JB, Chapman KD (2011) The seeds of green energy: expanding the contribution of plant oils as biofuels. Biochemist 33:34–38

Ohlrogge J, Allen D, Berguson B, Dellapenna D, Shachar-Hill Y, Stymne S (2009) Energy. Driving on biomass. Science 324:1019–1020

Otsuka H (1961) Changes of lipid and carbohydrate contents of *Chlorella* cells during the sulfur starvation, as studied by the technique of synchronous culture. J Gen Appl Microbiol 7:72–77

Park JJ, Wang H, Gargouri M, Deshpande RR, Skepper JN, Holguin FO, Juergens MT, Shachar-Hill Y, Hicks LM, Gang DR (2015) The response of *Chlamydomonas reinhardtii* to nitrogen deprivation: a systems biology analysis. Plant J 81:611–624

Petrie JR, Vanhercke T, Shrestha P, El Tahchy A, White A, Zhou XR, Liu Q, Mansour MP, Nichols PD, Singh SP (2012) Recruiting a new substrate for triacylglycerol synthesis in plants: the monoacylglycerol acyltransferase pathway. PLoS One 7, e35214

Pugh PR (1971) Changes in the fatty acid composition of *Coscinodiscus eccentricus* with culture-age and salinity. Mar Biol 11:118–124

Rabbani S, Beyer P, Lintig J, Hugueney P, Kleinig H (1998) Induced *β*-carotene synthesis driven by triacylglycerol deposition in the unicellular alga *Dunaliella bardawil*. Plant Physiol 116:1239–1248

Radakovits R, Jinkerson RE, Fuerstenberg SI, Tae H, Settlage RE, Boore JL, Posewitz MC (2012) Draft genome sequence and genetic transformation of the oleaginous alga *Nannochloropis gaditana*. Nat Commun 3:686

Reitan KI, Rainuzzo JR, Olsen Y (1994) Effect of nutrient limitation on fatty acid and lipid content of marine microalgae. J Phycol 30:972–979

Renaud SM, Thinh LV, Lambrinidis G, Parry DL (2002) Effect of temperature on growth, chemical composition and fatty acid composition of tropical Australian microalgae grown in batch cultures. Aquaculture 211:195–214

Riekhof RW, Benning C (2009) Glycerolipid biosynthesis. In: Stern DB, Harris EH (eds) The Chlamydomonas source book, vol 2, 2nd edn, Organellar and Metabolic Processes. Academic, Elsevier, Boston, pp 41–68

Rochaix JD (1995) *Chlamydomonas reinhardtii* as the photosynthetic yeast. Annu Rev Genet 29:209–230

Roessler PG (1988) Changes in the activities of various lipid and carbohydrate biosynthetic enzymes in the diatom *Cyclotella cryptica* in response to silicon deficiency. Arch Biochem Biophys 267:521–528

Ross B, Sanchez J, Millan F, Murphy DJ (1993) Differential presence of oleosins in oleogenic seed and mesocarp tissues in olive (*Olea europaea*) and avocado (*Persea americana*). Plant Sci 93:203–210

Roston RL, Wang K, Kuhn LA, Benning C (2014) Structural determinants allowing transferase activity in SENSITIVE TO FREEZING 2, classified as a family I glycosyl hydrolase. J Biol Chem 289:26089–26106

Sakaki T, Ohnishi J, Kondo N, Yamada M (1985) Polar and neutral lipid changes in spinach leaves with ozone fumigation: triacylglycerol synthesis from polar lipids. Plant Cell Physiol 26:253–262

Sakaki T, Kondo N, Yamada M (1990a) Free fatty acids regulate two galactosyltransferases in chloroplast envelope membranes isolated from spinach leaves. Plant Physiol 94:781–787

Sakaki T, Kondo N, Yamada M (1990b) Pathway for the synthesis of triacylglycerols from monogalactosyldiacylglycerols in ozone-fumigated spinach leaves. Plant Physiol 94:773–780

Sakaki T, Saito K, Kawaguchi A, Kondo N, Yamada M (1990c) Conversion of monogalactosyldiacylglycerols to triacylglycerols in ozone-fumigated spinach leaves. Plant Physiol 94:766–772

Sanjaya, Durrett TP, Weise SE, Benning C (2011) Increasing the energy density of vegetative tissues by diverting carbon from starch to oil biosynthesis in transgenic Arabidopsis. Plant Biotechnol J 9:874–883

Sanjaya, Miller R, Durrett TP, Kosma DK, Lydic TA, Muthan B, Koo AJ, Bukhman YV, Reid GE, Howe GA, Ohlrogge J, Benning C (2013) Altered lipid composition and enhanced nutritional value of *Arabidopsis* leaves following introduction of an algal diacylglycerol acyltransferase 2. Plant Cell 25:677–693

Santos-Mendoza M, Dubreucq B, Miquel M, Caboche M, Lepiniec L (2005) LEAFY COTYLEDON 2 activation is sufficient to trigger the accumulation of oil and seed specific mRNAs in *Arabidopsis* leaves. FEBS Lett 579:4666–4670

Santos-Mendoza M, Dubreucq B, Baud S, Parcy F, Caboche M, Lepiniec L (2008) Deciphering gene regulatory networks that control seed development and maturation in Arabidopsis. Plant J 54:608–620

Sato N, Murata N (1980) Temperature shift-induced responses in lipids in the blue-green alga, *Anabaena variabilis*: the central role of diacylmonogalactosylglycerol in thermo-adaptation. Biochim Biophys Acta 619:353–366

Schmollinger S, Muhlhaus T, Boyle NR, Blaby IK, Casero D, Mettler T, Moseley JL, Kropat J, Sommer F, Strenkert D, Hemme D, Pellegrini M, Grossman AR, Stitt M, Schroda M, Merchant SS (2014) Nitrogen-sparing mechanisms in *Chlamydomonas* affect the transcriptome, the proteome, and photosynthetic metabolism. Plant Cell 26:1410–1435

Shiratake T, Sato A, Minoda A, Tsuzuki M, Sato N (2013) Air-drying of cells, the novel conditions for stimulated synthesis of triacylglycerol in a green alga, *Chlorella kessleri*. PLoS One 8, e79630

Siaut M, Cuine S, Cagnon C, Fessler B, Nguyen M, Carrier P, Beyly A, Beisson F, Triantaphylides C, Li-Beisson Y, Peltier G (2011) Oil accumulation in the model green alga *Chlamydomonas reinhardtii*: characterization, variability between common laboratory strains and relationship with starch reserves. BMC Biotechnol 11:7

Simionato D, Block MA, La Rocca N, Jouhet J, Marechal E, Finazzi G, Morosinotto T (2013) The response of *Nannochloropsis gaditana* to nitrogen starvation includes *de novo* biosynthesis of triacylglycerols, a decrease of chloroplast galactolipids, and reorganization of the photosynthetic apparatus. Eukaryot Cell 12:665–676

Slocombe SP, Cornah J, Pinfield-Wells H, Soady K, Zhang Q, Gilday A, Dyer JM, Graham IA (2009) Oil accumulation in leaves directed by modification of fatty acid breakdown and lipid synthesis pathways. Plant Biotechnol J 7:694–703

Stephenson PG, Moore CM, Terry MJ, Zubkov MV, Bibby TS (2011) Improving photosynthesis for algal biofuels: toward a green revolution. Trends Biotechnol 29:615–623

Stoller EW, Weber EJ (1975) Differential cold tolerance, starch, sugar, protein, and lipid of yellow and purple nutsedge tubers. Plant Physiol 55:859–863

Stone SL, Braybrook SA, Paula SL, Kwong LW, Meuser J, Pelletier J, Hsieh TF, Fischer RL, Goldberg RB, Harada JJ (2008) *Arabidopsis* LEAFY COTYLEDON2 induces maturation traits and auxin activity: implications for somatic embryogenesis. Proc Natl Acad Sci U S A 105:3151–3156

Takagi M, Karseno, Yoshida T (2006) Effect of salt concentration on intracellular accumulation of lipids and triacylglyceride in marine microalgae *Dunaliella* cells. J Biosci Bioeng 101:223–226

Taleb A, Pruvost J, Legrand J, Marec H, Le-Gouic B, Mirabella B, Legeret B, Bouvet S, Peltier G, Li-Beisson Y, Taha S, Takache H (2015) Development and validation of a screening procedure of microalgae for biodiesel production: application to the genus of marine microalgae *Nannochloropsis*. Bioresour Technol 177:224–232

Thorlby G, Fourrier N, Warren G (2004) The *SENSITIVE TO FREEZING2* gene, required for freezing tolerance in *Arabidopsis thaliana*, encodes a β-glucosidase. Plant Cell 16:2192–2203

Traller JC, Hildebrand M (2013) High throughput imaging to the diatom *Cyclotella cryptica* demonstrates substantial cell-to-cell variability in the rate and extent of triacylglycerol accumulation. Algal Res 2:244–252

Tranbarger TJ, Dussert S, Joet T, Argout X, Summo M, Champion A, Cros D, Omore A, Nouy B, Morcillo F (2011) Regulatory mechanisms underlying oil palm fruit mesocarp maturation, ripening, and functional specialization in lipid and carotenoid metabolism. Plant Physiol 156:564–584

Tsai CH, Warakanont J, Takeuchi T, Sears BB, Moellering ER, Benning C (2014) The protein Compromised Hydrolysis of Triacylglycerols 7 (CHT7) acts as a repressor of cellular quiescence in Chlamydomonas. Proc Natl Acad Sci U S A 111:15833–15838

Urzica EI, Vieler A, Hong-Hermesdorf A, Page MD, Casero D, Gallaher SD, Kropat J, Pellegrini M, Benning C, Merchant SS (2013) Remodeling of membrane lipids in iron-starved *Chlamydomonas*. J Biol Chem 288:30246–30258

US Department of Energy (2007) Annual energy outlook 2007. Report number DOE/EIA-0383. Energy Information Administration, US Department of Energy, Washington, DC

Vanhercke T, El Tahchy A, Shrestha P, Zhou XR, Singh SP, Petrie JR (2013) Synergistic effect of WRI1 and DGAT1 coexpression on triacylglycerol biosynthesis in plants. FEBS Lett 587:364–369

Vanhercke T, El Tahchy A, Liu Q, Zhou XR, Shrestha P, Divi UK, Ral JP, Mansour MP, Nichols PD, James CN, Horn PJ, Chapman KD, Beaudoin F, Ruiz-Lopez N, Larkin PJ, de Feyter RC, Singh SP, Petrie JR (2014) Metabolic engineering of biomass for high energy density: oilseed-like triacylglycerol yields from plant leaves. Plant Biotechnol J 12:231–239

Venkata Mohan S, Rohit MV, Chiranjeevi P, Chandra R, Navaneeth B (2015) Heterotrophic microalgae cultivation to synergize biodiesel production with waste remediation: progress and perspectives. Bioresour Technol 184:169–178

Vieler A, Brubaker SB, Vick B, Benning C (2012a) A lipid droplet protein of *Nannochloropsis* with functions partially analogous to plant oleosins. Plant Physiol 158:1562–1569

Vieler A, Wu G, Tsai CH, Bullard B, Cornish AJ, Harvey C, Reca IB, Thornburg C, Achawanantakun R, Buehl CJ, Campbell MS, Cavalier D, Childs KL, Clark TJ, Deshpande R, Erickson E, Armenia Ferguson A, Handee W, Kong Q, Li X, Liu B, Lundback S, Peng C, Roston RL, Sanjaya, Simpson JP, Terbush A, Warakanont J, Zauner S, Farre EM, Hegg EL, Jiang N, Kuo MH, Lu Y, Niyogi KK, Ohlrogge J, Osteryoung KW, Shachar-Hill Y, Sears BB, Sun Y, Takahashi H, Yandell M, Shiu SH, Benning C (2012b) Genome, functional gene annotation, and nuclear transformation of the heterokont oleaginous alga *Nannochloropsis oceanica* CCMP1779. PLoS Genet 8, e1003064

Wang G, Lin Q, Xu Y (2007) *Tetraena mongolica* Maxim can accumulate large amounts of triacylglycerol in phloem cells and xylem parenchyma of stems. Phytochemistry 68:2112–2117

Wang ZT, Ullrich N, Joo S, Waffenschmidt S, Goodenough U (2009) Algal lipid bodies: stress induction, purification, and biochemical characterization in wild-type and starchless *Chlamydomonas reinhardtii*. Eukaryot Cell 8:1856–1868

Wang D, Ning K, Li J, Hu J, Han D, Wang H, Zeng X, Jing X, Zhou Q, Su X, Chang X, Wang A, Wang W, Jia J, Wei L, Xin Y, Qiao Y, Huang R, Chen J, Han B, Yoon K, Hill RT, Zohar Y, Chen F, Hu Q, Xu J (2014) *Nannochloropsis* genomes reveal evolution of microalgal oleaginous traits. PLoS Genet 10, e1004094

Winichayakul S, Scott RW, Roldan M, Hatier JH, Livingston S, Cookson R, Curran AC, Roberts NJ (2013) In vivo packaging of triacylglycerols enhances Arabidopsis leaf biomass and energy density. Plant Physiol 162:626–639

Work VH, Radakovits R, Jinkerson RE, Meuser JE, Elliott LG, Vinyard DJ, Laurens LM, Dismukes GC, Posewitz MC (2010) Increased lipid accumulation in the *Chlamydomonas reinhardtii sta7-10* starchless isoamylase mutant and increased carbohydrate synthesis in complemented strains. Eukaryot Cell 9:1251–1261

Xu C, Fan J, Froehlich JE, Awai K, Benning C (2005) Mutation of the TGD1 chloroplast envelope protein affects phosphatidate metabolism in *Arabidopsis*. Plant Cell 17:3094–3110

Yang Z, Ohlrogge JB (2009) Turnover of fatty acids during natural senescence of Arabidopsis, *Brachypodium*, and switchgrass and in Arabidopsis β-oxidation mutants. Plant Physiol 150:1981–1989

Yang Y, Yu X, Song L, An C (2011) ABI4 activates *DGAT1* expression in Arabidopsis seedlings during nitrogen deficiency. Plant Physiol 156:873–883

Yang ZK, Niu YF, Ma YH, Xue J, Zhang MH, Yang WD, Liu JS, Lu SH, Guan Y, Li HY (2013) Molecular and cellular mechanisms of neutral lipid accumulation in diatom following nitrogen deprivation. Biotechnol Biofuels 6:67

Yoon K, Han D, Li Y, Sommerfeld M, Hu Q (2012) Phospholipid:diacylglycerol acyltransferase is a multifunctional enzyme involved in membrane lipid turnover and degradation while synthesizing triacylglycerol in the unicellular green microalga *Chlamydomonas reinhardtii*. Plant Cell 24:3708–3724

Zhang HY, Hanna MA, Ali Y, Nan L (1996) Yellow nutsedge (*Cyperus esculentus* L) tuber oil as a fuel. Ind Crops Prod 5:177–181

Zhang R, Patena W, Armbruster U, Gang SS, Blum SR, Jonikas MC (2014) High-throughput genotyping of green algal mutants reveals random distribution of mutagenic insertion sites and endonucleolytic cleavage of transforming DNA. Plant Cell 26:1398–1409

Zhekisheva M, Boussiba S, Khozin-Goldberg I, Zarka A, Cohen Z (2002) Accumulation of oleic acid in *Haematococcus pluvialis* (Chlorophyceae) under nitrogen starvation or high light is correlated with that of astaxanthin esters. J Phycol 38:325–331

Chapter 9
Cellular Organization of Triacylglycerol Biosynthesis in Microalgae

Changcheng Xu, Carl Andre, Jilian Fan, and John Shanklin

Abstract Eukaryotic cells are characterized by compartmentalization and specialization of metabolism within membrane-bound organelles. Nevertheless, many fundamental processes extend across multiple subcellular compartments. Here, we describe and assess the pathways and cellular organization of triacylglycerol biosynthesis in microalgae. In particular, we emphases the dynamic interplay among the endoplasmic reticulum, lipid droplets and chloroplasts in acyl remodeling and triacylglycerol accumulation under nitrogen starvation in the model alga *Chlamydomonas reinhardtii*.

Keywords Triacylglycerol • Acyl remodeling • Oil droplet • Chloroplast • Endoplasmic reticulum

One of the defining features of photosynthetic eukaryotic organisms is the presence of the chloroplast. Beyond their role in photosynthesis, chloroplasts contribute to a wide array of fundamental functions that include the de novo synthesis of fatty acids, the building blocks of membrane lipids and storage triacylglycerol. Under normal growth conditions, the vast majority of de novo-synthesized fatty acids are used for membrane lipid assembly to support cell growth, organelle biogenesis and membrane proliferation. Under stress conditions such as nutrient deprivation, many microalgae can accumulate large amounts of triacylglycerol by diverting fatty acids from membrane lipid synthesis to the synthesis of TAG and/or by converting preformed membrane lipids to TAG. In most eukaryotic cells, storage lipids are packaged in simple structures termed lipid droplets (also referred to as oil bodies), which consist of a central core of TAG within a monolayer of membrane lipids with a small

C. Xu (✉) • C. Andre • J. Fan • J. Shanklin
Biology Department, Brookhaven National Laboratory, Upton, NY 11973, USA
e-mail: cxu@bnl.gov

Y. Nakamura, Y. Li-Beisson (eds.), *Lipids in Plant and Algae Development*, Subcellular Biochemistry 86, DOI 10.1007/978-3-319-25979-6_9

amount of embedded specific proteins. The current model of lipid body biogenesis favors formation of lipid droplets through budding from the ER (Thiele and Spandl 2008; Walther and Farese 2009; Chapman et al. 2012; Wilfling et al. 2014). According to this hypothesis, lipid droplets originate from specialized ER subdomains enriched with enzymes involved in TAG biosynthesis. Because the newly formed TAGs in these ER domains are unable to integrate into membrane bilayers due to lack of polar head groups, they accumulate in the hydrophobic region between the two leaflets of the ER membrane which leads to swelling of the membrane bilayer and eventually the budding of growing lipid droplets from ER into the cytosol. Recent years have seen rapid progress in identifying and characterizing molecular components of lipid droplets, revealing the genes and enzymes in lipid metabolic pathways and deciphering the regulatory networks controlling carbon metabolism and storage reserve accumulation in microalgae, particularly in the model alga *Chlamydomonas*. Here we review data on the pathways and cellular compartmentalization of glycerolipid biosynthesis in microalgae, focusing on interconnections among the endoplasmic reticulum, chloroplasts and lipid droplets in lipid remodeling and TAG biosynthesis in *Chlamydomonas*. Where appropriate, knowledge gained from higher plants, yeast and mammals will be discussed to highlight similarities and differences between these experimental systems.

Diacylglycerol and Glycerolipid Synthesis

Biosynthesis of diacylglycerol (DAG) and glycerolipids in plants and algae encompasses two parallel pathways involving multiple subcellular compartments. Fatty acid synthesis occurs almost exclusively in the plastid and is catalyzed by two large, evolutionarily conserved enzyme complexes: acetyl-CoA carboxylase (ACCase) and fatty acid synthase (Ohlrogge and Jaworski 1997). In most photosynthetic organisms, the major end products of fatty acid synthesis are 16:0- and 18:0 fatty acids esterified to acyl carrier protein (ACP), which are subsequently processed by stearoyl-ACP desaturase to generate 18:1-ACP and/or by acyl-ACP thioesterases to release fatty acids and ACP. A fraction of newly synthesized acyl-ACPs are used in plastids for the sequential acylation of glycerol-3-phosphate (G-3-P) catalyzed by plastidic acyltransferases, leading to the generation of phosphatidic acid (PA). PA is a key intermediate in the formation of the thylakoid membrane phospholipid phosphatidylglycerol. In addition, PA can be dephosphorylated by PA phosphohydrolases (PAHs) to generate DAG, which is normally used for assembly of thylakoid glycolipids including galactolipids, monogalactosyldiacylglycerol (MGDG), digalactosyldiacylglycerol (DGDG) and the sulfolipid sulfoquino vosyldiacylglycerol (SQDG) in the plastid envelope membranes. This sequence of reactions is commonly referred to as the prokaryotic pathway (Browse et al. 1986; Browse and Somerville 1991). In photosynthetic cells, two galactolipids MGDG and DGDG account for the bulk (i.e., up to 80 %) of thylakoid lipids (Dörmann et al. 1999).

Alternatively, fatty acids can be exported from the plastid to enter into glycerolipid biosynthetic pathways in the endoplasmic reticulum (ER), leading to the formation of DAG by sequential G-3-P acylation and PA dephosphorylation catalyzed by ER-resident acyltransferases and PAHs, respectively (Browse and Somerville 1991). The resulting DAG can be used to synthesize extraplastidic membrane lipids including phosphatidylcholine (PC), phosphatidylethanolamine (PE) and a non-phosporus-containing polar lipid diacylglycerol-*N,N,N*-trimethylhomoserine (DGTS) (Thompson 1996; Harwood and Guschina 2009). Some of the phospholipids assembled in the ER return to the plastid, thus providing the DAG moieties for the synthesis of thylakoid glycolipids in the plastid (Roughan and Slack 1982; Browse et al. 1986). This sequence of events is referred to as the eukaryotic pathway of thylakoid lipid synthesis. Because of the different substrate specificity of acyltransferases present in the plastid and endoplasmic reticulum, glycerolipids made via the prokaryotic and eukaryotic pathway are characterized by the presence of a 16-carbon (C16) or 18-carbon (C18) fatty acid at the sn-2 position of the glycerol backbone, respectively (Frentzen 1998).

Chlamydomonas chloroplast lipid biosynthesis is thought to be almost completely autonomous because it lacks thylakoid lipids derived from the eukaryotic pathway characterized by the presence of C18 acyl chains at the sn-2 position (Giroud et al. 1988; Giroud and Eichenberger 1989; Harwood and Guschina 2009). On the other hand, the galactolipids in the marine brown alga *Dictyopteris membranacea* (Hofmann and Eichenberger 1997) and several other brown algal species (Jones and Harwood 1992) are almost completely of the eukaryotic type. A few green algae such as *Chlorella kessleri* (Sato et al. 2003) and *Acetabularia mediterranea* (Thompson 1996) and some red and brown algae (Hofmann and Eichenberger 1997; Makewicz et al. 1997; Sato and Moriyama 2007) contain substantial amounts of ER-derived eukaryotic lipids in photosynthetic membranes and therefore employ two parallel pathways for chloroplast lipid assembly.

A characteristic feature of lipid metabolism in *Chlamydomonas* is the absence of PC (Sakurai et al. 2014). In higher plants, this phospholipid is known to play a critical role in supply of DAG moieties for the synthesis of eukaryotic thylakoid (Benning 2008, 2009). Therefore, the absence of the eukaryotic thylakoid lipid synthesis in *Chlamydomonas* has been assumed to be due to the lack of PC (Moellering et al. 2009; Riekhof and Benning 2009). However, radiotracer labeling studies in the marine brown alga *Dictyopteris membranacea* have shown that the absence of PC does not compromise the eukaryotic pathway of galactolipid biosynthesis (Hofmann and Eichenberger 1997). In addition, several brown algae contain almost exclusively eukaryotic thylakoid glycolipids despite the lack of PC (Jones and Harwood 1992). Together, these results suggest that PC is not absolutely required for the eukaryotic pathway of chloroplast lipid biosynthesis.

TAG synthesis shares the common precursor DAG produced by sequential G-3-P acylation and PA dephosphorylation reactions as described above with the synthesis of membrane lipids. Results from yeast, plant and mammals indicate that DAG partitioning between membrane lipids and TAG is, to a major extent, modulated by lipin family of PAHs (Harris and Finck 2011; Pascual and Carman 2013; Siniossoglou 2013; Fan et al. 2014). Disruption of the yeast lipin homolog impairs TAG synthesis (Han et al. 2006) and lipid droplet formation (Adeyo et al. 2011), causing an increase in phospholipid synthesis and a massive proliferation of ER and nuclear membranes (Santos-Rosa et al. 2005). Similarly, Arabidopsis lipins have been implicated in the regulation of TAG accumulation (Fan et al. 2014), phospholipid synthesis and ER membrane expansion (Eastmond et al. 2010). Mammalian lipins are bifunctional intracellular proteins, acting as regulators of DNA-bound transcription factors, in addition to catalyzing the dephosphorylation of PA (Harris and Finck 2011). A recent study showed that genetic modifications of PAHs result in altered TAG content in *Chlamydomonas* (Deng et al. 2013), suggesting that the role of lipin family of PAHs in TAG synthesis and regulation may be evolutionally conserved in eukaryotes ranging from yeast, algae to humans.

Acyl-CoA-Dependent and -Independent Pathways of TAG Synthesis

Microsomal membranes have long been recognized as the primary site of TAG assembly in yeast, plants and mammals. In support of the ER-localization of TAG biosynthesis, key enzymes involved in the final step of TAG synthesis are also associated with the ER network (Wakimoto et al. 2003; Shockey et al. 2006; Cao et al. 2007) and TAGs are packaged in oil droplets in the cytosol. Both acyl-CoA-dependent and –independent pathways contribute to the last step of TAG synthesis in yeast and higher plants and a similar situation may also apply to algae. The acyl-CoA-dependent TAG synthesis is catalyzed by diacylglycerol:acyl-CoA acyltransferases (DGATs) using acyl-CoA and DAG as substrates to form TAG. Two types of membrane-bound DGATs with distinct protein sequences, functionality, expression patterns and subcellular localization are known. The acyl-CoA-independent reaction is catalyzed by phospholipid:diacylglycerol acyltransferase (PDAT), which use phospholipids as acyl donor and DAG as acyl acceptor to produce TAG and lysophospholipids. The *Chlamydomonas* genome codes for five type 2 DGATs (DGTT1–DGTT5), one type 1 DGAT (DGAT1) and one PDAT (PDAT1). Among them, the functionality of DGTT1-DGTT3 and PDAT1 has been validated by heterologous complementation of a yeast *Δdga1Δlro1* mutant lacking both DGAT and PDAT activity (Sanjaya et al. 2013; Boyle et al. 2012; Yoon et al. 2012). In addition, DGTT4 has also been demonstrated to possess DGAT activity in vitro assays using recombinant proteins, while *DGTT5* is likely to be an inactive pseudogene (Sanjaya et al. 2013).

Studies in both yeast (Oelkers et al. 2000, 2002) and plants (Fan et al. 2013a, b) have indicated that PDAT plays a major role in TAG synthesis during stages of active cell growth and division, while DGATs appear to be more important in non-growing cells (Oelkers et al. 2002; Fan et al. 2013a, b). Similarly, A recent study has shown that artificial microRNA-mediated silencing of PDAT1 specifically compromises cell growth and TAG synthesis under favorable growth conditions in *Chlamydomonas* (Yoon et al. 2012). Taken together, these results point to evolutionarily conserved PDAT functions associated with rapid cell growth and membrane proliferation in yeast, microalgae and plants, and may suggest a role for TAG metabolism in membrane lipid homeostasis during cell growth.

The Origin of DAG for TAG Assembly

In plants, DAG used for TAG synthesis is mostly derived from PC (Bates et al. 2009) and TAG assembled in the ER is characterized by the exclusive presence of C18 fatty acids at the sn-2 position of the glycerol backbone in many plants (Mattson and Volpenhein 1963). In nitrogen-starved *Chlamydomonas* cells, however, as much as 90 % DAG used for TAG synthesis is of the prokaryotic type characterized by the presence of C16 fatty acids at the sn-2 position (Fan et al. 2011; Li et al. 2012; Urzica et al. 2013). In addition, while TAG is stored in cytosolic oil droplets in wild-type strains (Goodson et al. 2011), it is deposited in both the cytosol and the chloroplast in TAG hyperaccumulating *Chlamydomonas* starchless mutants (Fan et al. 2011; Goodson 2011). The use of the prokaryotic type of DAG for TAG assembly appears not to be restricted to *Chlamydomonas* cells, as TAGs from a wide variety of other algae were reported to be also highly enriched in C16 fatty acids (Suen et al. 1987; Yongmanitchai and Ward 1993; Davidi et al. 2014b). Our analysis of the fatty acid distribution of TAG isolated from N-starved *Nannochloropsis* sp. revealed that the sn-2 position of TAG is mostly occupied by 16:0 with less than 10 % of 14:0 and 18:0 (Fig. 9.1), whereas more than 20 % of acyl chains at the sn-1+3 positions are C18 fatty acids. Very similar results were obtained in the diatom *Phaeodactylum tricornutum*. TAG isolated from the N-starved, fresh water alga *Chlorella vulgaris* contains more than 70 % of C16 fatty acids at the sn-2 position, whereas the C16 and C18 fatty acids are equally present at the sn-1+3 positions, suggesting that the prokaryotic type of DAG may be the major precursor used for TAG assembly in many algal species.

We envision three pathways for accumulating the prokaryotic type of TAG in the cytosol in algae. In the first pathway, this prokaryotic type of TAG is assembled at the chloroplast envelope using DAG derived from the prokaryotic pathway and TAG is deposited in lipid droplets originating from the chloroplast envelope membranes. This hypothesis is supported by the presence of chloroplast envelope proteins such as TGD1, TGD2 and TGD3 in lipid droplets isolated from nitrogen-starved *Chlamydomonas* cells (Nguyen et al. 2011) and the association of TAG synthetic activities with chloroplast envelope membranes in plants (Martin and Wilson 1984; Kaup et al. 2002). A recent study has showed that phytyl ester synthases (PESs) from chloroplasts of Arabidopsis process diacylglycerol acyltransferase activities in addition to catalyzing the synthesis of fatty acid phytyl esters (Lippold et al. 2012). Homologs of Arabidopsis PESs were identified in *Chlamydomonas* (Moellering

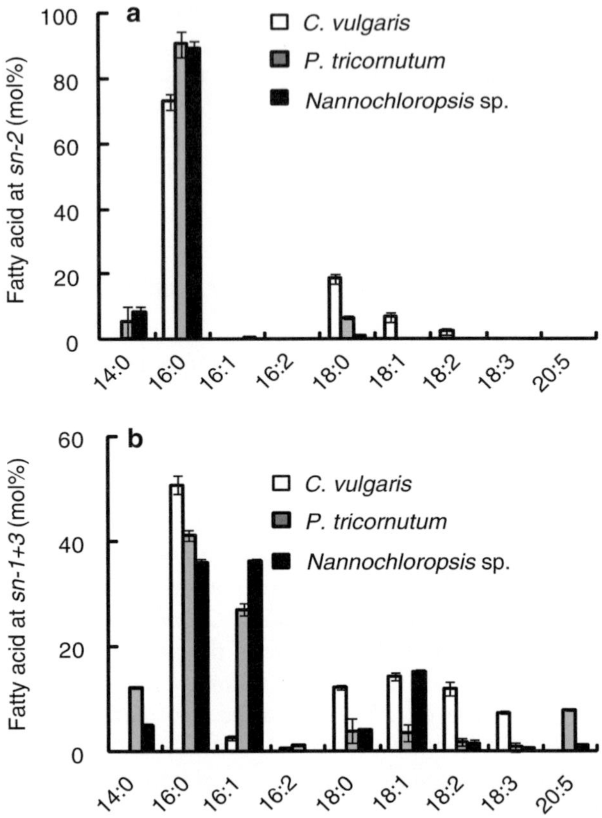

Fig. 9.1 Enrichment of 16-carbon fatty acids at the sn-2 position of TAG in algae. Fatty acid composition exclusively at the sn-2 (**a**) or sn-1+3 (**b**) positions of TAGs isolated from *C. vulgaris* (UTEX 259) *P. tricornutm* (UTEX 640) *or Nannochloropsis sp.* (CCMP 1779) grown in media lacking N for 1 day is shown. TAG positional analysis was done as described by Fan et al. (2011). Data are presented as the means and standard deviation of three or four replicates

and Benning 2010; Nguyen et al. 2011) and D*unaliella bardawil* (Davidi et al. 2014a) and these putative PES enzymes may be involved in TAG assembly in algal chloroplasts (Davidi et al. 2014a). Interestingly, a recent study has shown that PDAT1 from *Chlamydomonas* can catalyze transacylation reactions between galactolipids and the prokaryotic type of DAG, leading to generation of TAG and lysogalactolipids (Yoon et al. 2012), suggesting that PDAT1 may be important in generating the prokaryotic type of TAG in *Chlamydomonas*.

In the second pathway, TAG is assembled in the ER using DAG pools exported from the chloroplast and lipid droplets originate from the ER. In the third pathway, both DAG and TAG are assembled in the ER. This hypothesis requires the existence of acyltransferases in the ER that specifically incorporate C16 acyl chains into the sn-2 position of DAG used for TAG synthesis or selective channeling of C16 fatty acids toward the sn-2 position esterification catalyzed by ER-resident acyltransferases to generate "the prokaryotic form" of DAG in the ER. All three pathways are distinct from the relatively well known pathways present in higher plants and yeast. Testing these hypotheses awaits the identification and detailed characterization of acyltransferases from algae.

TAG Synthesis and Acyl Remodeling

In plants, fatty acids exported from the plastid are first incorporated into PC through acyl editing (remodeling) involving PC deacylation and reacylation reactions and acyl groups released from PC remodeling, rather than nascent fatty acids exported from the plastid, are used for the de novo-synthesis of membrane lipids and storage TAG in the ER (Bates and Browse 2011). In nitrogen-starved *Chlamydomonas* cells, a major fraction of fatty acids stored in TAG are derived from MGDG (Li et al. 2012). Both PDAT1 (Yoon et al. 2012) and a galactoglycerolipid lipase named PGD1 (Li et al. 2012) are implicated in the deacylation of MGDG, generating free fatty acids and lysoMGDG. Disruption of PDAT1 or PGD1 leads to a 20 or 50 % reduction in TAG content, respectively, but the exact subcellular localization of PDAT1- and PGD1-mediated deacylation reactions and the fate of resulting lyso MGDG remain uncertain.

Because the similarities in structure and biophysical properties between PC and DGTS (Sato and Murata 1991), it has been widely assumed that the betaine lipid DGTS may play some of the metabolic roles of PC in algae. Indeed, the presence of DGTS is often correlated with a decrease in the level of PC (Thompson 1996). Like PC, the betaine lipid DGTS was also found to be involved in lipid-linked fatty acid desaturation (Giroud and Eichenberger 1989), acyl remodeling and distribution in algal species *Dictyopteris membranacea* (Hofmann and Eichenberger 1997), *Fucus serratus* (Smith and Harwood 1984) and *Ascophyllum nodosum* and *Fucus vesiculosus* (Jones and Harwood 1993).

Consistent with a role of DGTS and other extraplastidic membrane lipids in supplying fatty acids for TAG synthesis, TAG isolated from nitrogen-starved cells also contains significant amounts of C18 polyunsaturated fatty acids specific to extraplastidic membrane lipids such as DGTS and PE at the sn-1 and sn-3 positions (Fan et al. 2011; Li et al. 2012; Sakurai et al. 2014). Additional evidence supporting the role of extraplasitidic membrane lipid turnover in TAG accumulation comes from pulse-chase labeling studies using radiolabeled free fatty acids. In these experiments, *Chlamydomonas* cells in the logarithmic growth phase were pulse-labeled with [1-^{14}C] labeled 16:0 or 18:1 for 16 h in complete growth medium and the movement of label was subsequently chased for 3 days. Because of the acyl asymmetrical distribution of fatty acids in glycerolipids, [1-^{14}C] 16:0 was reported to label mainly the *sn-2* position of chloroplast lipids such as MGDG and DGDG, but the *sn-1* position of extraplastidic lipids such as DGTS in *Chlamydomonas* (Giroud and Eichenberger 1989). As shown in Fig. 9.2a, immediately after the pulse, DGTS and SQDG/PI were the most strongly labeled lipids. During the chase in complete growth medium, DGTS, SQDG/PI and PG lost label over time. TAG contained minor amounts of label and stayed largely unchanged during the chase (Fig. 9.2c), whereas label in MGDG and DGDG increased significantly, suggesting that chloroplast membrane proliferation is the major sink for lipid intermediates derived from membrane lipid turnover in cells grown in nitrogen-replete medium. Under nitrogen starvation conditions, the labeled 16:0 fatty acid was rapidly decreased in DGTS,

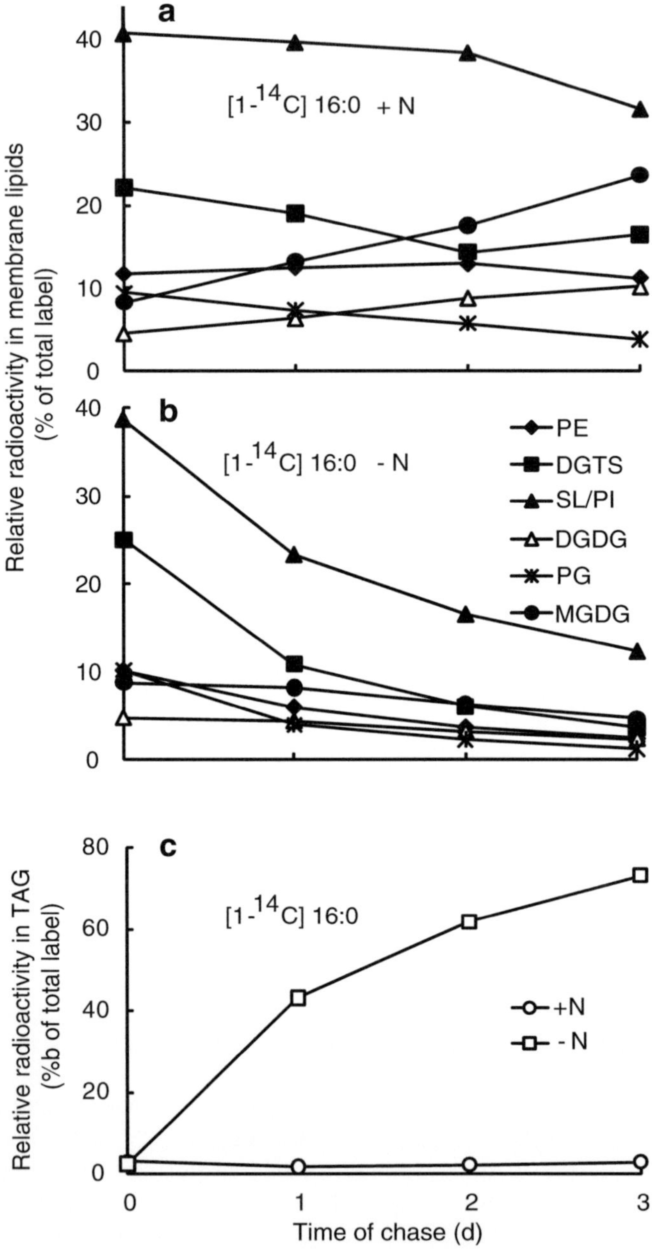

Fig. 9.2 Membrane lipid turnover contributes to TAG accumulation in response to N starvation. Cells were labeled for 16 h with [1-^{14}C] 16:0 in complete medium, thereafter shifted to unlabeled medium with (+N) or without (–N) N. Lipids were then extracted and separated by TLC at the indicated time points and radioactivity in individual polar lipids (**a** and **b**) and TAG (**c**) were determined by phosphor imaging. The experiment was repeated twice with similar results and a representative experiment is shown

SL/PI, PE and PG, but slowly in MGDG and DGDG (Fig. 9.2b), while the labeled TAG increased. At 2 days of chase, more than 60 % of total label was accumulated in TAG (Fig. 9.2c). Very similar results were obtained in pulse-chase experiments with [1-^{14}C] 18:1 (Fig. 9.3), which was previously shown to label mainly the *sn-1* position of MGDG and DGDG, but the *sn-2* position of DGTS (Giroud and Eichenberger 1989). Together, these results suggest an enhanced turnover of acyl groups at both the *sn-1* and *sn-2* positions of membrane lipids and the released fatty acids and/or other lipid intermediates are recycled into the TAG biosynthetic pathway under nitrogen starvation conditions. Because the losses in radioactivity in DGTS and PE (Figs. 9.2 and 9.3) are not accompanied by decreases in their mass under our growth conditions (Fan et al. 2011), the turnover of these membrane lipids is unlikely to be merely a consequence of net catabolism of membrane lipids but rather it may reflect the process of lipid remodeling involving deacylation and reacylation, a common mechanism widely conserved in yeasts (de Kroon 2007), plants (Bates et al. 2007, 2009) and mammals (Schmid et al. 1991). Thus, like PC in plants (Bates et al. 2007, 2009), the acyl remodeling of DGTS represents an important mechanism mediating the flux of fatty acids to TAG in *Chlamydomonas* under nitrogen starvation conditions.

The Role of Lipid Droplets in TAG Synthesis

Research in the past couple of decades has completely changed our perception of lipid droplets. It is now well recognized that rather than serving as inert globules of fats, lipid droplets are dynamic subcellular organelles that play vital roles in lipid metabolism, homeostasis and trafficking (Martin and Parton 2006; Beller et al. 2008; Guo et al. 2008; Olofsson et al. 2009; Chapman et al. 2012; Kohlwein et al. 2013; Wilfling et al. 2014). Thus, it is not surprising that many enzymes in lipid metabolic pathways are often found to be associated with lipid droplets (Sorger and Daum 2002; Rajakumari et al. 2008; Kohlwein 2010; Moessinger et al. 2011). Intriguingly, recent studies in mammalian cells have showed that the core machinery for TAG synthesis re-localizes from the ER to lipid droplets via membrane bridges between the organelles under conditions of fatty acid overload (Xu et al. 2012; Wilfling et al. 2013) and that such relocalization of TAG synthesis enzymes is essential for lipid droplet expansion (Wilfling et al. 2013). Similarly, although yeast DGAT2 is an integral ER membrane protein, it relocates to lipid droplets during stationary phase of growth, when rapid neutral lipid accumulation is occurring (Jacquier et al. 2011; Markgraf et al. 2014). These results suggest a coupling of TAG synthesis with lipid droplet expansion under conditions of lipid excess.

Recent functional genomic screens have identified as much as 1.5 % of all known genes functioning in oil-droplet formation and regulation in *Drosophila* (Guo et al. 2008) and 1.2 % in yeast (Szymanski et al. 2007). In plants, 25 oil droplet-associated proteins from mature *Brassica napus* seeds were identified in recent proteomic analysis (Jolivet et al. 2009), with oleosins being the most abundant proteins of seed

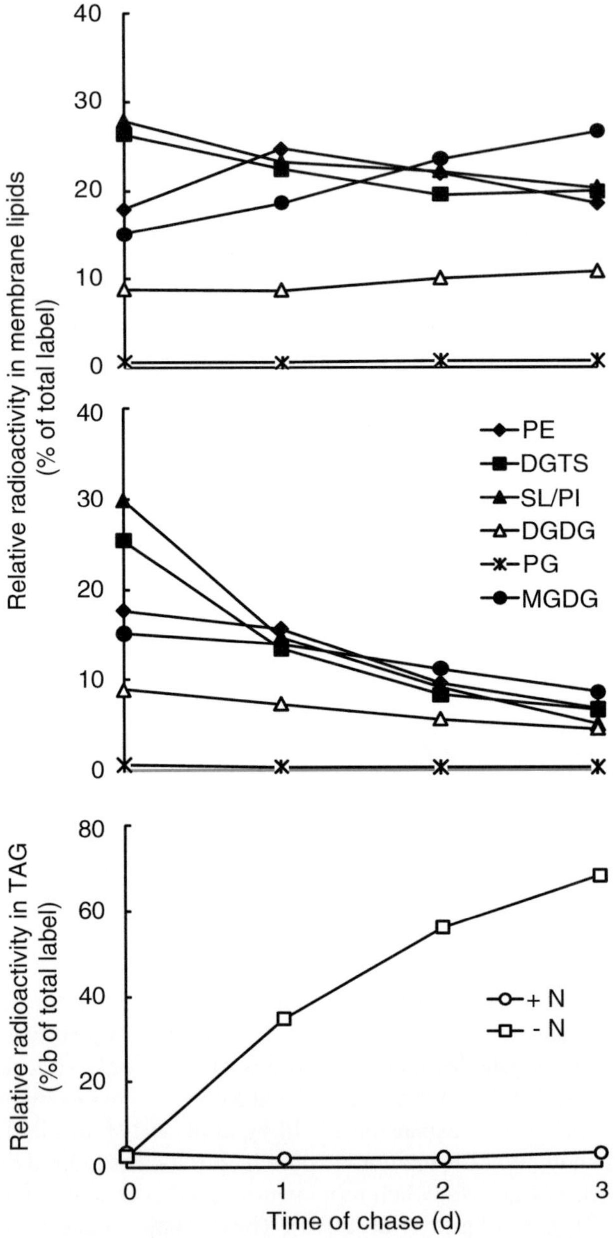

Fig. 9.3 Membrane lipid turnover contributes to TAG accumulation in response to N starvation. Cells were labeled for 16 h with [1-^{14}C] 18:1 in complete medium, thereafter shifted to unlabeled medium with (+N) or without (−N) N. Lipids were then extracted and separated by TLC at the indicated time points and radioactivity in individual polar lipids (**a** and **b**) and TAG (**c**) were determined by phosphor imaging. The experiment was repeated twice with similar results and a representative experiment is shown

oil droplets. Genetic studies showed that oleosins play an important role in regulating the size of lipid droplets in Arabidopsis seeds (Siloto et al. 2006). In addition, an OLEOSIN3 isoform from peanut cotyledons has been demonstrated to exhibit both a monoacylglycerol acyltransferase and a phospholipase activity, and overexpression of peanut OLEOSIN3 results in increased accumulation of DAG and TAG in yeast (Parthibane et al. 2012). Interestingly, a *Nannochloropsis* lipid droplet protein was recently shown to partially complement the OLEOSIN-deficiency with respect to lipid droplet size in Arabidopsis seeds (Vieler et al. 2012). Proteomic analysis yielded over 600 lipid droplet-associated proteins in *Chlamydomonas*, and many of them are likely involved in lipid metabolism (Moellering and Benning 2010; Nguyen et al. 2011). Notably, key enzymes in the TAG synthesis pathway including a glycerol-3-phosphate acyltransferase, a lysophosphatidic acid acyltransferase and a PDAT, along with proteins putatively involved in acyl remodeling, sterol synthesis and lipid trafficking were found in the recently reported *Chlamydomonas* oil droplets (Nguyen et al. 2011). Taken together, these results suggest that lipid droplets may play an evolutionary conserved role in many aspects of lipid metabolism including TAG synthesis in wide variety of organisms, ranging from microalgae to mammals and higher plants.

Conclusion

The study of biochemistry and cell biology of TAG metabolism in algae is still in its infancy but recent advances have been rapid, due primarily to a recent surge of interest in developing renewable fuels from microalgae. In particular, the discovery that TAG derived from the prokaryotic pathway accumulates in cytosolic lipid droplets in *Chlamydomonas* offers a new paradigm for interpreting TAG synthesis pathways and their organization in microalgae. In addition, the number of proteins involved in TAG synthesis, remodeling and regulation that are associated with lipid droplets has increased substantially and offers insights into the cellular compartmentalization of TAG metabolism. Future studies to exploit the molecular identity, biochemical properties, subcellular localization and dynamics of acyltransferases involved in DAG synthesis and the mechanisms underlying lipid droplet biogenesis and growth will be fundamental to our understanding of TAG metabolism and storage in microalgae and other organisms.

Acknowledgements We acknowledge the support of the Division of Chemical Sciences, Geosciences, and Biosciences, Office of Basic Energy Sciences, US Department of Energy (grant DOE KC0304000)

References

Adeyo O, Horn PJ, Lee SK, Binns DD, Chandrahas A, Chapman KD, Goodman JM (2011) The yeast lipin orthologue Pah1p is important for biogenesis of lipid droplets. J Cell Biol 192:1043–1055. doi:10.1083/jcb.201010111

Bates PD, Browse J (2011) The pathway of triacylglycerol synthesis through phosphatidylcholine in Arabidopsis produces a bottleneck for the accumulation of unusual fatty acids in transgenic seeds. Plant J 68:387–399. doi:10.1111/j.1365-313X.2011.04693

Bates PD, Ohlrogge JB, Pollard M (2007) Incorporation of newly synthesized fatty acids into cytosolic glycerolipids in pea leaves occurs via acyl editing. J Biol Chem 282:31206–31216

Bates PD, Durrett TP, Ohlrogge JB, Pollard M (2009) Analysis of acyl fluxes through multiple pathways of triacylglycerol synthesis in developing soybean embryos. Plant Physiol 150:55–72. doi:10.1104/pp. 109.137737

Beller M, Sztalryd C, Southall N, Bell M, Jackle H, Auld DS, Oliver B (2008) COPI complex is a regulator of lipid homeostasis. PLoS Biol 6:e292. doi:10.1371/journal.pbio.0060292

Benning C (2008) A role for lipid trafficking in chloroplast biogenesis. Prog Lipid Res 47:381–389. doi:10.1016/j.plipres.2008.04.001

Benning C (2009) Mechanisms of lipid transport involved in organelle biogenesis in plant cells. Ann Rev Cell Dev Biol 25:71–91. doi:10.1146/annurev.cellbio.042308.113414

Boyle NR, Page MD, Liu BS, Blaby IK, Casero D, Kropat J, Cokus SJ, Hong-Hermesdorf A, Shaw J, Karpowicz SJ, Gallaher SD, Johnson S, Benning C, Pellegrini M, Grossman A, Merchant SS (2012) Three acyltransferases and nitrogen-responsive regulator are implicated in nitrogen starvation-induced triacylglycerol accumulation in *Chlamydomonas*. J Biol Chem 287:15811–15825. doi:10.1074/jbc.M111.334052

Browse J, Somerville C (1991) Glycerolipid synthesis – biochemistry and regulation. Annu Rev Plant Physiol Plant Mol Biol 42:467–506

Browse J, Warwick N, Somerville CR, Slack CR (1986) Fluxes through the prokaryotic and eukaryotic pathways of lipid synthesis in the '16:3' plant Arabidopsis thaliana. Biochem J 235:25–31

Cao JS, Cheng L, Shi YG (2007) Catalytic properties of MGAT3, a putative triacylglycerol synthase. J Lipid Res 48:583–591

Chapman KD, Dyer JM, Mullen RT (2012) Biogenesis and functions of lipid droplets in plants. J Lipid Res 53:215–226. doi:10.1194/jlr.R021436

Davidi L, Levin Y, Ben-Dor S, Pick U (2014a) Proteome analysis of cytoplasmatic and of plastidic beta-carotene lipid droplets in *Dunaliella bardawil*. Plant Physiol 167:60–79. doi:10.1104/pp. 114.248450

Davidi L, Shimoni E, Khozin-Goldberg I, Zamir A, Pick U (2014b) Origin of beta-carotene-rich plastoglobuli in *Dunaliella bardawil*. Plant Physiol 164:2139–2156. doi:10.1104/pp. 113.235119

de Kroon AI (2007) Metabolism of phosphatidylcholine and its implications for lipid acyl chain composition in *Saccharomyces cerevisiae*. Biochim Biophys Acta 1771:343–352

Deng XD, Cai JJ, Fei XW (2013) Involvement of phosphatidate phosphatase in the biosynthesis of triacylglycerols in *Chlamydomonas reinhardtii*. J Zhejiang Univ Sci B 14:1121–1131. doi:10.1631/jzus.B1300180

Dörmann P, Balbo I, Benning C (1999) Arabidopsis galactolipid biosynthesis and lipid trafficking mediated by DGD1. Science 284:2181–2184

Eastmond PJ, Quettier AL, Kroon JTM, Craddock C, Adams N, Slabas AR (2010) PHOSPHATIDIC ACID PHOSPHOHYDROLASE1 and 2 regulate phospholipid synthesis at the endoplasmic reticulum in Arabidopsis. Plant Cell 22:4216–4216. doi:10.1105/tpc.109.071423

Fan JL, Andre C, Xu CC (2011) A chloroplast pathway for the de novo biosynthesis of triacylglycerol in *Chlamydomonas reinhardtii*. FEBS Lett 585:1985–1991. doi:10.1016/j.febslet.2011.05.018

Fan JL, Yan CS, Xu CC (2013a) Phospholipid: diacylglycerol acyltransferase-mediated triacylglycerol biosynthesis is crucial for protection against fatty acid-induced cell death in growing tissues of Arabidopsis. Plant J 76:930–942. doi:10.1111/tpj.12343

Fan JL, Yan CS, Zhang XB, Xu CC (2013b) Dual role for phospholipid: diacylglycerol acyltransferase: enhancing fatty acid synthesis and diverting fatty acids from membrane lipids to triacylglycerol in Arabidopsis leaves. Plant Cell 25:3506–3518. doi:10.1105/tpc.113.117358

Fan J, Yan C, Roston R, Shanklin J, Xu C (2014) Arabidopsis lipins, PDAT1 acyltransferase, and SDP1 triacylglycerol lipase synergistically direct fatty acids toward beta-oxidation, thereby maintaining membrane lipid homeostasis. Plant Cell 26:4119–4134. doi:10.1105/tpc.114.130377

Frentzen M (1998) Acyltransferases from basic science to modified seed oils. Fett Lipid 100:161–166

Giroud C, Eichenberger W (1989) Lipids of *Chlamydomonas reinhardtii* – incorporation of [C-14] acetate, [C-14] palmitate and [C-14] oleate into different lipids and evidence for lipid-linked desaturation of fatty acids. Plant Cell Physiol 30:121–128

Giroud C, Gerber A, Eichenberger W (1988) Lipids of *Chlamydomonas reinhardtii* – analysis of molecular species and intracellular sites of biosynthesis. Plant Cell Physiol 29:587–595

Goodson C, Roth R, Wang ZT, Goodenough U (2011) Structural correlates of cytoplasmic and chloroplast lipid body synthesis in *Chlamydomonas reinhardtii* and stimulation of lipid-body production with acetate-boost. Eukaryot Cell 10:1592–1606. doi:10.1128/EC.05242-11

Guo Y, Walther TC, Rao M, Stuurman N, Goshima G, Terayama K, Wong JS, Vale RD, Walter P, Farese RV (2008) Functional genomic screen reveals genes involved in lipid-droplet formation and utilization. Nature 453:657–661

Han GS, Wu WI, Carman GM (2006) The *Saccharomyces cerevisiae* lipin homolog is a Mg2+-dependent phosphatidate phosphatase enzyme. J Biol Chem 281:9210–9218

Harris TE, Finck BN (2011) Dual function lipin proteins and glycerolipid metabolism. Trends Endocrinol Metab 22:226–233. doi:10.1016/j.tem.2011.02.006

Harwood JL, Guschina IA (2009) The versatility of algae and their lipid metabolism. Biochimie 91:679–684. doi:10.1016/j.biochi.2008.11.004

Hofmann M, Eichenberger W (1997) Lipid and fatty acid composition of the marine brown alga *Dictyopteris membranacea*. Plant Cell Physiol 38:1046–1052

Jacquier N, Choudhary V, Mari M, Toulmay A, Reggiori F, Schneiter R (2011) Lipid droplets are functionally connected to the endoplasmic reticulum in *Saccharomyces cerevisiae*. J Cell Sci 124:2424–2437. doi:10.1242/jcs.076836

Jolivet P, Boulard C, Bellamy A, Larre C, Barre M, Rogniaux H, d'Andrea S, Chardot T, Nesi N (2009) Protein composition of oil bodies from mature *Brassica napus* seeds. Proteomics 9:3268–3284. doi:10.1002/pmic.200800449

Jones AL, Harwood JL (1992) Lipid composition of the brown algae *Fucus vesiculosus* and *Ascophyllum nodosum*. Phytochemistry 31:3397–3403

Jones AL, Harwood JL (1993) Lipid metabolism in the brown marine algae *Fucus vesiculosus* and *Ascophyllum nodosum*. J Exp Bot 44:1203–1210

Kaup MT, Froese CD, Thompson JE (2002) A role for diacylglycerol acyltransferase during leaf senescence. Plant Physiol 129:1616–1626

Kohlwein SD (2010) Triacylglycerol homeostasis: insights from yeast. J Biol Chem 285:15663–15667. doi:10.1074/jbc.R110.118356

Kohlwein SD, Veenhuis M, van der Klei IJ (2013) Lipid droplets and peroxisomes: key players in cellular lipid homeostasis or a matter of fat-store 'em up or burn 'em down. Genetics 193:1–50. doi:10.1534/genetics.112.143362

Li XB, Moellering ER, Liu BS, Johnny C, Fedewa M, Sears BB, Kuo MH, Benning C (2012) A galactoglycerolipid lipase is required for triacylglycerol accumulation and survival following nitrogen deprivation in *Chlamydomonas reinhardtii*. Plant Cell 24:4670–4686. doi:10.1105/tpc.112.105106

Lippold F, vom Dorp K, Abraham M, Holzl G, Wewer V, Yilmaz JL, Lager I, Montandon C, Besagni C, Kessler F, Stymne S, Dormann P (2012) Fatty acid phytyl ester synthesis in chloroplasts of Arabidopsis. Plant Cell 24:2001–2014. doi:10.1105/tpc.112.095588

Makewicz A, Gribi C, Eichenberger W (1997) Lipids of *Ectocarpus fasciculatus* (*Phaeophyceae*). Incorporation of [1-C-14] oleate and the role of TAG and MGDG in lipid metabolism. Plant Cell Physiol 38:952–960

Markgraf DF, Klemm RW, Junker M, Hannibal-Bach HK, Ejsing CS, Rapoport TA (2014) An ER protein functionally couples neutral lipid metabolism on lipid droplets to membrane lipid synthesis in the ER. Cell Rep 6:44–55. doi:10.1016/j.celrep.2013.11.046

Martin S, Parton RG (2006) Lipid droplets: a unified view of a dynamic organelle. Nat Rev Mol Cell Biol 7:373–378

Martin BA, Wilson RF (1984) Subcellular localization of triacylglycerol synthesis in Spinach leaves. Lipids 19:117–121

Mattson FH, Volpenhein RA (1963) Specific distribution of unsaturated fatty acids in triglycerides of plants. J Lipid Res 4:392–398

Moellering ER, Benning C (2010) RNA interference silencing of a major lipid droplet protein affects lipid droplet size in *Chlamydomonas reinhardtii*. Eukaryot Cell 9:97–106. doi:10.1128/EC.00203-09

Moellering ER, Miller R, Benning C (2009) Molecular genetics of lipid metabolism in the model green alga *Chlamydomonas reinhardtii*. In: Wada H, Murata M (eds) Lipids in photosynthesis: essential and regulatory functions. Springer, Dordrecht, pp 139–150

Moessinger C, Kuerschner L, Spandl J, Shevchenko A, Thiele C (2011) Human lysophosphatidylcholine acyltransferases 1 and 2 are located in lipid droplets where they catalyze the formation of phosphatidylcholine. J Biol Chem 286:21330–21339. doi:10.1074/jbc.M110.202424

Nguyen HM, Baudet M, Cuine S, Adriano JM, Barthe D, Billon E, Bruley C, Beisson F, Peltier G, Ferro M, Li-Beisson Y (2011) Proteomic profiling of oil bodies isolated from the unicellular green microalga *Chlamydomonas reinhardtii*: with focus on proteins involved in lipid metabolism. Proteomics 11:4266–4273. doi:10.1002/pmic.201100114

Oelkers P, Tinkelenberg A, Erdeniz N, Cromley D, Billheimer JT, Sturley SL (2000) A lecithin cholesterol acyltransferase-like gene mediates diacylglycerol esterification in yeast. J Biol Chem 275:15609–15612

Oelkers P, Cromley D, Padamsee M, Billheimer JT, Sturley SL (2002) The DGA1 gene determines a second triglyceride synthetic pathway in yeast. J Biol Chem 277:8877–8881

Ohlrogge JB, Jaworski JG (1997) Regulation of fatty acid synthesis. Annu Rev Plant Physiol Plant Mol Biol 48:109–136

Olofsson SO, Bostrom P, Andersson L, Rutberg M, Perman J, Boren J (2009) Lipid droplets as dynamic organelles connecting storage and efflux of lipids. Biochim Biophys Acta 1791:448–458. doi:10.1016/j.bbalip.2008.08.001

Parthibane V, Rajakumari S, Venkateshwari V, Iyappan R, Rajasekharan R (2012) Oleosin is bifunctional enzyme that has both monoacylglycerol acyltransferase and phospholipase activities. J Biol Chem 287:1946–1954. doi:10.1074/jbc.M111.309955

Pascual F, Carman GM (2013) Phosphatidate phosphatase, a key regulator of lipid homeostasis. Biochim Biophys Acta 1831:514–522. doi:10.1016/j.bbalip.2012.08.006

Rajakumari S, Grillitsch K, Daum G (2008) Synthesis and turnover of non-polar lipids in yeast. Prog Lipid Res 47:157–171

Riekhof W, Benning C (2009) Glycerolipid biosynthesis. In: Stern DB (ed) The chlamydomonas sourcebook: organellar and metabolic processes, 2nd edn. Academic, Oxford

Roughan PG, Slack CR (1982) Cellular organization of glycerolipid metabolism. Ann Rev Plant Physiol 33:97–132

Sakurai K, Moriyama T, Sato N (2014) Detailed identification of fatty acid isomers sheds light on the probable precursors of triacylglycerol accumulation in photoautotrophically grown *Chlamydomonas reinhardtii*. Eukaryot Cell 13:256–266. doi:10.1128/EC.00280-13

Sanjaya MR, Durrett TP, Kosma DK, Lydic TA, Muthan B, Koo AJ, Bukhman YV, Reid GE, Howe GA, Ohlrogge J, Benning C (2013) Altered lipid composition and enhance notional value of Arabidopsis leaves following the introduction of an algal diacylglycerol acyltransferase 2. Plant Cell 25:677–693. doi:10.1105/tpc.112.104752

Santos-Rosa H, Leung J, Grimsey N, Peak-Chew S, Siniossoglou S (2005) The yeast lipin Smp2 couples phospholipid biosynthesis to nuclear membrane growth. EMBO J 24:1931–1941

Sato N, Moriyama T (2007) Genomic and biochemical analysis of lipid biosynthesis in the unicellular rhodophyte *Cyanidioschyzon merolae*: lack of a plastidic desaturation pathway results in the coupled pathway of galactolipid synthesis. Eukaryot Cell 6:1006–1017

Sato N, Murata N (1991) Transition of lipid phase in aqueous dispersions of diacylglyceryltrimethylhomoserine. Biochim Biophys Acta 1082:108–111

Sato N, Tsuzuki M, Kawaguchi A (2003) Glycerolipid synthesis in *Chlorella kessleri* 11 h – I. Existence of a eukaryotic pathway. Biochim Biophys Acta 1633:27–34
Schmid PC, Johnson SB, Schmid HH (1991) Remodeling of rat hepatocyte phospholipids by selective acyl turnover. J Biol Chem 266:13690–13697
Shockey JM, Gidda SK, Chapital DC, Kuan JC, Dhanoa PK, Bland JM, Rothstein SJ, Mullen RT, Dyer JM (2006) Tung tree DGAT1 and DGAT2 have nonredundant functions in triacylglycerol biosynthesis and are localized to different subdomains of the endoplasmic reticulum. Plant Cell 18:2294–2313
Siloto RMP, Findlay K, Lopez-Villalobos A, Yeung EC, Nykiforuk CL, Moloney MM (2006) The accumulation of oleosins determines the size of seed oilbodies in Arabidopsis. Plant Cell 18:1961–1974
Siniossoglou S (2013) Phospholipid metabolism and nuclear function: roles of the lipin family of phosphatidic acid phosphatases. Biochim Biophys Acta 1831:575–581. doi:10.1016/j.bbalip.2012.09.014
Smith KL, Harwood JL (1984) Lipids and lipid metabolism in the brown alga, *Fucus serratus*. Phytochemistry 23:2469–2473
Sorger D, Daum G (2002) Synthesis of triacylglycerols by the acyl-coenzyme A: diacyl-glycerol acyltransferase Dga1p in lipid particles of the yeast *Saccharomyces cerevisiae*. J Bacteriol 184:519–524
Suen Y, Hubbard JS, Holzer G, Tornabene TG (1987) Total lipid production of the green-alga *Nannochloropsis* Sp Qii under different nitrogen regimes. J Phycol 23:289–296
Szymanski KM, Binns D, Bartz R, Grishin NV, Li WP, Agarwal AK, Garg A, Anderson RG, Goodman JM (2007) The lipodystrophy protein seipin is found at endoplasmic reticulum lipid droplet junctions and is important for droplet morphology. Proc Natl Acad Sci U S A 104:20890–20895
Thiele C, Spandl J (2008) Cell biology of lipid droplets. Curr Opin Cell Biol 20:378–385
Thompson GA (1996) Lipids and membrane function in green algae. Biochim Biophys Acta 1302:17–45
Urzica EI, Vieler A, Hong-Hermesdorf A, Page MD, Casero D, Gallaher SD, Kropat J, Pellegrini M, Benning C, Merchant SS (2013) Remodeling of membrane lipids in iron-starved *Chlamydomonas*. J Biol Chem 288:30246–30258. doi:10.1074/jbc.M113.490425
Vieler A, Brubaker SB, Vick B, Benning C (2012) A lipid droplet protein of *Nannochloropsis* with functions partially analogous to plant oleosins. Plant Physiol 158:1562–1569. doi:10.1104/pp.111.193029
Wakimoto K, Chiba H, Michibata H, Seishima M, Kawasaki S, Okubo K, Mitsui H, Torii H, Imai Y (2003) A novel diacylglycerol acyltransferase (DGAT2) is decreased in human psoriatic skin and increased in diabetic mice. Biochem Biophys Res Commun 310:296–302
Walther TC, Farese RV Jr (2009) The life of lipid droplets. Biochim Biophys Acta 1791:459–466. doi:10.1016/j.bbalip.2008
Wilfling F, Wang HJ, Haas JT, Krahmer N, Gould TJ, Uchida A, Cheng JX, Graham M, Christiano R, Frohlich F, Liu XR, Buhman KK, Coleman RA, Bewersdorf J, Farese RV, Walther TC (2013) Triacylglycerol synthesis enzymes mediate lipid droplet growth by relocalizing from the ER to lipid droplets. Dev Cell 24:384–399. doi:10.1016/j.devcel.2013.01.013
Wilfling F, Haas JT, Walther TC, Farese RV (2014) Lipid droplet biogenesis. Curr Opin Cell Biol 29:39–45. doi:10.1016/j.ceb.2014.03.008
Xu NY, Zhang SBO, Cole RA, McKinney SA, Guo FL, Haas JT, Bobba S, Farese RV, Mak HY (2012) The FATP1-DGAT2 complex facilitates lipid droplet expansion at the ER-lipid droplet interface. J Cell Biol 198:895–911. doi:10.1083/jcb.201201139
Yongmanitchai W, Ward OP (1993) Molecular species of triacylglycerols from the fresh-water diatom, *Phaeodactylum tricornutum*. Phytochemistry 32:1137–1139
Yoon K, Han DX, Li YT, Sommerfeld M, Hu Q (2012) Phospholipid:diacylglycerol acyltransferase is a multifunctional enzyme involved in membrane lipid turnover and degradation while synthesizing triacylglycerol in the unicellular green microalga *Chlamydomonas reinhardtii*. Plant Cell 24:3708–3724. doi:10.1105/tpc.112.100701

Chapter 10
High-Throughput Genetics Strategies for Identifying New Components of Lipid Metabolism in the Green Alga *Chlamydomonas reinhardtii*

Xiaobo Li and Martin C. Jonikas

Abstract Microalgal lipid metabolism is of broad interest because microalgae accumulate large amounts of triacylglycerols (TAGs) that can be used for biodiesel production (Durrett et al Plant J 54(4):593–607, 2008; Hu et al Plant J 54(4):621–639, 2008). Additionally, green algae are close relatives of land plants and serve as models to understand conserved lipid metabolism pathways in the green lineage. The green alga *Chlamydomonas reinhardtii* (Chlamydomonas hereafter) is a powerful model organism for understanding algal lipid metabolism. Various methods have been used to screen Chlamydomonas mutants for lipid amount or composition, and for identification of the mutated loci in mutants of interest. In this chapter, we summarize the advantages and caveats for each of these methods with a focus on screens for mutants with perturbed TAG content. We also discuss technical opportunities and new tools that are becoming available for screens of mutants altered in TAG content or perturbed in other processes in Chlamydomonas.

Keywords Triacylglycerol • Lipids • High-throughput genetics • Chlamydomonas • Algae • Nitrogen deprivation • Lipid analysis • Fluorescence-activated cell sorting • Lipid droplets

Chlamydomonas Is a Powerful Model Organism for Answering Key Questions in Microalgal Lipid Metabolism

Chlamydomonas is a unicellular green alga that can be used to study a broad spectrum of biological processes, including photosynthesis, the algal carbon concentrating mechanism, cell motility, light reception, phototaxis, nutrient stress responses and the cell cycle. Chlamydomonas has a number of useful features that make it

X. Li • M.C. Jonikas (✉)
Department of Plant Biology, Carnegie Institution for Science, Stanford, CA 94305, USA
e-mail: mjonikas@carnegiescience.edu

Y. Nakamura, Y. Li-Beisson (eds.), *Lipids in Plant and Algae Development*, Subcellular Biochemistry 86, DOI 10.1007/978-3-319-25979-6_10

convenient for genetic characterization of these processes. It is haploid during vegetative growth, so phenotypes of mutants can be observed without the need to generate homozygous lines (Harris et al. 2009). The nuclear (Merchant et al. 2007), chloroplast (Maul et al. 2002) and mitochondrial (GenBank accession number U03843) genomes have all been sequenced and can be transformed (Boynton et al. 1988; Kindle et al. 1989; Randolph-Anderson et al. 1993). A collection of selection markers, expression cassettes and fluorescent tags has been developed (Harris et al. 2009; Rasala et al. 2014). Large numbers of nuclear mutants can be generated by chemical or random insertional mutagenesis (Tam and Lefebvre 1993). Chlamydomonas has a well-characterized and rapid mating cycle, allowing combination of mutations and the analysis of segregation patterns by tetrad and random spore analyses (Harris et al. 2009).

Chlamydomonas is also a powerful model system for studies of lipid metabolism. Four qualities make it particularly useful for this application: (1) Chlamydomonas can be induced to accumulate TAGs in lipid droplets under nutrient-deprived conditions, for example under nitrogen (N) deprivation (Moellering and Benning 2010; Wang et al. 2009). Thus, Chlamydomonas can be used to understand how TAG is synthesized and assembled into lipid droplets and how algal lipid metabolism is regulated. (2) Its lipid metabolism is primarily regulated by environmental conditions, unlike in land plants, where lipid synthesis and turnover are also tissue-dependent (To et al. 2012). This simplifies the design, analysis and interpretation of experiments. (3) Chlamydomonas is unicellular, which enables higher throughput mutant screens. (4) Chlamydomonas is a member of the green lineage, with thylakoids and a chloroplast similar to those of land plants in structure and composition. Thus, Chlamydomonas can be used to understand processes that are conserved between green algae and land plants, such as the export of fatty acids from the chloroplast (Riekhof et al. 2005).

A significant fraction of current Chlamydomonas lipid studies is focused on the discovery of genes involved in TAG metabolism. Candidate genes with possible functions in TAG metabolism have been found by homology search (Riekhof et al. 2005), transcript profiling under normal growth conditions and N-deprived conditions (Blaby et al. 2013; Boyle et al. 2012; Goodenough et al. 2014; Miller et al. 2010; Schmollinger et al. 2014), and proteomic characterization of lipid droplets (Moellering and Benning 2010; Nguyen et al. 2011). However, the functions of only a handful of these candidate genes have been confirmed experimentally (Boyle et al. 2012; Li et al. 2012c; Yoon et al. 2012). Additionally, mutants with defects in lipids have been identified in screens, but the genes responsible for the lipid phenotypes have only been identified in a small fraction of cases (Cagnon et al. 2013; Xie et al. 2014; Yan et al. 2013; Li et al. 2012c). A comprehensive understanding of Chlamydomonas TAG metabolism will require additional unbiased screens for TAG mutants, combined with efficient methods for identifying the genes disrupted in the hits. Below, we review existing screening and genotyping tools, and present some tools that are in development.

High-Throughput Genetic Analyses of TAG Metabolism Traditionally Include Three Parts

A traditional forward genetic study consists of three major stages: mutagenesis, phenotypic screen and identification of mutated genes. Typically, a screen starts by randomly mutagenizing wild-type cells by radiation, chemicals or DNA insertions (Fig. 10.1). The TAG content in mutants is then analyzed by direct or indirect methods, in single cells or in a clonal population of each mutant. As there is usually a tradeoff between the throughput and the accuracy of phenotypic analyses, phenotyping is often performed in two stages: a high-throughput screen that rapidly identifies candidate mutants (called "primary screen" here) followed by a lower-throughput but higher accuracy analysis of the hits ("secondary screen"). Once mutants with abnormal TAG content are identified, mutated genes can be identified. We will first discuss screen growth conditions and methods of measuring TAG content. We will then discuss available approaches for generating mutants and for identifying mutated genes.

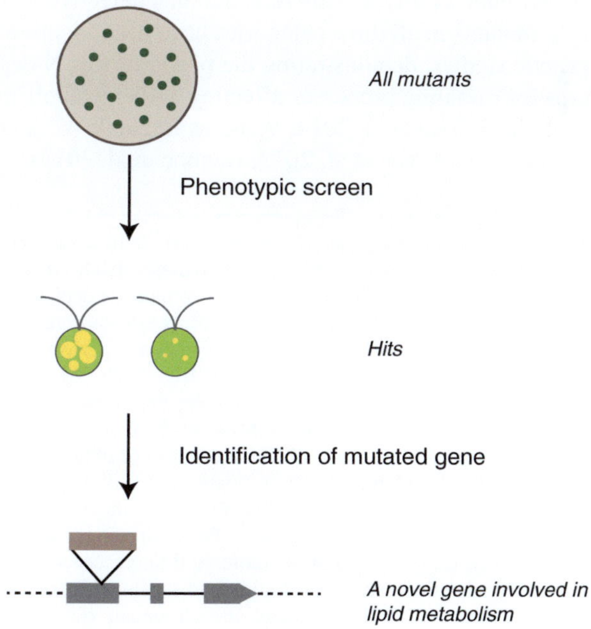

Fig. 10.1 A forward genetic study for mutants with abnormal TAG content consists of three steps. Mutants are generated from wild-type cells using random mutagenesis, followed by high-throughput phenotypic analysis (TAG content analysis in this case). Mutants with higher or lower amounts of TAG than wild-type cells, referred as "hits" in the figure and depicted as two cells with large and small lipid droplets, are isolated. The causative mutation in these mutants is then identified using map-based cloning, whole-genome sequencing or PCR-based methods. An example is given for an insertional mutant where a piece of exogenous DNA inserted into an exon of a gene results in perturbed TAG levels

Growth Conditions Can Affect TAG Content and the Classes of Mutants That Can Be Isolated

The most common environmental perturbation used for inducing TAG accumulation is N deprivation. Many other stress conditions such as high light and high salinity also induce TAG accumulation, but to a lesser extent (Fan et al. 2011). When N is replete, TAG is detectable but minimal. Under N deprivation, growth stops and TAG is gradually accumulated (Moellering and Benning 2010; Wang et al. 2009). Upon the resupply of N to N-deprived cells, growth resumes and TAG is gradually degraded or diluted by cell division until it reaches a level comparable with the level before N deprivation (Fig. 10.2a) (Cagnon et al. 2013; Li et al. 2012b, 2014; Tsai et al. 2014).

Broadly speaking, a N starvation and re-feeding experiment can reveal mutants that differ from wild-type in up to three stages: (1) Under N-replete conditions, some mutants may already accumulate excess or insufficient TAG (e.g. mutant with high steady state TAG, Fig. 10.2a); such mutants could have mis-regulated signal transduction or metabolism during vegetative growth. Considering the low levels of TAG in wild-type Chlamydomonas cells under N-replete conditions, it would be challenging to screen for mutants with lower steady-state TAG levels. However, this may be possible, as one mutant with such a phenotype was isolated with reverse genetic approaches (Yoon et al. 2012). (2) When N is deprived, mutants with defective or enhanced TAG synthesis or signaling pathways will accumulate less or more TAG, respectively, than wild-type cells (e.g. mutant with high TAG induction, Fig. 10.2a). (3) Finally, one could also identify mutants that degrade TAG faster or slower than wild-type cells (e.g. mutants defective in TAG turnover, Fig. 10.2a). In fact, mutants in all three categories have been identified through forward or reverse genetic studies, demonstrating the power of this N deprivation and re-feeding process for revealing processes affecting TAG metabolism (Li et al. 2012c; Terashima et al. 2014; Tsai et al. 2014; Wang et al. 2009; Work et al. 2010; Boyle et al. 2012; Xie et al. 2014; Yan et al. 2013; Cagnon et al. 2013).

Fig. 10.2 (continued) point. (**b**) In dot-thinFig. 10.2 (continued) layer chromatography (Dot-TLC), lipids are extracted and resolved radially. Each *circle* corresponds to a lipid species. A scheme is shown for eight mutants with two mutants depicted in *black*. Note that one of the two mutants over-accumulates one lipid species. (**c**) In gas chromatography-flame ionization detector (GC-FID) analysis, whole cell transesterification is performed, followed by gas chromatography separation of fatty acid methyl esters (FAMEs). Each FAME is depicted as one peak. FAME profiles are presented for two strains (the solid line depicts one strain; the dashed line depicts the other). Ratios between different FAMEs can be used to estimate the relative abundance of lipid species. The abundance of total FAMEs can also be obtained. (**d**) After staining with a lipid sensitive dye, cells of each mutant can be analyzed in bulk with the fluorescence quantified in 96-well format. (**e**) Flow cytometry allows the analysis of the lipid content of each cell of a mutant. Each *dot* in the graph represents a single cell. For each cell, the fluorescence emitted from a lipid sensitive dye is typically compared to chlorophyll fluorescence to offset the noise caused by extrinsic factors. Median fluorescence from the lipid sensitive dye across the whole population of cells is used as a metric of the TAG content of each mutant. (**f**) A mixed population of cells of different mutants is subjected to staining and fluorescence-activated cell sorting. Cells with high-TAG or low-TAG content (in the regions indicated by the *red triangles*) are isolated

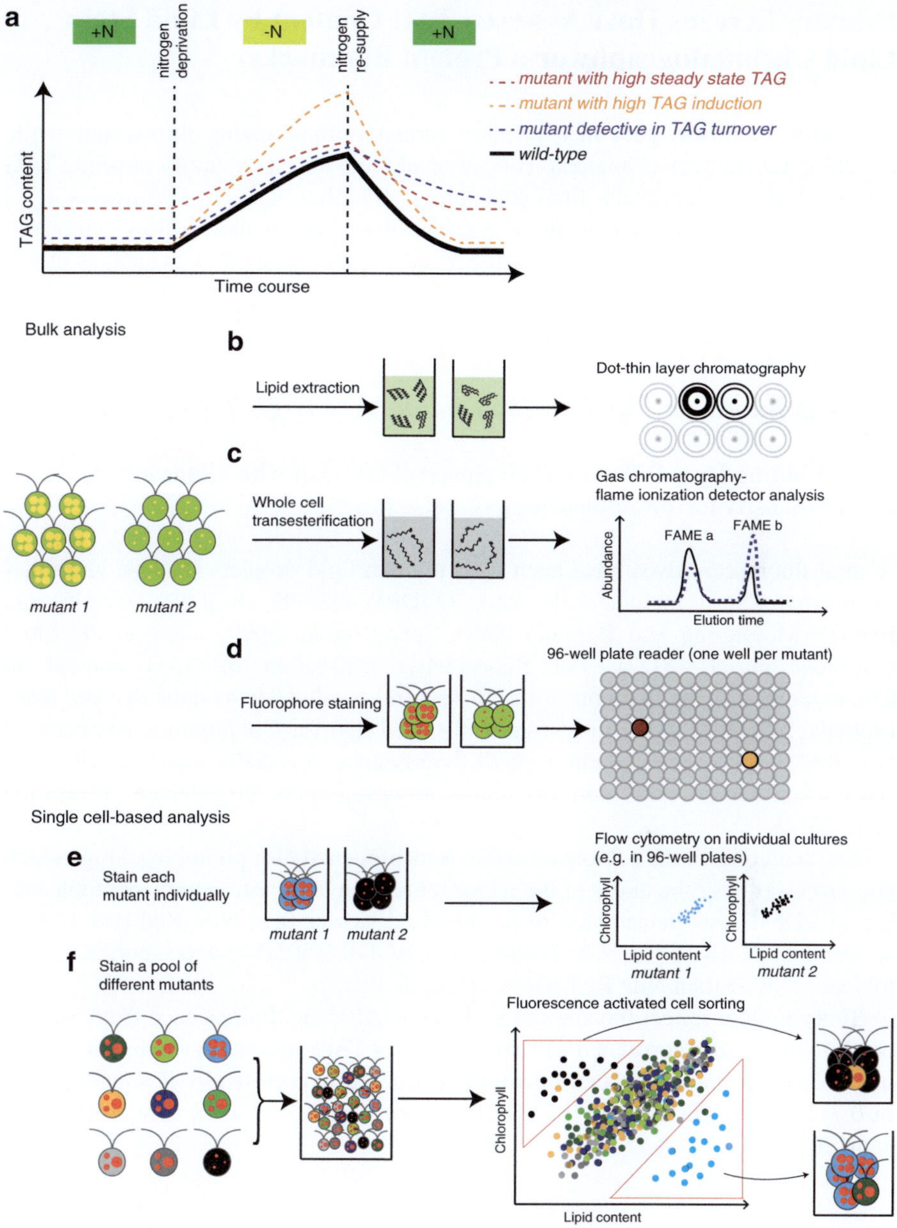

Fig. 10.2 Mutants can be screened for aberrations in the content of TAG or other lipids under several conditions with a variety of approaches. (**a**) Three examples of mutants with abnormal TAG contents in a certain stage within an N deprivation/resupply cycle are shown. One mutant accumulates a significant amount of TAG without nitrogen deprivation. Another mutant accumulates more TAG than wild-type under nitrogen deprivation. A third mutant has a defect in degrading the TAG accumulated when nitrogen is re-supplied. Note that this is not an exhaustive list of TAG phenotypes, and some mutants exhibit phenotypes under more than one condition. (**b**–**f**) Several methods are available for identifying mutants with perturbed lipid content at any time

Primary Screens Have Assessed TAG Content by Lipid Stains, Lipid Chromatography or a Protein Bio-marker

The major technical goal in a primary screen is maximizing throughput while enriching for mutants of interest. Biochemical methods can quantify absolute TAG amount but are relatively time-consuming. Staining with lipid-sensitive dyes enables high-throughput analysis with a microtiter plate reader or flow cytometer, although one must be cautious, as artifacts can be caused by factors such as inefficient dye penetration.

Screens Based on Lipid Stains Enable Very High Throughput

Lipid-Binding Dyes Differ in Their Spectral Overlap with Chlorophyll, and Propensity for Photobleaching

Several fluorescent dyes have been used to stain lipid droplets in algae and other organisms. These include Nile Red, BODIPY493/503, BODIPY505/515 and LD540 (Moellering and Benning 2010; Spandl et al. 2009; Wang et al. 2009; Govender et al. 2012). For fluorometric estimation of TAG content in Chlamydomonas, the first concern is the spectral overlap between the dye and chlorophylls. Nile Red is the most frequently used dye for Chlamydomonas because Nile Red fluorescence and chlorophyll fluorescence minimally interfere with each other when certain excitation and emission wavelengths are selected (Terashima et al. 2014).

Another concern for a fluorescent dye is its propensity for photobleaching, which compromises the robustness of the measurement. In an in vitro study, photobleaching of LD540 was determined to be threefold slower than Nile Red and 15-fold slower than BODIPY493/503 (Thiele 2011). BODIPY505/515 was reported to photobleach slower than Nile Red (Govender et al. 2012).

To accurately reflect the amount of TAG in cells, the fluorescence from the dye needs to be linearly correlated with the amount of TAG present. In Chlamydomonas, one staining protocol has been shown to yield a decent correlation (R-squared value of 0.7) between the intensity of Nile Red fluorescence and TAG content (Cagnon et al. 2013).

Staining Can Be Improved by Permeabilizing Cells to Facilitate Dye Entry

One major challenge for quantitative staining of lipid droplets is ensuring that the dye can efficiently enter cells. Several treatments, such as heating by microwaving (Chen et al. 2011), a low amount of detergents (Terashima et al. 2014), or presence of organic solvents (Doan and Obbard 2011) have been reported to enhance the staining efficiency in Chlamydomonas or other algae. The concentration of the dye also needs to be optimized, as a very high concentration can cause fluorescence from dye precipitates in addition to lipid droplets (Cirulis et al. 2012).

Lipid Content Can Be Quantified with a Plate Reader or by Flow Cytometry

Three general formats have been used for screening mutants stained with fluorescent lipid-sensitive dyes: a microtiter plate reader, flow cytometry of individual mutants, and flow cytometry of pools of mutants (Fig. 10.2b). Li et al. and Yan et al. stained mutants in 96-well plates and measured fluorescence emitted by lipid-sensitive dyes with a plate reader (Li et al. 2012c; Yan et al. 2013). Cagnon et al. also stained mutants in 96-well plates, but individually analyzed each mutant by flow cytometry (Cagnon et al. 2013). Deviation of the mutant population from the wild-type population in the lipid-sensitive dye fluorescence dimension was considered a phenotype. As flow cytometry measures the fluorescence of single cells, Xie et al. and Terashima et al. isolated high-TAG or low-TAG mutants from a mixed pool of mutants using fluorescence-activated cell sorting (FACS) (Terashima et al. 2014; Xie et al. 2014). Combined with high-throughput sequencing, FACS of pools of mutants has the potential to reveal all the genes involved in TAG metabolism in a single experiment. The throughput and other features of the above methods are summarized in Table 10.1.

Table 10.1 Properties of different TAG primary screen strategies in Chlamydomonas

Method	Throughput in 1 month	Properties	References
Dot-thin layer chromatography	~2,000 plants (Changcheng Xu, personal communication)	Can be used to estimate major lipid species	Xu et al. (2003)
Gas chromatography-flame ionization detector	~500 (Yonghua Li-Beisson, personal communication)	Can be used to estimate content of lipid species	Nguyen et al. (2013) and Pflaster et al. (2014)
Staining coupled with plate reader	~3,000 (Changcheng Xu, personal communication)	TAG specific	Li et al. (2012c) and Yan et al. (2013)
Staining coupled with flow cytometry	~1,000 (Yonghua Li-Beisson, personal communication)	TAG specific	Cagnon et al. (2013)
Staining coupled with FACS	60,000 for Terashima et al.	TAG specific	Xie et al. (2014) and Terashima et al. (2014)

Lipid Species Can Be Separated Physically in Medium-Throughput Screens

In contrast to fluorophore staining, analysis of extracted lipids can reveal differences in the abundances of individual lipid species. Transesterification of lipids into fatty acid methyl esters (FAMEs) followed by gas chromatography-flame ionization detector (GC-FID) analysis provides the composition as well as the amount of fatty acids contained in the lipid substrates used (Fig. 10.2b). For the purpose of high-throughput screens, direct transesterification with unprocessed algal cells has been developed. Fatty acid profile analysis has enabled the identification of the sole ω-3 fatty acid desaturase in Chlamydomonas (Nguyen et al. 2013; Pflaster et al. 2014). If the starting number of cells is known, this analysis yields the total amount of fatty acids per cell, which is often affected when TAG content is changed (Terashima et al. 2014). Additionally, certain fatty acid species are enriched in one lipid species compared to the others. The ratio of the abundance of a fatty acid species to the total abundance of all fatty acids serves as an indicator of the amount of the corresponding lipid. Because TAG is generally more saturated than membrane lipids, the ratio of C16:0 to C16:4 fatty acid was shown to correlate with TAG content (Liu et al. 2013).

Chlamydomonas studies could benefit from the adaptation of other high-throughput biochemical lipid screening techniques developed for other organisms. In Arabidopsis, Xu et al. screened for suppressors of a mutant deficient in synthesis of digalactosyldiacylglycerol (DGDG) by performing lipid extraction and TLC robotically (Xu et al. 2003). While lipids are resolved vertically in a traditional TLC, Xu et al. spotted lipid extracts on a TLC plate and applied a solvent mixture to each spot to separate lipid species radially outwards from the spot (Fig. 10.2b). This approach allows parallel analysis of 384 samples, and could be adapted to TAG mutant screens in algae, which are even more amenable to robotic operations.

Screens Based on Lipid Bio-markers Have Also Been Performed

High-throughput screens for lipid aberrations are also facilitated when a biomarker is available for the lipid of interest. The most abundant constituent of the Chlamydomonas lipid droplet proteome, Major Lipid Droplet Protein (MLDP), is highly correlated with TAG in abundance (Moellering and Benning 2010; Tsai et al. 2014). Tsai et al. screened for mutants with defects in TAG degradation by performing immuno-dot blots against MLDP in mutants arrayed in 96-well format. This allowed the identification of a mutant that exhibits a delay in TAG degradation (Tsai et al. 2014).

Raman Spectroscopy Has Potential for Single Cell-Based Screens

Since staining with fluorescent dyes can be biased by the cell penetration efficiency and provides little composition information on lipids, some groups have sought alternative approaches for rapid, single-cell lipid analysis. Single-cell Raman spectroscopy (SCRS) is a label-free technique for analyzing the in vivo chemical profiles of single cells (Li et al. 2012a). It detects the emitted light from light-excited molecules, with each wave number shift representing a distinct mode of vibration from a specific molecular structure. In contrast to infrared spectroscopy, Raman spectrometry is not compromised by water molecules thus is suitable for analyzing live biological samples. Recently, several groups have developed techniques based on Raman spectroscopy for analysis of lipids in single algal cells and obtained information on lipid amount and unsaturation levels (Wang et al. 2014; Wu et al. 2011). In addition to lipids, SCRS is also capable of quantitatively measuring the concentration of starch in individual microalgal cells such as *Chlamydomonas* spp. and *Chlorella* spp. (Ji et al. 2014). Furthermore, new technologies that isolate and sort single cells based on their Raman spectrum have recently been demonstrated (Wang et al. 2013; Zhang et al. 2014a, 2015). The SCRS approach is label-free, non-invasive, culture-independent, rapid, low-cost, and potentially able to simultaneously track multiple metabolites in individual live cells. As throughput, ease of deployment, cost, and multiplexing are all important considerations in design of assays for detecting and measuring intracellular compounds, such advantages can be exploited for development of many novel applications such as screening of microbial cell factories and optimization of bioprocesses.

Quantitative Analysis and Microscopy Are Used to Confirm TAG Phenotypes

GC-FID Can Be Used to Analyze the Fatty Acid Composition of TAG

The profile of the fatty acids contained in TAG is very useful for determining which pathway is responsible for the extra or missing fatty acids. For example, the *pgd1* mutant showed a major deficiency in the 18:1Δ9 fatty acid in TAG and other extraplastidic lipids, indicating a defect in exporting de novo synthesized fatty acids from the chloroplast rather than recycling of membrane lipid species (Li et al. 2012c). To analyze the fatty acid content of TAG specifically, instead of that of whole cells, lipids need to be extracted from cells and resolved, e.g. by TLC. The

TAG fraction can then be transesterified and analyzed by GC-FID. In this way, the molar amount of TAG as well as the fatty acid profile of TAG can be determined. A major limitation of GC-FID is that it cannot identify the molecular species of TAG, because the origins of FAMEs are not traceable.

Mass Spectroscopy (MS) Can Be Used to Identify and Quantify TAG Species

MS has been increasingly utilized to quantify TAG content in Chlamydomonas (Liu et al. 2013; Terashima et al. 2014; Wang et al. 2009; Yoon et al. 2012; Vieler et al. 2007). Direct infusion MS minimizes oxidation of polyunsaturated fatty acid species (Danielewicz et al. 2011) while liquid chromatography (LC) prior to MS enables separation and identification of positional isomers of TAG (Kind et al. 2012; MacDougall et al. 2011). With ultrahigh performance liquid chromatography–tandem mass spectrometry (UHPLC–MS/MS), 140 molecular species of TAG were identified for Chlamydomonas. Quantification with this method yielded a consistently lower amount for TAG compared to GC-FID performed on the same samples, possibly due to the difference in ionization efficiency between different TAG species (Liu et al. 2013).

Microscopy Can Be Used to Characterize the Size and Quantity of Lipid Droplets

Hyper or hypo-accumulation of TAG is expected to affect the number or size of lipid droplets. Disruption of different pathways may affect one metric over the other, depending on whether the pathway mainly affects the initial generation or the enlargement of lipid droplets (Wilfling et al. 2013). These parameters can be measured by microscopy on cells stained with lipid sensitive dyes. In a lipid droplet morphology mutant screen performed on a Drosophila cell line, five categories of abnormal lipid droplet morphology were observed (Guo et al. 2008). Microscopy has not yet been used for large-scale primary screens in Chlamydomonas, but it has been employed to visualize the phenotypes of specific mutants. For example, the over-accumulation of TAG in the *sta6* starchless mutant was initially observed as an increase in lipid droplet area by microscopy after Nile Red staining (Wang et al. 2009). Electron microscopy revealed a previously unknown association between lipid droplets and the chloroplast and allowed the discovery of large lipid droplets within the chloroplast in the *sta6* mutant (Fan et al. 2011; Goodson et al. 2011).

Point Mutations and Insertional Mutations Can Be Used to Disrupt Genes

Point Mutations Can Be Generated by Irradiation or Chemical Mutagens

Point mutations can produce gain-of-function and temperature-sensitive mutants. In Chlamydomonas, point mutations are commonly induced using ultraviolet (UV) radiation or chemical mutagens such as ethyl methane sulfonate (EMS). A general concern with these two approaches is the extent of mutation, which is correlated with the percentage of cells killed during mutagenesis (the "kill rate"). A low kill rate generally means that a larger number of cells need to be mutagenized to cover the genome. A very high kill rate indicates a higher density of point mutations, which may complicate the identification of the mutation causing the phenotype, as many mutations will be genetically linked to the phenotype. EMS mutagenesis has been employed by two groups for the purpose of obtaining high-TAG strains (Xie et al. 2014; Lee et al. 2014).

Traditionally, point mutations are identified through map-based cloning, which requires multiple rounds of crossing and a high-density molecular marker system, which is under-developed in Chlamydomonas compared with the plant model *A. thaliana* (Jander et al. 2002; Rymarquis et al. 2005). In recent years, next-generation sequencing technology has greatly facilitated the mapping of point mutations. Tulin and Cross were able to map 52 mutants by crossing each mutant to wild-type and performing whole-genome sequencing on pools of 10–12 progeny that showed the mutant phenotype. Candidate mutations were identified based on the expectation that only the causative point mutation or tightly linked ones would be uniform in the mutant progeny population (Tulin and Cross 2014).

Insertional Mutagenesis Is Broadly Used in Chlamydomonas

Transformation of an insertion cassette containing a resistance marker into Chlamydomonas cells disrupts the genome at random sites (Tam and Lefebvre 1993; Zhang et al. 2014b). Thus, insertional mutagenesis has been extensively employed to understand many processes including TAG metabolism (Cagnon et al. 2013; Li et al. 2012c; Terashima et al. 2014; Tsai et al. 2014; Yan et al. 2013). General considerations regarding the insertion cassette sequence and transformation method are summarized below.

Selection Is Facilitated with Antibiotic Resistance Markers or Auxotrophy Markers

Commonly used antibiotic resistance markers for Chlamydomonas nuclear genome transformation include *ble* for resistance to bleomycin, *aph7"* for resistance to hygromycin B and *aphVIII* for resistance to paromomycin (Berthold et al. 2002; Sizova et al. 2001; Stevens et al. 1996). *aph7"* and *aphVIII* each encode an aminoglycoside phosphotransferase but both can be used in the same strain without interfering with each other (Berthold et al. 2002). Recently, the chloroplast marker *AadA*, which confers resistance to spectinomycin, has been recoded for the nuclear genome and can be used as a shuttle marker between *E. coli* and Chlamydomonas (Meslet-Cladiere and Vallon 2012). Other drug-selectable markers include mutant ALS, encoding a mutated form of acetolactate synthase that confers resistance to sulfometuron methyl (Kovar et al. 2002), and *CRY1-1*, encoding a mutant ribosomal protein that confers resistance to emetine (Nelson et al. 1994).

Besides drug resistance markers, auxotrophy markers are also available for auxotrophic strains. Introduction of the *NIT1* gene (encoding the nitrate reductase) into a *nit1* background strain enables it to grow on nitrate as the nitrogen source (Kindle et al. 1989); the *ARG7* gene (encoding the argininosuccinate lyase) rescues the arginine auxotrophy of *arg7* mutants (Debuchy et al. 1989); NIC7 (encoding a quinolinate synthetase) rescues the nicotinamide auxotrophy of *nic7* mutants (Ferris 1995); and THI10 (encoding a hydroxyethylthiazole kinase) rescues thiamine auxotrophy of *thi10* mutants (Ferris 1995).

Most drug resistance markers are several-fold smaller than *NIT1* and *ARG7* in size. This makes drug resistance markers easier to manipulate and potentially less vulnerable to endonuclease cleavage during transformation (as discussed later).

Expression of Selection Marker Genes Can Be Enhanced by Strong Promoters or Introns

A strong promoter is desired for abundant production of resistance proteins and optimal transformation efficiency. The Chlamydomonas *PSAD* promoter (Fischer and Rochaix 2001), *RBCS2* promoter (Ribulose Bisphosphate Carboxylase/Oxygenase Small Subunit 2), β-2 tubulin promoter (Brunke et al. 1984; Kozminski et al. 1993) and a hybrid *HSP70A-RBCS2* promoter (Schroda et al. 2000) drive high expression of nuclear genes and are extensively used to drive selection markers.

Besides a strong promoter, the presence of the first intron of RBCS2 upstream or downstream of the start codon was demonstrated to enhance the abundance of the marker protein and the transformation efficiency (Berthold et al. 2002; Lumbreras and Purton 1998).

The Extent of Gene Disruption Is Affected by the Gene Feature of the Insertion Site

Insertions into exons typically disrupt gene functions. However, in the absence of a transcriptional terminator, insertions into introns can be spliced out. Commonly, a gene with an insertion in an intron will produce a normal protein but at a lower abundance (Wang 2008).

DNA Can Be Delivered into the Chlamydomonas Genome by Biolistics, Glass Bead Transformation, Electroporation or Agrobacterium-Mediated Gene Transfer

Early transformations of the Chlamydomonas nuclear genome were mostly achieved using biolistic particle delivery. Auxotrophic mutants were successfully complemented with native Chlamydomonas genes using this method (Debuchy et al. 1989; Kindle et al. 1989). However, the transformation efficiency of the biolistic method is typically much lower than the methods developed later. Moreover, it requires a sophisticated particle bombardment system. Soon afterwards, it was observed that exogenous DNA could be delivered into Chlamydomonas genome by agitating the cell/DNA mixture with glass beads. This method does not require specialized equipment but only yielded ~1,000 colonies with 1 μg DNA used in a cell wall-deficient strain (Kindle 1990). A protocol based on electroporation was later developed, enabling transformation efficiencies of up to 100,000 colonies per μg DNA (Shimogawara et al. 1998). More recently, *Agrobacterium tumefaciens*, which has been primarily used for transferring DNA into dicotyledonous plants, was also shown to introduce DNA into the Chlamydomonas genome (Kumar et al. 2004).

The Number of Insertions Needed to Obtain the Desired Genome Coverage Depends on the Number of Insertions per Mutant

Before mutagenesis, it is important to decide how many mutants are needed to cover a certain fraction of the genome. Results from a simulation suggest that 120,000 insertions are required to cover 80 % of Chlamydomonas genes with one or more allele for each gene (Zhang et al. 2014b). The number of mutants required to cover a certain fraction of the genome can be reduced by increasing the number of insertions per mutant, which can be achieved using a higher amount of transforming DNA per cell (Shimogawara et al. 1998).

Insertion Sites Can Be Identified Using Plasmid Rescue or Various PCR-Based Methods

Several methods have been applied in Chlamydomonas for identification of insertion loci (Table 10.2). If the insertion cassette contains *E. coli* plasmid sequences, it can be retrieved by restriction digestion of the mutant genome, circularization and

Table 10.2 Comparison of different strategies for insertion site identification in Chlamydomonas

Method	Advantage	Limitation	Time required	Application in chlamydomonas
Plasmid rescue	Can be applied to cases where the flanking sequences are hard to amplify	Requires an entire plasmid to be transformed	1 week for 10 mutants	Tam and Lefebvre (1993)
Nested PCR based methods	Relatively quick and simple	Sensitive to the distance to the annealing site of the genome side primer	1 week for 10 mutants	Dent et al. (2005), González-Ballester et al. (2005a, b), and Li et al. (2012)
3′ RACE	Relatively quick and simple	Requires an insertion cassette lacking a transcriptional terminator	1 week for 10 mutants	Meslet-Cladiere and Vallon (2012)
Chlamydomonas MmeI-based insertion site sequencing	Can be used for a pool of mutants	Requires digestion and ligation; short flanking sequence (20–21 bp)	2 weeks for 10,000 mutants	Zhang et al. (2014b)

transformation back into *E. coli* (Tam and Lefebvre 1993). Sequencing of the Chlamydomonas genomic regions in the resulting plasmid can reveal the insertion site. This method does not involve PCR and can be used when flanking regions are hard to amplify. However, it requires transformation of an entire plasmid, including bacterial replication origin and bacterial selection marker in addition to the Chlamydomonas selection marker.

Multiple methods developed later use nested PCR to amplify the junction between the insertion and the genome flanking sequence followed by sequencing of the PCR products. Examples include thermal asymmetric interlaced PCR (TAIL-PCR) (Dent et al. 2005), restriction enzyme site-directed amplification PCR (RESDA-PCR) (Gonzalez-Ballester et al. 2005a), and SiteFinding PCR (Li et al. 2012c).

In cases where the insertion cassette lacks a transcriptional terminator, a chimeric mRNA will be produced, containing both the sequence of the selection marker gene and the genomic sequence between the marker gene and the first native transcriptional terminator encountered in the genome. 3′ rapid amplification of cDNA ends (3′ RACE) can be applied to recover the sequence downstream of the marker gene (Meslet-Cladiere and Vallon 2012). Both 3′ RACE and nested PCR methods only require thermal cycling that can be easily performed on multiple microtiter tubes in parallel.

One challenge with 3′ RACE and nested PCR approaches is that the distance between the annealing location of the degenerate primer and the insertion junction varies between mutants. In some mutants, the distance may be too large for robust amplification. Additionally, differences in PCR efficiency between different sizes of DNA makes it difficult to adapt these methods to pools of mutants.

Recently, we developed a method for mapping insertion cassettes in pools of tens of thousands of Chlamydomonas mutants (Zhang et al. 2014b). This method, named Chlamydomonas *Mme*I-based insertion site sequencing (ChlaMmeSeq), relies on the presence of recognition sites for the type IIS restriction enzyme *Mme*I at the two ends of the cassette. To extract flanking sequences, the genome is digested with MmeI, which binds in the cassette and cleaves 20–21 base pairs into the flanking DNA. This produces a fragment containing 20–21 bases of flanking sequence attached to the cassette, onto which an adapter can be ligated to allow PCR amplification of the flanking sequence. When coupled with high-throughput sequencing, ChlaMmeSeq enables parallel retrieval of 20–21 bp of flanking sequences from more than 10,000 mutants. Approximately 70 % of short flanking sequences obtained can be aligned to a unique locus in the Chlamydomonas genome.

Insertion Sites Are Frequently Messy

The complex nature of many insertions can make it difficult to identify the insertion site and the gene whose disruption causes the observed phenotype. Insertions can contain large deletions, truncated cassettes, and insertions of genomic sequence between the cassette(s) and flanking genomic DNA (Fig. 10.3). In an ideal insertion site, the insertion cassette is inserted without any deletion in the cassette sequence or the genomic sequence at the insertion site (Fig. 10.3a). In this scenario, flanking sequences can be obtained from both sides of the inserted cassette, and a single gene will be directly disrupted by the insertion. However, insertion sites are rarely this clean.

Some insertions include a large deletion of genomic DNA that eliminates multiple genes. In such cases, one common strategy to identify the gene whose disruption causes the observed phenotype is to attempt complementation of the mutant with each of the disrupted genes (Fig. 10.3b) (Matsuo et al. 2008; Tsai et al. 2014).

Regardless of the presence of a deletion of genomic sequence, one or both sides of the insertion cassette can be truncated (Fig. 10.3c) (Zhang et al. 2014b). Such truncations can lead to the failure of flanking sequence extraction protocols, due to the loss of the necessary sequences, e.g. primer binding sites or sequences required for plasmid propagation in bacteria.

In addition to possible genomic deletions and cassette truncations, genomic DNA from a distant locus (referred as "junk DNA" here; Fig. 10.3d) can sometimes be found inserted between the cassette and the genomic sequence from the disrupted locus (Zhang et al. 2014b). The presence of junk DNA can make it challenging to identify the insertion site, as the DNA sequence flanking the cassette maps to the genome but does not indicate the true insertion site.

Last but not least, even when the insertion site can be correctly identified, the phenotype of a mutant could be caused by unmarked mutations, such as point mutations that occur during the transformation process (Fig. 10.3e). Thus, the causative relationship between the mutation and the phenotype should always be verified as suggested in the next part.

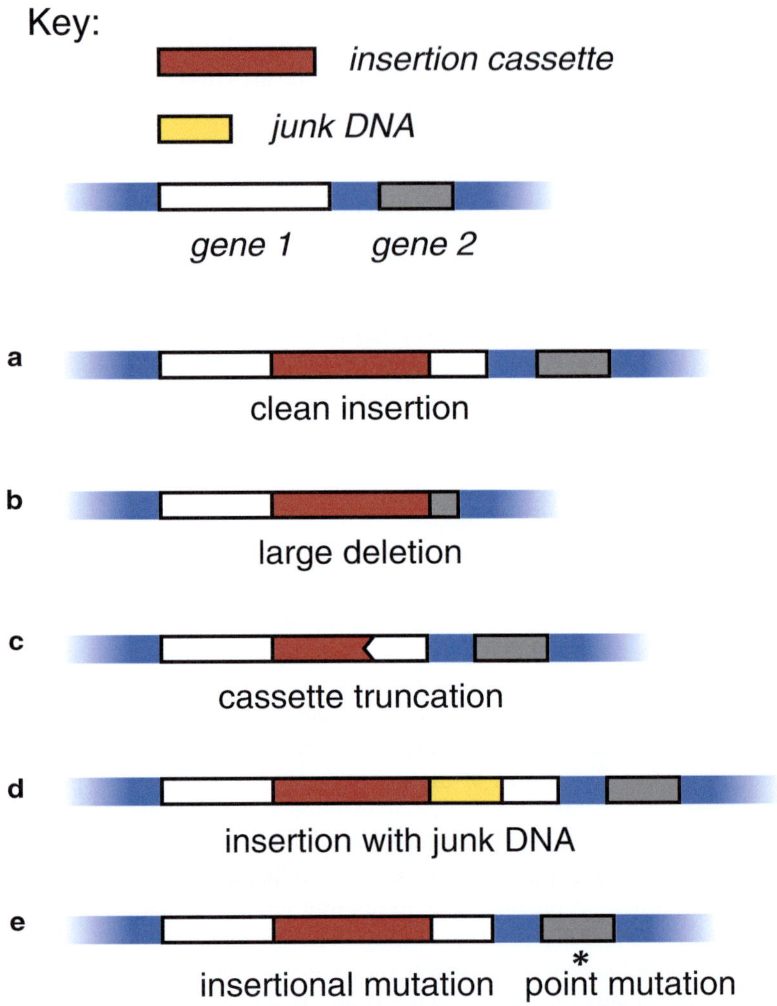

Fig. 10.3 Insertion sites are often complex in Chlamydomonas. Models of cassette insertions in genomic sequence are presented. (**a**) The cassette is inserted cleanly into the genome. (**b**) An insertion can be associated with a deletion of genomic DNA. (**c**) A cassette can be truncated on one or both ends. (**d**) Genomic DNA from a distant locus ("junk DNA", shown in *yellow*) can be inserted next to the cassette. (**e**) In addition to the mutation caused by cassette insertion, mutations can be present at one or more other loci. Note that a mutant may harbor a combination of these scenarios

A Mapped Insertional Mutant Library Will Be a Transformative Tool

Our lab is developing an indexed Chlamydomonas insertional mutant library in which the insertion site is known for each mutant. Mutants are maintained as colony arrays on solid medium, and also preserved cryogenically. This resource will

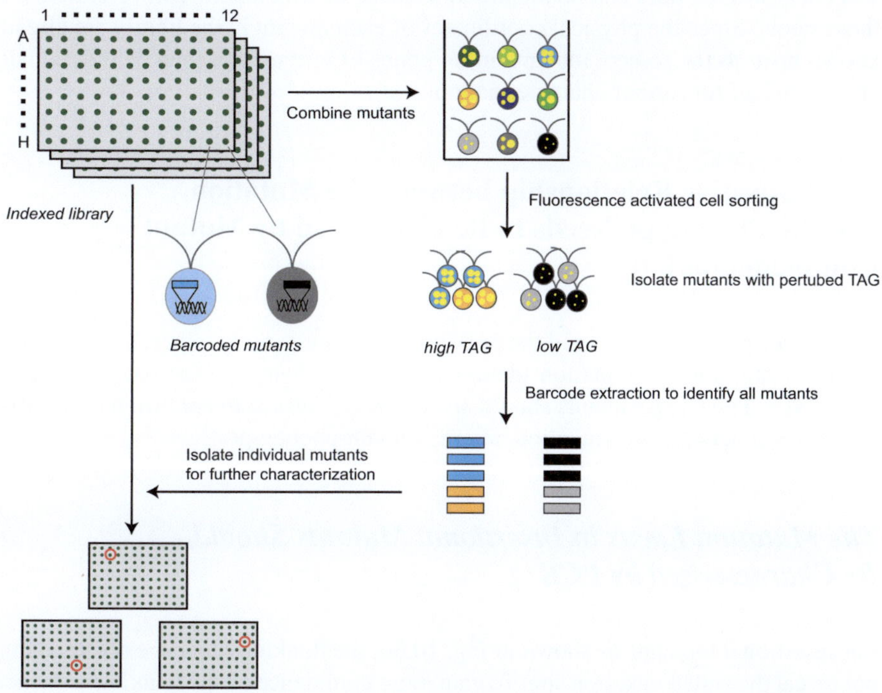

Fig. 10.4 An indexed library of insertional mutants will facilitate genetic screens for TAG phenotypes. Application of ChlaMmeSeq to each mutant produces a unique 20–21 bp genomic flanking sequence. This unique flanking sequence can serve as a barcode for detecting the presence or quantifying the abundance of the mutant in a pool. For a TAG mutant screen, the colonies from the library can be combined, grown and nitrogen deprived as a pooled sample. Fluorescence activated cell sorting (FACS) can be employed to sort out high-TAG mutants and low-TAG mutants as two pools of hits (depicted as cells with large and small lipid droplets, respectively). ChlaMmeSeq can then be applied to the two pooled samples to obtain insertion flanking sequences/barcodes. Individual mutants for genes of interest can be cherry-picked from the library for further studies

provide mutants in many genes of interest for reverse genetic studies. Additionally, this library will also be usable for various forward genetic screens, eliminating the time-consuming steps of mutant generation and mapping.

One scheme for a genome-saturating TAG mutant screen using this library takes advantage of a feature that enables tracking of the abundance of individual mutants in pools. The mutants in this library are generated with a cassette containing *Mme*I recognition sites at both ends, allowing the retrieval of a 20–21 bp flanking sequence from each end. If thousands of mutants are pooled together, the flanking sequence of each mutant can serve as a molecular barcode that indicates the abundance of that mutant in the pool (Fig. 10.4). To identify genes with a role in TAG metabolism, mutants can be combined into a single pool, and mutants with unusually high or low TAG content can be enriched by staining with a TAG-binding dye and Fluorescence-Activated Cell Sorting (FACS) (Terashima et al. 2014). Mutants enriched in a high-

TAG or low-TAG pool can be identified because their barcodes will be enriched in those pools. Since the physical coordinates of each mutant in the library are already known prior to the screen, mutants in hit genes of interest can then be individually cherry-picked for further characterization.

The Causative Relationship between the Mutation and the Phenotype Needs to Be Confirmed by Mutant Complementation

In the prior part, we mentioned the possibility that some mutations can be challenging to locate, and the mutation identified may not always be the one causing the phenotype. Thus, experiments should always be performed to confirm the causative relationship between the mutation and the mutant phenotype.

The Mutation Locus in Insertional Mutants Should Be Characterized by PCR

For insertional mutants, as shown in Fig. 10.3d, the flanking sequence obtained may not reveal the actual insertion site. To gain more confidence on the flanking sequence, one can attempt to retrieve longer flanking sequence beyond a possible junk DNA region. The challenge here is that this method can never tell conclusively that one has read into the actual flanking genomic sequence. A more robust approach is to obtain flanking sequences for both sides of the insertion and see whether they align to the same locus. If they do, it is extremely likely that the insertion is truly at that site. Finally, the most conclusive evidence that a region is disrupted is the absence of a wild-type product when the region spanning the insertion site is amplified by PCR.

Multiple Alleles Increase Confidence in a Genotype-Phenotype Link

When multiple independent mutants in the same gene exhibit the same phenotype, the likelihood that mutation of this gene causes the phenotype is greatly increased. Indeed, this principle is used extensively in Arabidopsis research (Krysan et al. 1999).

Genetic Linkage Analysis Can Be Used to Rule out Non-causative Mutations

In genetic linkage analysis, a mutant is crossed to the wild-type and the phenotypes and genotypes of progeny are determined. A non-causative mutation will typically be found in both progeny with and without the phenotype. However, co-segregation of the phenotype with a mutation of interest does not conclusively establish that the mutation of interest causes the phenotype, as the mutation of interest could merely be at a chromosomal location near the causative mutation.

Genetic Complementation Is Critical for Establishing the Causative Relationship between the Mutation and the Phenotype

The strongest evidence that a recessive mutation in a specific gene causes the observed phenotype is the observation that expression of the wild-type gene complements the mutant phenotype. To establish complementation robustly, multiple independent transformants of the wild-type gene into the mutant should be analyzed.

Investigators should be cautious as second site mutations can give the false appearance of complementation. During transformation, suppressor or reversion mutations can arise and revert the phenotype, giving the impression that the mutant has been complemented. To safeguard against this possibility, complementation transformations should include a parallel control transformation with a construct that is not expected to complement. If this latter construct produces colonies that revert the phenotype, reversion mutations are arising, and the investigator should perform multiple independent transformations and test whether the construct of interest yields a significantly greater number of wild-type-like transformants than the control construct.

Concluding Remarks

Algae are a promising platform for producing TAG for biodiesel. Many challenges remain to be overcome before algal biofuels are profitable, including our limited understanding of genes involved in algal TAG metabolism (Chisti 2007). The green alga Chlamydomonas is a powerful model organism for this process. However, due to difficulties in identifying the mutation sites, only a handful of genes have been identified with a role in TAG metabolism. Future efforts should focus on improving the throughput and robustness of screens and the success rate of mutation mapping.

More broadly, TAG serves as a universal storage lipid in eukaryotic organisms (Farese and Walther 2009), while glycerolipids are essential membrane components of all cells. Research on plant and animal lipid metabolism provides important insights into human nutrition and health. Yet, the multicellular nature of land plants and animals limits the throughput of genetic screens. For Chlamydomonas, in contrast, emerging tools including the genome-wide mutant library will enable high-throughput genome-saturating screens. We anticipate that future screens will reveal many conserved aspects of lipid metabolism.

Acknowledgements We thank Arthur Grossman, Robert Jinkerson, Bensheng Liu, Liz Freeman Rosenzweig and Jian Xu for critical reading of the manuscript. This work was supported by the Carnegie Institution for Science and a grant from the National Science Foundation (MCB-1146621).

References

Berthold P, Schmitt R, Mages W (2002) An engineered Streptomyces hygroscopicus aph 7″ gene mediates dominant resistance against hygromycin B in Chlamydomonas reinhardtii. Protist 153(4):401–412. doi:10.1078/14344610260450136

Blaby IK, Glaesener AG, Mettler T, Fitz-Gibbon ST, Gallaher SD, Liu B, Boyle NR, Kropat J, Stitt M, Johnson S, Benning C, Pellegrini M, Casero D, Merchant SS (2013) Systems-level analysis of nitrogen starvation-induced modifications of carbon metabolism in a Chlamydomonas reinhardtii starchless mutant. Plant Cell 25(11):4305–4323. doi:10.1105/tpc.113.117580

Boyle NR, Page MD, Liu B, Blaby IK, Casero D, Kropat J, Cokus SJ, Hong-Hermesdorf A, Shaw J, Karpowicz SJ, Gallaher SD, Johnson S, Benning C, Pellegrini M, Grossman A, Merchant SS (2012) Three acyltransferases and nitrogen-responsive regulator are implicated in nitrogen starvation-induced triacylglycerol accumulation in Chlamydomonas. J Biol Chem 287(19):15811–15825. doi:10.1074/jbc.M111.334052

Boynton JE, Gillham NW, Harris EH, Hosler JP, Johnson AM, Jones AR, Randolph-Anderson BL, Robertson D, Klein TM, Shark KB et al (1988) Chloroplast transformation in Chlamydomonas with high velocity microprojectiles. Science 240(4858):1534–1538

Brunke KJ, Anthony JG, Sternberg EJ, Weeks DP (1984) Repeated consensus sequence and pseudopromoters in the four coordinately regulated tubulin genes of Chlamydomonas reinhardi. Mol Cell Biol 4(6):1115–1124

Cagnon C, Mirabella B, Nguyen HM, Beyly-Adriano A, Bouvet S, Cuine S, Beisson F, Peltier G, Li-Beisson Y (2013) Development of a forward genetic screen to isolate oil mutants in the green microalga Chlamydomonas reinhardtii. Biotechnol Biofuels 6(1):178. doi:10.1186/1754-6834-6-178

Chen W, Sommerfeld M, Hu Q (2011) Microwave-assisted nile red method for in vivo quantification of neutral lipids in microalgae. Bioresour Technol 102(1):135–141. doi:10.1016/j.biortech.2010.06.076

Chisti Y (2007) Biodiesel from microalgae. Biotechnol Adv 25(3):294–306. doi:10.1016/j.biotechadv.2007.02.001

Cirulis JT, Strasser BC, Scott JA, Ross GM (2012) Optimization of staining conditions for microalgae with three lipophilic dyes to reduce precipitation and fluorescence variability. Cytometry A 81:618–626

Danielewicz MA, Anderson LA, Franz AK (2011) Triacylglycerol profiling of marine microalgae by mass spectrometry. J Lipid Res 52(11):2101–2108. doi:10.1194/jlr.D018408

Debuchy R, Purton S, Rochaix JD (1989) The argininosuccinate lyase gene of Chlamydomonas reinhardtii: an important tool for nuclear transformation and for correlating the genetic and molecular maps of the ARG7 locus. EMBO J 8(10):2803–2809

Dent RM, Haglund CM, Chin BL, Kobayashi MC, Niyogi KK (2005) Functional genomics of eukaryotic photosynthesis using insertional mutagenesis of Chlamydomonas reinhardtii. Plant Physiol 137(2):545–556. doi:10.1104/pp. 104.055244

Doan TY, Obbard JP (2011) Improved Nile Red staining of Nannochloropsis sp. J Appl Phycol 23(5):895–901. doi:10.1007/s10811-010-9608-5

Durrett TP, Benning C, Ohlrogge J (2008) Plant triacylglycerols as feedstocks for the production of biofuels. Plant J 54(4):593–607. doi:10.1111/j.1365-313X.2008.03442.x

Fan J, Andre C, Xu C (2011) A chloroplast pathway for the de novo biosynthesis of triacylglycerol in Chlamydomonas reinhardtii. FEBS Lett 585(12):1985–1991. doi:10.1016/j.febslet.2011.05.018

Farese RV Jr, Walther TC (2009) Lipid droplets finally get a little R-E-S-P-E-C-T. Cell 139(5):855–860. doi:10.1016/j.cell.2009.11.005

Ferris PJ (1995) Localization of the nic-7, ac-29 and thi-10 genes within the mating-type locus of Chlamydomonas reinhardtii. Genetics 141(2):543–549

Fischer N, Rochaix JD (2001) The flanking regions of PsaD drive efficient gene expression in the nucleus of the green alga Chlamydomonas reinhardtii. Mol Genet Genomics 265(5):888–894

Gonzalez-Ballester D, de Montaigu A, Galvan A, Fernandez E (2005a) Restriction enzyme site-directed amplification PCR: a tool to identify regions flanking a marker DNA. Anal Biochem 340(2):330–335. doi:10.1016/j.ab.2005.01.031

González-Ballester D, de Montaigu A, Higuera JJ, Galván A, Fernández E (2005b) Functional genomics of the regulation of the nitrate assimilation pathway in Chlamydomonas. Plant Physiol 137:522–533. doi:10.1104/pp.104.050914

Goodenough U, Blaby I, Casero D, Gallaher SD, Goodson C, Johnson S, Lee JH, Merchant SS, Pellegrini M, Roth R, Rusch J, Singh M, Umen JG, Weiss TL, Wulan T (2014) The path to triacylglyceride obesity in the sta6 strain of Chlamydomonas reinhardtii. Eukaryot Cell 13(5):591–613. doi:10.1128/EC.00013-14

Goodson C, Roth R, Wang ZT, Goodenough U (2011) Structural correlates of cytoplasmic and chloroplast lipid body synthesis in Chlamydomonas reinhardtii and stimulation of lipid body production with acetate boost. Eukaryot Cell 10(12):1592–1606. doi:10.1128/EC.05242-11

Govender T, Ramanna L, Rawat I, Bux F (2012) BODIPY staining, an alternative to the Nile Red fluorescence method for the evaluation of intracellular lipids in microalgae. Bioresour Technol 114:507–511. doi:10.1016/j.biortech.2012.03.024

Guo Y, Walther TC, Rao M, Stuurman N, Goshima G, Terayama K, Wong JS, Vale RD, Walter P, Farese RV (2008) Functional genomic screen reveals genes involved in lipid-droplet formation and utilization. Nature 453(7195):657–661. doi:10.1038/nature06928

Harris EH, Stern DB, Witman GB (2009) The Chlamydomonas sourcebook. Academic Press, Oxford

Hu Q, Sommerfeld M, Jarvis E, Ghirardi M, Posewitz M, Seibert M, Darzins A (2008) Microalgal triacylglycerols as feedstocks for biofuel production: perspectives and advances. Plant J 54(4):621–639. doi:10.1111/j.1365-313X.2008.03492.x

Jander G, Norris SR, Rounsley SD, Bush DF, Levin IM, Last RL (2002) Arabidopsis map-based cloning in the post-genome era. Plant Physiol 129(2):440–450. doi:10.1104/pp. 003533

Ji Y, He Y, Cui Y, Wang T, Wang Y, Li Y, Huang WE, Xu J (2014) Raman spectroscopy provides a rapid, non-invasive method for quantitation of starch in live, unicellular microalgae. Biotechnol J 9(12):1512–1518. doi:10.1002/biot.201400165

Kind T, Meissen JK, Yang D, Nocito F, Vaniya A, Cheng YS, Vandergheynst JS, Fiehn O (2012) Qualitative analysis of algal secretions with multiple mass spectrometric platforms. J Chromatogr A 1244:139–147. doi:10.1016/j.chroma.2012.04.074

Kindle KL (1990) High-frequency nuclear transformation of Chlamydomonas reinhardtii. Proc Natl Acad Sci U S A 87(3):1228–1232

Kindle KL, Schnell RA, Fernandez E, Lefebvre PA (1989) Stable nuclear transformation of Chlamydomonas using the Chlamydomonas gene for nitrate reductase. J Cell Biol 109(6 Pt 1):2589–2601

Kovar JL, Zhang J, Funke RP, Weeks DP (2002) Molecular analysis of the acetolactate synthase gene of Chlamydomonas reinhardtii and development of a genetically engineered gene as a dominant selectable marker for genetic transformation. Plant J 29(1):109–117

Kozminski KG, Diener DR, Rosenbaum JL (1993) High level expression of nonacetylatable alpha-tubulin in Chlamydomonas reinhardtii. Cell Motil Cytoskeleton 25(2):158–170. doi:10.1002/cm.970250205

Krysan PJ, Young JC, Sussman MR (1999) T-DNA as an insertional mutagen in Arabidopsis. Plant Cell 11(12):2283–2290

Kumar SV, Misquitta RW, Reddy VS, Rao BJ, Rajam MV (2004) Genetic transformation of the green alga—Chlamydomonas reinhardtii by Agrobacterium tumefaciens. Plant Sci 166(3):731–738. doi:10.1111/pbi.12190

Lee B, Choi GG, Choi YE, Sung M, Park MS, Yang JW (2014) Enhancement of lipid productivity by ethyl methane sulfonate-mediated random mutagenesis and proteomic analysis in Chlamydomonas reinhardtii. Korean J Chem Eng 31(6):1036–1042. doi:10.1007/s11814-014-0007-5

Li M, Xu J, Romero-Gonzalez M, Banwart SA, Huang WE (2012a) Single cell Raman spectroscopy for cell sorting and imaging. Curr Opin Biotechnol 23(1):56–63. doi:10.1016/j.copbio.2011.11.019

Li X, Benning C, Kuo MH (2012b) Rapid triacylglycerol turnover in Chlamydomonas reinhardtii requires a lipase with broad substrate specificity. Eukaryot Cell 11(12):1451–1462. doi:10.1128/EC.00268-12

Li X, Moellering ER, Liu B, Johnny C, Fedewa M, Sears BB, Kuo MH, Benning C (2012c) A galactoglycerolipid lipase is required for triacylglycerol accumulation and survival following nitrogen deprivation in Chlamydomonas reinhardtii. Plant Cell 24(11):4670–4686. doi:10.1105/tpc.112.105106

Li X, Umen JG, Jonikas MC (2014) Waking sleeping algal cells. Proc Natl Acad Sci U S A 111(44):15610–15611. doi:10.1073/pnas.1418295111

Liu B, Vieler A, Li C, Jones AD, Benning C (2013) Triacylglycerol profiling of microalgae Chlamydomonas reinhardtii and Nannochloropsis oceanica. Bioresour Technol 146:310–316. doi:10.1016/j.biortech.2013.07.088

Lumbreras V, Purton S (1998) Recent advances in chlamydomonas transgenics. Protist 149(1):23–27. doi:10.1016/S1434-4610(98)70006-9

MacDougall KM, McNichol J, McGinn PJ, O'Leary SJ, Melanson JE (2011) Triacylglycerol profiling of microalgae strains for biofuel feedstock by liquid chromatography-high-resolution mass spectrometry. Anal Bioanal Chem 401(8):2609–2616. doi:10.1007/s00216-011-5376-6

Matsuo T, Okamoto K, Onai K, Niwa Y, Shimogawara K, Ishiura M (2008) A systematic forward genetic analysis identified components of the Chlamydomonas circadian system. Genes Dev 22(7):918–930. doi:10.1101/gad.1650408

Maul JE, Lilly JW, Cui L, dePamphilis CW, Miller W, Harris EH, Stern DB (2002) The Chlamydomonas reinhardtii plastid chromosome: islands of genes in a sea of repeats. Plant Cell 14(11):2659–2679

Merchant SS, Prochnik SE, Vallon O, Harris EH, Karpowicz SJ, Witman GB, Terry A, Salamov A, Fritz-Laylin LK, Marechal-Drouard L, Marshall WF, Qu LH, Nelson DR, Sanderfoot AA, Spalding MH, Kapitonov VV, Ren QH, Ferris P, Lindquist E, Shapiro H, Lucas SM, Grimwood J, Schmutz J, Cardol P, Cerutti H, Chanfreau G, Chen CL, Cognat V, Croft MT, Dent R, Dutcher S, Fernandez E, Fukuzawa H, Gonzalez-Ballester D, Gonzalez-Halphen D, Hallmann A, Hanikenne M, Hippler M, Inwood W, Jabbari K, Kalanon M, Kuras R, Lefebvre PA, Lemaire SD, Lobanov AV, Lohr M, Manuell A, Meir I, Mets L, Mittag M, Mittelmeier T, Moroney JV, Moseley J, Napoli C, Nedelcu AM, Niyogi K, Novoselov SV, Paulsen IT, Pazour G, Purton S, Ral JP, Riano-Pachon DM, Riekhof W, Rymarquis L, Schroda M, Stern D, Umen J, Willows R, Wilson N, Zimmer SL, Allmer J, Balk J, Bisova K, Chen CJ, Elias M, Gendler K, Hauser C, Lamb MR, Ledford H, Long JC, Minagawa J, Page MD, Pan JM, Pootakham W, Roje S, Rose A, Stahlberg E, Terauchi AM, Yang PF, Ball S, Bowler C, Dieckmann CL, Gladyshev VN, Green P, Jorgensen R, Mayfield S, Mueller-Roeber B, Rajamani S, Sayre RT, Brokstein P, Dubchak I, Goodstein D, Hornick L, Huang YW, Jhaveri J, Luo YG, Martinez D, Ngau WCA, Otillar B, Poliakov A, Porter A, Szajkowski L, Werner G, Zhou KM, Grigoriev IV, Rokhsar DS, Grossman AR (2007) The Chlamydomonas genome reveals the evolution of key animal and plant functions. Science 318(5848):245–251. doi:10.1126/science.1143609

Meslet-Cladiere L, Vallon O (2012) A new method to identify flanking sequence tags in chlamydomonas using 3′-RACE. Plant Methods 8(1):21. doi:10.1186/1746-4811-8-21

Miller R, Wu G, Deshpande RR, Vieler A, Gartner K, Li X, Moellering ER, Zauner S, Cornish AJ, Liu B, Bullard B, Sears BB, Kuo MH, Hegg EL, Shachar-Hill Y, Shiu SH, Benning C (2010) Changes in transcript abundance in Chlamydomonas reinhardtii following nitrogen deprivation predict diversion of metabolism. Plant Physiol 154(4):1737–1752. doi:10.1104/pp. 110.165159

Moellering ER, Benning C (2010) RNA interference silencing of a major lipid droplet protein affects lipid droplet size in Chlamydomonas reinhardtii. Eukaryot Cell 9(1):97–106. doi:10.1128/EC.00203-09

Nelson JA, Savereide PB, Lefebvre PA (1994) The CRY1 gene in Chlamydomonas reinhardtii: structure and use as a dominant selectable marker for nuclear transformation. Mol Cell Biol 14(6):4011–4019

Nguyen HM, Baudet M, Cuine S, Adriano JM, Barthe D, Billon E, Bruley C, Beisson F, Peltier G, Ferro M, Li-Beisson Y (2011) Proteomic profiling of oil bodies isolated from the unicellular green microalga Chlamydomonas reinhardtii: with focus on proteins involved in lipid metabolism. Proteomics 11(21):4266–4273. doi:10.1002/pmic.201100114

Nguyen HM, Cuine S, Beyly-Adriano A, Legeret B, Billon E, Auroy P, Beisson F, Peltier G, Li-Beisson Y (2013) The green microalga Chlamydomonas reinhardtii has a single omega-3 fatty acid desaturase that localizes to the chloroplast and impacts both plastidic and extraplastidic membrane lipids. Plant Physiol 163(2):914–928. doi:10.1104/pp. 113.223941

Pflaster EL, Schwabe MJ, Becker J, Wilkinson MS, Parmer A, Clemente TE, Cahoon EB, Riekhof WR (2014) A high-throughput fatty acid profiling screen reveals novel variations in fatty acid biosynthesis in Chlamydomonas reinhardtii and related algae. Eukaryot Cell 13(11):1431–1438. doi:10.1128/EC.00128-14

Randolph-Anderson BL, Boynton JE, Gillham NW, Harris EH, Johnson AM, Dorthu MP, Matagne RF (1993) Further characterization of the respiratory deficient dum-1 mutation of Chlamydomonas reinhardtii and its use as a recipient for mitochondrial transformation. Mol Gen Genet 236(2–3):235–244

Rasala BA, Chao SS, Pier M, Barrera DJ, Mayfield SP (2014) Enhanced genetic tools for engineering multigene traits into green algae. PLoS One 9(4):e94028. doi:10.1371/journal.pone.0094028

Riekhof WR, Sears BB, Benning C (2005) Annotation of genes involved in glycerolipid biosynthesis in Chlamydomonas reinhardtii: discovery of the betaine lipid synthase BTA1Cr. Eukaryot Cell 4(2):242–252. doi:10.1128/EC.4.2.242-252.2005

Rymarquis LA, Handley JM, Thomas M, Stern DB (2005) Beyond complementation. Map-based cloning in Chlamydomonas reinhardtii. Plant Physiol 137(2):557–566. doi:10.1104/pp. 104.054221

Schmollinger S, Muhlhaus T, Boyle NR, Blaby IK, Casero D, Mettler T, Moseley JL, Kropat J, Sommer F, Strenkert D, Hemme D, Pellegrini M, Grossman AR, Stitt M, Schroda M, Merchant SS (2014) Nitrogen-sparing mechanisms in Chlamydomonas affect the transcriptome, the proteome, and photosynthetic metabolism. Plant Cell 26(4):1410–1435. doi:10.1105/tpc.113.122523

Schroda M, Blocker D, Beck CF (2000) The HSP70A promoter as a tool for the improved expression of transgenes in Chlamydomonas. Plant J 21(2):121–131

Shimogawara K, Fujiwara S, Grossman A, Usuda H (1998) High-efficiency transformation of Chlamydomonas reinhardtii by electroporation. Genetics 148(4):1821–1828

Sizova I, Fuhrmann M, Hegemann P (2001) A Streptomyces rimosus aphVIII gene coding for a new type phosphotransferase provides stable antibiotic resistance to Chlamydomonas reinhardtii. Gene 277(1–2):221–229

Spandl J, White DJ, Peychl J, Thiele C (2009) Live cell multicolor imaging of lipid droplets with a new dye, LD540. Traffic 10(11):1579–1584. doi:10.1111/j.1600-0854.2009.00980.x

Stevens DR, Rochaix JD, Purton S (1996) The bacterial phleomycin resistance gene ble as a dominant selectable marker in Chlamydomonas. Mol Gen Genet 251(1):23–30

Tam LW, Lefebvre PA (1993) Cloning of flagellar genes in Chlamydomonas reinhardtii by DNA insertional mutagenesis. Genetics 135(2):375–384

Terashima M, Freeman ES, Jinkerson RE, Jonikas MC (2014) A fluorescence-activated cell sorting-based strategy for rapid isolation of high-lipid Chlamydomonas mutants. Plant J. doi:10.1111/tpj.12682

Thiele C (2011) Fluorescence-based imaging and analysis of cells and cellular components using lipophilic dyes with improved specificity, spectral property and photostability. EP20090010408

To A, Joubes J, Barthole G, Lecureuil A, Scagnelli A, Jasinski S, Lepiniec L, Baud S (2012) WRINKLED transcription factors orchestrate tissue-specific regulation of fatty acid biosynthesis in Arabidopsis. Plant Cell 24(12):5007–5023. doi:10.1105/tpc.112.106120

Tsai CH, Warakanont J, Takeuchi T, Sears BB, Moellering ER, Benning C (2014) The protein Compromised Hydrolysis of Triacylglycerols 7 (CHT7) acts as a repressor of cellular quiescence in Chlamydomonas. Proc Natl Acad Sci U S A 111(44):15833–15838. doi:10.1073/pnas.1414567111

Tulin F, Cross FR (2014) A microbial avenue to cell cycle control in the plant superkingdom. Plant Cell 26(10):4019–4038. doi:10.1105/tpc.114.129312

Vieler A, Wilhelm C, Goss R, Suss R, Schiller J (2007) The lipid composition of the unicellular green alga Chlamydomonas reinhardtii and the diatom Cyclotella meneghiniana investigated by MALDI-TOF MS and TLC. Chem Phys Lipids 150(2):143–155. doi:10.1016/j.chemphyslip.2007.06.224

Wang YH (2008) How effective is T-DNA insertional mutagenesis in Arabidopsis? J Biochem Tech 1(1):11–20. doi:10.1111/pbi.12190

Wang ZT, Ullrich N, Joo S, Waffenschmidt S, Goodenough U (2009) Algal lipid bodies: stress induction, purification, and biochemical characterization in wild-type and starchless Chlamydomonas reinhardtii. Eukaryot Cell 8(12):1856–1868. doi:10.1128/EC.00272-09

Wang Y, Ji Y, Wharfe ES, Meadows RS, March P, Goodacre R, Xu J, Huang WE (2013) Raman activated cell ejection for isolation of single cells. Anal Chem 85(22):10697–10701. doi:10.1021/ac403107p

Wang T, Ji Y, Wang Y, Jia J, Li J, Huang S, Han D, Hu Q, Huang WE, Xu J (2014) Quantitative dynamics of triacylglycerol accumulation in microalgae populations at single-cell resolution revealed by Raman microspectroscopy. Biotechnol Biofuels 7:58. doi:10.1186/1754-6834-7-58

Wilfling F, Wang H, Haas JT, Krahmer N, Gould TJ, Uchida A, Cheng JX, Graham M, Christiano R, Frohlich F, Liu X, Buhman KK, Coleman RA, Bewersdorf J, Farese RV Jr, Walther TC (2013) Triacylglycerol synthesis enzymes mediate lipid droplet growth by relocalizing from the ER to lipid droplets. Dev Cell 24(4):384–399. doi:10.1016/j.devcel.2013.01.013

Work VH, Radakovits R, Jinkerson RE, Meuser JE, Elliott LG, Vinyard DJ, Laurens LM, Dismukes GC, Posewitz MC (2010) Increased lipid accumulation in the Chlamydomonas reinhardtii sta7-10 starchless isoamylase mutant and increased carbohydrate synthesis in complemented strains. Eukaryot Cell 9(8):1251–1261. doi:10.1128/EC.00075-10

Wu H, Volponi JV, Oliver AE, Parikh AN, Simmons BA, Singh S (2011) In vivo lipidomics using single-cell Raman spectroscopy. Proc Natl Acad Sci U S A 108(9):3809–3814. doi:10.1073/pnas.1009043108

Xie B, Stessman D, Hart JH, Dong H, Wang Y, Wright DA, Nikolau BJ, Spalding MH, Halverson LJ (2014) High-throughput fluorescence-activated cell sorting for lipid hyperaccumulating Chlamydomonas reinhardtii mutants. Plant Biotechnol J 12(7):872–882. doi:10.1111/pbi.12190

Xu C, Fan J, Riekhof W, Froehlich JE, Benning C (2003) A permease-like protein involved in ER to thylakoid lipid transfer in Arabidopsis. EMBO J 22(10):2370–2379. doi:10.1093/emboj/cdg234

Yan C, Fan J, Xu C (2013) Analysis of oil droplets in microalgae. Methods Cell Biol 116:71–82. doi:10.1016/B978-0-12-408051-5.00005-X

Yoon K, Han D, Li Y, Sommerfeld M, Hu Q (2012) Phospholipid:diacylglycerol acyltransferase is a multifunctional enzyme involved in membrane lipid turnover and degradation while synthesizing triacylglycerol in the unicellular green microalga Chlamydomonas reinhardtii. Plant Cell 24(9):3708–3724. doi:10.1105/tpc.112.100701

Zhang Q, Zhang P, Su Y, Mou C, Zhou T, Yang M, Xu J, Ma B (2014a) On-demand control of microfluidic flow via capillary-tuned solenoid microvalve suction. Lab Chip 14(24):4599–4603. doi:10.1039/c4lc00833b

Zhang R, Patena W, Armbruster U, Gang SS, Blum SR, Jonikas MC (2014b) High-throughput genotyping of green algal mutants reveals random distribution of mutagenic insertion sites and endonucleolytic cleavage of transforming DNA. Plant Cell 26(4):1398–1409. doi:10.1105/tpc.114.124099

Zhang P, Ren L, Zhang X, Shan Y, Wang Y, Ji Y, Yin H, Huang WE, Xu J, Ma B (2015) Raman-activated cell sorting based on dielectrophoretic single-cell trap and release. Anal Chem. doi:10.1021/ac503974e

Chapter 11
Plant Sphingolipid Metabolism and Function

Kyle D. Luttgeharm, Athen N. Kimberlin, and Edgar B. Cahoon

Abstract Sphingolipids, a once overlooked class of lipids in plants, are now recognized as abundant and essential components of plasma membrane and other endomembranes of plant cells. In addition to providing structural integrity to plant membranes, sphingolipids contribute to Golgi trafficking and protein organizational domains in the plasma membrane. Sphingolipid metabolites have also been linked to the regulation of cellular processes, including programmed cell death. Advances in mass spectrometry-based sphingolipid profiling and analyses of Arabidopsis mutants have enabled fundamental discoveries in sphingolipid structural diversity, metabolism, and function that are reviewed here. These discoveries are laying the groundwork for the tailoring of sphingolipid biosynthesis and catabolism for improved tolerance of plants to biotic and abiotic stresses.

Keywords Sphingolipid • Ceramide • Long-chain bases • Plasma membrane • Lipid-signaling • Lipid rafts • Programed cell death

Introduction

Sphingolipids were originally identified in the late nineteenth century by Johann Thudichum as an "enigmatic" major lipidic component of the brain (Thudichum 1884). Since this discovery, sphingolipids have been recognized as essential components of eukaryotic cells and have been extensively studied in humans due to their association with a number of lipid storage disorders, including Tay-Sachs disease and Niemann-Pick disease (Sandhoff 2013). Sphingolipids, however, were not identified in plants until the late 1950s (Carter et al. 1958) and for nearly four decades following this discovery, sphingolipid research in plants was limited mainly to structural and compositional analyses, including studies of sphingolipid compositional changes in response to abiotic stresses. Since the late 1990s, plant

K.D. Luttgeharm • A.N. Kimberlin • E.B. Cahoon (✉)
Center for Plant Science Innovation and Department of Biochemistry, University of Nebraska-Lincoln, E318 Beadle Center, 1901 Vine Street, Lincoln, NE 68588, USA
e-mail: ecahoon2@unl.edu

Y. Nakamura, Y. Li-Beisson (eds.), *Lipids in Plant and Algae Development*, Subcellular Biochemistry 86, DOI 10.1007/978-3-319-25979-6_11

sphingolipids have become an increasing research focus. Driving this heightened interest is the realization that sphingolipids are among the most abundant endomembrane lipids in plant cells and that they contribute not only to membrane structure and function that underlies abiotic and biotic stress resistance but also to the regulation of cellular processes. Recent advances in plant sphingolipid research have been spurred by development and application of advanced mass spectrometry methods that enable the rapid and quantitative measurement of molecular species of specific sphingolipid classes (Markham and Jaworski 2007). Coupling of these methods with the characterization of Arabidopsis mutants has resulted in advances in our fundamental understanding of plant sphingolipid metabolism and its regulation. The availability of Arabidopsis mutants has also led to discoveries of the involvement of sphingolipids in plant growth and responses to environmental stimuli. In addition, unexpected connections between sphingolipids and physiological processes such as programmed cell death and the related hypersensitive response that is important for plant resistance to bacterial and fungal pathogens have arisen from forward genetic studies of Arabidopsis. Recent research has also been directed at understanding the specialized functions of sphingolipids in plasma membrane microdomains that are believed to contribute to cell surface processes such as cell wall metabolism and ion and auxin transport. This chapter provides an overview of our recent understanding of plant sphingolipid structure, metabolism, and function and highlights unanswered questions in plant sphingolipid biology.

Sphingolipid Structure

Sphingolipids consist of hydrophobic ceramide backbones that are typically linked to polar sugar residues to form amphipathic lipid components of membranes (Lynch and Dunn 2004; Chen et al. 2010). The ceramide backbone contains a long chain amino alcohol referred to as a sphingoid long-chain base (LCB) linked through an amide bond to a fatty acid. LCBs are unique to sphingolipids. In plants, LCBs typically have chain lengths of 18 carbon atoms and can contain double bonds in the Δ4 or Δ8 positions (Fig. 11.1a). The Δ4 double bond is found only in the *trans* configuration, while the Δ8 double bond can be found in either the *trans* or *cis* configurations. Following its initial synthesis, a LCB has two hydroxyl groups at the C-1 and C-3 carbons (Lynch and Dunn 2004; Chen et al. 2010). These LCBs are referred to as dihydroxy LCBs. A third hydroxyl group can be enzymatically added at the C-4 carbon to form a trihydroxy LCB. In the short-hand nomenclature, a dihydroxy LCB with 18 carbons and 1 double bond is referred to as "d18:1", and a trihydroxy LCB with 18 carbons and 1 double bond is referred to as "t18:1". LCBs can be phosphorylated at the C-1 position to form LCB-phosphates (LCB-P). Free LCBs and their phosphorylated forms are typically in low abundance in plant cells (Markham and Jaworski 2007; Markham et al. 2006). Instead, the majority of LCBs are found linked to fatty acids in ceramides (Fig. 11.1b). The chain-lengths of plant ceramide fatty acids range from 16 to 26 carbon atoms, the majority of which contain an enzymatically added hydroxyl group at the C-2 or α-position (Lynch and

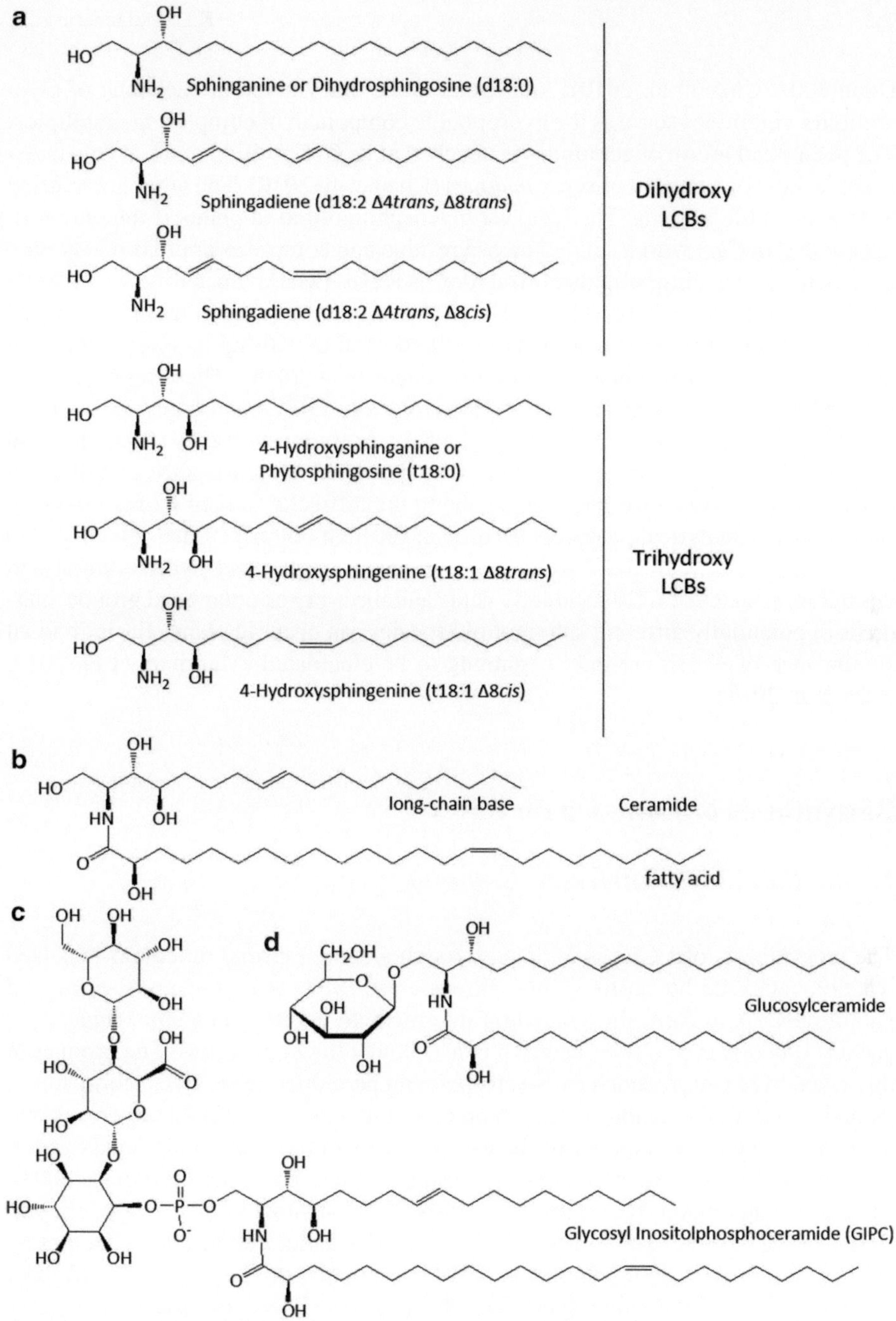

Fig. 11.1 Examples of long-chain bases (*LCB*) and sphingolipids found in plants. (**a**) Examples of LCB modifications found in plants. Shown are examples of dihydroxy and trihydroxy LCBs The nomenclature "d18:0" indicates that the LCB has two hydroxyl groups (d) and 18 carbon atoms and no double bonds, and the nomenclature "t18:0" indicates that the LCB has three hydroxyl groups (t) and 18 carbon atoms and no double bonds. (**b**) Hydoxyceramide composed of the LCB t18:1 Δ8*trans* and the fatty acid 24:1 ω9*cis* that is hydroxylated at the C-2 position. (**c**) Most abundant glycosyl inositolphosphoceramide (GIPC) found in Arabidopsis leaves. (**d**) Glucosylceramide

Dunn 2004; Chen et al. 2010). Analogous to the diacylglycerol backbone of glycerolipids, ceramides serve as the hydrophobic component of complex sphingolipids. The polar head group of ceramides is attached at its C-1 position and can be a phosphate residue or a variety of sugar residues (Chen et al. 2010). The latter are referred to as glycosphingolipids. The simplest glycosphingolipid in plants is the glucosylceramide (GlcCer) with a single glucose residue and comprises approximately one-third of the glycosphingolipids of Arabidopsis leaves (Markham and Jaworski 2007; Markham et al. 2006) (Fig. 11.1c). The most abundant glycosphingolipid in plants contains an inositol phosphate bound to the ceramide with up to seven additional hexose and pentose residues (Fig. 11.1c) (Cacas et al. 2013). These molecules are referred to as glycosyl inositolphosphoceramides or GIPCs and comprise approximately two-thirds of the glycosphingolipids of Arabidopsis leaves (Markham and Jaworski 2007; Markham et al. 2006). The quantitative significance of GIPCs in plants was overlooked for many years due to the difficulty in their extraction using standard lipid analytical protocols because of the high polarity of their glycosylated head groups. Between the different carbon chain-lengths and hydroxylation and unsaturation states of LCBs and fatty acids and the array of polar head groups, hundreds of potentially different sphingolipid species can occur in plants, the individual significance of which are only beginning to be elucidated (Markham et al. 2013; Bure et al. 2011).

Biosynthesis of Long-Chain Bases

Serine Palmitoyltransferase Complex

The biosynthesis of LCBs is initiated through an endoplasmic reticulum-localized reaction catalyzed by serine palmitoyltransferase (SPT) that condenses serine and palmitoyl-CoA to form the 18 carbon intermediate 3-ketosphinganine (Figs. 11.2 and 11.3) (Chen et al. 2006; Dietrich et al. 2008; Teng et al. 2008). The product of this reaction is then reduced by 3-ketosphinganine reductase (KSR) to form sphinganine or d18:0, the simplest long-chain base in plants and other eukaryotes (Chao et al. 2011). SPT is a member of the α-oxoamine synthase subfamily and is generally regarded as the main regulated step in sphingolipid biosynthesis (Hanada 2003). Similar to other eukaryotes, the Arabidopsis SPT functions as a heterodimer comprised of LCB1 and LCB2 subunits (Table 11.1) (Tamura et al. 2001; Chen et al. 2006; Dietrich et al. 2008; Teng et al. 2008). Although both LCB1 and LCB2 show similarity with α-oxoamine synthases, the catalytic lysine residue that forms a Schiff base with pyridoxal phosphate is found in the LCB2 subunit (Tamura et al. 2001; Hanada 2003). A third smaller subunit, termed the small subunit of SPT or ssSPT, also interacts with the LCB1/LCB2 subunits (Table 11.1) (Han et al. 2004; Kimberlin et al. 2013). ssSPT polypeptides in Arabidopsis contain only 56 amino acids that lack any predicted enzymatic activity but contain a single transmembrane

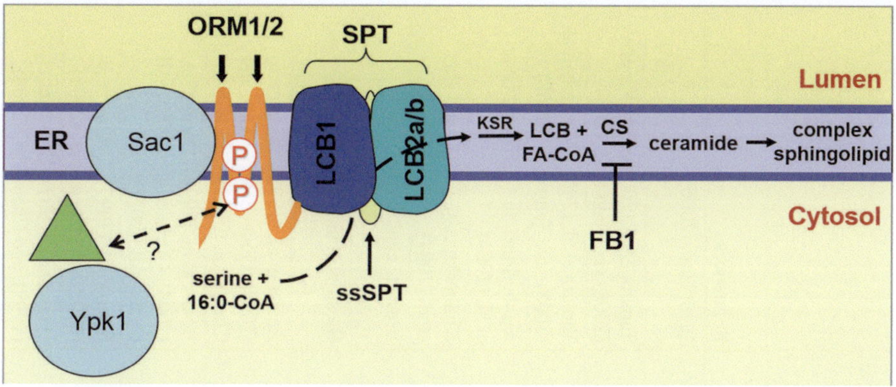

Fig. 11.2 Schematic of the serine palmitoyltransferase (*SPT*) complex. In plants, SPT exists as an ER-localized heterodimer of LCB1 and LCB2 subunits that interacts with several regulatory proteins. A stimulatory small subunit of SPT (*ssSPT*) binds to the core SPT heterodimer to increase SPT activity. Sac1 phosphatase and Ypk1 have been shown in *Saccharomyces cerevisiae* to catalyze reversible phosphorylation of the inhibitory regulator ORM in response to intracellular sphingolipid levels. The regulatory mechanism of ORM in plants remains to be elucidated. Long-chain bases (*LCBs*) produced by SPT and 3-ketosphinganine reductase (*KSR*) are used by ceramide synthase (*CS*) along with fatty acid acyl-CoAs to produce ceramides. Fumonisin B_1 (FB_1) is a mycotoxin that inhibits CS and triggers programmed cell death (Figure is derived from Kimberlin et al. 2013)

domain (Kimberlin et al. 2013). It is believed that the active site of SPT occurs at the interface of LCB1 and LCB2 with LCB1 and ssSPT acting to stabilize the complex (Gable et al. 2000, 2002). Although SPT can function as a heterodimer (LCB1 and LCB2) with minimal enzymatic activity, ssSPT enhances SPT activity to levels that produce LCBs in amounts that are sufficient to support cell viability in Arabidopsis (Kimberlin et al. 2013).

Studies of *LCB1*, *LCB2*, and *ssSPT* mutants have demonstrated that SPT activity is essential, and consequently, sphingolipids are required for the viability of plant cells (Kimberlin et al. 2013; Dietrich et al. 2008; Teng et al. 2008). In this regard, the *fbr11-2* mutant of the single Arabidopsis *LCB1* gene (At4g36480) displays male gametophytic lethality (Teng et al. 2008). Loss of pollen viability is also observed in double mutants of the two redundant LCB2 genes *LCB2a* (At5g23670) and *LCB2b* (At3g48780) (Dietrich et al. 2008) as well as in null mutants of *ssSPTa* (At1g06515), the more highly expressed of the two *ssSPT* genes in *Arabidopsis* (Kimberlin et al. 2013). Pollen deficient in sphingolipid synthesis in $lcb2a^{-/-}lcb2b^{-/+}$ mutants lack Golgi stacks and surrounding intine layer and have vesiculated ER, all of which are consistent with the contributions of sphingolipids to the structural and functional integrity of the endomembrane system (Dietrich et al. 2008).

The regulation of SPT is thought to occur primarily through ssSPT and a second class of interacting proteins termed orosomucoid or ORM polypeptides, rather than transcriptional mechanisms (Fig. 11.2 and Table 11.1) (Markham et al. 2013). The ssSPTs appear to be limiting and so modulation of their expression alters SPT

Table 11.1 Characterized and putative Arabidopsis genes involved in sphingolipid metabolism

Arabidopsis Gene (AGI code)	Designated gene symbol	Function (characterized or predicted)	*Saccharomyces cerevisiae* homolog	Mammalian homolog	References
At4g36480	*LCB1*	Subunit of serine palmitoyltransferase	*LCB1*	*LCB1*	Chen et al. (2006)
At5g23670	*LCB2a*	Subunit of serine palmitoyltransferase	*LCB2*	*LCB2*	Tamura et al. (2001), Dietrich et al. (2008) and Teng et al. (2008)
At3g48780	*LCB2b*	Subunit of serine palmitoyltransferase	*LCB2*	*LCB2*	Tamura et al. (2001), Dietrich et al. (2008), and Teng et al. (2008)
At1g06515	*ssSPTa*	Activator of serine palmitoyltransferase	*TSC3*	*ssSPT*	Kimberlin et al. (2013) and Han et al. (2004)
At2g30942	*ssSPTb*	Activator of serine palmitoyltransferase	*TSC3*	*ssSPT*	Kimberlin et al. (2013) and Han et al. (2004)
At1g01230	*ORM1*	Putative inhibitor of serine palmitoyltransferase	*ORM*	*ORMDL*	Breslow et al. (2010)
At5g42000	*ORM2*	Putative inhibitor of serine palmitoyltransferase	*ORM*	*ORMDL*	Breslow et al. (2010)
At3g06060	*KSR1*	3-ketosphinganine reductase	*TSC10*	*FVT1/KDSR*	Beeler et al. (1998), Dunn et al. (2004), Gupta et al. (2009), and Chao et al. (2011)
At5g19200	*KSR2*	3-ketosphinganine reductase	*TSC10*	*FVT1/KDSR*	Beeler et al. (1998), Dunn et al. (2004), Gupta et al. (2009), and Chao et al. (2011)
At1g69640	*SBH1*	LCB C-4 hydroxylase	*SUR2/SYR2*	None	Haak et al. (1997), Grilley et al. (1998), and Chen et al. (2008)
At1g14290	*SBH2*	LCB C-4 hydroxylase	*SUR2/SYR2*	None	Haak et al. (1997), Grilley et al. (1998), and Chen et al. (2008)

At3g61580	*SLD1*	LCB Δ8 desaturase	None	None	Sperling et al. (1998) and Ryan et al. (2007)
At2g46210	*SLD2*	LCB Δ8 desaturase	None	None	Sperling et al. (1998) and Ryan et al. (2007)
At4g04930	None	LCB Δ4 desaturase	None	*DES1/DEGS1*	Michaelson et al. (2009)
At2g34770	*FAH1*	Fatty acid α-hydroxylase	*FAH1/SCS7*	*FA2H*	Haak et al. (1997), Nagano et al. (2012), and Konig et al. (2012)
At4g20870	*FAH2*	Fatty acid α-hydroxylase	*FAH1/SCS7*	*FA2H*	Haak et al. (1997), Nagano et al. (2012), and Konig et al. (2012)
At3g25540	*LOH1*	Ceramide synthase	*LAC1/LAG1*	*CerS*	Markham et al. (2011), Brandwagt et al. (2000), and Ternes et al. (2011)
At3g19260	*LOH2*	Ceramide synthase	*LAC1/LAG1*	*CerS*	Markham et al. (2011), Brandwagt et al. (2000), and Ternes et al. (2011)
At1g13580	*LOH3*	Ceramide synthase	*LAC1/LAG1*	*CerS*	Markham et al. (2011), Brandwagt et al. (2000), and Ternes et al. (2011)
At2g37940	*IPCS1*	IPC synthase	*IPCS*	*SMS1*	Yang et al. (2005), Denny et al. (2006), Wang et al. (2008), and Mina et al. (2010)
At3g54020	*IPCS2*	IPC synthase	*IPCS*	*SMS1*	Yang et al. (2005), Denny et al. (2006), Wang et al. (2008), and Mina et al. (2010)
At2g29525	*IPCS3*	IPC synthase	*IPCS*	*SMS1*	Yang et al. (2005), Denny et al. (2006), Wang et al. (2008), and Mina et al. (2010)
At2g19880	*GCS*	Glucosylceramide synthase	*GCS*	*GCS*	Leipelt et al. (2001)
At5g18480	*IPUT1*	IPC glucuronosyltransferase	None	None	Rennie et al. (2014)

(continued)

Table 11.1 (continued)

Arabidopsis Gene (AGI code)	Designated gene symbol	Function (characterized or predicted)	*Saccharomyces cerevisiae* homolog	Mammalian homolog	References
At5g51290	*ACD5/AtCERK*	Ceramide kinase	*LCB4/LCB5*	*CerK*	Liang et al. (2003)
At5g23450	*AtLCBK1*	LCB kinase	*LCBK*	*LCBK*	Imai and Nishiura (2005) and Worrall et al. (2008)
At4g21540	*SPHK1*	LCB kinase	*LCBK*	*LCBK*	Imai and Nishiura (2005) and Worrall et al. (2008)
At2g46090	*SPHK2*	Putative LCB kinase	*LCBK*	*LCBK*	Imai and Nishiura (2005) and Worrall et al. (2008)
At1g27980	*AtDPL1*	LCB-1-P lyase	*DPL1*	*SPL*	Tsegaye et al. (2007) and Worrall et al. (2008)
At3g58490	*LCB-PP1*	LCB-1-P phosphatase	*LCB3/YSR3*	*SPP1*	Worrall et al. (2008)
At5g03080	*LCB-PP2*	LCB-1-P phosphatase	*LCB3/YSR3*	*SPP1*	Worrall et al. (2008)
At4g22330	*AtCES1*	Putative alkaline ceramidase	*YPC1/YDC1*	*aPHC*	Mao et al. (2000), Wu et al. (2015), and Chen et al. (2015)
At1g07380	None	Putative neutral ceramidase	None	*ASAH2*	Chen et al. (2010)
At2g38010	None	Putative neutral ceramidase	None	*ASAH2*	Pata et al. (2008) and Chen et al. (2010)
At5g58980	None	Putative neutral ceramidase	None	*ASAH2*	Chen et al. (2010)
At5g49900	None	Putative glucosylceramidase	None	*GBA2*	Boot et al. (2007) and Chen et al. (2010)
At1g33700	None	Putative glucosylceramidase	None	*GBA2*	Chen et al. (2010)
At4g10060	None	Putative glucosylceramidase	None	*GBA2*	Chen et al. (2010)
At3g24180	None	Putative glucosylceramidase	None	*GBA2*	Chen et al. (2010)
At4g29680	None	Putative inositolphosphoceramidase	*NPP1*	*ENPP7*	Chen et al. (2010) and Duan et al. (2003)
At4g29690	None	Putative inositolphosphoceramidase	*NPP1*	*ENPP7*	Chen et al. (2010)

At4g29700	None	Putative inositolphosphoceramidase	*NPP1*	*ENPP7*	Chen et al. (2010)
At3g06460	*AtELO1*	Putative condensing enzyme for VLCFA synthesis	*ELO1/ELO2/ELO3*	*ELOs*	Chen et al. (2010) and Lynch and Dunn (2004)
At3g06470	*AtELO2*	Putative condensing enzyme for VLCFA synthesis	*ELO1/ELO2/ELO3*	*ELOs*	Chen et al. (2010)
At4g36830	*AtELO3*	Putative condensing enzyme for VLCFA synthesis	*ELO1/ELO2/ELO3*	*ELOs*	Chen et al. (2010)
At1g75000	*AtELO4*	Putative condensing enzyme for VLCFA synthesis	*ELO1/ELO2/ELO3*	*ELOs*	Chen et al. (2010)
At1g01120	*KCS1*	3-ketoacyl-CoA synthase	None	None	Todd et al. (1999) and Joubes et al. (2008)
At1g04220	*KCS2*	3-ketoacyl-CoA synthase	None	None	Joubes et al. (2008) and Paul et al. (2006)
At1g07720	*KCS3*	3-ketoacyl-CoA synthase	None	None	Blacklock and Jaworski (2006) and Joubes et al. (2008)
At1g19440	*KCS4*	3-ketoacyl-CoA synthase	None	None	Blacklock and Jaworski (2006) and Joubes et al. (2008)
At1g25450	*KCS5/CER60*	3-ketoacyl-CoA synthase	None	None	Fiebig et al. (2000), Costaglioli et al. (2005), and Joubes et al. (2008)
At1g68530	*KCS6/CER6*	3-ketoacyl-CoA synthase	None	None	Fiebig et al. (2000), Costaglioli et al. (2005) and Joubes et al. (2008)
At1g71160	*KCS7*	3-ketoacyl-CoA synthase	None	None	Blacklock and Jaworski (2006); Joubes et al. (2008)
At2g15090	*KCS8*	3-ketoacyl-CoA synthase	None	None	Joubes et al. (2008)
At2g16280	*KCS9*	3-ketoacyl-CoA synthase	None	None	Paul et al. (2006), Kim et al. (2013), and Joubes et al. (2008)

(continued)

Table 11.1 (continued)

Arabidopsis Gene (AGI code)	Designated gene symbol	Function (characterized or predicted)	*Saccharomyces cerevisiae* homolog	Mammalian homolog	References
At2g26250	*KCS10/FDH*	3-ketoacyl-CoA synthase	None	None	Yephremov et al. (1999) and Joubes et al. (2008)
At2g26640	*KCS11*	3-ketoacyl-CoA synthase	None	None	Blacklock and Jaworski (2006) and Joubes et al. (2008)
At2g28630	*KCS12*	3-ketoacyl-CoA synthase	None	None	Joubes et al. (2008)
At2g46720	*KCS13/HIC*	3-ketoacyl-CoA synthase	None	None	Gray et al. (2000), Costaglioli et al. (2005), and Joubes et al. (2008)
At3g10280	*KCS14*	3-ketoacyl-CoA synthase	None	None	Joubes et al. (2008)
At3g52160	*KCS15*	3-ketoacyl-CoA synthase	None	None	Joubes et al. (2008)_
At4g34250	*KCS16*	3-ketoacyl-CoA synthase	None	None	Blacklock and Jaworski (2006) and Joubes et al. (2008)
At4g34510	*KCS17*	3-ketoacyl-CoA synthase	None	None	Trenkamp et al. (2004) and Joubes et al. (2008)
At4g34520	*KCS18*	3-ketoacyl-CoA synthase	None	None	Joubes et al. (2008)
At5g04530	*KCS19/FAE1*	3-ketoacyl-CoA synthase	None	None	Paul et al. (2006) and Joubes et al. (2008)
At5g43760	*KCS20*	3-ketoacyl-CoA synthase	None	None	Paul et al. (2006) and Joubes et al. (2008)
At5g49070	*KCS21*	3-ketoacyl-CoA synthase	None	None	Joubes et al. (2008)
At1g67730	*KCR*	3-ketoacyl-CoA reductase	*YBR159*	None	Beaudoin et al. (2002)
At1g24470	*KCR*	3-ketoacyl-CoA reductase	*YBR159*	None	Beaudoin et al. (2002)
At5g10480	*HACD*, *PAS2*	3-hydroxyacyl-CoA dehydratase	*PHS1*	*HACD*	Bach et al. (2008)
At5g59770	*HACD*	3-hydroxyacyl-CoA dehydratase	*PHS1*	*HACD*	Bach et al. (2008)
At3g55360	*ECR*, *CER10*	Enoyl-CoA reductase	*TSC13*	*SC2*	Gable et al. (2004) and Zheng et al. (2005)

activity. This has been shown in Arabidopsis with ssSPT over-expression leading to increased SPT activity and ssSPT RNAi suppression resulting in reduced SPT activity (Kimberlin et al. 2013). Whether alteration of ssSPT levels occurs naturally in response to intracellular cues to mediate sphingolipid homeostasis is unclear. In addition, no evidence currently exists for regulation of SPT activity through post-translational modifications of ssSPT polypeptides.

It is also notable that ssSPTs can dictate the acyl-CoA specificity of SPT (Han et al. 2009). In this regard, human ssSPTa and ssSPTb polypeptides have been shown to confer different acyl-CoA specificity when bound to SPT leading to the production of either C18 LCBs using palmitoyl (16:0)-CoA substrates or C20 LCBs using stearoyl (18:0)-CoA substrates (Han et al. 2009). These differences in acyl-CoA substrate specificities were shown to result from a single amino acid residue that is a Met in the human ssSPTa and Val in human ssSPTb (Han et al. 2009). The Arabidopsis ssSPTa and ssSPTb polypeptides both contain Met at the analogous position, but mutation of Met to Val results in the aberrant production of C20 LCBs when expressed in transgenic Arabidopsis (Kimberlin et al. 2013).

In yeast, ORM proteins have been shown to act as homeostatic negative regulators of SPT in response to intracellular sphingolipid levels (Roelants et al. 2011; Han et al. 2010; Breslow et al. 2010). The regulation involves TORC2-dependent phosphorylation of ORM to gradually relieve ORM suppression of SPT to enhance LCB synthesis in response to sub-optimal intracellular sphingolipid levels and reversible dephosphorylation of ORM by SAC1 phosphatase to engage ORM suppression of SPT activity in response to excess intracellular sphingolipid levels (Muir et al. 2014; Roelants et al. 2011). Arabidopsis has two homologs of the yeast *ORM* genes, *ORM1* (At1g01230) and *ORM2* (At5g42000). Although these proteins have not yet been characterized, RNAi suppression of *ORM* genes in rice results in temperature sensitivity and pollen abnormalities (Chueasiri et al. 2014).

Regulation of SPT by ORM proteins in plants and other eukaryotes appears to be more complex than ORM phosphorylation/dephosphorylation in yeast. As described above, the primary regulatory mechanism of ORM in yeast occurs through TORC2 dependent YPK1 phosphorylation of ORM that relieves inhibition of SPT that can be reversed by SAC1 phosphatase activity that restores inhibition of SPT (Han et al. 2010; Muir et al. 2014). This mechanism is adjustable and dependent on intracellular sphingolipid levels and has been shown to be coordinated with ceramide synthase activity (Muir et al. 2014). An N-terminal extension of approximately 80 amino acids in yeast ORM was found to contain several Ser residues that are responsible for this phosphorylation mechanism (Roelants et al. 2011). This phosphorylation domain, however, is absent in plant and mammalian ORM homologs (Roelants et al. 2011). As a result, it is unclear if ORM phosphorylation/dephosphorylation also regulates SPT activity in plants and mammals. Overall, homeostatic regulation of SPT in plants and mammals remains an open and active area of research.

3-Ketosphinganine Reductase

In the second step of LCB synthesis, the SPT product 3-ketosphinganine is reduced by the enzyme 3-ketosphinganine reductase (KSR) to form sphinganine (d18:0), the simplest LCB found in plants (Fig. 11.3 and Table 11.1). KSR is encoded by two genes in *Arabidopsis thaliana*, *KSR-1* (At3g06060) and *KSR-2* (At5g19200). Both genes are essential and contribute to the reductase activity (Chao et al. 2011), although *KSR-1* is more highly expressed throughout the plant (Chao et al. 2011). *KSR-1* and *KSR-2* are functionally redundant, but *KSR-1* is the primary contributor to the reductase activity (Chao et al. 2011). Loss-of-function mutants of *KSR-1* are viable but display greatly reduced reductase activity (Chao et al. 2011). These

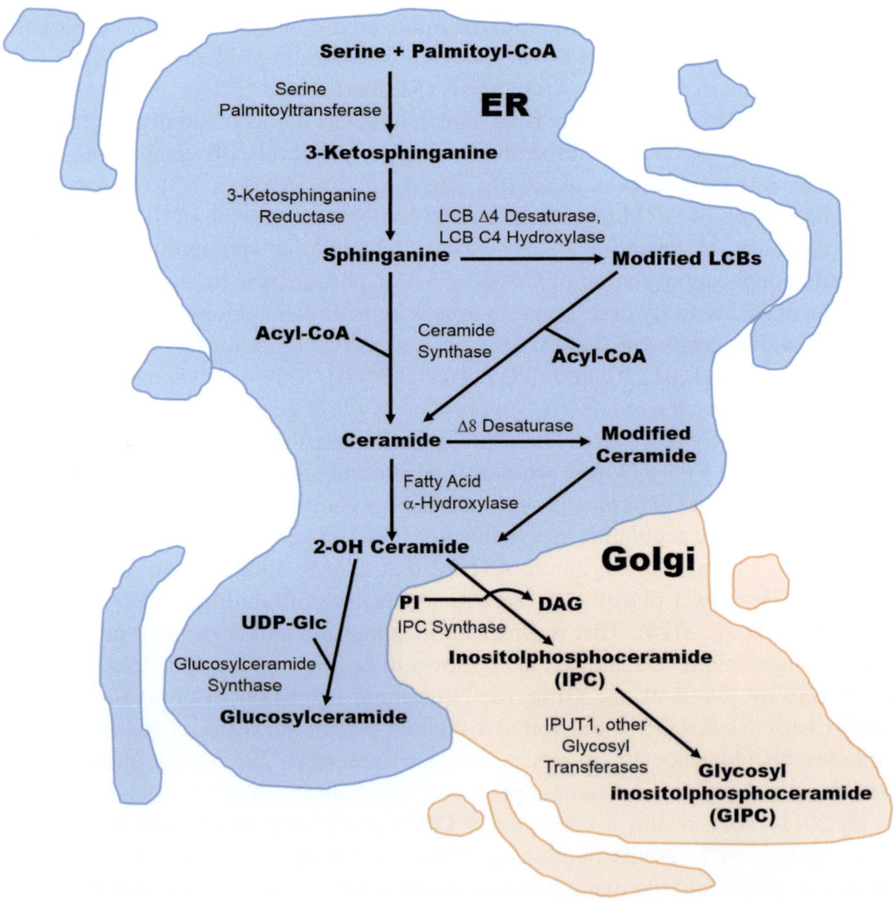

Fig. 11.3 Abbreviated plant sphingolipid biosynthetic pathway. Abbreviations: *LCB* long-chain base, *Glc* glucose, *PI* phosphatidylinositol, *DAG* diacylglycerol, *IP* inositolphosphate, *GIPCase* glycosyl inositolphosphoceramidase, *IPUT1* inositol phosphorylceramide glucuronosyltransferase 1

mutants also display an altered leaf ionome that is associated with increased root suberization, altered root morphology, and altered root iron homeostasis (Chao et al. 2011). The sphinganine (d18:0) produced from the combined activities of SPT and KSR can be used directly by ceramide synthase or modified by hydroxylation or desaturation at the C-4 position prior to use for ceramide synthesis.

LCB Modifications

LCB C-4 Hydroxylation

The d18:0 LCB resulting from the sequential activities of SPT and KSR can undergo combinations of three modification reactions to generate trihydroxylated and unsaturated LCBs (Figs. 11.1a, 11.3, and Table 11.1). In Arabidopsis leaves, ~90 % of the total LCBs contain three hydroxyl groups and Δ8 unsaturation. The third hydroxyl group of these LCBs occurs at the C-4 position and is introduced by a LCB C-4 hydroxylase (Chen et al. 2008; Sperling et al. 2001). This enzyme is a di-iron oxo protein with homology to desaturases and hydroxylases (Sperling et al. 2001). The two genes that encode the LCB C-4 hydroxylase in Arabidopsis are designated *SPHINGOID BASE HYDROXYLASE (SBH)1* (At1g69640) and *2* (At1g14290). Expression of these genes in mutants of the *Saccharomyces cerevisiae SUR2* gene (Haak et al. 1997) that encodes a related LCB C-4 hydroxylase restores trihydroxy LCB synthesis (Chen et al. 2008; Sperling et al. 2001). It is presumed that the Arabidopsis LCB C-4 hydroxylase uses a free dihydroxy LCB as its substrate, in part, because of the prevalence of trihydroxy LCBs in the free LCB pool (Markham and Jaworski 2007).

LCB Δ8 Desaturation

LCBs with Δ8 unsaturation, either in the dihydroxy or trihydroxy form, are also abundant in sphingolipids of most plant species (Lynch and Dunn 2004) (Figs. 11.1a and 11.3). Like the LCB C-4 hydroxylase, LCB Δ8 desaturases are di-iron oxo enzymes (Shanklin and Cahoon 1998). The plant Δ8 LCB desaturase was originally identified in sunflower as a desaturase-like enzyme that also contains an N-terminal cytochrome b_5 domain and shown to confer production of Δ8 unsaturated LCBs when expressed in *Saccharomyces cerevisiae* (Sperling et al. 1995). Notably, the LCB Δ8 desaturase is not found in mammals and *Saccharomyces cerevisiae*, but is present in plants and filamentous or dimorphic fungi such as *Pichia patoris* and *Yarrowia lipolytica*. Two homologs, *SLD1* (At3g61580) and *SLD2* (At2g46210), were identified in Arabidopsis and confirmed to be Δ8 desaturases through yeast and *in planta* studies (Sperling et al. 1998; Chen et al. 2012). To further add to the structural diversity found in LCBs, the Δ8 double bond can be introduced in either the *cis* or *trans* configuration (Markham et al. 2006), which likely results from presentation of LCB substrates in alternative conformations relative to the di-iron oxo atoms in the active site of these enzymes (Beckmann et al. 2002). Though evidence to date cannot preclude that at least a portion of LCB Δ8 desaturation uses free LCBs as substrates, it is presumed that these enzymes largely use LCBs bound in ceramides as substrates (Beckmann et al. 2002; Sperling et al. 1998).

LCB Δ4 Desaturation

Long-chain bases (LCBs) with Δ4 unsaturation are also prevalent in sphingolipids in many plant species. LCB Δ4 unsaturation occurs almost entirely in combination with LCB Δ8 unsaturation in dihydroxy LCBs. These di-unsaturated, dihydroxy LCBs (d18:2) also are found almost exclusively in ceramides of GlcCer, but absent from ceramides of GIPCs (Markham and Jaworski 2007; Markham et al. 2006; Sperling et al. 2005) (Figs. 11.1a, c, 11.3, and Table 11.1). Arabidopsis contains one Δ4 desaturase gene (At4g04930) that was identified by homology to analogous genes in filamentous fungi and mammals (Ternes et al. 2002). In contrast to the LCB Δ8 desaturase, the Δ4 desaturase introduces double bonds exclusively in the *trans* configuration, most likely using free LCBs as substrates (Ternes et al. 2002). As a result, two d18:2 isomers occur in plants: d18:2-*trans*Δ4, *trans*Δ8 and d18:2-*trans*Δ4, *cis*Δ8. It is notable that LCB C-4 hydroxylases and LCB Δ4 desaturase can both use d18:0 as substrates. As a result, C-4 hydroxylation precludes Δ4 desaturation, and conversely, Δ4 desaturation prevents C-4 hydroxylation. In Arabidopsis and likely other Brassicaceae, the LCB Δ4 desaturase gene has little or no expression in leaves (Michaelson et al. 2009). Instead, expression is limited almost entirely to flowers and, specifically, pollen, which is consistent with the occurrence of d18:2 in Arabidopsis organs (Michaelson et al. 2009). In most species outside of the Brassicaceae family, LCB Δ4 desaturation, as evidenced by d18:2 production, occurs throughout the plant, and in species such as tomato and soybean, d18:2 is the most abundant LCB in GlcCer (Markham et al. 2006; Sperling et al. 2005).

Influence of Long-Chain Base Hydroxylation and Desaturation on Metabolic Routing

Evidence that has emerged from sphingolipid compositional profiling of Arabidopsis mutants indicates that LCB hydroxylation and desaturation affect metabolic outcomes in sphingolipid biosynthesis. LCB C-4 hydroxylase mutants, for example, accumulate high levels of sphingolipids with ceramide backbones containing C16 fatty acids and dihydroxy LCBs, rather than the more typical ceramides with very long-chain fatty acids and trihydroxy LCBs (Chen et al. 2008). As discussed below, this metabolic phenotype arises from the substrate preferences of ceramide synthases. In addition, Arabidopsis *sld1sld2* double mutants lacking LCB Δ8 unsaturation have a 50 % reduction of GlcCer, perhaps due to the substrate specificity of GlcCer synthase (Chen et al. 2012). Similarly, Arabidopsis mutants for the LCB Δ4 desaturase, have a 50 % reduction in GlcCer in pollen (Michaelson et al. 2009). This phenotype is more extreme in LCB Δ4 mutants of the yeast *Pichia pastoris* (Michaelson et al. 2009). Disruption of this LCB Δ4 desaturase in *P. pastoris* results in nearly complete loss of GlcCer, likely due to a ceramide synthase that exclusively uses LCBs with Δ4 unsaturation and is dedicated to GlcCer synthesis (Michaelson et al. 2009).

In addition to metabolic alterations, loss of LCB C-4 hydroxylation and Δ8 desaturation affects plant performance. LCB C-4 hydroxylase mutants have impaired growth and constitutive up-regulation of PCD (Chen et al. 2008). In addition, Arabidopsis mutants lacking LCB Δ8 unsaturation are more sensitive to low temperature (Chen et al. 2012), and the relative ratio of *cis-trans* Δ8 unsaturation affects resistance of Arabidopsis to aluminum (Ryan et al. 2007). However, Arabidopsis LCB Δ4 desaturase mutants have no detectable impairment of pollen viability or germination or other growth phenotypes, despite the reduction of GlcCer levels (Michaelson et al. 2009). These findings bring into question the quantitative significance of GlcCer for plant performance.

Sphingolipid Fatty Acid Synthesis and Structural Modifications

Carbon chain-length, unsaturation, and hydroxylation of fatty acids also contribute to the structural diversity of the ceramide backbone of sphingolipids (Table 11.1). In plants, the fatty acid component ranges from 16 to 26 carbon atoms (Markham and Jaworski 2007), including small amounts of odd-chain fatty acids with 21, 23, and 25 carbon atoms (Cahoon and Lynch 1991). In Arabidopsis leaves, C16, C24, and C26 fatty acids predominate (Markham and Jaworski 2007; Markham et al. 2006). The C16 fatty acids of ceramides arise from palmitic acid formed by de novo fatty acid synthesis, whereas the very long-chain fatty acids or VLCFAs (i.e., fatty acids with ≥C20) of sphingolipids arise from the ER-localized reactions involving the two-carbon sequential elongation of fatty acids produced de novo in plastids (Smith et al. 2013). Each two carbon elongation cycle involves the four successive reactions catalyzed by 3-ketoacyl-CoA synthase (KCS), 3-ketoacyl-CoA reductase (KCR), hydroxyacyl-CoA dehydratase (HACD), and enoyl-CoA reductase (ECR) (Smith et al. 2013). Interestingly, KCS enzymes do not occur in *Saccharomyces cerevisiae* (Paul et al. 2006). Instead, fatty acid elongation is initiated by ELO enzymes that are structurally unrelated, but functionally equivalent to KCS (Paul et al. 2006). Four ELO homologs occur in Arabidopsis, but no findings to date link these enzymes to the synthesis of sphingolipid VLCFAs (Haslam and Kunst 2013). Arabidopsis mutants of the *PAS2* gene (At5g10480) encoding HACD are defective in VLCFA synthesis and have demonstrated the importance of sphingolipid VLCFAs for cellular function. Partial *PAS2* mutants are defective in growth and phragmoplast (or cell plate) formation resulting in impaired cell division, and null *PAS2* mutants display embryo lethality (Bach et al. 2008, 2011).

Sphingolipid VLCFAs are typically saturated in the plant kingdom, but monounsaturated VLCFAs occur in sphingolipids of Brassicaceae and some Poaceae species as well as selected species from other families (Cahoon and Lynch 1991; Lynch and Dunn 2004; Markham et al. 2006; Sperling et al. 2005). The double bond in sphingolipid VLCFAs of these species is at the ω-9 position (Imai et al. 2000). In Arabidopsis, this double bond is introduced by an enzyme encoded by *ADS2* (At2g31360) that has homology to acyl-CoA desaturases (Smith et al. 2013). The *ADS2* gene is induced by low temperatures and *ADS2* null mutants display chilling sensitivity, indicating a link between sphingolipid structure and low temperature performance, as also shown for the LCB Δ8 desaturase (Chen and Thelen 2013). Notably, *ADS2*-encoded enzymes do not occur in Poaceae and other monocots (unpublished observation), suggesting that a distinct pathway has evolved for biosynthesis of monounsaturated VLCFA synthesis in Poaceae.

Fatty acids in ceramides of glycosphingolipids occur almost entirely with C-2 or α-hydroxylation (Lynch and Dunn 2004). The C-2 hydroxyl group is introduced by a di-iron-oxo enzyme related to the *Saccharomyces cerevisiae* fatty acid C-2 hydroxylase encoded by the *FAH1* or *SCS7* gene (Haak et al. 1997; Mitchell and Martin 1997). The Arabidopsis homologs AtFAH1 (encoded by At2g34770) and AtFAH2 (encoded by At4g20870) notably lack the N-terminal cytochrome b_5 domain that is found in the *Saccharomyces cerevisiae* enzyme (Konig et al. 2012; Mitchell and Martin 1997; Nagano et al. 2012). Based on phenotypes in T-DNA insertion mutants and RNAi suppression lines, AtFAH1 appears to be primarily associated with hydroxylation of VLCFAs, and AtFAH2 appears to be primarily associated with hydroxylation of C16 fatty acids *in planta* (Nagano et al. 2012). It is presumed that AtFAH1 and AtFAH2 use fatty acids in ceramides rather than free or CoA esters of fatty acids as substrates, given that a substantial portion of fatty acids in the free ceramide pool lack C-2 hydroxylation, even though hydroxylated fatty acids predominate in glycosphingolipid ceramide backbones (Markham and Jaworski 2007). Double mutants of the At*FAH1* and At*FAH2* genes have elevated levels of ceramides but ~25 % reduction in glucosylceramide level (Konig et al. 2012). These results suggest that ceramides with C-2 hydroxylated fatty acids are important for metabolic channeling of ceramides to form glycosphingolipids, due possibly to the substrate preference of enzymes such as glucosylceramide synthase (see below). Suppression of PCD by ER-associated Bax inhibitor-1 protein in Arabidopsis has been shown to be dependent on functional fatty acid C-2 hydroxylases, and overexpression of the Bax inhibitor 1 gene increases fatty acid C-2 hydroxylation of ceramides through direct interaction with cytochrome b_5 (Nagano et al. 2009, 2012). From these findings, it has been speculated that accumulation of ceramides with fatty acids lacking the C-2 hydroxyl group initiates PCD, whereas this response is reduced when the fatty acids of these ceramides are hydroxylated (Nagano et al. 2012).

Ceramide Synthesis

Ceramides are synthesized by the condensation of a long-chain base and fatty acyl-CoA through an acyltransferase-type reaction catalyzed by ceramide synthase (Fig. 11.3 and Table 11.1). Three ceramide synthases have been identified in Arabidopsis through homology with the yeast ceramide synthase encoded by *LAG1* (*LONGEVITY ASSURANCE GENE1*). These enzymes are designated Lag One Homolog (LOH)-1, -2, and -3 and correspond to genes encoded by *LOH1*, At3g25440; *LOH2*, At3g19260; and *LOH3*, At1g13580, respectively (Ternes et al. 2011; Markham et al. 2011). Homologs of these three enzymes are found throughout the plant kingdom and appear to form two distinct evolutionary branches, LOH1/LOH3-related isoforms and LOH2-related isoforms (Markham et al. 2011; Ternes et al. 2011). Arabidopsis LOH1 and LOH3 share approximately 90 % amino acid sequence identity, while LOH2 shares approximately 60 % identity with LOH1 and LOH3 (Ternes et al. 2011; Markham et al. 2011). In other mammals, multiple ceramide synthases occur that have distinct specificity for fatty acyl-CoAs and/or long-chain bases (Venkataraman et al. 2002; Laviad et al. 2008; Mizutani et al. 2005, 2006; Riebeling et al. 2003). Studies of Arabidopsis LCB C-4 hydroxylase mutants initially pointed to the likelihood that two functional classes of ceramide synthases occur in plants (Chen et al. 2008). Loss of or reduced LCB C-4 hydroxylation has been shown to result in the aberrant accumulation of high levels of sphingolipids with ceramides containing C16 fatty acids bound to dihydroxy LCBs (Chen et al. 2008). Based on this observation, it was proposed that Arabidopsis has one class of ceramide synthase that links C16 fatty acyl-CoAs with dihydoxy LCBs (termed "Class I"), and a second class ("Class II") that primarily links very long-chain fatty acyl CoAs with trihydroxy LCBs (Chen et al. 2008) (Fig. 11.4). This prediction was supported by the identification, biochemical and genetic characterization of LOH1, LOH2, and LOH3 in Arabidopsis. Studies using yeast complementation showed that LOH2 prefers C16 acyl-CoAs, similar to the predicted Class I ceramide synthase (Ternes et al. 2011). Similarly, Arabidopsis *LOH2* mutants were found to be deficient in sphingolipids with ceramide backbones containing C16 fatty acids and dihydroxy fatty acids (Markham et al. 2011). Consistent with the substrate properties of Class II ceramide synthase, partial knock-out mutants of *LOH1* and *LOH3* contained reduced amounts of ceramides with very long-chain fatty acids and trihydroxy LCBs (Markham et al. 2011). It is notable that under ideal growth conditions, null mutants of *LOH2* are viable, suggesting that the Class I ceramide synthase and hence ceramides with C16 fatty acids and dihydroxy LCBs are not essential in Arabidopsis (Markham et al. 2011). Conversely, double null mutants of *LOH1* and *LOH3* were not recoverable, indicating that the Class II ceramide synthase and ceramides with very long-chain fatty acids and trihydroxy LCBs are essential (Markham et al. 2011).

Ceramide synthases are known targets for competitive inhibition by sphinganine analog mycotoxins (SAMs) such as fumonisin B_1 or FB_1 produced by a variety of *Fusarium* species and AAL toxin produced by *Alternaria alternata* f. sp. *lycoper-*

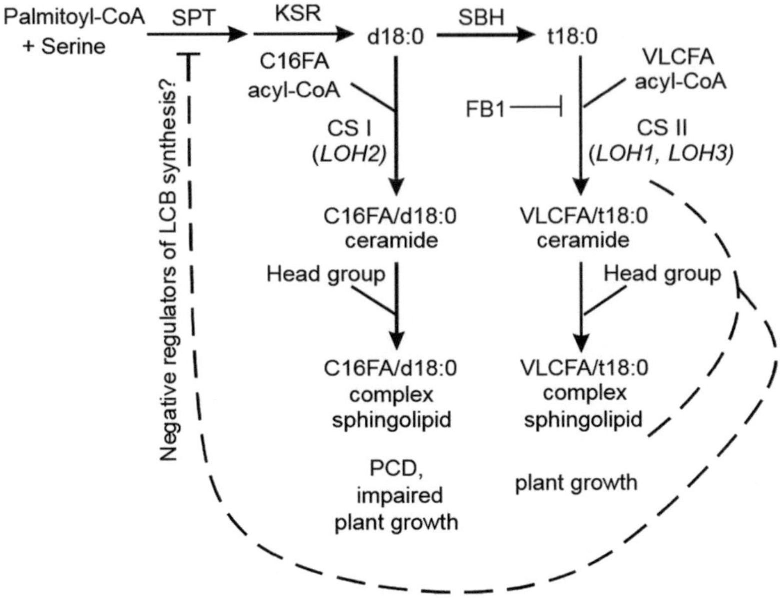

Fig. 11.4 Model of ceramide synthase mediated long-chain base (*LCB*) and fatty acid routing. The Arabidopsis gene names are shown as reference. As indicated, Class I ceramide synthase (*CSI*) encoded by *LOH2* displays strict substrate specificity of C16 fatty acid acyl-CoAs and dihydroxy LCBs, and Class II ceramide synthase (*CSII*) encoded by *LOH1* or *LOH3* display strict substrate specificity for very long-chain fatty acyl-CoAs and trihydroxy LCBs. One or more products of the CSII pathway appear to negatively regulate serine palmitoyltransferase (*SPT*) activity. In addition, sphingolipids with ceramides from the CSI pathway do not support growth, while those from the CSII pathway are essential for plant growth. The mycotoxin fumonisin B_1 (FB_1) appears to preferentially inhibit CSII enzymes. KSR, 3-ketosphinganine reductase; SBH, LCB C-4 hydroxylase

sici (Abbas et al. 1994). These compounds, particularly FB_1, have been widely used as tools for induction of programmed cell death (PCD) in plants, presumably due to the accumulation of cytotoxic LCBs from their inhibition of ceramide synthases (Stone et al. 2000). Recent evidence using FB_1 treatment of Arabidopsis ceramide synthase mutants has suggested that FB_1 is a more potent inhibitor of Class II ceramide synthases (i.e. LOH1 and LOH3 ceramide synthases) (Markham et al. 2011). Interestingly, in addition to accumulation of free LCBs, elevated levels of ceramides with C16 fatty acids and dihydroxy LCBs formed by Class I ceramide synthases (i.e. LOH2 ceramide synthase) are detectable following treatment of Arabidopsis with FB_1 (Markham et al. 2011). These results suggest that FB_1 cytotoxicity and PCD induction may be triggered by accumulated ceramides rather than or in addition to accumulated LCBs. FB_1 has also been used as a tool to study sphingolipid homeostasis in plants based on the observation that down-regulation of serine palmitoyltransferase (SPT) activity reduces FB_1 cytotoxicity and up-regulation of SPT activity enhances sensitivity of plants to FB_1 (Kimberlin et al. 2013; Shi et al. 2007).

Glucosylceramide Synthesis

Following its synthesis by Class I or Class II ceramide synthases, the ceramide backbone can be glycosylated at its C-1 OH to form either of two classes of glycosphingolipids: glucosylceramides (GlcCer) or glycosyl inositolphosphoceramides (GIPCs) (Fig. 11.3 and Table 11.1). GlcCer are the simplest glycosphingolipid and occur broadly in eukaryotes, with the notable exception of *Saccharomyces cerevisiae* (Lynch and Dunn 2004). GlcCer consist of a glucose bound to the ceramide backbone by a 1,4-glycosidic linkage and are formed by the condensation of a ceramide substrate with UDP-glucose (Leipelt et al. 2001). This reaction is catalyzed by GlcCer synthase, an ER-localized enzyme in Arabidopsis that is encoded by At2g19980 (Melser et al. 2010). Compared to GIPCs, GlcCer are more enriched in ceramides with C16 fatty acids and dihydroxy LCBs (Markham et al. 2006; Sperling et al. 2005). In plants such as tomato and soybean, ceramides with C16 fatty acids and the LCB d18:2 predominate (Markham et al. 2006; Sperling et al. 2005). Based on this composition, it appears that a large portion of the GlcCer ceramide backbone is channeled from Class I-type ceramide synthases that have substrate preference for C16 fatty acids and dihydroxy LCBs (Markham et al. 2011). Although it is an abundant glycosphingolipid in plants, null mutants of the LCB $\Delta4$ deasaturase in Arabidopsis have 30 % reductions in GlcCer levels in flowers (Michaelson et al. 2009) and 50 % reduction in GlcCer levels in pollen (Luttgeharm et al. 2015) without any apparent effect on flower and pollen physiology and function (Michaelson et al. 2009). GlcCer synthase is potently inhibited by d,l-*threo*-1-phenyl-2-decanoylamino-3-morpholino-1-propanol (PDMP) (Melser et al. 2010). Treatment of Arabidopsis roots with PDMP results in altered Golgi morphology, including reduced numbers of Golgi stacks, and defects in endomembrane trafficking (Melser et al. 2010, 2011). PDMP application to Arabidopsis root cells has also been shown to result in rapid vacuolar fusion and altered vacuole morphology including the appearance of vacuolar invaginations (Kruger et al. 2013). Arabidopsis GlcCer synthase mutants devoid of GlcCer have yet to be described. Such mutants will clarify whether GlcCer are essential in plants, which is an open question because viable fungal cells can be recovered that lack GlcCer (Michaelson et al. 2009; Rittenour et al. 2011).

Inositolphosphoceramide Synthesis

As an alternative fate to GlcCer synthesis, ceramides can be used for the production of GIPCs. GIPCs, which are approximately twofold more abundant in Arabidopsis leaves than GlcCer, are typically enriched in ceramides with VLCFAs and trihydroxy LCBs that arise from Class II ceramide synthases (Markham et al. 2006). The first step in GIPC synthesis occurs by the transfer of the inositolphosphoryl head group of phosphatidylinositol (PI) onto ceramide to form inositolphosphoceramides (IPCs) (Mina et al. 2010; Wang et al. 2008) (Fig. 11.3 and Table 11.1). This activity is catalyzed by IPC synthase, a phosphatidic acid phosphatase-2 (PAP2)-related enzyme, that is encoded by three genes in Arabidopsis: *IPCS1* (or *ERH1*), At2g37940; *IPCS2*, At2g37940; *IPCS3*, At2g29525 (Mina et al. 2010; Wang et al. 2008). In contrast to the ER localization of GlcCer synthase, IPC synthases predominantly occur in Golgi bodies of Arabidopsis. Plant IPC synthases are most closely related to analogous enzymes in the protozoa *Leishmania major* and *Trypanosoma brucei* than to the *Saccharomyces cerevisiae* IPC synthase (encoded by the *AUR1* gene) (Mina et al. 2010; Wang et al. 2008). Despite this, the three Arabidopsis IPC synthase genes are able to rescue lethality associated with the loss of IPC production in the *Saccharomyces cerevisiae* AUR1 mutant (Mina et al. 2010; Wang et al. 2008). Although triple mutants of the three Arabidopsis IPC synthase genes have not been reported, it is presumed that IPC biosynthesis is essential, as the three genes are likely partially redundant.

Following the synthesis of IPC, up to seven additional sugar residues can be added to the inositolphosphoryl head group to form an array of different GIPCs (Bure et al. 2011; Cacas et al. 2013). The first residue added to the inositophosphoryl head group is a glucuronic acid moiety (Rennie et al. 2014). This reaction, which uses a UDP-glucuronic acid substrate, was recently shown to be catalyzed by a glycosyltransferase encoded by *IPUT1* (At5g18480) in Arabidopsis (Rennie et al. 2014). T-DNA null mutants of *IPUT1* are not transmitted through pollen, indicating that this gene is essential in Arabidopsis (Rennie et al. 2014). The remaining glycosyltransferases associated with GIPC synthesis have yet to be identified. Interestingly, a Golgi lumen-localized GDP-mannose transporter encoded by *GOLGI-LOCALIZED NUCLEOTIDE SUGAR TRANSPORTER 1* (*GONST1*, At2g13650) in Arabidopsis was found to be deficient in GIPCs with glycosylation beyond the glucuronic acid introduced by the *IPUT1*-encoded glycosyltransferase (Mortimer et al. 2013). This suggests that like *Saccharomyces cerevisiae*, Arabidopsis GIPCs contain mannose, presumably bound to the glucuronic acid moiety. Interestingly, *GONST1* mutants display a dwarfed phenotype, constitutive induction of the hypersensitive response, and elevated salicylic acid levels, consistent with a connection between GIPC synthesis and plant pathogen defense (Mortimer et al. 2013).

LCB and Ceramide Phosphorylation/Dephosphorylation

In addition to their occurrence in free form and in ceramides, LCBs are also detectable in low levels as phosphate derivatives that have been attributed to triggers of physiological responses, such as ABA-dependent guard cell closure (Coursol et al. 2003; Ng et al. 2001) (Figs. 11.5, 11.6 and Table 11.1). Phosphorylation of LCBs at their C-1 hydroxyl group is catalyzed by LCB kinases (often referred to sphingosine kinases or SPHKs). To date three LCB kinases have been identified in Arabidopsis:

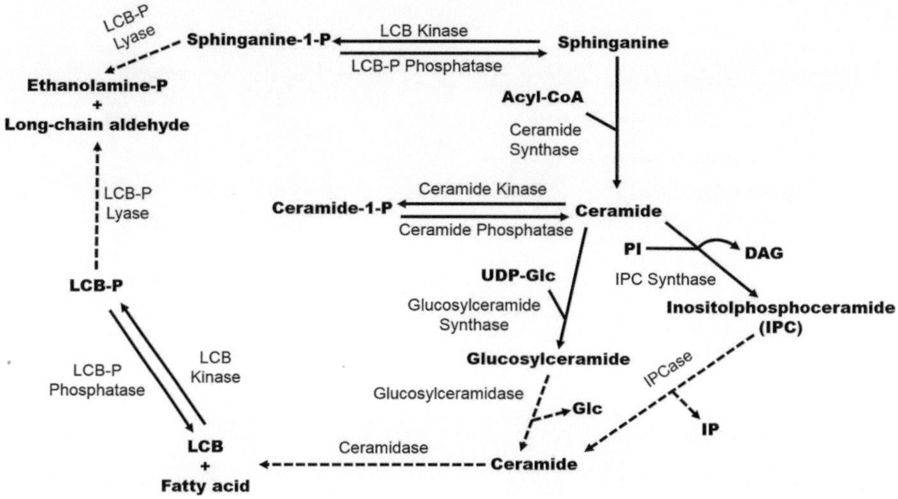

Fig. 11.5 Sphingolipid catabolic and ceramide and long-chain base phosphorylation/dephosphorylation pathways. *Dashed arrows* represent enzymatic steps involved in catabolism. Abbreviations: *LCB* long-chain base, *LCB-P* long-chain base-1-phosphate, *Glc* glucose, *PI* phosphatidylinositol, *DAG* diacylglycerol, *IP* inositolphosphate, *IPCase* inositolphosphoceramidase

SPHK1 (At5g23450), *SPHK2* (At2g46090), and *AtLCBK1* (At5g23450) (Imai and Nishiura 2005; Worrall et al. 2008; Guo et al. 2012). Release of the phosphate group from LCB-P molecules is catalyzed by the enzyme LCB-P phosphatase, which are encoded by two genes in Arabidopsis (At3g58490 and At5g03080) (Nakagawa et al. 2012; Worrall et al. 2008). As described below, the interplay between LCB kinases and LCB-P phosphatases are believed to be important for signaling pathways in plants (Nakagawa et al. 2012; Worrall et al. 2008).

Similar to LCBs, ceramides can also be found in phosphorylated forms (Figs. 11.5, 11.6. and Table 11.1). Although ceramide-1-phosphates are believed to be of low abundance in plants, they have proven difficult to measure by recently developed mass spectrometry-based protocols. Mutants of the proposed ceramide kinase (encoded by At5g51290), termed *accelerated death 5* or *acd5* display spontaneous onset of programmed cell death or PCD in late development (Greenberg et al. 2000; Liang et al. 2003). This is accompanied by enhanced accumulation of ceramides (Greenberg et al. 2000; Liang et al. 2003). This observation led to the hypothesis, now accepted as dogma, that elevation of ceramide levels triggers PCD in plants (Greenberg et al. 2000; Liang et al. 2003) via accumulation of mitochondrial-derived hydrogen peroxide (Bi et al. 2014). A ceramide-1-phosphate phosphatase that would convert ceramide-1-phosphates to their free form has yet to be identified in plants.

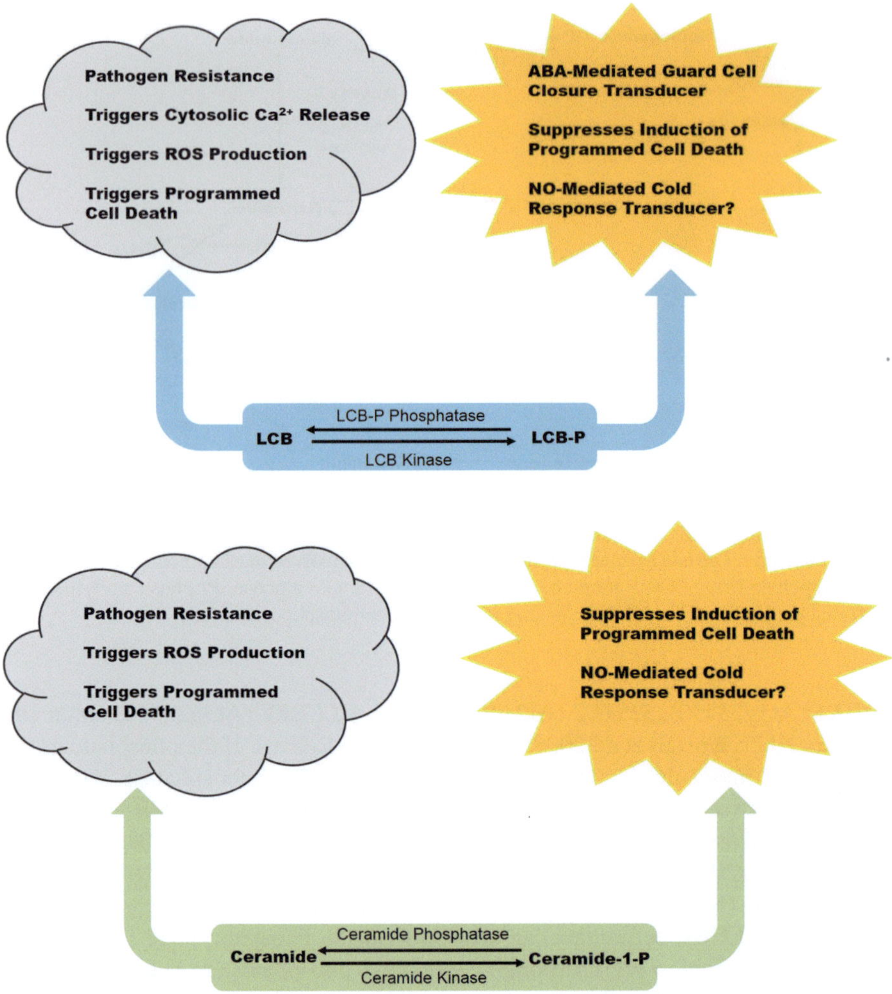

Fig. 11.6 Phosphorylated/dephosphorylated long-chain bases (*LCBs*) and ceramides serve as mediators of physiological processes in plants. The interplay between LCBs and ceramides and their phosphorylated forms regulates cellular process and responses to environmental stimuli. Abbreviations: *LCB* long-chain base, *LCB-P* long-chain base-1-phosphate, *ABA* Abscisic acid, *ROS* Reactive oxygen species, *NO* nitrous oxide

Sphingolipid Turnover

The net content and composition of sphingolipids in membranes are determined by rates of synthesis and turnover. Little is currently known about rates of sphingolipid turnover, and the contributions of sphingolipid catabolism to membrane function and plant responses to altered environmental conditions. Also unexplored to date in plants are enzymes associated with removal of glycosphingolipid head groups, although candidate genes have been proposed (Chen et al. 2010). More is known about ceramide turnover. Enzymes referred to as ceramidases convert ceramides to free LCBs and fatty acids. Ceramidases are classified into three distinct forms based upon their optimal pH preferences in *in vitro* assays: acid, neutral, and alkaline ceramidases (Mao and Obeid 2008). Three predicted neutral ceramidase genes and one predicted alkaline ceramidase have been identified by homology with human ceramidase genes (Chen et al. 2010; Wu et al. 2015). The Arabidopsis alkaline ceramidase homolog, AtACER (At4g22330), has been shown to function as a ceramidase with mutant, and RNAi suppression lines for this gene had elevated ceramide levels and increased salt sensitivity and enhanced susceptibility to a bacterial pathogen (Wu et al. 2015). A second gene *TOD1* corresponding to At5g46220 was recently shown to encode a polypeptide with alkaline ceramidase activity, but notably, lacked close homology to known alkaline ceramidases (Chen et al. 2015). Based on mutant phenotypes, this enzyme was linked to control of turgor pressure in pollen tubes and silique guard cells (Chen et al. 2015). A neutral ceramidase has also been cloned from rice and confirmed *in vitro* to be a member of the neutral ceramidase subclass (Pata et al. 2008) (Fig. 11.5 and Table 11.1).

LCBs, including those released by ceramidase activity, can be degraded following phosphorylation by LCB kinases. This process is catalyzed by LCB-P lyase (often referred to as DPL1, based on homology to the yeast enzyme), which generates C16-fatty aldehyde and phosphoethanolamine from a C18 LCB-P. Arabidopsis contains only a single *DPL1* gene (At1g27980) (Nishikawa et al. 2008; Tsegaye et al. 2007; Worrall et al. 2008) that is constitutively expressed and strongly upregulated by senescence (Tsegaye et al. 2007). Null mutants of *DPL1* displayed small increases in accumulation of the LCB-P t18:1-P, but surprisingly no obvious growth phenotypes (Tsegaye et al. 2007) (Fig. 11.5 and Table 11.1).

Sphingolipids and Membrane Structure-Function

Membrane Function

Sphingolipids compose an estimated ~40 % of the total lipids in plasma membrane of plants (Sperling et al. 2005), where they are enriched in the outer leaflet (van Genderen et al. 1991; Burger et al. 1996; Tjellstrom et al. 2010). Sphingolipids are also abundant lipid components of other endomembranes in the plant, including ER, Golgi, and tonoplast (Bayer et al. 2014; Mongrand et al. 2004; Sperling et al. 2005; Verhoek et al. 1983). GlcCer was first identified in the plasma membrane and tonoplast of plant cells in a number of studies conducted in the 1980s (Cahoon and Lynch 1991; Lynch and Steponkus 1987; Verhoek et al. 1983; Yoshida et al. 1986). In these membranes, GlcCer was reported to compose between 7 and 30 % of plasma membrane and tonoplast, depending on the plant species and tissue type analyzed (Cahoon and Lynch 1991; Lynch and Steponkus 1987; Uemura et al.

1995; Uemura and Steponkus 1994; Verhoek et al. 1983; Yoshida et al. 1986). More recently, it has been shown that GIPCs, rather than GlcCer, are the more abundant glycosphingolipid in plants (Markham and Jaworski 2007; Markham et al. 2006). The quantitative importance of GIPCs was largely overlooked until very recently because their highly glycosylated head groups offer challenges for extraction using typical organic solvents, such as mixtures of chloroform and methanol (Markham et al. 2006). Initially, by quantification of long-chain bases of GIPCs and GlcCer and later by LC-MS/MS analysis, GIPCs were found to be nearly twofold more abundant than GlcCer in Arabidopsis leaves, whereas amounts of GIPCs and GlcCer were nearly the same in tomato leaves (Markham and Jaworski 2007; Markham et al. 2006). GIPCs have subsequently been identified as one of the most abundant lipids of plant plasma membrane and are also enriched in detergent resistant membranes (DRMs) derived from isolated plasma membrane and in plasmodesmata (Grison et al. 2015; Cacas et al. 2012). Given their abundance in plasma membrane and tonoplast, it is likely that the content and composition of sphingolipids affect the ability of plants to respond to abiotic stress, particularly osmotic stresses such as freezing, drought, and salinity. For example, GlcCer concentrations were shown to decrease by nearly half in plasma membrane during cold acclimation of rye and Arabidopsis (Lynch and Steponkus 1987; Uemura et al. 1995; Uemura and Steponkus 1994). More recently, it was reported that GIPCs increase and GlcCer decrease in response to chilling of Arabidopsis (Nagano et al. 2014). Although this is likely an adaptive response to low temperatures, the impact of such adjustments in relative amounts of GIPCs and GlcCer on plant performance has not yet been established. In addition, the fatty acid and long-chain base composition of sphingolipids also affects plant resistance to abiotic stress. For example, Arabidopsis mutants lacking LCB Δ8 and ceramide fatty acid unsaturation display sensitivity to low temperature growth and alterations in the relative amounts of LCB cis-trans Δ8 unsaturation affects resistance of plants to aluminum (Chen et al. 2012; Chen and Thelen 2013; Ryan et al. 2007).

The unique structural components of sphingolipid hydrophobic ceramide backbones include VLCFAs and an abundance of hydroxyl groups distributed between the LCB and fatty acid moieties. Through these structural features, sphingolipids confer rigidity to membranes. In addition, the hydroxyl groups enable the formation of extensive hydrogen bonding networks that result in elevated phase transition temperatures and reduced ion permeability (Lunden et al. 1977; Pascher 1976). The rigidity and high phase transition temperatures of sphingolipid micelles is moderated by interactions with other lipids, including sterols (Curatolo 1987). Sphingolipids have been shown to cluster with sterol in membrane microdomains or lipid rafts (Cacas et al. 2012). Lipid microdomains have long been hypothesized to be present in membranes (Karnovsky et al. 1982) with sphingolipids potentially aiding in the sorting of membrane proteins, such as GPI-anchored proteins, by forming lipid domains that slow lateral protein diffusion (Simons and van Meer 1988; van Meer and Simons 1988; Brown and Rose 1992; Simons and Ikonen 1997;

de Almeida et al. 2003). Indeed, pure sphingolipid membranes form a 'solid gel' phase with little lateral movement that is fluidized by the presence of sterols (van Meer et al. 2008; Estep et al. 1980; Roche et al. 2008; Grosjean et al. 2015). The co-localization of sterols and sphingolipids in the membrane may be due to sphingolipids complex sugar head group's ability to shield the non-polar sterol from the bulk solvent, much like an umbrella (Huang and Feigenson 1999; Ali et al. 2006). Raft formation is also dependent on the various sphingolipid modifications, including fatty acid chain length and fatty acid and LCB hydroxylation (Klose et al. 2010).

Proteomic analysis of DRMs prepared from plant plasma membrane has revealed an enrichment of proteins in lipid rafts related to signaling, responses to biotic and abiotic stress, cellular trafficking, auxin transport, and cell wall synthesis and degradation (Brodersen et al. 2002; Lefebvre et al. 2007; Lin et al. 2008; Morel et al. 2006), suggesting that raft regions contribute to these cellular functions. Similar to DRMs, plasmodesmata have also recently been shown to be enriched in sphingolipids and sterols and contain specific GPI-anchored proteins (Bayer et al. 2014; Grison et al. 2015).

Endomembrane Trafficking

Given their abundance in the endomembrane system, sphingolipids are presumed to play a major role in ER export and Golgi-mediated trafficking of proteins through the secretory system. Consistent with this, Arabidopsis pollen deficient in sphingolipids have been shown to have vesiculated ER and lack Golgi stacks (Dietrich et al. 2008). Consistent with defects in Golgi-trafficking to the plasma membrane, the sphingolipid-deficient pollen lacked a surrounding intine layer (Dietrich et al. 2008). Similarly, chemical inhibition of GlcCer synthesis has been shown to alter Golgi morphology and impair Golgi-mediated trafficking of secretory proteins to the plasma membrane (Melser et al. 2010, 2011). Recent studies using an Arabidiopsis *KSR* mutant and GlcCer synthase and ceramide synthase inhibitors demonstrated the importance of sphingolipids in trafficking of ATP-binding cassette B19 (ABCB19) auxin transporter to the Golgi, *trans*-Golgi network, and plasma membrane (Yang et al. 2012). Similar studies targeting sterols indicated that sterols have greater significance for post-Golgi transport of ABCB19 from *trans*-Golgi to the plasma membrane (Yang et al. 2012). Of particular importance for trafficking of proteins through the secretory system is the presence of the very long-fatty acid (VLCFA) component of sphingolipids, which are enriched in GIPCs. Mutants defective in VLCFA synthesis or that have reduced activity of Class II ceramide synthases that incorporate VLCFAs in ceramides have impaired trafficking of secretory proteins, including PIN1 and AUX1 that are required for auxin transport and plant growth (Bach et al. 2008, 2011; Markham et al. 2011; Zheng et al. 2005).

Sphingolipids as Physiological Mediators

ABA-Dependent Guard Cell Closure

In addition to the contributions of the glycosphingolipids GIPCs and GlcCer to membrane structure and function, the less abundant sphingolipid biosynthetic intermediates LCBs, LCBPs, ceramides, and ceramide-1-phosphates have been linked to the mediation of numerous, diverse physiological processes in plant cells. An important contributor to the formation of these physiological mediators are kinase and phosphatase reactions that convert LCBs and ceramides between their phosphorylated and free forms, as described above. The phosphorylation status of LCBs and ceramides are key to the particular physiological process that they regulate (Fig. 11.6).

One of the first links between sphingolipids and control of cellular processes was the observation that the LCB-P sphingosine-1-phosphate or S1P participates in the ABA-mediated signaling pathway that controls stomatal aperture by elevating cytosolic Ca^{2+} levels which, in turn, activates ion channels in guard cell membranes, with the resulting K^{+} efflux causing loss of guard cell turgor pressure and stomatal closing (Kim et al. 2010). After drought treatment of the plant *Commelina communis*, S1P levels were found to increase in leaves, and when applied exogenously, S1P resulted in a $Ca2^{+}$ spike followed by stomatal closure (Ng et al. 2001). Phytosphingosine-1-phosphate elicits the same response, although sphinganine-1-phosphate does not, indicating some level of LCB specificity in the mediation of guard cell closure (Coursol et al. 2003). Treatment of Arabidopsis plants with ABA was found to activate LCB kinase, and this activity was sensitive to the mammalian LCB kinase inhibitor, N,N-dimethylsphingosine (Coursol et al. 2003). As in mammals, the target of S1P in plants is presumed to be a G-protein coupled receptor, and Arabidopsis mutants lacking the G-protein α-subunit (GPA1) did not respond upon exogenous S1P application (Coursol et al. 2005; Strub et al. 2010). More recently, a connection between phospholipids and sphingolipids in the signaling pathway for ABA-dependent guard cell closure has been proposed. In this regard, ABA and phosphatidic acid (PA) produced by phospholipase Dα1 (PLDα1) have been shown to activate sphingosine kinase (SPK) to promote production of LCB-P (Guo et al. 2012; Guo and Wang 2012; Worrall et al. 2008). PA enhancement of SPK activity was found to occur by direct interaction of PA with this enzyme (Guo et al. 2012; Guo and Wang 2012). Given that LCB-P induction of guard cell closure requires a functional PLDα1, it was proposed that LCB-P functions upstream of PLDα1 in this signaling pathway (Guo et al. 2012; Guo and Wang 2012).

Programmed Cell Death

Sphingolipids, primarily in the form of ceramides and LCBs, have also been strongly implicated in mediation of programmed cell death (PCD) in plants. As described above, an initial indication of the role of ceramides as PCD triggers was obtained from the Arabidopsis *acd5* mutant that is defective in a proposed ceramide kinase (Greenberg et al. 2000; Liang et al. 2003). This mutant accumulates enhanced levels of free ceramides and displays early onset of PCD relative to wild-type controls (Greenberg et al. 2000; Liang et al. 2003), resulting in part to the enhanced release of mitochondrial reactive oxygen species (Bi et al. 2014). PCD induction in the *acd11* mutant has also been linked to ceramide accumulation associated with defects in ceramide-1-phosphate transport in this mutant (Simanshu et al. 2014). Similar findings have been obtained by treatment of Arabidopsis cell cultures with C2 ceramide at a concentration of 50 μM (Townley et al. 2005). This treatment induces a transient increase in cytosolic $Ca^{2+}$and hydrogen peroxide production, followed by cell death, which was reversed by inhibition of Ca^{2+} release (Townley et al. 2005). These findings implicate Ca^{2+} as an essential component of ceramide induction of PCD. Notably, C2 ceramides containing 2- or α-hydroxylated fatty acids were not effective in PCD induction in Arabidopsis cell cultures (Townley et al. 2005). Consistent with this observation, the ability of Bax inhibitor-1 (BI-1) to suppress cell death in Arabidopsis is dependent on 2-hydroxylation of ceramide VLCFAs (Nagano et al. 2012).

Similar to results with ceramides, application of the free LCBs d18:1, d18:0, and t18:0 to Arabidopsis leaves also induces PCD, albeit at concentrations lower than that observed with ceramides (Shi et al. 2007). This induction of PCD was also dependent on ROS generation, but was suppressed by application of LCB-P along with free LCBs (Alden et al. 2011; Shi et al. 2007). These findings suggest that the ratio of free LCB to LCB-P, mediated by LCB kinases and LCB-P phosphatases, is an important "rheostat" for regulation of PCD (Fig. 11.6) (Alden et al. 2011; Shi et al. 2007). This is analogous to the dependence of PCD induction on relative levels of ceramides and ceramide-1-phosphates (Greenberg et al. 2000; Liang et al. 2003). The transduction pathway for elicitation of PCD by free LCBs has been shown to be dependent in Arabidopsis on mitogen-activated protein kinase 6 (MPK6) (Saucedo-García et al. 2011) as well as 14-3-3 protein phosphorylation by calcium-dependent kinase 3 (CPK3) that is activated by LCB-triggered release of cytosolic Ca^{2+} (Lachaud et al. 2013).

Fungal-derived sphingosine-analog mycotoxins (SAMs) including fumonisin B_1 produced by *Fusarium* species and AAL toxin produced by *Alternaria alternata lycopersici* are also potent triggers of PCD in plants. These molecules competitively inhibit ceramide synthases leading to the accumulation of free LCBs that, in turn, elicit PCD (Abbas et al. 1994). Consistent with this, reduction of LCB synthesis by chemical inhibition of serine palmitoyltransferase (SPT) activity enhances resistance of plants to SAMs (Spassieva et al. 2002). Increased resistance to FB_1-triggered PCD induction has also been observed in an Arabidopsis LCB1 mutant and in small subunit of SPT RNAi suppression lines that have reduced SPT activity (Kimberlin et al. 2013; Shi et al. 2007). Recent evidence has also emerged that FB_1 not only increases levels of free LCBs in plant cells but also elevates levels of ceramides containing C16 fatty acids (Markham et al. 2011; Ternes et al. 2011). This finding suggests that FB_1 most effectively inhibits Class II ceramide synthases (i.e., LOH1, LOH3) that produce ceramides with VLCFAs and are less effective inhibitors of Class I ceramide synthase (i.e., LOH2) that produces ceramides with C16 fatty acids (Markham et al. 2011; Ternes et al. 2011). These findings suggest that the potency of SAMs for PCD-induction is due to their ability to enhance accumulation of LCBs and ceramides.

Pathogen Resistance

The hypersensitive response (HR) is an important process for resistance to bacterial and fungal pathogens that is characterized by localized induction of PCD that reduces or prevents the spread of pathogens in plants. Given the importance of LCBs and ceramides to PCD induction, a considerable body of research has emerged linking sphingolipids to bacterial and fungal pathogen resistance as described in a recent review (Berkey et al. 2012). Notably, ceramide accumulation in *acd5* and *acd11* mutants has been shown to be associated with salicylic acid (SA)-dependent upregulation of HR-type PCD and pathogen-resistance genes, including genes for PR1, ERD11, and chitinase (Brodersen et al. 2002; Greenberg et al. 2000). More recently, Arabidopsis mutants defective in 2-hydroxylation of ceramide fatty acids were found to have elevated LCB and ceramide levels, as well as, increased levels of free and glycosylated SA and constitutive induction of *PR1* and *PR2* genes (Konig et al. 2012). These mutants also displayed enhanced resistance to the biotrophic fungal pathogen *Golovinomyces cichoracearum* (Konig et al. 2012). In addition, infection of Arabidopsis with the bacterial pathogen *Pseudomonas syringae* was accompanied by transient increases in the LCB phytosphingosine (t18:0) and induction of ROS and cell death (Fig. 11.6) (Peer et al. 2010; Bach et al. 2011). Furthermore, resistance to the bacterial pathogen *Pseudomonas cichorii* was compromised in tobacco upon chemical inhibition of SPT and an accompanying reduction in LCB synthesis (Takahashi et al. 2009). This resistance appears to be mediated by MPK6, as FB_1-elicited Arabidopsis *mpk6* mutants displayed reduced resistance to the bacterial pathogen *Pseudomonas syringae* pv. *tomato avrRpm1* due to compromised induction of PCD in this mutant (Saucedo-García et al. 2011).

Cold Stress Signaling

Sphingolipids as abundant components of plasma membrane and tonoplast contribute to the ability of plants to resist chilling and freezing stresses. As evidence of this, Arabidopsis mutants lacking LCB $\Delta 8$ unsaturation have increased sensitivity to prolonged exposure to low, non-freezing temperatures (Chen et al. 2012). In addition to their roles as membrane components, recent studies have implicated sphingolipids in cold stress signaling pathways (Cantrel et al. 2011; Guillas et al. 2011, 2013). Exposure of Arabidopsis plants to 4 °C resulted in accumulation of PA and nitrous oxide (NO). In addition, within 5 min of this cold treatment amounts of the LCB-P phytosphingosine phosphate and ceramide-1-phosphate increased by ~50 %

(Fig. 11.6) (Cantrel et al. 2011). This increase was negatively regulated by NO, as chemical inhibition of NO production enhanced the accumulation of these molecules but chemically-induced enhancement of NO levels reduced accumulation of the phosphorylated LCB and ceramides (Cantrel et al. 2011). From these findings, it was suggested that NO may regulate the relative levels of phosphorylated and dephosphorylated LCBs and ceramides as part of a rapid signaling response pathway to low, non-freezing temperatures (Cantrel et al. 2011; Guillas et al. 2011). The mechanistic details of this potential signaling pathway remain uncharacterized.

Conclusions

As outlined here, significant advances have been made in plant sphingolipid biology during the past 15 years. Nearly all of the genes involved in sphingolipid synthesis have now been identified and numerous insights have been made in understanding the role of sphingolipids in membrane function and physiological mediation pathways. In addition, the combination of mass spectrometry-based sphingolipid compositional profiling or sphingolipidomics and the characterization of Arabidopsis mutants have revealed the unexpected importance of LCB modifications in regulating flux into glycosphingolipids due, in part, to the narrow substrate specificities of the two functional classes of ceramide synthases. Furthermore, the biological importance of sphingolipids in processes such as programmed cell death and pathogen resistance have been fortuitously discovered by forward genetic screens. Despite the large amount of progress to date, many open questions and challenges remain for plant sphingolipid research. For example, sphingolipid biosynthesis and its regulation have been mostly examined in isolation, and metabolic networks that supply the serine and fatty acid precursors to sphingolipid synthesis have not been examined. We also have yet to understand the interconnections of sphingolipid metabolic and phospholipid and sterol metabolic pathways, particularly in response to environmental stimuli, such as altered growth temperature, that together lead to adjustments in membrane properties. In addition, details of sphingolipid homeostatic regulation in plants have yet to be fully resolved. Also missing in our knowledge is the contributions of sphingolipid turnover and the identities of many of the catabolic enzymes that contribute to mediation of sphingolipid levels in plants. Furthermore, the exact sugar residues, their linkages, and the enzymes that introduce these residues in GIPC head groups have yet to be fully identified. Moreover, it is not currently understood if the distinctive ceramide compositions in GlcCer and GIPCs arise from substrate preferences of GlcCer and IPC synthases or from differential partitioning of specific ceramides to the GlcCer synthase in the ER and to IPC synthase in the Golgi. Similarly, the distinct roles of GlcCer and GIPCs in plants have yet to be elucidated. Finally, an understanding of sphingolipid-associated signaling pathways and their possible interconnections await further resolution. Ultimately, with this information, it will be possible to more precisely genetically tailor sphingolipid metabolism for improved plant performance, particularly in response to environmental perturbations.

Acknowledgements Research that contributed to this chapter was supported by the U.S. National Science Foundation (Grant MCB-1158500).

References

Abbas HK, Tanaka T, Duke SO, Porter JK, Wray EM, Hodges L, Sessions AE, Wang E, Merrill AH Jr, Riley RT (1994) Fumonisin- and AAL-toxin-induced disruption of sphingolipid metabolism with accumulation of free sphingoid bases. Plant Physiol 106(3):1085–1093

Alden KP, Dhondt-Cordelier S, McDonald KL, Reape TJ, Ng CK, McCabe PF, Leaver CJ (2011) Sphingolipid long chain base phosphates can regulate apoptotic-like programmed cell death in plants. Biochem Biophys Res Commun 410(3):574–580

Ali MR, Cheng KH, Huang J (2006) Ceramide drives cholesterol out of the ordered lipid bilayer phase into the crystal phase in 1-palmitoyl-2-oleoyl-sn-glycero-3-phosphocholine/cholesterol/ceramide ternary mixtures. Biochemistry 45(41):12629–12638

Bach L, Michaelson LV, Haslam R, Bellec Y, Gissot L, Marion J, Da Costa M, Boutin JP, Miquel M, Tellier F, Domergue F, Markham JE, Beaudoin F, Napier JA, Faure JD (2008) The very-long-chain hydroxy fatty acyl-CoA dehydratase *PASTICCINO2* is essential and limiting for plant development. Proc Natl Acad Sci U S A 105(38):14727–14731

Bach L, Gissot L, Marion J, Tellier F, Moreau P, Satiat-Jeunemaitre B, Palauqui JC, Napier JA, Faure JD (2011) Very-long-chain fatty acids are required for cell plate formation during cytokinesis in *Arabidopsis thaliana*. J Cell Sci 124(19):3223–3234

Bayer EM, Mongrand S, Tilsner J (2014) Specialised membrane domains of plasmodesmata, plant intercellular nanopores. Front Plant Sci 5:507

Beaudoin F, Gable K, Sayanova O, Dunn T, Napier JA (2002) A Saccharomyces cerevisiae gene required for heterologous fatty acid elongase activity encodes a microsomal β-keto-reductase. J Biol Chem 277(13):11481–11488

Beckmann C, Rattke J, Oldham NJ, Sperling P, Heinz E, Boland W (2002) Characterization of a Δ8-sphingolipid desaturase from higher plants: a stereochemical and mechanistic study on the origin of E, Z isomers. Angew Chem Int Ed 41(13):2298–2300

Beeler T, Bacikova D, Gable K, Hopkins L, Johnson C, Slife H, Dunn T (1998) The *Saccharomyces cerevisiae TSC10/YBR265w* gene encoding 3-ketosphinganine reductase is identified in a screen for temperature-sensitive suppressors of the Ca^{2+}-sensitive *csg2*Δ mutant. J Biol Chem 273(46):30688–30694

Berkey R, Bendigeri D, Xiao S (2012) Sphingolipids and plant defense/disease: the "death" connection and beyond. Front Plant Sci 3:68

Bi FC, Liu Z, Wu JX, Liang H, Xi XL, Fang C, Sun TJ, Yin J, Dai GY, Rong C, Greenberg JT, Su WW, Yao N (2014) Loss of ceramide kinase in Arabidopsis impairs defenses and promotes ceramide accumulation and mitochondrial H2O2 bursts. Plant Cell 26(8):3449–3467

Blacklock BJ, Jaworski JG (2006) Substrate specificity of Arabidopsis 3-ketoacyl-CoA synthases. Biochem Biophys Res Commun 346(2):583–590

Boot RG, Verhoek M, Donker-Koopman W, Strijland A, van Marle J, Overkleeft HS, Wennekes T, Aerts JMFG (2007) Identification of the non-lysosomal glucosylceramidase as β-glucosidase 2. J Biol Chem 282(2):1305–1312

Brandwagt BF, Mesbah LA, Takken FLW, Laurent PL, Kneppers TJA, Hille J, Nijkamp HJJ (2000) A longevity assurance gene homolog of tomato mediates resistance to *Alternaria alternata* f. sp lycopersici toxins and fumonisin B_1. Proc Natl Acad Sci U S A 97(9):4961–4966

Breslow DK, Collins SR, Bodenmiller B, Aebersold R, Simons K, Shevchenko A, Ejsing CS, Weissman JS (2010) Orm family proteins mediate sphingolipid homeostasis. Nature 463(7284):1048–1053

Brodersen P, Petersen M, Pike H, Olszak B, Skov S, Odum N, Jørgensen L, Brown R, Mundy J (2002) Knockout of Arabidopsis *accelerated-cell-death11* encoding a sphingosine transfer protein causes activation of programmed cell death and defense. Genes Dev 16(4):490–502

Brown DA, Rose JK (1992) Sorting of GPI-anchored proteins to glycolipid-enriched membrane subdomains during transport to the apical cell surface. Cell 68(3):533–544

Bure C, Cacas JL, Wang F, Gaudin K, Domergue F, Mongrand S, Schmitter JM (2011) Fast screening of highly glycosylated plant sphingolipids by tandem mass spectrometry. Rapid Commun Mass Spectrom 25(20):3131–3145

Burger KN, van der Bijl P, van Meer G (1996) Topology of sphingolipid galactosyl transferases in ER and Golgi: transbilayer movement of monohexosyl sphingolipids is required for higher glycosphingolipid biosynthesis. J Cell Biol 133(1):15–28

Cacas JL, Furt F, Le Guedard M, Schmitter JM, Bure C, Gerbeau-Pissot P, Moreau P, Bessoule JJ, Simon-Plas F, Mongrand S (2012) Lipids of plant membrane rafts. Prog Lipid Res 51(3):272–299

Cacas JL, Bure C, Furt F, Maalouf JP, Badoc A, Cluzet S, Schmitter JM, Antajan E, Mongrand S (2013) Biochemical survey of the polar head of plant glycosylinositolphosphoceramides unravels broad diversity. Phytochemistry 96:191–200

Cahoon EB, Lynch DV (1991) Analysis of glucocerebrosides of rye (*Secale cereale* L. cv Puma) leaf and plasma membrane. Plant Physiol 95(1):58–68

Cantrel C, Vazquez T, Puyaubert J, Reze N, Lesch M, Kaiser WM, Dutilleul C, Guillas I, Zachowski A, Baudouin E (2011) Nitric oxide participates in cold-responsive phosphosphingolipid formation and gene expression in *Arabidopsis thaliana*. New Phytol 189(2):415–427

Carter HE, Gigg RH, Law JH, Nakayama T, Weber E (1958) Biochemistry of the sphingolipides: structure of phytoglycolipide. J Biol Chem 233(6):1309–1314

Chao DY, Gable K, Chen M, Baxter I, Dietrich CR, Cahoon EB, Guerinot ML, Lahner B, Lu S, Markham JE, Morrissey J, Han G, Gupta SD, Harmon JM, Jaworski JG, Dunn TM, Salt DE (2011) Sphingolipids in the root play an important role in regulating the leaf ionome in *Arabidopsis thaliana*. Plant Cell 23(3):1061–1081

Chen M, Thelen JJ (2013) *ACYL-LIPID DESATURASE2* is required for chilling and freezing tolerance in Arabidopsis. Plant Cell 25(4):1430–1444

Chen M, Han G, Dietrich CR, Dunn TM, Cahoon EB (2006) The essential nature of sphingolipids in plants as revealed by the functional identification and characterization of the Arabidopsis LCB1 subunit of serine palmitoyltransferase. Plant Cell 18(12):3576–3593

Chen M, Markham JE, Dietrich CR, Jaworski JG, Cahoon EB (2008) Sphingolipid long-chain base hydroxylation is important for growth and regulation of sphingolipid content and composition in Arabidopsis. Plant Cell 20(7):1862–1878

Chen M, Cahoon E, Saucedo-García M, Plasencia J, Gavilanes-Ruíz M (2010) Plant sphingolipids: structure, synthesis and function. In: Wada H, Murata N (eds) Lipids in photosynthesis, vol 30, Advances in Photosynthesis and Respiration. Springer, Dordrecht, p 77

Chen M, Markham JE, Cahoon EB (2012) Sphingolipid Δ8 unsaturation is important for glucosylceramide biosynthesis and low-temperature performance in Arabidopsis. Plant J 69(5):769–781

Chen LY, Shi DQ, Zhang WJ, Tang ZS, Liu J, Yang WC (2015) The Arabidopsis alkaline ceramidase TOD1 is a key turgor pressure regulator in plant cells. Nat Commun 6:6030

Chueasiri C, Chunthong K, Pitnjam K, Chakhonkaen S, Sangarwut N, Sangsawang K, Suksangpanomrung M, Michaelson LV, Napier JA, Muangprom A (2014) Rice ORMDL controls sphingolipid homeostasis affecting fertility resulting from abnormal pollen development. PLoS One 9(9):e106386

Costaglioli P, Joubes K, Garcia C, Stef M, Arveiler B, Lessire R, Garbay B (2005) Profiling candidate genes involved in wax biosynthesis in *Arabidopsis thaliana* by microarray analysis. BBA-Mol Cell Biol Lipids 1734(3):247–258

Coursol S, Fan LM, Le Stunff H, Spiegel S, Gilroy S, Assmann SM (2003) Sphingolipid signalling in Arabidopsis guard cells involves heterotrimeric G proteins. Nature 423(6940):651–654

Coursol S, Le Stunff H, Lynch D, Gilroy S, Assmann S, Spiegel S (2005) Arabidopsis sphingosine kinase and the effects of phytosphingosine-1-phosphate on stomatal aperture. Plant Physiol 137(2):724–737

Curatolo W (1987) The physical properties of glycolipids. Biochim Biophys Acta 906(2):111–136 de Almeida RF, Fedorov A, Prieto M (2003) Sphingomyelin/phosphatidylcholine/cholesterol phase diagram: boundaries and composition of lipid rafts. Biophys J 85(4):2406–2416

Denny PW, Shams-Eldin H, Price HP, Smith DF, Schwarz RT (2006) The protozoan inositol phosphorylceramide synthase – a novel drug target that defines a new class of sphingolipid synthase. J Biol Chem 281(38):28200–28209

Dietrich CR, Han G, Chen M, Berg RH, Dunn TM, Cahoon EB (2008) Loss-of-function mutations and inducible RNAi suppression of Arabidopsis *LCB2* genes reveal the critical role of sphingolipids in gametophytic and sporophytic cell viability. Plant J 54(2):284–298

Duan RD, Bergman T, Xu N, Wu J, Cheng Y, Duan JX, Nelander S, Palmberg C, Nilsson A (2003) Identification of human intestinal alkaline sphingomyelinase as a novel ecto-enzyme related to the nucleotide phosphodiesterase family. J Biol Chem 278(40):38528–38536

Dunn TM, Lynch DV, Michaelson LV, Napier JA (2004) A post-genomic approach to understanding sphingolipid metabolism in *Arabidopsis thaliana*. Ann Bot 93(5):483–497

Estep TN, Calhoun WI, Barenholz Y, Biltonen RL, Shipley GG, Thompson TE (1980) Evidence for metastability in stearoylsphingomyelin bilayers. Biochemistry 19(1):20–24

Fiebig A, Mayfield JA, Miley NL, Chau S, Fischer RL, Preuss D (2000) Alterations in *CER6*, a gene identical to *CUT1*, differentially affect long-chain lipid content on the surface of pollen and stems. Plant Cell 12(10):2001–2008

Gable K, Slife H, Bacikova D, Monaghan E, Dunn TM (2000) Tsc3p is an 80-amino acid protein associated with serine palmitoyltransferase and required for optimal enzyme activity. J Biol Chem 275(11):7597–7603

Gable K, Han G, Monaghan E, Bacikova D, Natarajan M, Williams R, Dunn TM (2002) Mutations in the yeast *LCB1* and *LCB2* genes, including those corresponding to the hereditary sensory neuropathy type I mutations, dominantly inactivate serine palmitoyltransferase. J Biol Chem 277(12):10194–10200

Gable K, Garton S, Napier JA, Dunn TM (2004) Functional characterization of the Arabidopsis thaliana orthologue of Tsc13p, the enoyl reductase of the yeast microsomal fatty acid elongating system. J Exp Bot 55(396):543–545

Gray JE, Holroyd GH, van der Lee FM, Bahrami AR, Sijmons PC, Woodward FI, Schuch W, Heterington AM (2000) The HIC signalling pathway links CO_2 perception to stomatal development. Nature 408(6813):713–716

Greenberg JT, Silverman FP, Liang H (2000) Uncoupling salicylic acid-dependent cell death and defense-related responses from disease resistance in the Arabidopsis mutant *acd5*. Genetics 156(1):341–350

Grilley MM, Stock SD, Dickson RC, Lester RL, Takemoto JY (1998) Syringomycin action gene *SYR2* is essential for sphingolipid 4-hydroxylation in *Saccharomyces cerevisiae*. J Biol Chem 273(18):11062–11068

Grison MS, Brocard L, Fouillen L, Nicolas W, Wewer V, Dormann P, Nacir H, Benitez-Alfonso Y, Claverol S, Germain V, Boutte Y, Mongrand S, Bayer EM (2015) Specific membrane lipid composition is important for plasmodesmata function in Arabidopsis. Plant Cell 27(4):1228–1250

Grosjean K, Mongrand S, Beney L, Simon-Plas F, Gerbeau-Pissot P (2015) Differential effect of plant lipids on membrane organization: specificities of phytosphingolipids and phytosterols. J Biol Chem 290(9):5810–5825

Guillas I, Zachowski A, Baudouin E (2011) A matter of fat: interaction between nitric oxide and sphingolipid signaling in plant cold response. Plant Signal Behav 6(1):140–142

Guillas I, Guellim A, Reze N, Baudouin E (2013) Long chain base changes triggered by a short exposure of Arabidopsis to low temperature are altered by AHb1 non-symbiotic haemoglobin overexpression. Plant Physiol Biochem 63:191–195

Guo L, Wang X (2012) Crosstalk between phospholipase D and sphingosine kinase in plant stress signaling. Front Plant Sci 3:51

Guo L, Mishra G, Markham JE, Li M, Tawfall A, Welti R, Wang X (2012) Connections between sphingosine kinase and phospholipase D in the abscisic acid signaling pathway in Arabidopsis. J Biol Chem 287(11):8286–8296

Gupta SD, Gable K, Han GS, Borovitskaya A, Selby L, Dunn TM, Harmon JM (2009) Tsc10p and FVT1: topologically distinct short-chain reductases required for long-chain base synthesis in yeast and mammals. J Lipid Res 50(8):1630–1640

Haak D, Gable K, Beeler T, Dunn T (1997) Hydroxylation of *Saccharomyces cerevisiae* ceramides requires Sur2p and Scs7p. J Biol Chem 272(47):29704–29710

Han G, Gable K, Yan L, Natarajan M, Krishnamurthy J, Gupta SD, Borovitskaya A, Harmon JM, Dunn TM (2004) The topology of the Lcb1p subunit of yeast serine palmitoyltransferase. J Biol Chem 279(51):53707–53716

Han G, Gupta SD, Gable K, Niranjanakumari S, Moitra P, Eichler F, Brown RH Jr, Harmon JM, Dunn TM (2009) Identification of small subunits of mammalian serine palmitoyltransferase that confer distinct acyl-CoA substrate specificities. Proc Natl Acad Sci U S A 106(20):8186–8191

Han S, Lone MA, Schneiter R, Chang A (2010) Orm1 and Orm2 are conserved endoplasmic reticulum membrane proteins regulating lipid homeostasis and protein quality control. Proc Natl Acad Sci U S A 107(13):5851–5856

Hanada K (2003) Serine palmitoyltransferase, a key enzyme of sphingolipid metabolism. Biochim Biophys Acta 1632(1–3):16–30

Haslam TM, Kunst L (2013) Extending the story of very-long-chain fatty acid elongation. Plant Sci 210:93–107

Huang J, Feigenson GW (1999) A microscopic interaction model of maximum solubility of cholesterol in lipid bilayers. Biophys J 76(4):2142–2157

Imai H, Nishiura H (2005) Phosphorylation of sphingoid long-chain bases in Arabidopsis: functional characterization and expression of the first sphingoid long-chain base Kinase gene in plants. Plant Cell Physiol 46(2):375–380

Imai H, Yamamoto K, Shibahara A, Miyatani S, Nakayama T (2000) Determining double-bond positions in monoenoic 2-hydroxy fatty acids of glucosylceramides by gas chromatography-mass spectrometry. Lipids 35(2):233–236

Joubes J, Raffaele S, Bourdenx B, Garcia C, Laroche-Traineau J, Moreau P, Domergue F, Lessire R (2008) The VLCFA elongase gene family in *Arabidopsis thaliana*: phylogenetic analysis, 3D modelling and expression profiling. Plant Mol Biol 67(5):547–566

Karnovsky MJ, Kleinfeld AM, Hoover RL, Dawidowicz EA, McIntyre DE, Salzman EA, Klausner RD (1982) Lipid domains in membranes. Ann N Y Acad Sci 401:61–75

Kim T-H, Böhmer M, Hu H, Nishimura N, Schroeder JI (2010) Guard cell signal transduction network: advances in understanding abscisic acid, CO_2, and Ca^{2+} signaling. Annu Rev Plant Biol 61(1):561–591

Kim J, Jung JH, Lee SB, Go YS, Kim HJ, Cahoon R, Markham JE, Cahoon EB, Suh MC (2013) Arabidopsis 3-ketoacyl-coenzyme a synthase9 is involved in the synthesis of tetracosanoic acids as precursors of cuticular waxes, suberins, sphingolipids, and phospholipids. Plant Physiol 162(2):567–580

Kimberlin AN, Majumder S, Han G, Chen M, Cahoon RE, Stone JM, Dunn TM, Cahoon EB (2013) Arabidopsis 56-amino acid serine palmitoyltransferase-interacting proteins stimulate sphingolipid synthesis, are essential, and affect mycotoxin sensitivity. Plant Cell 25(11):4627–4639

Klose C, Ejsing CS, Garcia-Saez AJ, Kaiser HJ, Sampaio JL, Surma MA, Shevchenko A, Schwille P, Simons K (2010) Yeast lipids can phase-separate into micrometer-scale membrane domains. J Biol Chem 285(39):30224–30232

Konig S, Feussner K, Schwarz M, Kaever A, Iven T, Landesfeind M, Ternes P, Karlovsky P, Lipka V, Feussner I (2012) Arabidopsis mutants of sphingolipid fatty acid α-hydroxylases accumulate ceramides and salicylates. New Phytol 196(4):1086–1097

Kruger F, Krebs M, Viotti C, Langhans M, Schumacher K, Robinson DG (2013) PDMP induces rapid changes in vacuole morphology in Arabidopsis root cells. J Exp Bot 64(2):529–540

Lachaud C, Prigent E, Thuleau P, Grat S, Da Silva D, Briere C, Mazars C, Cotelle V (2013) 14-3-3-Regulated Ca^{2+}-dependent protein kinase CPK3 is required for sphingolipid-induced cell death in Arabidopsis. Cell Death Differ 20(2):209–217

Laviad EL, Albee L, Pankova-Kholmyansky I, Epstein S, Park H, Merrill AH Jr, Futerman AH (2008) Characterization of ceramide synthase 2: tissue distribution, substrate specificity, and inhibition by sphingosine 1-phosphate. J Biol Chem 283(9):5677–5684

Lefebvre B, Furt F, Hartmann M-A, Michaelson LV, Carde J-P, Sargueil-Boiron F, Rossignol M, Napier JA, Cullimore J, Bessoule J-J, Mongrand S (2007) Characterization of lipid rafts from *Medicago truncatula* root plasma membranes: a proteomic study reveals the presence of a raft-associated redox system. Plant Physiol 144(1):402–418

Leipelt M, Warnecke D, Zahringer U, Ott C, Muller F, Hube B, Heinz E (2001) Glucosylceramide synthases, a gene family responsible for the biosynthesis of glucosphingolipids in animals, plants, and fungi. J Biol Chem 276(36):33621–33629

Liang H, Yao N, Song JT, Luo S, Lu H, Greenberg JT (2003) Ceramides modulate programmed cell death in plants. Genes Dev 17(21):2636–2641

Lin S-S, Martin R, Mongrand S, Vandenabeele S, Chen K-C, Jang I-C, Chua N-H (2008) RING1 E3 ligase localizes to plasma membrane lipid rafts to trigger FB_1-induced programmed cell death in Arabidopsis. Plant J 56(4):550–561

Lunden BM, Lofgren H, Pascher I (1977) Accommodation of hydroxyl groups and their hydrogen bond system in a hydrocarbon matrix. Chem Phys Lipids 20(4):263–271

Luttgeharm KD, Kimberlin AN, Cahoon RE, Cerny RL, Napier JA, Markham JE, Cahoon EB (2015) Sphingolipid metabolism is strikingly different between pollen and leaf in Arabidopsis as revealed by compositional and gene expression profiling. Phytochemistry 115:121–129

Lynch DV, Dunn TM (2004) An introduction to plant sphingolipids and a review of recent advances in understanding their metabolism and function. New Phytol 161(3):677–702

Lynch DV, Steponkus PL (1987) Plasma membrane lipid alterations associated with cold acclimation of winter rye seedlings (*Secale cereale* L. cv Puma). Plant Physiol 83(4):761–767

Mao C, Obeid LM (2008) Ceramidases: regulators of cellular responses mediated by ceramide, sphingosine, and sphingosine-1-phosphate. Biochim Biophys Acta 1781(9):424–434

Mao CG, Xu RJ, Bielawska A, Obeid LM (2000) Cloning of an alkaline ceramidase from *Saccharomyces cerevisiae* – an enzyme with reverse (CoA-independent) ceramide synthase activity. J Biol Chem 275(10):6876–6884

Markham JE, Jaworski JG (2007) Rapid measurement of sphingolipids from *Arabidopsis thaliana* by reversed-phase high-performance liquid chromatography coupled to electrospray ionization tandem mass spectrometry. Rapid Commun Mass Spectrom 21(7):1304–1314

Markham JE, Li J, Cahoon EB, Jaworski JG (2006) Separation and identification of major plant sphingolipid classes from leaves. J Biol Chem 281(32):22684–22694

Markham JE, Molino D, Gissot L, Bellec Y, Hematy K, Marion J, Belcram K, Palauqui JC, Satiat-Jeunemaitre B, Faure JD (2011) Sphingolipids containing very-long-chain fatty acids define a secretory pathway for specific polar plasma membrane protein targeting in Arabidopsis. Plant Cell 23(6):2362–2378

Markham JE, Lynch DV, Napier JA, Dunn TM, Cahoon EB (2013) Plant sphingolipids: function follows form. Curr Opin Plant Biol 16(3):350–357

Melser S, Batailler B, Peypelut M, Poujol C, Bellec Y, Wattelet-Boyer V, Maneta-Peyret L, Faure JD, Moreau P (2010) Glucosylceramide biosynthesis is involved in Golgi morphology and protein secretion in plant cells. Traffic 11(4):479–490

Melser S, Molino D, Batailler B, Peypelut M, Laloi M, Wattelet-Boyer V, Bellec Y, Faure JD, Moreau P (2011) Links between lipid homeostasis, organelle morphodynamics and protein trafficking in eukaryotic and plant secretory pathways. Plant Cell Rep 30(2):177–193

Michaelson LV, Zauner S, Markham JE, Haslam RP, Desikan R, Mugford S, Albrecht S, Warnecke D, Sperling P, Heinz E, Napier JA (2009) Functional characterization of a higher plant sphingolipid Δ4-desaturase: defining the role of sphingosine and sphingosine-1-phosphate in Arabidopsis. Plant Physiol 149(1):487–498

Mina JG, Okada Y, Wansadhipathi-Kannangara NK, Pratt S, Shams-Eldin H, Schwarz RT, Steel PG, Fawcett T, Denny PW (2010) Functional analyses of differentially expressed isoforms of the Arabidopsis inositol phosphorylceramide synthase. Plant Mol Biol 73(4–5):399–407

Mitchell AG, Martin CE (1997) Fah1p, a Saccharomyces cerevisiae cytochrome b_5 fusion protein, and its Arabidopsis thaliana homolog that lacks the cytochrome b_5 domain both function in the α-hydroxylation of sphingolipid-associated very long chain fatty acids. J Biol Chem 272(45):28281–28288

Mizutani Y, Kihara A, Igarashi Y (2005) Mammalian Lass6 and its related family members regulate synthesis of specific ceramides. Biochem J 390:263–271

Mizutani Y, Kihara A, Igarashi Y (2006) *LASS3* (*longevity assurance homologue 3*) is a mainly testis-specific (dihydro)ceramide synthase with relatively broad substrate specificity. Biochem J 398:531–538

Mongrand S, Morel J, Laroche J, Claverol S, Carde JP, Hartmann MA, Bonneu M, Simon-Plas F, Lessire R, Bessoule JJ (2004) Lipid rafts in higher plant cells: purification and characterization of Triton X-100-insoluble microdomains from tobacco plasma membrane. J Biol Chem 279(35):36277–36286

Morel J, Claverol S, Mongrand S, Furt F, Fromentin J, Bessoule JJ, Blein JP, Simon-Plas F (2006) Proteomics of plant detergent-resistant membranes. Mol Cell Proteomics 5(8):1396–1411

Mortimer JC, Yu XL, Albrecht S, Sicilia F, Huichalaf M, Ampuero D, Michaelson LV, Murphy AM, Matsunaga T, Kurz S, Stephens E, Baldwin TC, Ishii T, Napier JA, Weber APM, Handford MG, Dupree P (2013) Abnormal glycosphingolipid mannosylation triggers salicylic acid-mediated responses in arabidopsis. Plant Cell 25(5):1881–1894

Muir A, Ramachandran S, Roelants FM, Timmons G, Thorner J (2014) TORC2-dependent protein kinase Ypk1 phosphorylates ceramide synthase to stimulate synthesis of complex sphingolipids. Elife 3:e03779

Nagano M, Ihara-Ohori Y, Imai H, Inada N, Fujimoto M, Tsutsumi N, Uchimiya H, Kawai-Yamada M (2009) Functional association of cell death suppressor, Arabidopsis Bax inhibitor-1, with fatty acid 2-hydroxylation through cytochrome b_5. Plant J 58(1):122–134

Nagano M, Takahara K, Fujimoto M, Tsutsumi N, Uchimiya H, Kawai-Yamada M (2012) Arabidopsis sphingolipid fatty acid 2-hydroxylases (AtFAH1 and AtFAH2) are functionally differentiated in fatty acid 2-hydroxylation and stress responses. Plant Physiol 159(3):1138–1148

Nagano M, Ishikawa T, Ogawa Y, Iwabuchi M, Nakasone A, Shimamoto K, Uchimiya H, Kawai-Yamada M (2014) Arabidopsis Bax inhibitor-1 promotes sphingolipid synthesis during cold stress by interacting with ceramide-modifying enzymes. Planta 240(1):77–89

Nakagawa N, Kato M, Takahashi Y, Shimazaki K, Tamura K, Tokuji Y, Kihara A, Imai H (2012) Degradation of long-chain base 1-phosphate (LCBP) in Arabidopsis: functional characterization of LCBP phosphatase involved in the dehydration stress response. J Plant Res 125(3):439–449

Ng CKY, Carr K, McAinsh MR, Powell B, Hetherington AM (2001) Drought-induced guard cell signal transduction involves sphingosine-1-phosphate. Nature 410(6828):596–599

Nishikawa M, Hosokawa K, Ishiguro M, Minamioka H, Tamura K, Hara-Nishimura I, Takahashi Y, Shimazaki K, Imai H (2008) Degradation of sphingoid long-chain base 1-phosphates (LCB-1Ps): functional characterization and expression of AtDPL1 encoding LCB-1P lyase involved in the dehydration stress response in Arabidopsis. Plant Cell Physiol 49(11):1758–1763

Pascher I (1976) Molecular arrangements in sphingolipids. Conformation and hydrogen bonding of ceramide and their implication on membrane stability and permeability. Biochim Biophys Acta 455(2):433–451

Pata MO, Wu BX, Bielawski J, Xiong TC, Hannun YA, Ng CKY (2008) Molecular cloning and characterization of OsCDase, a ceramidase enzyme from rice. Plant J 55(6):1000–1009

Paul S, Gable K, Beaudoin F, Cahoon E, Jaworski J, Napier JA, Dunn TM (2006) Members of the Arabidopsis FAE1-like 3-ketoacyl-CoA synthase gene family substitute for the Elop proteins of *Saccharomyces cerevisiae*. J Biol Chem 281(14):9018–9029

Peer M, Stegmann M, Mueller MJ, Waller F (2010) *Pseudomonas syringae* infection triggers *de novo* synthesis of phytosphingosine from sphinganine in *Arabidopsis thaliana*. FEBS Lett 584(18):4053–4056

Rennie EA, Ebert B, Miles GP, Cahoon RE, Christiansen KM, Stonebloom S, Khatab H, Twell D, Petzold CJ, Adams PD, Dupree P, Heazlewood JL, Cahoon EB, Scheller HV (2014) Identification of a sphingolipid α-glucuronosyltransferase that is essential for pollen function in Arabidopsis. Plant Cell 26(8):3314–3325

Riebeling C, Allegood JC, Wang E, Merrill AH, Futerman AH (2003) Two mammalian *longevity assurance gene* (*LAG1*) family members, *trh1* and *trh4*, regulate dihydroceramide synthesis using different fatty acyl-CoA donors. J Biol Chem 278(44):43452–43459

Rittenour WR, Chen M, Cahoon EB, Harris SD (2011) Control of glucosylceramide production and morphogenesis by the Bar1 ceramide synthase in *Fusarium graminearum*. PLoS One 6:e19385

Roche Y, Gerbeau-Pissot P, Buhot B, Thomas D, Bonneau L, Gresti J, Mongrand S, Perrier-Cornet JM, Simon-Plas F (2008) Depletion of phytosterols from the plant plasma membrane provides evidence for disruption of lipid rafts. FASEB J 22(11):3980–3991

Roelants FM, Breslow DK, Muir A, Weissman JS, Thorner J (2011) Protein kinase Ypk1 phosphorylates regulatory proteins Orm1 and Orm2 to control sphingolipid homeostasis in *Saccharomyces cerevisiae*. Proc Natl Acad Sci U S A 108(48):19222–19227

Ryan PR, Liu Q, Sperling P, Dong B, Franke S, Delhaize E (2007) A higher plant Δ8 sphingolipid desaturase with a preference for (Z)-isomer formation confers aluminum tolerance to yeast and plants. Plant Physiol 144(4):1968–1977

Sandhoff K (2013) Metabolic and cellular bases of sphingolipidoses. Biochem Soc Trans 41(6):1562–1568

Saucedo-García M, Guevara-García A, González-Solís A, Cruz-García F, Vázquez-Santana S, Markham J, Lozano-Rosas M, Dietrich C, Ramos-Vega M, Cahoon E, Gavilanes-Ruíz M (2011) MPK6, sphinganine and the *LCB2a* gene from serine palmitoyltransferase are required in the signaling pathway that mediates cell death induced by long chain bases in Arabidopsis. New Phytol 191(4):943–957

Shanklin J, Cahoon EB (1998) Desaturation and related modifications of fatty acids. Annu Rev Plant Physiol Plant Mol Biol 49:611–641

Shi L, Bielawski J, Mu J, Dong H, Teng C, Zhang J, Yang X, Tomishige N, Hanada K, Hannun YA, Zuo J (2007) Involvement of sphingoid bases in mediating reactive oxygen intermediate production and programmed cell death in Arabidopsis. Cell Res 17(12):1030–1040

Simanshu DK, Zhai X, Munch D, Hofius D, Markham JE, Bielawski J, Bielawska A, Malinina L, Molotkovsky JG, Mundy JW, Patel DJ, Brown RE (2014) Arabidopsis *accelerated cell death 11*, *ACD11*, is a ceramide-1-phosphate transfer protein and intermediary regulator of phytoceramide levels. Cell Rep 6(2):388–399

Simons K, van Meer G (1988) Lipid sorting in epithelial cells. Biochemistry 27(17):6197–6202

Simons K, Ikonen E (1997) Functional rafts in cell membranes. Nature 387(6633):569–572

Smith MA, Dauk M, Ramadan H, Yang H, Seamons LE, Haslam RP, Beaudoin F, Ramirez-Erosa I, Forseille L (2013) Involvement of Arabidopsis *ACYL-COENZYME A DESATURASE-LIKE2* (At2g31360) in the biosynthesis of the very-long-chain monounsaturated fatty acid components of membrane lipids. Plant Physiol 161(1):81–96

Spassieva SD, Markham JE, Hille J (2002) The plant disease resistance gene *Asc-1* prevents disruption of sphingolipid metabolism during AAL-toxin-induced programmed cell death. Plant J 32(4):561–572

Sperling P, Schmidt H, Heinz E (1995) A cytochrome-b containing fusion protein similar to plant acyl lipid desaturases. Eur J Biochem 232(3):798–805

Sperling P, Zahringer U, Heinz E (1998) A sphingolipid desaturase from higher plants. Identification of a new cytochrome b_5 fusion protein. J Biol Chem 273(44):28590–28596

Sperling P, Ternes P, Moll H, Franke S, Zahringer U, Heinz E (2001) Functional characterization of sphingolipid C4-hydroxylase genes from *Arabidopsis thaliana*. FEBS Lett 494(1–2):90–94

Sperling P, Franke S, Luthje S, Heinz E (2005) Are glucocerebrosides the predominant sphingolipids in plant plasma membranes? Plant Physiol Biochem 43(12):1031–1038

Stone JM, Heard JE, Asai T, Ausubel FM (2000) Simulation of fungal-mediated cell death by fumonisin B_1 and selection of *fumonisin B_1-resistant* (*fbr*) Arabidopsis mutants. Plant Cell 12(10):1811–1822

Strub GM, Maceyka M, Hait NC, Milstien S, Spiegel S (2010) Extracellular and intracellular actions of sphingosine-1-phosphate. Adv Exp Med Biol 688:141–155

Takahashi Y, Berberich T, Kanzaki H, Matsumura H, Saitoh H, Kusano T, Terauchi R (2009) Serine palmitoyltransferase, the first step enzyme in sphingolipid biosynthesis, is involved in nonhost resistance. Mol Plant Microbe Interact 22(1):31–38

Tamura K, Mitsuhashi N, Hara-Nishimura I, Imai H (2001) Characterization of an Arabidopsis cDNA encoding a subunit of serine palmitoyltransferase, the initial enzyme in sphingolipid biosynthesis. Plant Cell Physiol 42(11):1274–1281

Teng C, Dong H, Shi L, Deng Y, Mu J, Zhang J, Yang X, Zuo J (2008) Serine palmitoyltransferase, a key enzyme for de novo synthesis of sphingolipids, is essential for male gametophyte development in Arabidopsis. Plant Physiol 146(3):1322–1332

Ternes P, Franke S, Zahringer U, Sperling P, Heinz E (2002) Identification and characterization of a sphingolipid Δ4-desaturase family. J Biol Chem 277(28):25512–25518

Ternes P, Feussner K, Werner S, Lerche J, Iven T, Heilmann I, Riezman H, Feussner I (2011) Disruption of the ceramide synthase LOH1 causes spontaneous cell death in *Arabidopsis thaliana*. New Phytol 192(4):841–854

Thudichum JLW (1884) A treatise on the chemical constitution of the brain. Baillière, Tindall and Cox, London

Tjellstrom H, Hellgren LI, Wieslander A, Sandelius AS (2010) Lipid asymmetry in plant plasma membranes: phosphate deficiency-induced phospholipid replacement is restricted to the cytosolic leaflet. FASEB J 24(4):1128–1138

Todd J, Post-Beittenmiller D, Jaworski Jan G (1999) *KCS1* encodes a fatty acid elongase 3-ketoacyl-CoA synthase affecting wax biosynthesis in *Arabidopsis thaliana*. Plant J 17(2):119–130

Townley HE, McDonald K, Jenkins GI, Knight MR, Leaver CJ (2005) Ceramides induce programmed cell death in Arabidopsis cells in a calcium-dependent manner. Biol Chem 386(2):161–166

Trenkamp S, Martin W, Tietjen K (2004) Specific and differential inhibition of very-long-chain fatty acid elongases from Arabidopsis thaliana by different herbicides. Proc Natl Acad Sci U S A 101(32):11903–11908

Tsegaye Y, Richardson CG, Bravo JE, Mulcahy BJ, Lynch DV, Markham JE, Jaworski JG, Chen M, Cahoon EB, Dunn TM (2007) Arabidopsis mutants lacking long chain base phosphate lyase are fumonisin-sensitive and accumulate trihydroxy-18:1 long chain base phosphate. J Biol Chem 282(38):28195–28206

Uemura M, Steponkus PL (1994) A contrast of the plasma membrane lipid composition of oat and rye leaves in relation to freezing tolerance. Plant Physiol 104(2):479–496

Uemura M, Joseph RA, Steponkus PL (1995) Cold acclimation of *Arabidopsis thaliana*. effect on plasma membrane lipid composition and freeze-induced lesions. Plant Physiol 109(1):15–30

van Genderen IL, van Meer G, Slot JW, Geuze HJ, Voorhout WF (1991) Subcellular localization of Forssman glycolipid in epithelial MDCK cells by immuno-electron microscopy after freeze-substitution. J Cell Biol 115(4):1009–1019

van Meer G, Simons K (1988) Lipid polarity and sorting in epithelial cells. J Cell Biochem 36(1):51–58 van Meer G, Voelker DR, Feigenson GW (2008) Membrane lipids: where they are and how they behave. Nat Rev Mol Cell Biol 9(2):112–124

Venkataraman K, Riebeling C, Bodennec J, Riezman H, Allegood JC, Sullards MC, Merrill AH Jr, Futerman AH (2002) Upstream of growth and differentiation factor 1 (uog1), a mammalian homolog of the yeast *longevity assurance gene 1* (*LAG1*), regulates N-stearoyl-sphinganine (C18-(dihydro)ceramide) synthesis in a fumonisin B_1-independent manner in mammalian cells. J Biol Chem 277(38):35642–35649

Verhoek B, Haas R, Wrage K, Linscheid M, Heinz E (1983) Lipids and enzymatic-activities in vacuolar membranes isolated via protoplasts from oat primary leaves. Z Naturforsch 38(9–10):770–777

Wang W, Yang X, Tangchaiburana S, Ndeh R, Markham JE, Tsegaye Y, Dunn TM, Wang GL, Bellizzi M, Parsons JF, Morrissey D, Bravo JE, Lynch DV, Xiao S (2008) An inositolphosphorylceramide synthase is involved in regulation of plant programmed cell death associated with defense in Arabidopsis. Plant Cell 20(11):3163–3179

Worrall D, Liang YK, Alvarez S, Holroyd GH, Spiegel S, Panagopulos M, Gray JE, Hetherington AM (2008) Involvement of sphingosine kinase in plant cell signalling. Plant J 56(1):64–72

Wu J-X, Li J, Liu Z, Yin J, Chang Z-Y, Rong C, Wu J-L, Bi F-C, Yao N (2015) The Arabidopsis ceramidase AtACER functions in disease resistance and salt tolerance. Plant J 81(5):767–780

Yang Z, Jean-Baptiste G, Khoury C, Greenwood MT (2005) The mouse *sphingomyelin synthase 1* (*SMS1*) gene is alternatively spliced to yield multiple transcripts and proteins. Gene 363:123–132

Yang H, Richter GL, Wang X, Mlodzinska E, Carraro N, Ma G, Jenness M, Chao DY, Peer WA, Murphy AS (2012) Sterols and sphingolipids differentially function in trafficking of the Arabidopsis ABCB19 auxin transporter. Plant J 74(1):37–47

Yephremov A, Wisman E, Huijser P, Huijser C, Wellesen K, Saedler H (1999) Characterization of the *FIDDLEHEAD* gene of Arabidopsis reveals a link between adhesion response and cell differentiation in the epidermis. Plant Cell 11(11):2187–2201

Yoshida S, Kawata T, Uemura M, Niki T (1986) Isolation and characterization of tonoplast from chilling-sensitive etiolated seedlings of *Vigna radiata* L. Plant Physiol 80(1):161–166

Zheng H, Rowland O, Kunst L (2005) Disruptions of the Arabidopsis enoyl-CoA reductase gene reveal an essential role for very-long-chain fatty acid synthesis in cell expansion during plant morphogenesis. Plant Cell 17(5):1467–1481

Chapter 12
Plant Surface Lipids and Epidermis Development

Camille Delude, Steven Moussu, Jérôme Joubès, Gwyneth Ingram, and Frédéric Domergue

Abstract The epidermis has a strategic position at the interface between the plant and the environment. In order to control exchanges with the environment as well as to protect the plant from external threats, the epidermis synthesises and secretes surface lipids to form a continuous, transparent and hydrophobic layer known as the cuticle. Cuticle formation is a strictly epidermal property in plants and all aerial epidermal cells produce some sort of cuticle on their surface. Conversely, all cuticularized plant surfaces are of epidermal origin. This seemingly anodyne observation has surprisingly profound implications in terms of understanding the function of the plant cuticle, since it underlies in part, the difficultly of functionally separating epidermal cell fate specification from cuticle biogenesis.

Keywords Cuticle • Epidermis • Regulation • Waxes • Development • Arabidopsis

Introduction

The lipid metabolism of the epidermis is highly specialized because this cell monolayer is located at the interface between the plant and its surrounding environment. Due to this highly strategic positioning, one important function of the epidermis is to form a hydrophobic barrier over aerial surfaces to permit the regulation of water and gas exchange as well as protecting the plant from both biotic and abiotic stresses. In order to achieve this role, the epidermis synthesises and secretes surface lipids to form a continuous transparent and hydrophobic layer on the outside of leaves, stems and flowers. In contrast to the lipid metabolism of most cortical cells,

C. Delude • J. Joubès • F. Domergue (✉)
Laboratoire de Biogenèse Membranaire, UMR 5200, CNRS, Université de Bordeaux, 71 av. E. Bourleaux, CS 20032, 33140 Villenave d'Ornon, France
e-mail: frederic.domergue@u-bordeaux.fr

S. Moussu • G. Ingram
Laboratoire Reproduction et Développement des Plantes, UMR 5667, CNRS/INRA/UCBL/ENS Lyon, 46, allée d'Italie, 69364 Lyon, France

Y. Nakamura, Y. Li-Beisson (eds.), *Lipids in Plant and Algae Development*, Subcellular Biochemistry 86, DOI 10.1007/978-3-319-25979-6_12

which is mainly channelled towards the generation of thylakoid membranes for photosynthesis, that of epidermal cells is therefore principally devoted to the biosynthesis of surface lipids.

All aerial parts of the plants as well as parts of the root system are delimitated by a layer of epidermal cells. Above ground epidermal cells include a "basal" cell type, known as pavement cells, which are abundant and show a highly organ dependent morphology, as well as trichomes, also known as leaf hairs, and the guard cells of stomata. Underground, young primary roots possess an epidermis layer, but this tissue, together with the endodermis and the cortex, disappears during secondary development, when a periderm originating from the pericycle develops. Aerial and mature root surfaces are covered by lipid layers known as the cuticle and suberin, respectively. Over the last 20 years, the biosynthetic pathways producing the lipids of these apoplastic diffusion barriers have been well described, especially in the plant model *Arabidopsis thaliana*, where many mutants have been isolated and characterised (Nawrath et al. 2013). None of the suberin mutants identified to date have been associated with root epidermal or peridermal defects, but this may be due to technical difficulties in studying the precise morphology of these underground tissues. In contrast, the fact that several mutants affected in their cuticle composition and/or content also display defects in epidermal development, such as organ fusions, abnormal trichome morphology or decreased stomatal index, suggests a link between surface lipids and the differentiation of epidermal cells (Bird and Gray 2003). Furthermore, two studies have highlighted a potential role of the epidermis in restricting plant growth (Savaldi-Goldstein et al. 2007; Nobusawa et al. 2013a). Interestingly, cutinase treatments in rapidly growing internodes of deepwater rice suggest that the cuticle acts as a growth-limiting structure, and could thus be a key epidermal element in keeping the underlying parenchyma cells under compression (Hoffmann-Benning and Kende 1994).

In this chapter, we will present our current knowledge of the biosynthesis of surface lipids and describe in more detail how certain mutants with defects in their surface lipid content and/or composition are also affected in their epidermal development. We will then discuss the complex transcriptional regulation of plant surface lipid biosynthesis and epidermis development. Finally, we will take advantage of recent advances in our understanding of the establishment of the protocuticule in Arabidopsis embryos to present the inter-cellular signaling pathway monitoring the formation of a functional embryonic cuticle and enabling the physical separation of the embryo from the endosperm.

Biosynthesis of the Cuticular Barrier by Epidermal Cells

The plant cuticle is a continuous lipophilic layer, of which the main roles are to limit non-stomatal water loss, control gaseous exchanges, protect the plant from both biotic and abiotic environmental stresses, provide mechanical strength, and prevent organ fusion during development. The cuticle is composed of a fatty acid derived polymer, known as cutin, which is impregnated with, and covered by, a mixture of very long-chain (VLC) aliphatic compounds, called cuticular waxes (Yeats and Rose 2013).

Cuticular waxes consist of a mixture of very-long chain aliphatic compounds, containing from 22 to 48 carbon atoms, which are produced in the endoplasmic reticulum (ER) (Bernard and Joubès 2013). Their biosynthesis relies on the elongase complex, which produces VLC-acyl-CoAs (and free VLC fatty acids; VLCFAs), the alcohol-forming pathway, which converts these into even-numbered primary alcohols and alkyl esters, and the alkane-forming pathway, which yields aldehydes and odd-numbered alkanes, secondary alcohols and ketones (Fig. 12.1). The cutin polymer is mainly made of interesterified hydroxy fatty acids, but also contains some glycerol and dicarboxilic acids, as well as phenolic compounds. Cutin biosynthesis starts in the ER where hydroxyl fatty acids are produced and esterified to mono-acyl glycerol (MAG), and ends in the apoplasm where polymerization takes place (Fig. 12.1). A key step in surface lipid biosynthesis is therefore the transport of the cutin monomers and cuticular waxes across both the plasma membrane and the hydrophilic cell wall (Fig. 12.1).

The Fatty Acid Elongase Complex Produces Very Long Chain Fatty Acids

Free C_{16} and C_{18} fatty acids, issuing from *de novo* fatty acid synthesis in the plastids, are used as precursors for fatty acid elongation (Haslam and Kunst 2013). Recent reports revealed the important role of at least three Long-Chain-Acyl-CoA Synthetase (LACS) iso-enzymes, which activate free fatty acids into CoA thioesters, in Arabidopsis wax synthesis (Lü et al. 2009; Weng et al. 2010; Jessen et al. 2011). Each cycle of the ER-bound multi-enzymatic fatty acid elongase (FAE) consists of four successive reactions that generate an acyl-chain extended by two carbons. The reactions involve (1) formation of β-ketoacyl-CoA by condensation of malonyl-CoA with an a C_n-acyl-CoA catalysed by a β-ketoacyl-CoA synthase (KCS), (2) reduction to β-hydroxyacyl-CoA by a β-ketoacyl-CoA reductase (KCR), (3) dehydration to enoyl-CoA by a β-hydroxyacyl-CoA dehydratase (HCD) and (4) a final reduction by a enoyl-CoA reductase (ECR) yielding a C_{n+2}-acyl-CoA. Biochemical studies and the co-existence of 21 putative KCS-encoding genes annotated in the Arabidopsis genome (Joubès et al. 2008), led to the idea that multiple elongase complexes with distinct chain-length specificities perform sequential and parallel reactions to produce the wide range of chain-lengths found in plant VLCFAs (von Wettstein-Knowles 1982).

Over the last decade, major advances in understanding VLCFA synthesis in yeast have permitted the identification of the ECR, KCR and HCD enzymes of Arabidopsis. Sequence similarity searches with ECR from yeast revealed one single locus in Arabidopsis. Complementation of the Arabidopsis *cer10* (*cer* standing for *eceriferum*, i.e. wax-less) mutant by expression of *AtECR* demonstrated that *CER10*

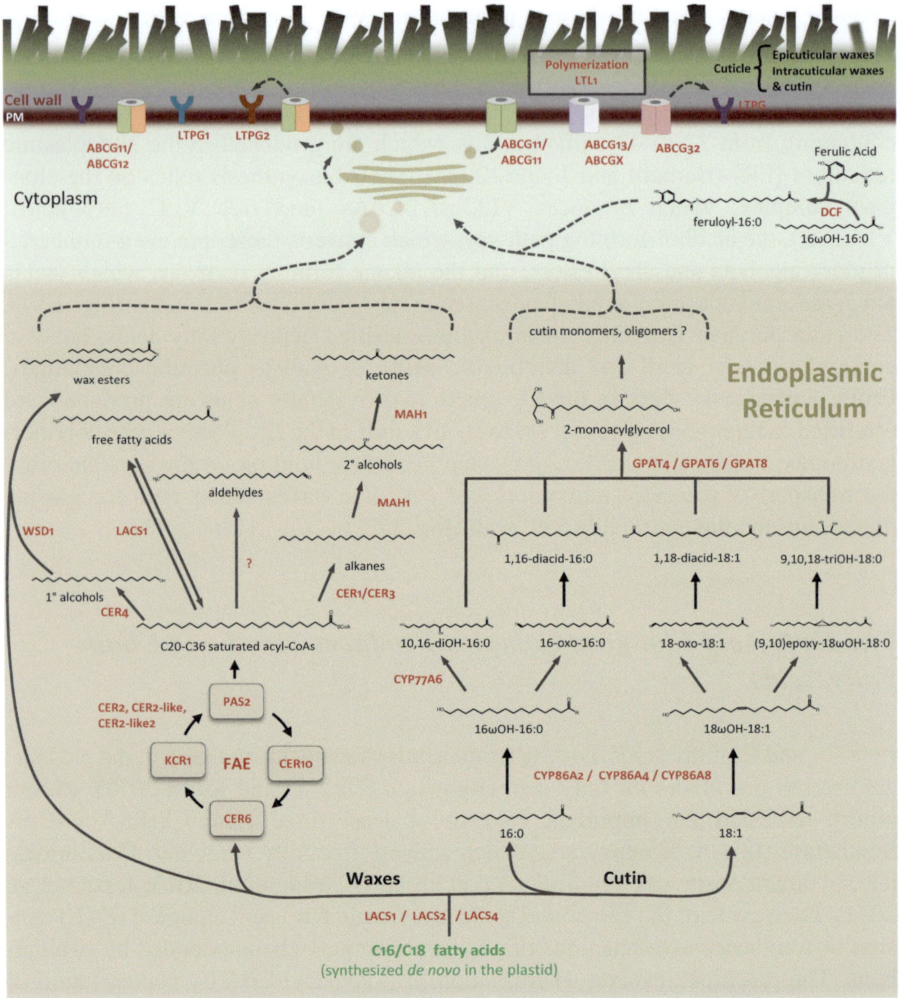

Fig. 12.1 Cuticle biosynthetic pathways. The genes in *red* are those discovered in *Arabidopsis thaliana*. Wax biosynthetic pathways: Very long chain fatty acyl-CoAs are synthesized by the Fatty Acid Elongase (FAE) multienzymatic complex from C16-C18 fatty acids. VLCFAs are then modified via two distinct biosynthetic pathways to generate the aliphatic compounds of waxes; the alcohol forming pathway resulting in the formation of primary alcohols (1° alcohols) and wax esters, and the alkane forming pathway which leads to the production of aldehydes, alkanes, secondary alcohols (2° alcohols) and ketones. Cutin biosynthetic pathways: Addition of a hydroxyl group at the terminal or mid-chain position of C16 and C18 is catalyzed by members of the cytochrome P450 family (CYPs). Formation of dicarboxylic acids (DCA) from ω-hydroxyacids (ω-OH) may involve oxydoreductases. Intermediates are annotated "COOR", where R could be H, CoA or glycerol, since the extact substrate of P450 and order of reaction remains unclear. Esterification of ω-OH and DCA to glycerol-3-phosphate by glycerol-3-phosphate acyl-CoA transferase (GPATs) produces *sn*-2-monoacylglycerol. Incorporation of phenolic components into the cutin polymer requires BAHD-type acyltransferases. Extracellular polymerisation are performed by cutin synthases. Export of precursors to the cuticular matrix: Cuticle precursors produced in the endoplasmic reticulum (ER) reach the plasma membrane (PM) via the Golgi apparatus through the secretory pathway. Export of cuticle compounds from the plasma membrane is carried out by ATP-Binding Cassette (ABCs) transporters and glycosylphosphatidylinositol-anchored Lipid Transfer Proteins (LTPGs)

encodes a functional ECR activity (Zheng et al. 2005). Likewise, two Arabidopsis loci were identified as encoding potential HCDs and two loci as encoding potential KCRs (Bach et al. 2008; Beaudoin et al. 2009). Surprisingly, complementation assays in corresponding yeast mutants revealed that only *PASTICCINO2* encodes a functional HCD and only *KCR1* has KCR activity. Although several *KCS* mutants show defects in their cuticular waxes, *CER6* is the only one considered to be strictly wax-specific (Millar et al. 1999; Hooker et al. 2002), indicating a preponderant role in wax synthesis. The recent functional characterization of CER2 and related proteins suggests that these proteins are necessary for elongation up to C_{34}, even though the precise function of these proteins remains to be elucidated (Haslam et al. 2012, 2015; Pascal et al. 2013).

The VLC-acyl-CoAs produced by the FAEs are either used as precursors in the alcohol- and alkane-forming pathways (below), or simply converted by thioesterase(s) to free VLCFAs, a significant quantity of which are detected in surface lipids.

The Alcohol Forming Pathway Produces Primary Alcohols and Alkyl Esters

The first biochemical studies of primary alcohol formation suggested a two-step reaction in which fatty acyl reductase (FAR) reduces VLCFAs to aldehydes, which are further reduced to primary alcohol by an aldehyde reductase (Kolattukudy 1971). However, biochemical studies on jojoba seeds and pea leaves, as well as expression of genes encoding FAR alcohol-forming activities in heterologous systems, revealed that a single enzyme produced fatty alcohols, with the potential intermediate aldehyde remaining bound to the enzyme (Rowland and Domergue 2012). In Arabidopsis, the *cer4* mutant shows a severe reduction of primary alcohols and wax esters, suggesting that *CER4* could play a role in this biosynthetic pathway (Jenks et al. 1995). Expression of *CER4* in yeast resulted in the production of VLC-primary alcohols with C_{24} and C_{26} chain lengths, confirming the FAR activity of CER4 (Rowland et al. 2006). However, the activity of CER4 on C_{28} and C_{30} fatty acids has never been demonstrated, indicating that another FAR or functionally related enzyme, is likely responsible for C_{30} primary alcohol formation (Rowland and Domergue 2012). Detailed analysis of wax ester chain-lengths from the stems of Arabidopsis *cer4* mutants indicated that primary alcohols formed by CER4 act as substrates for subsequent alkyl ester formation (Lai et al. 2007). Wax Synthase (WS) enzymes catalyze the esterification of primary alcohols with acyl-CoAs in higher plants, mammals and bacteria (Lardizabal et al. 2000; Cheng and Russell 2004; Stoveken et al. 2005). In Arabidopsis, sequence similarity with jojoba seed WS and bi-functional WS/Diacylglycerol acyltransferases (DGATs) from *Acinetobacter calcoaceticus*, revealed 12 and 11 similar sequences respectively. Analysis of a putative WS/DGAT encoding gene (*WSD1*), highly expressed in the epidermis, subsequently confirmed its involvement as the major WS in cuticular wax synthesis (Li et al. 2008).

The Alkane Forming Pathway Produces Aldehydes and Odd-Numbered Alkanes, Secondary Alcohols and Ketones

Analyses of *cer* mutants and biochemical experiments established a putative alkane forming pathway in which VLCFAs are used as precursors to form alkanes via a potential aldehyde intermediate (Bernard and Joubès 2013). Several *cer* mutants with a decreased alkane load have been biochemically characterized (Jenks et al. 1995). The *cer3* mutant showed a dramatic reduction in aldehydes, alkanes, secondary alcohols and ketones, while the *cer1* mutant exhibits a drastic decrease in alkanes and a near abolition of secondary alcohol and ketone production, accompanied by a slight increase in aldehyde content (Aarts et al. 1995; Chen et al. 2003; Kurata et al. 2003; Bourdenx et al. 2011). It has been proposed from these phenotypes that *CER3* could encode a potential VLCFA reductase producing aldehydes, whereas *CER1* could encode the alkane-forming enzyme, catalyzing the presumed decarbonylation of aldehydes to alkanes. Consistent with a role in wax associated processes, the expression of *CER1* and *CER3* was found to be restricted to the aerial epidermis, and to be up-regulated in drought conditions when wax synthesis is particularly active (Kurata et al. 2003; Bourdenx et al. 2011). Additionally, wax analyses of *CER1* overexpressors revealed a specific increase in alkanes with chain-lengths between 27 and 33 carbon atoms, consistent with *CER1* encoding an alkane-forming activity with strict substrate specificity for compounds containing more than 27 carbon atoms (Bourdenx et al. 2011). Recently, the proof that CER1 and CER3 act as an enzymatic complex catalyzing the synthesis of VLC-alkanes was provided by co-expression of the two proteins in yeast (Bernard et al. 2012).

Whereas alkanes are the end products of the alkane-forming pathway in Arabidopsis leaves, they can also be further modified by consecutive oxidation to produce secondary alcohols, and subsequently ketones, in stems. By looking for genes up-regulated in the epidermis and encoding proteins potentially involved in lipid oxidation, a cytochrome P_{450} encoding gene, *CYP96A15*, was pinpointed as a candidate for a catalytic role in secondary alcohol and ketone formation (Greer et al. 2007). Its ectopic expression resulted in the production of secondary alcohols and ketones in leaves, suggesting that CYP96A15 functions as a mid-chain alkane hydroxylase (MAH1).

Acyl-Oxydation, Esterification to Glycerol and Extracellular Polymerization Generate the Insoluble Cutin Polymer

Although the major steps involved in the biosynthesis of cutin have been elucidated over the last 10 years, its exact three-dimensional structure remains unknown. Cutin isolation followed by depolymerization reactions have shown that it is a polyester mostly composed of C16 and C18 ω-hydroxyacid, polyhydroxyacid and α,ω-dicarboxylic acid (DCA), as well as of glycerol and small amounts of phenylpropanoids (Beisson et al. 2012). In Arabidopsis, the cutin of leaves and stems in enriched in DCA, while that of flowers is dominated by polyhydroxyacids, suggesting that the composition of cutin varies between different organs of the same plant species (Bonaventure et al. 2004; Franke et al. 2005). The oxidative reactions converting fatty acids to ω-hydroxyacids are catalyzed by members of the CYP86 cytochrome P450 family. The characterization of the *lcr* and *att1* mutants showed that LCR (CYP86A8) and ATT1 (CYP86A2) are the two major cytochrome P450s involved in the acyl-oxydation of cutin precursors in Arabidopsis leaves and stems (Wellesen et al. 2001; Xiao et al. 2004). The mechanism underlying further conversion to DCA is still unclear but could involve oxydoreductases such as HOTHEAD (HTH), even though HTH is not specifically expressed in the epidermis (Krolikowski et al. 2003; Kurdyukov et al. 2006a). In Arabidopsis flowers, CYP86A4 and CYPA77 respectively are the ω-hydroxylase and in-chain hydroxylase responsible for the formation of 10,16-dihydroxypalmitate, the main floral cutin precursor (Li-Beisson et al. 2009). Although the sequential order of reactions is still a matter of debate, substrate specificities tested in vitro suggest that fatty acyl oxidation is most probably followed by acyl activation before transfer to glycerol.

Two long chain acyl-CoA acyltransferases, LACS1 and LACS2 have been shown to be important for cutin synthesis in Arabidopsis (Lü et al. 2009; Schnurr et al. 2004). Once activated as acyl-CoAs, hydroxylated fatty acyls are most probably esterified to the *sn-2* position of glycerol by glycerol-3-phosphate:acyl-CoA acyltransferases such as GPAT4 and GPAT8 in leaves and stems (Li et al. 2007) and GPAT6 in flowers (Li-Beisson et al. 2009). These enzymes also possess phosphatase activity so that end products are 2-monoacylglycerol instead of 2-acyl-lysophos-phatidic acids. This additional activity could be instrumental in the separation of glycerolipid precursors for cutin and membrane lipid biosynthesis (Yang et al. 2012; Rautengarten et al. 2012). Incorporation of phenylpropanoids into the cutin polyester requires BAHD acyltransferases such as Deficient in Cutin Ferulate (DCF) and possibly Defective in Cuticular Ridges (DCR) which transfer ferulic acid to ω-hydroxyacids (Rautengarten et al. 2012; Panikashvili et al. 2009).

Using fluorescent fusion proteins, all the reactions described above, with the exception of those catalyzed by BAHD-type proteins which could be cytosolic, were shown to take place in the endoplasmic reticulum. Thus precursors must be transported across the plasma membrane and through the cell wall (see below) before polymerization occurs. Although this process is the least well understood part of cutin biosynthesis, the recent identification of a extracellular tomato cutin synthase (GDSL1/CD1), has provided a major step forward (Yeats et al. 2012; Girard et al. 2012). The *cd1* mutant shows a dramatic reduction in polymerized cutin, and in vitro assays showed that CD1 catalyzes the formation of polyester oligomers from 2-monoacylglycerol (Girard et al. 2012). Loss of function of Li-tolerant lipase 1 (LTL1), which encodes an Arabidopsis ortholog of CD1, causes typical characteristics of cutin mutants, while LTL1 can also catalyze cutin polymerization in vitro (Yeats et al. 2014).

ABC Transporters and LTPs Are Involved in the Export of Precursors to the Cuticular Matrix

Although most cuticle biosynthetic enzymes have been shown to be localized in the ER, little is known about the intracellular trafficking of the cuticular precursors before their export out of epidermal cells. Nevertheless, the recent use of mutants defective in vesicle trafficking and protein secretion suggests that the transfer of these hydrophobic molecules through the hydrophilic cytoplasm involves vesicles, and that they transit through the ER-Golgi interface and the trans-Golgi network to deliver cargo to the plasma membrane (PM) (McFarlane et al. 2014). Once the cuticle components have reached the PM, their export is carried out by ABC transporters. The gene encoding the first ABC transporter identified as required for wax transport was found to be allelic to *CER5*, and was named *ABCG12* (Pighin et al. 2004). A search for ABC protein-encoding genes with an expression pattern similar to *ABCG12* revealed *ABCG11* as a potential candidate for a role in wax export (Bird et al. 2007; Panikashvili et al. 2007). The *abcg11* mutant and the double mutant *abcg11 cer5* show similar wax composition, suggesting that ABCG11 and ABCG12 act in the same pathway or ABC transporter unit. Furthermore, *abcg11* showed organ fusions, defects in cuticle permeability and reduced cutin load, indicating that ABCG11 is also involved in cutin monomer export in vegetative organs (Panikashvili et al. 2007) as well as in flowers, siliques and seedcoats (Panikashvili et al. 2010). However, residual export of waxes and cutin monomers onto the plant surface in the absence of ABCG11 and ABCG12 indicates that other ABC transporters might also export these compounds. Interestingly, ABCG11 and ABCG12 are half transporters, and ABCG11/ABCG12 heterodimers have a function in wax export, while ABCG11 can also homodimerize or heterodimerize with other unknown component(s) to transport cutin monomers (McFarlane et al. 2010). Recently, two other ABCG transporters have been characterized. ABCG13, which is closely related to ABCG11 and ABCG12, contributes to cutin formation in flowers (Panikashvili et al. 2011). In contrast, ABCG32/PEC1, which is a full length transporter, is required for hydroxylated fatty acid transport in leaves and flowers (Bessire et al. 2011).

Based on transcriptome analysis of Arabidopsis stems and stem epidermal cells, seven candidate LTPs, which could play a role in wax and/or cutin monomer transport were isolated (Suh et al. 2005). The analysis of the cuticle phenotype of *ltpg1* and *ltpg2* mutants suggests that both of these glycosylphosphatidylinositol-anchored LTPs (LTPGs) could be involved in cuticle formation (DeBono et al. 2009; Lee et al. 2009; Kim et al. 2012). However, the very specific transport shown to be affected in *ltpg1* or *ltpg2* mutants has given rise to the proposition that multiple specialized LTPs, possibly with overlapping functions, are required to deliver the whole diversity of cuticle compounds to the epidermal surface.

Cuticle Mutants with Epidermal Defects

As illustrated above, numerous Arabidopsis lines with mutations in genes involved in cuticle metabolism have been characterized by reverse and forward genetics. Besides their defects in surface lipid composition and/or content, several also present abnormalities in the development of their epidermis, the most common being post-genital organ fusions. The severity of these phenotypes seems to vary dramatically depending on the enzymatic step affected, as summarized in Table 12.1.

Mutations in Genes of the Fatty Elongation Complex Affect Epidermal Development to Various Degrees

Loss of KCR or HCD, two of the three activities of the elongation complex that are encoded by a single functional gene, causes embryo lethality. Corresponding knock-down mutants (*AtKCR-RNAi* lines or *pas2*-1, a weak allele of *HCD*) display strong developmental defects, such as dwarfism, spontaneous organ fusions and abnormal epidermal cell morphology. For example, *pas2-1* lines show ectopic cell divisions and loss of cell adhesion (Faure et al. 1998), while *AtKCR-RNAi* lines have abnormally short trichomes with fused branches, and produce swollen pavement cells with adhesion defects (Beaudoin et al. 2009). Knock-out mutants in the other activity of the elongation complex encoded by a single functional gene (*ECR*) are viable because of a partial functional complementation by a yet unidentified enzyme. Nevertheless, *ecr* mutants exhibit severe morphological abnormalities such as size reduction in aerial organs, and flower buds with fused tissues (Zheng et al. 2005). However, because the elongation complex is also necessary for the production of the VLCFAs present in sphingolipids, it is not possible to distinguish whether epidermal defects are linked to abnormal surface lipid composition or to impaired sphingolipid metabolism and/or plasma membrane abnormality.

Among the *KCS* multigene family, loss of function alleles of four genes show epidermal abnormalities, amongst which *kcs10/fiddlehead/fdh* is the most strongly affected. *fdh* mutant lines are dwarf, show fusions in both floral organs and leaves, have an increased cuticular permeability, allow pollen to germinate on leaves, and show a 50 % reduction in trichome density (Lolle et al. 1997; Yephremov et al. 1999.). Although *FDH* is specifically expressed in the epidermis and appears to affect epidermal cell interactions (Pruitt et al. 2000), no strong alterations in the lipid composition of *fdh* has been reported, even though mutant cell walls are enriched in one particular C24-VLCFA (Lolle et al. 1997). Similarly, the *kcs13/hic* mutant shows no major modification in its lipid profiles, but exhibits a 20–25 % increase in stomatal index at elevated atmospheric CO_2 concentrations (Gray et al. 2000). Consistent with this phenotype, *KCS13* is specifically expressed in guard cells. KCS6/CER6/CUT1 is the most important condensing enzyme for cuticular wax biosynthesis since *cer6* mutants show a 48 and 81 % decreases in leaf and stem

Table 12.1 Epidermal defects observed in Arabidopsis cuticular mutants

			General features				Organ fusion		Epidermal cell defects				
Mutant	Locus	Pathway affected	Dwarfism	Male Sterility	Glossy stems	Curly leaves	Flowers	Leaf	Pavement cells morphology	Trichome morphology	Stomatal density	Defective cuticular permeability[b]	References
kcr[a]	At1g67730	VLCFA biosynthesis	+	+	+	+	+	+	+	+		+	Beaudoin et al. (2009)
hcd	At5g10480		+	+			+	+					Bach et al. (2008) and Faure et al. (1998)
cer10[a]	At3g55360		+	+	+	+	+		+	+			Zheng et al. (2005)
cer6	At1g68530			+	+						+		Bird and Gray (2003), Millar et al. (1999), and Hooker et al. (2002)
fdh	At2g26250		+			+	+	+		+		+	Bird and Gray (2003), Lolle et al. (1997), Yephremov et al. (1999), and Pruitt et al. (2000)
hic	At2g46720										+		Bird and Gray (2003) and Gray et al. (2000)
kcs1	At1g01120		+	+									Bird and Gray (2003) and Todd et al. (1999)
cer1	At1g02205	Wax biosynthesis		+	+						±		Aarts et al. (1995), Bourdenx et al. (2011), and Gray et al. (2000)
cer3	At5g57800			+	+		+	+		+	+	+	Chen et al. (2003), Kurata et al. (2003), Ariizumi et al. (2003), and Rowland et al. (2007)

Mutant	Locus	Pathway affected	General features				Organ fusion		Epidermal cell defects			Defective cuticular permeability[b]	References
			Dwarfism	Male Sterility	Glossy stems	Curly leaves	Flowers	Leaf	Pavement cells morphology	Trichome morphology	Stomatal density		
wbc11	At1g17840	Cutin biosynthesis	+	+		+	+	+	+	+	+	+	Bird et al. (2007) and Panikashvili et al. (2007)
abcg13	At1g51460						+					+	Panikashvili et al. (2011)
lcr	At2g45970		+			+	+	+		+			Wellesen et al. (2001)
lacs2	At1g49430		+	+		+			+	+	+	+	Lü et al. (2009), Weng et al. (2010), and Schnurr et al. (2004)
hth	At1g72970						+						Krolikowski et al. (2003) and Kurdyukov et al. (2006a)
gpat4 *gpat8*	At1g01610/ At4g00400		+						+		+	+	Li et al. (2007) and Yang et al. (2010, 2012)
dcr	At3g48720						+		+	+			Panikashvili et al. (2009)
gpat6	At2g38110						+		+			+	Li-Beisson et al. (2009) and Yang et al. (2010, 2012)
bdg	At1g64670		+			+		+		+		+	Kurdyukov et al. (2006b)

[a]Embryo lethal knock-out

[b]Assayed using toluidine blue, chlorophyll leaching, weight loss or pollen germination on leaf test

wax loads, respectively (Millar et al. 1999; Fiebig et al. 2000). Despite these dramatic reductions in wax load, *cer6* mutants do not display any major epidermal defect, with the exception of a 30 % increase in stomatal index (Gray et al. 2000). The defects observed in *kcs1* mutants depend on the relative humidity (RH) used for growth. At high RH, *kcs1-1* plants resemble wild-type segregants but have 25 % fewer flowers, a 35 % reduction in silique length and very thin stems. In contrast, at low RH, 2 week old seedlings are very small and have a mortality rate of over 90 % (Todd et al. 1999). Finally, no epidermal defects were reported in plants with a mutation in *KCS9* or any CER2-like protein, although the corresponding mutants are affected in fatty acid elongation and/or the chain length of epicuticular wax compounds (Haslam et al. 2012; Pascal et al. 2013; Kim et al. 2013).

Knock-Out Mutations in Cuticular Wax Biosynthesis Have no Major Epidermal Defects

The *cer4* and *wsd1* mutants, which are affected in the reduction pathway producing fatty alcohols and wax esters, are phenotypically similar to wild-type plants, and show normal epidermis development (Rowland et al. 2006; Li et al. 2008). A mutation in *CER1*, which is responsible for the synthesis of the major wax components in Arabidopsis (alkanes), results in conditional male sterility (at low RH) despite the fact that the pollen coat (tryphine) of the mutant looks like that of wild-type (Aarts et al. 1995). Contradictory results have been reported concerning the stomatal index of *cer1* plants: Gray et al. (2000) report an increase, but in contrast Bourdenx et al. (2011) report no difference compared to wild-type, suggesting that in a *cer1* background, growth conditions could affect the development of stomata. Conversely, mutations in *CER3*, which is also involved in the biosynthesis of alkanes, cause a reduced stomatal index but also result in male sterility at low RH (Chen et al. 2003). Depending on the study, *cer3* mutants either have no other epidermal defect (Ariizumi et al. 2003; Rowland et al. 2007), show smaller trichomes and aerial organ adhesions (Kurata et al. 2003), or can even display postgenital organ fusions in leaves and flower buds (Chen et al. 2003). Although this last study attributed fusions to a putative role for CER3 in cutin formation, lipid analysis later showed that CER3 has no role in the biosynthesis of cutin, at least in leaves (Rowland et al. 2007).

Cutin Mutants Often Present Pleiotropic Defects in Epidermal Development

In contrast to cuticular wax mutants in which development and morphology is often indistinguishable from that of wild-type plants, many cutin mutants present extremely severe phenotypes with multiple epidermal defects, especially in reproductive organs. Mutations in the ABC transporters responsible for the export of cuticle precursors best exemplify this discrepancy. The *cer5* mutant, which is affected in wax load (54 % less wax on stems) but not in cutin content, does not present any visible phenotype (Pighin et al. 2004; Bird et al. 2007). In contrast, the *wbc11* mutant, which is affected in both wax and cutin loads, shows many surface defects which are typical of an abnormal epidermal development (Bird et al. 2007; Panikashvili et al. 2007). Mutant plants present a strong growth retardation at the vegetative stage, and multiple thin-stemmed short inflorescences at the reproductive stage (loss of apical dominance). Numerous fusion events between leaves or between leaves and flower buds are also observed in this background, as are the production of misshapen rosette leaves, an increase in permeability to solutes and the appearance of pavement cell abnormalities. In addition the production of asymmetric stomata, shorter trichomes with irregular branching and frequent trichome collapse has been reported. Finally flower morphology is affected in the *wbc11* mutant where the fertility of pollen and the functionality of stigmatic papillae are reduced, and siliques are shorter and contain fewer seeds than those of wild-type plants. Consistent with the strong expression of *ABCG13* in flowers, the *abcg13* mutant also shows inter-organ post-genital fusions in inflorescences, but shows no defects during the vegetative phase (Panikashvili et al. 2011). The analysis of *abcg13* flower surface lipids showed that although wax load is normal, cutin load is reduced by 50 %. Petals are most strongly affected, with a loss of the typical conical shape of the epidermal cells and an absence of cuticular ridges (Panikashvili et al. 2011). An identical floral phenotype was observed in the *gpat6*, *cyp77a6*, and *dcr* mutants, which all show a reduction in 10,16-dihydroxypalmitate, the main floral cutin precursor, of over 90 % (Li-Beisson et al. 2009; Panikashvili et al. 2009).

Although mutants with reduced floral organ cutin always show strong epidermal defects such as post genital organ fusions, phenotypes associated with a reduction in leaf cutin content can vary dramatically. For example, the *lcr* mutant has multiple developmental abnormalities whereas the *att1* mutant looks morphologically normal, although both genes are thought to code for P450 hydroxylases involved in fatty acid oxidation (Wellesen et al. 2001; Xiao et al. 2004). Unfortunately, lipid analyses were only conducted for *att1*, where the authors did not report any epidermal defects although the stem cutin load was decreased by 70 % (Xiao et al. 2004). Similar to the situation for *lcr*, the effect of a mutation in *LACS2* on cutin content and/or composition has not been reported, even though epidermis development is strongly affected in *lacs2* mutants (Schnurr et al. 2004). In particular leaf pavement cells, which are shaped like jigsaw puzzle pieces in wild-type, have fewer lobes or even collapse in *lacs2* mutants. This phenotype is reminiscent to that of *dcr* mutants, in which the pavement cell patterning is disrupted and trichomes often collapse (Panikashvili et al. 2009). In contrast, mutation in the *DCF* gene, encoding another BAHD, does not result in any epidermal defect even though the leaf cutin of *dcf* is devoid of ferulate. Finally, *gapt4gpat8* double mutants, containing 65 % less cutin in their stems, have deformed pavement cells and stomata lacking cuticular edges, while *bdg* mutants, in which cutin load is 1.2–3 times higher than that of wild type, have deformed leaves with flattened trichomes that often collapse and adhere to neighboring pavement cells (Kurdyukov et al. 2006b). It should nevertheless be noted that growing conditions can apparently strongly modulate the consequences of cutin defects on epidermis development, since *lacs1lacs2* double mutants were described as unaffected in size, growth and development by Lü and coworkers (Lü et al. 2009), but severely disrupted with temporary organ fusion in flowers, by Weng and coworkers (Weng et al. 2010).

Together, these results suggest that epidermal developmental defects are related to modifications in the composition and/or content of cutin, rather than waxes. This idea is further supported by the fact that similar epidermal defects have been observed in Arabidopsis transgenic plants expressing a fungal cutinase (Sieber et al. 2000) or an Arabidopsis putative cutinase (Takahashi et al. 2010). In addition, the primary role of cuticular waxes is widely considered to be related to the waterproof function of the cuticle, consistent with their having a much stronger hydrophobic character than cutin monomers. In contrast, cutin monomers or derivatives have been shown to function as elicitors of plant defense suggesting that plants actively monitor cutin integrity (Schweizer et al. 1996; Fauth et al. 1998). Intriguingly, an increased permeability of the cuticle to pathogen-derived lipid-signals eliciting plant defenses was proposed as an explanation for the enhanced biotic resistance observed in certain cutin-deficient mutants (Reina-Pinto and Yephremov 2009). Similarly, Bird and Grey (2003) suggested that a lipid molecule, most probably a VLCFA (or derivative) produced in guard cells, could diffuse in the cutin matrix and prevent surrounding cells from differentiating in guard cells, thereby participating in the fine-tuning of stomatal patterning. The control of epidermal cell fate by non-autonomous VLCFA-derivative signals is further supported by the *fdh* mutant which presents higher levels of a cell-wall localized C24:0 fatty acid derivative and strong developmental defects (Lolle et al. 1997), as well as by the recent demonstration that synthesis of VLCFAs in the epidermis can restrict cell proliferation in the vasculature (Nobusawa et al. 2013b). The cutin polymer could therefore function to restrict the mobility of lipid signaling molecules within the cuticle, thus fine tuning the development of the epidermis.

Transcriptional Regulation of Cuticular Lipid Biosynthesis and Epidermis Specification

The epidermis-specific expression of many genes involved in the biosynthesis and transport of cuticle components suggests that these processes must be under strict transcription regulation. Large-scale screens of Arabidopsis overexpressor/activation tagged lines identified the AP2/EREBP transcription factor SHN1/WIN1 as a major positive regulator of cuticle production (Broun et al. 2004; Aharoni et al. 2004). Plants over expressing SHN1/WIN1, or one the two other members of this small AP2/EREBP clade (SHN2 and SHN3), display clear visual phenotypes including dwarfism, shiny leaves that curl, smaller and more shiny siliques, and flowers with affected morphology, especially in petals. In addition, the differentiation of their leaf epidermal cells was clearly altered as trichomes were less abundant and often single branched, pavement cells were enlarged, and stomatal density was decreased by two thirds (Aharoni et al. 2004). The first published lipid analyses conducted on these plants indicated that their wax load was several fold increased, consistent with their shiny appearance (Broun et al. 2004; Aharoni et al. 2004). However, further analyses showed that cutin production was the primary target of regulation (resulting in a 2–3 times higher cutin load in leaves), with *LACS2* identified as a direct target of WIN1, while wax deposition was only up-regulated in a second later phase (Kannangara et al. 2007).

The redundant functions of SHN1, 2 and 3 in shaping the surface and morphology of Arabidopsis flowers was shown using microRNA technology. Plants silenced simultaneously for all three genes displayed flowers with organs fusions and lacking the typical nanoridges on the adaxial epidermis of petals (Shi et al. 2011). Transcriptomic analyses further showed that downstream targets are either related to cutin biosynthesis or to cell wall metabolism, suggesting that SHN transcription factors not only regulate the deposition of cuticular lipids, but also modulate cell wall pectins and epidermal cell elongation, adhesion and separation, to prevent organ fusion. The relation between cutin metabolism and epidermal development is further supported by a study conducted on *SlSHN3* in tomato. In fruits from *SlSHN3*-silenced lines, a 60 % decrease in cutin load was correlated with strong modification of the morphology and patterning of the fruit epidermal cells. Most interestingly, among the putative SlSHN3 target genes were several encoding transcription factors associated with epidermal cell patterning. These included *SlSHN2*, the R2R3-MYB-encoding *SlMIXTA*, and three HDZIPIV genes (*SlGL2*, *SlHDG11a* and *SlANL2c*). In Arabidopsis, the MIXTA-like MYB transcription factor MYB106, which was known to regulate the formation and branching of trichomes, was recently shown to act as a positive regulator of WIN1/SHNs, supporting the link between cuticle and epidermal cell differentiation (Oshima et al. 2013). Similarly, Arabidopsis HDG1, a member of the HDZIP IV family of homeodomain transcription factors, was shown to bind the promoter of the cutin-related genes *BDG* and *FDH*, and HDG1 chimeric repressor plants were shown to have a defective cuticle resulting in leaf fusion events (Wu et al. 2011).

In Arabidopsis, the *HDZIP IV* gene family comprises 16 members, of which several have been shown to be involved in both epidermal specification and differentiation (Abe et al. 2003; Nakamura et al. 2006; Marks et al. 2009; Vernoud et al. 2009; Depege-Fargeix et al. 2011; Javelle et al. 2011a; Nadakuduti et al. 2012; Roeder et al. 2012; Takada et al. 2013; San-Bento et al. 2014). This gene family encodes plant-specific transcription factors which play a key role in regulating the intimately intertwined processes of epidermal fate specification and cuticle biogenesis. Loss of function of two closely related family members, ATML1 and PDF2, leads to a complete loss of epidermal cell fate specification during early embryogenesis (Abe et al. 2003; San-Bento et al. 2014). Conversely, the ectopic expression of *ATML1*, and/or of the closely related *HDG2* gene leads to the ectopic formation of stomata in internal positions within developing organs, suggesting that these proteins are not only necessary, but also sufficient for the specification of some epidermal cell types (Takada et al. 2013; Peterson et al. 2013). Transcriptional analyses of *ATML1* and *HDG2* ectopic expression lines show an over-expression (probably also ectopic) of other members of the *HDZIPIV* gene family, such as PDF2.

Interestingly however, over-expression of either *ATML1* or *PDF2* specifically within the epidermis leads to a repression of the endogenous *ATML1* and *PDF2* genes during post-embryonic development, suggesting that the level of expression of these two genes, and probably of other family members, is tightly regulated by complex feedback mechanisms. In support of this hypothesis, it has been shown that ATML1 and PDF2 are capable of binding to their own, and each other's, promoters. Furthermore, the normal expression of the ATML1 gene is at least in part dependent upon the presence of an L1 box, the cognate binding site for HDZIPIV proteins (Sessions et al. 1999; Abe et al. 2001; Nakamura et al. 2006; Takada and Jurgens 2007). Interestingly, ATML1 and PDF2 are capable of both homodimerization and heterodimerization *in planta*, the latter both with each other and with other HDZIPIV proteins (our unpublished results) and indeed with other transcription factor families (San-Bento et al. 2014; Rombola-Caldentey et al. 2014). Taken together, these observations suggest a complex picture in which the production of multiple different heterodimeric complexes could fine-tune both feedback regulation and target gene activation by this protein family.

Because of their important role in specifying epidermal cell fate it could be argued that the effects of altering HDZIPIV activity on cuticle composition are largely indirect. Nonetheless, several pieces of evidence support a more direct role for members of the HDZIPIV family in the control of cuticle production. Ectopic overexpression of the maize HDZIPIV protein OCL1 leads to changes in cuticle composition associated with regulation of several genes involved in cuticle biogenesis. In the case of at least one gene, *ZmWBC11a*, this regulation requires an L1 box (Javelle et al. 2010). More evidence for regulation of cuticle biogenesis by HDZIP proteins, albeit potentially indirect, comes from *Arabidopsis* lines ectopically overexpressing *ATML1* and *HDG2*, which show strong up-regulation of both *FDH* and *WBC12/CER5* (Takada et al. 2013). Furthermore the defective trichomes of *hdg2* mutants show reduced cuticle load and reduced expression of *CYP94C*, a gene required for the production of dicarboxylic acids potentially involved in cutin biosynthesis (Marks et al. 2009). Finally, defects in both the tomato HDZIPIV encoding *CUTIN DEFICIENT2* gene, and its *Arabidopsis* orthologue *ANTHOCYANINLESS2* (*ANL2*) reduce cuticle loads (Nadakuduti et al. 2012). At a more global level, both co-expression data and the presence of L1 boxes in the promoters of a subset of genes with important roles in cuticle production support a link between HDZIPIV expression and cuticle biogenesis (unpublished results), although how this is integrated with the activity of other key cuticle-regulating transcription factors, such as members of the WIN/SHINE clade, remains rather unclear.

Protocuticle and Epidermal Cell Fate

Early plant embryogenesis involves the sequential specification of key cell types including the epidermis. In Arabidopsis, where this process has perhaps been most extensively studied, epidermal cell fate is generally considered to be "fixed" at the dermatogen stage of embryogenesis. At this stage, the eight roughly geometrically equivalent cells derived from the original apical cell of the two celled embryo by two perpendicular longitudinal and one latitudinal rounds of division, each undergo a periclinal cell division to give 16 cells, of which 8 form the skin (protoderm) of the embryo. Interestingly, markers of epidermal cell fate, including the *ATML1* gene, are expressed as early as the two cell embryo stage suggesting that "epidermis" or at least "protoderm" may be a default state, at least of the embryo proper. Our unpublished results suggest that at this stage in embryo development, and indeed during early preglobular stages, there is no detectable trace of cuticle deposition within the cell wall separating the membrane of the embryo from that of the surrounding syncytial endosperm. However, in silico transcriptomic data issuing from painstaking laser microdissection and microarray analysis of Arabidopsis seed development (Le et al. 2010), shows that many key components of the cuticle biosynthesis pathway including *LACS2*, *FDH*, and *BDG1* are already strongly expressed in the pre-globular embryo proper, suggesting that cuticle biogenesis and deposition is likely initiated very early in the developing embryo. Both *FDH* and *BDG1* contain L1 boxes upstream their transcription start sites, consistent with a potential regulation by HDZIPIV proteins (Abe et al. 2001, 2003; Nakamura et al. 2006). However, although it is clear that well before embryo maturity cotyledons are covered in a layer of cuticular material (Tanaka et al. 2001), frustratingly little is known about the biogenesis of the embryonic cuticle.

In terms of cuticle biogenesis, plant embryos are unique from two important points of view. Firstly, because they arise from gametophyte cells, which in turn are derived from non-epidermal cells buried within the ovule nucellus, their cuticle, like the epidermal identity of the cells on which it appears, arises de novo. In contrast, clonal analyses in several species suggests that after early embryogenesis the overwhelming majority of, if not all plant epidermal cells arise from anticlinal divisions of existing epidermal cells, and are thus "pre-equipped" with a cuticle bearing surface (Javelle et al. 2011b). Secondly, not only does the embryonic epidermis and therefore its cuticle arise de novo, but it also arises in a position where it is not juxtaposed by another cuticularized surface. In angiosperms, the embryo develops in an invagination of the endosperm, which initially takes the form of a rapidly growing coenocytic single cell (commonly called the syncytial endosperm). This cell subsequently undergoes cellularization in a wave, starting in the zone surrounding the developing embryo. There is no evidence in the literature or from TEM data (our unpublished results) that the endosperm produces a cuticle at its junction with the developing embryo. The embryonic cuticle, and thus the apoplastic separation of the embryo from the endosperm, therefore arises in developmentally unique circumstances since in all other aerial plant organs the outer, cuticle bearing surfaces of all epidermal cells juxtapose the equivalent surfaces of other epidermal cells. The juxtaposed cuticles of these cells play a seminal role in preventing post genital fusion of epidermal surfaces, no matter how tightly they are packed together in developing buds. The developmental importance of this "separation" function is manifest in the many cuticle-related mutants where this function is impacted, and in which organs fuse, impacting growth, fertility and viability (see previous section).

The unique circumstances surrounding angiosperm embryonic cuticle formation might be predicted to demand the deployment of specific molecular and cellular mechanisms which permit the correct definition and positioning of the nascent cuticle, and the successful separation of the embryo from surrounding endosperm tissues. Consistent with this idea, recent research has uncovered a seed-specific signaling pathway, involving both endosperm and embryonically expressed components, which is necessary for the formation of a functional embryonic cuticle, and for physical separation of the embryo and endosperm, in *Arabidopsis*. In the interests of conciseness, we will refer to this pathway as the Embryo Cuticle Functionality (ECF) pathway. To date, five critical components of this pathway have been identified. The first two comprise two very distantly related bHLH transcription factors: ZHOUPI (ZOU)/RGE1 which is expressed specifically in the embryo surrounding endosperm (Kondou et al. 2008; Yang et al. 2008; Xing et al. 2013), and ICE1/SCREAM which acts as a heterodimer with ZOU (Denay et al. 2014), but is also involved in several other developmental and physiological processes (Shirakawa et al. 2014; Kanaoka et al. 2008; Chinnusamy et al. 2003; Lee et al. 2005; Miura et al. 2007). Both *zou* and *ice1* mutants fail to form a functional embryonic cuticle, and their cotyledons are highly permeable to toluidine blue (Denay et al. 2014). In addition mutant embryos fail to physically separate from the surrounding endosperm (Yang et al. 2008). Interestingly however, cuticle components appear to be present at the surface of *zou* mutant embryos, and, importantly, the expression of neither epidermal markers (Yang et al. 2008; Xing et al. 2013; Denay et al. 2014) nor genes involved in cuticle biosynthesis (our unpublished results) is strongly affected in either mutant background. The third gene involved in the ECF pathway is again expressed specifically in the embryo-surrounding endosperm, but in this case encodes a Subtilisin-like serine protease called ABNORMAL LEAF SHAPE1 (ALE1), which is necessary for normal embryonic surface formation and endosperm/embryo separation (Tanaka et al. 2001). ALE1 is predicted to act in the secretory pathway or apoplast, and therefore represents a non-cell autonomous component of the pathway. The expression of *ALE1* is almost completely abolished in *zou* and *ice1* mutants, and reintroduction of *ALE1* expression into *zou* mutants partially complements their cuticle defects (Yang et al. 2008; Xing et al. 2013; Denay et al. 2014). Consistent with the fact that *ZOU* and *ALE1* act in the same genetic pathway their mutant phenotypes show no additivity in double mutants (Yang et al. 2008). Finally, two genes encoding the closely related receptor kinases GASSHO1 (GSO1) and GSO2, and which show strong expression in developing embryos, act redundantly to promote the production of a functional embryonic cuticle and embryo/endosperm separation (Tsuwamoto et al. 2008). Double *gso1 gso2* mutant shows remarkably similar phenotypes to *ale1* mutants and non-additive genetic interactions with both *ale1* and *zou* mutants, confirming its likely participation in a signaling processes involving communication between the developing embryo and endosperm necessary for normal embryonic cuticle biogenesis (Xing et al. 2013; Waters et al. 2013).

Despite their defective cuticle phenotypes, like *zou* mutants, neither *gso1 gso2* double mutants nor *ale1* mutants show significant changes in the expression of epi-

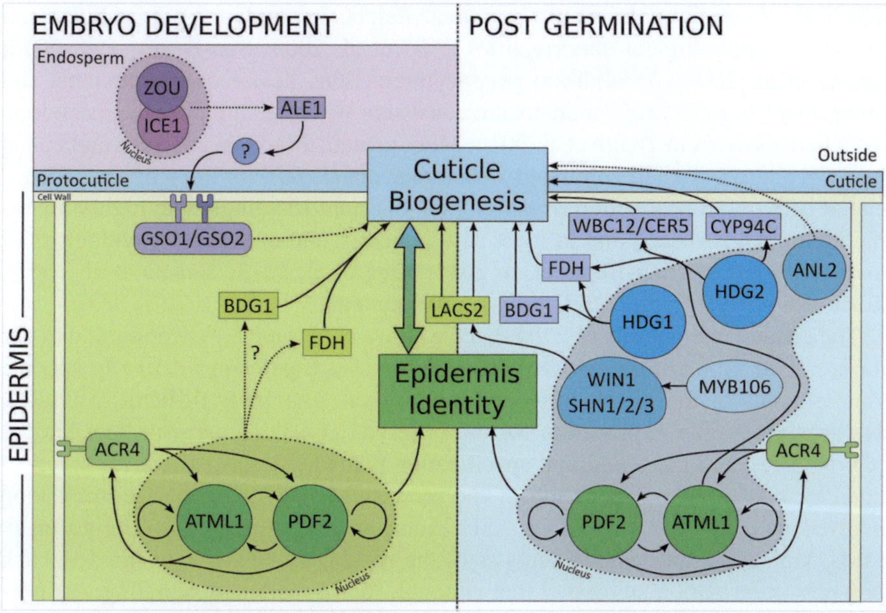

Fig. 12.2 Transcriptional control of cuticle biosynthesis and epidermis specification during development. During plant development, epidermis identity is thought to be specified by the HDZIPIV transcription factors ATML1 and PDF2. Cell-cell signaling, for example mediated by the ACR4 receptor, generates a feedback loop allowing identity maintenance. During zygotic embryogenesis, cuticle biogenesis is affected by two different genetic pathways. The first involves an intercompartmental signaling pathway between the endosperm and the embryo. The second is poorly understood but may be regulated by ATML1, PDF2 and other HDZIPIV transcription factors, and thus be intricately linked to epidermis identity. Later during development, many transcription factors from different classes, including HDZIPIV transcription factors have been shown to regulate the expression of different effectors involved in cuticle biosynthesis. Similar transcription factors could be involved in cuticle biogenesis during embryogenesis, although evidence to support this is currently lacking. Transcription factors are represented by circles, effectors by rectangles and receptor by round-cornered boxes. *Left* panel shows an epidermal cell and its environment during early embryogenesis. *Right* panel shows the same cell after germination

dermal marker genes such as *ATML1* and *PDF2* (San-Bento et al. 2014). Consistent with this observation, the formation of an epidermal cell layer does not appear to be significantly affected in these mutants suggesting that the defects in cuticle integrity which they display are not linked to a fundamental defect in epidermal cell fate specification. Interestingly however, mutants in the ECF pathway show very strong synergistic interactions with single mutants in either *ATML1*, *PDF2* and with mutant alleles of *ACR4*, a receptor-kinase encoding gene, homologues of which are required for normal epidermal development in several species (Roeder et al. 2012; Pu et al. 2012; Becraft et al. 1996; Gifford et al. 2005; Tanaka et al. 2004, 2007) and which has recently been shown to be both directly regulated by, and necessary for the maintained expression of, ATML1 and PDF2 (San-Bento et al. 2014) (Fig. 12.2). Although *atml1 pdf2* mutants show early embryo lethality, single mutant *atml1*, *pdf2* or *acr4* seedlings show only very weak defects in cotyledon permeability, and no major morphological abnormalities (Abe et al. 2003; San-Bento et al. 2014; Tanaka et al. 2004). In addition genetic interactions between *acr4* mutants and either *atml1* or *pdf2* single mutants are consistent with the signaling feedback loop described above (San-Bento et al. 2014; Moussu et al. 2013). In contrast, mutants in the ECF pathway, when combined with either *atml1*, *pdf2* or *acr4* mutant alleles, lead at best to the production of extremely abnormal seedlings with highly abnormal epidermal surfaces, and in most cases to early embryo lethality and a loss of epidermal identity (San-Bento et al. 2014; Xing et al. 2013; Tanaka et al. 2007), effectively phenocopying *atml1 pdf2* double mutants.

The genetic interactions described in the above paragraph are complex and rather non-intuitive. In addition, the fact that they involve a pathways active in specific compartments of the developing seed makes them extremely difficult to dissect. However a possible explanation for the observed genetic synergism between the ECF pathway and the epidermal specification pathway described above, was provided by mathematical modeling of the predicted behavior of the feedback loop involved in epidermal cell fate specification and maintenance (San-Bento et al. 2014). Molecular and genetic analysis of the regulatory behavior of this feedback loop post germination suggested that although ACR4 mediated signaling was necessary for maintaining the expression levels of ATML1 and PDF2, these two proteins fed back negatively on the expression of the *ACR4*, *ATML1* and *PDF2* genes. In mathematical models, this scenario gives a robust maintenance of *ATML1* and *PDF2* expression (and thus epidermal identity) over a wide range of possible values for ACR4 signaling activity (which can be considered to be equivalent to the presence of the, as yet unidentified, ligand of ACR4). Interestingly however, data concerning the regulation of elements of the epidermal specification feedback loop in developing seeds suggested that during early embryogenesis the net regulation of *ACR4* expression, and likely the expression of *ATML1* and *PDF2* by HDZIP IV class proteins, is likely to be positive rather than negative (San-Bento et al. 2014; Takada et al. 2013; Takada and Jurgens 2007). Mathematical modeling of this scenario leads to an unstable scenario, in which modestly decreasing the level of ACR4 signaling activity can lead to a sudden and irreversible loss of expression of *ATML1* and *PDF2* (San-Bento et al. 2014). A plausible scenario then is that the production of a functional embryonic cuticle, mediated by the ECF pathway, is necessary for the maintenance of ACR4 signaling activity, possibly via the restriction of a diffusible apoplastically localized ligand to the embryo. In an otherwise wild-type background, defects in the ECF pathway are not sufficient to destabilize the epidermal specification feedback loop. However in backgrounds where this loop is defective, cuticle defects are able to “tip the balance” leading to a sudden and catastrophic loss of epidermal identity. Interestingly, and in accord with this hypothesis, loss of epidermal cell fate in ECF component mutant combinations with mutants in elements of the epidermal feedback loop, occurs at the early heart stage, at the time when endosperm dramatically increases the potential for apoplastic diffusion in the embryo-surrounding endosperm. However, until more is known about when and where the cuticle is formed during embryogenesis, it will be difficult to address its function as a developmentally important diffusion barrier in more detail.

Concluding Remarks

The differentiation of a distinct, continuous epidermal cell layer containing highly specialized cells such as stomata, and the appearance of a cuticle on the surface of epidermal cells, represent key adaptive responses developed by the green phyla to cope with dry conditions when plants colonized non-aqueous environments about 450 million years ago. Ever since, epidermal differentiation and the biosynthesis of surface lipids, and their respective regulation, have co-evolved to allow plants to successfully colonize an astounding range of terrestrial habitats. Recent studies focusing on the transcriptional regulation of epidermal characters have started to unravel a highly complex network and have highlighted interplay in the regulation of key epidermal traits. Major cuticle regulators interact with proteins necessary for epidermal cell differentiation, which in turn can control the expression of genes coding for proteins involved in surface lipid biosynthesis. Tantalizingly, HDZIPIV transcription factors, which play a central role in specifying epidermal cell fate and in the differentiation of various epidermal cell types, contain a START domain that has the potential to bind regulatory lipids, while VLCFA derivatives have been identified as candidate signaling lipids involved in fine-tuning the development of both the epidermis and the cuticle in response to environmental clues. Together, these data hint at mechanisms underlying a complex cross-talk between cuticle formation and epidermis differentiation both at the transcriptional and metabolic level.

Acknowledgments This work was supported by the Agence Nationale de la Recherche through the INASEED grant (ANR-13-BSV2-0002 to G.I.).

References

Aarts MGM, Keijzer CJ, Stiekema WJ et al (1995) Molecular characterization of the *CER1* gene of Arabidopsis involved in epicuticular wax biosynthesis and pollen fertility. Plant Cell 7:2115–2127

Abe M, Takahashi T, Komeda Y (2001) Identification of a cis-regulatory element for L1 layer-specific gene expression, which is targeted by an L1-specific homeodomain protein. Plant J 26:487–494

Abe M, Katsumata H, Komeda Y et al (2003) Regulation of shoot epidermal cell differentiation by a pair of homeodomain proteins in Arabidopsis. Development 130:635–643

Aharoni A, Dixit S, Jetter R et al (2004) The SHINE clade of AP2 domain transcription factors activates wax biosynthesis, alters cuticle properties, and confers drought tolerance when over-expressed in Arabidopsis. Plant Cell 16:2463–2480

Ariizumi T, Hatakeyama K, Hinata K et al (2003) A novel male-sterile mutant of Arabidopsis thaliana, faceless pollen-1, produces pollen with a smooth surface and an acetolysis-sensitive exine. Plant Mol Biol 53:107–116

Bach L, Michaelson LV, Haslam R et al (2008) The very-long-chain hydroxy fatty acyl-CoA dehydratase PASTICCINO2 is essential and limiting for plant development. Proc Natl Acad Sci U S A 105:14727–14731

Beaudoin F, Wu X, Li F et al (2009) Functional characterization of the Arabidopsis β-ketoacyl-Coenzyme A reductase candidates of the fatty acid elongase. Plant Physiol 150:1174–1191

Becraft PW, Stinard PS, McCarty DR (1996) CRINKLY4: a TNFR-like receptor kinase involved in maize epidermal differentiation. Science 273:1406–1409
Beisson F, Li-Beisson Y, Pollard M (2012) Solving the puzzles of cutin and suberin polymer biosynthesis. Curr Opin Plant Biol 15:329–337
Bernard A, Joubès J (2013) Arabidopsis cuticular waxes: advances in regulation, synthesis, export and functions. Prog Lipid Res 52:110–129
Bernard A, Domergue F, Pascal S et al (2012) Reconstitution of plant alkane biosynthesis in yeast demonstrates that Arabidopsis ECERIFERUM1 and ECERIFERUM3 are core components of a very-long-chain alkane synthesis complex. Plant Cell 24:3106–3118
Bessire M, Borel S, Fabre G et al (2011) A member of the PLEIOTROPIC DRUG RESISTANCE family of ATP binding cassette transporters is required for the formation of a functional cuticle in Arabidopsis. Plant Cell 23:1958–1970
Bird SM, Gray JE (2003) Signals from the cuticle affect epidermal cell differentiation. New Phytol 157:9–23
Bird D, Beisson F, Brigham A et al (2007) Characterization of Arabidopsis ABCG11/WBC11, an ATP binding cassette (ABC) transporter that is required for cuticular lipid secretion. Plant J 52:485–498
Bonaventure G, Beisson F, Ohlrogge J et al (2004) Analysis of the aliphatic monomer composition of polyesters associated with Arabidopsis epidermis: occurrence of octadeca-cis-6, cis-9-diene-1,18-dioate as the major component. Plant J 40:920–930
Bourdenx B, Bernard A, Domergue F et al (2011) Overexpression of Arabidopsis ECERIFERUM1 promotes wax very-long-chain alkane biosynthesis and influences plant response to biotic and abiotic stresses. Plant Physiol 156:29–45
Broun P, Poindexter P, Osborne E et al (2004) WIN1, a transcriptional activator of epidermal wax accumulation in Arabidopsis. Proc Natl Acad Sci U S A 101:4706–4711
Chen X, Goodwin SM, Boroff VL et al (2003) Cloning and characterization of the WAX2 gene of Arabidopsis involved in cuticle membrane and wax production. Plant Cell 15:1170–1185
Cheng JB, Russell DW (2004) Mammalian wax biosynthesis-I: identification of two fatty acyl-Coenzyme A reductases with different substrate specificities and tissue distributions. J Biol Chem 279:37789–37797
Chinnusamy V, Ohta M, Kanrar S et al (2003) ICE1: a regulator of cold-induced transcriptome and freezing tolerance in Arabidopsis. Genes Dev 17:1043–1054
DeBono A, Yeats TH, Rose JKC et al (2009) Arabidopsis LTPG is a glycosylphosphatidylinositol-anchored lipid transfer protein required for export of lipids to the plant surface. Plant Cell 21:1230–1238
Denay G, Creff A, Moussu S et al (2014) Endosperm breakdown in Arabidopsis requires heterodimers of the basic helix-loop-helix proteins ZHOUPI and INDUCER OF CBP EXPRESSION 1. Development 141:1222–1227
Depege-Fargeix N, Javelle M, Chambrier P et al (2011) Functional characterization of the HD-ZIP IV transcription factor OCL1 from maize. J Exp Bot 62:293–305
Faure JD, Vittorioso P, Santoni V et al (1998) The PASTICCINO genes of Arabidopsis thaliana are involved in the control of cell division and differentiation. Development 125:909–918
Fauth M, Schweizer P, Buchala A et al (1998) Cutin monomers and surface wax constituents elicit H_2O_2 in conditioned cucumber hypocotyl segments and enhance the activity of other H_2O_2 elicitors. Plant Physiol 117:1373–1380
Fiebig A, Mayfield JA, Miley NL et al (2000) Alterations in CER6, a gene identical to CUT1, differentially affect long-chain lipid content on the surface of pollen and stems. Plant Cell 12:2001–2008
Franke R, Briesen I, Wojciechowski T et al (2005) Apoplastic polyesters in Arabidopsis surface tissues – a typical suberin and a particular cutin. Phytochemistry 66:2643–2658
Gifford ML, Robertson FC, Soares DC et al (2005) ARABIDOPSIS CRINKLY4 function, internalization, and turnover are dependent on the extracellular crinkly repeat domain. Plant Cell 17:1154–1166

Girard L, Mounet F, Lemaire-Chamley M et al (2012) Tomato GDSL1 is required for cutin deposition in the fruit cuticle. Plant Cell 24:3119–3134

Gray JE, Holroyd GH, van der Lee FM et al (2000) The HIC signalling pathway links CO_2 perception to stomatal development. Nature 408:713–716

Greer S, Wen M, Bird D et al (2007) The cytochrome P450 enzyme CYP96A15 is the midchain alkane hydroxylase responsible for formation of secondary alcohols and ketones in stem cuticular wax of Arabidopsis. Plant Physiol 145:653–667

Haslam TM, Kunst L (2013) Extending the story of very-long-chain fatty acid elongation. Plant Sci 210:93–107

Haslam TM, Manas-Fernandez A, Zhao L et al (2012) Arabidopsis ECERIFERUM2 is a component of the fatty acid elongation machinery required for fatty acid extension to exceptional lengths. Plant Physiol 160:1164–1174

Haslam TM, Haslam R, Thoraval D et al (2015) CER2-LIKE proteins have unique biochemical and physiological functions in very-long-chain fatty acid elongation. Plant Physiol 167:682–692

Hoffmann-Benning S, Kende H (1994) Cuticle biosynthesis in rapidly growing internodes of deepwater rice. Plant Physiol 104:719–723

Hooker TS, Millar AA, Kunst L (2002) Significance of the expression of the CER6 condensing enzyme for cuticular wax production in Arabidopsis. Plant Physiol 129:1568–1580

Javelle M, Vernoud V, Depège-Fargeix N et al (2010) Overexpression of the epidermis-specific homeodomain-leucine zipper IV transcription factor Outer Cell Layer1 in maize identifies target genes involved in lipid metabolism and cuticle biosynthesis. Plant Physiol 154:273–286

Javelle M, Klein-Cosson C, Vernoud V et al (2011a) Genome-wide characterization of the HD-ZIP IV transcription factor family in maize: preferential expression in the epidermis. Plant Physiol 157:790–803

Javelle M, Vernoud V, Rogowsky PM et al (2011b) Epidermis: the formation and functions of a fundamental plant tissue. New Phytol 189:17–39

Jenks MA, Tuttle HA, Eigenbrode SD et al (1995) Leaf epicuticular waxes of the eceriferum mutants in Arabidopsis. Plant Physiol 108:369–377

Jessen D, Olbrich A, Knüfer J et al (2011) Combined activity of LACS1 and LACS4 is required for proper pollen coat formation in Arabidopsis. Plant J 68:715–726

Joubès J, Raffaele S, Bourdenx B et al (2008) The VLCFA elongase gene family in Arabidopsis thaliana: phylogenetic analysis, 3D modelling and expression profiling. Plant Mol Biol 67:547–566

Kanaoka MM, Pillitteri LJ, Fujii H et al (2008) SCREAM/ICE1 and SCREAM2 specify three cell-state transitional steps leading to arabidopsis stomatal differentiation. Plant Cell 20:1775–1785

Kannangara R, Branigan C, Liu Y et al (2007) The transcription factor WIN1/SHN1 regulates cutin biosynthesis in Arabidopsis thaliana. Plant Cell 19:1278–1294

Kim H, Lee SB, Kim HJ et al (2012) Characterization of glycosylphosphatidylinositol-anchored lipid transfer protein 2 (LTPG2) and overlapping function between LTPG/LTPG1 and LTPG2 in cuticular wax export or accumulation in *Arabidopsis thaliana*. Plant Cell Physiol 53:1391–1403

Kim J, Jung JH, Lee SB et al (2013) Arabidopsis 3-ketoacyl-coenzyme a synthase9 is involved in the synthesis of tetracosanoic acids as precursors of cuticular waxes, suberins, sphingolipids, and phospholipids. Plant Physiol 162:567–580

Kolattukudy PE (1971) Enzymatic synthesis of fatty alcohols in Brassica oleracea. Arch Biochem Biophys 142:701–709

Kondou Y, Nakazawa M, Kawashima M et al (2008) RETARDED GROWTH OF EMBRYO1, a new basic helix-loop-helix protein, expresses in endosperm to control embryo growth. Plant Physiol 147:1924–1935

Krolikowski K, Victor JL, Wagler TN et al (2003) Isolation and characterization of the Arabidopsis organ fusion gene HOTHEAD. Plant J 35:501–511

Kurata T, Kawabata-Awai C, Sakuradani E et al (2003) The YORE-YORE gene regulates multiple aspects of epidermal cell differentiation in Arabidopsis. Plant J 36:55–66

Kurdyukov S, Faust A, Trenkamp S et al (2006a) Genetic and biochemical evidence for involvement of HOTHEAD in the biosynthesis of long-chain a-, x-dicarboxylic fatty acids and formation of extracellular matrix. Planta 224:315–329

Kurdyukov S, Faust A, Nawrath C et al (2006b) The epidermis-specific extracellular BODYGUARD controls cuticle development and morphogenesis in Arabidopsis. Plant Cell 18:321–339

Lai C, Kunst L, Jetter R (2007) Composition of alkyl esters in the cuticular wax on inflorescence stems of Arabidopsis thaliana cer mutants. Plant J 50:189–196

Lardizabal KD, Metz JG, Sakamoto T et al (2000) Purification of a jojoba embryo wax synthase, cloning of its cDNA, and production of high levels of wax in seeds of transgenic Arabidopsis. Plant Physiol 122:645–656

Le BH, Cheng C, Bui AQ et al (2010) Global analysis of gene activity during Arabidopsis seed development and identification of seed-specific transcription factors. Proc Natl Acad Sci U S A 107:8063–8070

Lee BH, Henderson DA, Zhu JK (2005) The Arabidopsis cold-responsive transcriptome and its regulation by ICE1. Plant Cell 17:3155–3175

Lee SB, Go YS, Bae H-J et al (2009) Disruption of glycosylphosphatidylinositol-anchored lipid transfer protein gene altered cuticular lipid composition, increased plastoglobules, and enhanced susceptibility to infection by the fungal pathogen *Alternaria brassicicola*. Plant Physiol 150:42–54

Li Y, Beisson F, Koo AJK et al (2007) Identification of acyltransferases required for cutin biosynthesis and production of cutin with suberin-like monomers. Proc Natl Acad Sci U S A 104:18339–18344

Li F, Wu X, Lam P et al (2008) Identification of the wax ester synthase/acyl-CoenzymeA:diacylglycerol acyltransferase WSD1 required for stem wax ester biosynthesis in Arabidopsis. Plant Physiol 148:97–107

Li-Beisson Y, Pollard M, Sauveplane V et al (2009) Nanoridges that characterize the surface morphology of flowers require the synthesis of cutin polyester. Proc Natl Acad Sci U S A 106:22008–22013

Lolle SJ, Berlyn GP, Engstrom EM et al (1997) Developmental regulation of cell interactions in the Arabidopsis fiddlehead-1 mutant: a role for the epidermal cell wall and cuticle. Dev Biol 189:311–321

Lü S, Song T, Kosma DK et al (2009) Arabidopsis CER8 encodes LONG-CHAIN ACYL-COA SYNTHETASE 1 (LACS1) that has overlapping functions with LACS2 in plant wax and cutin synthesis. Plant J 59:553–564

Marks MD, Wenger JP, Gilding E et al (2009) Transcriptome analysis of Arabidopsis wild-type and gl3-sst sim trichomes identifies four additional genes required for trichome development. Mol Plant 2009:803–822

McFarlane HE, Shin JJH, Bird DA et al (2010) Arabidopsis ABCG transporters, which are required for export of diverse cuticular lipids, dimerize in different combinations. Plant Cell 22:3066–3075

McFarlane HE, Watanabe Y, Yang W et al (2014) Golgi- and trans-Golgi network-mediated vesicle trafficking is required for wax secretion from epidermal cells. Plant Physiol 164:1250–1260

Millar AA, Clemens S, Zachgo S et al (1999) CUT1, an Arabidopsis gene required for cuticular wax biosynthesis and pollen fertility, encodes a very-long-chain fatty acid condensing enzyme. Plant Cell 11:825–838

Miura K, Jin JB, Lee J et al (2007) SIZ1-mediated sumoylation of ICE1 controls CBF3/DREB1A expression and freezing tolerance in Arabidopsis. Plant Cell 19:1403–1414

Moussu S, San-Bento R, Galletti R et al (2013) Embryonic cuticle establishment: the great (apoplastic) divide. Plant Signal Behav 8, e27491

Nadakuduti SS, Pollard M, Kosma DK et al (2012) Pleiotropic phenotypes of the sticky peel mutant provide new insight into the role of CUTIN DEFICIENT2 in epidermal cell function in tomato. Plant Physiol 159:945–960

Nakamura M, Katsumata H, Abe M et al (2006) Characterization of the class IV homeodomain-Leucine Zipper gene family in Arabidopsis. Plant Physiol 141:1363–1375

Nawrath C, Schreiber L, Franke RB et al (2013) Apoplastic diffusion barriers in Arabidopsis. Arabidopsis Book. doi:10.1199/tab.0167

Nobusawa T, Okushima Y, Nagata N et al (2013a) Synthesis of very-long-chain fatty acids in the epidermis controls plant organ growth by restricting cell proliferation. PLoS Biol 11(4), e1001531

Nobusawa T, Okushima Y, Nagata N et al (2013b) Synthesis of very-long-chain fatty acids in the epidermis controls plant organ growth by restricting cell proliferation. PLoS Biol. doi:10.1371/journal.pbio.1001531

Oshima Y, Shikata M, Koyama T et al (2013) MIXTA-like transcription factors and WAX INDUCER1/SHINE1 coordinately regulate cuticle development in Arabidopsis and Torenia fournieri. Plant Cell 25:1609–1624

Panikashvili D, Savaldi-Goldstein S, Mandel T et al (2007) The Arabidopsis DESPERADO/AtWBC11 transporter is required for cutin and wax secretion. Plant Physiol 145:1345–1360

Panikashvili D, Shi JX, Schreiber L et al (2009) The Arabidopsis DCR encoding a soluble BAHD acyltransferase is required for cutin polyester formation and seed hydration properties. Plant Physiol 151:1773–1789

Panikashvili D, Shi JX, Bocobza S et al (2010) The Arabidopsis DSO/ABCG11 transporter affects cutin metabolism in reproductive organs and suberin in roots. Mol Plant 3:563–575

Panikashvili D, Shi JX, Schreiber L et al (2011) The Arabidopsis ABCG13 transporter is required for flower cuticle secretion and patterning of the petal epidermis. New Phytol 190:113–124

Pascal S, Bernard A, Sorel M et al (2013) The Arabidopsis cer26 mutant, like the cer2 mutant, is specifically affected in the very-long-chain fatty acid elongation process. Plant J 73:733–746

Peterson KM, Shyu C, Burr CA et al (2013) Arabidopsis homeodomain-leucine zipper IV proteins promote stomatal development and ectopically induce stomata beyond the epidermis. Development 140:1924–1935

Pighin JA, Zheng H, Balakshin LJ et al (2004) Plant cuticular lipid export requires an ABC transporter. Science 306:702–704

Pruitt RE, Vielle-Calzada JP, Ploense SE et al (2000) FIDDLEHEAD, a gene required to suppress epidermal cell interactions in Arabidopsis, encodes a putative lipid biosynthetic enzyme. Proc Natl Acad Sci U S A 97:1311–1316

Pu CX, Ma Y, Wang J et al (2012) Crinkly4 receptor-like kinase is required to maintain the interlocking of the palea and lemma, and fertility in rice, by promoting epidermal cell differentiation. Plant J 70:940–953

Rautengarten C, Ebert B, Ouellet M et al (2012) Arabidopsis deficient in cutin ferulate encodes a transferase required for feruloylation of ω-hydroxy fatty acids in cutin polyester. Plant Physiol 158:654–665

Reina-Pinto JJ, Yephremov A (2009) Surface lipids and plant defenses. Plant Physiol Biochem 47:540–549

Roeder AH, Cunha A, Ohno CK et al (2012) Cell cycle regulates cell type in the Arabidopsis sepal. Development 139:4416–4427

Rombola-Caldentey B, Rueda-Romero P, Iglesias-Fernández R et al (2014) Arabidopsis DELLA and two HD-ZIP transcription factors regulate GA signaling in the epidermis through the L1 box cis-element. Plant Cell 26:2905–2919

Rowland O, Domergue F (2012) Plant fatty acyl reductases: enzymes generating fatty alcohols for protective layers with potential for industrial applications. Plant Sci 193–194:28–38

Rowland O, Zheng H, Hepworth SR et al (2006) CER4 encodes an alcohol-forming fatty acyl-coenzyme A reductase involved in cuticular wax production in Arabidopsis. Plant Physiol 142:866–877

Rowland O, Lee R, Franke R et al (2007) The CER3 wax biosynthetic gene from Arabidopsis thaliana is allelic to WAX2/YRE/FLP1. FEBS Lett 581:3538–3544
San-Bento R, Farcot E, Galletti R et al (2014) Epidermal identity is maintained by cell-cell communication via a universally active feedback loop in Arabidopsis thaliana. Plant J 77:46–58
Savaldi-Goldstein S, Peto C, Chory J (2007) The epidermis both drives and restricts plant shoot growth. Nature 446:199–202
Schnurr J, Shockey J, Browse J (2004) The acyl-CoA synthetase encoded by LACS2 is essential for normal cuticle development in Arabidopsis. Plant Cell 16:629–642
Schweizer P, Felix G, Buchala A et al (1996) Perception of free cutin monomers by plant cells. Plant J 10:331–341
Sessions A, Weigel D, Yanofsky MF (1999) The Arabidopsis thaliana MERISTEM LAYER 1 promoter specifies epidermal expression in meristems and young primordia. Plant J 20:259–263
Shi JX, Malitsky S, De Oliveira S et al (2011) SHINE transcription factors act redundantly to pattern the archetypal surface of Arabidopsis flower organs. PLoS Genet 7, e1001388
Shirakawa M, Ueda H, Nagano AJ et al (2014) FAMA is an essential component for the differentiation of two distinct cell types, Myrosin cells and guard cells, in Arabidopsis. Plant Cell 26:4039–4052
Sieber P, Schorderet M, Ryser U et al (2000) Transgenic Arabidopsis plants expressing a fungal cutinase show alterations in the structure and properties of the cuticle and postgenital organ fusions. Plant Cell 12:721–738
Stoveken T, Kalscheuer R, Malkus U et al (2005) The wax ester synthase/acyl coenzyme A:diacylglycerol acyltransferase from Acinetobacter sp. Strain ADP1: characterization of a novel type of acyltransferase. J Bacteriol 187:1369–1376
Suh MC, Samuels AL, Jetter R et al (2005) Cuticular lipid composition, surface structure, and gene expression in Arabidopsis stem epidermis. Plant Physiol 139:1649–1665
Takada S, Jurgens G (2007) Transcriptional regulation of epidermal cell fate in the Arabidopsis embryo. Development 134:1141–1150
Takada S, Takada N, Yoshida A (2013) ATML1 promotes epidermal cell differentiation in Arabidopsis shoots. Development 140:1919–1923
Takahashi K, Shimada T, Kondo M et al (2010) Ectopic expression of an esterase, which is a candidate for the unidentified plant cutinase, causes cuticular defects in Arabidopsis thaliana. Plant Cell Physiol 51:123–131
Tanaka H, Onouchi H, Kondo M et al (2001) A subtilisin-like serine protease is required for epidermal surface formation in Arabidopsis embryos and juvenile plants. Development 128:4681–4689
Tanaka T, Tanaka H, Machida C et al (2004) A new method for rapid visualization of defects in leaf cuticle reveals five intrinsic patterns of surface defects in Arabidopsis. Plant J 37:139–146
Tanaka H, Watanabe M, Sasabe M et al (2007) Novel receptor-like kinase ALE2 controls shoot development by specifying epidermis in Arabidopsis. Development 134:1643–1652
Todd J, Post-Beittenmiller D, Jaworski JG (1999) KCS1 encodes a fatty acid elongase 3-ketoacyl-CoA synthase affecting wax biosynthesis in Arabidopsis thaliana. Plant J 17:119–130
Tsuwamoto R, Fukuoka H, Takahata Y (2008) GASSHO1 and GASSHO2 encoding a putative leucine-rich repeat transmembrane-type receptor kinase are essential for the normal development of the epidermal surface in Arabidopsis embryos. Plant J 54:30–42
Vernoud V, Laigle G, Rozier F et al (2009) The HD-ZIP IV transcription factor OCL4 is necessary for trichome patterning and anther development in maize. Plant J 59:883–894
von Wettstein-Knowles P (1982) Elongase and epicuticular wax biosynthesis. Physiol Vég 20:797–809
Waters A, Creff A, Goodrich J et al (2013) "What we've got here is failure to communicate": zou mutants and endosperm cell death in seed development. Plant Signal Behav 8, e24368

Wellesen K, Durst F, Pinot F et al (2001) Functional analysis of the LACERATA gene of Arabidopsis provides evidence for different roles of fatty acid ω-hydroxylation in development. Proc Natl Acad Sci U S A 98:9694–9699

Weng H, Molina I, Shockey J et al (2010) Organ fusion and defective cuticle function in a *lacs1lacs2* double mutant of Arabidopsis. Planta 231:1089–1100

Wu R, Li S, He S et al (2011) CFL1, a WW domain protein, regulates cuticle development by modulating the function of HDG1, a class IV homeodomain transcription factor, in rice and Arabidopsis. Plant Cell 23:3392–3411

Xiao F, Goodwin SM, Xiao Y et al (2004) Arabidopsis CYP86A2 represses Pseudomonas syringae type III genes and is required for cuticle development. EMBO J 23:2903–2913

Xing Q, Creff A, Waters A et al (2013) ZHOUPI controls embryonic cuticle formation via a signalling pathway involving the subtilisin protease ABNORMAL LEAF-SHAPE1 and the receptor kinases GASSHO1 and GASSHO2. Development 140:770–779

Yang S, Johnston N, Talideh E et al (2008) The endosperm-specific ZHOUPI gene of Arabidopsis thaliana regulates endosperm breakdown and embryonic epidermal development. Development 135:3501–3509

Yang W, Pollard M, Li-Beisson Y et al (2010) A distinct type of glycerol-3-phosphate acyltransferase with sn-2 preference and phosphatase activity producing 2-monoacylglycerol. Proc Natl Acad Sci U S A 107:12040–12045

Yang W, Simpson JP, Li-Beisson Y et al (2012) A land-plant-specific glycerol-3-phosphate acyltransferase family in Arabidopsis: substrate specificity, sn-2 preference, and evolution. Plant Physiol 160:638–652

Yeats TH, Rose JKC (2013) The formation and function of plant cuticles. Plant Physiol 163:5–20

Yeats TH, Martin LBB, Viart HM-F et al (2012) The identification of cutin synthase: formation of the plant polyester cutin. Nat Chem Biol 8:609–611

Yeats TH, Huang W, Chatterjee S et al (2014) Tomato cutin deficient 1 (CD1) and putative orthologs comprise an ancient family of cutin synthase-like (CUS) proteins that are conserved among land plants. Plant J 77:667–675

Yephremov A et al (1999) Characterization of the FIDDLEHEAD gene of Arabidopsis reveals a link between adhesion response and cell differentiation in the epidermis. Plant Cell 11:2187–2201

Zheng H, Rowland O, Kunst L (2005) Disruptions of the Arabidopsis enoyl-CoA reductase gene reveal an essential role for very-long-chain fatty acid synthesis in cell expansion during plant morphogenesis. Plant Cell 17:1467–1481

Chapter 13
Role of Lipid Metabolism in Plant Pollen Exine Development

Dabing Zhang, Jianxin Shi, and Xijia Yang

Abstract Pollen plays important roles in the life cycle of angiosperms plants. It acts as not only a biological protector of male sperms but also a communicator between the male and the female reproductive organs, facilitating pollination and fertilization. Pollen is produced within the anther, and covered by the specialized outer envelope, pollen wall. Although the morphology of pollen varies among different plant species, the pollen wall is mainly comprised of three layers: the pollen coat, the outer exine layer, and the inner intine layer. Except the intine layer, the other two layers are basically of lipidic nature. Particularly, the outer pollen wall layer, the exine, is a highly resistant biopolymer of phenylpropanoid and lipidic monomers covalently coupled by ether and ester linkages. The precise molecular mechanisms underlying pollen coat formation and exine patterning remain largely elusive. Herein, we summarize the current genetic, phenotypic and biochemical studies regarding to the pollen exine development and underlying molecular regulatory mechanisms mainly obtained from monocot rice (*Oryza sativa*) and dicot *Arabidopsis thaliana*, aiming to extend our understandings of plant male reproductive biology. Genes, enzymes/proteins and regulatory factors that appear to play conserved and diversified roles in lipid biosynthesis, transportation and modification during pollen exine formation, were highlighted.

Keywords ABC transporter • Exine • Lipid transport protein • Sporopollenin • Tapetum

Introduction

The alternation of the life cycle of angiosperms relies on the male reproductive development, which forms the haploid male gametes within pollen grains via meiosis and mitosis. During the pollen development, pollination and fertilization, the

D. Zhang (✉) • J. Shi • X. Yang
School of Life Sciences and Biotechnology, Shanghai Jiao Tong University–University of Adelaide Joint Centre for Agriculture and Health,
Shanghai Jiao Tong University, Dongchuan Road 800, Shanghai 200240, China
e-mail: zhangdb@sjtu.edu.cn; jianxin.shi@sjtu.edu.cn; yangxijia1986@163.com

Y. Nakamura, Y. Li-Beisson (eds.), *Lipids in Plant and Algae Development*,
Subcellular Biochemistry 86, DOI 10.1007/978-3-319-25979-6_13

pollen wall, a lipidic structure surrounding pollen grains, provides not only a protective role for male gametes (Scott et al. 2004), but also functions in pollen-stigma communication (Zinkl et al. 1999). Disruption of pollen wall structure frequently causes pollen degradation and/or abortion, leading to partial or complete male sterility (Wilson and Zhang 2009; Li and Zhang 2010). In plant kingdom, pollen grains exhibit remarkable biodiversity in shape and surface patterning. For example, the pollen grains of model dicot plant *Arabidopsis thaliana* display ellipsoidal epidermis covered by reticulate exine, while those of the model monocot plant rice (*Oryza sativa*) exhibit globular epidermis with smooth surface without reticulate cavities (Zhang et al. 2011). The diversity of plant pollen surface results mainly from the structure of the exine (the outer pollen wall), which is also likely associated with pollination, adhesion to vectors and stigmas during pollen-stigma contact, facilitating communication between pollen and stigma and subsequent hydration response during germination (Heslop-Harrison 1979; Murphy 2006). Due to its biological importance, the mechanism underlying the biosynthesis of pollen wall, particularly exine, is an interesting biological question to be addressed for plant biologists.

Structure and Development of Pollen Wall Structure

Although the morphology of pollen grains varies among different species, the pollen wall is mainly comprised of three layers: the pollen coat, the outer exine, and the inner intine (Blackmore et al. 2007; Zinkl et al. 1999; Edlund et al. 2004). The pollen coat, the outermost stratum of the pollen surface, is extremely hydrophobic, and can be eluted from the exine using organic solvents. It is mainly composed of lipids, and ever-expending list of proteins, pigments, aromatic substances, and other unknown compounds. The pollen coat fills in the depressions or the interspace of the pollen exine, protecting male gametophytes from dehydration and facilitating subsequent pollen-stigma communication and adhesion (Piffanelli et al. 1998; Edlund et al. 2004). Current evidence indicates that mono- and di-cot plants displayed diverse pollen coat proteins (Murphy 2006; Wu et al. 2015), even though limited is known about the lipids in the pollen coat. The report that sterol esters are the major components of pollen coat, indicating that pollen coat is different from membrane lipid, sharing similarity to oil droplet. The intine, an innermost bilayered structure of pollen wall, is mainly composed of exintine and the endintine. The main components of intine are pectin, cellulose, hemicellulose, hydrolytic enzymes and hydrophobic proteins, are required for the maintaining the structural integrity of pollen grains, pollen germination and pollen tube growing into the stigma (Scott et al. 2004; Knox and Heslop-Harrison 1971).

Exine, the most complex layer of the pollen wall, contains two layers, the inner nexine and the outer sexine. The nexine is a bilayer structure, consisting of nexine I (foot layer) and nexine II (endexine) (Ariizumi and Toriyama 2011), and the sexine is composed of tectum and bacula (Zhang et al. 2011; Li and Zhang 2010). Exine has extremely physical and chemical stability resistant to biotic and abiotic stresses such as high temperature, desiccation, ultraviolet (UV) irradiation and mechanical damage (Scott et al. 2004) due to its major component, sporopollenin, a highly resistant biopolymer of phenylpropanoid and lipidic monomers covalently coupled by ether and ester linkages (Ahlers et al. 2000). Sporopollenin provides the rigid and sculptured framework of the exine, functions to encapsulate and protect the pollen contents, and to aid in stigmatic capture. Despite the evolutionary history of sporopollenin in land plants is not clear, sporopollenin has been found in ferns, moss, fossilized green algae, and even in fungi, suggesting that it has remained evolutionarily conserved since the initial colonization of land (Scott 1994). Since sporopollenin is one of the most durable substances in nature, the biochemical mechanism on the biosynthesis of sporopollenin are still largely uncharacterized (Blackmore et al. 2007). Recent metabolomic analysis with MALDI-TOF mass spectrometer on lipids revealed that plant pollen contains also membrane-associated phospholipids (phosphatidylserine and phosphatidylcholine) and diacylglycerol (Liang et al. 2013).

Although the fundamental structure of pollen wall shares similarity among species, morphological analysis reveals that the pollen wall of rice is distinct from that of *Arabidopsis* (Li and Zhang 2010). *Arabidopsis* exine has thin nexine layer, semi-open tectum layer and longer baculum, with an abundant pollen coat deposited in the sculptured cavities of exine (Zhang et al. 2011; Edlund et al. 2004). Compared with *Arabidopsis*, rice exine has thicker tectum and nexine, and higher density of bacula, with much less pollen coat filled in the space between tectum and nexine (Li and Zhang 2010; Zhang et al. 2011; Ariizumi and Toriyama 2011). In addition, the outer surface of *Arabidopsis* pollen grains is fully reticulate while that of rice is relatively smooth. This morphological difference may be associated with different pollination methods between rice and *Arabidopsis*. The pollen grains of *Arabidopsis* are mainly distributed by insects, while rice pollen grains are transmitted by wind (Ariizumi and Toriyama 2011).

Exine Development

Current evidence shows that most plant species share similar biological processes in exine development (Fig. 13.1) even though different plants exhibit diversity of pollen grain morphology (Zhang et al. 2011; Ariizumi and Toriyama 2011; Li and Zhang 2010; Blackmore et al. 2007; Wilson and Zhang 2009). Morphological analysis indicates that the development of pollen wall starts from the formation of callose wall before meiosis (Blackmore et al. 2007; Ariizumi and Toriyama 2011; Scott et al. 2004). The callose wall consists of the linear β-1,3-glucan polymer, and it was assumed to act as a temporary cell wall between the anther wall and the plasma membrane from the meiosis stage to the end of the tetrad stage. After the formation of tetrads, the surrounding callose wall is digested by β-1,3-glucanase secreted from the tapetum (Ariizumi and Toriyama 2007; Piffanelli et al. 1998; Verma and Hong 2001; Worrall et al. 1992).

Several genes, such as *AtMYB103* (Zhang et al. 2007), *Callose Synthase* (*CALS5*) (Dong et al. 2005), *AUXIN RESPONSE FACTOR17* (*ARF17*) (Yang et al. 2013) and *CYCLIN-DEPENDENT KINASE G1* (*CDKG1*) (Huang et al. 2013b), have been

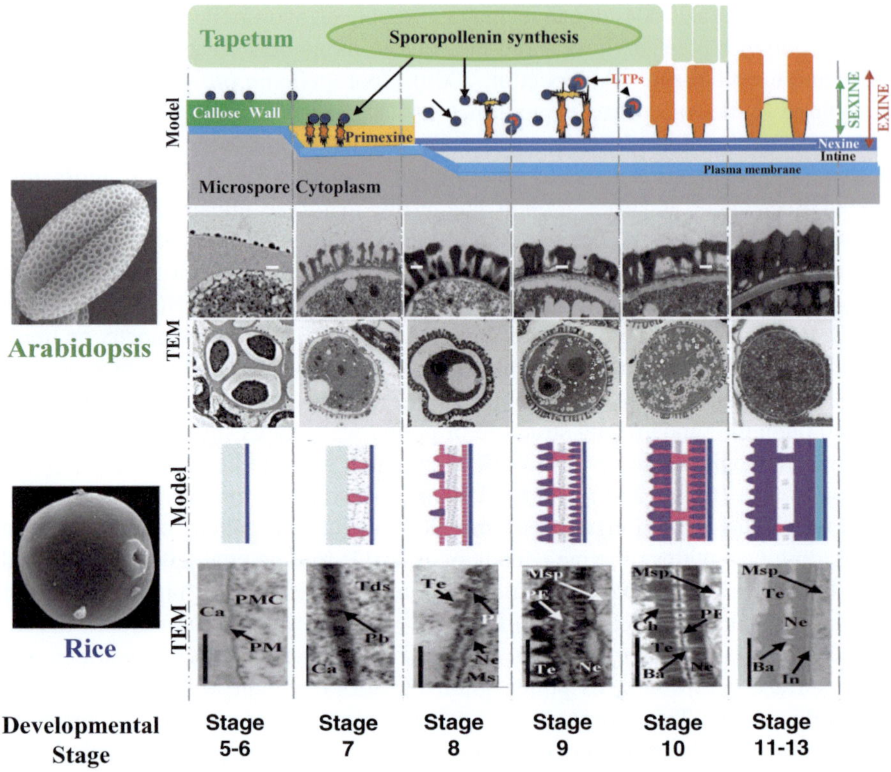

Fig. 13.1 Current model of pollen wall formation in monocot rice and dicot Arabidopsis (Adopted from Ariizumi and Toriyama 2011; Xu et al. 2014a with modification). Developmental stages for rice and Arabidopsis are defined by Zhang et al. (2011) and Blackmore et al. (2007), respectively

reported to be required for callose formation or dissolution in *Arabidopsis*. Mutations of these genes causes defective pollen exine formation and patterning. It is proposed that the callose wall around the tetrad may recruit the primexine and provide a structural support for the exine formation. During the meiosis, pollen mother cells develop into the tetrad, each containing four newly formed haploid microspores enclosed by callose, and the primexine with the low electron density is synthesized around each microspore, between the callose wall and the plasma membrane (Ariizumi and Toriyama 2011; Blackmore et al. 2007; Paxson-Sowders et al. 1997). Chemical staining experiments show that the primexine is a microfibrillar matrix mainly composed of cellulose components (Paxson-Sowders et al. 1997). Some reports also suggest that the primexine may serve as a template for initial sporopollenin deposition and following polymerization and patterning because of a kind of enzymes called sporopollenin acceptor particles (SAPs), which exist in the primexine and initiate the sporopollenin accumulation and the appearance of the exine pattern (Gabarayeva et al. 2009; Gabarayeva and Grigorjeva 2004) In *Arabidopsis*, there are some genes known to be involved in the primexine formation including *DEX1* (Paxson-Sowders et al. 2001), *RUPTURED POLLEN GRAIN1* (*RPG1*) (Guan et al. 2008; Sun et al. 2013),

NOEXINEFORMATION1 (*NEF1*) (Ariizumi et al. 2004) and *NO PRIMEXINE AND PLASMA MEMBRANE UNDULATION* (*NPU*) (Ariizumi et al. 2004; Chang et al. 2012; Guan et al. 2008; Paxson-Sowders et al. 2001; Sun et al. 2013). But there is no gene identified to be involved in primexine development in rice.

At stage 8 during the tetrad formation, according to the classification of rice anther development stage (Zhang et al. 2011), the probaculum is formed and deposited onto the primexine (Paxson-Sowders et al. 1997). At stage 9, when microspores are released from the tetrad, massive sporopollenin precursors putatively secreted by the tapetum are deposited onto the primexine to form a very thin tectum (sexine) (Bedinger 1992; Li and Zhang 2010). Meanwhile, the formation of nexine is seen on the surface of microspores. With the development of microspores, the sporopollenin gradually accumulates to thicken and consolidate the exine structure, a two-layer high density structure with low electron density intervals or channels across the exine is mostly completed by the late stage 10, when the microspores become vacuolated (Li and Zhang 2010). Subsequently, during the first mitosis each microspore generates one generative cell and one vegetative cell, and the microspore starts the synthesis of intine below the exine (Paxson-Sowders et al. 1997; Owen and Makaroff 1995). Finally, at stage 13, after the second mitosis of microspores to form mature pollen grains with two sperm cells and one vegetative nucleus, the pollen exine shape is almost established by the accumulation of sporopollenin compositions and pollen coat deposition (Li and Zhang 2010; Owen and Makaroff 1995). It is notably that different from the exine development, which appears to be actively produced and deposited by intact tapetal cells at the early stages of pollen development, the components of lipidic pollen coat are produced and deposited by degenerating tapetal cells at the late stages of pollen development (Murgia et al. 1991).

Additionally, all abovementioned information associated with the developmental course of pollen wall and anther wall layers are almost obtained from the common light and electron microscopy in which the chemical fixation is used. Increasing evidence demonstrated that the tapetum is a transitory tissue, which is very sensitive to chemical fixation, so do those early emerging ultra-structures of developing pollens. In a recent study, cryo-fixation combined with transmission electron microscopy were employed to examine the sequence of developmental events in anther and pollen development in *Arabidopsis*, and revealed novel information on middle layer, locule fluid and the morphology of tapetal cells (Quilichini et al. 2014). The new features of tapetal cells observed by Quilichini et al. (2014) included: tapetal cell lacks a cell wall as early as the tetrad stage; tapetal cell ultrastructure does not show the sign of active endomembrane secretion system; tapetal cell is distinguishable at the tricellular pollen stage; tapetosomes and elaioplastes, two specific organelles within tapetal cells during the late stage are the major contributors to pollen coat development (Quilichini et al. 2014). In addition, unlike to previous reports from chemical fixation, the middle layer cell persists after the release of pollen and into the late stage of uninucleate microscope development. Furthermore, this analysis revealed a clear and dense presence of locule fluid from the tetrad stage to the early bicellular pollen stage, after which it becomes less dense or absent when locule is filled with tricellular pollen grains, indicating its involvement in the traffic of sporopollenin from tapetal cells to microspores, because at this stage, the locule fluid is abundant with various lipid transfer proteins (LTPs) (Huang et al. 2013a; Zhang et al. 2010). These morphological features in developing anthers of plants rather than *Arabidopsis* need to be performed to update current knowledge regarding pollen wall formation.

Tapetum: The Supportive Tissue for Exine Formation

The anther, the male organ of flowering plants, consists of four somatic cell layers, the epidermis, endothecium, middle layer, and tapetum from the surface to interior, surrounding the developing microspores inside. Pollen development relies heavily on the surrounding tissues of the anther wall layers, especially the tapetum, which has the most intimate contact with the microspores and plays an secretory role by supporting developing microspores with enzymes, nutrients, metabolites, sporopollenin precursors and pollen coat components during the exine and pollen coat formation (Blackmore et al. 2007; Pacini and Hesse 1984; Shivanna et al. 1997). The tapetal cytoplasm is usually filled with ribosomes, plastids, mitochondria, Golgi bodies, and both rough and smooth endoplasmic reticulum (ER), and plastids and ER are the places for fatty acid synthesis and modification [i.e., unsaturation, hydroxylation, fatty acid chain elongation, or diversification of aliphatic compounds of very long-chain fatty acid (VLCFAs)] (Yang et al. 2014a, b), respectively, suggesting the essential function of tapetal cells in lipid biogenesis.

The anther primordium in angiosperms is composed initially of three layers, L1, L2, and L3 from the outer to the inner (Poethig 1987), with the L1 developing into the epidermis, and the L3 forming the connective cells and vascular tissues (Ma 2005; Zhang et al. 2011). At the same time, the L2 layer differentiates into the peripheral L2-derived (L2-d) cells dividing and forming the endothecium, the middle layer and the tapetum, and the central L2-d cells developing into the sporogenous cells (Kelliher and Walbot 2011, 2012; Zhang and Yang 2014; Zhang et al. 2011).

When microspore mother cells (MMCs) are formed, the tapetal cell with a single nucleus keep integrity has clear boundaries from MMCs. When entering meiosis stage, the tapetal cells undergo a series of changes to obtain the unique cytological features consisted with the biological function, such as exuberantly synthetized proteins and RNAs, high levels of DNA, and abundant mitochondria, plastids, ERs and other organelles, these characteristics are different from those of other sporophytic cells (Gunning and Steer 1996). At the end of meiosis I, the tapetum cytoplasm becomes condensed and the nuclear DNA becomes fragmented, indicating the tapetum degradation initiated by programmed cell death (PCD) (Li et al. 2006, 2011; Varnier et al. 2005). During this stage, the condensed cytoplasm of tapetal cells contains abundant mitochondria, ribosomes, varisized vacuoles, ERs, and plastids with high electron density substance. After meiosis II, when the tetrads are formed, the tapetum becomes more condensed and vacuolated, but there are still abundant organelles, of which the highly developed ER is the most obvious (Li et al. 2006).

The metabolism in the tapetum is most active during the tetrad stage of anther development, and the β-1,3-glucanase produced by the tapetum breaks down the callose wall to release the microspores into the locule (Ariizumi and Toriyama 2011; Verma and Hong 2001). With the degradation of tapetum cells after microspores released from tetrads, the tapetum-synthesized nutrition and structural matters, such as sporopollenin precursors, flavonoids and proteins, are provided for pollen development including exine formation. When the trinucleated pollen grains are formed, the tapetum is almost completely degraded, and the proteins and lipids that compose the tryphine as the last materials supplied by the tapetum are deposited on the pollen exine (Edlund et al. 2004; Piffanelli et al. 1998). Throughout the whole process of pollen development, the tapetum, as a source for synthesis, storage and transport of various nutrients and structural compositions, is an important layer of anther wall, and its abnormal structure and function is closely related to pollen abortion (Ariizumi and Toriyama 2011; Wilson and Zhang 2009; Zhang et al. 2011).

Increasing evidence indicates that exine synthesis and patterning rely on the tapetal PCD, which has been shown to be controlled by regulators such as basic helix-loop-helix (bHLH) transcription factors, MYB transcription factors and PHD-finger proteins (Aya et al. 2009; Fu et al. 2014; Ji et al. 2013; Li et al. 2006; Wilson and Zhang 2009; Xu et al. 2010, 2014a; Yang et al. 2007; Zhang et al. 2011). Loss of function of these regulators frequently causes defective tapetal PCD and abnormal exine formation. Emerging evidence suggests that many other genetic factors such as pollen specific mRNA-like non-coding RNA gene (Song et al. 2013) and cysteine protease encoding genes (Lee et al. 2004; Yang et al. 2014b; Zhang et al. 2014) also play important roles in tapetum and microspore development in plants. These findings suggest the possible close link between tapetal PCD and nutritive support for exine biosynthesis. However, it is not clear whether these two processes function in the same pathway or in parallel.

Biosynthesis of Exine

As the major component of pollen exine, the chemical compositions of sporopollenin are still elusive due to its highly insoluble and resistant to various degradations (Scott 1994). However, with the application of a series of elaborate physical and chemical techniques, such as nuclear magnetic resonance (NMR) spectroscopy (Bubert et al. 2002) and thiocarbamate herbicides treatment (Wilmesmeier and Wiermann 1995), our knowledge about the chemical compounds in sporopollenin is substantially updated. Increasing evidence indicates that the sporopollenin mainly consists of complex aliphatic monomers including VLCFAs and their polyhydroxylated derivatives, and phenolic compounds, which means the lipid metabolism is critical for sporopollenin biosynthesis (Kim and Douglas 2013; Scott 1994; Piffanelli et al. 1998). Comparison with pollen transcriptomes from lilly (Lang et al. 2015), *Arabidopsis* (Honys and Twell 2003), soybean (Haerizadeh et al. 2009), rice (Wei et al. 2010) and tobacco (Hafidh et al. 2012) clearly shows that similar protein/transcript categories including lipid metabolism and transport are highly expressed as compared with vegetative tissues, indicating the roles of lipid metabolism in pollen development (Fig. 13.2).

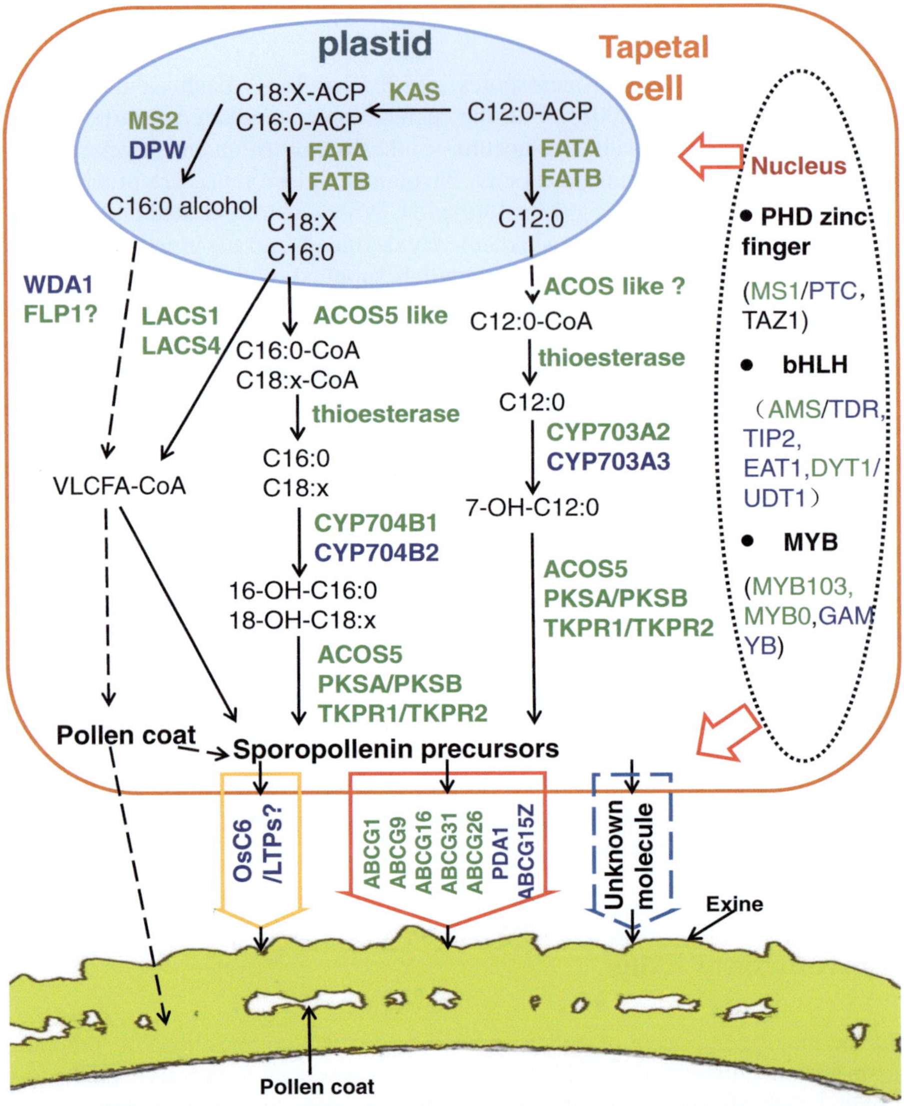

Fig. 13.2 The metabolic network of pollen wall development in monocot rice and dicot Arabidopsis (Adopted from Yang et al. 2014a with modification). In the endoplasmic reticulum (ER) of tapetal cells, plastid produced fatty acids (C12, C16, and C18) conversed from their corresponding CoAs by thioesterase are hydroxylated by CYP703/704 proteins, and resulting hydroylated fatty acids are further converted to phenolic compounds via pathway involving ACOS/like, PKS/like and TKPR/like proteins. In ER, VLCFA-CoA pool contributes significantly to aliphatic components of spropopollenin precursors and perhaps pollen coat compositions as well. VLCFA-CoA pool is produced either from fatty acids produced in plastid by the joint activity of LACS proteins or from the likely condensation of fatty alcohols, generated by the plastidic fatty acid reductase activity of MS2/DPW protein, by WDA1 like proteins. The synthesized spropopollenin precursors are transported by several ABCG transporters, LTPs and unknown molecules, and the transportation of pollen coat components remains unknown. There are at least three transcription factor families that are involved in the regulation of exine development and pollen coat formation individually or interactively. *FATA/B* fatty acyl thioesterase A/B, *KAS* ketoacyl-ACP synthase, *VLCFA* very long chain fatty acid. Genes in *blue* and *green* color are ortholog ones from rice and Arabidopsis, respectively

Fatty Acid Synthesis

Biochemical and genetic investigations have recently revealed a number of enzymes catalyzing the metabolism of fatty acids and their derivatives, required for sporopollenin biosynthesis (Aarts et al. 1997; Ariizumi and Toriyama 2011; Chang et al. 2012; Chen et al. 2011b; de Azevedo Souza et al. 2009; Dobritsa et al. 2009, 2010; Grienenberger et al. 2010; Jung et al. 2006; Morant et al. 2007; Shi et al. 2011; Tang et al. 2009). In heterotrophic eukaryotes, the synthesis of fatty acids usually occurs in the cytosol, whereas plants de novo fatty acids are produced in plastids (Li-Beisson et al. 2010, 2013; Ohlrogge et al. 1979), and the elongation (>C18) and modification (i.e., unsaturation, hydroxylation) of fatty acids usually occur in the ER or cytosol (Kunst and Samuels 2003; Li-Beisson et al. 2013).

De Novo *Fatty Acid Synthesis in Plastids*

Simply, the fatty acid synthesis is a complex cyclic process, and during each cycle, there are four reactions: condensation, reduction, dehydration, and reduction, catalyzed by fatty acid synthase (FAS), a multi-subunit complex composed of monofunctional enzymes (Brown et al. 2006; Li-Beisson et al. 2013). The plastidial acetyl-coenzyme A (CoA), a two-carbon molecule rapidly generated by the plastidial pyruvate dehydrogenase complex (PDHC) (Johnston et al. 1997), is used as a building block for the process. During the fatty acid synthesis, acetyl-CoA is the starting unit, and malonyl-ACP, catalyzed from acetyl-CoA (Konishi et al. 1996), is used as a carbon donor in each cycle of elongation (Li-Beisson et al. 2013). The condensation reaction is catalyzed by 3-ketoacyl-ACP synthases (KAS) to mediate the combination of two-carbon units provided by malonyl-ACP and the acetyl-CoA (subsequently acyl-ACP acceptors) and release one CO_2 (Li-Beisson et al. 2013). There are three KAS isoforms identifies: the KAS I catalyzes the subsequent condensations (up to 16:0-ACP), the KAS II catalyzes the final condensation of 16:0-ACP to 18:0-ACP, and the KAS III catalyzes the condensation of malonyl-ACP and acetyl-CoA at the first step of the fatty acid synthesis (Pidkowich et al. 2007). To complete a whole cycle of fatty acid elongation, additional two reductases and a dehydrase are necessary. At last, fatty acids or alcohols with the carbon chain length up to C18 are exported from the plastids to the ER for further modification or elongation and compose the sporopollenin precursors.

MALE STERILITY2 (*MS2*) is the first gene identified to participate in the exine development, as evidenced by the aborted pollen grains without exine in the *ms2* mutant (Aarts et al. 1997). Rice *DEFECTIVE POLLEN WALL* (*DPW*), the ortholog gene of *MS2*, is capable of complementing the *ms2* mutant. *dpw* mutant displays completely male sterility and defective pollen exine (Shi et al. 2011), suggesting that DPW/MS2 represents a conserved pathway in pollen wall formation. Subsequent studies revealed that DPW and MS2, expressed in both tapetal cells and microspores, encode a class of plastid-localized novel fatty acid reductases that catalyze fatty acyl carrier proteins (ACPs) into fatty alcohols, with a strong preference for C16:0-ACP, uncovering a conserved step in fatty alcohols biosynthesis in the plastid for anther cuticle and pollen sporopollenin biosynthesis in monocots and dicots (Chen et al. 2011b; Shi et al. 2011). Rice Wax-Deficient Anther1 (WDA1) (Jung et al. 2006) and its ortholog *FACELESS POLLEN-1* (*FLP1*) (Ariizumi et al. 2003) in *Arabidopsis* are both the orthologs of *ECERIFERUM1* (*CER1*) (Bourdenx et al. 2011) in *Arabidopsis*, which was predicted to encode an enzyme involved in the decarbonylation pathway of alkane biosynthesis. The mutation of both *WDA1* and *FLP1* caused the abnormal sporopollenin deposition, indicating their possible roles in the sporopollenin biosynthesis. Further gene expression and biochemical analysis in *wda1* showed that WDA1 probably participate in the biosynthesis of VLCFAs and affect the exine formation directly or indirectly (Bourdenx et al. 2011).

Among the cytochrome P450 family, members of CYP703A and CYP704B subfamilies play essential roles in hydroxylating fatty acids in ER for sporopollenin biosynthesis (Dobritsa et al. 2009, 2010; Li et al. 2010; Li and Zhang 2010; Morant et al. 2007; Yang et al. 2014a; Yi et al. 2010). CYP703As catalyze the in-chain hydroxylation of fatty acids. The *cyp703a2* mutants can produce partial fertile pollen grains with abnormally developed exine, on which sporopollenin deposited. The heterologous CYP703A2 protein catalyzes in-chain hydroxylase of fatty acids with the chain length from C10 to C16, and the productions are thought to be sporopollenin precursors (Morant et al. 2007). By contrast, the *cyp703a3* mutant is completely male sterile with defective exine formation (Aya et al. 2009; Yang et al. 2014a). *In vitro* enzyme activity assay showed that CYP703A3 catalyzes the in-chain hydroxylation of lauric acid specifically, to produce 7OH-C12 fatty acid (Yang et al. 2014a). While CYP703As encode in-chain hydroxylases with substrates preference for lauric acid, CYP704Bs catalyze the ω-hydroxylation of long-chain fatty acids (Yang et al. 2014a; Dobritsa et al. 2009, 2010; Li et al. 2010; Morant et al. 2007). *Arabidopsis* CYP704B1 (Dobritsa et al. 2009) and rice CYP704B2 (Li et al. 2010) have similar biochemical functions, which can catalyze the ω-hydroxylation of C16 and C18 fatty acids to produce the important precursors involved in sporopollenin synthesis. The *cyp704b1* and *cyp704b2* mutants produce aborted pollens with abnormal exine (Dobritsa et al. 2009, 2010; Li et al. 2010). The mutants of *BnMS1* and *BnMS2*, the orthologs of *CYP704B1* in *Brassica napus*, display delayed tapetum PCD and defective exine without sporopollenin accumulation (Yi et al. 2010). The hydroxylated fatty acids by CYP703A and CYP704B can be further catalyzed by cytoplasm-localized fatty acyl-CoA synthetase (ACOS) to generate CoA esters for sporopollenin synthesis (de Azevedo Souza et al. 2009).

Synthesis of Phenolic Compounds

The other important sporopollenin precursors are the phenolic compounds derived from the phenylpropanoid pathway, involving several key enzymes encoded by the tapetum-expressed Acyl-CoA Synthetase5 (ACOS5) (de Azevedo Souza et al.

2009), LAP6 (LESS ADHESIVE POLLEN)/PKSA (POLYKETIDE SYNTHASE A) and LAP5/PKSB (Dobritsa et al. 2010; Grienenberger et al. 2010), and TETRAKETIDE α-PYRONE REDUCTASE1 (TKPR1) and TKPR2 (Grienenberger et al. 2010) in *Arabidopsis*. ACOS5 functions *in vitro* as a fatty acyl-CoA synthetase taking medium- to long-chain fatty acids including hydroxylated fatty acids as the substrates. The fatty acyl-CoA esters are considered as a central intermediate during the sporopollenin precursor synthesis, and the CoA esters are likely essential for the fatty acids traveling across the membrane (de Azevedo Souza et al. 2009). Loss of function mutant of ACOS5 displays defective pollen exine without sporopollenin (de Azevedo Souza et al. 2009). The products of ACOS5, fatty acyl-CoA esters, are further utilized as the substrates of ER-localized LAP6/PKSA and LAP5/PKSB to yield tri- and tetra-ketide alpha-pyrones required for sporopollenin synthesis (de Azevedo Souza et al. 2009; Dobritsa et al. 2010; Kim et al. 2010). The *pksa/pksb* double mutant lacks pollen exine and sporopollenin deposition (Kim et al. 2010). Next, the reductase encoded by *TKPR1* and *TKPR2* takes tri- and tetra-ketide α-pyrones generated by PKSA/B as the substrates to produce the hydroxylated alpha-pyrone for sporopollenin synthesis (Grienenberger et al. 2010). In accordance with the direct upstream and downstream relations in the common sporopollenin synthesis pathway, *ACOS5*, *PKSA/B* and *TKPR1* (Grienenberger et al. 2010) are tightly co-expressed as predicted, which suggests the possibility that these enzymes function as a metabolon for sporopollenin biosynthesis. Interestingly, it was recently reported that the ACOS5, PKSA/B and TKPR1 are all localized on the ER of the tapetal cells and can interact together *in vivo*, which strongly supports the existence of sporopollenin metabolon (Lallemand et al. 2013). Notably, none of the orthologs of these *Arabidopsis* synthetic genes of phenolic compounds has been reported in rice.

As compared with exine biosynthesis, studies on pollen coat biosynthesis are much less. It is well known that in *Arabidopsis*, the biosynthesis of pollen coat is derived from the tapetum degenerative debris, which shares the same fatty acid CoA pools as those of VLCFAs, in which long-chain acyl-CoA synthetases (LACS), such as LACS1 and LACS4, play important roles. It is reported that the biosynthesis of pollen coat depends on the combined activity of both LACS1 and LACS4 proteins (Jessen et al. 2011). It is not clear whether other genes affecting long chain fatty acid CoA pools in plants affect pollen coat development in plants or not.

Potential Carriers for Sporopollenin Precursors

How tapetum derived chemicals are exported into the locule for the assembly of sporopollenin is a mysterious question for biologists. Up to date, ABC transporters, LTPs and other molecules (Zhang and Li 2014) have been proved to be responsible for cargo transport of exine constituents (Fig. 13.2).

ABC Transporters

Plant ABC transporters have been shown to be involved in a wide range of cellular activities, such as hormone homeostasis, detoxification, antibiotics resistance and secondary metabolism (Kang et al. 2011). Among these ABC transporters, some G subfamily members such as *Arabidopsis ABCG26* (*WBC27*), ABCG9 and ABCG31, and ABCG1 and ABCG16 are reported to be responsible for transport of sporopollenin precursors from tapetal cells into microspore surface for exine development. Knockdown of either one of those ABCG proteins leads to the aborted exine formation (Choi et al. 2011, 2014; Dou et al. 2011; Kuromori et al. 2010; Quilichini et al. 2010; Xu et al. 2010; Yadav et al. 2014). These findings suggest that the ABC transporter plays a key role in exine formation, likely secreting the sporopollenin precursors from the tapetum to developing microspores.

Recently, rice ABCG15 (also called *Post-meiotic Deficient Anther 1*, *PDA1*), an ortholog of Arabidopsis *ABCG26*, has been shown to be required for exine development (Hu et al. 2010; Niu et al. 2013a; Qin et al. 2013; Wu et al. 2014; Zhu et al. 2013). Mutation of the tapetum-expressed *ABCG15/PDA1* leads to aborted microscope without exine formation and vanished orbicules on tapetal cells (Niu et al. 2013a; Qin et al. 2013; Wu et al. 2014; Zhu et al. 2013). Besides, chemical analysis data shows that *ABCG15*/PDA1-mdeidated transport of sporepollenin precursors in turn affects the synthesis of lipidic components. Consistently, the expression of two pollen exine biosynthetic genes *CYP704B2* and *CYP703A3* was decreased in *pda1* anthers (Zhu et al. 2013). These findings suggest that the tapetum-expressed ABC transporters play a conserved role in secreting lipidic sporopollenin precursors from the tapetum to developing microspores. However, the biochemical evidence on how ABC transporters translocate sporopollenin precursors and what are the exact substrates remain to be investigated. Future biochemical function assay of ABCG26/ABCG15 may help us in better understanding of the roles of ABC transporters in plant pollen exine development.

Lipid Transfer Proteins

Beside the possible role of transmembrane ABCG transporters in actively delivering the lipid molecules across the membranes, other lipid transporters such as low molecular weight (LTPs) have also been shown to be involved in lipid trafficking (Huang et al. 2013a, b; Hu et al. 2010; Yeats and Rose 2008; Kader 1996). LTPs were termed as they have the capacity to transfer phospholipids and fatty acids between membranes *in vitro* (Kader 1996). Expression analysis showed that some tapetal PCD regulators affect the expression of LTPs (Aya et al. 2009; Ariizumi and Toriyama 2011; Brown et al. 2006; Jung et al. 2005), implying that LTPs regulate the pollen development. Nevertheless, CaMF2, an anther-specific LTP gene, affects pollen development in *Capsicum annuum* L (Chen et al. 2011a). *OsC6*, a rice LTP, was shown to function as lipidic transporter from tapetal cells to anther wall layers and microspore surface (Zhang et al. 2010). OsC6 has the typical eight-Cys motif, and its transcription is highly detectable in tepetal cell. However, the localization of OsC6 protein is seen in tapetal cells, anther locule, anther epidermis, as well as the extracellular space between the tapetum and the middle layer, indicating that the tapetum expressed OsC6 can be secreted among anther cells. In addition, the recombinant OsC6 protein has the bind ability of fatty acid molecules (Zhang et al. 2010), suggesting that OsC6 is involved in distribution of lipidic molecules for male reproduction. Furthermore, knock-down of *OsC6* caused aborted exine and orbicule development in rice. This work demonstrates the key role of LTPs in secreting lipidic molecule from the tapetum for exine development. Recently, Huang et al. (2013a) also demonstrated that the tapetum-expressed type III LTP plays a role for the lipid molecular trafficking from tapetum into the pollen exine in *Arabidopsis*, confirming the novel vesicular trafficking mechanism for pollen exine development (Huang et al. 2013a). However, the substrates and molecular mechanisms of LTPs remain to be analyzed in the future.

In addition to ABCG transporters and LTPs, multidrug and toxic efflux (MATE) transporters are also involved in trafficking of sporopollenin precursors across tapetal cells for exine development in plants (reviewed by Zhang and Li 2014).

Regulation Network

Pollen wall development is one of the most important events of plant organ development, involving hundreds of genes and proteins (Honys and Twell 2003, 2004; Xu et al. 2010, 2014a). The transcription factor is the regulatory point of many genes. In higher plants, a number of conserved transcription factors have been shown to be associated with tapetal function and pollen development. So far the known transcription factors involved in pollen development were fallen into following families, forming specific and crosstalk regulatory networks governing the microspore development in plants, which will continue to be the hotpot of this research area (Fig. 13.2).

MALE STERILITY 1 (*MS1*), is a tapetum-expressed gene encoding a plant PHD-zinc finger transcription factor (Ito et al. 2007; Ito and Shinozaki 2002; Wilson et al. 2001). Loss of function of *MS1* results in severe vacuolation of tapetal cells, exine structural defects and microspore degradation. By analyzing the microarray data of *MS1*, over 260 genes, the majority of which are required for tapetum development and exine formation like the genes encoding cysteine proteases and *LTP*s, are directly or indirectly regulated by *MS1* (Ito et al. 2007; Ito and Shinozaki 2002; Wilson et al. 2001). *PERSISTENT TAPETAL CELL1* (*PTC1*), an ortholog of *MS1* in rice, encodes a PHD-zinc finger transcription factor that is specifically expressed in tapetum and microspores (Li et al. 2011). The *ptc1* mutant exhibited many similar phenotypes to those of *ms1*, such as lack of DNA fragmentation in tapetal cells, delayed tapetum PCD and pollen exine defect. In addition, *ptc1* mutants also showed uncontrolled proliferation of tapetal cells and abnormal orbicules (Li et al. 2011). It is reported that mutation of *PTC1* significantly alters the expression of 2,417 genes involved mainly in tapetum development and exine formation (Li et al. 2011). In addition, mutation in *TAPETUM DEVELOPMENT ZINC FINGER PROTEIN1* (*TAZ1*), an anther-specific TFIIIA-type zinc finger transcription factor in petunia, leads to defective in exine formation due to precocious degeneration of tapetum (Kapoor et al. 2002), which is similar to *cp51* (Yang et al. 2014b).

ABORTED MICROSPORE (*AMS*) in Arabidopsis, encoding a bHLH transcription factor, affects the tapetum development and PCD (Sorensen et al. 2003). The *ams* mutants display enlarged tapetal cells and aborted microspore development. In-depth analysis on *AMS* revealed that *AMS* plays a central role in tapetal PCD and pollen exine formation by directly regulating target genes involved in different developmental events of pollen formation, those affected developmental processes include the separation of the microspore mother cell, callose dissociation, and sporopollenin precursor biosynthesis and secretion at the transcription level, such as *ABCG26* (Xu et al. 2010, 2014a). *Tapetum Degeneration Retardation* (*TDR*/*bHLH5*) (Li et al. 2006), the ortholog of *AMS* in rice, is preferentially expressed in the tapetum from the meiosis stage to the young microspore stage (Li et al. 2006), which functions as a key regulator in tapetum PCD and pollen exine development (Li et al. 2006; Zhang et al. 2008). The pollen wall was severely defective in the *tdr* mutant. Similarly, the expression of a group of genes encoding putative enzymes involved in lipid metabolism is altered in *tdr*, including *OsC6*, *CYP703A3*, both are confirmed to be directly targeted by *TDR* (Li et al. 2006; Yang et al. 2014a; Zhang et al. 2010). In addition, TDR can affect the pollen development by regulating other transcription factors. Rice *ETERNAL TAPETUM 1* (*EAT1*), another bHLH transcription factor, is also a direct target gene of *TDR* (Niu et al. 2013b). *eat1* mutant showed delayed tapetum PCD and abnormal pollen development. The third rice bHLH transcription factor, *TDR INTERACTING PROTEIN2* (*TIP2*), functions as an upstream regulator of *TDR* and *EAT1*, and determines the differentiation of the inner three layers of anther wall and tapetum PCD (Fu et al. 2014). Recently, it is reported that TDR controls tapetal cell development and pollen formation via triggering the possible ADF-mediated proteolysis pathway via its target gene *OsADF*, an anther development F-Box protein (Li et al. 2015). Those data demonstrate that TDR play important roles in the regulation of tapetal and pollen development in rice.

Moreover, *AMS* directly regulates the expression of *ABCG26* via binding to *ABCG26* promoter and coordinates the synthesis of aliphatic molecules in the tapetum (Xu et al. 2010). *Delayed Tapetum Degeneration* (DTD), the rice bHLH 141 transcription factor (EAT1), functions coordinately with *TDR* to regulate lipidic orbicules and exine synthesis, through *OsC6* and *OsC4*, two LTPs, to affect pollen development (Ji et al. 2013; Niu et al. 2013b). In addition, the *bHLH142* coordinates with *TDR* to modulate the expression of *EAT1* and regulate pollen development in rice (Ko et al. 2014). Recently, it is reported that *MS10*[35] encodes a tomato ortholog of *AtDYT1* and *OsUDT1* (*bHLH164*) (Jung et al. 2005), positively regulates genes for lipid metabolism, cell wall modification/degradation, pollen wall/coat proteins, and transporters. There was neither orbicules nor normal sporopollenin structures in *ms10*^{35} and the expression of sporopollenin biosynthesis-related genes, such as Solyc12g010920.1.1 (LCFA reductases) and Solyc04g081780.2.1 (lipase), was also reduced in *ms10*^{35} (Jeong et al. 2014).

AtMYB103, a R2R3 MYB family member, is important for tapetum and pollen development. Cytological observation showed that the mutation of AtMYB103 caused the defective tapetum development and blocked callose degradation, further leading to aborted microspores without exine. RT-PCR analysis showed that *AtMYB103* regulates the *A6* gene relative to the callose degradation (Zhang et al. 2007). Rice *GAMYB* plays an important role in pollen development (Aya et al. 2009; Kaneko et al. 2004). Loss-of-function of *GAMYB* results in defective tapetum PCD and exine development, and abnormal orbicules (Kaneko et al. 2004). It is reported that cytochrome P450 hydroxylase *CYP703A3* was directly regulated by *GAMYB*, indicating an essential role of *GAMYB* in sporopollenin biosynthesis (Aya et al. 2009). In addition, MYB80, a known MYB protein required for the regulation of tapetal programmed cell death and pollen development in Arabidopsis (Phan et al. 2011), is conserved in its structure and function in all tested monocot and dicot species (Xu et al. 2014b).

Notably, interactions between different group transcription factors play essential roles in the regulation of lipid metabolism during pollen development, In Arabidopsis, a recent study shows that DYT1, a putative bHLH transcription factor, plays a role during the synthesis and transport of pollen wall lipidic materials via TDF1, a putative R2R3 MYB transcription factor, by forming a heterodimer to bind to the E-box motif of the TDF1 promoter (Gu et al. 2014). Further investigation into the correlation network of different regulatory transcription factors governing the lipid metabolism during pollen development would expend greatly our understanding of the molecular interactive network for gametogenesis and reproductive development.

Evolutionary

The colonization of land by plants was a highly significant event in Earth's history (Beerling 2007), during which key adaptations include rooting structures, conducting tissues, cuticle, stomata, and sex organs such as gametangia and spores/pollen (Menand et al. 2007), play important roles in the successful colonization. Another important innovation was the development of a durable spore wall structure capable of withstanding physical abrasion, desiccation and UV-B radiation environment encountered on land (Wellman 2004). The major component of the spore/pollen wall proposed to be of primary importance in enabling resistance to the conditions described above is the highly resistant biopolymer sporopollenin (Cronk and Cronk 2009; Ito et al. 2007). It seems reasonable to hypothesize that colonization of the land by plants was not possible prior to the evolution of the sporopollenin spore wall, and this adaptation is considered to be a synapomorphy of the embryophytes. Additionally, spore walls are not present in the green algae (Wellman 2004). However, the production of sporopollenin is highly likely to be pre-adaptive as it is present in a number of different algal groups. Fossil green algae dating back to the Devonian period have been shown to contain sporopollenin (Wall 1962), and there are reports that sporopollenin also occurs in fungi (Shaw 1970), indicating an origin predating the appearance of plants.

Phylogenetic studies and fossil evidence have shown that the most basal living land plants are the paraphyletic 'bryophytes' (Kenrick and Crane 1997). They comprise the liverworts, mosses and hornworts. The moss Physcomitrella patens is the first 'bryophyte' genome to be sequenced. This genome, through comparisons with angiosperm genomes, is proved to be a valuable tool in experimental studies that attempt to reconstruct genome evolution during the colonization of land (Quatrano et al. 2007; Rensing et al. 2008). Bioinformatic studies have suggested that genes implicated in pollen wall development in angiosperms are also present in moss and lycopsids, and may therefore be involved in spore wall development in basal plants (Wallace et al. 2011). This suggests that the molecular genetics of sporopollenin biosynthesis are highly conserved among all land plants, despite the large morphological and functional differences between spores and pollens. Comparative phylogenetic and genomic analyses support this hypothesis, as putative orthologs of ACOS5, PKSA, PKSB, CYP703A2 and CYP704B1 are broadly distributed in flowering plants, and in the moss Physcomitrella patens (Colpitts et al. 2011; de Azevedo Souza et al. 2009; Dobritsa et al. 2010; Grienenberger et al. 2010; Kim et al. 2010). These genes are apparently absent from the green algal lineage, supporting a key role for acquisition of these genes in the progression of plant life onto land (Colpitts et al. 2011; de Azevedo Souza et al. 2009; Dobritsa et al. 2010; Grienenberger et al. 2010; Kim et al. 2010; Morant et al. 2007). Recently, reverse genetic analysis on moss MS2 ortholog gene revealed a core component of the biochemical and developmental pathway required for angiosperm pollen wall development was recruited early in land plant evolution but the continued increase in pollen wall complexity observed in angiosperms has been accompanied by divergence in MS2 gene function. Knock out of moss *MS2* homolog results in defective spore wall and extremely compromised germination. Nevertheless, the moss *MS2* gene could not rescue the *Arabidopsis ms2* phenotype (Wallace et al. 2015). The hypothesis that a highly conserved biochemical pathway leading to sporopollenin biosynthesis exists among land plants has been further supported by genetic and biochemical approaches used to study a variety of plant species (Quilichini et al. 2014). In particular, *Physcomitrella patens* ASCL encodes an enzyme with *in vitro* preference for hydroxy fatty acyl-CoA esters that is capable of hydroxyalkylpyrone synthase activity, suggesting that PpASCL is a functional ortholog of Arabidopsis PKSA and that the pathway to sporopollenin may be conserved among land plants (Colpitts et al. 2011). The similar functions and enzymatic activities of Arabidopsis MS2, CYP704B1, CYP703A2 and there rice orthologs DPW, CYP704B2 and CYP703A3 support the conservation of sporopollenin synthesis among monocotyledonous and dicotyledonous plants (Aya et al. 2009; Chen et al. 2011b; de Azevedo Souza et al. 2009; Dobritsa et al. 2009; Morant et al. 2007; Shi et al. 2011; Yang et al. 2014a).

Perspective

Lipidic and phenolic molecules secreted from tapetum cell are considered to be the key compounds determining the framework for pollen exine. Even though this process is essential for reproduction success in flowering plants, many questions remain to be addressed. For instance, how about the link between tapetal PCD and the supply of sporopollen precursors? How about the signaling between the microspore and tapetal cells for coordinating the developmental programs, and how about the complex transcriptional regulation of tapetal cell PCD and pollen development? Furthermore, during the microspore development, the microspore always attaches to the inner side of tapetal cells, whether the liquid in the anther locule plays a short-distance transport role for exine formation? Moreover, the substrates of different transport machineries/molecules during exine formation remain to be elucidated. As the advance of the tools for cell biology, biochemistry, in-depth understandings will be revealed on the mechanism underlying the synthesis and transport of sporopollenin precursors during plant exine formation.

In addition, tapetal cell layer is a key player in forming the highly sculptured sporopollenin wall of the exine. However, the dynamic cellular structures of tapetum and developing pollen walls are not fully understood due to the limitation of current microscopy examination. It is reported that those fine structures are sensitive to the chemical fixation (Ariizumi et al. 2003). Combination with the advance in living scanning technology and tapetal cell-specific expression system would facilitate our fully understanding of the dynamic molecular and ultra-structural changes of this specific cell layer occurring along the pollen development in general and of the role of lipids in pollen development in particular.

Finally, although pollen development relies heavily on the tapetum, other cell layers of the anther, namely middle layer, endothecium and epidermal cell layer, may also play important roles in pollen development, particularly the epidermis. Mutation in *WDA1*, an epidermal cell specific expressed gene in rice, causes defective exine development, indicating that *WDA1* might affect pollen development via affecting biosynthesis of lipidic components of sporopollenin.

References

Aarts M, Hodge R, Kalantidis K et al (1997) The Arabidopsis MALE STERILITY 2 protein shares similarity with reductases in elongation/condensation complexes. Plant J 12(3):615–623

Ahlers F, Bubert H, Steuernagel S et al (2000) The nature of oxygen in sporopollenin from the pollen of Typha angustifolia L. Z Naturforsch C 55(3–4):129–136

Ariizumi T, Toriyama K (2007) Pollen exine pattern formation is dependent on three major developmental processes in Arabidopsis thaliana. Int J Plant Dev Biol 1:106–115

Ariizumi T, Toriyama K (2011) Genetic regulation of sporopollenin synthesis and pollen exine development. Annu Rev Plant Biol 62:437–460

Ariizumi T, Hatakeyama K, Hinata K et al (2003) A novel male-sterile mutant of Arabidopsis thaliana, faceless pollen-1, produces pollen with a smooth surface and an acetolysis-sensitive exine. Plant Mol Biol 53(1–2):107–116

Ariizumi T, Hatakeyama K, Hinata K et al (2004) Disruption of the novel plant protein NEF1 affects lipid accumulation in the plastids of the tapetum and exine formation of pollen, resulting in male sterility in Arabidopsis thaliana. Plant J 39(2):170–181

Aya K, Ueguchi-Tanaka M, Kondo M et al (2009) Gibberellin modulates anther development in rice via the transcriptional regulation of GAMYB. Plant Cell 21(5):1453–1472

Bedinger P (1992) The remarkable biology of pollen. Plant Cell 4(8):879–887

Beerling D (2007) The emerald planet: how plants changed earth's history. Oxford University Press, New York

Blackmore S, Wortley A, Skvarla J et al (2007) Pollen wall development in flowering plants. New Phytol 174(3):483–498. doi:10.1111/j.1469-8137.2007.02060.x

Bourdenx B, Bernard A, Domergue F et al (2011) Overexpression of Arabidopsis ECERIFERUM1 promotes wax very-long-chain alkane biosynthesis and influences plant response to biotic and abiotic stresses. Plant Physiol 156(1):29–45

Brown AP, Affleck V, Fawcett T et al (2006) Tandem affinity purification tagging of fatty acid biosynthetic enzymes in Synechocystis sp. PCC6803 and Arabidopsis thaliana. J Exp Bot 57(7):1563–1571

Bubert H, Lambert J, Steuernagel S et al (2002) Continuous decomposition of sporopollenin from pollen of Typha angustifolia L. by acidic methanolysis. Z Naturforsch C 57(11–12):1035–1041

Chang HS, Zhang C, Chang YH et al (2012) No primexine and plasma membrane undulation is essential for primexine deposition and plasma membrane undulation during microsporogenesis in Arabidopsis. Plant Physiol 158(1):264–272

Chen C, Chen G, Hao X et al (2011a) CaMF2, an anther-specific lipid transfer protein (LTP) gene, affects pollen development in Capsicum annuum L. Plant Sci 181(4):439–448

Chen W, Yu XH, Zhang K et al (2011b) Male Sterile2 encodes a plastid-localized fatty acyl carrier protein reductase required for pollen exine development in Arabidopsis. Plant Physiol 157(2):842–853

Choi H, Jin JY, Choi S et al (2011) An ABCG/WBC-type ABC transporter is essential for transport of sporopollenin precursors for exine formation in developing pollen. Plant J 65(2):181–193

Choi H, Ohyama K, Kim Y-Y et al (2014) The role of Arabidopsis ABCG9 and ABCG31 ATP binding cassette transporters in pollen fitness and the deposition of steryl glycosides on the pollen coat. Plant Cell 26(1):310–324

Colpitts CC, Kim SS, Posehn SE et al (2011) PpASCL, a moss ortholog of anther-specific chalcone synthase-like enzymes, is a hydroxyalkylpyrone synthase involved in an evolutionarily conserved sporopollenin biosynthesis pathway. New Phytol 192(4):855–868

Cronk Q, Cronk Q (2009) The molecular organography of plants. Oxford University Press, Oxford

de Azevedo Souza C, Kim SS, Koch S et al (2009) A novel fatty Acyl-CoA synthetase is required for pollen development and sporopollenin biosynthesis in Arabidopsis. Plant Cell 21(2):507–525

Dobritsa AA, Shrestha J, Morant M et al (2009) CYP704B1 is a long-chain fatty acid omega-hydroxylase essential for sporopollenin synthesis in pollen of Arabidopsis. Plant Physiol 151(2):574–589

Dobritsa A, Lei Z, Nishikawa S et al (2010) LAP5 and LAP6 encode anther-specific proteins with similarity to chalcone synthase essential for pollen exine development in Arabidopsis. Plant Physiol 153(3):937–955

Dong X, Hong Z, Sivaramakrishnan M et al (2005) Callose synthase (CalS5) is required for exine formation during microgametogenesis and for pollen viability in Arabidopsis. Plant J 42(3):315–328

Dou XY, Yang KZ, Zhang Y et al (2011) WBC27, an adenosine tri-phosphate-binding cassette protein, controls pollen wall formation and patterning in Arabidopsis. J Integr Plant Biol 53(1):74–88

Edlund AF, Swanson R, Preuss D (2004) Pollen and stigma structure and function: the role of diversity in pollination. Plant Cell 16(Suppl):S84–S97

Fu Z, Yu J, Cheng X et al (2014) The rice basic helix-loop-helix transcription factor TDR INTERACTING PROTEIN2 is a central switch in early anther development. Plant Cell 26(4):1512–1524

Gabarayeva NI, Grigorjeva VV (2004) Exine development in Encephalartos altensteinii (Cycadaceae): ultrastructure, substructure and the modes of sporopollenin accumulation. Rev Palaeobot Palynol 132(3–4):175–193

Gabarayeva N, Grigorjeva V, Rowley JR et al (2009) Sporoderm development in Trevesia burckii (Araliaceae). I. Tetrad period: further evidence for the participation of self-assembly processes. Rev Palaeobot Palynol 156(1–2):211–232

Grienenberger E, Kim SS, Lallemand B et al (2010) Analysis of TETRAKETIDE alpha-PYRONE REDUCTASE function in Arabidopsis thaliana reveals a previously unknown, but conserved, biochemical pathway in sporopollenin monomer biosynthesis. Plant Cell 22(12):4067–4083

Gu J, Zhu J, Yu Y et al (2014) DYT1 directly regulates the expression of TDF1 for tapetum development and pollen wall formation in Arabidopsis. Plant J 80(6):1005–1013

Guan Y, Huang X, Zhu J et al (2008) RUPTURED POLLEN GRAIN1, a member of the MtN3/saliva gene family, is crucial for exine pattern formation and cell integrity of microspores in Arabidopsis. Plant Physiol 147(2):852–863

Gunning B, Steer M (1996) Plant cell biology: structure and function. Jones & Bartlett Learning, Burlington

Haerizadeh F, Wong C, Bhalla P et al (2009) Genomic expression profiling of mature soybean (Glycine max) pollen. BMC Plant Biol 9:25

Hafidh S, Breznenová K, Růžička P et al (2012) Comprehensive analysis of tobacco pollen transcriptome unveils common pathways in polar cell expansion and underlying heterochronic shift during spermatogenesis. BMC Plant Biol 12(1):24

Heslop-Harrison J (1979) Pollen wall as adaptive systems. Ann Mo Bot Gard 66(4):813–829

Honys D, Twell D (2003) Comparative analysis of the Arabidopsis pollen transcriptome. Plant Physiol 132(2):640–652

Honys D, Twell D (2004) Transcriptome analysis of haploid male gametophyte development in Arabidopsis. Genome Biol 5(11):R85

Hu L, Tan H, Liang W et al (2010) The Post-meiotic Deficient Anther1 (PDA1) gene is required for post-meiotic anther development in rice. J Genet Genomics 37(1):37–46

Huang M, Chen T, Huang A (2013a) Abundant type III lipid transfer proteins in Arabidopsis tapetum are secreted to the locule and become a constituent of the pollen exine. Plant Physiol 163:1218–1229

Huang X, Niu J, Sun M et al (2013b) CYCLIN-DEPENDENT KINASE G1 is associated with the spliceosome to regulate CALLOSE SYNTHASE5 splicing and pollen wall formation in Arabidopsis. Plant Cell 25(2):637–648

Ito T, Shinozaki K (2002) The MALE STERILITY1 gene of Arabidopsis, encoding a nuclear protein with a PHD-finger motif, is expressed in tapetal cells and is required for pollen maturation. Plant Cell Physiol 43(11):1285–1292

Ito T, Nagata N, Yoshiba Y et al (2007) Arabidopsis MALE STERILITY1 encodes a PHD-type transcription factor and regulates pollen and tapetum development. Plant Cell 19(11):3549–3562

Jeong H, Kang J, Zhao M et al (2014) Tomato Male sterile 10^{35} is essential for pollen development and meiosis in anthers. J Exp Bot 65(22):6693–6709, eru389

Jessen D, Olbrich A, Knüfer J et al (2011) Combined activity of LACS1 and LACS4 is required for proper pollen coat formation in Arabidopsis. Plant J 68(4):715–726

Ji C, Li H, Chen L et al (2013) A novel rice bHLH transcription factor, DTD, acts coordinately with TDR in controlling tapetum function and pollen development. Mol Plant 6(5):1715–1718

Johnston M, Luethy M, Miernyk J et al (1997) Cloning and molecular analyses of the Arabidopsis thaliana plastid pyruvate dehydrogenase subunits. Biochim Biophys Acta 1321(3):200–206

Jung K, Han M, Lee Y et al (2005) Rice Undeveloped Tapetum1 is a major regulator of early tapetum development. Plant Cell 17(10):2705–2722

Jung K, Han M, Lee Y et al (2006) Wax-deficient anther1 is involved in cuticle and wax production in rice anther walls and is required for pollen development. Plant Cell 18(11):3015–3032
Kader J (1996) Lipid-transfer proteins in plants. Ann Rev Plant Biol 47(1):627–654
Kaneko M, Inukai Y, Ueguchi-Tanaka M et al (2004) Loss-of-function mutations of the rice GAMYB gene impair α-amylase expression in aleurone and flower development. Plant Cell 16(1):33–44
Kang J, Park J, Choi H, et al (2011) Plant ABC transporters. The Arabidopsis Book. No. 9. Am Soc Plant Biol. doi:10.1199/tab.0153
Kapoor S, Kobayashi A, Takatsuji H (2002) Silencing of the tapetum specific zinc finger gene TAZ1 causes premature degeneration of tapetum and pollen abortion in Petunia. Plant Cell 14:2353–2367
Kelliher T, Walbot V (2011) Emergence and patterning of the five cell types of the Zea mays anther locule. Dev Biol 350(1):32–49
Kenrick P, Crane P (1997) The origin and early diversification of land plants. A cladistic study. Smithsonian Institute Press, Washington, DC
Kim SS, Douglas CJ (2013) Sporopollenin monomer biosynthesis in Arabidopsis. J Plant Biol 56(1):1–6
Kim S, Grienenberger E, Lallemand B et al (2010) LAP6/POLYKETIDE SYNTHASE A and LAP5/POLYKETIDE SYNTHASE B encode hydroxyalkyl alpha-pyrone synthases required for pollen development and sporopollenin biosynthesis in Arabidopsis thaliana. Plant Cell 22(12):4045–4066
Knox R, Heslop-Harrison J (1971) Pollen-wall proteins: the fate of intine-held antigens on the stigma in compatible and incompatible pollinations of Phalaris tuberosa L. J Cell Sci 9(1):239–251
Ko S, Li M, Ku M et al (2014) The bHLH142 transcription factor coordinates with TDR1 to modulate the expression of EAT1 and regulate pollen development in rice. Plant Cell 26(6):2486–2504
Konishi T, Shinohara K, Yamada K et al (1996) Acetyl-CoA carboxylase in higher plants: most plants other than gramineae have both the prokaryotic and the eukaryotic forms of this enzyme. Plant Cell Physiol 37(2):117–122
Kunst L, Samuels A (2003) Biosynthesis and secretion of plant cuticular wax. Prog Lipid Res 42(1):51–80
Kuromori T, Miyaji T, Yabuuchi H et al (2010) ABC transporter AtABCG25 is involved in abscisic acid transport and responses. PNAS 107(5):2361–2366
Lallemand B, Erhardt M, Heitz T et al (2013) Sporopollenin biosynthetic enzymes interact and constitute a metabolon localized to the endoplasmic reticulum of tapetum cells. Plant Physiol 162(2):616–625
Lang V, Usadel B, Obermeyer G (2015) De novo sequencing and analysis of the lily pollen transcriptome: an open access data source for an orphan plant species. Plant Mol Biol 87(1–2):69–80
Lee S, Jung KH, An G et al (2004) Isolation and characterization of a rice cysteine protease gene, OsCP1, using T-DNA gene trap system. Plant Mol Biol l54:755–765
Li H, Zhang D (2010) Biosynthesis of anther cuticle and pollen exine in rice. Plant Signal Behav 5(9):1121–1123
Li N, Zhang D, Liu H et al (2006) The rice tapetum degeneration retardation gene is required for tapetum degradation and anther development. Plant Cell 18(11):2999–3014
Li H, Pinot F, Sauveplane V et al (2010) CYP704B2 catalyzing the ω-hydroxylation of fatty acids is required for anther cutin biosynthesis and pollen exine formation in rice. Plant Cell 22:173–190
Li H, Yuan Z, Vizcay-Barrena G et al (2011) PERSISTENT TAPETAL CELL1 encodes a PHD-finger protein that is required for tapetal cell death and pollen development in rice. Plant Physiol 156(2):615–630
Li L, Li Y, Song S et al (2015) An anther development F-box (ADF) protein regulated by tapetum degeneration retardation (TDR) controls rice anther development. Planta 241(1):157–166

Liang M, Zhang P, Shu X et al (2013) Characterization of pollen by MALDI-TOF lipid profiling. Int J Mass Spectrom 334:13–18
Li-Beisson Y, Shorrosh B, Beisson F et al (2010) Acyl-lipid metabolism. Arabidopsis Book Am Soc Plant Biologist 8:e0133
Li-Beisson Y, Shorrosh B, Beisson F et al (2013) Acyl-lipid metabolism. The Arabidopsis Book, No.11. Am Soc Plant Biol. doi:10.1199/tab.0133
Ma H (2005) Molecular genetic analyses of microsporogenesis and microgametogenesis in flowering plants. Annu. Rev. Plant Biol 56:393–434
Menand B, Yi K, Jouannic S et al (2007) An ancient mechanism controls the development of cells with a rooting function in land plants. Science 316(5830):1477–1480
Morant M, Jorgensen K, Schaller H et al (2007) CYP703 is an ancient cytochrome P450 in land plants catalyzing in-chain hydroxylation of lauric acid to provide building blocks for sporopollenin synthesis in pollen. Plant Cell 19(5):1473–1487
Murgia M, Charzynska M, Rougier M et al (1991) Secretory tapetum of Brassica oleracea L.: polarity and ultrastructural features. Sex. Plant Reprod 4:28–35
Murphy D (2006) The extracellular pollen coat in members of the Brassicaceae: composition, biosynthesis, and functions in pollination. Protoplasma 228(1–3):31–39
Niu B, He R, He M et al (2013a) The ATP-binding cassette transporter OsABCG15 is required for anther development and pollen fertility in rice. J Integr Plant Biol 55:710–720
Niu N, Liang W, Yang X et al (2013b) EAT1 promotes tapetal cell death by regulating aspartic proteases during male reproductive development in rice. Nat Commun 4:1445
Ohlrogge J, Kuhn D, Stumpf P (1979) Subcellular localization of acyl carrier protein in leaf protoplasts of Spinacia oleracea. PNAS 76(3):1194–1198
Owen H, Makaroff C (1995) Ultrastructure of microsporogenesis and microgametogenesis in Arabidopsis thaliana (L.) Heynh. ecotype Wassilewskija (Brassicaceae). Protoplasma 185(1–2):7–21
Pacini E, Hesse M (1984) The tapetum: its form, function, and possible phylogeny in Embryophyta. Plant Syst Evol 149(3–4):155–185
Paxson-Sowders D, Owen H, Makaroff C (1997) A comparative ultrastructural analysis of exine pattern development in wild-type Arabidopsis and a mutant defective in pattern formation. Protoplasma 198(1–2):53–65
Paxson-Sowders D, Dodrill C, Owen H et al (2001) DEX1, a novel plant protein, is required for exine pattern formation during pollen development in Arabidopsis. Plant Physiol 127(4):1739–1749
Phan H, Iacuone S, Li S et al (2011) The MYB80 transcription factor is required for pollen development and the regulation of tapetal programmed cell death in Arabidopsis thaliana. Plant Cell 23(6):2209–2224
Pidkowich M, Nguyen H, Heilmann I et al (2007) Modulating seed β-ketoacyl-acyl carrier protein synthase II level converts the composition of a temperate seed oil to that of a palm-like tropical oil. PNAS 104(11):4742–4747
Piffanelli P, Ross J, Murphy D (1998) Biogenesis and function of the lipidic structures of pollen grains. Sex Plant Reprod 11(2):65–80
Poethig RS (1987) Clonal analysis of cell lineage patterns in plant development. Am J Bot 74(4):581–594
Qin P, Tu B, Wang Y et al (2013) ABCG15 encodes an ABC transporter protein, and is essential for post-meiotic anther and pollen exine development in rice. Plant Cell Physiol 54(1):138–154
Quatrano R, McDaniel S, Khandelwal A et al (2007) Physcomitrella patens: mosses enter the genomic age. Curr Opin Plant Biol 10(2):182–189
Quilichini T, Friedmann M, Samuels A et al (2010) ATP-binding cassette transporter G26 is required for male fertility and pollen exine formation in Arabidopsis. Plant Physiol 154(2):678–690
Quilichini T, Grienenberger E, Douglas CJ (2014) The biosynthesis, composition and assembly of the outer pollen wall: a tough case to crack. Phytochemistry. doi:10.1016/j.phytochem.2014.05.002

Rensing S, Lang D, Zimmer A et al (2008) The Physcomitrella genome reveals evolutionary insights into the conquest of land by plants. Science 319(5859):64–69
Scott R (1994) Pollen exine-the sporopollenin enigma and the physics of pattern. In: Seminar series-society for experimental biology. Cambridge University Press, pp 49–49
Scott RJ, Spielman M, Dickinson HG (2004) Stamen structure and function. Plant Cell 16(Suppl):S46–S60
Shaw G (1970) Chemistry of sporopollenin. In: Symposium on sporopollenin
Shi J, Tan H, Yu XH et al (2011) Defective pollen wall is required for anther and microspore development in rice and encodes a fatty acyl carrier protein reductase. Plant Cell 23(6):2225–2246
Shivanna K, Cresti M, Ciampolini F et al (1997) Pollen development and pollen-pistil interaction. In: Shivann KR, Sawhney VK (eds) Pollen biotechnology for crop production and improvement. Cambridge University Press, Cambridge, pp 15–39
Song J, Cao J, Wang C (2013) BcMF11, a novel non-coding RNA gene from Brassica campestris, is required for pollen development and male fertility. Plant Cell Rep 32(1):21–30
Sorensen A, Krober S, Unte U et al (2003) The Arabidopsis ABORTED MICROSPORES (AMS) gene encodes a MYC class transcription factor. Plant J Cell Mol Biol 33(2):413–423
Sun M, Huang X, Yang J et al (2013) Arabidopsis RPG1 is important for primexine deposition and functions redundantly with RPG2 for plant fertility at the late reproductive stage. Plant Reprod 26(2):83–91
Tang L, Chu H, Yip W et al (2009) An anther-specific dihydroflavonol 4-reductase-like gene (DRL1) is essential for male fertility in Arabidopsis. New Phytol 181(3):576–587
Varnier A, Mazeyrat-Gourbeyre F, Sangwan R et al (2005) Programmed cell death progressively models the development of anther sporophytic tissues from the tapetum and is triggered in pollen grains during maturation. J Struct Biol 152(2):118–128
Verma DP, Hong Z (2001) Plant callose synthase complexes. Plant Mol Biol 47(6):693–701
Wall D (1962) Evidence from recent plankton regarding the biological affinities of Tasmanites Newton 1875 and Leiosphaeridia Eisenack 1958. Geol Mag 99(04):353–362
Wallace S, Fleming A, Wellman CH et al (2011) Evolutionary development of the plant and spore wall. AoB Plants plr027. doi:10.1093/aobpla/plr027
Wallace S, Chater C, Kamisugi Y et al (2015) Conservation of Male Sterility 2 function during spore and pollen wall development supports an evolutionarily early recruitment of a core component in the sporopollenin biosynthetic pathway. New Phytol 205(1):390–401
Wei L, Xu W, Deng Z et al (2010) Genome-scale analysis and comparison of gene expression profiles in developing and germinated pollen in Oryza sativa. BMC Genomics 11(1):338
Wellman C (2004) Origin, function and development of the spore wall in early land plants. In: Hemsley AR, Poole I (eds) The evolution of plant physiology. Elsevier Academic Press, pp 43–64
Wilmesmeier S, Wiermann R (1995) Influence of EPTC (S-Ethyl-Dipropyl-Thiocarbamate) on the composition of surface waxes and sporopollenin structure in Zea mays. J Plant Physiol 146(1):22–28
Wilson Z, Zhang D (2009) From Arabidopsis to rice: pathways in pollen development. J Exp Bot 60(5):1479–1492
Wilson Z, Morroll S, Dawson J et al (2001) The Arabidopsis MALE STERILITY1 (MS1) gene is a transcriptional regulator of male gametogenesis, with homology to the PHD-finger family of transcription factors. Plant J Cell Mol Biol 28(1):27–39
Worrall D, Hird DL, Hodge R et al (1992) Premature dissolution of the microsporocyte callose wall causes male sterility in transgenic tobacco. Plant Cell 4(7):759–771
Wu L, Guan Y, Wu Z et al (2014) OsABCG15 encodes a membrane protein that plays an important role in anther cuticle and pollen exine formation in rice. Plant Cell Rep 33:1881–1899
Wu X, Cai G, Gong F et al (2015) Proteome profiling of maize pollen coats reveals novel protein components. Plant Mol Biol Rep 33:975–986

Xu J, Yang C, Yuan Z et al (2010) The ABORTED MICROSPORES regulatory network is required for postmeiotic male reproductive development in Arabidopsis thaliana. Plant Cell 22(1):91–107

Xu J, Ding Z, Vizcay-Barrena G et al (2014a) ABORTED MICROSPORES acts as a master regulator of pollen wall formation in Arabidopsis. Plant Cell 26(4):1544–1556

Xu Y, Iacuone S, Li S et al (2014b) MYB80 homologues in Arabidopsis, cotton and Brassica: regulation and functional conservation in tapetal and pollen development. BMC Plant Biol 14(1):278

Yadav V, Molina I, Ranathunge K et al (2014) ABCG transporters are required for suberin and pollen wall extracellular barriers in Arabidopsis. Plant Cell 26:3569–3588

Yang C, Vizcay-Barrena G, Conner K et al (2007) MALE STERILITY1 is required for tapetal development and pollen wall biosynthesis. Plant Cell 19(11):3530–3548

Yang J, Tian L, Sun MX et al (2013) AUXIN RESPONSE FACTOR17 is essential for pollen wall pattern formation in Arabidopsis. Plant Physiol 162(2):720–731

Yang X, Wu D, Shi J et al (2014a) Rice CYP703A3, a cytochrome P450 hydroxylase, is essential for development of anther cuticle and pollen exine. J Integr Plant Biol 56(10):979–994

Yang Y, Dong C, Yu J et al (2014b) Cysteine Protease 51 (CP51), an anther-specific cysteine protease gene, is essential for pollen exine formation in Arabidopsis. Plant Cell Tissue Organ Cult 119(2):383–397

Yeats TH, Rose JK (2008) The biochemistry and biology of extracellular plant lipid-transfer proteins (LTPs). Protein Sci 17(2):191–198

Yi B, Zeng F, Lei S et al (2010) Two duplicate CYP704B1-homologous genes BnMs1 and BnMs2 are required for pollen exine formation and tapetal development in Brassica napus. Plant J Cell Mol Biol 63(6):925–938

Zhang D, Li H (2014) Exine export in pollen. In: Plant ABC transporters. Springer International Publishing, pp 49–62

Zhang D, Yang L (2014) Specification of tapetum and microsporocyte cells within the anther. Curr Opin Plant Biol 17C:49–55

Zhang Z, Zhu J, Gao J et al (2007) Transcription factor AtMYB103 is required for anther development by regulating tapetum development, callose dissolution and exine formation in Arabidopsis. Plant J 52(3):528–538

Zhang D, Liang W, Yuan Z et al (2008) Tapetum degeneration retardation is critical for aliphatic metabolism and gene regulation during rice pollen development. Mol Plant 1(4):599–610

Zhang D, Liang W, Yin C et al (2010) OsC6, encoding a lipid transfer protein, is required for postmeiotic anther development in rice. Plant Physiol 154(1):149–162

Zhang D, Luo X, Zhu L (2011) Cytological analysis and genetic control of rice anther development. J Genet Genomics 38(9):379–390

Zhang D, Liu D, Lv X et al (2014) The cysteine protease CEP1, a key executor involved in tapetal programmed cell death, regulates pollen development in Arabidopsis. Plant Cell 26(7):2939–2961

Zhu L, Shi J, Zhao G et al (2013) Post-meiotic deficient anther1 (PDA1) encodes an ABC transporter required for the development of anther cuticle and pollen exine in rice. J Plant Biol 56(1):59–68

Zinkl G, Zwiebel B, Grier D et al (1999) Pollen-stigma adhesion in Arabidopsis: a species-specific interaction mediated by lipophilic molecules in the pollen exine. Development 126(23):5431–5440

Chapter 14
Long-Distance Lipid Signaling and its Role in Plant Development and Stress Response

Allison M. Barbaglia and Susanne Hoffmann-Benning

Abstract Lipids are important signaling compounds in plants. They can range from small lipophilic molecules like the dicarboxylic acid Azelaic acid to complex phosphoglycerolipids and regulate plant development as well as the response to biotic and abiotic stress. While their intracellular function is well described, several lipophilic signals are known to be found in the plant phloem and can, thus, also play a role in long-distance signaling. Mostly, they play a role in the pathogen response and systemic acquired resistance. This is particularly true for oxylipins, dehydroabietinal, and azelaic acid. However, several phospholipids have now been described in phloem exudates. Their intracellular function as well as implications and a model for long-distance signaling are discussed in this chapter.

Keywords Lipid signaling • Phloem lipids • Lipid-binding proteins • Abiotic and biotic stress signaling

Introduction

The world population is projected to increase by 2.4 billion, from 7.2 billion in 2013 to 9.6 billion in 2050, with most of the increase occurring within less developed regions (www.un.com). Combined with the encroachment of cities onto arable land this necessitates an increase in agricultural yields. When faced with similar issues in the 1940s, the "green revolution," led by Norman Borlaug, provided solutions through the development of high-yielding varieties of cereal grains, the modernization of management techniques and irrigation systems, as well as distribution of hybrid seeds, synthetic fertilizers, and pesticides to farmers. Today, the amount of arable land is limited and often, there is now a competition between food and fuel crops. In addition, changes in the global climate may impact future yields. Unlike

A.M. Barbaglia • S. Hoffmann-Benning (✉)
Department of Biochemistry and Molecular Biology, Michigan State University, East Lansing, MI, USA
e-mail: hoffma16@cns.msu.edu

Y. Nakamura, Y. Li-Beisson (eds.), *Lipids in Plant and Algae Development*, Subcellular Biochemistry 86, DOI 10.1007/978-3-319-25979-6_14

animals, plants cannot move to escape adverse conditions. Plants had to develop mechanisms to detect changes in their environment, communicate these changes to different organs, and adjust development accordingly (Thomas and Vince-Prue 1997). These responses can occur within one cell or tissue; however, often we see a systemic response at a location distal from the detection of those environmental changes. One path to move these long-distance signals is the plant phloem. Its main conductive system is composed of large tubular cells, the sieve elements, which connect to form sieve tubes. The length of the phloem can reach up to 100 m in tall trees. To allow for an unobstructed flow, the sieve elements, cells through which transport occurs, have lost their nuclei, ribosomes, and most organelles during development. Cell walls at the interphase between two sieve elements contain large sieve pores. Sieve elements are left with the plasma membrane and a thin cytoplasm containing ER, phloem-specific plastids, and a few dilated mitochondria (VanBel and Knoblauch 2000; Turgeon and Wolfe 2009). The residual ER is found near the plasmodesmata which connect the sieve elements with the companion cells (Lucas et al. 2013). Molecules found within the sieve element are largely synthesized in the companion cells. For long-distance movement, these molecules need to be transported into the sieve element via plasmodesmata, moved along the sieve tube, and are either perceived at their target location or transported out of the phloem. Transport of photoassimilates as well as signaling molecules is thought to occur from source (photosynthetically active mature leaves) to sink (immature leaves, roots, fruits, flowers, etc.) in a mechanism driven by the osmotic gradient ("Pressure flow hypothesis"; Münch 1930; for a review see Froehlich et al. 2011; Lucas et al. 2013). The view of the phloem function has evolved from that of simple assimilate transport to a complex trafficking system for stress signals and developmental regulators (Zeevaart 1976; Wu et al. 2003; Ding et al. 2003; Ayre and Turgeon 2004; Haywood et al. 2005) in the form of small molecules (Corbesier et al. 2003; Chen et al. 2001), peptides/proteins (Fisher et al. 1992; Kühn et al. 1997; Marentes and Grusak 1998; Kehr et al. 1999; Xoconostle-Cazares et al. 1999; Haebel and Kehr 2001; Hoffmann-Benning et al. 2002; Giavalisco et al. 2006; Lin et al. 2009; Benning et al. 2012), nucleic acids (Citovsky and Zambryski 2000; Haywood et al. 2005; Ding et al. 2003; Ruiz-Medrano et al. 1999; Yoo et al. 2004), and more recently, lipids (Madey et al. 2002; Behmer et al. 2010; Guelette et al. 2012; Benning et al. 2012; Tetyuk et al. 2013). Thus, the study of signaling compounds within the phloem is essential for our understanding of the transmission of environmental cues throughout the plant. While there is a large focus on the function of proteins, mRNAs and microRNAs in long-distance signaling, less attention has been paid to small lipophilic metabolites, and even less to more complex lipids. Yet, both, small lipophilic molecules as well as several phospholipids, di- and triacylglycerols (DAG and TAG, respectively) have been found in phloem exudates (Table 14.1). The former have already been shown to play an important role in long-distance signaling. Most play a role in the response to biotic stress and systemic-acquired resistance (SAR). Others are plant hormones with functions ranging from pathogen response to plant development. Their known function as well as possible roles of phospholipids will be discussed below.

Table 14.1 Hydrophobic molecules in the phloem and their possible function

Name	Category	Long-distance function	Example-structures
Indole-acetic acid (IAA)	Aromatic ring with carboxylic acid	Root growth; apical dominance; vascular bundle development	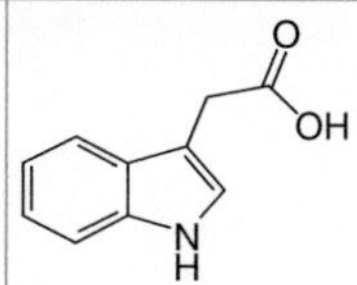
Abscisic acid (ABA)	Isoprenoid (Sesquiterpene)	Abiotic stress signal in response to drought, cold, salt stress	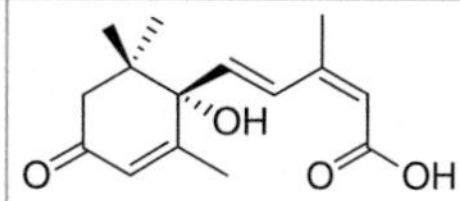
Gibberellins (GA)	Isoprenoid (Pentacyclic diterpene)	Germination; plant growth	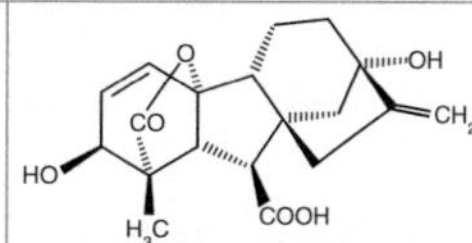
Cytokinin	Adenine derivatives; often isoprenylated	Vascular bundle development; root development; cell division/differentiation	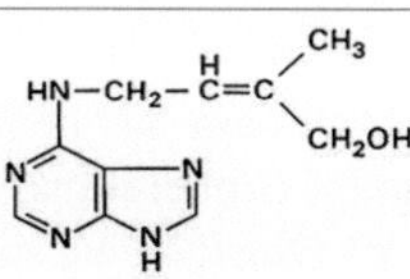
Salicylic acid (SA)	Monohydroxybenzoic acid	Pathogen defense; systemic acquired resistance (SAR)	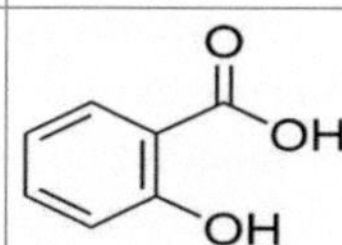
Dehydroabietinal (DA)	Isoprenoid (Abietane diterpenoid)	SAR; flowering?	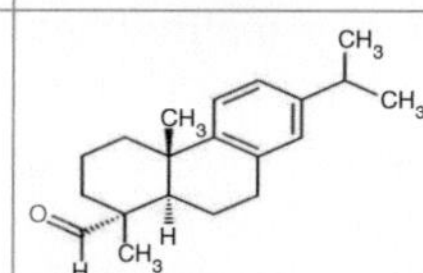
Azelaic acid	Dicarboxylic acid	SAR	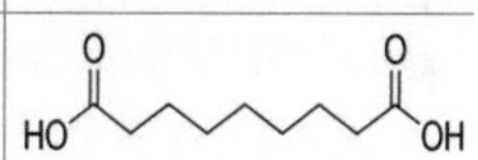
Glycerol-3-phosphate*	Unknown G-3-P-derivative	SAR	
Jasmonic acid/ oxylipins	Oxylipins	SAR; wounding; senescence; drought response	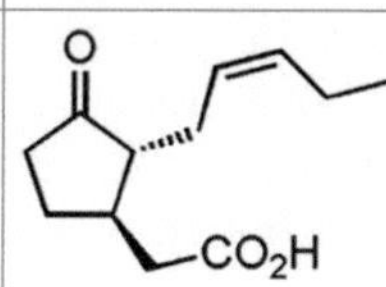
Diacylglycerols	Diacylglycerols	Precursor for PtdOH through the PLC-DGK pathway	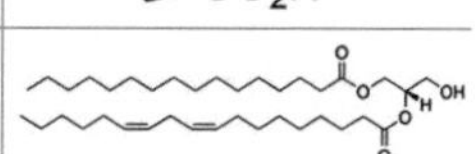
Triacylglycerols	Triacylglycerols	Storage lipid	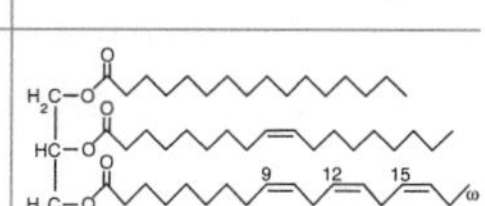

(continued)

Table 14.1 (continued)

Name	Category	Long-distance function	Example-structures
Phosphatidyl choline	Phosphoglyerolipid	Precursor for PtdOH through PLD pathway	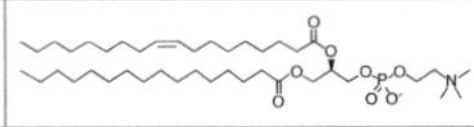
Phosphatidyl inositol	Phosphoglyerolipid	Abiotic stress; light- and gravitropic signaling; vesicle trafficking	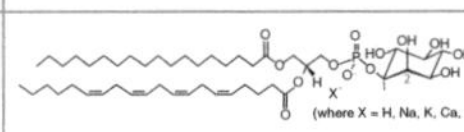
Phosphatidic acid	Phosphoglyerolipid	Drought, salt, osmotic, wounding, pathogen?	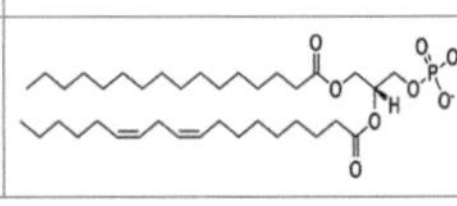

Lipophilic Compounds in the Phloem

Plant Hormones

The phloem has been shown to contain several "lipid" plant hormones including Auxin, Gibberellins, Cytokinins, Jasmonate, and Abscisic acid (ABA; Hoad 1995), all of which signal different aspects of plant development and, in the case of ABA, certain abiotic stresses. Auxins, Cytokinins and ABA are found in the xylem as well. ***Auxins*** have been detected in the phloem of over 14 species and shown to be mobile. Whether auxin is synthesized in the sieve elements or imported from other cells is, as of yet, unknown (Hoad 1995). Auxin and Cytokinin play an important role in signaling/regulating the development of both the xylem and the phloem (see Lucas et al. 2013). Auxin is thought to integrate the phosphate status with the plant growth response by changing root architecture (López-Bucio et al. 2002; Chiou and Lin 2011). It is also implied in systemic acquired resistance (Truman et al. 2010). ***ABA*** is transported in the phloem of many woody and herbaceous species in response to many stresses including drought. A function in phloem loading and unloading has also been proposed (Hoad 1995). It will be discussed further below. Plant hormones such as systemin, ***salicylic acid*** (*SA*) and ***jasmonic acid*** **(JA)**, which provoke a pathogen response, have been identified and transported within the phloem in response to pathogen infection (Schilmiller and Howe 2005; Truman et al. 2007). In addition, the phloem contains methyl salicylate (2-hydroxy benzoic acid), which plays a role in thermogenesis (Raskin 1992) and plant defense/systemic acquired resistance:

Small Lipophilic Metabolites and Systemic Acquired Resistance (SAR)

Systemic acquired resistance is a response to pathogen infection that puts distal, pathogen-free parts of a plant into a state of increased preparedness against further infections. As a result, defense molecules are induced faster and stronger during subsequent infections. This response can be passed on through several generations in an epigenetic mechanism (Luna and Ton 2012). However, its origin lies within one plant and is mediated by proteins and small metabolites. An essential component of SAR is the movement of a SAR-signal from the infection site to distal organs of the plant, most likely through the phloem (Tuzun and Kuc 1985; Maldonado et al. 2002; Chaturvedi et al. 2008; Jung et al. 2009; Chanda et al. 2011). Candidates for such a SAR-signal include several small lipophilic molecules: methyl salicylate (Chaturvedi and Shah 2007), dehydroabietinal (Chaturvedi et al. 2012), a Glycerol-3-phosphate-dependent factor (Chanda et al. 2011; Mandal et al. 2011), and azelaic acid (Jung et al. 2009). All four are increased in leaves infected with pathogens.

Salicylic acid (SA) is considered one of the key factors in SAR. Both, SA as well as its glucoside are increased in tissues infected by a variety of pathogens (Shah et al. 2014). There they lead to the subsequent expression of SA-responsive genes like PR1 (pathogenesis-related 1) and a local defense response. In addition to this localized response, SA is converted to methyl salicylate (Me-SA) and elicits a systemic response as well. While SA has been found in phloem exudates, its movement is still debated and not sufficient to explain the increase of SA in distal leaves (Malamy et al. 1990; Métraux et al. 1990). One possibility is an increased SA biosynthesis in the distal leaves in response to additional mobile signals. This includes Me-SA, which is produced from SA in Tobacco-Mosaic-Virus (TMW)-infected leaves, and moves through the phloem to elicit a SAR response in the distal (non-infected) tobacco leaves (Park et al. 2007). A second important factor in SA accumulation and SAR is the abietane diterpenoid ***dehydroabietinal*** (DA; Chaturvedi et al. 2012): DA was purified and identified as a small hydrophobic factor in SAR. Picomolar applications of DA to leaves led to its movement to distal leaves as well as SA accumulation and a SAR response. While the DA level does not increase in petiole exudates from infected plants, ^{2}H-DA can rapidly translocate from the point of application to the rest of the plant, suggesting a mobile DA-derivative (Chaturvedi et al. 2012). Interestingly, free dehydroabietic acid was found in phloem exudates from non-infected plants (Guelette et al. 2012). DA, on the other hand, is found in a pathogen-induced high-molecular-weight complex (HMW), which also includes DIR1 (see Shah et al. 2014). While this complex is likely not phloem mobile, it is possible that DIR 1 facilitates the long-distance movement of DA. An additional factor in systemic signaling is a ***Glycerol-3-phosphate-derivative*** of unknown structure. Glycerol-3-phosphate (G3P) in itself is not a lipid; yet, it forms the backbone of glycerolipids, which include membrane-, storage-, and signaling lipids. During SAR, G3P-levels increase in the infected leaf, petiole exudates, and the distal, non-infected leaves, however, when radiolabeled G3P was applied to leaves, it did not move. Instead, the radioactivity was recovered in an unknown compound within the distal leaf (Chanda et al. 2011). This suggests that rather than G3P, an as of yet, unidentified G3P derivative (G3P*) may be the mobile signal. ***Azelaic acid*** (AzA), a small dicarboxylic acid, was found in phloem exudates of infected plants (Jung et al. 2009). It is derived through the oxidation of C18 unsaturated fatty acids in a

ROS-mediated process. Only a small fraction of the labeled AzA that had been applied to a leaf could be recovered in a distal leaf making long-distance movement questionable. It may, however, act in SAR by increasing the production of G3P (Yu et al. 2013).

In addition to these four lipophilic molecules, the SAR response is mediated by two proteins: DIR1 (Defective in induced resistance 1) and AZI1 (Azelaic acid induced 1). DIR1 is a lipid transfer protein that has been found within the phloem (Maldonado et al. 2002; Mitton et al. 2009) and was shown to move systemically throughout the plant (Champigny et al. 2013; Yu et al. 2013); DIR1 can interact with itself as well as with AZI1. The proposed networking between these proteins and G3P, DA, and AzA has been summarized comprehensively in Shah et al. 2014.

Oxylipins/Jasmonic Acid

Additional, more complex lipophilic signals moving through the phloem are the oxylipins. Oxylipins are oxygenated polyunsaturated fatty acids derived from the polyunsaturated acyl groups in chloroplast membrane galactolipids (Howe and Schilmiller 2002; Feussner and Wasternack 2002; Yan et al. 2013). Jasmonates are formed though a LOX-catalyzed peroxidation of the trienoic fatty acids to form 13-hydroperoxide. They are further modified through the allene oxide synthase pathway (AOS) or the hydrodyperoxide lyase pathway (HPL), which partition oxylipin biosynthesis into two pools of molecules with varying functions (Chehab et al. 2008). In the AOS pathway, 12-OPDA (*cis*12-oxo-phytodienoic acid) and JA are produced, while the HPL pathway produces aldehydes and their volatile relatives (green leaf volatiles; GLV). Jasmonates include Jasmonic acid (JA), methyl-jasmonate (Me-JA), jasmonate-isoleucine (JA-Ile), and OPDA. They play a role in the regulation of plant growth and development as well as in the response to biotic and abiotic stress. 12-OPDA is not just a JA precursor but a signaling molecule itself (Dave and Graham 2012). In response to wounding or herbivory, JA is synthesized and moves throughout the plant to elicit a (systemic) defense response (Truman et al. 2007; Thorpe et al. 2007). This response is not limited to herbivores but extends to bacterial and fungal pathogens as well (Yan et al. 2012). In many cases jasmonates display a cooperative response with other plant hormones – it can act synergistically with ethylene, salicylic acid, ABA, and gibberellins (Yan et al. 2013). Different environmental stresses produce different oxylipins profiles: Savchenko et al. (2014) found that while wounding induces JA and its precursor 12-OPDA, drought only induces 12-OPDA. This induction is correlated with a differential increase in plant ABA levels under the same conditions. More importantly, external application of ABA and/or 12-OPDA could overcome the defect in stomatal closure in ABA- and 12-OPDA biosynthesis mutants, suggesting ABA/OPDA crosstalk during drought signaling (Savchenko et al. 2014). Interestingly, this mirrors findings that OPDA plays a role in seed development and germination, which are cellular processes that are also mediated by ABA (Dave et al. 2011).

(Phospho-)Glycerolipids in the Phloem: A Possible Model of their Transport

In analogy to animal systems, lipids and lipid-binding proteins may play an important role in stress and developmental signaling in plants. Lipids, specifically phospholipids, are mainly found within cell membranes to which they provide both structure and act as mediators that regulate various aspects of plant development and environmental interactions (Wang and Chapman 2013; Wang 2004; Xue et al. 2009). Signaling lipids include glycerolipids, sphingolipids, fatty acids, oxylipins, and sterols. Phospholipid signaling cascades are normally grouped according to the phospholipase that catalyzes the formation of other messengers such as PtdOH, DAG, DAG pyrophosphate (DGPP), lysophospholipids, free fatty acids, oxylipins, phosphoinositides, and inositol phosphates (Zhu 2002; Wang 2004; Wang et al. 2007).

Unlike structural lipids, signaling lipids occur only in minute amounts, but increase rapidly and transiently in response to certain stimuli. The resulting signal is quickly downregulated. Consequently, signaling lipids typically have a higher turnover than other structural lipids (Testerink and Munnik 2005). Lipids such as phosphatidic acid (PtdOH), phosphatidyl inositol (PtdIns), and its phosphates (PtdInsPs) are well-known secondary messengers within plant cells (Wang 2004; Munnik and Testerink 2009), but essentially nothing is known about their involvement in long-distance signaling. A recent analysis of phloem exudates of *Arabidopsis thaliana*, led to the identification of several glycerolipids within the phloem including phosphatidic acid (PtdOH), phosphatidyl choline (PtdCho), phosphatidyl inositol (PtdIns), di- and triacylglycerols (DAG and TAG, respectively; see Guelette et al. 2012; Benning et al. 2012; Tetyuk et al. 2013). Similarly, lipids have been detected in canola phloem as well (Madey et al. 2002). Several possibilities exist for the function of these lipids in the phloem:

I. Lipids could be transported through the phloem as energy carriers, assimilates, or building blocks. However, most tissues including seeds are capable of making their own (storage) lipids, hence, transport as energy carriers or building blocks would be inefficient (Ohlrogge and Browse 1995).
II. Lipids could be the product of regular membrane turnover within the sieve element (Lucas et al. 2013). When we compared leaf and phloem lipids, the lipid profiles were distinct, with several lipids being unique to the phloem (Guelette et al. 2012). It does not exclude the possibility though, that they are generated from other membrane lipids to serve as anchors for proteins or precursors for signaling molecules. For example, one of the PtdCho species found in phloem exudates contains a 18:3 acyl group. Polyunsaturated lipids are known precur-

sors to signaling lipids, such as oxylipins, which are phloem mobile. However, the first step of oxylipins biosynthesis occurs in the chloroplast, a plastid that is not present in sieve elements. Moreover, Buseman et al. (2006) had proposed that galactolipids, which are plastidal lipids, are the substrate for the production of OPDA. This suggests that phloem lipids have phloem-specific functions and are likely not products of non-specific membrane turnover. Similarly, as part of the seasonal senescence-related mobilization, cellular components including lipids are metabolized and moved. However, this mobilization tends to occur in the form of sucrose rather than lipids (Thomas and Stoddard 1980).

III. Phloem lipids could serve as long distance signals. Long-distance transport of hydrophobic compounds does occur in aqueous biological systems: For example, lipids in the blood are transported while bound to proteins. These lipid-protein complexes are transported to other tissues for storage, modification, or degradation (e.g. cholesterol; Glatz et al. 1995; Charbonneau et al. 2009), others transport vitamins (e.g., vitamin A: Blaner 1989) or serve as messengers and modulate transcription factor activity (Tontonez et al. 1994; Nagy and Szanto 2005; Musille et al. 2013). While these lipid-protein mechanisms are key to mammalian development their possible importance in plants is virtually unexplored.

All three of the above mechanisms would require the assistance/binding of proteins. As a result, we and others proposed that phloem (phospho-) lipids could act as <u>long-distance developmental signals</u> or in response to abiotic stress and that phloem lipid-binding proteins participate in different aspects of this signaling cascade (Benning et al. 2012; Zheng et al. 2012):

(i) Proteins could release the lipid into the phloem, either by generating it in within the sieve element, possibly from the membrane and through a lipase-like activity; or they could transport the lipid into the phloem via plasmodesmata.

(ii) Proteins could solubilize the lipid and transport it to its target organ. This would imply that the lipid is the mobile signal. Alternatively, the protein could be a signal which needs a bound lipid for functionality. An example for such a mechanism would be the Frizzled-Wnt interaction. Wnt is an important developmental regulator in animal systems, which binds to its receptor, Frizzled. It had long been known that Wnt can only solicit the downstream signaling cascade when bound to a lipid, palmitoleic acid. In 2012, crystallographic analysis showed that palmitoleic acid binding to Wnt is essential for Wnt-Frz interaction (Janda et al. 2012). A similar mechanism is conceivable in plants.

(iii) Proteins either transport the lipid out of the sieve element or are part of a receptor that binds the signaling lipid and leads to a change in development (Fig. 14.1).

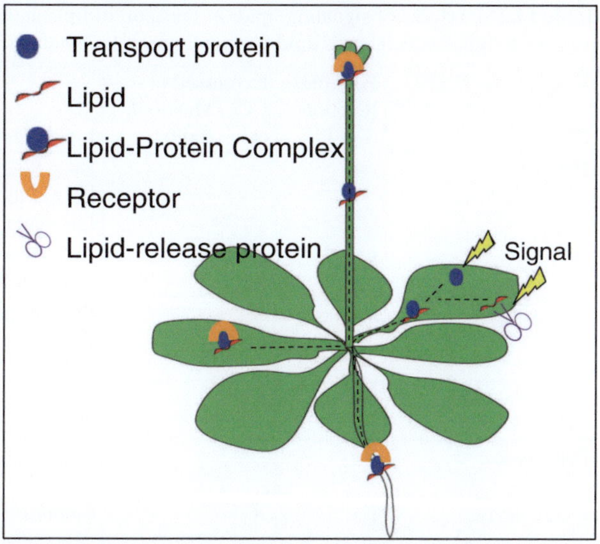

Fig. 14.1 Model of possible lipid signaling (derived from Hoffmann-Benning 2015)

To identify possible proteinaceous components of long-distance lipid signaling and transport in *Arabidopsis* we had used a proteomics (LC-ESI-MS/MS) approach. This allowed us to identify phloem proteins with predicted lipid-binding sites (Guelette et al. 2012). These findings together with the analysis of other phloem proteome databases led to the identification of 13 putative lipid-binding proteins (LBPs; Guelette et al. 2012; Benning et al. 2012) in *Arabidopsis* and additional plant species such as rice, canola, cucurbits, lupine, and *Perilla* (Hayashi et al. 2000; Hoffmann-Benning et al. 2002; Barnes et al. 2004; Giavalisco et al. 2006; Lin et al. 2009; Guelette et al. 2012). Genes encoding most of these proteins are expressed in companion cells (CC; Mustroph et al. 2009). References describing their first description in the phloem as well as their function according to the model above are summarized in Table 14.2.

The question arises whether any of the glycerolipids identified in phloem exudates are binding to any of the phloem lipid-binding proteins (LBPs)? And could they be signaling molecules? The answer is a definite yes. Six of these predicted phloem lipid-binding proteins have been studied further: **DIR1**, a lipid transfer protein (LTP) plays a role in systemic acquired resistance in *Arabidopsis* and tomato (Maldonado et al. 2002; Lascombe et al. 2008; Mitton et al. 2009; see Shah et al. 2014). It has been discussed above. **Annexin** binds phospholipids and is involved in intracellular Ca^{2+}-signaling as well as callose formation (Andrawis et al. 1993; Mortimer et al. 2008). **FT** signals the induction of flowering and seasonal leaf abscission (Corbesier et al. 2007; Tamaki et al. 2007; Lin et al. 2007). It binds PtdChos, however, whether this interaction occurs in the phloem/SE, leaf mesophyll cells or apical meristem, remains to be elucidated (Nakamura et al. 2014). **14-3-3 proteins** are a component of the receptor system detecting the presence of FT in the

Table 14.2 i: release of signaling lipid; ii: transport of lipid signal; iii: receptor for a lipid signal; iv: protein signal, which needs lipid-partner for receptor binding similar to the Frz – wnt interaction

Protein name	Possible function (i–iv)	Expressed in CCs (Mustroph et al. 2009)	Lipid identified	First report of phloem localization
Dir1/LTP	i–iii		DA?; Shah et al. 2014; PtdCho: Lascombe et al. 2008	Maldonado et al. (2002)
ACBP	i	X	ACBP6: PC; Chen et al. (2008)	Hayashi et al. (2000)
Annexin	ii	X	Phospholipids (PL); Rescher and Gerke (2004)	Barnes et al. (2004)
GRP17/oleosin	ii	X	PL; Tzen and Huang (1992)	Guelette et al. (2012)
Put. lipase	i			Guelette et al. (2012)
GDSL-lipase	i	X	(unpub)	Guelette et al. (2012)
PLAFP	i–iv	X	PtdOH; Benning et al. (2012)	Guelette et al. (2012)
Aspartic protease	i	X		Guelette et al. (2012)
14-3-3 protein	iii	X		Guelette et al. (2012)
Major latex protein	ii, iv	X		Giavalisco et al. (2006)
Bet v1 allergen	ii, iv	X		Guelette et al. (2012)
PIG-P-like protein	iii		(unpub)	Guelette et al. (2012)
Flowering locus T	ii, iv	X	PtdCho; Nakamura et al. (2014)	Giavalisco et al. (2006)

phloem near the apical meristem (Taoka et al. 2011). While lipid-binding of these proteins has been shown, no connection has been made between these proteins and phloem lipids. **Acyl-CoA-binding proteins** (**ACBPs**) have been identified in phloem exudates from rice (Hayashi et al. 2000) and associated with vascular bundles in *Arabidopsis* (Chen et al. 2008; Zheng et al. 2012). ACBPs will be discussed in a separate chapter in this book. **PLAFP**, is a 20 kDa protein with a PLAT/LH2 domain (Polycystin-1, Lipoxygenase, Alpha-Toxin-domain). This domain is thought to mediate interaction with lipids or membrane-bound proteins (Bateman and Sandford 1999). Proteins containing this domain are typically induced by stress. PLAFP binds PtdOH, one of the lipid species found in phloem exudates (Benning et al. 2012). PLAFP and its possible function in long-distance lipid signaling will be discussed below.

The Role of Phospholipases in Phospholipid Signaling

Phospholipids are more than just membrane lipids. Phospholipid signaling cascades are normally grouped according to the phospholipase that catalyzes their formation (Zhu 2002; Wang 2004, ; Wang et al. 2007): phospholipase D (PLD), phospholipase C (PLC), phospholipase A1 (PLA_1), and phospholipase A2 (PLA_2). PLC and PLD cleave at the first and the second phosphodiester bond, respectively, while the PLAs cleave the acyl chains from the membrane lipid. Plant phospholipases participate in different cellular and physiological processes, which are classified into three categories: cellular regulation, membrane lipid remodeling, and lipid degradation; furthermore, some phospholipases can participate in the biosynthesis of storage lipids (Wang et al. 2012). Phospholipase D catalyzes the hydrolysis of glycerophospholipids into phosphatidic acid (PtdOH) and a free head group, such as choline or serine. *Arabidopsis* encodes 12 distinct PLDs: PLDα(1,2,3), PLDβ(1,2), PLDγ(1,2,3), PLDδ, PLDε, and PLDζ (Wang et al. 2012). Each of these subgroups is based on molecular and biochemical properties. Additionally, PLD can be classified into two groups based on their lipid binding domains; those with both pleckstrin homology (PH) and phox homology (PX) domains belong to the PLDζ class. The majority of PLDs fall into a class known as C2, which contain a C2 (calcium- and lipid- binding) domain. Phospholipase C cleaves phospholipids generating diacylglycerol (DAG) and a phosphorylated head group. Plants possess two distinct PLC families: phosphoinositide-specific PLC (PtdIns-PLC) and nonspecific PLC (NPC). PtdIns-PLC facilitates the hydrolysis of phosphatidylinositol 4, 5-bisphosphate ($PtdInP_2$) to generate 1, 4, 5-triphosphate ($InsP_3$) and DAG. The resulting DAG remains bound to the membrane, while $InsP_3$ is released into the cytosol to act as a mediator (Wang et al. 2012). NPC functions in cleaving other membrane phospholipids, such as PtdCho and PtdEtn. PhospholipaseAs hydrolyze phospholipids leading to the production of lysophospholipids and free fatty acids. PLA_1 and PLA_2 cleave the fatty acid from sn-1 and sn-2 positions, respectively. In plants, there are four families of PLAs: the PC-hydrolyzing PLA_1 (PLA_1), the PA-preferring PLA_1 (PA-PLA_1), the secretory low molecular weight PLA_2 ($sPLA_2$), and the patatin-like PLA (pPLA; Wang et al. 2012).

In our phloem sap analysis, we detected PtdChos, PtdIns, PtdOHs, TAGs, and DAGs. While all of those play a role in intracellular signaling, PtdCho and PtdIns are likely precursors to the signaling molecules PtdOH and inositol phosphates. Their role in intracellular signaling and its possible implications for long-distance signaling are discussed below:

Phosphatidyl Inositol and Its Phosphates

Early findings in animal systems had suggested the existence of a phosphoinositide (PtdIns)/phospholipase C (PLC) intracellular signaling system: upon receptor stimulation, PLC hydrolyses phosphatidyl inositol 4,5-bisphosphate into the second messengers diacylglycerol and inositol- 1,4,5-trisphosphate ($InsP_3$) and leads to a downstream change in developmental processes. The scenario is different in plants due to their low $PtdInsP_2$ concentration (Munnik and Testerink 2009). Plant PLCs will hydrolyze phosphatidylinositol-4-phosphate but can also use phosphatidylinositol (PtdIns) as a substrate (Munnik et al. 1998). In addition, so far no $InsP_3$ receptor has been found. However, $InsP_3$ is an intermediate in the lipid-mediated biosynthesis as well as in the new lipid-independent biosynthesis of $InsP_6$, a known signaling molecule in plants (Gillaspy 2013).

Phosphoinositides, the phosphorylated versions of PtdIns play an important role in lipid signaling, membrane signaling (through regulating ion channels and pumps), and vesicle trafficking. They appear to be localized in discrete signaling regions within the membrane (Boss et al. 2006). The lipid-soluble phosphatidylinositol phosphates and water-soluble inositolphosphates (InsPs) are interrelated in that $InsP_3$ and ***diacylglycerol*** (DAG) are produced from $PtdInsP_2$ through the action of PLCs (Berridge 1993). Some PLCs have the capability to hydrolyze PtdIns(4)P or PtdIns (Munnik et al. 1998). DAG remains in the membrane where, in animals, it can activate a kinase C localized near the membrane and lead to the phosphorylation of target proteins (PKC; Newton 2010). However, so far, DAG could not be shown to activate any of the plant PKC homologues (Gillaspy 2013). In plants, the resulting DAG is further converted to phosphatidic acid (PtdOH; Arisz et al. 2009) while $InsP_3$ can be increasingly phosphorylated up to $InsP_6$ (hexakisphosphate). $InsP_6$ is a high-energy storage compound in plants, but inositol polyphosphates can also function as a signaling molecule, regulate plant hormone receptors like those for IAA (Tan et al. 2007) and JA (Mosblech et al. 2011), and affect trafficking (Nagy et al. 2009). In addition, it can be converted to inositol pyrophosphates ($InsP_7$, $InsP_8$; Bennett et al. 2006; Burton et al. 2009). These high-energy molecules have been linked to energy sensing in yeast and animal systems. In plants, their role is yet to be determined. The phosphoinositide pathway and inositol-1,4,5-triphosphate ($InsP_3$) were shown to be involved in the response to many abiotic stresses (Im et al. 2007), however, more recent findings suggest a role in the signaling of biotic stress as well (Hung et al. 2014): One of the early responses in both pathogen- and effector-triggered immunity (PTI and ETI, respectively) is a transient increase in cytosolic Ca^{2+}, likely mediated by an influx through a plasma membrane localized Ca^{2+} -channel/pump or release of intracellular stores of Ca^{2+}. This is likely due to the PLC-dependent generation of $InsP_3$ and $InsP_6$: *Arabidopsis* plants expressing the human type I inositol polyphosphate 5-phosphatase (InsP5-ptase), which specifically hydrolyzes soluble inositol phosphates and thus, interferes with $InsP_3$-mediated signaling, exhibit reduced $InsP_3$ and $InsP_6$ levels as well as a reduced response to gravity and impaired Ca signaling following salt and cold-stress (Perera et al. 2008). Hung et al. (2014) further showed that InsP5-ptase expressing plants lacked the pathogen-induced Ca^{2+} increase in response to the elicitor flagellin 22 (flg22). Expression of pathogenesis related genes, SA levels and SAR in response to *P. syringae* were reduced as well. These findings suggest that reduced levels of $InsP_3$ (or $InsP_6$) lead to a reduced intracellular Ca^{2+} release and impaired downstream responses. Interestingly, both PtdIns, as well as one of the PLC-cleavage products,

DAG, have been found in the phloem exudates of *Arabidopsis* (Guelette et al. 2012). While similar responses within the sieve elements or the companion cells are possible, so far, no experimental evidence points at their long-distance function in biotic stress.

The same holds true for the role of PtdInsPs in abiotic stress: drought, ABA, salt stress, cold, anoxia, light, or gravitropic signals all lead to the production of, albeit different, InsPs from PtdInsPs. Plants contain a variety of PLCs, all of which appear to be regulated by different environmental stimuli (Dowd and Gilroy 2010) and can be, at least in response to flg22, post-translationally modified. These increases in InsPs, however, may not solely be due to an increase in PLC activity, but can also be founded in the increase of PtdInsP_2 production. Contrary to animals, plants have a lower PtdIns(4,5)P_2 to PtdIns(4)P ratio (1:20 instead of 1:2). It has been suggested that PtdIns(4)P may have a dual function as membrane anchor for proteins as well as PtdIns(4,5)P_2 precursor. As the former, they may anchor proteins to the membrane and thus remove them from their signaling path. In addition, PtdInsPs may function as second messengers themselves (see Gillaspy 2013).

A recent report suggested that the regulatory functions of **PtdIns(4,5)P_2** in plant cells is dependent on the phosphatidylinositol 4-phosphate 5-kinases isoform which generates the lipid (PtdIns4P 5-kinases/PIP5K; Stenzel et al. 2012). In plants, PtdIns4P 5-kinases belong to a family of 11 proteins, which act in pairs of closely related proteins (Müller-Röber and Pical 2002). PtdIns(4,5)P_2 is membrane localized and appears to play a role in vesicle trafficking. For example it is necessary for the polarized distribution of the PIN1 auxin efflux carrier and thus, indirectly for the directional auxin transport and gravitropism in root and shoot (Salinas-Mondragon et al. 2010; Thole and Nielsen 2008; Heilmann 2009). It was proposed that PIN cycling is impaired in *Arabidopsis pi5k2* mutants possibly due to interference with clathrin-mediated endocytosis (König et al. 2008; Mei et al. 2012). Indeed, Ischebeck et al. (2013) showed that *pip5K1/pip5K2* double mutants showed significantly reduced PtdIns(4,5)P_2 levels, which was accompanied by an attenuated gravitropic response and reduced shoot and root growth. The lack of PtdIns(4,5)P_2 in *pip5K1/pip5K2* double mutants affected PIN2 localization through the perturbation of clathrin-mediated endocytic trafficking.

While PtdIns has been identified in phloem exudates, the methodology employed could not unambiguously identify phosphoinositides. Hence, any possible function in long-distance signaling may be indirect, for example through the enhancement of trafficking of signaling molecules to the sieve element or through the conversion of PtdIns to DAG to PtdOH as signaling molecules.

Phosphatidic Acid

Phosphatidic acid (PtdOH) is an important intermediate in lipid biosynthesis, a membrane component, and a signaling molecule. As membrane component it may affect the membrane curvature, and, as a consequence, can regulate trafficking and membrane biogenesis (Kooijman et al. 2003; Wang 2004). In addition, it participates in signaling pathways, often by tethering components of these pathways to the membrane, thus alternating their location and function.

PtdOH is rapidly and transiently generated in response to several biotic and abiotic stresses, such as pathogen infection, drought, salinity, wounding, cold, cell death, and oxylipin production by either the PLC-DGK pathway or directly by PLD (Munnik and Testerink 2009; Xue et al. 2009; Wang et al. 2007; Hong et al. 2010; Kim et al. 2013; Testerink and Munnik 2011; Wang et al. 2014; Julkowska et al. 2015). PLD cleaves structural phospholipids such as phosphatidylcholine (PtdCho) and phosphatidylethanolamine (PtdEtn) in order to generate PtdOH and a free head group. PLC, on the other hand, hydrolyzes PtdIns(4,5)P_2 into Ins(1,4,5)P_3 and DAG (Julkowska et al. 2015). InsP_3 diffuses into the cytosol where it triggers Ca^{2+} release from intracellular stores, while DAG remains within the membrane and is phosphorylated to PtdOH by DAG kinase (DGK) (Testerink and Munnik 2005, 2011).

Different stresses elicit different PtdOH production paths. For example, PtdOH produced in response to cold stress is generated via the PLC-DGK pathway. However, mutations in individual DGK genes did not affect this increase in PtdOH, likely due to redundancy (Arisz et al. 2013). Similarly, pathogens appear to induce the PLC-DGK pathway: an infection-mimicking treatment with xylanase inducers leads to an increased transcription of *StPLC5* and *StPLC2* in tomato cell suspension cultures (Gonorazky et al. 2014). On the other hand, the responses to drought, nutrient deficiency, and salt stress are largely mediated by PLD pathways (McLoughlin and Testerink 2013): PLDζ2 is involved in the response to salt and phosphate stress by not only affecting PtdOH production but PIN redistribution and, as a result, root curvature and root growth (Li and Xue 2007; Oropeza-Aburto et al. 2011; Galvan-Ampudia et al. 2013). PLDα1 and PLDδ mediate drought/ABA-induced and ROS-induced PtdOH production, respectively (Uraji et al. 2012). They appear to play a role in freezing tolerance as well (Welti et al. 2002). The drought response will be discussed further below.

PtdOH binds a myriad of proteins, including transcription factors, protein kinases, lipid kinases, protein phosphatases, as well as proteins involved in vesicular trafficking and cytoskeletal rearrangements (Guo et al. 2011). Examples of PtdOH -interacting proteins include abscisic acid insensitive 1 (ABI1), phosphoinositide-dependent protein kinase (PDK1), constitutive triple response 1 (CTR1), trigalactosyldiacylglycerol 2 (TGD2), and NADPH oxidase (Guo et al. 2011). When PtdOH binds to ABI1, CTR1, and phosphoethanolamine N-methyltransferase (PEAMT), it prevents their phosphatase or kinase activity. On the other hand, through direct interaction, PtdOH stimulates the catalytic activity of PDK1, NADPH oxidases (RESPIRATORY BURST OXIDASE HOMOLOG D/F; RbohD/F), sphingosine kinases (SPHK1/2), mitogen activated protein kinase 6 (MPK6), and SNF1-related kinase 2 and glyceraldehyde-3-phosphate dehydrogenase (G3PDH; SnRK2; Kim et al. 2013; McLoughlin and Testerink 2013).

In addition to regulating protein activity through their active site, PtdOH has been shown to tether some of these proteins to the membrane. A well-known example is the function of PtdOH in the response to drought/ABA-signaling cascade: In response to ABA, PtdOH is produced by PLDα1. This phospholipase, in addition to PLDδ, participates in mediating the plant's response to ABA as well as promoting stomatal closure, although the two lipases regulate different areas of the signaling pathway. PLDδ facilitates ROS response, while PLDα1 acts upstream of PLDδ and promotes ROS production (Uraji et al. 2012; Lu et al. 2013). Both, PLDα1 and PLDδ positively regulate ABA-induced stomatal closure and are involved in ABA-induced ROS and NO production. It appears that PLDα1 functions mainly under moderate environmental conditions while PLDα1 and PLDδ, together work under more severe environmental stress conditions (Uraji et al. 2012). Under drought stress, ABA binds to the pyrabactin resistance/pyr1-like protein/regulatory components of its receptor (PYR/PYL/RCAR), which results the inhibition of ABI1, a type 2C phosphatase (PP2C) and a negative regulator, allowing SNF1-related kinase 2 (SnRK2) to be activated, which mediates downstream signaling such as the phosphorylation of transcription factors and ABA-induced gene expression (Julkowska et al. 2015; Dorosh et al. 2013). In turn, PtdOH is produced by PLDα1 and interacts with SPHK (sphingosine kinase). Increased phyto-S1P (a phosphorylated long-chain base-1-phosphate) activates more PLDα1, leading to an increase in PtdOH levels. PtdOH regulates proteins such as ABI1, NADPH oxidase, and ion channels to facilitate stomatal closure (Guo et al. 2012). As part of this response, PtdOH binds and tethers ABI1 to the plasma membrane which prevents ABI1 from translocating to the nucleus thus preventing its binding to ATHB6, a transcription factor that negatively regulates ABA responses (Klingler et al. 2010; Kim et al. 2013; Zhang et al. 2013). A recent study noted that under drought conditions, PtdOH binds the MYB transcription factor WEREWOLF, and may tether it to the nuclear membrane, which would allow for the localization of the proteins from the cytosol to the nucleus (Yao et al. 2013). This could aid in further transcriptional regulation of other drought-responsive genes.

In addition to its effect on stomatal closure, PtdOH is important for normal root development and thus normal water and nutrient uptake in plants: PLDα knock-down mutants display shorter roots and reduced lateral roots (*pld*α3) as well as an altered PtdOH accumulation in response to water deficit and ABA (*pld*α1 and *pld*α3; Wang 2005; Singh et al. 2012). This suggests that under standard conditions, PtdOH regulates normal root development, possibly via the action of PIN proteins (Gao et al. 2013).

Outlook

So far, the described role of PtdOH in stress responses has solely been of <u>intracellular nature</u>. Recently, we have been able to show that PtdOH is present in phloem exudates (Guelette et al. 2012). Its function there could be to tether a signaling protein to the membrane and thus change its activity, it could tether a receptor to the membrane, or it could participate in the movement of essential components of ABA/drought-related signaling from the companion cell into the sieve element. In addition, it could be part of the signal itself, similar to the Wnt-palmitoleic acid interaction (Janda et al. 2012). We have identified at least one phloem-localized PtdOH-binding protein, PLAFP (phloem-localized lipid-associated protein), a 20 kDa uncharacterized protein with a PLAT/LH2 domain. In humans and animals, PLAT domains (Polycystin-1, Lipoxygenase, Alpha-Toxin-domains) are found in a variety of lipid- or membrane-associated proteins. The domain contains eight beta-strands and forms a beta-sandwich composed of two sheets of four strands each. While structurally similar, the PLAT domain of a group of PLAFPs is very different from that of humans: the plant-specific PLAT_plant_stress domain (cd01754) is loosely related but not identical to lipase and lipoxygenase domains. Homology to human lipoxygenases is less than 30 %. Plant proteins with this domain, on the contrary, are highly conserved in different plant species. Most importantly, these PLAFP proteins do not contain the lipoxygenase catalytic domain but only their PLAT domain with a role in protein-protein interaction. Contrary to larger, usually polymeric lipoxygenases, their size of only 20 kDa would allow for movement through plasmodesmata and through the phloem. Thus, PLAFP could function in the transport of lipid-signals into, out, or through the phloem sieve element, the detection of lipid-signals, or be part of a signal itself. PtdOH is an important signaling compound, yet the characteristics of the PtdOH-binding site in proteins remain elusive. SnRK2.4, a known PtdOH-binding protein, contains a 42-amino acid PtdOH-binding domain with five conserved basic amino acids (Julkowska et al. 2015). These conserved amino acids are essential for PtdOH binding but not sufficient to mediate salt-induced re-localization of the protein. Hence, additional factors may be necessary. If PtdOH anchors a protein to the membrane, a conserved hydrophilic pocket with basic amino acids as found in SnrRK.4 could interact with the head group while the remainder of the lipid remains integrated in the membrane. However, if the PtdOH-binding specificity extends beyond the binding of the head group and protein–lipid interaction is also based on the characteristics of the acyl groups, it would suggest that the protein binds the entire lipid, in which case the latter may not remain integrated in the membrane but instead is embedded in a hydrophobic groove on the surface of the protein . This is particularly interesting in the context of findings by other labs that different stresses are involving distinct PtdOH species (Guo et al. 2012). A dynamic modeling approach (Visual Molecular Dynamics – VMD; http://www.ks.uiuc.edu/Research/vmd/) using a ligand obtained from the RCSB-protein data bank (http://www.rcsb.org/; (R)-2-(formyloxy)-3-(phosphonooxy) propyl pentanoate) as approximation for an, albeit short-chain-, PtdOH suggested the presence of a hydrophobic grove on the surface of PLAFP bordered by amino acids with basic side chains (Fig. 14.2). These features would facilitate not only the interaction of the protein with the phosphatidic acid head group but with the acyl chains as well, suggesting interaction with the entire length of the molecule and thus possible PtdOH movement away from the membrane and intracellular or long-distance transport. While this model, at this point, is entirely speculative, it provides further avenues for studying possible mechanisms of phospholipid movement between membranes and throughout the plant.

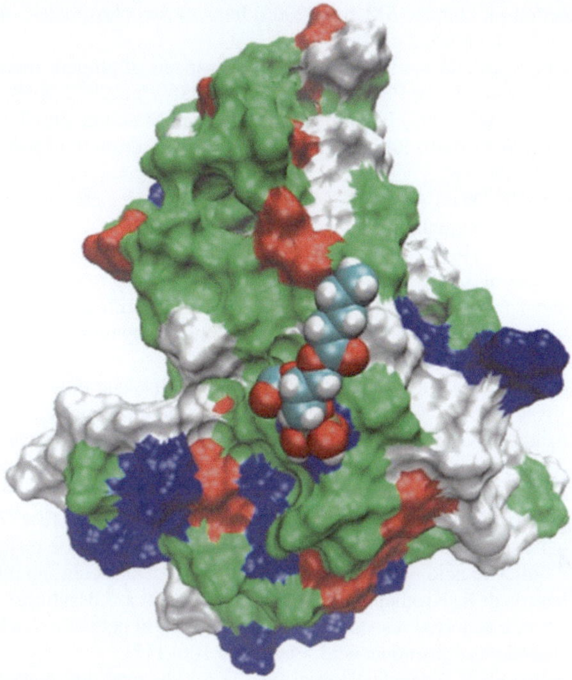

Fig. 14.2 Predicted ligand binding to PLAFP using vmd. C5-PA was used as a ligand. Glycerol backbone and phosphate headgroup are adjacent to hydrophilic (*green*) and basic (*blue*) amino acid side chains, respectively. The acyl groups extend into a hydrophobic area (*grey*) of the protein. A longer, monounsaturated acyl group could extend into the hydrophobic region at the top of the protein.

Acknowledgements The model in Fig. 14.2 was generated by Michael O'Keefe with the help of Michael Feig. This work was funded by NSF grant #1144391 and USDA-NIFA hatch project # MICL02233 to SHB. Lipid structures were obtained from Wikipedia.org, lipidlibrary.aocs.org, www.bio-protocol.org.

References

Andrawis A, Solomon M, Delmer DP (1993) Cotton fiber annexins – a potential role in the regulation of callose synthase. Plant J 3:763–772

Arisz SA, Testerink C, Munnik T (2009) Plant PA signaling via diacylglycerol kinase. Biochim Biophys Acta 1791(9):869–875

Arisz SA, van Wijk R et al (2013) Rapid phosphatidic acid accumulation in response to low temperature stress in Arabidopsis is generated through diacylglycerol kinase. Front Plant Sci 4:1. doi:10.3389/fpls.2013.00001. eCollection 2013

Ayre BG, Turgeon R (2004) Graft transmission of a floral stimulant derived from CONSTANS. Plant Physiol 135:2271–2278

Barnes A, Bale J et al (2004) Determining protein identity from sieve element sap in *Ricinus communis* L. by quadrupole time of flight (Q-TOF) mass spectrometry. J Exp Bot 55:1473–1481

Bateman A, Sandford R (1999) The PLAT domain: a new piece in the PKD1 puzzle. Curr Biol 9:R588–R590. doi:10.1016/S0960-9822(99)80380-7

Behmer ST, Grebenok RJ, Douglas AE (2010) Plant sterols and host plant suitability for a phloem feeding insect. Funct Ecol 25:484–491

Bennett M, Onnebo SM et al (2006) Inositol pyrophosphates: metabolism and signaling. Cell Mol Life Sci 63:552–564
Benning UF, Tamot B et al (2012) New aspects of phloem-mediated long-distance lipid signaling in plants. Front Plant Sci 3:53. doi:10.3389/fpls.2012.0053
Berridge MJ (1993) Inositol trisphosphate and calcium signaling. Nature 361:315–325
Blaner WS (1989) Retinol-binding protein: the serum transport protein for vitamin A. Endocr Rev 10:308–316
Boss WF, Davis AJ et al (2006) Phosphoinositide metabolism: towards an understanding of subcellular signaling. Subcell Biochem 39:181–205
Burton A, Hu X, Saiardi A (2009) Are inositol pyrophosphates signalling molecules? J Cell Physiol 220:8–15
Buseman CM, Tamura P et al (2006) Wounding stimulates the accumulation of glycerolipids containing oxophytodienoic acid and dinor-oxophytodienoic acid in *Arabidopsis* leaves. Plant Physiol 142:28–39
Champigny MJ, Isaacs M et al (2013) Long distance movement of DIR1 and investigation of the role of DIR1-like during systemic acquired resistance in Arabidopsis. Front Plant Sci 4:230
Chanda B, Xia Y et al (2011) Glycerol-3-phosphate is a critical mobile inducer of systemic immunity in plants. Nat Genet 43:421–427. doi:10.1038/ng.798. Epub 2011 Mar 27
Charbonneau D, Beauregard M, Tajmir-Riahi HA (2009) Structural analysis of human serum albumin complexes with cationic lipids. J Phys Chem B 113:1777–1781
Chaturvedi R, Shah J (2007) Salicylic acid in plant disease resistance. In: Hayat S, Ahmad A (eds) Salicylic acid – a plant hormone. Springer, Dordrecht, pp 335–370
Chaturvedi R, Krothapalli K et al (2008) Plastid x-3 desaturase-dependent accumulation of a systemic acquired resistance inducing activity in petiole exudates of Arabidopsis thaliana is independent of jasmonic acid. Plant J 54:106–117
Chaturvedi R, Venables B et al (2012) An abietane diterpenoid is a potent activator of systemic acquired resistance. Plant J 71:161–172
Chehab EW, Kaspi R et al (2008) Distinct roles of jasmonates and aldehydes in plant-defense responses. PLoS One 3, e1904
Chen S, Petersen BL et al (2001) Long-distance phloem transport of glucosinolates in Arabidopsis. Plant Physiol 127:194–201
Chen QF, Xiao S, Chye ML (2008) Overexpression of the Arabidopsis 10-kilodalton acyl-coenzyme A-binding protein ACBP6 enhances freezing tolerance. Plant Physiol 148:304–315
Chiou TJ, Lin SI (2011) Signaling network in sensing phosphate availability in plants. Annu Rev Plant Biol 62:185–206
Citovsky V, Zambryski P (2000) Systemic transport of RNA in plants. Trends Plant Sci 5:52
Corbesier L, Prinsen E et al (2003) Cytokinin levels in leaves, leaf exudate and shoot apical meristem of *Arabidopsis thaliana* during floral transition. J Exp Bot 54:2511–2517
Corbesier L, Vincent C et al (2007) FT protein movement contributes to long-distance signaling in floral induction of Arabidopsis. Science 316:1030–1033
Dave A, Graham IA (2012) Oxylipin signalling: a distinct role for the Jasmonic acid precursor 12-oxo-phytodienoic acid (OPDA). Front Plant Sci 3:42
Dave A, Hernández ML et al (2011) 12-Oxo-phytodienoicacid accumulation during seed development represses seed germination in *Arabidopsis*. Plant Cell 23:583–599
Ding B, Itaya A, Qi YJ (2003) Symplasmic protein and RNA traffic: regulatory points and regulatory factors. Curr Opin Plant Biol 6:596–602
Dorosh L, Kharenko OA et al (2013) Molecular mechanisms in the activation of abscisic acid receptor PYR1. PLoS Comput Biol 9:e1003114. doi:10.1371/journal.pcbi.1003114. Epub 2013 Jun 27
Dowd PE, Gilroy S (2010) The emerging roles of phospholipase C in plant growth and development. In: Munnik T (ed) Lipid signaling in plants. Springer, Berlin
Feussner I, Wasternack C (2002) The lipoxygenase pathway. Annu Rev Plant Biol 53:275–297
Fisher DB, Wu Y, Ku MSB (1992) Turnover of soluble-proteins in the wheat sieve tube. Plant Physiol 100:1433–1441

Froehlich DR, Mullendore DL et al (2011) Phloem ultrastructure and pressure flow: sieve-element-occlusion-related agglomerations do not affect translocation. Plant Cell 23:4428–4445

Galvan-Ampudia CS, Julkowska MM et al (2013) Halotropism is a response of plant roots to avoid a saline environment. Curr Biol 23:2044–2050. doi:10.1016/j.cub.2013.08.042. Epub 2013 Oct 3

Gao HB, Chu YJ, Xue HW (2013) Phosphatidic acid (PA) binds PP2AA1 to regulate PP2A activity and PIN1 polar localization. Mol Plant 6(5):1692–1702. doi:10.1093/mp/sst076. Epub 2013 May 17

Giavalisco P, Kapitza K et al (2006) Towards the proteome of *Brassica napus* phloem sap. Proteomics 6:896–909

Gillaspy GE (2013) The role of phosphoinositides and inositol phosphates in plant cell signaling. In: Capelluto GS (ed) Lipid-mediated protein signaling. Springer, Dordrecht, pp 141–157

Glatz JFC, Boerchers T et al (1995) Fatty acids in cell signalling: modulation by lipid binding proteins. Prostaglandins Leukot Essent Fatty Acids 52:121

Gonorazky G, Ramirez L et al (2014) The tomato phosphatidylinositol-phospholipase C2 (*Sl*PLC2) is required for defense gene induction by the fungal elicitor xylanase. J Plant Physiol 171:959–965. doi:10.1016/j.jplph.2014.02.008. Epub 2014 Mar 12

Guelette BS, Benning UF, Hoffmann-Benning S (2012) Identification of lipids and lipid-binding proteins in phloem exudates from Arabidopsis thaliana. J Exp Bot 63:3603–3616. doi:10.1093/jxb/ers028

Guo L, Mishra G, Taylor K, Wang X (2011) Phosphatidic acid binds and stimulates *Arabidopsis* sphingosine kinases. J Biol Chem 286(15):13336–13345

Guo L, Mishra G et al (2012) Connections between sphingosine kinase and phospholipase D in the abscisic acid signaling pathway in Arabidopsis. J Biol Chem 287:8286–8296

Haebel S, Kehr J (2001) Matrix-assisted laser desorption/ionization time of flight mass spectrometry peptide mass fingerprints and post source decay: a tool for the identification and analysis of phloem proteins from Cucurbita maxima Duch. separated by two-dimensional polyacrylamide gel electrophoresis. Planta 214:332

Hayashi H, Fukuda A et al (2000) Proteins in the sieve element-companion cell complexes: their detection, localization and possible functions. Aust J Plant Physiol 27:489–496

Haywood V, Yu TS et al (2005) Phloem long-distance trafficking of gibberellic acid-insensitive RNA regulates leaf development. Plant J 42:49–68

Heilmann I (2009) Using genetic tools to understand plant phosphoinositide signalling. Trends Plant Sci 14:171–179

Hoad GV (1995) Transport of hormones in the phloem of higher plants. Plant Growth Reg 16:173–182

Hoffmann-Benning S, Gage DA et al (2002) Comparison of peptides in the phloem sap of flowering and non-flowering *Perilla* and lupine plants using microbore HPLC followed by matrix-assisted laser desorption/ionization time-of-flight mass spectrometry. Planta 216:140–147

Hoffmann-Benning S (2015) https://bmb.natsci.msu.edu/faculty/susanne-hoffmann-benning-assistant-professor/current-research

Hong Y, Zhang W, Wang X (2010) Phospholipase D and phosphatidic acid signaling in plant response to drought and salinity. Plant Cell Environ 33:627–635

Howe GA, Schilmiller AL (2002) Oxylipin metabolism in response to stress. Curr Opin Plant Biol 5:230–236

Hung CY, Aspesi P Jr et al (2014) Phosphoinositide-signaling is one component of a robust plant defense response. Front Plant Sci 5:267. doi:10.3389/fpls.2014.00267

Im YJ, Perera IY et al (2007) Increasing plasma membrane phosphatidylinositol (4,5)bisphosphate biosynthesis increases phosphoinositide metabolism in *Nicotiana tabacum*. Plant Cell 19:1603–1616

Ischebeck T, Werner S et al (2013) Phosphatidylinositol 4,5-bisphosphate influences PIN polarization by controlling clathrin-mediated membrane trafficking in Arabidopsis. Plant Cell 25:4894–4911. doi:10.1105/tpc.113.116582. Epub 2013 Dec 10

Janda CY, Waghray D et al (2012) Structural basis of Wnt recognition by frizzled. Science 337:59–64

Julkowska MM, McLoughlin F et al (2015) Identification and functional characterization of the *Arabidopsis* Snf1-related protein kinase SnRK2.4 phosphatidic acid-binding domain. Plant Cell Environ 38:614–624. doi:10.1111/pce.12421. Epub 2014 Sep 1

Jung HW, Tschaplinski TJ et al (2009) Priming in systemic plant immunity. Science 324:89–91

Kehr J, Haebel S et al (1999) Analysis of phloem protein patterns from different organs of *Cucurbita maxima* Duch. by matrix-assisted laser desorption/ionization time of flight mass spectroscopy combined with sodium dodecyl sulfate-polyacrylamide gel electrophoresis. Planta 207:612–619

Kim SC, Guo L, Wang X (2013) Phosphatidic acid binds to cytosolic glyceraldehyde-3-phosphate dehydrogenase and promotes its cleavage in *Arabidopsis*. J Biol Chem 288:11834–11844. doi:10.1074/jbc.M112.427229

Klingler JP, Batelli G, Zhu J-K (2010) ABA receptors: the START of a new paradigm in phytohormone signaling. J Exp Bot 61:3199–3210

König S, Ischebeck T et al (2008) Salt-stress-induced association of phosphatidylinositol 4,5-bisphosphate with clathrin-coated vesicles in plants. Biochem J 415:387–399

Kooijman EE, Chupin V et al (2003) Modulation of membrane curvature by phosphatidic acid and lysophosphatidic acid. Traffic 4:162–174

Kühn C, Franceschi VR et al (1997) Macromolecular trafficking indicated by localization and turnover of sucrose transporters in enucleate sieve elements. Science 275:1298–1300

Lascombe MB, Bakan B et al (2008) The structure of "defective in induced resistance" protein of Arabidopsis thaliana, DIR1, reveals a new type of lipid transfer protein. Protein Sci 17:1522–1530

Li G, Xue HW (2007) *Arabidopsis* PLD zeta2 regulates vesicle trafficking and is required for auxin response. Plant Cell 19:281–295. doi:10.1105/tpc.106.041426

Lin MK, Belanger H et al (2007) Flowering locus T protein may act as the long-distance florigenic signal in the cucurbits. Plant Cell 19:1488–1506

Lin MK, Lee YJ et al (2009) Analysis of the pumpkin phloem proteome provides functional insights into angiosperm sieve tube function. Mol Cell Proteomics 8:343–356

López-Bucio J, Hernández-Abreu E et al (2002) Phosphate availability alters architecture and causes changes in hormone sensitivity in the *Arabidopsis* root system. Plant Physiol 129:244–256

Lu S, Bahn SC et al (2013) Increased expression of phospholipase Dα1 in guard cells decreases water loss with improved seed production under drought in Brassica napus. Plant Biotechnol J 11:380–389

Lucas WJ, Groover A et al (2013) The plant vascular system: evolution, development and functions. J Integr Plant Biol 55:294–388. doi:10.1111/jipb.12041

Luna E, Ton J (2012) The epigenetic machinery controlling transgenerational systemic acquired resistance. Plant Signal Behav 7:615–618

Madey E, Nowack LM, Thompson JE (2002) Isolation and characterization of lipid in phloem sap of canola. Planta 214:625–634

Malamy J, Carr JP et al (1990) Salicylic acid: a likely endogenous signal in the resistance response of tobacco to viral infection. Science 250:1002–1004

Maldonado AM, Doerner P et al (2002) A putative lipid transfer protein involved in systemic acquired resistance signaling in Arabidopsis. Nature 419:399–403

Mandal M, Chanda B et al (2011) Glycerol-3-phosphate and systemic immunity. Plant Signal Behav 6:1871–1874

Marentes E, Grusak MA (1998) Mass determination of low-molecular-weight proteins in phloem sap using matrix-assisted laser desorption/ionization time-of-flight mass spectrometry. J Exp Bot 49:903–911

McLoughlin F, Testerink C (2013) Phosphatidic acid, a versatile water-stress signal in roots. Front Plant Sci 4:525. doi:10.3389/fpls.2013.00525

Mei Y, Jia WJ et al (2012) Arabidopsis phosphatidylinositol monophosphate 5-kinase 2 is involved in root gravitropism through regulation of polar auxin transport by affecting the cycling of PIN proteins. Cell Res 22:581–597

Métraux JP, Signer H et al (1990) Increase in salicylic acid at the onset of systemic acquired resistance in cucumber. Science 250:1004–1006

Mitton F, Pinedo ML, de la Canal L (2009) Phloem sap of tomato plants contains a DIR1 putative ortholog. J Plant Physiol 166:543–547

Mortimer JC, Laohavisit A et al (2008) Annexins: multifunctional components of growth and adaptation. J Exp Bot 59:533–544

Mosblech A, Thurow C et al (2011) Jasmonic acid perception by COI1 involves inositol polyphosphates in *Arabidopsis thaliana*. Plant J 65(6):949–957

Mueller-Roeber B, Pical C (2002) Inositol phospholipid metabolism in Arabidopsis. Characterized and putative isoforms of inositol phospholipid kinase and phosphoinositide-specific phospholipase C. Plant Physiol 130:22–46

Münch E (1930) Material flow in plants. Translated 2003 by Milburn JA and Kreeb KH, University of Bremen, Germany. Gustav Fischer Verlag, Jena

Munnik T, Testerink C (2009) Plant phospholipid signaling: "in a nutshell". J Lipid Res 50:S.260–S265

Munnik T, Irvine RF, Musgrave A (1998) Phospholipid signalling in plants. Biochim Biophys Acta 1389:222–272

Musille PM, Kohn JA, Ortlund EA (2013) Phospholipid-driven gene regulation. FEBS Lett 587(8):1238–1246

Mustroph A, Zanetti ME et al (2009) Profiling translatomes of discrete cell populations resolves altered cellular priorities during hypoxia in *Arabidopsis*. Proc Natl Acad Sci U S A 106:18843–18848

Nagy L, Szanto A (2005) Roles for lipid activated transcription factors. Mol Nutr Food Res 49:1072–1074

Nagy R, Grob H et al (2009) The *Arabidopsis* ATP-binding cassette protein AtMRP5/AtABCC5 is a high affinity inositol hexakisphosphate transporter involved in guard cell signaling and phytate storage. J Biol Chem 284:33614–33622

Nakamura Y, Andrés F et al (2014) Arabidopsis florigen FT binds to diurnally oscillating phospholipids that accelerate flowering. Nat Commun 5:3553

Newton AC (2010) Protein kinase C: poised to signal. Am J Physiol Endocrinol Metab 298(3):E395–E402

Ohlrogge J, Browse J (1995) Lipid biosynthesis. Plant Cell 7:957–970

Oropeza-Aburto A, Cruz-Ramirez A et al (2011) Functional analysis of the *Arabidopsis* PLDZ2 promoter reveals an evolutionarily conserved low-Pi- responsive transcriptional enhancer element. J Exp Bot 63:2189–2202. doi:10.1093/jxb/err446

Park SW, Kaimoyo E et al (2007) Methyl salicylate is a critical mobile signal for plant systemic acquired resistance. Science 318:113–116

Perera IY, Hung CY et al (2008) Transgenic Arabidopsis plants expressing the type1 inositol 5-phosphatase exhibit increased drought tolerance and altered abscisic acid signaling. Plant Cell 20:2876–2893. doi:10.1105/tpc.108.061374

Raskin I (1992) Role of salicylic acid in plants. Annu Rev Plant Physiol Plant Mol Biol 43:439–463

Rescher U, Gerke V (2004) Annexins – unique membrane binding proteins with diverse functions. J Cell Sci 117:2631–2639

Ruiz-Medrano R, Xoconostle-Cázares B, Lucas WJ (1999) Phloem long-distance transport of CmNACP mRNA: implications for supracellular regulation in plants. Development 126:4405–4419

Salinas-Mondragon RE, Kajla JD et al (2010) Role of inositol 1,4,5-trisphosphate signaling in gravitropic and phototropic gene expression. Plant Cell Envrion 33:2041–2055

Savchenko T, Kolla VA et al (2014) Functional convergence of oxylipin and abscisic acid pathways controls stomatal closure in response to drought. Plant Physiol 164:1151–1160

Schilmiller AL, Howe GA (2005) Systemic signaling in the wound response. Curr Opin Plant Biol 8:369–377

Shah J, Chaturvedi R et al (2014) Signaling by small metabolites in systemic acquired resistance. Plant J 79:645–658

Singh A, Pandey et al (2012) Comprehensive expression analysis of rice phospholipase D gene family during abiotic stresses and development. Plant Signal Behav 7:847–855

Stenzel I, Ischebeck T et al (2012) Variable regions of PI4P 5-kinases direct PtdIns(4,5)P2 toward alternative regulatory functions in tobacco pollen tubes. Front Plant Sci 2:114

Tamaki S, Matsuo S et al (2007) Hd3a protein is a mobile flowering signal in rice. Science 316:1033–1036

Tan X, Calderon-Villalobos LI et al (2007) Mechanism of auxin perception by the TIR1 ubiquitin ligase. Nature 446:640–645

Taoka K et al (2011) 14-3-3 proteins act as intracellular receptors for rice Hd3a florigen. Nature 476:332–335

Testerink C, Munnik T (2005) Phosphatidic acid: a multifunctional stress signaling lipid in plants. Trends Plant Sci 10:368–375

Testerink C, Munnik T (2011) Molecular, cellular, and physiological responses to phosphatidic acid formation in plants. J Exp Bot 62(7):2349–2361

Tetyuk O, Benning UF, Hoffmann-Benning S (2013) Collection and analysis of Arabidopsis phloem exudates using the EDTA-facilitated method. J Vis Exp 80, e51111. doi:10.3791/51111. Movie and Text

Thole JM, Nielsen E (2008) Phosphoinositides in plants: novel functions in membrane trafficking. Curr Opin Plant Biol 11:620–631

Thomas B, Vince-Prue D (1997) Photoperiodism in plants. Academic, London

Thomas H, Stoddard JL (1980) Leaf senescence. Annu Rev Plant Physiol 31:83–111. doi:10.1146/annurev.pp.31.060180.000503

Thorpe MR, Ferrieri AP et al (2007) (11)C-imaging: methyl jasmonate moves in both phloem and xylem, promotes transport of jasmonate, and of photoassimilate even after proton transport is decoupled. Planta 226:541–551

Tontonez P, Hu E, Spiegelman BM (1994) Stimulation of adipogenesis in fibroblasts by PPARg2, a lipid-activated transcription factor. Cell 79:1147–1156

Truman W, Bennett MH et al (2007) Arabidopsis systemic immunity uses conserved defense signaling pathways and is mediated by jasmonates. Proc Natl Acad Sci U S A 104:1075–1080

Truman W, Bennett MH et al (2010) *Arabidopsis* auxin mutants are compromised in systemic acquired resistance and exhibit aberrant accumulation of various indolic compounds. Plant Physiol 152:1562–1573

Turgeon R, Wolfe S (2009) Phloem transport: cellular pathways, molecular trafficking. Annu Rev Plant Biol 60:207–221

Tuzun S, Kuc J (1985) Movement of a factor in tobacco infected with Peranospora tabacina Adam which systemically protects against blue mold. Physiol Plant Pathol 26:321–330

Tzen JTC, Huang AHC (1992) Surface structure and properties of plant seed oil bodies. J Cell Biol 117:327–335

Uraji M, Katagiri T et al (2012) Cooperative function of PLD delta and PLD alpha1 in ABA-induced stomatal closure in *Arabidopsis*. Plant Physiol 159:450–460. doi:10.1104/pp.112.195578

van Bel AJE, Knoblauch M (2000) Sieve element and companion cell: the story of the comatose patient and the hyperactive nurse. Aust J Plant Physiol 27:477–487

Wang X (2004) Lipid signaling. Curr Opin Plant Biol 7:329–336

Wang X (2005) Regulatory functions of phospholipase D and phosphatidic acid in plant growth, development, and stress responses. Plant Physiol 139(2):566–573

Wang X, Zhang W, et al (2007) Phospholipid signaling in plant response to drought and salt stress. In Jenks MA et al (eds) Advances in molecular breeding toward drought and salt tolerant crops. Springer, Netherlands, pp 183–192

Wang G, Ryu S, Wang X (2012) Plant phospholipases: an overview. Methods Mol Biol 861:123–137

Wang X, Chapman KD (2013) Lipid signaling in plants. Front Plant Sci 4:1–2

Wang X, Guo L, Wang G, Li M (2014) PLD: phospholipase Ds in plant signaling phospholipases in plant signaling. Springer-Verlag Berlin Heidelberg, pp 3–26

Welti R, Li W et al (2002) Profiling membrane lipids in plant stress responses. Role of phospholipase D alpha in freezing-induced lipid changes in Arabidopsis. J Biol Chem 277:31994–32002. doi:10.1074/jbc.M205375200

Wu X, Dinneny JR et al (2003) Modes of intercellular transcription factor movement in the *Arabidopsis* apex. Development 130:3735–3745

Xoconostle-Cazares B, Yu X et al (1999) Plant paralog to viral movement protein that potentiates transport of mRNA into the phloem. Science 283:94–98

Xue H-W, Chen X, Mei Y (2009) Function and regulation of phospholipid signalling in plants. Biochem J 421:145–156

Yan Y, Christensen S et al (2012) Disruption of *OPR7* and *OPR8* reveals the versatile functions of JA in maize development and defense. Plant Cell 24:1420–1436

Yan Y, Borrego E, Kolomiets MV (2013) Jasmonate biosynthesis, perception and function in plant development and stress responses. In: Baez RV (ed) Lipid metabolism. Intech, pp 393–442. http://dx.doi.org/10.5772/52675

Yao H, Wang G et al (2013) Phosphatidic acid interacts with a MYB transcription factor and regulates its nuclear localization and function in Arabidopsis. The Plant Cell 25:5030–5042

Yoo BC, Kragler F et al (2004) A systemic small RNA signaling system in plants. Plant Cell 16:1979–2000

Yu K, Soares JM et al (2013) A feedback regulatory loop between G3P and lipid transfer proteins DIR1 and AZI1 mediates azelaic-acid-induced systemic immunity. Cell Rep 3:1266–1278

Zeevaart JAD (1976) Physiology of flower formation. Annu Rev Plant Physiol Plant Mol Biol 27:321–348

Zhang X, Jiang L et al (2013) Structural insights into the abscisic acid stereospecificity by the ABA receptors PYR/PYL/RCAR. PLoS One 8, e67477

Zheng SX, Xiao S, Chye ML (2012) The gene encoding Arabidopsis Acyl-CoA-binding protein 3 is pathogen-inducible and subject to circadian regulation. J Exp Bot 63:2985–3000

Zhu J-K (2002) Salt and drought stress signal transduction in plants. Annu Rev Plant Biol 53:247–273

Chapter 15
Acyl-CoA-Binding Proteins (ACBPs) in Plant Development

Shiu-Cheung Lung and Mee-Len Chye

Abstract Acyl-CoA-binding proteins (ACBPs) play a pivotal role in fatty acid metabolism because they can transport medium- and long-chain acyl-CoA esters. In eukaryotic cells, ACBPs are involved in intracellular trafficking of acyl-CoA esters and formation of a cytosolic acyl-CoA pool. In addition to these ubiquitous functions, more specific non-redundant roles of plant ACBP subclasses are implicated by the existence of multigene families with variable molecular masses, ligand specificities, functional domains (e.g. protein-protein interaction domains), subcellular locations and gene expression patterns. In this chapter, recent progress in the characterization of ACBPs from the model dicot plant, *Arabidopsis thaliana*, and the model monocot, *Oryza sativa*, and their emerging roles in plant growth and development are discussed. The functional significance of respective members of the plant ACBP families in various developmental and physiological processes such as seed development and germination, stem cuticle formation, pollen development, leaf senescence, peroxisomal fatty acid β-oxidation and phloem-mediated lipid transport is highlighted.

Keywords Acyl-CoA esters • Fatty acids • Lipid trafficking • Phospholipid metabolism • Transporters

Introduction

In eukaryotic cells, the metabolism of fatty acids (FAs) takes place at various subcellular locations and requires extensive metabolite exchange *via* intracellular lipid trafficking. Plant FAs are synthesized *de novo* predominantly in the stroma of plastids, which houses acetyl-CoA carboxylase for acetyl-CoA carboxylation as the first committed step and FA synthetase for subsequent acyl-chain extension (Rawsthorne 2002). These nascent FAs are either utilized directly for plastidial lipid biosynthesis *via* the prokaryotic pathway (Roughan et al. 1980; Sparace and Mudd 1982; Heinz

S.-C. Lung • M.-L. Chye (✉)
School of Biological Sciences, The University of Hong Kong, Pokfulam, Hong Kong, China
e-mail: mlchye@hkucc.hku.hk

Y. Nakamura, Y. Li-Beisson (eds.), *Lipids in Plant and Algae Development*,
Subcellular Biochemistry 86, DOI 10.1007/978-3-319-25979-6_15

and Roughan 1983) or exported into the cytosol after activation into CoA-esters by acyl-CoA synthetase at the outer membrane of the plastidial envelope (Andrews and Keegstra 1983; Block et al. 1983). The plant mitochondria are also capable of limited FA synthesis, mainly for the production of lipoic acid (Wada et al. 1997; Gueguen et al. 2000). At the endoplasmic reticulum (ER) membrane, the cytosolic acyl-CoA pool fuels the assembly of triacylglycerols (TAGs) and extraplastidial membrane lipids such as phosphatidylcholine (PC) *via* the eukaryotic pathway (Slack and Roughan 1975; Simpson and Williams 1979; Dubacq et al. 1983). During this process, an acyl group is released from PC by the reverse reaction of acyl-CoA:lyso-PC acyltransferase (LPCAT) or phospholipase A, and the resulting lyso-PC can be reacylated by the forward reaction of LPCAT (Stymne and Stobart 1984; Wang 2001). The net outcome of this deacylation-reacylation cycle, often known as "acyl editing", is an exchange of acyl groups between the cytosolic acyl-CoA pool and PC at the ER (Bates et al. 2007; Bates and Browse 2012). PC, after the removal of its phosphocholine headgroup, can return to the diacylglycerol (DAG) pool at the ER membrane for TAG assembly, or it can enter the DAG pool at the inner membrane of plastid envelope for the assembly of thylakoid glycerolipids (Browse et al. 1986; Benning 2009). In the ER of epidermal cells, an acyl-CoA elongase catalyzes the formation of very-long-chain (VLC; C22 or longer) acyl-CoA esters (Lessire et al. 1985; Bessoule et al. 1989), which can be further metabolized to wax constituents *via* the primary alcohol and alkane pathways (Rowland et al. 2006; Greer et al. 2007). The wax molecules are delivered from the plasma membrane into the apoplast for cuticle formation (Samuels et al. 2008). During seed germination, FAs are liberated from TAGs by lipase activities on oil bodies, translocated across the single-membraned glyoxysomes (i.e. specialized peroxisomes) in the form of free FAs or activated CoA-esters, and eventually β-oxidized to acetyl-CoA molecules (Beisson et al. 2001; Fulda et al. 2002, 2004). In view of the subcellular nature of plant FA metabolism in the compartmentation of discrete pathways, highly dynamic lipid transfer mechanisms are essential to route the substrates, intermediates and products to destinations including the endomembrane system, the cytosol, plastids, mitochondria, peroxisomes and the extracellular space.

Due to their hydrophobic nature and lack of sorting signals in contrast to nascent polypeptides, lipids and their derivatives are delivered intracellularly in a facilitated manner. In vesicular trafficking, lipophilic metabolites are sorted between organelles *via* vesicle budding and fusion. Movement of a shorter distance (i.e. a few nm) across organellar membranes can be mediated by transbilayer flip-flop mechanisms or direct contact between the membranes (Raggers et al. 2000; Sprong et al. 2001; Holthuis and Levine 2005; van Meer et al. 2008; Toulmay and Prinz 2011). In a membrane-independent mode of transfer, lipids and their derivatives rely on protein facilitators or transporters such as ATP-binding cassette (ABC) transporters (Zolman et al. 2001; Kim et al. 2013), lipid transfer proteins (Kader 1997; Yeats and Rose 2008) and acyl-CoA-binding proteins (ACBPs; Xiao and Chye 2011a). For an in-depth discussion of the various mechanisms of lipid transfer within the plant cell and how they affect organelle biogenesis, readers are referred to recent review articles (Benning 2009; Hurlock et al. 2014). This chapter focuses on new progress in the biochemical and functional characterization of plant ACBPs and their emerging roles in various facets of plant growth and development.

Conservation of ACBP Families in Eukaryotes

ACBPs, virtually found in all eukaryotic species and a few pathogenic prokaryotes, represent a multigene family conserved phylogenetically and functionally (Burton et al. 2005; Faergeman et al. 2007). ACBPs bind acyl-CoA esters (C12–C26) with high specificities and affinities in a non-covalent, reversible manner (Rasmussen et al. 1993; Faergeman and Knudsen 1997; Chye 1998; Chye et al. 2000; Knudsen et al. 2000; Burton et al. 2005; Leung et al. 2004, 2006; Hsiao et al. 2014a; Xue et al. 2014). The prototype form in ACBP families is represented by highly-conserved, low molecular mass (ca. 10-kDa) cytosolic proteins, which may be accompanied by larger variants identified previously from the majority of eukaryotic phyla except the fungi (Kragelund et al. 1999; Faergeman et al. 2007). The yeast (*Saccharomyces cerevisiae*) genome encodes a single-copy ACBP homolog, and its depletion adversely affected FA elongation as well as membrane assembly and organization (Gaigg et al. 2001), whilst its overexpression significantly altered acyl-CoA pool size and composition (Mandrup et al. 1993; Knudsen et al. 1994). The first characterized ACBP was purified from rat brain as a neurotransmitter termed diazepam-binding inhibitor (Guidotti et al. 1983), which shares sequence homology with the ACBPs subsequently purified from rat and bovine livers (Mogensen et al. 1987; Knudsen et al. 1989).

Following the discoveries of mammalian ACBPs, the prototype 10-kDa ACBP homologs were identified from a number of plant species, including *Brassica napus* (oilseed rape; Hills et al. 1994; Brown et al. 1998), *Ricinus communis* (castor bean; van de Loo et al. 1995), *Arabidopsis thaliana* (thale cress; Engeseth et al. 1996), *Gossypium hirsutum* (cotton; Reddy et al. 1996), *Digitalis lanata* (foxglove; Metzner et al. 2000), *Oryza sativa* (rice; Suzui et al. 2006) and *Vernicia fordii* (tung tree; Pastor et al. 2013). In addition to the 10-kDa ACBP isoforms, the acyl-CoA-binding (ACB) domain is conserved in larger multi-domain proteins from *A. thaliana* (Chye 1998; Chye et al. 2000; Leung et al. 2004, 2006; Xiao and Chye 2009), *Agave americana* (century plant; Guerrero et al. 2006), *O. sativa* (Meng et al. 2011, 2014; Meng and Chye 2014), *V. fordii* (Pastor et al. 2013) and *Vitis vinifera* (grape; Takato et al. 2013). A recent BLASTP search using sequences of Arabidopsis paralogs retrieved 84 ACB domain-containing sequences comprising 66–682 amino acid residues from 14 plant species and 2 green algae (Meng et al. 2011). The majority of these proteins possess a conserved *N*-terminal ACB domain accompanied by highly dissimilar *C*-terminal regions (Meng et al. 2011). Based on molecular mass and domain architecture, Meng et al. (2011) classified plant ACBPs into four subgroups (Fig. 15.1; Table 15.1), namely small ACBPs (Class I), ankyrin-repeat ACBPs (Class II), large

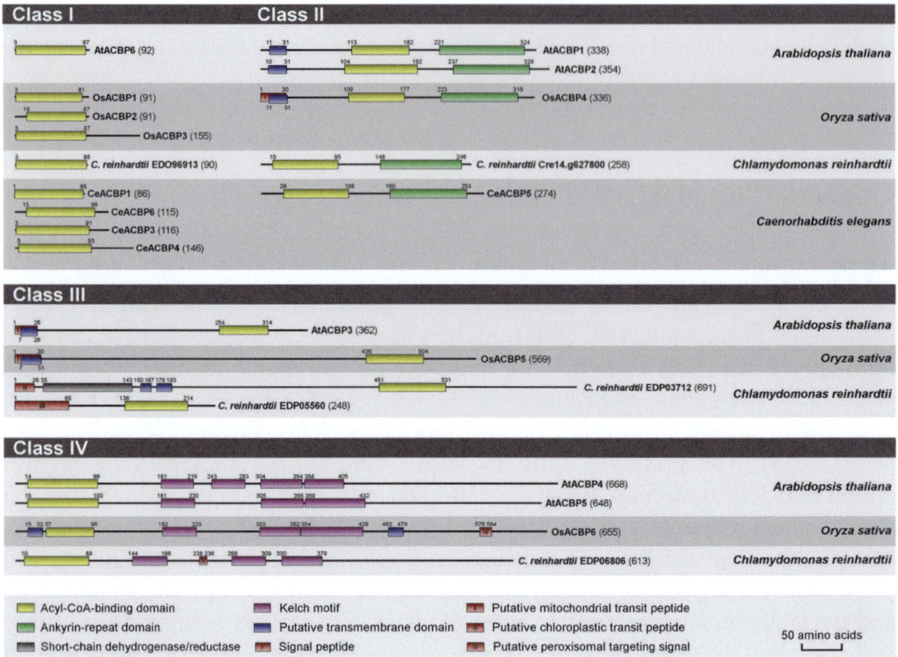

Fig. 15.1 Domain architecture of Classes I–IV ACBP members from plant and non-plant species. Representative members from Arabidopsis (*Arabidopsis thaliana*; At), rice (*Oryza sativa*; Os), green alga (*Chlamydomonas reinhardtii*; Cr) and the nematode worm (*Caenorhabditis elegans*; Ce). Classes III and IV ACBPs were not identified from *C. elegans*. Protein queries were submitted to NCBI's protein BLAST search (http://blast.ncbi.nlm.nih.gov/Blast.cgi) to scan for the presence of conserved domains and their intervals were annotated by alignment with the Conserved Domain Database (CDD) collection including acyl-CoA-binding domain (cd00435), ankyrin-repeat domain (cd00204), short-chain dehydrogenase/reductase (pfam01113) and kelch motif (pfam01344, pfam07646, pfam13415, pfam13418 and pfam13854). Amino acid residues are indicated in parentheses. AtACBPs, OsACBPs and CeACBPs are numbered according to Xiao and Chye (2009), Meng et al. (2011), and Elle et al. (2011), respectively. Their sequence data can be found under the following GenBank accession number: AtACBP1 (AED96361), AtACBP2 (AEE85391), AtACBP3 (AEE84874), AtACBP4 (AEE74237), AtACBP5 (AED93708), AtACBP6 (AEE31396), OsACBP1 (BAG86980), OsACBP2 (BAG86809), OsACBP3 (ABF97253), OsACBP4 (BAF16206), OsACBP5 (BAG93201), OsACBP6 (ABF99748), CeACBP1 (CCD67306), CeACBP3 (CAA91987), CeACBP4 (CCD61830), CeACBP5 (CAB03343) and CeACBP6 (CAA19448)

ACBPs (Class III) and kelch-ACBPs (Class IV), which are each represented in 12 of the 13 higher plant species analyzed (except *Picea sitchensis*) using genome sequence information available to date. The indispensability of the diverse subclasses of plant ACBPs implicates their functional non-redundancy with respect to their specific roles for normal plant growth and development.

Over the years, a combination of techniques including molecular genetics, microscopy, cell biology, biochemistry and mass spectrometry have been utilized in this laboratory to unravel the discrete functions of the six ACBP members in

Table 15.1 Properties of four classes of ACBPs from various plant sources

Class	Member	Locus ID	Protein size		pI	Tissue specificity[a]	Subcellular localization	Binding preferences			References
								Acyl-CoA			
			Residues	M_r (kDa)				By Lipidex assays	By quantitative assays[b] (K_d)	Phospholipid	
I	AtACBP6	At1g31812	92	10.4	4.91	High in leaves, stems, developing seeds, siliques and flowers (esp. microspores, tapetum, pollen)	Cytosol	18:2 > 20:4 > 18:1 ≈ 16:0 > 18:3	16:0 (35.9 nM) 18:2 (36.4 nM) 14:0 (38.7 nM) 18:1 (45.3 nM) 18:0 (60.6 nM) 18:3 (84.1 nM)	di 16:0-PC di 18:0-PC di 18:1-PC di 18:2-PC	Engeseth et al. (1996), Chen et al. (2008) and Hsiao et al. (2014a, 2015)
	OsACBP1	Os08g06550	91	10.2	4.86	High in leaves and roots		18:1 > 18:2 > 18:3 > 16:0	N.D.	di 18:0-PA/PC di 18:1-PA/PC di 18:2-PC	Meng et al. (2011, 2014)
	OsACBP2	Os06g02490		10.3	4.69	High in leaves, roots and soft dough seeds		18:3			
	OsACBP3	Os03g37960	155	17.7	9.05	High in leaves, roots and seeds at anthesis	Cytosol and others[c]				
	BnACBP	Bna.2377	92	10.2	5.50	High in developing embryos, flowers and cotyledons	Cytosol	18:2 > 1 8:1 > 18:3		N.D.	Hills et al. (1994), Engeseth et al. (1996) and Yurchenko et al. (2009)

(continued)

Table 15.1 (continued)

Class	Member	Locus ID	Protein size		pI	Tissue specificity[a]	Subcellular localization	Binding preferences			References
								Acyl-CoA			
			Residues	M_r (kDa)				By Lipidex assays	By quantitative assays[b] (K_d)	Phospholipid	
II	AtACBP1	At5g53470	338	37.5	4.32	High in developing seeds, siliques, vasculature, root primordia, flowers (not in petals/stamens), stem epidermis and trichomes	ER and PM	20:4 > 18:2 > 18:3 > 18:1 > 16:0	18:3 (0.44 μM) 18:1 (0.76 μM) 18:2 (0.83 μM) 25:0 (1.69 μM) 26:0 (1.94 μM) 24:0 (2.14 μM)	di 16:0-PA di 18:0-PA di 18:1-PA/PC di 18:2-PC	Chye (1998), Chye et al. (1999), Leung et al. (2006), Xiao et al. (2008b), Gao et al. (2009), Chen et al. (2010), Du et al. (2010a, 2013a) and Xue et al. (2014)
	AtACBP2	At4g27780	354	38.5	4.22	High in roots (esp. vasculature), pollen, developing embryos, siliques and guard cells		18:2 > 18:3 > 20:4 ≈ 16:0 > 18:1	N.D.	16:0-lysoPC di 18:1-PC di 18:2-PC	Chye et al. (2000), Li and Chye (2003), Kojima et al. (2007), Gao et al. (2009, 2010a), Chen et al. (2010) and Du et al. (2013b)
	OsACBP4	Os04g58550	336	36.0		High in leaves, roots and seeds at anthesis	ER	18:2 > 16:0 > 18:3		di 16:0-PA di 18:0-PA/PC di 18:1-PA/PC di 18:2-PC	Meng et al. (2011, 2014) and Meng and Chye (2014)

III	AtACBP3	At4g24230	362	39.3	3.96	High in shoots, roots (not in root hair/tip), vasculature and flowers (esp. stigma)	ER and apoplast	20:4 > 18:3 > 18:2 > 18:1 ≈ 16:0	22:0 (29.9 nM) 20:0 (66.0 nM) 24:0 (80.6 nM) 18:2 (788 nM)	di 16:0-PC/PE di 18:0-PC/PE di 18:1-PC/PE di 18:2-PC/PE	Leung et al. (2004, 2006), Xiao et al. (2010), Zheng et al. (2012) and Xie et al. (2015)
	OsACBP5	Os03g14000	569	61.2	3.98	High in leaves and seeds at all stages	ER	18:3 > 16:0	N.D.	di 18:0-PA/PC di 18:1-PA/PC di 18:2-PC	Meng et al. (2011, 2014) and Meng and Chye (2014)
IV	AtACBP4	At3g05420	668	73.2	4.95	High in leaves, roots, siliques and flowers (esp. pollen grains; not in microspores/ tapetum)	Cytosol	18:1 > 16:0 > 18:2 > 18:3	18:0 (2.7 μM) 16:0 (23.5 μM) 14:0 (65.3 μM) 18:2 (73.8 μM) 18:1 (95.2 μM) 18:3 (189 μM)	di 18:1-PC di 18:2-PC	Leung et al. (2004), Xiao et al. (2008a, 2009a) and Hsiao et al. (2014a, 2015)
	AtACBP5	At5g27630	648	71.0	5.86	High in leaves, roots and flowers (esp. microspores, tapetum; not in pollen grains)			18:0 (35.4 μM) 16:0 (35.7 μM) 14:0 (41.9 μM) 18:2 (64.1 μM) 18:1 (74.4 μM) 18:3 (92.4 μM)		
	OsACBP6	Os03g61930	655	71.4	5.05	High in leaves, roots, developing and germinating seeds	Peroxisomes	18:3 > 18:2	18:3 (32.3 μM)	di 18:0-PA/PC di 18:1-PA/PC di 18:2-PC	Meng et al. (2011, 2014)

ER endoplasmic reticulum, M_r relative molecular mass, *N.D.* not determined, *PA* phosphatidic acid, *PC* phosphatidylcholine, *PE* phosphatidylethanolamine, *PM* plasma membrane

[a]All ACBPs were detectable in all tissues tested albeit at varying expression levels

[b]The dissociation constants (K_d) were determined by microscale thermophoresis (for rAtACBP3) or isothermal titration calorimetry (otherwise)

[c]Irregular membranous and punctate structures

Arabidopsis, constituting the first comprehensive study on an ACBP family in the Plant Kingdom (Xiao and Chye 2009). Our recent work on the rice ACBP family has also provided insights into similarities and differences between the representative dicot (Arabidopsis) and monocot (rice) ACBP families (Meng et al. 2011, 2014). The important features and specific roles of the Arabidopsis and rice ACBP subclasses in plant growth and development are described herein.

Arabidopsis ACBPs

The Arabidopsis genome contains six *ACBP* paralogs (i.e. *AtACBP1–AtACBP6*), all of which encode functional proteins as demonstrated *in vitro* through the binding of *Escherichia coli*-expressed recombinant (r) ACBPs to acyl-CoA esters (Engeseth et al. 1996; Chye 1998; Chye et al. 2000; Leung et al. 2004, 2006; Gao et al. 2009; Xiao et al. 2009a, 2010; Xue et al. 2014; Hsiao et al. 2014a; Xie et al. 2015). In addition to the prototype 10-kDa AtACBP6 (Class I), five other AtACBPs ranging from 37.5 to 73.2 kDa exist (Xiao and Chye 2009). Class II members, AtACBP1 and AtACBP2, are highly homologous sharing 71 % amino acid sequence identity, and are transmembrane proteins that are targeted to the ER and plasma membrane (Chye 1998; Chye et al. 1999, 2000; Li and Chye 2003). Both proteins harbor a *C*-terminal domain of ankyrin repeats (Li and Chye 2004; Gao et al. 2009, 2010a; Du et al. 2013a), a potential domain for protein-protein interactions (Michaely and Bennett 1992; Bork 1993), subsequently substantiated for AtACBP2 (Gao et al. 2009, 2010a). AtACBP3 (Class III) contains an *N*-terminal cleavable signal peptide which directs the protein to the endomembrane system and the apoplastic space, as verified using confocal laser scanning microscopy of green fluorescent protein (GFP)-tagged proteins and subcellular fractionation experiments followed by Western blot analysis using AtACBP3-specific antibodies (Leung et al. 2006; Xiao et al. 2010). The location of an ACB domain within the *C*-terminal region of AtACBP3 contrasts with the more common appearance of this domain at the *N*-termini in the other ACBPs (Meng et al. 2011; Xiao and Chye 2011a). Besides the prototype AtACBP6, Class IV AtACBP4 and AtACBP5 are the two other cytosolic isoforms, and they represent the largest (i.e. 73.2 and 71.0 kDa, respectively) members in the AtACBP family (Leung et al. 2004; Xiao et al. 2008a). The core regions of AtACBP4 and AtACBP5 invariably feature kelch motifs (Leung et al. 2004; Li et al. 2008), that constitute potential sites for protein-protein interactions (Adams et al. 2000).

Sequence alignment of the ACB domains from the six AtACBPs and other plant ACBPs shows high conservation of the "YKQA" and "KWDAW" motifs (Xiao and Chye 2011a), which correlates well with the X-ray and nuclear magnetic resonance structures of ligand-bound ACBP complexes from cow, man, *Plasmodium falciparum* and yeast (Kragelund et al. 1993; van Aalten et al. 2001; Faergeman et al. 2007; Taskinen et al. 2007). Site-directed mutagenesis of either one of these two motifs affected the binding of rAtACBP2 to palmitoyl-CoA ester (Chye et al.

2000; Leung et al. 2004) and rAtACBP4 and rAtACBP5 to oleoyl-CoA ester (Leung et al. 2004), but not rAtACBP1 or rAtACBP3 to arachidonyl-CoA ester, which was instead disrupted by amino acid substitution of an arginine residue upstream of the "YKQA" motif (Leung et al. 2006). Thus, the amino acid determinants of ACB functions in AtACBPs appear to be isoform- and ligand-dependant. In fact, the six AtACBPs, which show more or less sequence variations within their ACB domains, exhibit varying ligand specificities. Both rAtACBP1 and rAtACBP2 displayed preferences for linoleoyl- and linolenoyl-CoA esters (Gao et al. 2009), and recent data from isothermal titration calorimetry (ITC) also revealed high affinities of rAtACBP1 to VLC acyl-CoA esters (Xue et al. 2014). Also, rAtACBP3 bound VLC acyl-CoA esters (Xie et al. 2015), and exhibited higher affinity for arachidonyl-CoA ester than linoleoyl- and linolenoyl-CoA esters (Leung et al. 2006; Xiao et al. 2010). Recently, the thermodynamics of interactions between the three cytosolic AtACBPs and long-chain (C16–C18) acyl-CoA esters were quantitatively compared using ITC (Hsiao et al. 2014a), and the binding affinities of rAtACBP4 and rAtACBP5 (dissociation constants K_d in the micromolar range) were reported to be much weaker than that of rACBP6 (K_d in the nanomolar range). Other than acyl-CoA esters, AtACBPs exhibited affinities for phospholipids, as typified by the binding of all six rAtACBPs to PC *in vitro* (Chen et al. 2008, 2010; Xiao et al. 2009a, 2010). In addition, rAtACBP1 and rAtACBP2 bind phosphatidic acid (PA; Du et al. 2010a) and lyso-PC (Gao et al. 2010a), respectively, whilst rAtACBP3 binds phosphatidylethanolamine (PE; Xiao et al. 2010).

While the differential subcellular localization and ligand specificities of AtACBPs for different acyl-CoA esters and (lyso-)phospholipids *in vitro* are in favor of the hypothesis that individual ACBPs play non-redundant roles, studies of protein interactome have provided insights on their functions (Du and Chye 2013). Bimolecular fluorescence complementation (BiFC) and yeast 2-hybrid (Y2H) assays indicated that one of the protein partners of the ankyrin repeat-containing isoform, AtACBP1, is phospholipase Dα1 (PLDα1; Du et al. 2013a), which hydrolyzes structural phospholipids to produce PA as a second messenger in abscisic acid (ABA) signaling (Jacob et al. 1999; Zhang et al. 2004; Li et al. 2009). The studies of Arabidopsis AtACBP1-overexpressors (AtACBP1-OEs) and the *acbp1* knockout mutant revealed the potential of AtACBP1 in promoting PLDα1 activity related to ABA signaling during seed germination and seedling development (Du et al. 2013a; see Sect. AtACBPs in the Regulation of Seed Germination and Seedling Development for details). Using AtACBP2, another ankyrin repeat-containing isoform, as bait in Y2H screening of an Arabidopsis cDNA library identified a member of the ethylene response factor (ERF) family, termed ethylene-responsive element-binding protein (AtEBP) or RELATED TO APETALA2.3 (RAP2.3), as a protein interactor (Li and Chye 2004). The AtACBP2-AtEBP interaction was confirmed by an *in vitro* pull-down assay, and their fluorescent fusion proteins were found to co-localize at the plasma membrane (Li and Chye 2004). The same interactor AtEBP/RAP2.3 was also found in another Y2H screen using AtACBP4 as bait, and this interaction was confirmed by co-immunoprecipitation and fluorescence resonance energy transfer, the latter of which revealed that they interact predominantly in the cytosol (Li et al.

2008). In Northern blot analysis, the concomitant induction of *AtACBP4* and *AtEBP* expression by the treatment with the ethylene precursor 1-aminocyclopropane-1-carboxylic acid (ACC), methyl jasmonate (MeJA) or the pathogen *Botrytis cinerea* suggested that AtACBP4 may play a role in AtEBP-mediated defense response *via* ethylene and/or jasmonate (JA) signaling (Li et al. 2008). AtACBP1 and AtACBP2 were also demonstrated to interact with another related ERF member, RAP2.12, using BiFC and Y2H assays (Licausi et al. 2011). Under aerobic conditions, AtACBP1 and AtACBP2 sequester the soluble RAP2.12 at the plasma membrane and protect it from degradation *via* the ubiquitin-dependent *N*-end rule pathway, whereas low oxygen concentration triggers the release of RAP2.12 from the plasma membrane into the nucleus for activation of hypoxia-responsive gene expression (Licausi et al. 2011). More recently, a role for AtACBP3 was identified in VLCFA-mediated response to hypoxic stress in conjugation with salicylic acid (SA) and ethylene signaling (Xie et al. 2015).

AtACBP2 also interacts at the plasma membrane with other stress-responsive proteins including a farnesylated protein AtFP6 (Gao et al. 2009) and a lyso-phospholipase AtLYSOPL2 (Gao et al. 2010a), both of which have roles in conferring plant tolerance to heavy metal stress. AtFP6 is a member of the isoprenylated protein family (i.e. farnesylated proteins) featuring at least one characteristic metal-binding motif (core sequence: M/LXCXXC; Dykema et al. 1999). In roots, *AtFP6* mRNA expression was induced by cadmium and zinc (Gao et al. 2009), whereas *AtACBP2* expression was lead-inducible (Xiao et al. 2008b). The *in vitro* binding of AtFP6 and AtACBP2 to lead, cadmium and copper, in conjunction with the enhanced cadmium-tolerance of Arabidopsis overexpressing either protein, led to the assumption that the AtACBP2-AtFP6 interaction at the plasma membrane of root cells facilitates the transport of these heavy metal ions (Xiao et al. 2008b; Gao et al. 2009). Another AtACBP2-interactor, AtLYSOPL2, is one of the few known plant LYSOPLs to degrade lyso-phospholipids into FAs and glycerolphosphate derivatives (Wang and Dennis 1999), representing an important step in the detoxification of lyso-phospholipids, the intermediates of phospholipid metabolism post-stress (Flieger et al. 2002; Ryu 2004). In Arabidopsis, the overexpression of *AtLYSOPL2* or *AtACBP2* improved plant tolerance to cadmium and hydrogen peroxide (H_2O_2), whilst *lysopl2* knockout mutants were more sensitive to zinc and H_2O_2 (Gao et al. 2009, 2010a), suggesting a protective role of AtLYSOPL2 and AtACBP2 against metal-induced oxidative stress (Gao et al. 2010a). Considering the binding activity of rAtACBP2 to linoleoyl-CoA ester, linolenoyl-CoA ester, PC and lyso-PC (Gao et al. 2010a), and the tight association of AtACBP2 with two metal-inducible proteins AtFP6 and AtLYSOPL2, it is possible that AtACBP2 has a role in phospholipid repair following lipid peroxidation upon heavy metal-induced stress (Gao et al. 2009, 2010a, b). In another study, quantitative real-time PCR (qRT-PCR) further identified the up-regulation of *AtACBP1* and *AtACBP4*, besides *AtACBP2*, mRNAs in Arabidopsis shoots and roots upon lead (II) treatment (Du et al. 2015). The overexpression of *AtACBP1* or *AtACBP4* in transgenic *Brassica juncea* culminated in an enhanced accumulation of lead (II) in the cytosol of root cells without increasing lead (II) tolerance (Du et al. 2015).

Other than hypoxia, heavy metal and oxidative stresses, some members of the AtACBP family function in responses to other abiotic stresses including low temperature and drought. Chen et al. (2008) indicated relatively higher mRNA and protein levels of AtACBP6 in rosettes of wild-type Arabidopsis upon cold treatment. The freezing-tolerant phenotype of AtACBP6-OEs was attributed to the up-regulated expression of *PLDδ* in rosettes, leading to the elevation of PA and the decline in PC (Chen et al. 2008), but the freezing-tolerant flowers of the same AtACBP6-OEs were recently shown to be directly associated with elevations of PC and monogalactosyldiacylglycerol and decline in PA (Liao et al. 2014). On the other hand, the overexpression of AtACBP1 rendered transgenic Arabidopsis more sensitive to freezing stress accompanied by a decrease in PC and an increase in PA (Du et al. 2010a). Reciprocally, the Arabidopsis *acbp1* mutant plants exhibited freezing tolerance accompanied by PC accumulation and PA reduction in rosette leaves, possibly due to suppression in hydrolysis of PC to PA as evident from the down-regulated expression of *PLDα1* (Du et al. 2010a). In spite of the high homology to AtACBP1, AtACBP2 was not associated with the freezing response (Du et al. 2010b). Instead, it has been ascribed a role in drought response (Du et al. 2013b). Du et al. (2013b) showed that the overexpression of *AtACBP2* in Arabidopsis altered the expression profiles of ABA signaling genes, including the up-regulation of Respiratory Burst Oxidase Homolog D (*AtRBOHD*) and *AtRBOHF* and the down-regulation of *HYPERSENSITIVE TO ABA1*. The over-production of ABA-mediated reactive oxygen species in the guard cells induced stomatal closure and reduced transpiration loss leading to an enhanced drought tolerance in *AtACBP2-OEs*, in comparison with *acbp2* mutant plants which were more sensitive to drought than the wild type (Du et al. 2013b).

In addition to abiotic stresses, AtACBPs have been linked to defense responses against pathogens. Expression of both *AtACBP3* and *AtACBP4* was shown to be inducible by treatment with the defense-related phytohormones including MeJA and ACC and necrotrophic fungus *B. cinerea* (Li et al. 2008; Xiao and Chye 2011b). *AtACBP3* expression was also up-regulated by treatment with SA, arachidonic acid (a fungal elicitor) and bacterial pathogen *Pseudomonas syringae* pv *tomato* DC3000 (Xiao and Chye 2011b). These phytohormones and pathogens induced β-glucuronidase (*GUS*) expression in transgenic Arabidopsis expressing *GUS* from *AtACBP3* (Zheng et al. 2012). Electrophoretic mobility shift assays (EMSA) further revealed that an S-box of AT-rich sequence (−516/−512) binds nuclear extracts from pathogen-infected leaves (Zheng et al. 2012). In Arabidopsis AtACBP3-OEs, the constitutive activation of pathogenesis-related (*PR*) gene expression and the induction of cell death and hydrogen peroxide production likely account for the enhanced resistance to *P. syringae* DC3000 (Xiao and Chye 2011b). Conversely, Arabidopsis *acbp3* knock-out mutant and *AtACBP3* RNA interference (RNAi) transgenic plants were more susceptible to *P. syringae* attack and exhibited lower mRNAs of the defense-related *PR1*, *PR2* and *PR5* (Xiao and Chye 2011b). Consistently, transgenic Arabidopsis expressing an *AtACBP3* homolog from grape (i.e. *VvACBP*) showed resistance to *P. syringae* DC3000 and a hemibiotrophic fungus, *Colletotrichum higginsianum* (Takato et al. 2013).

Besides the stress-responsiveness of *AtACBPs*, the expression of some is subject to diurnal oscillation during light/dark cycles (Xiao et al. 2009a, b, 2010; Zheng et al. 2012; Hsiao et al. 2014b). In Arabidopsis rosettes, the expression patterns of *AtACBP4* and *AtACBP5* in light/dark cycling (Xiao et al. 2009a, b) correlated well with that of *FAD7* which encodes an Arabidopsis plastidial ω-3 FA desaturase (Nishiuchi et al. 1995). All three genes are expressed at higher levels during the day than night, and their expression was rapidly silenced in continuous darkness (Xiao et al. 2009a). On the other hand, an opposing pattern was observed for *AtACBP3* and *AtACBP6*, both of which were up-regulated in the dark (Xiao et al. 2009a). Similar RNA gel blot analysis indicated that *AtACBP3* was highly expressed in continuous darkness (Xiao et al. 2010). By using transgenic Arabidopsis expressing GUS driven by various deletions of the *AtACBP3* promoter, the *cis*-elements responsible for the dark-induced expression of *AtACBP3* were mapped to a 160-bp 5′-flanking region (−434/−274), within which a DNA-BINDING WITH ONE FINGER (DOF) box and a GT-1 motif were shown to interact specifically with nuclear proteins from dark-treated Arabidopsis leaves in EMSA and DNase I footprinting assays (Zheng et al. 2012). In Arabidopsis seedlings, the biological clock regulates *AtACBPs* transcriptionally, amongst which *AtACBP3* most diurnally oscillates (Hsiao et al. 2014b). The diurnal patterns of all six *AtACBPs* appeared slightly altered in an arrhythmic line (i.e. *cca1lhy* double mutant) with defects in internal clock functions (Hsiao et al. 2014b).

Rice ACBPs

As our studies on AtACBPs and their stress-responsive protein interactors have opened new possibilities in protecting plant against abiotic stresses and phytopathogens (Xiao and Chye 2011a; Chen et al. 2014), we sought to investigate if genetic engineering strategies could be extended to agronomically important species including those taxonomically distant from the eudicot Arabidopsis. In this regard, rice represents an excellent candidate as a monocot model plant that is arguably the most important staple crop in Asia. The completion of the rice genome sequence (Goff et al. 2002; Yu et al. 2002) facilitated the retrieval of homologous sequences from available rice genome databases. BLASTP searches using the amino acid sequences of AtACBP1, AtACBP3, AtACBP4 and AtACBP6 as query identified six rice ACBP homologs designated as OsACBP1 to OsACBP6, respectively (Meng et al. 2011). Meng et al. (2011) studied the phylogeny, gene expression and biochemical analyses of the rice ACBP family. The characterization of the six rice ACBPs in comparison to AtACBPs has revealed similarities and differences (Meng et al. 2011, 2014; Meng and Chye 2014), as summarized below.

Resembling the prototype AtACBP6 (Class I), OsACBP1, OsACBP2 and OsACBP3 are of relatively low molecular masses (i.e. 10.2–17.7 kDa) (Meng et al.

2011). In contrast to Class I in which three members exist in rice, the remaining three classes (Classes II, III and IV) of OsACBPs are each represented by a single member (Meng et al. 2011). The 36.0-kDa OsACBP4 resembles the ankyrin-repeat-containing Class II AtACBP1 and AtACBP2. The 61.2-kDa OsACBP5 belongs to Class III (large ACBPs), whilst the largest (71.4-kDa) kelch-motif-containing OsACBP6 falls under Class IV.

Analysis by qRT-PCR on organ-specificity and stress-responsiveness of the rice *ACBP* family provided the initial clues on their possible functions (Meng et al. 2011). In terms of spatial expression, transcripts of the six rice *ACBPs* were detectable in all organs tested, in descending order of relative expression levels from leaves, roots to stems in 7-week-old plants (Meng et al. 2011). The expression patterns of the six rice *ACBP*s varied during seed development (Meng et al. 2011). Both *OsACBP1* and *OsACBP5* were stably expressed throughout the entire reproductive phase, but *OsACBP3* and *OsACBP4* expression were elevated at anthesis, and *OsACBP2* expression peaked at the more mature soft dough-staged seeds (Meng et al. 2011). On the other hand, *OsACBP6* expression was higher in anthesis- and soft dough- than milk-staged seeds (Meng et al. 2011). The highest expression of Class III *OsACBP5* occurred during seed development, resembling Class II *AtACBP1* and *AtACBP2* (Chen et al. 2010), and deviated from *AtACBP3* (Class III) which was expressed at a low level in siliques (Zheng et al. 2012). When the regulation of rice *ACBP* expression in response to abiotic and biotic stresses was tested by qRT-PCR (Meng et al. 2011), they appeared to be consistent with microarray data (http://www.ricearray.org/). *OsACBP4* and *OsACBP5* were induced by high-salinity treatment and *OsACBP4* was also induced by drought (Meng et al. 2011). Upon cold stress, *OsACBP6* expression slightly declined after 24 h, whilst down-regulation of the other five *OsACBPs* occurred within 12 h. After wounding, the transcripts of the three Class I *OsACBPs* and Class II *OsACBP4* declined but *OsACBP5* and *OsACBP6* were induced at 30 min post-treatment followed by a significant decrease below the basal level. Inoculation with the rice blast fungus *Magnaporthe grisea* up-regulated *OsACBP5* but reduced the expression of the other five *OsACBPs* (Meng et al. 2011). These varying expression patterns of respective rice *ACBPs* in response to environmental and biotic stresses appeared to differ from their Arabidopsis counterparts within the same subclasses (see Sect. Arabidopsis ACBPs). Differential spatial and stress-responsive expression patterns between rice and Arabidopsis *ACBPs* indicate functional divergence of ACBPs following gene duplication during the evolution of monocot and eudicot genomes (Meng et al. 2011).

In vitro binding assays confirmed that the six *OsACBP*s encode functional proteins (Meng et al. 2011, 2014). In rice, as in Arabidopsis, individual ACBP homologs exhibited differential binding preferences for acyl-CoA species (Meng et al. 2011), implicating that rice ACBPs play non-redundant physiological roles. Meng et al. (2011) tested the binding of rOsACBPs to four selected acyl-CoA species representing saturated (i.e. palmitoyl-CoA ester), monounsaturated (i.e. oleoyl-CoA ester) and polyunsaturated (i.e. linoleoyl- and linolenoyl-CoA esters) fatty acyl-CoA esters. While only rOsACBP1 binds oleoyl-CoA ester, all six rOsACBPs bind linolenoyl-CoA ester (Meng et al. 2011). Also, rOsACBP1, rOsACBP4 and rOsACBP5 showed binding activities to palmitoyl-CoA ester (Meng et al. 2011). Linoleoyl-CoA ester was bound by rOsACBP1, rOsACBP4 and rOsACBP6 (Meng et al. 2011). In addition to acyl-CoA esters, the interactions of rOsACBPs with phospholipids were tested (Meng et al. 2014). While all rOsACBPs bind 18:0-PC, 18:1-PC and 18:2-PC, none bind 14:0-PC, 16:0-PC or lyso-PC (Meng et al. 2014). All rOsACBPs also bind 18:0-PA and 18:1-PA but only rOsACBP4 binds 16:0-PA (Meng et al. 2014). Although rAtACBP3 binds PE (Xiao et al. 2010) and rAtACBP2 binds lyso-PC (Gao et al. 2010a), no rOsACBP emerged to bind PE, lyso-PC or phosphatidylserine (PS) (Meng et al. 2014).

In terms of subcellular localization, differences were also observed between Arabidopsis and rice ACBPs within the same subclasses (Meng et al. 2014; Meng and Chye 2014). The Class I ACBPs from rice (i.e. OsACBP1, OsACBP2 and OsACBP3) were localized to the cytosol of agroinfiltrated tobacco leaf epidermal cells as well as the cotyledonary and root cells of transgenic Arabidopsis using fluorescent protein fusions (Meng et al. 2014). In addition, GFP-tagged OsACBP3 which has a 64-amino acid *C*-terminal extension was localized to some irregular membranous structures and randomly-scattered punctate structures in transgenic Arabidopsis (Meng et al. 2014). The ER localization of Class II OsACBP4 resembled AtACBP1 and AtACBP2 belonging to the same class, but OsACBP4, unlike its Arabidopsis counterparts, was not sorted to the plasma membrane. However, Class III OsACBP5 was confined to the ER and was not targeted extracellularly (Meng et al. 2014), unlike Class III AtACBP3 which was localized to the endomembranes and the apoplastic space (Leung et al. 2006; Xiao et al. 2010). In transgenic Arabidopsis seedlings, both OsACBP4 and OsACBP5 were visualized at the peripheral tubular ER, whereas OsACBP4 was also observed at the central and peripheral cisternal ER structures (Meng et al. 2014; Meng and Chye 2014), which are specialized for ribosome binding and cotranslational translocation of proteins (Friedman and Voeltz 2011). In addition, both OsACBP4 and OsACBP5 were localized to the membranes of ER bodies (Meng and Chye 2014), which have been identified as ribosome-surrounded, spindle-shaped ER structures in healthy seedlings or wounded/JA-treated rosette leaves of Arabidopsis (Hayashi et al. 2001; Matsushima et al. 2003). Interestingly, OsACBP6 represented the first plant ACBP demonstrated to be targeted to the peroxisomes (Meng et al. 2014). OsACBP6::GFP fusion proteins were confirmed to colocalize with the peroxisomal marker DsRed::SKL in primary roots and leaves of transgenic Arabidopsis as well as particle-bombarded rice sheath cells (Meng et al. 2014). Proteomic analysis of rat liver and mouse kidney peroxisomes had previously identified 56.6-kDa ACBP homologs (Kikuchi et al. 2004; Wiese et al. 2007).

Roles of ACBPs in Plant Development

Lipids are vital macromolecules acting as a major constituent of biological membranes and other structural components, fuel for many metabolic processes and key mediator of signal transduction (Wang 2004). Apart from the involvement of plant

ACBPs in different stress responses to environmental and biotic cues (Xiao and Chye 2011a), accumulating evidence suggests that ACBPs are crucial for normal plant growth and development (Table 15.2). The current knowledge of the physiological roles of individual plant ACBP homologs in plant development is discussed.

ACBPs in Seed Oil Biosynthesis

In oleaginous plants, TAG is the major form of seed storage lipids. Its synthesis involves the sequential incorporation of three FA moieties onto a glycerol backbone *via* a series of acyl-CoA-dependent acylation steps in the Kennedy pathway (Kennedy 1961). This pathway starts from the acylation of glycerol-3-phosphate to form lyso-PA by the action of glycerol-3-phosphate acyltransferase (GPAT) (Kennedy 1961). Lyso-PA acyltransferase (LPAAT) catalyzes the acylation of lyso-PA to PA, which is dephosphorylated into DAG by the action of phosphatidate phosphatase (Kennedy 1961). The final acylation of DAG to form TAG is catalyzed by DAG acyltransferase (DGAT), although acyl-CoA-independent reactions also exist (Lung and Weselake 2006). During the acyl transfer reactions, ACBP facilitates TAG assembly by maintaining a cytosolic acyl-CoA pool as a source of FAs, while the free acyl-CoA concentration in the cytosol is typically maintained below 10 nM (Knudsen et al. 1999). The ACBP:acyl-CoA ratio regulates the feedback inhibitory effect of substrates on enzyme activities (Rasmussen et al. 1993). ACBP also protects long-chain acyl-CoA esters from degradation by microsomal acyl-hydrolases (Rasmussen et al. 1993; Engeseth et al. 1996; Jolly et al. 2000). The tendency of ACBP to associate with acyl-CoA esters on the membrane periphery suggests its potential role in intermembrane acyl-CoA transport (Simonsen et al. 2003), and in the protection of membranes and membrane-associated enzymes from the deleterious detergent effect of amphiphilic acyl-CoA esters (Rasmussen et al. 1990). More specifically, mammalian ACBPs are known to modulate the activities of GPAT (Jolly et al. 2000; Kannan et al. 2003; Huang et al. 2005), acyl-CoA:lysophospholipid acyltransferase (Fyrst et al. 1995) and acyl-CoA:cholesterol acyltransferase (Kerkhoff et al. 1997; Chao et al. 2003).

In *B. napus*, the activities of TAG biosynthetic enzymes are also regulated by a 10-kDa cytosolic ACBP (Brown et al. 1998; Hobbs and Hills 2000; Yurchenko and Weselake 2011). GPAT activity was stimulated *in vitro* in the presence of rBnACBP if its concentration exceeded that of the [^{14}C]-oleoyl-CoA substrate; otherwise GPAT activity was inhibited (Brown et al. 1998). In fact, the level of BnACBP was elevated 12-fold during seed maturation coinciding with peak TAG accumulation, implicating an important role of ACBP in seed oil biosynthesis (Engeseth et al. 1996). In contrast, BnACBP was near undetectable in desiccating seeds at maturity (Brown et al. 1998). In addition, rBnACBP stimulated the activity of BnDGAT, when expressed in *B. napus* microspore-derived cell suspension cultures or in a

Table 15.2 Summary of ACBP functions in plant development

Function	Species	ACBP(s) involved	Binding partner(s) involved	Effect of knockout mutation, silencing (RNAi) and overexpression in Arabidopsis			References
				Line	Phenotypic effects	Regulation of associated genes	
Seed oil biosynthesis	*Brassica napus*	BnACBP	LC acyl-CoAs	*phaP::BnACBP*	Higher 18:2 and 18:3 at the expense of MUFAs and SFAs in seed oils at maturity	N.D.	Yurchenko et al. (2009, 2014)
				phaP::OleoH3P-BnACBP	Higher 18:2 at the expense of MUFAs and SFAs in seed oils at maturity		
	Arabidopsis thaliana	AtACBP4; AtACBP5; AtACBP6	18:1-CoA	*acbp6*	Higher 18:1-CoA in cotyledonary-staged embryos		Hsiao et al. (2014a)
				acbp4acbp5acbp6	Lower seed weight compared to double mutants and wild type		
Peroxisomal fatty acid β-oxidation	*Oryza sativa*	OsACBP6	18:3-CoA	*35S::OsACBP6/pxa1*	Restored IBA sensitivity, normal root length, hypocotyl elongation and wound-induced JA production compared to *pxa1*	Up-regulation of JA-responsive gene *VSP1* upon wounding	Meng et al. (2014)

Seed germination and seedling development	*Arabidopsis thaliana*	AtACBP1	PA; PLDα1 protein	*acbp1*	Lower ABA sensitivity during seed germination and seedling development	Down-regulation of *ABA2* and of *PLDα1* and *AREB1* under ABA treatment	Du et al. (2013a)
				35S::AtACBP1	Higher ABA sensitivity during seed germination and seedling development; higher PA at the expense of PC in rosettes	Up-regulation of *PLDα1* and down-regulation of *ABI1* under ABA treatment; up-regulation of ABA signaling genes (i.e. *AREB1*, *RD29A*, *MYC2*, *PLDα1*, *NCED3*, *ABA2*, *AtrbohD*, *AtrbohF*) in seedlings	
		AtACBP4; AtACBP5; AtACBP6	LC acyl-CoAs	*acbp6*	Higher 18:1- and 18:2-CoA in 5-days-old seedlings	N.D.	Hsiao et al. (2014a)
				acbp4acbp5	Hypersensitivity to ABA during seed germination		
				acbp5acbp6			
				acbp4acbp6			
				acbp4acbp5acbp6			

(continued)

Table 15.2 (continued)

Function	Species	ACBP(s) involved	Binding partner(s) involved	Effect of knockout mutation, silencing (RNAi) and overexpression in Arabidopsis			References
				Line	Phenotypic effects	Regulation of associated genes	
Embryo development	*Arabidopsis thaliana*	AtACBP1; AtACBP2	PC	*acbp1*	Higher MGDG at the expense of PC, PI & PS and higher levels of most polyunsaturated species of phospholipids in siliques	Up-regulation of *AtACBP2*	Chen et al. (2010)
				acbp2	No phenotype	Up-regulation of *AtACBP1*	
				acbp1acbp2	Embryo-lethal	N.D.	
Pollen development	*Arabidopsis thaliana*	AtACBP4; AtACBP5; AtACBP6	N.D.	*acbp4acbp6*	Aborted and vacuolated pollen; reduced seed number	N.D.	Hsiao et al. (2015)
				acbp5acbp6	*Ditto*		
				acbp4acbp5acbp6	*Ditto*; aberrant exine formation; fewer oil bodies in pollen; reduced pollen activity		

Cuticle formation	*Arabidopsis thaliana*	AtACBP1; AtACBP3; AtACBP4; AtACBP6	LC & VLC acyl-CoAs	*acbp1*	Fewer epicuticular wax crystals, ruptured cuticle membrane and altered cuticular wax composition in stems; more susceptible to *Botrytis cinerea* infection	Down-regulation of wax (i.e. *CER8*, *KCR1*, *ECR*, *CUT1*/*KCS6*) & cutin (i.e. *LACS2*, *CYP86A2*/*4*, *GPAT8*) biosynthetic genes	Xue et al. (2014)
				acbp6	Leaky leaf cuticle; faster water loss under drought; aberrant cuticle morphology; altered cuticular wax composition in leaves; more susceptible to *Pseudomonas syringae* infection; failed to generate SAR signals	N.D.	Xia et al. (2012)
				acbp3	*Ditto*; reduced levels of cutin monomers in leaves		
				acbp4	*Ditto*		

(continued)

Table 15.2 (continued)

Function	Species	ACBP(s) involved	Binding partner(s) involved	Effect of knockout mutation, silencing (RNAi) and overexpression in Arabidopsis			References
				Line	Phenotypic effects	Regulation of associated genes	
Leaf senescence	*Arabidopsis thaliana*	AtACBP3	PE	*acbp3*	Delayed dark-induced leaf senescence; lower PE in rosettes	N.D.	Xiao et al. (2010) and Xiao and Chye (2010)
				AtACBP3-RNAi	*Ditto*		
				35S::AtACBP3	Early age-dependent and dark/N_2-starvation-induced leaf senescence; higher PE and PI at the expense of PC and PI in rosettes; higher oxylipin-containing polar lipids; inhibited autophagosome formation	Up-regulation of senescence-associated genes (i.e. *SAG12*, *SEN1*, *SAG101*, *PLDδ*, *PLDγ*, *PLDζ2*)	

ABA abscisic acid, *IBA* indole-3-butyric acid, *LC* long-chain, *MUFAs* monounsaturated fatty acids, *N.D.* not determined, *PA* phosphatidic acid, *PC* phosphatidylcholine, *PE* phosphatidylethanolamine, *PI* phosphatidylinositol, *PS* phosphatidylserine, *SAR* systemic acquired resistance, *SFAs* saturated fatty acids, *VLC* very-long-chain yeast knockout mutant, at an ACBP:oleoyl-CoA ratio of 1:3, but DGAT activity declined if the ratio exceeded 1:1 (Yurchenko and Weselake 2011). Also, rBnACBP showed a subtle stimulatory effect on AtDGAT activity when expressed in insect cells at all (5, 15 or 30 μM) oleoyl-CoA ester concentrations tested (Hobbs and Hills 2000).

Apart from being incorporated into the glycerol backbone during glycerolipid production, palmitoyl, stearoyl and oleoyl-CoA esters exported from the plastids are subject to further modifications at the ER membrane *via* two major pathways for VLCFA and polyunsaturated FA (PUFA) syntheses (Harwood 1988). While FA elongases use acyl-CoA substrates (Pollard et al. 1979; Pollard and Stumpf 1980a, b), the newly synthesized oleic acid for desaturation into linoleic acid and α-linolenic acid must first be incorporated into PC (Slack et al. 1979; Citharel et al. 1983; Stymne et al. 1983; Murphy et al. 1985), predominantly at the *sn*-2 position (Bates et al. 2007, 2009; Tjellström et al. 2012). The resulting PUFAs may be released from PC by the reverse reaction of LPCAT, and its forward reaction can then incorporate another FA during the next round of acyl editing, leading to the enrichment of polyunsaturated acyl-CoA esters in the cytosolic pool for TAG assembly (Stymne and Stobart 1984; Wang 2001). In transgenic Arabidopsis, this acyl exchange between acyl-CoA esters and PC was stimulated by the expression of the cytosolic 10-kDa BnACBP under the control of seed-specific phaseolin promoter, resulting in a higher PUFA (linoleic and linolenic acids) seed content at the expense of eicosenoic acid and saturated FAs (Yurchenko et al. 2009). Similar acyl compositional change was reported in transgenic lines expressing an ER-targeted BnACBP, despite lower linolenic acid contents in both the acyl-CoA pool and seed oil compared with the cytosolic BnACBP lines (Yurchenko et al. 2014). As rBnACBP stimulated AtLPCAT activity *in vitro,* it has been proposed that a higher number of available ACB sites in the cytosol or on the ER may favor the partitioning of oleoyl-CoA ester into PC for subsequent desaturation rather than elongation (Yurchenko et al. 2009, 2014). To support this hypothesis, protein-protein interactions of BnACBP with acyl-CoA metabolic enzymes have been proposed for future investigations (Yurchenko and Weselake 2011).

In Arabidopsis, the three cytosolic isoforms (i.e. AtACBP4, AtACBP5 and AtACBP6) are potential candidates as cytosolic acyl-CoA pool formers that can facilitate acyl transfer from plastids to the ER for glycerolipid synthesis (Xiao and Chye 2009). ITC revealed that all three proteins bind palmitoyl-, oleoyl-, linoleoyl- and linolenoyl-CoA esters *in vitro,* although rAtACBP6 showed much stronger affinities than rAtACBP4 and rAtACBP5 (Hsiao et al. 2014a). The cotyledonary-staged embryos of the *acbp6* mutant accumulated a higher level of oleoyl-CoA ester (Hsiao et al. 2014a), in agreement with its reduction in developing seeds (20 days after flowering) in *BnACBP*-overexpressing Arabidopsis in comparison with the wild type (Yurchenko et al. 2014). Given that GUS was also strongly expressed in the cotyledonary-staged embryos of transgenic *AtACBP6pro::GUS* lines, AtACBP6 appears to have a potential role in seed oil biosynthesis (Hsiao et al. 2014a). On the other hand, the possibility that AtACBP4 and AtACBP5 play overlapping roles with AtACBP6 cannot be ruled out. *AtACBP4* and *AtACBP5* expression, coincided with that of *FAD7* (Nishiuchi et al. 1995), and were up-regulated during the day (Xiao et al. 2009a, b), consistent with the light/dark cycles of acetyl-CoA production from carbon fixation that fuels plastidial FA biosynthesis (Sasaki et al. 1997; Harmer et al. 2000). In fact, the depletion of all three cytosolic AtACBPs in a triple mutant resulted in the most notable reduction in seed weight when compared with the double mutants, implicating the involvement of AtACBP4, AtACBP5 and AtACBP6 in seed development (Hsiao et al. 2014a).

OsACBP6 in Peroxisomal FA β-Oxidation

A novel ACBP function in plant lipid catabolism has been proposed recently after observations that OsACBP6 is targeted to the peroxisomes (Meng et al. 2014). In higher plants, the peroxisomes are microbodies specialized for a variety of metabolic functions, including β-oxidative breakdown of FAs into acetyl-CoA units which can be further metabolized into succinate through the glyoxylate cycle, hence converting FAs into carbohydrates (Tolbert and Essner 1981; Olsen 1998; Graham and Eastmond 2002). Despite the presence of peroxisomal FA β-oxidation in most plant tissues (Gerhardt 1983), it is most active during seed germination and early seedling growth when FAs are liberated from TAGs in oil bodies and activated into CoA-esters as substrates for β-oxidation (Fulda et al. 2004). The import of FAs into the peroxisomes, despite the uncertainty whether it is in the form of free acids or CoA esters (Fulda et al. 2004; van Roermund et al. 2012; De Marcos Lousa et al. 2013), is mediated by a peroxisomal membrane-localized ABC transporter encoded by a single gene, *COMATOSE* (*CTS*; Footitt et al. 2002), which is also named *PEROXISOMAL ABC TRANSPORTER1* (*PXA1*; Zolman et al. 2001) and *PEROXISOME DEFICIENT3* (*PED3*; Hayashi et al. 2002).

In the Arabidopsis *pxa1* mutant, carbon reserves from seed oil cannot be mobilized into sugars owing to the defects in peroxisomal FA β-oxidation, resulting in the arrest of seedlings during post-germinative growth if no sucrose were to be supplemented (Zolman et al. 2001). In fact, the *ped3* mutants cannot degrade seed reserve lipids for gluconeogenesis as verified by its unaltered lipid content 5 days post-germination (Hayashi et al. 2002). In addition, root elongation of the *pxa1* mutant was not inhibited by the auxin precursor indole-3-butyric acid (IBA) because it cannot be converted into active indole-3-acetic acid when peroxisomal β-oxidation activities are impaired (Zolman et al. 2000, 2001). Similarly, the *ped3* mutants were resistant to the herbicide precursor 2,4-dichlorophenoxybutyric acid as it could not be β-oxidized into toxic 2,4-dichlorophenoxyacetic acid (Hayashi et al. 1998, 2002). On the other hand, the overexpression of *OsACBP6* complemented the *pxa1* mutant phenotypes (Meng et al. 2014). In the *OsACBP6-OE/pxa1* lines, hypocotyl elongation of seedlings was restored in a sucrose-free medium, and the number and diameter of rosettes from 3-week-old plants appeared similar to the wild type (Meng et al. 2014). Root elongation of the *OsACBP6-OE/pxa1* lines was inhibited significantly by 10 μM IBA, in contrast to the subtly-affected root length of the *pxa1* mutant (Meng et al. 2014).

In addition to acyl-CoA esters, 12-oxo-phytodienoic acid (OPDA) is another potential import substrate of CTS (Theodoulou et al. 2005). In chloroplasts, the JA biosynthetic pathway starts with the phospholipase-catalyzed hydrolysis of membrane lipids to release linolenic acid, which is oxidized and further modified to form OPDA (Vick and Zimmerman 1983). In the peroxisomes, OPDA is subject to several rounds of β-oxidation in JA formation (Vick and Zimmerman 1983). Despite its membrane-permeable nature, OPDA enters the peroxisomes in a transporter-regulated manner (Stenzel et al. 2003). CTS is believed to facilitate the transfer of OPDA as the basal and wound-induced JA levels in the *cts* mutant were detected to be lower than the wild type (Theodoulou et al. 2005). The expression of a JA-responsive gene, *VEGETATIVE STORAGE PROTEIN1* (*VSP1*), was also down-regulated by nearly 50 % in flowers of the *cts* mutant in comparison with the wild type (Theodoulou et al. 2005). On the other hand, the overexpression of *OsACBP6* in the *pxa1* mutant led to the recovery of induced JA production 1.5 h after wound-treatment (Meng et al. 2014). Concomitantly, *VSP1* expression in the *OsACBP6-OE/pxa1* lines was rapidly up-regulated 0.5 h post-wounding (Meng et al. 2014).

ITC was further used to test if rOsACBP6 binds potential CTS substrates including linolenoyl-CoA ester, OPDA and IBA (Meng et al. 2014). Although rOsACBP6 binds linolenoyl-CoA ester in ITC consistent with previous data from Lipidex 1000 binding assays (Meng et al. 2011, 2014), it did not bind IBA or OPDA (Meng et al. 2014). Hence, the possibility of IBA and OPDA being transported into the peroxisomes in the form of CoA-thioesters cannot be ruled out (Hu et al. 2012). Especially, given the potential existence of an alternative CTS-independent route which accounts for the low but detectable JA levels in the *cts* mutant (Theodoulou et al. 2005; Dave et al. 2011), OsACBP6 may emerge to be such a candidate that regulates peroxisomal FA β-oxidation (Meng et al. 2014).

AtACBPs in the Regulation of Seed Germination and Seedling Development

While a new role has been ascribed to OsACBP6 (Class IV) in lipid catabolism during seed germination and seedling growth, no such peroxisome-localized homolog was identified in Arabidopsis. However, another ACBP isoform, AtACBP1 (Class II), is linked to ABA signaling during seed germination and seedling development in Arabidopsis (Du et al. 2013a). ABA is an osmotic stress-related phytohormone that plays important roles in adaptive responses to abiotic and biotic stresses (Hirayama and Shinozaki 2007; Ton et al. 2009; Cutler et al. 2010), as well as in plant growth and development such as seed maturation and dormancy (Cheng et al. 2002; Fujii and Zhu 2009; Nakashima et al. 2009). The ABA receptors, REGULATORY COMPONENTS OF ABA RECEPTOR/PYRABACTIN

RESISTANCE1/PYRABACTIN RESISTANCE1-LIKE, have been recently identified as members of the START domain superfamily (Ma et al. 2009; Park et al. 2009; Klingler et al. 2010; Nishimura et al. 2010). Upon ABC transporter-mediated uptake into the cells (Kang et al. 2010; Kuromori et al. 2010; Boursiac et al. 2013), ABA binds to the cytosolic receptors, which in turn suppress phosphatase activity of type 2C protein phosphatase (PP2C), a negative regulator of ABA responses (Ma et al. 2009; Park et al. 2009; Santiago et al. 2009; Szostkiewicz et al. 2010). PP2C inhibition in the presence of ABA triggers the release of SNF1-related protein kinases 2, which phosphorylate and regulate the activity of downstream transcription factors, ABA-RESPONSIVE ELEMENT-BINDING PROTEIN/FACTOR (Nakashima and Yamaguchi-Shinozaki 2013). During seed germination, PA is produced as an important second messenger in ABA responses (Katagiri et al. 2005). The signaling PA molecules are generated by the PLD-catalyzed hydrolysis of structural phospholipids or by the sequential action of phospholipase C and DAG kinase (Testerink and Munnik 2005). It has been previously shown that PLD and its product, PA, inhibit the function of the negative regulator ABA INSENSITIVE1 (ABI1; Zhang et al. 2004), a member of the PP2C family (Gosti et al. 1999).

In Arabidopsis seedlings, *ACBP1* is highly expressed and its expression is inducible by ABA or drought treatment (Du et al. 2013a). In addition, the overexpression of *AtACBP1* up-regulated *PLDα1* and three ABA/stress-responsive genes including *AREB1*, *RESPONSE TO DESICCATION29A* (*RD29A*) and *bHLH-TRANSCRIPTION FACTOR MYC2* (*MYC2*) with or without ABA treatment, whilst *PLDα1* and *AREB1* were suppressed in the *acbp1* mutant upon ABA treatment (Du et al. 2013a). These observations are consistent with the higher accumulation of PA in the ABA-treated germinating seeds of AtACBP1-OEs and its lower level in the *acbp1* mutant (Du et al. 2013a). Functionally, the overexpression of *AtACBP1* rendered the freshly-harvested seeds more dormant than the wild type, whereas the *acbp1* mutant seeds became less dormant (Du et al. 2013a). The dry after-ripened seeds from AtACBP1-OEs were more sensitive to inhibition by exogenous ABA during germination in comparison with the wild type, while the *acbp1* mutant seeds became more resistant (Du et al. 2013a). Thus, AtACBP1 likely plays a role in ABA signaling during seed dormancy, germination and seedling development. As rAtACBP1 is the sole member of the AtACBP family that binds both PC and PA *in vitro* (Du et al. 2010a, 2013a), and that AtACBP1 and PLDα1 proteins were identified to interact at the plasma membrane in BiFC assays (Du et al. 2013a), AtACBP1 probably promotes PLDα1 activities *via* protein-protein interactions during ABA-mediated regulation of seed germination and seedling development in Arabidopsis.

Recently, Hsiao et al. (2014a) reported the accumulation of oleoyl- and linoleoyl-CoA esters in Arabidopsis *acbp6* seedlings in comparison to the wild type. During seed germination in the presence of exogenous ABA, double mutants of *acbp4acbp5*, *acbp4acbp6* and *acbp5acbp6* and the *acbp4acbp5acbp6* triple mutant exhibited a lower germination rate than the wild type, implicating an overlapping role of the three cytosolic AtACBPs in seed germination (Hsiao et al. 2014a).

AtACBP1 and AtACBP2 in Embryo Development

In the attempt to unravel the physiological functions of AtACBPs using a reverse genetics approach, embryo lethality of the *acbp1acbp2* double mutant represents the only severe phenotype observed thus far (Chen et al. 2010). On the other hand, neither *acbp1* nor *acbp2* single mutant plants exhibited morphological anomaly during seed development or seedling growth (Chen et al. 2010). It is plausible that AtACBP1 and AtACBP2 share an overlapping function in lipid metabolism during embryogenesis as both proteins are endomembrane-localized and highly conserved with 71 % sequence identity (Chye 1998; Chye et al. 1999, 2000; Li and Chye 2003). Microarray data also suggested that the temporal expression pattern of *AtACBP1* during seed development coincides with *AtACBP2* (Chen et al. 2010), and immunohistochemical staining confirmed that both proteins accumulated in cotyledonary-staged embryos (Chye et al. 1999; Chen et al. 2010). The ovules of *acbp1acbp2* double mutants aborted during early embryo development at either zygotic, two-cell or eight-cell stages, preceding the active phase of FA and TAG accumulation (Chen et al. 2010). The aborted embryos could not be induced to form calli, confirming that the embryos were arrested at the very early stages of development (Chen et al. 2010). Lipid profiling further revealed that siliques of the *acbp1* single mutant accumulated more galactolipid monogalactosyldiacylglycerol and stearoyl-CoA ester than the wild type, but most of the polyunsaturated species of phospholipids, including PC, PE, PS and phosphatidylinositol (PI), were notably reduced (Chen et al. 2010). As rAtACBP1 binds both acyl-CoA esters and phospholipids (PC and PA) *in vitro* (Chye 1998; Gao et al. 2009; Du et al. 2010a, 2013a; Xue et al. 2014), AtACBP1 possibly contributes to acyl-CoA metabolism and membrane phospholipid biogenesis during embryo development. AtACBP1- and AtACBP2-mediated acyl exchange between the acyl-CoA and phospholipid pools is one such possibility (Napier and Haslam 2010).

The importance of normal lipid metabolism during seed development can be reflected by the embryo-lethality phenotypes observed in several loss-of-function mutants that have impaired acyl-CoA-dependent enzymes in Arabidopsis. For instance, embryo development was arrested at the heart-torpedo stage by a T-DNA insertion within the *LPAAT1* gene that encodes a plastid LPAAT for the acylation of lyso-PA into PA (Kim and Huang 2004). In addition, depletion of the 3-hydroxy-acyl-CoA dehydratase PASTICCINO2 (PAS2), a component of the microsomal fatty acyl elongase complex, affected cotyledon development as visualized at the torpedo stage (Bach et al. 2008; Bach and Faure 2010). Multiple allelic mutants in the *PAS3/ACC1* gene that encodes an acetyl-CoA carboxylase also resulted in defective embryo morphogenesis (Baud et al. 2003, 2004). In the *pas2* and *acc1* mutants, the VLCFA content significantly declined or was barely detectable in seed storage TAGs, respectively (Baud et al. 2003; Bach et al. 2008). Similarly, insertional mutants in *AtKCR1* which encodes β-ketoacyl-CoA reductase in the fatty acyl elongase complex also resulted in embryo lethality (Beaudoin et al. 2009). Collectively, reverse genetics studies indicate a link between embryogenesis and the synthesis of VLCFAs, which constitute a significant proportion (i.e. 13–21 %) of total FAs in Arabidopsis seeds (O'Neill et al. 2003). Although the modified content of VLCFAs in the TAG pool may not severely influence plant development, their altered compositions in phospholipids and complex sphinogolipids may lead to profound consequences in plant development affecting embryogenesis (Bach and Faure 2010). With respect to the phospholipid profile of siliques from the *acbp1* mutant (Chen et al. 2010), the levels of VLCFAs esterified to PS (i.e. C40:3- and C40:4-PS) were particularly reduced (Napier and Haslam 2010). Given our recent evidence that rAtACBP1 binds VLC acyl-CoA esters *in vitro* (Xue et al. 2014), an alternative role of AtACBP1 in VLCFA formation in seed development may now be proposed. However, it is noteworthy that the differences in lipid profile and acyl-CoA content between the *acbp2* mutant and the wild type were statistically insignificant (Chen et al. 2010). Whether AtACBP2 acts concertedly with AtACBP1 in binding VLC acyl-CoA esters that may account for the embryo lethality of the *acbp1acbp2* double mutant remains to be investigated.

AtACBPs in Cuticle Formation

In terrestrial plants, the cuticle is a highly hydrophobic layer that covers the surface of all aerial organs as a protective barrier against non-stomatal water loss (Riederer and Schreiber 2001) and pathogens (Barthlott and Neinhuis 1997), and as an interface for interaction with the environment in stress responses and adaptations (Bernard and Joubès 2013). It also has a role in contact-mediated cell-cell interactions such as pollen-stigma interaction and in the prevention of fusions between different plant organs (Pruitt et al. 2000; Sieber et al. 2000). The cuticle is composed of cutin and wax, both of which are synthesized exclusively in the epidermal cells (Bernard and Joubès 2013; Lee and Suh 2013). Cutin forms the structural skeleton of the cuticle and is a polyester built from monomers of hydroxylated and epoxy-hydroxylated C16 and C18 FAs, glycerol and minute amounts of phenolic compounds (Heredia 2003; Nawrath 2006; Samuels et al. 2008). Intracuticular waxes are embedded within the cutin in the cuticle membrane which is covered by an outermost layer of epicuticular waxes (Jenks et al. 2002; Kunst and Samuels 2003). On the ER membrane, cuticular waxes are produced by the extension of leucoplast-synthesized C16 and C18 FAs as CoA-esters to form VLCFAs, which are reduced into primary alcohols and wax esters *via* acyl reduction or modified into aldehydes, alkanes, secondary alcohols and ketones *via* the decarbonylation pathway (Jenks et al. 2002; Nawrath 2006; Samuels et al. 2008; Lee and Suh 2013).

The binding specificities of rAtACBP1 to C18 and VLC acyl-CoA esters *in vitro* coinciding with strong GUS expression in stem epidermis from the *AtACBP1pro::GUS* transformants led to investigations to address whether ER-localized AtACBP1 facilitates cutin and wax biosyntheses from such acyl-CoA precursors (Xue et al. 2014). This novel role of AtACBP1 was supported by phenotypic abnormalities of its loss-of-function mutant (Xue et al. 2014). Scanning and transmission electron microscopy showed fewer epicuticular wax crystals and a ruptured cuticle membrane in Arabidopsis *acbp1* mutant stems (Xue et al. 2014). Gas chromatography-mass spectrometry (GC-MS) further revealed an aberrant composition of cuticular wax in *acbp1* mutant stems with significantly reduced levels of C29 alkane, C28 and C30 primary alcohols, and C29 secondary alcohol and ketone (Xue et al. 2014). GC-MS also indicated lower levels of C18:1 and C18:2 ω-hydroxyl and dicarboxylic FAs, which are the constituents of cutin monomers (Xue et al. 2014). Concomitantly, the expression of several wax and cutin biosynthetic genes was down-regulated in stems of the *acbp1* mutant in comparison with the wild type and *acbp1*-complemented lines (Xue et al. 2014). Although AtACBP1 may not play a direct role in leaf wax synthesis as evident from the lack of *AtACBP1pro::GUS* signal in leaf epidermis, the wax content was reduced in *acbp1* mutant leaves, which became more susceptible to *B. cinerea* infection (Xue et al. 2014). As cuticular wax biosynthesis is tightly controlled by a number of transcription factors in response to diverse external cues (Aharoni et al. 2004; Zhang et al. 2007; Cominelli et al. 2008; Lü et al. 2009; Seo et al. 2011), defective stem cuticle formation in the *acbp1* mutant may possibly alter the status of the plant as a whole and exert an indirect effect on leaf wax synthesis (Xue et al. 2014).

In another study, AtACBP3, AtACBP4 and AtACBP6 were shown to be crucial for normal leaf cuticle development as knockout mutant plants exhibited higher water loss under drought stress, increased folding on the adaxial leaf surface, irregularity in cuticular appearance and permeability of leaves to toluidine blue staining (Xia et al. 2012). The aberrant cuticle structures of knockout mutant plants were linked to higher levels of several components of cuticular wax including FA (C16:0, C18:0), alkanes (C29, C31 and C33) and primary alcohols (C28 and C32), and reduced levels of cutin monomers including C16:0, C18:1 and C18:2 dicarboxylic FAs (Xia et al. 2012). The defective cuticles of the *acbp3*, *acbp4* and *acbp6* mutant plants also account for their higher susceptibility to *P. syringae* (both virulent DC3000 and avirulent *avrRpt2* strains) and, occasionally, to *B. cinerea* and *C. higginsianum* (Xia et al. 2012). When petiole exudates from *avrRpt2*-challenged leaves of the wild type were infiltrated into *acbp3*, *acbp4* and *acbp6*, the mutant plants were protected against virulent bacteria inoculated at distal leaves away from the exudate-infiltrated leaves (Xia et al. 2012). On the other hand, the petiole exudates from *acbp3*, *acbp4* and *acbp6* did not confer comparable protection of the wild type against virulent bacteria in similar experimental set-ups (Xia et al. 2012). Accordingly, it was concluded that AtACBP3, AtACBP4 and AtACBP6 are essential for the formation of an intact cuticle and the generation of mobile systemic acquired resistance (SAR) signals in response to microbial pathogens but not for the perception of SAR signals (Xia et al. 2012).

Cytosolic AtACBPs in Pollen Development

Lipids form a structurally and metabolically essential constituent in pollen, which contain extracellular lipidic deposits (i.e. tryphine) covering the exine wall surface (primarily composed of the chemically-stable polymer sporopollenin) and intracellular lipid pools as storage oil bodies and extensive membranous structures (Evans et al. 1991, 1992; Piffanelli et al. 1997; Murphy 2006). The importance of pollen lipids is evident from aberrant pollen development resulting by the manipulation of gene expression affecting acyl lipid metabolism, including intracellular storage lipid accumulation (Zhang et al. 2009), extracellular pollen lipid secretion (Zheng et al. 2003), wax deposition (Aarts et al. 1995, 1997; Millar et al. 1999; Ariizumi et al. 2003; Chen et al. 2003; Kurata et al. 2003) and membrane lipid catabolism (Kim et al. 2011). As potential intracellular acyl transporters involved in lipid metabolic processes, the three cytosolic AtACBPs (i.e. AtACBP4, AtACBP5 and AtACBP6) were recently investigated with respect to their roles in pollen lipid formation (Hsiao et al. 2015).

Normal pollen development in single mutants of the three cytosolic AtACBPs has been attributed to the overlapping roles of AtACBP4 and AtACBP5 with AtACBP6, as *AtACBP6pro*::*GUS* signals coincided with *AtACBP5pro*::*GUS* in microspores and tapetal cells, and with *AtACBP4pro*::*GUS* in pollen grains at later stages (Hsiao et al. 2015). On the other hand, the *acbp4acbp6* and *acbp5acbp6* double mutants, but not *acbp4acbp5*, exhibited morphological aberrance in floral development, such as reduction in the number of normal pollen, appearance of aborted pollen, reduction in seed number per silique and more prominent cytoplasmic vacuolation in mature pollen grains (Hsiao et al. 2015). The morphology of flowers and siliques appeared to be the most affected in the *acbp4acbp5acbp6* triple mutant (Hsiao et al. 2015). Scanning and transmission electron microscopy further confirmed defects in the exine of the triple mutant in a smoother pollen surface and an irregular arrangement between the bacula and tryphine in comparison with the wild type, whilst lower abundance of oil bodies in its pollen was in agreement with the decreased pollen activity (Hsiao et al. 2015).

Although the composition of the sporopollenin polymer and its biosynthetic pathways in the tapetum remain elusive, the inter-organellar transport (e.g. from plastids to ER) and some modifications (e.g. acyl reduction) of the acyl-lipid intermediates require their esterification into CoA-esters, as mutation of Arabidopsis *ACYL-COA SYNTHETASE5* led to defective pollen exine formation (de Azevedo Souza et al. 2009; Grienenberger et al. 2010; Ariizumi and Toriyama 2011). The subsequent secretion of lipidic sporopollenin precursors across the plasma membrane of tapetal cells onto the surface of developing microspore walls is mediated by ABC TRANSPORTER G26 (ABCG26), and *abcg26* mutants showed reduced male fertility and exine-less microspores (Quilichini et al. 2010; Choi et al. 2011). The comparable but less severe phenotypic changes in pollen of the *acbp4acbp5acbp6* triple mutant implicate overlapping roles for the three cytosolic AtACBPs in sporopollenin biosynthesis, potentially by intracellular transport of the acyl-lipid precursors (Hsiao et al. 2015).

AtACBP3 in Leaf Senescence

During senescence and nutrient starvation, dysfunctional or unwanted organelles and cellular constituents are rapidly recycled to remobilize the supply of carbon and nitrogen for sustaining growth and survival in a turnover process known as autophagy (Li and Vierstra 2012). The sophisticated machinery for autophagy is now known to be conserved from yeast to higher eukaryotes including plants (He and Klionsky 2009; Liu and Bassham 2012). Autophagy is mediated by the sequestration of intracellular components in double-membrane autophagosomes, which subsequently fuse with lysosomes or plant vacuoles for breakdown and recycling of degradation products (He and Klionsky 2009). The biogenesis of autophagosomes requires two ubiquitination-like conjugation systems. The ubiquitin-like proteins ATG8 and ATG12 conjugate to PE and ATG5, respectively (Geng and Klionsky 2008; Liu and Bassham 2012). ATG8-PE constitutes a scaffold for membrane expansion of autophagosomes, the size of which correlates with the amount of ATG8 (Geng and Klionsky 2008; Xie et al. 2008). The lipidation of ATG8 is facilitated by the ATG12-ATG5 conjugate, and is reversible by the action of PLD (Fujioka et al. 2008; Chung et al. 2009, 2010).

In Arabidopsis, the ATG8-PE association during starvation-induced and age-dependent leaf senescence may also be modulated by AtACBP3, which has been demonstrated to bind PE *in vitro* and its expression was up-regulated in continuous darkness and senescing rosettes (Xiao et al. 2010). The overexpression of *AtACBP3* led to higher accumulation of PE in rosettes, whereas its level decreased in the *acbp3* mutant and *AtACBP3-RNAi* transgenic lines (Xiao et al. 2010). AtACBP3-OEs exhibited faster leaf senescence upon dark treatment or nitrogen starvation, and they also showed early senescence of 5-week-old rosettes in a SA signaling-dependent manner under normal light/dark condition (Xiao et al. 2010). These stress-inducible and age-dependent senescence phenotypes of AtACBP3-OEs resemble those of Arabidopsis *atg2*, *atg5*, *atg7*, *atg10* and *atg18a* mutants defective in the autophagic pathway (Doelling et al. 2002; Thompson et al. 2005; Xiong et al. 2005; Phillips et al. 2008; Yoshimoto et al. 2009). Consistently, the *acbp3* mutant and *AtACBP3-RNAi* transgenic lines showed delayed dark-induced leaf senescence (Xiao et al. 2010). The root cells of AtACBP3-OE seedlings had fewer autophagosome structures as visualized by the expression of the autophagy marker GFP-ATG8e or in monodansylcadaverine staining in comparison with the wild type (Xiao et al. 2010). As protein degradation of GFP-ATG8e was also stimulated in an AtACBP3-OE background, AtACBP3 may regulate autophagosome formation by controlling ATG8-PE lipidation, possibly *via* competition for a common ligand, PE (Xiao et al. 2010; Xiao and Chye 2010). Apart from the modulation of ATG8 stability, AtACBP3 may affect membrane lipid metabolism during leaf senescence as membrane phospholipids (e.g. PC and PI) were lost and their degradative and oxidative products (e.g. PA, lyso-phospholipids, arabidopsides) increased in dark-treated and premature senescing AtACBP3-OE plants (Xiao et al. 2010; Xiao and Chye 2010).

ACBPs in Systemic Transport via the Phloem

In vascular plants, the phloem conducts long-distance translocation of phytohormones, inorganic ions and assimilates such as sugars, amino acids and organic acids. A number of soluble mobile proteins have also been identified from the phloem sap in earlier studies (Fisher et al. 1992; Ishiwatari et al. 1995; Hayashi et al. 2000), prompting subsequent comprehensive analyses of the protein complements in phloem exudates by peptide sequencing and tandem mass spectrometry-based proteomic studies (Walz et al. 2004; Suzui et al. 2006; Guelette et al. 2012). Partial peptide sequences from the phloem exudates of cucumber (*Cucumis sativus*) and pumpkin (*Cucurbita maxima*) were highly homologous to ACBP and some other defense-related proteins (Walz et al. 2004). In another study, ACBP was identified as a major phloem sap protein in rice, and its existence was also confirmed in the phloem exudates of winter squash (*C. maxima*), *B. napus* and coconut palm (*Cocos nucifera*) by Western blot analysis using an anti-BnACBP antibody (Suzui et al. 2006). More recently, proteomic analysis of the Arabidopsis phloem exudate revealed the identities of 65 proteins, 11 of which were lipid-binding proteins (Guelette et al. 2012). The same study also identified several FAs and lipids, some of which function as signaling compounds (e.g. PA and phosphatidyl bisphosphate), leading to a hypothesis that some lipid-binding proteins in the phloem sap may play a role in long-distance transport of lipid signals (Guelette et al. 2012). As to whether the ACBPs present in phloem exudates play a role in systemic lipid trafficking remains to be determined. In Arabidopsis, *AtACBP3*, which was expressed in the phloem as indicated by promoter-*GUS* fusions in transgenic Arabidopsis (Zheng et al. 2012), presents a promising candidate for future investigations.

Conclusion

Lipids have widespread biological roles in energy storage and provision, cellular structure formation and signal transduction. ACBPs, as “lipid chaperones”, are emerging to be versatile proteins that contribute to a number of different processes in plant growth and development, besides their elucidated functions in stress responses. In addition to the binding of acyl-CoA esters as prototypical ligands highlighting the pertinence of ACBPs to acyl-CoA-dependent biosyntheses of lipids (e.g. seed oils and cuticle), the phospholipid-binding affinities of ACBPs further diversify their functions in the regulation of specific plant developmental events such as seed germination, seedling development and leaf senescence. Whilst the vast majority of novel ACBP functions were identified in the eudicot model plant Arabidopsis, recent studies in the monocot rice have unravelled substantial differences between the two ACBP families in terms of ligand specificities, domain architecture, subcellular localization patterns and stress-responsiveness in gene expression. The possible involvement of the peroxisome-localized OsACBP6 in

FA catabolism represents the first example of a unique rice ACBP function not previously reported in Arabidopsis, marking a contrast between rice and Arabidopsis. Further dissection of the functions of rice ACBP homologs and phloem-mobile ACBPs from other species will provide a more thorough understanding of the developmental roles of plant ACBPs in the future.

Acknowledgements MLC is grateful to the Wilson and Amelia Wong Endowment Fund and the Research Grants Council of the Hong Kong Special Administrative Region, China (project no. HKU17105615M & HKU765813M) for supporting her research on plant ACBPs. SCL is supported by a postdoctoral fellowship and the Committee on Research and Conference Grants (Small Project Funding- 201309176052) from the University of Hong Kong.

References

Aarts MGM, Keijzer CJ, Stiekema WJ, Pereira A (1995) Molecular characterization of the *CER1* gene of Arabidopsis involved in epicuticular wax biosynthesis and pollen fertility. Plant Cell 7:2115–2127

Aarts MGM, Hodge R, Kalantidis K, Florack D, Wilson ZA, Mulligan BJ, Stiekema WJ, Scott R, Pereira A (1997) The *Arabidopsis MALE STERILITY 2* protein shares similarity with reductases in elongation/condensation complexes. Plant J 12:615–623

Adams J, Kelso R, Cooley L (2000) The kelch repeat superfamily of proteins: propellers of cell function. Trends Cell Biol 10:17–24

Aharoni A, Dixit S, Jetter R, Thoenes E, van Arkel G, Pereira A (2004) The SHINE Clade of AP2 domain transcription factors activates wax biosynthesis, alters cuticle properties, and confers drought tolerance when overexpressed in Arabidopsis. Plant Cell 16:2463–2480

Andrews J, Keegstra K (1983) Acyl-CoA synthetase is located in the outer membrane and acyl-CoA thioesterase in the inner membrane of pea chloroplast envelopes. Plant Physiol 72:735–740

Ariizumi T, Toriyama K (2011) Genetic regulation of sporopollenin synthesis and pollen exine development. Annu Rev Plant Biol 62:437–460

Ariizumi T, Hatakeyama K, Hinata K, Sato S, Kato T, Tabata S, Toriyama K (2003) A novel male-sterile mutant of *Arabidopsis thaliana*, *faceless pollen-1*, produces pollen with a smooth surface and an acetolysis-sensitive exine. Plant Mol Biol 53:107–116

Bach L, Faure JD (2010) Role of very-long-chain fatty acids in plant development, when chain length does matter. C R Biol 333:361–370

Bach L, Michaelson LV, Haslam R, Bellec Y, Gissot L, Marion J, Costa MD, Boutin JP, Miquel M, Tellier F et al (2008) The very-long-chain hydroxy fatty acyl-CoA dehydratase PASTICCINO2 is essential and limiting for plant development. Proc Natl Acad Sci U S A 105:14727–14731

Barthlott W, Neinhuis C (1997) Purity of the sacred lotus, or escape from contamination in biological surfaces. Planta 202:1–8

Bates PD, Browse J (2012) The significance of different diacylgycerol synthesis pathways on plant oil composition and bioengineering. Front Plant Sci 3:147

Bates PD, Ohlrogge JB, Pollard M (2007) Incorporation of newly synthesized fatty acids into cytosolic glycerolipids in pea leaves occurs via acyl editing. J Biol Chem 282:31206–31216

Bates PD, Durrett TP, Ohlrogge JB, Pollard M (2009) Analysis of acyl fluxes through multiple pathways of triacylglycerol synthesis in developing soybean embryos. Plant Physiol 150:55–72

Baud S, Guyon V, Kronenberger J, Wuillème S, Miquel M, Caboche M, Lepiniec L, Rochat C (2003) Multifunctional acetyl-CoA carboxylase 1 is essential for very long chain fatty acid elongation and embryo development in *Arabidopsis*. Plant J 33:75–86

Baud S, Bellec Y, Miquel M, Bellini C, Caboche M, Lepiniec L, Faure JD, Rochat C (2004) *gurke* and *pasticcino3* mutants affected in embryo development are impaired in acetyl-CoA carboxylase. EMBO Rep 5:515–520

Beaudoin F, Wu X, Li F, Haslam RP, Markham JE, Zheng H, Napier JA, Kunst L (2009) Functional characterization of the Arabidopsis β-ketoacyl-coenzyme A reductase candidates of the fatty acid elongase. Plant Physiol 150:1174–1191

Beisson F, Ferté N, Bruley S, Voultoury R, Verger R, Arondel V (2001) Oil-bodies as substrates for lipolytic enzymes. Biochim Biophys Acta 1531:47–58

Benning C (2009) Mechanisms of lipid transport involved in organelle biogenesis in plant cells. Annu Rev Cell Dev Biol 25:71–91

Bernard A, Joubès J (2013) Arabidopsis cuticular waxes: advances in synthesis, export and regulation. Prog Lipid Res 52:110–129

Bessoule JJ, Lessire R, Cassagne C (1989) Partial purification of the Acyl-CoA elongase of *Allium porrum* leaves. Arch Biochem Biophys 268:475–484

Block MA, Dorne AJ, Joyard J, Douce R (1983) The acyl-CoA synthetase and acyl-CoA thioesterase are located on the outer and inner membrane of the chloroplast envelope, respectively. FEBS Lett 153:377–381

Bork P (1993) Hundreds of ankyrin-like repeats in functionally diverse proteins: mobile modules that cross phyla horizontally? Proteins 17:363–374

Boursiac Y, Léran S, Corratgé-Faillie C, Gojon A, Krouk G, Lacombe B (2013) ABA transport and transporters. Trends Plant Sci 18:325–333

Brown AP, Johnson P, Rawsthorne S, Hills MJ (1998) Expression and properties of acyl-CoA binding protein from *Brassica napus*. Plant Physiol Biochem 36:629–635

Browse J, Warwick N, Somerville CR, Slack CR (1986) Fluxes through the prokaryotic and eukaryotic pathways of lipid synthesis in the'16:3' plant *Arabidopsis thaliana*. Biochem J 235:25–31

Burton M, Rose TM, Faergeman NJ, Knudsen J (2005) Evolution of the acyl-CoA binding protein (ACBP). Biochem J 392:299–307

Chao H, Zhou M, McIntosh A, Schroeder F, Kier AB (2003) ACBP and cholesterol differentially alter fatty acyl CoA utilization by microsomal ACAT. J Lipid Res 44:72–83

Chen X, Goodwin SM, Boroff VL, Liu X, Jenks MA (2003) Cloning and characterization of the *WAX2* gene of Arabidopsis involved in cuticle membrane and wax production. Plant Cell 15:1170–1185

Chen QF, Xiao S, Chye ML (2008) Overexpression of the Arabidopsis 10-kilodalton acyl-coenzyme A-binding protein ACBP6 enhances freezing tolerance. Plant Physiol 148:304–315

Chen QF, Xiao S, Qi W, Mishra G, Ma J, Wang M, Chye ML (2010) The *Arabidopsis acbp1acbp2* double mutant lacking acyl-CoA-binding proteins ACBP1 and ACBP2 is embryo lethal. New Phytol 186:843–855

Chen MX, Lung SC, Du ZY, Chye ML (2014) Engineering plants to tolerate abiotic stresses. Biocatal Agric Biotechnol 3:81–87

Cheng WH, Endo A, Zhou L, Penney J, Chen HC, Arroyo A, Leon P, Nambara E, Asami T, Seo M et al (2002) A unique short-chain dehydrogenase/reductase in Arabidopsis glucose signaling and abscisic acid biosynthesis and functions. Plant Cell 14:2723–2743

Choi H, Jin JY, Choi S, Hwang JU, Kim YY, Suh MC, Lee Y (2011) An ABCG/WBC-type ABC transporter is essential for transport of sporopollenin precursors for exine formation in developing pollen. Plant J 65:181–193

Chung T, Suttangkakul A, Vierstra RD (2009) The ATG autophagic conjugation system in maize: ATG transcripts and abundance of the ATG8-lipid adduct are regulated by development and nutrient availability. Plant Physiol 149:220–234

Chung T, Phillips AR, Vierstra RD (2010) ATG8 lipidation and ATG8-mediated autophagy in Arabidopsis require ATG12 expressed from the differentially controlled *ATG12A* AND *ATG12B* loci. Plant J 62:483–493

Chye ML (1998) *Arabidopsis* cDNA encoding a membrane-associated protein with an acyl-CoA binding domain. Plant Mol Biol 38:827–838

Chye ML, Huang BQ, Zee SY (1999) Isolation of a gene encoding *Arabidopsis* membrane-associated acyl-CoA binding protein and immunolocalization of its gene product. Plant J 18:205–214

Chye ML, Li HY, Yung MH (2000) Single amino acid substitutions at the acyl-CoA-binding domain interrupt ^{14}[C]palmitoyl-CoA binding of ACBP2, an *Arabidopsis* acyl-CoA-binding protein with ankyrin repeats. Plant Mol Biol 44:711–721

Citharel B, Oursel A, Mazliak P (1983) Desaturation of oleoyl and linoleoyl residues linked to phospholipids in growing roots of yellow lupin. FEBS Lett 161:251–256

Cominelli E, Sala T, Calvi D, Gusmaroli G, Tonelli C (2008) Over-expression of the Arabidopsis *AtMYB41* gene alters cell expansion and leaf surface permeability. Plant J 53:53–64

Cutler SR, Rodriguez PL, Finkelstein RR, Abrams SR (2010) Abscisic acid: emergence of a core signaling network. Annu Rev Plant Biol 61:651–679

Dave A, Hernández ML, He Z, Andriotis VME, Vaistij FE, Larson TR, Graham IA (2011) 12-Oxo-phytodienoic acid accumulation during seed development represses seed germination in *Arabidopsis*. Plant Cell 23:583–599

de Azevedo Souza C, Kim SS, Koch S, Kienow L, Schneider K, McKim SM, Haughn GW, Kombrink E, Douglas CJ (2009) A novel fatty acyl-CoA synthetase is required for pollen development and sporopollenin biosynthesis in *Arabidopsis*. Plant Cell 21:507–525

De Marcos Lousa C, van Roermund CWT, Postis VLG, Dietrich D, Kerr ID, Wanders RJA, Baldwin SA, Baker A, Theodoulou FL (2013) Intrinsic acyl-CoA thioesterase activity of a peroxisomal ATP binding cassette transporter is required for transport and metabolism of fatty acids. Proc Natl Acad Sci U S A 110:1279–1284

Doelling JH, Walker JM, Friedman EM, Thompson AR, Vierstra RD (2002) The APG8/12-activating enzyme APG7 is required for proper nutrient recycling and senescence in *Arabidopsis thaliana*. J Biol Chem 277:33105–33114

Du ZY, Chye ML (2013) Interactions between Arabidopsis acyl-CoA-binding proteins and their protein partners. Planta 238:239–245

Du ZY, Xiao S, Chen QF, Chye ML (2010a) Depletion of the membrane-associated acyl-coenzyme A-binding protein ACBP1 enhances the ability of cold acclimation in Arabidopsis. Plant Physiol 152:1585–1597

Du ZY, Xiao S, Chen QF, Chye ML (2010b) Arabidopsis acyl-CoA-binding proteins ACBP1 and ACBP2 show different roles in freezing stress. Plant Signal Behav 5:607–609

Du ZY, Chen MX, Chen QF, Xiao S, Chye ML (2013a) Arabidopsis acyl-CoA-binding protein ACBP1 participates in the regulation of seed germination and seedling development. Plant J 74:294–309

Du ZY, Chen MX, Chen QF, Xiao S, Chye ML (2013b) Overexpression of Arabidopsis acyl-CoA-binding protein ACBP2 enhances drought tolerance. Plant Cell Environ 36:300–314

Du ZY, Chen MX, Chen QF, Gu JD, Chye ML (2015) Expression of *Arabidopsis* acyl-CoA-binding proteins AtACBP1 and AtACBP4 confers Pb(II) accumulation in *Brassica juncea* roots. Plant Cell Environ 38:101–117

Dubacq JP, Drapier D, Trémoliéres A (1983) Polyunsaturated fatty acid synthesis by a mixture of chloroplasts and microsomes from spinach leaves: evidence for two distinct pathways of the biosynthesis of trienoic acids. Plant Cell Physiol 24:1–9

Dykema PE, Sipes PR, Marie A, Biermann BJ, Crowell DN, Randall SK (1999) A new class of proteins capable of binding transition metals. Plant Mol Biol 41:139–150

Elle IC, Simonsen KT, Olsen LCB, Birck PK, Ehmsen S, Tuck S, Le TT, Faergeman NJ (2011) Tissue- and paralogue-specific functions of acyl-CoA-binding proteins in lipid metabolism in *Caenorhabditis elegans*. Biochem J 437:231–241

Engeseth NJ, Pacovsky RS, Newman T, Ohlrogge JB (1996) Characterization of an acyl-CoA-binding protein from *Arabidopsis thaliana*. Arch Biochem Biophys 331:55–62

Evans DE, Taylor PE, Singh MB, Knox RB (1991) Quantitative analysis of lipids and protein from the pollen of *Brassica napus* L. Plant Sci 73:117–126

Evans DE, Taylor PE, Singh MB, Knox RB (1992) The interrelationship between the accumulation of lipids, protein and the level of acyl carrier protein during the development of *Brassica napus* L. pollen. Planta 186:343–354

Faergeman NJ, Knudsen J (1997) Role of long-chain fatty acyl-CoA esters in the regulation of metabolism and in cell signalling. Biochem J 323:1–12

Faergeman NJ, Wadum M, Feddersen S, Burton M, Kragelund BB, Knudsen J (2007) Acyl-CoA binding proteins; structural and functional conservation over 2000 MYA. Mol Cell Biochem 299:55–65

Fisher DB, Wu Y, Ku MSB (1992) Turnover of soluble proteins in the wheat sieve tube. Plant Physiol 100:1433–1441

Flieger A, Neumeister B, Cianciotto NP (2002) Characterization of the gene encoding the major secreted lysophospholipase A of *Legionella pneumophila* and its role in detoxification of lysophosphatidylcholine. Infect Immun 70:6094–6106

Footitt S, Slocombe SP, Larner V, Kurup S, Wu Y, Larson T, Graham I, Baker A, Holdsworth M (2002) Control of germination and lipid mobilization by *COMATOSE*, the *Arabidopsis* homologue of human ALDP. EMBO J 21:2912–2922

Friedman JR, Voeltz GK (2011) The ER in 3-D: a multifunctional dynamic membrane network. Trends Cell Biol 21:709–717

Fujii H, Zhu JK (2009) Arabidopsis mutant deficient in 3 abscisic acid-activated protein kinases reveals critical roles in growth, reproduction, and stress. Proc Natl Acad Sci U S A 106:8380–8385

Fujioka Y, Noda NN, Fujii K, Yoshimoto K, Ohsumi Y, Inagaki (2008) *In vitro* reconstitution of plant Atg8 and Atg12 conjugation systems essential for autophagy. J Biol Chem 283:1921–1928

Fulda M, Shockey J, Werber M, Wolter FP, Heinz E (2002) Two long-chain acyl-CoA synthetases from *Arabidopsis thaliana* involved in peroxisomal fatty acid β-oxidation. Plant J 32:93–103

Fulda M, Schnurr J, Abbadi A, Heinz E, Browse J (2004) Peroxisomal acyl-CoA synthetase activity is essential for seedling development in *Arabidopsis thaliana*. Plant Cell 16:394–405

Fyrst H, Knudsen J, Schott MA, Lubin BH, Kuypers FA (1995) Detection of acyl-CoA-binding protein in human red blood cells and investigation of its role in membrane phospholipid renewal. Biochem J 306:793–799

Gaigg B, Neergaard TBF, Schneiter R, Hansen JK, Faergeman NJ, Jensen NA, Andersen JR, Friis J, Sandhoff R, Schrøder HD et al (2001) Depletion of acyl-coenzyme A-binding protein affects sphingolipid synthesis and causes vesicle accumulation and membrane defects in *Saccharomyces cerevisiae*. Mol Biol Cell 12:1147–1160

Gao W, Xiao S, Li HY, Tsao SW, Chye ML (2009) *Arabidopsis thaliana* acyl-CoA-binding protein ACBP2 interacts with heavy-metal-binding farnesylated protein AtFP6. New Phytol 181:89–102

Gao W, Li HY, Xiao S, Chye ML (2010a) Acyl-CoA-binding protein 2 binds lysophospholipase 2 and lysoPC to promote tolerance to cadmium-induced oxidative stress in transgenic Arabidopsis. Plant J 62:989–1003

Gao W, Li HY, Xiao S, Chye ML (2010b) Protein interactors of acyl-CoA-binding protein ACBP2 mediate cadmium tolerance in Arabidopsis. Plant Signal Behav 5:1025–1027

Geng J, Klionsky DJ (2008) The Atg8 and Atg12 ubiquitin-like conjugation systems in macroautophagy. EMBO Rep 9:859–864

Gerhardt B (1983) Localization of *β*-oxidation enzymes in peroxisomes isolated from nonfatty plant tissues. Planta 159:238–246

Goff SA, Ricke D, Lan TH, Presting G, Wang R, Dunn M, Glazebrook J, Sessions A, Oeller P, Varma H et al (2002) A draft sequence of the rice genome (*Oryza sativa* L. ssp. *japonica*). Science 296:92–100

Gosti F, Beaudoin N, Serizet C, Webb AAR, Vartanian N, Giraudat J (1999) ABI1 protein phosphatase 2C is a negative regulator of abscisic acid signaling. Plant Cell 11:1897–1909

Graham IA, Eastmond PJ (2002) Pathways of straight and branched chain fatty acid catabolism in higher plants. Prog Lipid Res 41:156–181

Greer S, Wen M, Bird D, Wu X, Samuels L, Kunst L, Jetter R (2007) The cytochrome P450 enzyme CYP96A15 is the midchain alkane hydroxylase responsible for formation of secondary alcohols and ketones in stem cuticular wax of Arabidopsis. Plant Physiol 145:653–667

Grienenberger E, Kim SS, Lallemand B, Geoffroy P, Heintz D, de Azevedo Souza C, Heitz T, Douglas CJ, Legrand M (2010) Analysis of *TETRAKETIDE α-PYRONE REDUCTASE* function in *Arabidopsis thaliana* reveals a previously unknown, but conserved, biochemical pathway in sporopollenin monomer biosynthesis. Plant Cell 22:4067–4083

Gueguen V, Macherel D, Jaquinod M, Douce R, Bourguignon J (2000) Fatty acid and lipoic acid biosynthesis in higher plant mitochondria. J Biol Chem 275:5016–5025

Guelette BS, Benning UF, Hoffmann-Benning S (2012) Identification of lipids and lipid-binding proteins in phloem exudates from *Arabidopsis thaliana.* J Exp Bot 63:3603–3616

Guerrero C, Martín-Rufián M, Reina JJ, Heredia A (2006) Isolation and characterization of a cDNA encoding a membrane bound acyl-CoA binding protein from *Agave americana* L. epidermis. Plant Physiol Biochem 44:85–90

Guidotti A, Forchetti CM, Corda MG, Konkel D, Bennett CD, Costa E (1983) Isolation, characterization, and purification to homogeneity of an endogenous polypeptide with agonistic action on benzodiazepine receptors. Proc Natl Acad Sci U S A 80:3531–3535

Harmer SL, Hogenesch JB, Straume M, Chang HS, Han B, Zhu T, Wang X, Kreps JA, Kay SA (2000) Orchestrated transcription of key pathways in *Arabidopsis* by the circadian clock. Science 290:2110–2113

Harwood JL (1988) Fatty acid metabolism. Annu Rev Plant Physiol Plant Mol Biol 39:101–138

Hayashi M, Toriyama K, Kondo M, Nishimura M (1998) 2,4-dichlorophenoxybutyric acid-resistant mutants of Arabidopsis have defects in glyoxysomal fatty acid β-oxidation. Plant Cell 10:183–195

Hayashi H, Fukuda A, Suzui N, Fujimaki S (2000) Proteins in the sieve element-companion cell complexes: their detection, localization and possible functions. Aust J Plant Physiol 27:489–496

Hayashi Y, Yamada K, Shimada T, Matsushima R, Nishizawa NK, Nishimura M, Hara-Nishimura I (2001) A proteinase-storing body that prepares for cell death or stresses in the epidermal cells of Arabidopsis. Plant Cell Physiol 42:894–899

Hayashi M, Nito K, Takei-Hoshi R, Yagi M, Kondo M, Suenaga A, Yamaya T, Nishimura M (2002) Ped3p is a peroxisomal ATP-binding cassette transporter that might supply substrates for fatty acid β-oxidation. Plant Cell Physiol 43:1–11

He C, Klionsky DJ (2009) Regulation mechanisms and signaling pathways of autophagy. Annu Rev Genet 43:67–93

Heinz E, Roughan PG (1983) Similarities and differences in lipid metabolism of chloroplasts isolated from 18:3 and 16:3 plants. Plant Physiol 72:273–279

Heredia A (2003) Biophysical and biochemical characteristics of cutin, a plant barrier biopolymer. Biochim Biophys Acta 1620:1–7

Hills MJ, Dann R, Lydiate D, Sharpe A (1994) Molecular cloning of a cDNA from *Brassica napus* L. for a homologue of acyl-CoA-binding protein. Plant Mol Biol 25:917–920

Hirayama T, Shinozaki K (2007) Perception and transduction of abscisic acid signals: keys to the function of the versatile plant hormone ABA. Trends Plant Sci 12:343–351

Hobbs DH, Hills MJ (2000) Expression and characterization of diacylglycerol acyltransferase from *Arabidopsis thaliana* in insect cell cultures. Biochem Soc Trans 28:687–689

Holthuis JCM, Levine TP (2005) Lipid traffic: floppy drives and a superhighway. Nat Rev Mol Cell Biol 6:209–220

Hsiao AS, Haslam RP, Michaelson LV, Liao P, Chen QF, Sooriyaarachchi S, Mowbray SL, Napier JA, Tanner JA, Chye ML (2014a) *Arabidopsis* cytosolic acyl-CoA-binding proteins ACBP4,

ACBP5 and ACBP6 have overlapping but distinct roles in seed development. Biosci Rep 34:e00165
Hsiao AS, Haslam RP, Michaelson LV, Liao P, Napier JA, Chye ML (2014b) Gene expression in plant lipid metabolism in Arabidopsis seedlings. PLOS ONE 9:e107372
Hsiao AS, Yeung EC, Ye ZW, Chye ML (2015) The Arabidopsis cytosolic acyl-CoA-binding proteins play combinatory roles in pollen development. Plant Cell Physiol 56:322–333
Hu J, Baker A, Bartel B, Linka N, Mullen RT, Reumann S, Zolman BK (2012) Plant peroxisomes: biogenesis and function. Plant Cell 24:2279–2303
Huang H, Atshaves BP, Frolov A, Kier AB, Schroeder F (2005) Acyl-coenzyme A binding protein expression alters liver fatty acyl-coenzyme A metabolism. Biochemistry 44:10282–10297
Hurlock AK, Roston RL, Wang K, Benning C (2014) Lipid trafficking in plant cells. Traffic 15:915–932
Ishiwatari Y, Honda C, Kawashima I, Nakamura S, Hirano H, Mori S, Fujiwara T, Hayashi H, Chino M (1995) Thioredoxin h is one of the major proteins in rice phloem sap. Planta 195:456–463
Jacob T, Ritchie S, Assmann SM, Gilroy S (1999) Abscisic acid signal transduction in guard cells is mediated by phospholipase D activity. Proc Natl Acad Sci U S A 96:12192–12197
Jenks MA, Eigenbrode SD, Lemieux B (2002) Cuticular waxes of Arabidopsis. Arabidopsis Book 1:e0016
Jolly CA, Wilton DC, Schroeder F (2000) Microsomal fatty acyl-CoA transacylation and hydrolysis: fatty acyl-CoA species dependent modulation by liver fatty acyl-CoA binding proteins. Biochim Biophys Acta 1483:185–197
Kader JC (1997) Lipid-transfer proteins: a puzzling family of plant proteins. Trends Plant Sci 2:66–70
Kang J, Hwang JU, Lee M, Kim YY, Assmann SM, Martinoia E, Lee Y (2010) PDR-type ABC transporter mediates cellular uptake of the phytohormone abscisic acid. Proc Natl Acad Sci U S A 107:2355–2360
Kannan L, Knudsen J, Jolly CA (2003) Aging and acyl-CoA binding protein alter mitochondrial glycerol-3-phosphate acyltransferase activity. Biochim Biophys Acta 1631:12–16
Katagiri T, Ishiyama K, Kato T, Tabata S, Kobayashi M, Shinozaki K (2005) An important role of phosphatidic acid in ABA signaling during germination of *Arabidopsis thaliana*. Plant J 43:107–117
Kennedy EP (1961) Biosynthesis of complex lipids. Fed Proc Am Soc Exp Biol 20:934–940
Kerkhoff C, Beuck M, Threige-Rasmussen J, Spener F, Knudsen J, Schmitz G (1997) Acyl-CoA binding protein (ACBP) regulates acyl-CoA:cholesterol acyltransferase (ACAT) in human mononuclear phagocytes. Biochim Biophys Acta 1346:163–172
Kikuchi M, Hatano N, Yokota S, Shimozawa N, Imanaka T, Taniguchi H (2004) Proteomic analysis of rat liver peroxisome: presence of peroxisome-specific isozyme of Lon protease. J Biol Chem 279:421–428
Kim HU, Huang AHC (2004) Plastid lysophosphatidyl acyltransferase is essential for embryo development in Arabidopsis. Plant Physiol 134:1206–1216
Kim HJ, Ok SH, Bahn SC, Jang J, Oh SA, Park SK, Twell D, Ryu SB, Shin JS (2011) Endoplasmic reticulum- and Golgi-localized phospholipase A_2 plays a critical roles in *Arabidopsis* pollen development and germination. Plant Cell 23:94–110
Kim S, Yamaoka Y, Ono H, Kim H, Shim D, Maeshima M, Martinoia E, Cahoon EB, Nishida I, Lee Y (2013) AtABCA9 transporter supplies fatty acids for lipid synthesis to the endoplasmic reticulum. Proc Natl Acad Sci U S A 110:773–778
Klingler JP, Batelli G, Zhu JK (2010) ABA receptors: the START of a new paradigm in phytohormone signalling. J Exp Bot 61:3199–3210
Knudsen J, Højrup P, Hansen HO, Hansen HF, Roepstorff P (1989) Acyl-CoA-binding protein in the rat. Biochem J 262:513–519

Knudsen J, Faergeman NJ, Skøtt H, Hummel R, Børsting C, Rose TM, Andersen JS, Højrup P, Roepstorff P, Kristiansen K (1994) Yeast acyl-CoA-binding protein: acyl-CoA-binding affinity and effect on intracellular acyl-CoA pool size. Biochem J 302:479–485

Knudsen J, Jensen MV, Hansen JK, Faergeman NJ, Neergaard TBF, Gaigg B (1999) Role of acyl-CoA binding protein in acylCoA transport, metabolism and cell signaling. Mol Cell Biochem 192:95–103

Knudsen J, Neergaard TBF, Gaigg B, Jensen MV, Hansen JK (2000) Role of acyl-CoA binding protein in acyl-CoA metabolism and acyl-CoA-mediated cell signaling. J Nutr 130:294S–298S

Kojima M, Casteel J, Miernyk JA, Thelen JJ (2007) The effects of down-regulating expression of *Arabidopsis thaliana* membrane-associated acyl-CoA binding protein 2 on acyl-lipid composition. Plant Sci 172:36–44

Kragelund BB, Andersen KV, Madsen JC, Knudsen J, Poulsen FM (1993) Three-dimensional structure of the complex between acyl-coenzyme A binding protein and palmitoyl-coenzyme A. J Mol Biol 230:1260–1277

Kragelund BB, Knudsen J, Poulsen FM (1999) Acyl-coenzyme A binding protein (ACBP). Biochim Biophys Acta 1441:150–161

Kunst L, Samuels AL (2003) Biosynthesis and secretion of plant cuticular wax. Prog Lipid Res 42:51–80

Kurata T, Kawabata-Awai C, Sakuradani E, Shimizu S, Okada K, Wada T (2003) The *YORE-YORE* gene regulates multiple aspects of epidermal cell differentiation in *Arabidopsis*. Plant J 36:55–66

Kuromori T, Miyaji T, Yabuuchi H, Shimizu H, Sugimoto E, Kamiya A, Moriyama Y, Shinozaki K (2010) ABC transporter AtABCG25 is involved in abscisic acid transport and responses. Proc Natl Acad Sci U S A 107:2361–2366

Lee SB, Suh MC (2013) Recent advances in cuticular wax biosynthesis and its regulation in *Arabidopsis*. Mol Plant 6:246–249

Lessire R, Juguelin H, Moreau P, Cassagne C (1985) Elongation of acyl-CoAs by microsomes from etiolated leek seedlings. Phytochemistry 24:1187–1192

Leung KC, Li HY, Mishra G, Chye ML (2004) ACBP4 and ACBP5, novel *Arabidopsis* acyl-CoA-binding proteins with kelch motifs that bind oleoyl-CoA. Plant Mol Biol 55:297–309

Leung KC, Li HY, Xiao S, Tse MH, Chye ML (2006) *Arabidopsis* ACBP3 is an extracellularly targeted acyl-CoA-binding protein. Planta 223:871–881

Li HY, Chye ML (2003) Membrane localization of *Arabidopsis* acyl-CoA binding protein ACBP2. Plant Mol Biol 51:483–492

Li HY, Chye ML (2004) *Arabidopsis* acyl-CoA-binding protein ACBP2 interacts with an ethylene-responsive element-binding protein, AtEBP, *via* its ankyrin repeats. Plant Mol Biol 54:233–243

Li F, Vierstra RD (2012) Autophagy: a multifaceted intracellular system for bulk and selective recycling. Trends Plant Sci 17:526–537

Li HY, Xiao S, Chye ML (2008) Ethylene- and pathogen-inducible *Arabidopsis* acyl-CoA-binding protein 4 interacts with an ethylene-responsive element binding protein. J Exp Bot 59:3997–4006

Li M, Hong Y, Wang X (2009) Phospholipase D- and phosphatidic acid-mediated signaling in plants. Biochim Biophys Acta 1791:927–935

Liao P, Chen QF, Chye ML (2014) Transgenic Arabidopsis flowers overexpressing acyl-CoA-binding protein ACBP6 are freezing tolerant. Plant Cell Physiol 55:1055–1071

Licausi F, Kosmacz M, Weits DA, Giuntoli B, Giorgi FM, Voesenek LACJ, Perata P, van Dongen JT (2011) Oxygen sensing in plants is mediated by an N-end rule pathway for protein destabilization. Nature 479:419–422

Liu Y, Bassham DC (2012) Autophagy: pathways for self-eating in plant cells. Annu Rev Plant Biol 63:215–237

Lü S, Song T, Kosma DK, Parsons EP, Rowland O, Jenks MA (2009) Arabidopsis *CER8* encodes LONG-CHAIN ACYL-COA SYNTHETASE 1 (LACS1) that has overlapping functions with LACS2 in plant wax and cutin synthesis. Plant J 59:553–564

Lung SC, Weselake RJ (2006) Diacylglycerol acyltransferase: a key mediator of plant triacylglycerol synthesis. Lipids 41:1073–1088

Ma Y, Szostkiewicz I, Korte A, Moes D, Yang Y, Christmann A, Grill E (2009) Regulators of PP2C phosphatase activity function as abscisic acid sensors. Science 324:1064–1068

Mandrup S, Jepsen R, Skøtt H, Rosendal J, Højrup P, Kristiansen K, Knudsen J (1993) Effect of heterologous expression of acyl-CoA-binding protein on acyl-CoA level and composition in yeast. Biochem J 290:369–374

Matsushima R, Hayashi Y, Yamada K, Shimada T, Nishimura M, Hara-Nishimura I (2003) The ER body, a novel endoplasmic reticulum-derived structure in *Arabidopsis*. Plant Cell Physiol 44:661–666

Meng W, Chye ML (2014) Rice acyl-CoA-binding proteins OsACBP4 and OsACBP5 are differentially localized in the endoplasmic reticulum of transgenic *Arabidopsis*. Plant Signal Behav 9:e29544

Meng W, Su YCF, Saunders RMK, Chye ML (2011) The rice acyl-CoA-binding protein gene family: phylogeny, expression and functional analysis. New Phytol 189:1170–1184

Meng W, Hsiao AS, Gao C, Jiang L, Chye ML (2014) Subcellular localization of rice acyl-CoA-binding proteins (ACBPs) indicates that OsACBP6::GFP is targeted to the peroxisomes. New Phytol 203:469–482

Metzner M, Ruecknagel KP, Knudsen J, Kuellertz G, Mueller-Uri F, Diettrich B (2000) Isolation and characterization of two acyl-CoA-binding proteins from proembryogenic masses of *Digitalis lanata* Ehrh. Planta 210:683–685

Michaely P, Bennett V (1992) The ANK repeat: a ubiquitous motif involved in macromolecular recognition. Trends Cell Biol 2:127–129

Millar AA, Clemens S, Zachgo S, Giblin EM, Taylor DC, Kunst L (1999) *CUT1*, an Arabidopsis gene required for cuticular wax biosynthesis and pollen fertility, encodes a very-long-chain fatty acid condensing enzyme. Plant Cell 11:825–838

Mogensen IB, Schulenberg H, Hansen HO, Spener F, Knudsen J (1987) A novel acyl-CoA-binding protein from bovine liver. Effect on fatty acid synthesis. Biochem J 241:189–192

Murphy DJ (2006) The extracellular pollen coat in members of the Brassicaceae: composition, biosynthesis, and functions in pollination. Protoplasma 228:31–39

Murphy DJ, Woodrow IE, Mukherjee KD (1985) Substrate specificities of the enzymes of the oleate desaturase system from photosynthetic tissue. Biochem J 225:267–270

Nakashima K, Yamaguchi-Shinozaki K (2013) ABA signaling in stress-response and seed development. Plant Cell Rep 32:959–970

Nakashima K, Fujita Y, Kanamori N, Katagiri T, Umezawa T, Kidokoro S, Maruyama K, Yoshida T, Ishiyama K, Kobayashi M et al (2009) Three Arabidopsis SnRK2 protein kinases, SRK2D/SnRK2.2, SRK2E/SnRK2.6/OST1 and SRK2I/SnRK2.3, involved in ABA signaling are essential for the control of seed development and dormancy. Plant Cell Physiol 50:1345–1363

Napier JA, Haslam RP (2010) As simple as ACB – new insights into the role of acyl-CoA-binding proteins in Arabidopsis. New Phytol 186:783–785

Nawrath C (2006) Unraveling the complex network of cuticular structure and function. Curr Opin Plant Biol 9:281–287

Nishimura N, Sarkeshik A, Nito K, Park SY, Wang A, Carvalho PC, Lee S, Caddell DF, Cutler SR, Chory H et al (2010) PYR/PYL/RCAR family members are major *in-vivo* ABI1 protein phosphatase 2C-interacting proteins in Arabidopsis. Plant J 61:290–299

Nishiuchi T, Nakamura T, Abe T, Kodama H, Nishimura M, Iba K (1995) Tissue-specific and light-responsive regulation of the promoter region of the *Arabidopsis thaliana* chloroplast ω-3 fatty acid desaturase gene (*FAD7*). Plant Mol Biol 29:599–609

O'Neill CM, Gill S, Hobbs D, Morgan C, Bancroft I (2003) Natural variation for seed oil composition in *Arabidopsis thaliana*. Phytochemistry 64:1077–1090

Olsen LJ (1998) The surprising complexity of peroxisome biogenesis. Plant Mol Biol 38:163–189

Park SY, Fung P, Nishimura N, Jensen DR, Fujii H, Zhao Y, Lumba S, Santiago J, Rodrigues A, Chow TF et al (2009) Abscisic acid inhibits type 2C protein phosphatases via the PYR/PYL family of START proteins. Science 324:1068–1071

Pastor S, Sethumadhavan K, Ullah AHJ, Gidda S, Cao H, Mason C, Chapital D, Scheffler B, Mullen R, Dyer J et al (2013) Molecular properties of the class III subfamily of acyl-coenzyme A binding proteins from tung tree (*Vernicia fordii*). Plant Sci 203–204:79–88

Phillips AR, Suttangkakul A, Vierstra RD (2008) The ATG12-conjugating enzyme ATG10 is essential for autophagic vesicle formation in *Arabidopsis thaliana*. Genetics 178:1339–1353

Piffanelli P, Ross JHE, Murphy DJ (1997) Intra- and extracellular lipid composition and associated gene expression patterns during pollen development in *Brassica napus*. Plant J 11:549–562

Pollard MR, Stumpf PK (1980a) Long chain (C_{20} and C_{22}) fatty acid biosynthesis in developing seeds of *Tropaeolum majus*. Plant Physiol 66:641–648

Pollard MR, Stumpf PK (1980b) Biosynthesis of C_{20} and C_{22} fatty acids by developing seeds of *Limnanthes alba*. Plant Physiol 66:649–655

Pollard MR, McKeon T, Gupta LM, Stumpf PK (1979) Studies on biosynthesis of waxes by developing jojoba seed. II. The demonstration of wax biosynthesis by cell-free homogenates. Lipids 14:651–662

Pruitt RE, Vielle-Calzada JP, Ploense SE, Grossniklaus U, Lolle SJ (2000) *FIDDLEHEAD*, a gene required to suppress epidermal cell interactions in *Arabidopsis*, encodes a putative lipid biosynthetic enzyme. Proc Natl Acad Sci U S A 97:1311–1316

Quilichini TD, Friedmann MC, Samuels AL, Douglas CJ (2010) ATP-binding cassette transporter G26 is required for male fertility and pollen exine formation in Arabidopsis. Plant Physiol 154:678–690

Raggers RJ, Pomorski T, Holthuis JCM, Kälin N, van Meer G (2000) Lipid traffic: the ABC of transbilayer movement. Traffic 1:226–234

Rasmussen JT, Börchers T, Knudsen J (1990) Comparison of the binding affinities of acyl-CoA-binding protein and fatty-acid-binding protein for long-chain acyl-CoA esters. Biochem J 265:849–855

Rasmussen JT, Rosendal J, Knudsen J (1993) Interaction of acyl-CoA binding protein (ACBP) on processes for which acyl-CoA is a substrate, product or inhibitor. Biochem J 292:907–913

Rawsthorne S (2002) Carbon flux and fatty acid synthesis in plants. Prog Lipid Res 41:182–196

Reddy AS, Ranganathan B, Haisler RM, Swize MA (1996) A cDNA encoding acyl-CoA-binding protein from cotton (accession no. U35015). Plant Physiol 111:348 (PGR 96-028)

Riederer M, Schreiber L (2001) Protecting against water loss: analysis of the barrier properties of plant cuticles. J Exp Bot 52:2023–2032

Roughan PG, Holland R, Slack CR (1980) The role of chloroplasts and microsomal fractions in polar-lipid synthesis from [1-^{14}C]acetate by cell-free preparations from spinach (*Spinacia oleracea*) leaves. Biochem J 188:17–24

Rowland O, Zheng H, Hepworth SR, Lam P, Jetter R, Kunst L (2006) *CER4* encodes an alcohol-forming fatty acyl-Coenzyme A reductase involved in cuticular wax production in Arabidopsis. Plant Physiol 142:866–877

Ryu SB (2004) Phospholipid-derived signaling mediated by phospholipase A in plants. Trends Plant Sci 9:229–235

Samuels L, Kunst L, Jetter R (2008) Sealing plant surfaces: cuticuliar wax formation by epidermal cells. Annu Rev Plant Biol 59:683–707

Santiago J, Dupeux F, Round A, Antoni R, Park SY, Jamin M, Cutler SR, Rodrigeuz PL, Márquez JA (2009) The abscisic acid receptor PYR1 in complex with abscisic acid. Nature 462:665–668

Sasaki Y, Kozaki A, Hatano M (1997) Link between light and fatty acid synthesis: thioredoxin-linked reductive activation of plastidic acetyl-CoA carboxylase. Proc Natl Acad Sci U S A 94:11096–11101

Seo PJ, Lee SB, Suh MC, Park MJ, Go YS, Park CM (2011) The MYB96 transcription factor regulates cuticular wax biosynthesis under drought conditions in *Arabidopsis*. Plant Cell 23:1138–1152

Sieber P, Schorderet M, Ryser U, Buchala A, Kolattukudy P, Métraux JP, Nawrath C (2000) Transgenic Arabidopsis plants expressing a fungal cutinase show alternations in the structure and properties of the cuticle and postgenital organ fusions. Plant Cell 12:721–737

Simonsen AC, Jensen UB, Faergeman NJ, Knudsen J, Mouritsen OG (2003) Acyl-coenzyme A organizes laterally in membranes and is recognized specifically by acyl-coenzyme A binding protein. FEBS Lett 552:253–258

Simpson EE, Williams JP (1979) Galactolipid synthesis in *Vicia faba* leaves: IV. Site(s) of fatty acid incorporation into the major glycerolipids. Plant Physiol 63:674–676

Slack CR, Roughan PG (1975) The kinetics of incorporation *in vivo* of [^{14}C]acetate and [^{14}C]carbon dioxide into the fatty acids of glycerolipids in developing leaves. Biochem J 152:217–228

Slack CR, Roughan PG, Browse J (1979) Evidence for an oleoyl phosphatidylcholine desaturase in microsomal preparations from cotyledons of safflower (*Carthamus tinctorius*) seed. Biochem J 179:649–656

Sparace SA, Mudd JB (1982) Phosphatidylglycerol synthesis in spinach chloroplasts: characterization of the newly synthesized molecule. Plant Physiol 70:1260–1264

Sprong H, van der Sluijs P, van Meer G (2001) How proteins move lipids and lipids move proteins. Nat Rev Mol Cell Biol 2:504–513

Stenzel I, Hause B, Miersch O, Kurz T, Maucher H, Weichert H, Ziegler J, Feussner I, Wasternack C (2003) Jasmonate biosynthesis and the allene oxide cyclase family of *Arabidopsis thaliana*. Plant Mol Biol 51:895–911

Stymne S, Stobart AK (1984) Evidence for the reversibility of the acyl-CoA:lysophosphatidylcholine acyltransferase in microsomal preparations from developing safflower (*Carthamus tinctorius* L.) cotyledons and rat liver. Biochem J 223:305–314

Stymne S, Stobart AK, Glad G (1983) The role of the acyl-CoA pool in the synthesis of polyunsaturated 18-carbon fatty acids and triacylglycerol production in the microsomes of developing safflower seeds. Biochim Biophys Acta 752:198–208

Suzui N, Nakamura S, Fujiwara T, Hayashi H, Yoneyama T (2006) A putative acyl-CoA-binding protein is a major phloem sap protein in rice (*Oryza sativa* L.). J Exp Bot 57:2571–2576

Szostkiewicz I, Richter K, Kepka M, Demmel S, Ma Y, Korte A, Assaad FF, Christmann A, Grill E (2010) Closely related receptor complexes differ in their ABA selectivity and sensitivity. Plant J 61:25–35

Takato H, Shimidzu M, Ashizawa Y, Takei H, Suzuki S (2013) An acyl-CoA-binding protein from grape that is induced through ER stress confers morphological changes and disease resistance in *Arabidopsis*. J Plant Physiol 170:591–600

Taskinen JP, van Aalten DM, Knudsen J, Wierenga RK (2007) High resolution crystal structures of unliganded and liganded human liver ACBP reveal a new mode of binding for the acyl-CoA ligand. Proteins 66:229–238

Testerink C, Munnik T (2005) Phosphatidic acid: a multifunctional stress signaling lipid in plants. Trends Plant Sci 10:368–375

Theodoulou FL, Job K, Slocombe SP, Footitt S, Holdsworth M, Baker A, Larson TR, Graham IA (2005) Jasmonic acid levels are reduced in COMATOSE ATP-binding cassette transporter mutants. Implications for transport of jasmonate precursors into peroxisomes. Plant Physiol 137:835–840

Thompson AR, Doelling JH, Suttangkakul A, Vierstra RD (2005) Autophagic nutrient recycling in Arabidopsis directed by the ATG8 and ATG12 conjugation pathways. Plant Physiol 138:2097–2110

Tjellström H, Yang Z, Allen DK, Ohlrogge JB (2012) Rapid kinetic labeling of Arabidopsis cell suspension cultures: implications for models of lipid export from plastids. Plant Physiol 158:601–611

Tolbert NE, Essner E (1981) Microbodies: peroxisomes and glyoxysomes. J Cell Biol 91:271s–283s
Ton J, Flors V, Mauch-Mani B (2009) The multifaceted role of ABA in disease resistance. Trends Plant Sci 14:310–317
Toulmay A, Prinz WA (2011) Lipid transfer and signaling at organelle contact sites: the tip of the iceberg. Curr Opin Cell Biol 23:458–463
van Aalten DMF, Milne KG, Zou JY, Kleywegt GJ, Bergfors T, Ferguson MAJ, Knudsen J, Jones TA (2001) Binding site differences revealed by crystal structures of *Plasmodium falciparum* and bovine acyl-CoA binding protein. J Mol Biol 309:181–192
van de Loo FJ, Turner S, Somerville C (1995) Expressed sequence tags from developing castor seeds. Plant Physiol 108:1141–1150
van Meer G, Voelker DR, Feigenson GW (2008) Membrane lipids: where they are and how they behave. Nat Rev Mol Cell Biol 9:112–124
van Roermund CWT, IJlst L, Majczak W, Waterham HR, Folkerts H, Wanders RJA, Hellingwerf KJ (2012) Peroxisomal fatty acid uptake mechanism in *Saccharomyces cerevisiae*. J Biol Chem 287:20144–20153
Vick BA, Zimmerman DC (1983) The biosynthesis of jasmonic acid: a physiological role for plant lipoxygenase. Biochem Biophys Res Commun 111:470–477
Wada H, Shintani D, Ohlrogge J (1997) Why do mitochondria synthesize fatty acids? Evidence for involvement in lipoic acid production. Proc Natl Acad Sci U S A 94:1591–1596
Walz C, Giavalisco P, Schad M, Juenger M, Klose J, Kehr J (2004) Proteomics of curcurbit phloem exudate reveals a network of defence proteins. Phytochemistry 65:1795–1804
Wang X (2001) Plant phospholipases. Annu Rev Plant Physiol Plant Mol Biol 52:211–231
Wang X (2004) Lipid signaling. Curr Opin Plant Biol 7:329–336
Wang A, Dennis EA (1999) Mammalian lysophospholipases. Biochim Biophys Acta 1439:1–16
Wiese S, Gronemeyer T, Ofman R, Kunze M, Grou CP, Almeida JA, Eisenacher M, Stephan C, Hayen H, Schollenberger L et al (2007) Proteomics characterization of mouse kidney peroxisomes by tandem mass spectrometry and protein correlation profiling. Mol Cell Proteomics 6:2045–2057
Xia Y, Yu K, Gao QM, Wilson EV, Navarre D, Kachroo P, Kachroo A (2012) Acyl CoA binding proteins are required for cuticle formation and plant responses to microbes. Front Plant Sci 3:224
Xiao S, Chye ML (2009) An *Arabidopsis* family of six acyl-CoA-binding proteins has three cytosolic members. Plant Physiol Biochem 47:479–484
Xiao S, Chye ML (2010) The *Arabidopsis thaliana* ACBP3 regulates leaf senescence by modulating phospholipid metabolism and ATG8 stability. Autophagy 6:802–804
Xiao S, Chye ML (2011a) New roles for acyl-CoA-binding proteins (ACBPs) in plant development, stress responses and lipid metabolism. Prog Lipid Res 50:141–151
Xiao S, Chye ML (2011b) Overexpression of Arabidopsis ACBP3 enhances NPR1-dependent plant resistance to *Pseudomonas syringe* pv *tomato* DC3000. Plant Physiol 156:2069–2081
Xiao S, Li HY, Zhang JP, Chan SW, Chye ML (2008a) Arabidopsis acyl-CoA-binding proteins ACBP4 and ACBP5 are subcellularly localized to the cytosol and ACBP4 depletion affects membrane lipid composition. Plant Mol Biol 68:571–583
Xiao S, Gao W, Chen QF, Ramalingam S, Chye ML (2008b) Overexpression of membrane-associated acyl-CoA-binding protein ACBP1 enhances lead tolerance in Arabidopsis. Plant J 54:141–151
Xiao S, Chen QF, Chye ML (2009a) Light-regulated Arabidopsis ACBP4 and ACBP5 encode cytosolic acyl-CoA-binding proteins that bind phosphatidylcholine and oleoyl-CoA ester. Plant Physiol Biochem 47:926–933
Xiao S, Chen QF, Chye ML (2009b) Expression of ACBP4 and ACBP5 proteins is modulated by light in Arabidopsis. Plant Signal Behav 4:1063–1065
Xiao S, Gao W, Chen QF, Chan SW, Zheng SX, Ma J, Wang M, Welti R, Chye ML (2010) Overexpression of Arabidopsis acyl-CoA binding protein ACBP3 promotes starvation-induced and age-dependent leaf senescence. Plant Cell 22:1463–1482

Xie Z, Nair U, Klionsky DJ (2008) Atg8 controls phagophore expansion during autophagosome formation. Mol Biol Cell 19:3290–3298

Xie LJ, Yu LJ, Chen QF, Wang FZ, Huang L, Xia FN, Zhu TR, Wu JX, Yin J, Liao B et al (2015) Arabidopsis acyl-CoA-binding protein ACBP3 participates in plant response to hypoxia by modulating very-long-chain fatty acid metabolism. Plant J 81:53–67

Xiong Y, Contento AL, Bassham DC (2005) AtATG18a is required for the formation of autophagosomes during nutrient stress and senescence in *Arabidopsis thaliana*. Plant J 42:535–546

Xue Y, Xiao S, Kim J, Lung SC, Chen L, Tanner JA, Suh MC, Chye ML (2014) *Arabidopsis* membrane-associated acyl-CoA-binding protein ACBP1 is involved in stem cuticle formation. J Exp Bot 65:5473–5483

Yeats TH, Rose JKC (2008) The biochemistry and biology of extracellular plant lipid-transfer proteins (LTPs). Protein Sci 17:191–198

Yoshimoto K, Jikumaru Y, Kamiya Y, Kusano M, Consonni C, Panstruga R, Ohsumi Y, Shirasu K (2009) Autophagy negatively regulates cell death by controlling NPR1-dependent salicylic acid signaling during senescence and the innate immune response in *Arabidopsis*. Plant Cell 21:2914–2927

Yu J, Hu S, Wang J, Wong GKS, Li S, Liu B, Deng Y, Dai L, Zhou Y, Zhang X et al (2002) A draft sequence of the rice genome (*Oryza sativa* L. ssp. *indica*). Science 296:79–92

Yurchenko OP, Weselake RJ (2011) Involvement of low molecular mass soluble acyl-CoA-binding protein in seed oil biosynthesis. New Biotechnol 28:97–109

Yurchenko OP, Nykiforuk CL, Moloney MM, Ståhl U, Banaś A, Stymne S, Weselake RJ (2009) A 10-kDa acyl-CoA-binding protein (ACBP) from *Brassica napus* enhances acyl exchange between acyl-CoA and phosphatidylcholine. Plant Biotechnol J 7:602–610

Yurchenko O, Singer SD, Nykiforuk CL, Gidda S, Mullen RT, Moloney MM, Weselake RJ (2014) Production of a *Brassica napus* low-molecular mass acyl-coenzyme A-binding protein in Arabidopsis alters the acyl-coenzyme A pool and acyl composition of oil in seeds. Plant Physiol 165:550–560

Zhang W, Qin C, Zhao J, Wang X (2004) Phospholipase Dα1-derived phosphatidic acid interacts with ABI1 phosphatase 2C and regulates abscisic acid signaling. Proc Natl Acad Sci U S A 101:9508–9513

Zhang JY, Broeckling CD, Sumner LW, Wang ZY (2007) Heterologous expression of two *Medicago truncatula* putative ERF transcription factor genes, *WXP1* and *WXP2*, in *Arabidopsis* led to increased leaf wax accumulation and improved drought tolerance, but differential response in freezing tolerance. Plant Mol Biol 64:265–278

Zhang M, Fan J, Taylor DC, Ohlrogge JB (2009) *DGAT1* and *PDAT1* acyltransferases have overlapping functions in *Arabidopsis* triacylglycerol biosynthesis and are essential for normal pollen and seed development. Plant Cell 21:3885–3901

Zheng Z, Xia Q, Dauk M, Shen W, Selvaraj G, Zou J (2003) Arabidopsis *AtGPAT1*, a member of the membrane-bound glycerol-3-phosphate acyltransferase gene family, is essential for tapetum differentiation and male fertility. Plant Cell 15:1872–1887

Zheng SX, Xiao S, Chye ML (2012) The gene encoding *Arabidopsis* acyl-CoA-binding protein 3 is pathogen inducible and subject to circadian regulation. J Exp Bot 63:2985–3000

Zolman BK, Yoder A, Bartel B (2000) Genetic analysis of indole-3-butyric acid responses in *Arabidopsis thaliana* reveals four mutant classes. Genetics 156:1323–1337

Zolman BK, Silva ID, Bartel B (2001) The Arabidopsis *pxa1* mutant is defective in an ATP-binding cassette transporter-like protein required for peroxisomal fatty acid β-oxidation. Plant Physiol 127:1266–1278

Chapter 16
The Rise and Fall of Jasmonate Biological Activities

Thierry Heitz, Ekaterina Smirnova, Emilie Widemann, Yann Aubert, Franck Pinot, and Rozenn Ménard

Abstract Jasmonates (JAs) constitute a major class of plant regulators that coordinate responses to biotic and abiotic threats and important aspects of plant development. The core biosynthetic pathway converts linolenic acid released from plastid membrane lipids to the cyclopentenone *cis*-oxo-phytodienoic acid (OPDA) that is further reduced and shortened to jasmonic acid (JA) in peroxisomes. Abundant pools of OPDA esterified to plastid lipids also occur upon stress, mainly in the *Arabidopsis* genus. Long thought to be the bioactive hormone, JA only gains its pleiotropic hormonal properties upon conjugation into jasmonoyl-isoleucine (JA-Ile). The signaling pathway triggered when JA-Ile promotes the assembly of COI1-JAZ (Coronatine Insensitive 1-JAsmonate Zim domain) co-receptor complexes has been the focus of most recent research in the jasmonate field. In parallel, OPDA and several other JA derivatives are recognized for their separate activities and contribute to the diversity of jasmonate action in plant physiology. We summarize in this chapter the properties of different bioactive JAs and review elements known for their perception and signal transduction. Much progress has also been gained on the enzymatic processes governing JA-Ile removal. Two JA-Ile catabolic pathways, operating through ω-oxidation (cytochromes P450) or conjugate cleavage (amido hydrolases) shape signal dynamics to allow optimal control on defense. JA-Ile turnover not only participates in signal attenuation, but also impact the homeostasis of the entire JA metabolic pathway.

Keywords Jasmonic acid • JA-Ile • OPDA • Biological activity • Catabolism • CYP94 • Hormone homeostasis

---

T. Heitz (✉) • E. Smirnova • E. Widemann • Y. Aubert • F. Pinot • R. Ménard
Institut de Biologie Moléculaire des Plantes CNRS-UPR 2357, associé à l'Université de Strasbourg, 12 rue du Général Zimmer, 67084 Strasbourg, France
e-mail: thierry.heitz@ibmp-cnrs.unistra.fr

Y. Nakamura, Y. Li-Beisson (eds.), *Lipids in Plant and Algae Development*, Subcellular Biochemistry 86, DOI 10.1007/978-3-319-25979-6_16

Introduction

Plants perform their lifecycle and adapt to changing environmental conditions under the complex control of hormonal compounds endowed with signaling properties. Among the wide class of active fatty acid-derived compounds known as phytooxylipins (Blée 2002; Feussner and Wasternack 2002), the best studied regulators are members of the jasmonate family. These compounds are collectively referred to as jasmonates (JAs), a name originating from their founding member, jasmonic acid (JA). JAs are cyclopente(a)none compounds formed under developmental or external cues via a cascade of reactions involving lipoxygenase oxidation of plastidial polyinsaturated fatty acids. An abundant literature has documented the widespread occurrence and diverse activities of JAs in examined plant species (for review, see Wasternack and Hause 2013). This knowledge is dominated by the master control exerted on defense and development by jasmonic acid (JA), and more precisely by the conjugate jasmonoyl-isoleucine (JA-Ile) with the elucidation of mechanistic insights into the COI1-JAZ signaling module.

JAs have been repeatedly associated with induced defense against attacks by herbivores and a subclass of microbial pathogens, generally of the necrotrophic type. JA-defective lines are highly susceptible to such attackers, and JA-controlled defense against herbivores is not limited to leaf-chewing insects but extends to other arthropods, nematodes, slugs, or even vertebrate herbivores (Campos et al. 2014). The JA pathway also seems to modulate further types of plant interactions with microorganisms, including growth-promoting bacteria or mycorrhiza (Wasternack and Hause 2013). Widespread genetic evidence supports a central role of JA promoting arrays of chemical and morphological defenses that will curtail fitness of attackers. JA-triggered plant immunity encompasses most major classes of secondary metabolites including alcaloids, phenylpropanoids and terpenoids, as well as antimicrobial and anti-nutritive proteins (Campos et al. 2014). Such chemicals are frequently sequestered into specialized structures like glandular trichomes or nectaries. Substantial cross-talk of the JA pathway with other hormonal pathways offers a wealth of synergistic or antagonistic signaling combinations that result in the induction of a stimulus-specific set of defenses (Pieterse et al. 2009).

While defense-eliciting properties of JA along with initial evidence for *in planta* JA accumulation were established in the early 1990s, mechanistic insights and a unifying signaling model were only elucidated by molecular genetics approaches in the late 2000s. This central JA mode of action is not limited to defense-related responses, as mutants defective in the core JA biosynthetic or perception pathway were known to be affected in specific development processes such as root growth inhibition, light signaling, reproduction or senescence. Particularly, Arabidopsis JA-deficient lines are male sterile due to defects in stamen filament elongation, anther dehiscence and pollen maturation. JA is embedded in a complex hormone cascade where it controls late steps in stamen development (Song et al. 2013). JA requirement in reproduction seems to be a general feature across plant species, but JAs promotes distinct developmental programs underlying female fertility in tomato (Goetz et al. 2012) or sex determination in maize (Acosta et al. 2009).

Besides this core pathway directing conserved development and adaptation programs in all examined plant species, more specialized jasmonates are accumulated specifically in some plant species or organs. Their study offers potential insight into original metabolic routes and modes of action in the control of specialized responses.

This review will focus on the diversity of biologically active jasmonates (Fig. 16.1), the ways they are generated, and their perception and turnover systems when described. The chapter will also illustrate how co-regulation tools applied to a simple biological model – mechanical leaf wounding – allowed to decipher novel

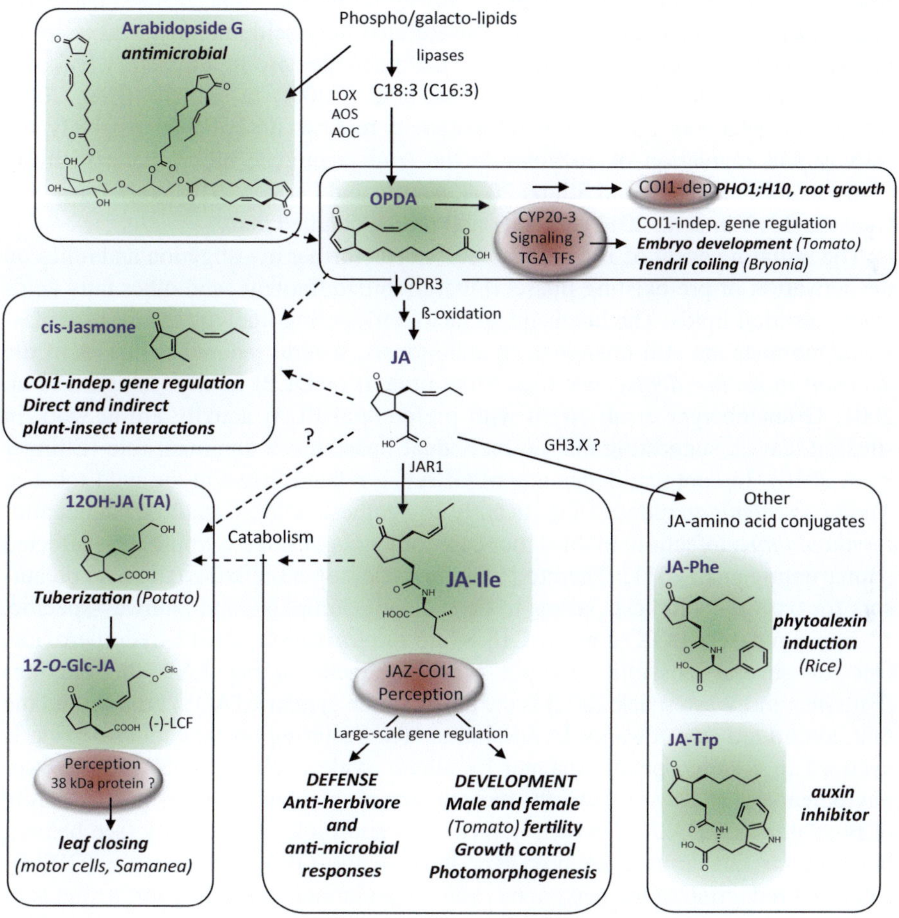

Fig. 16.1 Diversity of structures and biological activities in the jasmonate family. Major biological activities characterized along the jasmonate pathway are depicted and range from direct antifungal properties to a large array of developmental and defensive processes. Most responses are mediated by JA-Ile through COI1-JAZ signaling, but other sensing systems for distinct jasmonates are emerging (*red background*). *TF* transcription factor, *LCF* leaf-closing factor. Enzyme and other protein acronyms are detailed in main text. Plant name is indicated when effect was specifically evidenced in a given species

catabolic pathways that orchestrate the dynamics of JA-Ile and provide a renewed view of JA homeostasis.

The Core Jasmonate Biosynthetic and Signaling Pathway

Jasmonate Biosynthesis

Jasmonates are synthetized through a pathway termed as "octadecanoid pathway" because the main precursor is the polyinsaturated fatty acid linolenic acid (C18:3), in reference to the "eicosanoid pathway" that generates pro-inflammatory leukotrienes and prostaglandins from arachidonic acid (C20:4) in animals (Blée 2002; Kazan and Manners 2008). Several excellent reviews described extensively the activity and regulation of enzymes in the core jasmonate biosynthetic pathway (Browse 2009; Schaller and Stintzi 2009; Acosta and Farmer 2010; Wasternack and Hause 2013; Yan et al. 2013) and that will not be detailed here.

The initiating event of JA synthesis still needs further investigation and relies on the activation of pre-existing lipase(s) that mobilize linolenic and other fatty acids from plastidial lipids. The biochemical factors triggering lipolytic catalysis at plastid membranes are still unknown. In arabidopsis, several redundant lipases of the *deficient in anther dehiscence* (*dad1*-like protein or DLP) family (Ishiguro et al. 2001; Grienenberger et al. 2010) with preferential PLA_1 activity are at work in stressed leaves, suggesting that no individual lipase has a dominant role (Ellinger et al. 2010). In contrast, deficiency of GLA1, a related lipase in the wild tobacco *Nicotiana attenuata* reduced significantly JA synthesis after wounding, but not after *Phythophthora* infection; in this latter case, other classes of oxylipins were affected (Bonaventure et al. 2011). Therefore, the lipolytic step mobilizing fatty acid precursors for oxylipin synthesis seems to be species-, stimulus-, and pathway-specific. 13-lipoxygenases (LOX) convert C18:3 into the formation of a 13-hydroperoxide. One of several possible oxylipin branches metabolizing FA-hydroperoxides (Feussner and Wasternack 2002) is the allene oxide synthase (AOS) branch, and the first committed step towards JA formation. The resulting unstable allene oxide is cyclized in a stereospecific manner by allene oxide cyclase (AOS) into 12-oxo-phytodienoic acid (*cis*-OPDA), the most upstream member of the jasmonate family of regulators. OPDA has its own homeostasis and biological activity (see below), but a major fate is as a JA precursor through the OPDA reductase 3 (OPR3)-catalyzed reduction of the pentenone cycle to pentanone, preceding three rounds of ß-oxidation to generate jasmonic acid (JA). Among many possible subsequent JA modifications, the enzyme Jasmonate Resistant 1 (JAR1) catalyzes JA conjugation to the amino acid isoleucine, generating the active hormonal signal jasmonoyl-isoleucine (JA-Ile).

The Canonical Jasmonate Signaling Pathway

The jasmonate signaling pathway has been the topic of numerous reviews (Katsir et al. 2008; Browse 2009; Fonseca et al. 2009a; Wasternack and Hause 2013) and only major elements needed for understanding the general concept will be mentioned here. In 2007, a breakthrough by several laboratories has connected some previous findings on JAR1-catalyzed JA conjugation with COI1-based perception, and has provided a mechanistic frame to explain JA action at the molecular level. The so-called "relief of repression model" is widespread in plant hormone perception, and was initially discovered for auxin signaling (Santner and Estelle 2009). JA-Ile was identified as a specific ligand promoting assembly of a multiprotein co-receptor complex whose minimal elements are the F-box protein COI1 and members of the family of JAZ repressors (Chini et al. 2007; Thines et al. 2007; Yan et al. 2007). Further studies have identified the associated co-repressors TOPLESS and NINJA (Perez and Goossens 2013). Under non-induced conditions with low or no JA-Ile, JAZ proteins repress JA-regulated genes by blocking the action of specific transcription factors (TF). Upon stress or in specific developmental situations, JA-Ile builds-up and triggers the ubiquitination and subsequent proteolysis of JAZ proteins by the 26S proteasome, relieving active transcription of JA-responsive genes. In different organs/tissues/conditions, JAZ protein elimination allows JA-Ile control over target TFs and corresponding responses (Pauwels and Goossens 2011). JAZ proteins also constitute an important interface to modulate crosstalk with other hormones (Song et al. 2014) and JAZ are also targeted by microbial effectors to subvert host defense pathways. Most *JAZ* genes are strongly stress-inducible and their transcriptional response provides a rapid supply of JAZ proteins on promoter sites and allows the return of target genes to the repressed state.

From Structural Fatty Acids to Hormonal Properties: Distinct Biological Activities Pave the Jasmonate Pathway

JA has given the generic name to the family because of its powerful and diverse biological properties upon exogenous application, and because it is the hub of a metabolic wheel that leads to a large number of derivatives. JAs modified by hydroxylation, decarboxylation, glucosylation or conjugation have been identified in early investigations in many species and include dihydro-JA, JA-Glucose, 11OH-JA, 12OH-JA, 6-OH-JA, cucurbic acid, and JA conjugated to the ethylene precursor JA-ACC (Miersch et al. 2008; Wasternack and Hause 2013). For most of these derivatives, precise biosynthetic pathways or biological activities, if any, are poorly described. In many studies, JA was considered as the final compound of the pathway and the active hormone, mostly because of the plethora of transcriptional, biochemical and physiological responses this compound triggers.

More recently, significant advances have been obtained on the formation, biological properties and inactivation mechanisms of JA-amino acid conjugates, in particular the hormone JA-Ile. The following sections will review specific JAs that are endowed with well-defined biological activities, their perception/transduction mechanism when available, and current knowledge of active mechanisms for their controled turnover.

cis-*Oxo-Phytodienoic Acid (OPDA)*

Besides its position as a metabolic intermediate, OPDA has long been recognized as bearing distinct signaling properties from JA. OPDA triggers faster/stronger responses in a number of bioassays or transcriptional surveys (Böttcher and Pollmann 2009; Wasternack and Hause 2013). These effects include asymetric growth leading to tendril coiling in *Bryonia* (Weiler et al. 1993) or thigmomorphogenesis in *Phaseolus* (Stelmach et al. 1998). In these cases, increases in OPDA but not JA preceded the onset of the growth responses. These findings may underlie a more widespread role of OPDA in touch-induced growth adaptation (Chehab et al. 2012). In fact, OPDA and JA have ovelapping yet distinct activities, and accumulating literature suggests that their respective requirement can vary significantly depending on plant species, stimulus, or developmental process.

Beyond correlative studies, a major advance was the generation of Arabidopsis and tomato mutants deficient in OPDA Reductase 3 (OPR3), that accumulate OPDA but are unable to convert it into JA. OPDA cannot substitute for JA in Arabidopsis male fertility (Stintzi and Browse 2000), but the male sterile *opr3* mutant retained resistance to some pathogens and herbivores (Stintzi et al. 2001), in line with the powerful elicitation of phytoalexin induction by OPDA in *Escholtzia* cell cultures (Blechert et al. 1995). Comparative transcriptome analysis after OPDA or JA or Methyl-Jasmonate (MJ) treatment revealed partial overlap but also a number of OPDA-specific genes that are not induced by JA, including both COI1-independent and COI1-dependent stress response-regulated genes (Taki et al. 2005). OPDA negatively regulates Arabidopsis seed germination along with ABA (Dave et al. 2011), but a positive role in tomato embryo development and seed set was determined (Goetz et al. 2012). In this latter species, regulation of foliar jasmonate defense signaling appeared even more complex as determined with OPR3-silenced plants. JA-deficiency rendered plants more attractive to insect larvae in two-choice feeding or oviposition assays and this preference was associated with impaired trichome formation and terpene content, and increased *cis*-3-hexenal emission (Bosch et al. 2014). However, despite of these features, resistance to the specialist herbivore *Manduca sexta* was as WT in OPR3-silenced plants, but strongly reduced in the jasmonate-insensitive (*jai1*) mutant affected in the tomato COI1 homolog. This shows that JA/JA-Ile biosynthesis is dispensable for insect resistance but that COI1 signaling is not. Furthermore, systemic wound-induction of the typical defense marker proteinase inhibitor II was lost in OPR3-silenced scions grafted on WT rootstocks, indicating that OPDA is competent for local, but not systemic defense, and confirming an earlier study on ß-oxidation requirement for long distance signaling in tomato (Li et al. 2005). In an independent study, Scalschi et al. (2014) reported that similar OPR3-deficient plants that displayed reduced OPDA and nearly abolished JA-Ile levels were impaired in callose deposition and were more susceptible to *B. cinerea* infection. Unexpectedly, only OPDA but not JA could restore basal resistance and callose deposition in silenced plants. An OPDA-specific function was also evidenced by AOC but not OPR3 overexpression in rice defense against a piercing-sucking insect (Guo et al. 2014).

The knowledge of the OPDA perception/transduction system is still fragmentory. Very few OPDA-specific responses were found to be COI1-dependent, and include the induction of *PHO1;H10*, a member of a phosphate transporter gene family (Ribot et al. 2008), or root growth responses (Stotz et al. 2013). Wether these effects rely or not on JAZ destabilization is unknown. COI1-independent action of OPDA may be due to its distinct structural features, including the presence of an α,ß-unsaturated carbonyl group that confers a particular chemical reactivity as strong electrophile (Almeras et al. 2003). This property is shared with phytoprostanes, a class of oxylipins generated non enzymatically under oxidative or biotic stress, and with some lipid breakdown compounds, all being collectively designated as reactive electrophile species (RES) (Farmer and Mueller 2013). RES signaling does not proceed through COI1-JAZ interaction, and accordingly, structural analysis showed that OPDA cannot fit into the COI1-JAZ co-receptor binding pocket (Sheard et al. 2010). RES have been proposed to bind free thiol groups and modify cellular proteins. RES formation is generally accompanied by increases in glutathione-S-transferase activity, and accordingly, OPDA-GSH adducts have been identified in elicited tissues (Davoine et al. 2006) a class of compounds that can be further transported to the vacuole (Ohkama-Ohtsu et al. 2011). Whether this participates in the modulation of OPDA levels or signaling properties is not known. Recently, a signaling module was characterized where OPDA promotes the interaction between cyclophilin 20-3 (CYP20-3) and serine acyltransferase, a component of the cysteine synthase complex, thereby ensuring redox-dependent transmission of OPDA signal (Park et al. 2013). Consistently, induction of OPDA-responsive genes is dependent on the function of the transcription factors TGA2, 5 and 6, that are themselves deregulated in redox-unbalanced mutants (Stotz et al. 2013). Of note, these transcription factors also mediate responses to phytoprostanes (Mueller et al. 2008).

From an evolutionary point of view, OPDA is an ancient signal present in the moss *Physcomitrella patens* that, interestingly, seems to have a truncated pathway limited to OPDA, as no JA or derivatives were found. Mutations in PpAOC lead to reduced fertility and altered sporophyte morphology, again pointing to the early evolution of active OPDA signaling (Stumpe et al. 2010). In higher plants, OPDA accumulation may also largely be affected by its turnover by downstream enzymes, including conversion to JA and derivatives. However, specific OPDA elimination/modification routes may exist, although no direct OPDA oxidative catabolic pathway has been described in any plant. Interestingly, *cis*-OPDA is isomerized into iso-OPDA by an enzyme in the insect's gut, a strategy believed to help in disarming the plant defense or alternatively to prevent OPDA, a prostaglandin-related compound, from affecting insect vital functions (Schulze et al. 2007).

Arabidopsides

While studying the canonical JA biosynthetic pathway initiated from free FAs, came as an unexpected finding that some JAs can occur abundantly as acyl parts of plastid galactolipids (Stelmach et al. 2001), exceeding largely the abundance of free JAs. A number of different variants known as "arabidopsides" have successively been characterized, and contain generally one to three OPDA or dinor-OPDA moieties esterified to the *sn*-1 or *sn*-2 glycerol position, or bound to the galactose headgroup itself (Fig. 16.1; Hisamatsu et al. 2005; Andersson et al. 2006; Buseman et al. 2006; Böttcher and Weiler 2007). Several of these compounds accumulate to very high levels in vegetative tissues after wounding or during the hypersensitive response to an ectopically-expressed bacterial avirulence protein (Andersson et al. 2006), and this accumulation depends on an intact JA signaling pathway (Kourtchenko et al. 2007). Whether free JAs are re-esterified to galactolipids or directly formed on non-hydrolyzed lipids was a matter of debate. Recently, support to the latter hypothesis was provided based on stable isotope labelling, showing that arabidopside formation does not require prior fatty acid release (Nilsson et al. 2012). The presumed biological function of these bound JAs is in direct defense. Consistently with their abundant stress-induced accumulation, arabidopside E and G inhibit growth of the fungus *Botrytis cinerea* (Kourtchenko et al. 2007). Recently, the diversity of arabidopsides has been extended to more lipid classes, including phospholipids (Nilsson et al. 2014). Finally, while arabidopsides were so far limited to Arabidopsis and some closely related species (Böttcher and Weiler 2007), their presence was recently detected in *Circium arvense*, a plant from the more distant *Asteracea* family, in response to endophytic fungal growth (Hartley et al. 2015). These compounds are therefore believed to fulfill a dual function, either by directly antagonizing agressors, or as a source for delayed OPDA release.

cis-*Jasmone*

Cis-jasmone (CJ) is a major component emitted in floral scent in several species like jasmine, and that regulates the interaction with insect pollinators or indirect defense by attracting herbivore parasitoids (Birkett et al. 2000). CJ is also emitted by vegetative tissues after herbivory, in response to insect oral secretions. Two possible biosynthetic routes leading to CJ have been reported. CJ can derive from JA by decarboxylation (Koch et al. 1997). Non-exclusively, CJ may also be formed from OPDA when this latter is isomerized to the planar *iso*-OPDA and subsequently ß-oxidized (Dabrowska and Boland 2007). CJ directly acts as a repellent for some pests like aphids, and has attractant activity for antagonistic insects like ladybirds (Birkett et al. 2000). In addition, CJ signals transcriptional changes in plants. In a microarray-based study, Matthes et al. (2010) have shown that CJ treatment induces a distinct set of genes than MJ through a pathway independent of COI1 and JAR1, and that involves similar TGA transcription factors than OPDA or phytoprostanes. This response stimulates the synthesis of secondary metabolites including volatile terpenes, reinforcing indirect defenses in multitrophic interactions (Birkett et al. 2000; Bruce et al. 2008; Matthes et al. 2010). Neither the mode of perception of CJ in plants nor potential inactivation pathways are known.

Jasmonic Acid (JA) and Methyl Jasmonate (MJ)

JA and methyl jasmonate (MJ) were considered for a long time to be bioactive forms because of the pleiotrophic effects these compounds induce in various species and JA abundance globally correlates with timing and amplitude of developmental and defensive responses (Creelman and Mullet 1997). Arabidopsis lines engineered for overexpressing a JA-methyltransferase (JMT) initially provided evidence that MJ could be an active signal (Seo et al. 2001). MJ concentrations are usually much lower than those of JA in wild-type tissues and most MJ is released as a volatile. This has fed the hypothesis that MJ could function as a mediator of interplant communication, but the ecological relevance of such signalling could not be established when using *N. attenuata* as indicator plants growing beside sagebrush emitting MJ in a natural habitat (Preston et al. 2004).

The 2007 breakthrough revealing JA-Ile as a specific ligand of a COI1-JAZ coreceptor complex and further findings pointed to the evidence that most JA and MJ effects are mediated via conversion to JA-Ile and COI1-JAZ-based perception (Browse 2009). A more detailed study examined JA homeostasis and consequences of overexpressing the homologous JMT in *N. attenuata* (Stitz et al. 2011). In such JMT-OE plants, the JA profile was largely shifted towards massive MJ accumulation with reduced JA and JA-Ile bursts, identifying MJ formation as a metabolic sink that subtracts free JA from being conjugated into bioactive JA-Ile. In addition to provoking reproductive defects, this shift was associated with enhanced susceptibility to different classes of native herbivores due to impaired direct and indirect defense responses (Stitz et al. 2011). This demonstrates that MJ is not per se an active signal for defense activation.

Tuberonic Acid (TA) and Glucosylated or Sulfated Derivatives

12OH-JA (and its glucoside 12-*O*-Glc-JA, Fig. 16.1) was initially isolated from potato tubers and shown to induce tuberization of roots, hence its name tuberonic acid or TA (Yoshihara et al. 1989). Later, it was found that TA was not specific to tuberizing plants but occurs widely in many species, sometimes more abundantly than JA itself (Miersch et al. 2008). No biosynthetic pathway was known until recently (see below). In addition to 12-*O*-Glc-JA, TA can also be modified to the sulfated derivative $12HSO_4$-JA by the ST2a sulfotransferase (Gidda et al. 2003), and their formation was proposed to participate in switching-off JA responses (Miersch et al. 2008).

However, there may be more functional information in such compounds, revealed by genus- and organ-specific bioactivity. For example, such a specialized function has been discovered for 12-*O*-Glc-JA as a leaf closing factor (LCF) in motor cells of nyctinastic plants like *Samanea* (rain tree) or *Albizia* (Nakamura et al. 2011). Structure-function study has shown that only the specific enantiomer, (-)LCF is active, and that stereochemistry of the glucone moiety is also important (Ueda et al. 2015). The LCF response was not triggered by other jasmonates or the JA-Ile mimick coronatine, suggesting that LCF triggers motor cell shrinkage independently from the COI1-JAZ module (Nakamura et al. 2011). By using fluorescent probes, authors showed that active but not inactive LCF labels a 38 kDa membrane protein that may be involved in LCF signal transduction (Nakamura et al. 2008). 12-*O*-Glc-JA is a widely distributed jasmonate, but its formation is not fully elucidated. Genes encoding a glucosyltransferase active on TA and on salicylic acid (Seto et al. 2009) and a TA-glucoside glucosidase (Wakuta et al. 2010) have been isolated from rice, but their physiological importance in controlling specifically reversible TA/12-*O*-Glc-JA conversions awaits genetic characterization.

Jasmonoyl-Isoleucine (JA-Ile) and Other Bioactive JA-Amino Acid Conjugates

JA-amino acid conjugates were described in pioneering investigations, and in particular, the strong gene-inducing activity of JA-Ile was identified in barley well before the JA-Ile/COI1/JAZ perception paradigm had been elucidated and could thus not be mechanistically explained (Kramell et al. 1997). Molecular genetic and biochemical evidence supports the concept that JA-Ile mediates JA-triggered responses through COI1-JAZ-dependent de-repression of target genes. This specific activity of an amino acid conjugate is unique in plant hormone signalling because hormone conjugation generally leads to their inactivation (Santner and Estelle 2009). The requirement of JA-Ile in Arabidopsis JA signalling is based on structure-function and biochemical studies where stereoisomer purification and COI1-JAZ cristallization identified (+)-7-*iso*-JA-Ile as the genuine ligand of the co-receptor and the endogenous bioactive jasmonate (Fonseca et al. 2009b; Sheard et al. 2010). JAR1 conjugates different amino acids to JA, but only JA-Ile is known to promote assembly of co-receptor complexes (Thines et al. 2007). Arabidopsis *jar1* mutants exhibit near WT wound responses (Suza and Staswick 2008), despite of reduced JA-Ile formation, and are therefore not well suited to dissect JA-Ile effects in leaves. *jar1* was initially isolated phenotypically by its reduced root sensitivity to MJ-induced growth inhibition (Staswick et al. 1992), indicating that JAR1-dependent JA-Ile synthesis is needed for JA responses in roots. In support to organ-specific effects, JAR1 was shown to be critical for JA signaling in roots after cotyledon wounding (Acosta et al. 2013). In comparison, the *coi1* mutant is strongly impaired in most developmental and defensive JA responses in different organs, identifying COI1 as an obligate point of convergence of most JA responses.

The biological impact of JA-Ile signaling was also extensively studied in the wild tobacco *N. attenuata*. Independently silencing two JAR members impairs levels of trypsin proteinase inhibitors but not nicotine in wounded leaves (Wang et al. 2007). In addition, a comprehensive analysis of JA homeostasis in *N. attenuata* flowers revealed that JA-Ile coordinates metabolic networks required for anthesis and pollinator attraction (Stitz et al. 2014). Emerging data in rice highlight the importance of JA-Ile as a signal in monocots. Although OsJAR1 and OsJAR2 have overlapping JA-Ile-producing activity in vitro (Wakuta et al. 2011), *osjar1* single mutant is deficient in photomorphogenesis (Riemann et al. 2008), fertility and optimal seed development (Fukumoto et al. 2013). This mutant lacks JA-Ile accumulation, induction of the flavonoid sakuranetin, but not of diterpenoid phytoalexins and displays enhanced susceptibility to rice blast fungus (Shimizu et al. 2013). Finally, the central role of JAs in maize reproductive development and immunity to pathogens and insects was demonstrated using *opr7opr8* mutants, but the specific involvement of JA-Ile is not yet formally identified in this species (Yan et al. 2012).

How and why JA-Ile in particular has evolved as a master hormonal signal is unknown as is the question of whether other conjugates can form COI1-JAZ co-receptors has not yet been systematically explored. For example, JA-Gln accumulates to much higher levels than JA-Ile in wounded *N. attenuata* leaves and is formed by distinct enzyme(s), but no signaling activity was associated so far (Wang et al. 2007).

Several other conjugates are present at different levels in plant tissues such as JA-Tyr in broad bean flowers (Kramell et al. 2005), and may represent storage forms. However, in addition to JA-Ile, JA-Phe also elicits accumulation of phytoalexin synthesis in rice (Tamogami et al. 1997), and JA-Trp behaves as an endogenous auxin inhibitor by causing agravitropic root growth in Arabidopsis (Staswick 2009).

Anti-cancer Properties of JAs

Although JAs do not occur in mammalian cells, growing evidence suggests that JAs can be active against different types of cancer cells, without affecting normal cells (Cohen and Flescher 2009; Cesari et al. 2014). MJ activity on such cell lines is multifacetted, and includes (1) cell cycle arrest, leading to cell growth and proliferation inhibition, (2) cell death through the intrinsic/extrinsic proapoptotic, p53-independent apoptotic, and nonapoptotic (necrosis) pathways, (3) disruption of mitochondrial and reactive oxygen species-mediated functions, (4) MAPK signaling, (5) inhibition of cell migration and antimetastatic activities. Combination of MJ with chemotherapeutic drugs resulted in highly additive cytotoxic effects on different cell lines. Most data derive from assays with MJ, the most accessible jasmonate, but the full potential as anticancer agents of the diverse JA family present in plants has not yet been explored. Raw plant material whose processing as food implies mechanical stress like chopping or peeling may be enriched in oxidized JAs (*vide infra*) and provide potentially beneficial effects to consumer. The properties of such catabolites have not been explored and deserve more attention.

The Dynamic Leaf Wound Response to Decipher JA-Ile Catabolism

The relevance of JA pathway induction for induced resistance to insects and pathogens has been recognized for two decades (Creelman and Mullet 1997; Campos et al. 2014) and implies for most biochemical steps a rapid transcriptional gene activation. Plant responses induced after mechanical wounding largely but not totally overlap with responses to insect attacks as evidenced in multiple microarray-based studies (Howe and Jander 2008). Obviously, insect feeding is a more subtle stimulus than mechanical tissue disruption and differences are believed to arise from sophisticated feeding patterns generated by mouth parts of herbivores, and also the presence of active elicitor and suppressor compounds in insect oral secretions that shape multilayered interactions. Mechanical wounding as practiced in lab experiments inflicts a synchronous stimulus that triggers a robust JA burst (Chung et al. 2008; Glauser et al. 2008; Chauvin et al. 2013), providing access to the dynamics of metabolic conversions between JAs. Figure 16.2a shows the temporal evolution of 8 distinct conjugated and unconjugated JAs in Arabidopsis leaves within 4 h

a

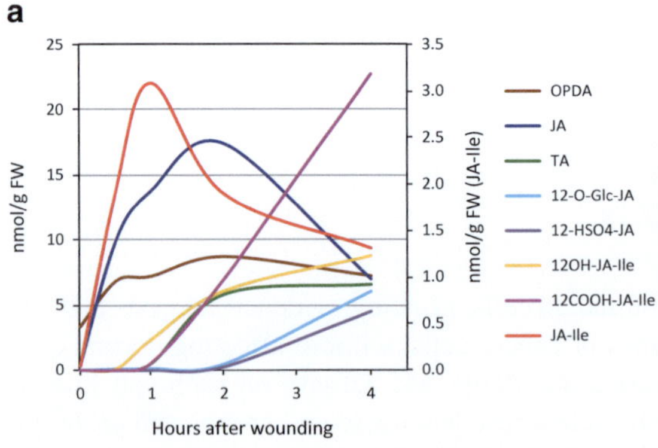

b

Pathogen Set	
score	Name
(bait) 1	CYP94C1
0.931	OPR3
0.925	Inosit tri-Pkin
0.91	LOX3
0.907	OPCL1
0.901	ILL6
0.895	AOC3
0.886	JAZ6
0.832	JAZ5
0.823	JAZ9
0.818	JAZ7

Hormone Set	
score	Name
(bait) 1	CYP94C1
0.913	JAZ7
0.814	JAZ2
0.789	JAZ10
0.78	JAZ8
0.776	MYC2
0.764	JAZ3
0.74	JAZ6
0.735	JAZ1
0.709	OPR3
0.684	OPCL1
0.674	JAZ9
0.658	JAZ5
0.652	ILL6
0.633	CYP94B1

Fig. 16.2 Genome-wide co-regulation analysis of pathway genes identifies novel actors involved in metabolic jasmonate conversions. (**a**) Kinetic analysis of the abundance of eight distinct jasmonates upon wounding in *Arabidopsis* leaves. The right y axis scale applies to JA-Ile only. (**b**) CYP94C1, previously characterized as a fatty-acid hydroxylase, was found co-regulated with numerous genes in the JA biosynthesis and signaling pathway. ILL6, a member of the ILL1-related amido-hydrolase family is also picked

after wounding. Each individual compound presents a specific kinetic/abundance profile. JA-Ile, the hormonally relevant (and also the least abundant) form, displays a rapid and transient peak, whereas oxidized derivatives undergo a slower build-up. This complex pattern is reflective of active metabolic/catabolic conversions between JAs and also highlights tight control of the wound-induced JA-Ile pulse.

JA-Ile Catabolism by CYP94 Enzymes

Contrary to other plant hormones for which mechanisms and impact of active compound removal are relatively well described (Mizutani and Ohta 2010), knowledge of signal extinction mechanisms in the JA pathway lagged behind, mostly because their search focused on free JA rather than on JA-Ile turnover. In 2008, the report of the accumulation of the derivatives 12OH-JA-Ile and 12COOH-JA-Ile in wounded Arabidopsis leaves (Glauser et al. 2008) provided a strong lead as to the existence of an oxidative JA-Ile disposal pathway. CYP94C1, a wound- and MJ-inducible cytochrome P450 oxidase has been previously characterized, and performs in vitro the sequential oxidation of the terminal (ω) carbon of linear fatty acids, generating 12-hydroxy- and 12-carboxy- derivatives (Kandel et al. 2007). Many genes in the core JA pathway are strongly co-regulated across a wide spectrum of physiological situations involving JA reponses. This allowed to mine large-scale Arabidopsis expression data sets to identify unknown/candidate genes linked by high co-expression scores to well-characterized genes. Once believed to provide diacid precursors for cutin synthesis, *CYP94C1* in fact picked many known genes of the JA biosynthetic and signaling pathway when used as a bait in coregulation screens (Fig. 16.2b). When different expression datasets are mined, the expression association of CYP94C1 with the JA pathway was extended to CYP94B1 and CYP94B3, two other inducible members of the 6-gene *CYP94* subfamily in Arabidopsis.

Functional analysis by three laboratories established that CYP94 enzymes are in fact JA-Ile oxidases that turnover the active hormone, but interestingly, are inactive on unconjugated JA (Kitaoka et al. 2011; Koo et al. 2011; Heitz et al. 2012). Recombinant expression in yeast demonstrated the sequential action on JA-Ile, with CYP94B3 generating mostly 12OH-JA-Ile when CYP94C1 catalyzes further oxidation to 12COOH-JA-Ile (Heitz et al. 2012). Furthermore, detailed analysis of oxydation products revealed that CYP94C1 also formed the aldehyde intermediate 12-oxo-jasmonoyl-isoleucine (12CHO-JA-Ile) (Widemann et al. 2015a) and likely acts on other JA conjugates (Widemann et al. 2015b). Genetic analysis in single *cyp94b3*, *cyp94c1* or double *cyp94b3c1* mutants clearly established that these two enzymes define a major oxidative turnover pathway accounting for the transient JA-Ile accumulation in WT plants. CYP94B1 was recently characterized as being catalytically similar to CYP94B3 (Koo et al. 2014), but its much lower expression makes it a minor contributor to leaf JA-Ile catabolism. Interestingly, despite of much persistant wound-induced JA-Ile levels, CYP93B3-deficient plants only exhibit a moderate and transient gain in defense marker induction (Heitz et al. 2012), associated with enhanced *JAZ* gene expression. This can be interpreted as a negative feed-back by JAZ repressors to prevent overinduction of defense under excessive JA-Ile accumulation. In contrast, ectopic CYP94B3 and CYP94C1 over-expression strongly reduced JA-Ile levels (Koo et al. 2011; Heitz et al. 2012), and these plants displayed phenotypes indicative of deficient JA-Ile signaling, such as reduced male fertility, reduced sensitivity to jasmonate-induced growth inhibition and susceptibility to insect attack. Consistent with CYP94B3 in vivo impact, 12OH-JA-Ile was less efficient than JA-Ile to promote formation of COI1-JAZ receptor complexes in in vitro assays (Koo et al. 2011). Together, these data demonstrate that JA-Ile oxidation by CYP94 enzymes is a key event for attenuation of JA responses. This signal switch-off is likely to function also in other situations of JA-Ile signaling, like microbial infection.

JA-Ile Catabolism by Amido-Hydrolases

The atypical situation in the JA pathway that the bioactive form is generated by JA conjugation to isoleucine offers the possibility that JA-Ile could be readily eliminated by simple conjugate cleavage in a JAR1-reverse reaction. Consistently, coregulated gene lists comprise genes encoding ILL1-related family of amido-hydrolases, such as *ILL6* (Fig. 16.2b). Protein members of this family, that also include IAR3, have long been known as cleaving auxin-amino acid conjugates, thereby releasing active auxin (LeClere et al. 2002). In view of their JA-regulated expression, recent examination revealed that recombinant IAR3 as well as the newly characterized ILL6 cleave JA-Ile in vitro (Widemann et al. 2013). Furthermore, *iar3* and *ill6* mutants have altered wound-induced JA profiles, consistent with a role in JA-Ile turnover. Although increased JA-Ile levels were specific to *iar3* mutants, strong increases in 12OH-JA-Ile and 12COOH-JA-Ile levels were recorded in both *iar3*- and *ill6*-deficient lines. A possible explanation is that excessive JA-Ile accumulating due to impaired cleavage capacity is channeled into the oxidation pathway, as suggested by hyperinduction of *CYP94* genes in amido hydrolase mutants (Widemann et al. 2013). This JA-Ile-cleaving pathway was previously shown to be biologically relevant in the wild species *N. attenuata*. NaJIH, an herbivore-inducible IAR3 homolog, was found to equally hydrolyze JA-Ile and IAA-Ala in vitro. However, its silencing extended JA-Ile half-life after herbivory, increased herbivore predator attraction due to enhanced JA-Ile-regulated volatile emission, and increased overall insect resistance (Woldemariam et al. 2012). These data demonstrated that JA-Ile cleavage in WT plants participates in the attenuation of JA-Ile accumulation and signalling. JA-Ile is therefore catabolized by an irreversible, CYP94-mediated ω-oxidation pathway and also by a reversible hydrolytic pathway, that antagonizes the action of the conjugating enzyme JAR1. The exact impact of JA-Ile-cleaving activity on JA pools is difficult to assess, because the steady-state JA levels measured are the result of its synthesis and consumption by several enzymatic activities. The collaborative action of the oxidative and hydrolytic pathways for JA-Ile disposal, consistent with their simultaneous transcriptional upregulation, has recently been demonstrated in *cyp94b3ill6* double mutants (Koo et al. 2014).

JA-Ile Catabolic Pathways Redefine the Jasmonate Metabolic Grid

While characterizing JA profiles in CYP94-modified plants, we detected unexpected changes in unconjugated JAs. For example, 12OH-JA levels, for which no biosynthetic route was described, were reduced in wounded plants impaired in CYP94-encoded JA-Ile oxidating activity (Widemann et al. 2013). Additionally, we found that the *jar1* mutant only displays half WT 12OH-JA levels, and residual level of all conjugates, including 12OH-JA-Ile. This contradicts the supposed –but not demonstrated- 12OH-JA formation by direct JA hydroxylation, because excess JA in *jar1* should provide more substrate for such a direct oxidation. An indirect metabolic route via conjugate intermediates was suspected and could be established biochemically and genetically. Evidence for 12OH-JA-Ile cleavage was obtained in vitro for recombinant IAR3 only, but both *iar3* and *ill6* KO lines displayed 50 % reduced 12OH-JA accumulation. This established that an indirect 12OH-JA biosynthetic route operates in wounded leaves by successive JA conjugation, JA-Ile ω-hydroxylation, and 12OH-JA-Ile cleavage (Widemann et al. 2013). Further analysis revealed that 12OH-JA generated through this pathway feeds the synthesis of its two derivatives $12HSO_4$-JA and 12-*O*-Glc-JA (Fig. 16.3). In contrast, infection with the necrotrophic fungus *Botrytis cinerea* triggered a quantitatively different JA profile, and 12OH-JA accumulation seems less dependent on this indirect route (Aubert et al. 2015). This suggests the possible existence of several 12OH-JA biosynthetic routes whose relative contributions may depend on the type of stimulus. Collectively, these findings highlight the dual role of JA-Ile catabolic pathways: the coordinated action of oxidative and hydrolytic conversions ensure tight control of hormonal JA-Ile half-life and dynamics, but also provide new connections between conjugated and unconjugated compounds and as such redefine the jasmonate conversion pathway as a metabolic grid (Fig. 16.3). JA-Ile not only acts as the master signal, but also stands as a metabolic hub in the pathway. Quantitative analysis along with genetic tools has shown that specific enzymatic control over the least abundant compound, JA-Ile, affects larger pools of precursors and catabolites, suggesting an important flux through JA-Ile. Functional IAR3/ILL6 redundancy in 12OH-JA generation will need to be addressed, as well as the possible existence of a conjugate-independent biosynthesis route.

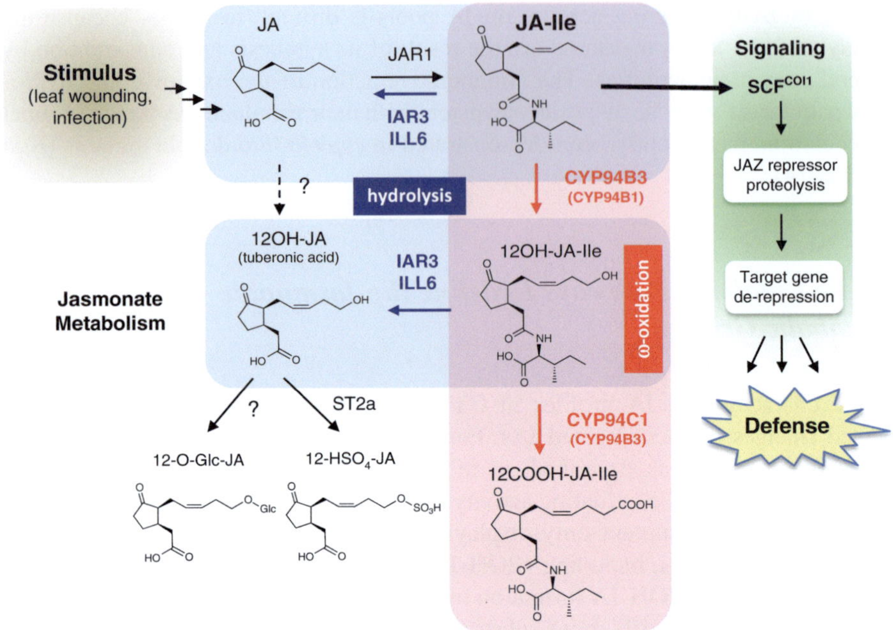

Fig. 16.3 JA-Ile catabolic pathways redefine the jasmonate metabolic grid. JA-Ile hormone generated through JAR1 activity signals through the SCF[COI1] E3 ubiquitin ligase that directs JAZ repressor proteolytic removal and de-repression of target defense genes. Two biochemical pathways divert JA-Ile to attenuate defense through conjugate oxidation and hydrolysis. The names of characterized enzymes for jasmonate conversion are given. The coordinated action of cytochrome P450 (CYP94) and amido-hydrolase enzymes in JA-Ile turnover impacts the homeostasis of conjugated and unconjugated jasmonates and provides an indirect route for the production of tuberonic acid and its derivatives

Conclusions

Recent advances have enriched our knowledge on diverse biological activities associated with jasmonate compounds. OPDA synthesis and signaling seems to have arisen very early in evolution, and acts through redox changes, similarly to phytoprostane effects. The JA-Ile/COI1/JAZ signaling module is a sophisticated system to mediate broad jasmonate actions and its functional conservation has been confirmed recently in a number of plant species. JA-Ile seems to be a generic signal coordinating major defensive or developmental responses, but other conjugates have been associated with more specific biological activities. The question whether different JAZ-COI1 combinations may bind different jasmonate ligands remains open, as is the particular function of individual JAZ repressors. Besides JA-Ile, a number of more specialized biological activities have evolved within unconjugated JA derivatives in a species- or organ-specific manner. These two aspects may be interconnected as the recently-identified JA-Ile catabolic pathways have implications on global JA homeostasis. More progress is needed to understand and exploit the full repertoire of biosynthetic routes, signaling pathways, and targets of the rich world of jasmonates.

Acknowledgements This work was partly funded by the Agence Nationale de la Recherche (contract ANR-12-BSV8-005) (YA and ES) and a doctoral fellowship from the Ministère de l'Enseignement Supérieur et de la Recherche to EW.

References

Acosta IF, Farmer EE (2010) Jasmonates. Arabidopsis Book 8:e0129

Acosta IF, Laparra H, Romero SP, Schmelz E, Hamberg M, Mottinger JP, Moreno MA, Dellaporta SL (2009) tasselseed1 is a lipoxygenase affecting jasmonic acid signaling in sex determination of maize. Science 323:262–265

Acosta IF, Gasperini D, Chetelat A, Stolz S, Santuari L, Farmer EE (2013) Role of NINJA in root jasmonate signaling. Proc Natl Acad Sci U S A 110:15473–15478

Almeras E, Stolz S, Vollenweider S, Reymond P, Mene-Saffrane L, Farmer EE (2003) Reactive electrophile species activate defense gene expression in Arabidopsis. Plant J 34:205–216

Andersson MX, Hamberg M, Kourtchenko O, Brunnstrom A, McPhail KL, Gerwick WH, Gobel C, Feussner I, Ellerstrom M (2006) Oxylipin profiling of the hypersensitive response in Arabidopsis thaliana. Formation of a novel oxo-phytodienoic acid-containing galactolipid, arabidopside E. J Biol Chem 281:31528–31537

Aubert Y, Widemann E, Miesch L, Pinot F, Heitz T (2015) CYP94-mediated jasmonoyl-isoleucine hormone oxidation shapes jasmonate profiles and attenuates defense responses to *Botrytis cinerea* infection. J Exp Bot 66:3879–3892

Birkett MA, Campbell CA, Chamberlain K, Guerrieri E, Hick AJ, Martin JL, Matthes M, Napier JA, Pettersson J, Pickett JA, Poppy GM, Pow EM, Pye BJ, Smart LE, Wadhams GH, Wadhams LJ, Woodcock CM (2000) New roles for cis-jasmone as an insect semiochemical and in plant defense. Proc Natl Acad Sci U S A 97:9329–9334

Blechert S, Brodschelm W, Holder S, Kammerer L, Kutchan TM, Mueller MJ, Xia ZQ, Zenk MH (1995) The octadecanoic pathway: signal molecules for the regulation of secondary pathways. Proc Natl Acad Sci U S A 92:4099–4105

Blee E (2002) Impact of phyto-oxylipins in plant defense. Trends Plant Sci 7:315–322

Bonaventure G, Schuck S, Baldwin IT (2011) Revealing complexity and specificity in the activation of lipase-mediated oxylipin biosynthesis: a specific role of the Nicotiana attenuata GLA1 lipase in the activation of jasmonic acid biosynthesis in leaves and roots. Plant Cell Environ 34:1507–1520

Bosch M, Wright LP, Gershenzon J, Wasternack C, Hause B, Schaller A, Stintzi A (2014) Jasmonic acid and its precursor 12-oxophytodienoic acid control different aspects of constitutive and induced herbivore defenses in tomato. Plant Physiol 166:396–410

Böttcher C, Pollmann S (2009) Plant oxylipins: plant responses to 12-oxo-phytodienoic acid are governed by its specific structural and functional properties. FEBS J 276:4693–4704

Böttcher C, Weiler EW (2007) Cyclo-oxylipin-galactolipids in plants: occurrence and dynamics. Planta 226:629–637

Browse J (2009) Jasmonate passes muster: a receptor and targets for the defense hormone. Annu Rev Plant Biol 60:183–205

Bruce TJ, Matthes MC, Chamberlain K, Woodcock CM, Mohib A, Webster B, Smart LE, Birkett MA, Pickett JA, Napier JA (2008) cis-Jasmone induces Arabidopsis genes that affect the chemical ecology of multitrophic interactions with aphids and their parasitoids. Proc Natl Acad Sci U S A 105:4553–4558

Buseman CM, Tamura P, Sparks AA, Baughman EJ, Maatta S, Zhao J, Roth MR, Esch SW, Shah J, Williams TD, Welti R (2006) Wounding stimulates the accumulation of glycerolipids containing oxophytodienoic acid and dinor-oxophytodienoic acid in Arabidopsis leaves. Plant Physiol 142:28–39

Campos ML, Kang JH, Howe GA (2014) Jasmonate-triggered plant immunity. J Chem Ecol 40:657–675

Cesari IM, Carvalho E, Figueiredo Rodrigues M, Mendonca Bdos S, Amoedo ND, Rumjanek FD (2014) Methyl jasmonate: putative mechanisms of action on cancer cells cycle, metabolism, and apoptosis. Int J Cell Biol 2014:572097

Chauvin A, Caldelari D, Wolfender JL, Farmer EE (2013) Four 13-lipoxygenases contribute to rapid jasmonate synthesis in wounded Arabidopsis thaliana leaves: a role for lipoxygenase 6 in responses to long-distance wound signals. New Phytol 197:566–575

Chehab EW, Yao C, Henderson Z, Kim S, Braam J (2012) Arabidopsis touch-induced morphogenesis is jasmonate mediated and protects against pests. Curr Biol 22:701–706

Chini A, Fonseca S, Fernandez G, Adie B, Chico JM, Lorenzo O, Garcia-Casado G, Lopez-Vidriero I, Lozano FM, Ponce MR, Micol JL, Solano R (2007) The JAZ family of repressors is the missing link in jasmonate signalling. Nature 448:666–671

Chung HS, Koo AJ, Gao X, Jayanty S, Thines B, Jones AD, Howe GA (2008) Regulation and function of Arabidopsis JASMONATE ZIM-domain genes in response to wounding and herbivory. Plant Physiol 146:952–964

Cohen S, Flescher E (2009) Methyl jasmonate: a plant stress hormone as an anti-cancer drug. Phytochemistry 70:1600–1609

Creelman RA, Mullet JE (1997) Biosynthesis and action of jasmonates in plants. Annu Rev Plant Physiol Plant Mol Biol 48:355–381

Dabrowska P, Boland W (2007) Iso-OPDA: an early precursor of cis-jasmone in plants? Chembiochem 8:2281–2285

Dave A, Hernandez ML, He Z, Andriotis VM, Vaistij FE, Larson TR, Graham IA (2011) 12-oxo-phytodienoic acid accumulation during seed development represses seed germination in Arabidopsis. Plant Cell 23:583–599

Davoine C, Falletti O, Douki T, Iacazio G, Ennar N, Montillet JL, Triantaphylides C (2006) Adducts of oxylipin electrophiles to glutathione reflect a 13 specificity of the downstream lipoxygenase pathway in the tobacco hypersensitive response. Plant Physiol 140:1484–1493

Ellinger D, Stingl N, Kubigsteltig II, Bals T, Juenger M, Pollmann S, Berger S, Schuenemann D, Mueller MJ (2010) DONGLE and DEFECTIVE IN ANTHER DEHISCENCE1 lipases are not essential for wound- and pathogen-induced jasmonate biosynthesis: redundant lipases contribute to jasmonate formation. Plant Physiol 153:114–127

Farmer EE, Mueller MJ (2013) ROS-mediated lipid peroxidation and RES-activated signaling. Annu Rev Plant Biol 64:429–450

Feussner I, Wasternack C (2002) The lipoxygenase pathway. Annu Rev Plant Biol 53:275–297

Fonseca S, Chico JM, Solano R (2009a) The jasmonate pathway: the ligand, the receptor and the core signalling module. Curr Opin Plant Biol 12:539–547

Fonseca S, Chini A, Hamberg M, Adie B, Porzel A, Kramell R, Miersch O, Wasternack C, Solano R (2009b) (+)-7-iso-Jasmonoyl-L-isoleucine is the endogenous bioactive jasmonate. Nat Chem Biol 5:344–350

Fukumoto K, Alamgir K, Yamashita Y, Mori IC, Matsuura H, Galis I (2013) Response of rice to insect elicitors and the role of OsJAR1 in wound and herbivory-induced JA-Ile accumulation. J Integr Plant Biol 55:775–784

Gidda SK, Miersch O, Levitin A, Schmidt J, Wasternack C, Varin L (2003) Biochemical and molecular characterization of a hydroxyjasmonate sulfotransferase from Arabidopsis thaliana. J Biol Chem 278:17895–17900

Glauser G, Grata E, Dubugnon L, Rudaz S, Farmer EE, Wolfender JL (2008) Spatial and temporal dynamics of jasmonate synthesis and accumulation in Arabidopsis in response to wounding. J Biol Chem 283:16400–16407

Goetz S, Hellwege A, Stenzel I, Kutter C, Hauptmann V, Forner S, McCaig B, Hause G, Miersch O, Wasternack C, Hause B (2012) Role of cis-12-oxo-phytodienoic acid in tomato embryo development. Plant Physiol 158:1715–1727

Grienenberger E, Geoffroy P, Mutterer J, Legrand M, Heitz T (2010) The interplay of lipid acyl hydrolases in inducible plant defense. Plant Signal Behav 5:1181–1186

Guo HM, Li HC, Zhou SR, Xue HW, Miao XX (2014) Cis-12-oxo-phytodienoic acid stimulates rice defense response to a piercing-sucking insect. Mol Plant 7:1683–1692

Hartley SE, Eschen R, Horwood JM, Gange AC, Hill EM (2015) Infection by a foliar endophyte elicits novel arabidopside-based plant defence reactions in its host, Cirsium arvense. New Phytol 205:816–827

Heitz T, Widemann E, Lugan R, Miesch L, Ullmann P, Desaubry L, Holder E, Grausem B, Kandel S, Miesch M, Werck-Reichhart D, Pinot F (2012) Cytochromes P450 CYP94C1 and CYP94B3 catalyze two successive oxidation steps of plant hormone Jasmonoyl-isoleucine for catabolic turnover. J Biol Chem 287:6296–6306

Hisamatsu Y, Goto N, Sekiguchi M, Hasegawa K, Shigemori H (2005) Oxylipins arabidopsides C and D from Arabidopsis thaliana. J Nat Prod 68:600–603

Howe GA, Jander G (2008) Plant immunity to insect herbivores. Annu Rev Plant Biol 59:41–66

Ishiguro S, Kawai-Oda A, Ueda J, Nishida I, Okada K (2001) The DEFECTIVE IN ANTHER DEHISCIENCE gene encodes a novel phospholipase A1 catalyzing the initial step of jasmonic acid biosynthesis, which synchronizes pollen maturation, anther dehiscence, and flower opening in Arabidopsis. Plant Cell 13:2191–2209

Kandel S, Sauveplane V, Compagnon V, Franke R, Millet Y, Schreiber L, Werck-Reichhart D, Pinot F (2007) Characterization of a methyl jasmonate and wounding-responsive cytochrome P450 of Arabidopsis thaliana catalyzing dicarboxylic fatty acid formation in vitro. FEBS J 274:5116–5127

Katsir L, Chung HS, Koo AJ, Howe GA (2008) Jasmonate signaling: a conserved mechanism of hormone sensing. Curr Opin Plant Biol 11:428–435

Kazan K, Manners JM (2008) Jasmonate signaling: toward an integrated view. Plant Physiol 146:1459–1468

Kitaoka N, Matsubara T, Sato M, Takahashi K, Wakuta S, Kawaide H, Matsui H, Nabeta K, Matsuura H (2011) Arabidopsis CYP94B3 encodes jasmonyl-L-isoleucine 12-hydroxylase, a key enzyme in the oxidative catabolism of jasmonate. Plant Cell Physiol 52:1757–1765

Koch T, Bandemer K, Boland W (1997) Biosynthesis of *cis*-jasmone: a pathway for the inactivation and the disposal of the plant stress hormone jasmonic acid to the gas phase ? Helv Chem Acta 80:838–850

Koo AJ, Cooke TF, Howe GA (2011) Cytochrome P450 CYP94B3 mediates catabolism and inactivation of the plant hormone jasmonoyl-L-isoleucine. Proc Natl Acad Sci U S A 108:9298–9303

Koo AJ, Thireault C, Zemelis S, Poudel AN, Zhang T, Kitaoka N, Brandizzi F, Matsuura H, Howe GA (2014) Endoplasmic reticulum-associated inactivation of the hormone jasmonoyl-L-isoleucine by multiple members of the cytochrome P450 94 family in Arabidopsis. J Biol Chem 289:29728–29738

Kourtchenko O, Andersson MX, Hamberg M, Brunnstrom A, Gobel C, McPhail KL, Gerwick WH, Feussner I, Ellerstrom M (2007) Oxo-phytodienoic acid-containing galactolipids in Arabidopsis: jasmonate signaling dependence. Plant Physiol 145:1658–1669

Kramell R, Miersch O, Hause B, Ortel B, Parthier B, Wasternack C (1997) Amino acid conjugates of jasmonic acid induce jasmonate-responsive gene expression in barley (Hordeum vulgare L.) leaves. FEBS Lett 414:197–202

Kramell R, Schmidt J, Herrmann G, Schliemann W (2005) N-(jasmonoyl)tyrosine-derived compounds from flowers of broad beans (Vicia faba). J Nat Prod 68:1345–1349

LeClere S, Tellez R, Rampey RA, Matsuda SP, Bartel B (2002) Characterization of a family of IAA-amino acid conjugate hydrolases from Arabidopsis. J Biol Chem 277:20446–20452

Li C, Schilmiller AL, Liu G, Lee GI, Jayanty S, Sageman C, Vrebalov J, Giovannoni JJ, Yagi K, Kobayashi Y, Howe GA (2005) Role of beta-oxidation in jasmonate biosynthesis and systemic wound signaling in tomato. Plant Cell 17:971–986

Matthes MC, Bruce TJ, Ton J, Verrier PJ, Pickett JA, Napier JA (2010) The transcriptome of cis-jasmone-induced resistance in Arabidopsis thaliana and its role in indirect defence. Planta 232:1163–1180

Miersch O, Neumerkel J, Dippe M, Stenzel I, Wasternack C (2008) Hydroxylated jasmonates are commonly occurring metabolites of jasmonic acid and contribute to a partial switch-off in jasmonate signaling. New Phytol 177:114–127

Mizutani M, Ohta D (2010) Diversification of P450 genes during land plant evolution. Annu Rev Plant Biol 61:291–315

Mueller S, Hilbert B, Dueckershoff K, Roitsch T, Krischke M, Mueller MJ, Berger S (2008) General detoxification and stress responses are mediated by oxidized lipids through TGA transcription factors in Arabidopsis. Plant Cell 20:768–785

Nakamura Y, Miyatake R, Ueda M (2008) Enantiodifferential approach for the detection of the target membrane protein of the jasmonate glycoside that controls the leaf movement of Albizzia saman. Angew Chem Int Ed Engl 47:7289–7292

Nakamura Y, Mithofer A, Kombrink E, Boland W, Hamamoto S, Uozumi N, Tohma K, Ueda M (2011) 12-hydroxyjasmonic acid glucoside is a COI1-JAZ-independent activator of leaf-closing movement in Samanea saman. Plant Physiol 155:1226–1236

Nilsson AK, Fahlberg P, Ellerstrom M, Andersson MX (2012) Oxo-phytodienoic acid (OPDA) is formed on fatty acids esterified to galactolipids after tissue disruption in Arabidopsis thaliana. FEBS Lett 586:2483–2487

Nilsson AK, Johansson ON, Fahlberg P, Steinhart F, Gustavsson MB, Ellerstrom M, Andersson MX (2014) Formation of oxidized phosphatidylinositol and 12-oxo-phytodienoic acid containing acylated phosphatidylglycerol during the hypersensitive response in Arabidopsis. Phytochemistry 101:65–75

Ohkama-Ohtsu N, Sasaki-Sekimoto Y, Oikawa A, Jikumaru Y, Shinoda S, Inoue E, Kamide Y, Yokoyama T, Hirai MY, Shirasu K, Kamiya Y, Oliver DJ, Saito K (2011) 12-oxo-phytodienoic acid-glutathione conjugate is transported into the vacuole in Arabidopsis. Plant Cell Physiol 52:205–209

Park SW, Li W, Viehhauser A, He B, Kim S, Nilsson AK, Andersson MX, Kittle JD, Ambavaram MM, Luan S, Esker AR, Tholl D, Cimini D, Ellerstrom M, Coaker G, Mitchell TK, Pereira A, Dietz KJ, Lawrence CB (2013) Cyclophilin 20-3 relays a 12-oxo-phytodienoic acid signal during stress responsive regulation of cellular redox homeostasis. Proc Natl Acad Sci U S A 110:9559–9564

Pauwels L, Goossens A (2011) The JAZ proteins: a crucial interface in the jasmonate signaling cascade. Plant Cell 23:3089–3100

Perez AC, Goossens A (2013) Jasmonate signalling: a copycat of auxin signalling? Plant Cell Environ 36:2071–2084

Pieterse CM, Leon-Reyes A, Van der Ent S, Van Wees SC (2009) Networking by small-molecule hormones in plant immunity. Nat Chem Biol 5:308–316

Preston CA, Laue G, Baldwin IT (2004) Plant-plant signaling: application of trans- or cis-methyl jasmonate equivalent to sagebrush releases does not elicit direct defenses in native tobacco. J Chem Ecol 30:2193–2214

Ribot C, Zimmerli C, Farmer EE, Reymond P, Poirier Y (2008) Induction of the Arabidopsis PHO1;H10 gene by 12-oxo-phytodienoic acid but not jasmonic acid via a CORONATINE INSENSITIVE1-dependent pathway. Plant Physiol 147:696–706

Riemann M, Riemann M, Takano M (2008) Rice JASMONATE RESISTANT 1 is involved in phytochrome and jasmonate signalling. Plant Cell Environ 31:783–792

Santner A, Estelle M (2009) Recent advances and emerging trends in plant hormone signalling. Nature 459:1071–1078

Scalschi L, Sanmartin M, Camanes G, Troncho P, Sanchez-Serrano JJ, Garcia-Agustin P, Vicedo B (2014) Silencing of OPR3 in tomato reveals the role of OPDA in callose deposition during the activation of defense responses against Botrytis cinerea. Plant J 81:304–315
Schaller A, Stintzi A (2009) Enzymes in jasmonate biosynthesis – structure, function, regulation. Phytochemistry 70:1532–1538
Schulze B, Dabrowska P, Boland W (2007) Rapid enzymatic isomerization of 12-oxophytodienoic acid in the gut of lepidopteran larvae. Chembiochem 8:208–216
Seo HS, Song JT, Cheong JJ, Lee YH, Lee YW, Hwang I, Lee JS, Choi YD (2001) Jasmonic acid carboxyl methyltransferase: a key enzyme for jasmonate-regulated plant responses. Proc Natl Acad Sci U S A 98:4788–4793
Seto Y, Hamada S, Matsuura H, Matsushige M, Satou C, Takahashi K, Masuta C, Ito H, Matsui H, Nabeta K (2009) Purification and cDNA cloning of a wound inducible glucosyltransferase active toward 12-hydroxy jasmonic acid. Phytochemistry 70:370–379
Sheard LB, Tan X, Mao H, Withers J, Ben-Nissan G, Hinds TR, Kobayashi Y, Hsu FF, Sharon M, Browse J, He SY, Rizo J, Howe GA, Zheng N (2010) Jasmonate perception by inositol-phosphate-potentiated COI1-JAZ co-receptor. Nature 468:400–405
Shimizu T, Miyamoto K, Miyamoto K, Minami E, Nishizawa Y, Iino M, Nojiri H, Yamane H, Okada K (2013) OsJAR1 contributes mainly to biosynthesis of the stress-induced jasmonoyl-isoleucine involved in defense responses in rice. Biosci Biotechnol Biochem 77:1556–1564
Song S, Qi T, Huang H, Xie D (2013) Regulation of stamen development by coordinated actions of jasmonate, auxin, and gibberellin in Arabidopsis. Mol Plant 6:1065–1073
Song S, Qi T, Wasternack C, Xie D (2014) Jasmonate signaling and crosstalk with gibberellin and ethylene. Curr Opin Plant Biol 21:112–119
Staswick PE (2009) The tryptophan conjugates of jasmonic and indole-3-acetic acids are endogenous auxin inhibitors. Plant Physiol 150:1310–1321
Staswick PE, Su W, Howell SH (1992) Methyl jasmonate inhibition of root growth and induction of a leaf protein are decreased in an Arabidopsis thaliana mutant. Proc Natl Acad Sci U S A 89:6837–6840
Stelmach BA, Muller A, Hennig P, Laudert D, Andert L, Weiler EW (1998) Quantitation of the octadecanoid 12-oxo-phytodienoic acid, a signalling compound in plant mechanotransduction. Phytochemistry 47:539–546
Stelmach BA, Muller A, Hennig P, Gebhardt S, Schubert-Zsilavecz M, Weiler EW (2001) A novel class of oxylipins, sn1-O-(12-oxophytodienoyl)-sn2-O-(hexadecatrienoyl)-monogalactosyl Diglyceride, from Arabidopsis thaliana. J Biol Chem 276:12832–12838
Stintzi A, Browse J (2000) The Arabidopsis male-sterile mutant, opr3, lacks the 12-oxophytodienoic acid reductase required for jasmonate synthesis. Proc Natl Acad Sci U S A 97:10625–10630
Stintzi A, Weber H, Reymond P, Browse J, Farmer EE (2001) Plant defense in the absence of jasmonic acid: the role of cyclopentenones. Proc Natl Acad Sci U S A 98:12837–12842
Stitz M, Baldwin IT, Gaquerel E (2011) Diverting the flux of the JA pathway in Nicotiana attenuata compromises the plant's defense metabolism and fitness in nature and glasshouse. PLoS One 6:e25925
Stitz M, Hartl M, Baldwin IT, Gaquerel E (2014) Jasmonoyl-l-isoleucine coordinates metabolic networks required for anthesis and floral attractant emission in wild tobacco (Nicotiana attenuata). Plant Cell 26:3964–3983
Stotz HU, Mueller S, Zoeller M, Mueller MJ, Berger S (2013) TGA transcription factors and jasmonate-independent COI1 signalling regulate specific plant responses to reactive oxylipins. J Exp Bot 64:963–975
Stumpe M, Gobel C, Faltin B, Beike AK, Hause B, Himmelsbach K, Bode J, Kramell R, Wasternack C, Frank W, Reski R, Feussner I (2010) The moss Physcomitrella patens contains cyclopentenones but no jasmonates: mutations in allene oxide cyclase lead to reduced fertility and altered sporophyte morphology. New Phytol 188:740–749
Suza WP, Staswick PE (2008) The role of JAR1 in Jasmonoyl-L: -isoleucine production during Arabidopsis wound response. Planta 227:1221–1232

Taki N, Sasaki-Sekimoto Y, Obayashi T, Kikuta A, Kobayashi K, Ainai T, Yagi K, Sakurai N, Suzuki H, Masuda T, Takamiya K, Shibata D, Kobayashi Y, Ohta H (2005) 12-oxo-phytodienoic acid triggers expression of a distinct set of genes and plays a role in wound-induced gene expression in Arabidopsis. Plant Physiol 139:1268–1283

Tamogami S, Rakwal R, Kodama O (1997) Phytoalexin production by amino acid conjugates of jasmonic acid through induction of naringenin-7-O-methyltransferase, a key enzyme on phytoalexin biosynthesis in rice (Oryza sativa L.). FEBS Lett 401:239–242

Thines B, Katsir L, Melotto M, Niu Y, Mandaokar A, Liu G, Nomura K, He SY, Howe GA, Browse J (2007) JAZ repressor proteins are targets of the SCF(COI1) complex during jasmonate signalling. Nature 448:661–665

Ueda M, Yang G, Nukadzuka Y, Ishimaru Y, Tamura S, Manabe Y (2015) Functional importance of the sugar moiety of jasmonic acid glucoside for bioactivity and target affinity. Org Biomol Chem 13:55–58

Wakuta S, Hamada S, Ito H, Matsuura H, Nabeta K, Matsui H (2010) Identification of a beta-glucosidase hydrolyzing tuberonic acid glucoside in rice (Oryza sativa L.). Phytochemistry 71:1280–1288

Wakuta S, Suzuki E, Saburi W, Matsuura H, Nabeta K, Imai R, Matsui H (2011) OsJAR1 and OsJAR2 are jasmonyl-L-isoleucine synthases involved in wound- and pathogen-induced jasmonic acid signalling. Biochem Biophys Res Commun 409:634–639

Wang L, Halitschke R, Kang JH, Berg A, Harnisch F, Baldwin IT (2007) Independently silencing two JAR family members impairs levels of trypsin proteinase inhibitors but not nicotine. Planta 226:159–167

Wasternack C, Hause B (2013) Jasmonates: biosynthesis, perception, signal transduction and action in plant stress response, growth and development. An update to the 2007 review in Annals of Botany. Ann Bot 111:1021–1058

Weiler E, Albrecht T, Groth B, Xia Z-Q, Luxem M, Liß H, Andert L, Spengler P (1993) Evidence for the involvement of jasmonates and their octadecanoid precursors in the tendril coiling response of *Bryonia dioica*. Phytochemistry 32:591–600

Widemann E, Miesch L, Lugan R, Holder E, Heinrich C, Aubert Y, Miesch M, Pinot F, Heitz T (2013) The amidohydrolases IAR3 and ILL6 contribute to jasmonoyl-isoleucine hormone turnover and generate 12-hydroxyjasmonic acid upon wounding in Arabidopsis leaves. J Biol Chem 288:31701–31714

Widemann E, Grausem B, Renault H, Pineau E, Heinrich C, Lugan R, Ullmann P, Miesch L, Aubert Y, Miesch M et al (2015a) Sequential oxidation of Jasmonoyl-Phenylalanine and Jasmonoyl-Isoleucine by multiple cytochrome P450 of the CYP94 family through newly identified aldehyde intermediates. Phytochemistry 117:388–399

Widemann E, Heitz T, Miesch L, Miesch M, Heinrich C, Pinot F, Lugan R (2015b) Identification of the 12-oxojasmonoyl-isoleucine, a new intermediate of jasmonate metabolism in Arabidopsis, by combining chemical derivatization and LC–MS/MS analysis. Metabolomics. 11:991–997. doi:10.1007/s11306–014-0754-7

Woldemariam MG, Onkokesung N, Baldwin IT, Galis I (2012) Jasmonoyl-L-isoleucine hydrolase 1 (JIH1) regulates jasmonoyl-L-isoleucine levels and attenuates plant defenses against herbivores. Plant J 72:758–767

Yan Y, Stolz S, Chetelat A, Reymond P, Pagni M, Dubugnon L, Farmer EE (2007) A downstream mediator in the growth repression limb of the jasmonate pathway. Plant Cell 19:2470–2483

Yan Y, Christensen S, Isakeit T, Engelberth J, Meeley R, Hayward A, Emery RJ, Kolomiets MV (2012) Disruption of OPR7 and OPR8 reveals the versatile functions of jasmonic acid in maize development and defense. Plant Cell 24:1420–1436

Yan Y, Borrego E, Kolomiets M (2013) Jasmonate biosynthesis, perception and function in plant development and stress responses. InTech, www.intechopen.com

Yoshihara T, Omer E-LA, Koshino H, Sakamura S, Kikuta Y, Koda Y (1989) Structure of a tuber-inducing stimulus from potato leaves (*Solanum tuberosum* L.). Agric Biol Chem 53:2835–2837

Chapter 17
Green Leaf Volatiles in Plant Signaling and Response

Kenji Matsui and Takao Koeduka

Abstract Most 'green' plants form green leaf volatiles (GLVs). GLVs are a familiar plant secondary metabolite, but knowledge of their physiological and ecological functions is limited. GLV formation is tightly suppressed when plant tissues are intact, but upon mechanical wounding, herbivore attack, or abiotic stresses, GLVs are formed rapidly, within seconds or minutes. Thus, this may be an important system for defense responses, allowing plants to protect themselves from damage as soon as possible. Because GLV formation in the natural environment is roughly related to the degree of stress in the plant life, sensing the amount of GLVs in the atmosphere might allow plants to recognize their surroundings. Because some plants respond to GLVs, they may communicate with GLVs. GLVs that contain α,β-unsaturated carbonyl groups might activate signaling systems regulated under the redox state of plant cells. Plasma membranes would also be targets of interactions with GLVs. Additionally, the metabolism of GLVs in plant cells after absorption from the atmosphere could also be classified as a plant–plant interaction.

Keywords Green leaf volatile • Hexenal • Reactive electrophilic species • Plant–plant communication

Introduction

Lipids, in general, have three important roles in living organisms. First, they are essential components of the cell membrane, which can separate the cells from an abiotic environment. Lipids are also important nutrients for many organisms, and the catabolism through β-oxidation of lipid constituents yields large amounts of energy.

K. Matsui (✉)
Department of Applied Molecular Bioscience, Graduate School of Medicine, Yamaguchi University, Yamaguchi 753-8515, Japan
e-mail: matsui@yamaguchi-u.ac.jp

T. Koeduka
Department of Biological Chemistry, Faculty of Agriculture, Yamaguchi University, Yamaguchi 753-8515, Japan

Y. Nakamura, Y. Li-Beisson (eds.), *Lipids in Plant and Algae Development*, Subcellular Biochemistry 86, DOI 10.1007/978-3-319-25979-6_17

Lipids can also be used as signal molecules. Lipid mediators in mammalian cells, such as prostaglandins, leukotrienes, or platelet-activating factors, have various biological functions in inflammation, blood clotting, and immune systems (Murakami 2011). This is also the case with plants, even though the structures and functions of the lipid mediators found in plants are usually different from those in mammals.

Jasmonates are typical and important lipid mediators in plants, as described in detail elsewhere of this book. Volatile compounds with six carbon (C6) backbones, which are collectively called green leaf volatiles (GLVs) (Fig. 17.1), are also formed in plants. They use a biosynthetic pathway similar to that for jasmonates, i.e., using lipoxygenases to form fatty acid hydroperoxides, followed by the homolytic cleavage of the hydroperoxide by cytochrome P450s (Grechkin et al. 2006). GLVs consist of C6 volatile compounds containing aldehyde, alcohol, and esters. Because almost all green leaves on the Earth form GLVs, human beings correlate the olfactory sensation emitted by GLVs with green leaves, and thus, we usually sense the volatiles as a green note.

One GLV, (*E*)-2-hexenal (leaf aldehyde), was isolated in 1912 by Curtius and Franzen as a component in the essential oil prepared from 600 kg of hornbeam leaves by steam distillation (Curtius and Franzen 1912). Since then, because of their peculiar aromatic properties and common occurrence in plants, GLVs have been studied in flavor chemistry. However, their functions in ecological systems have

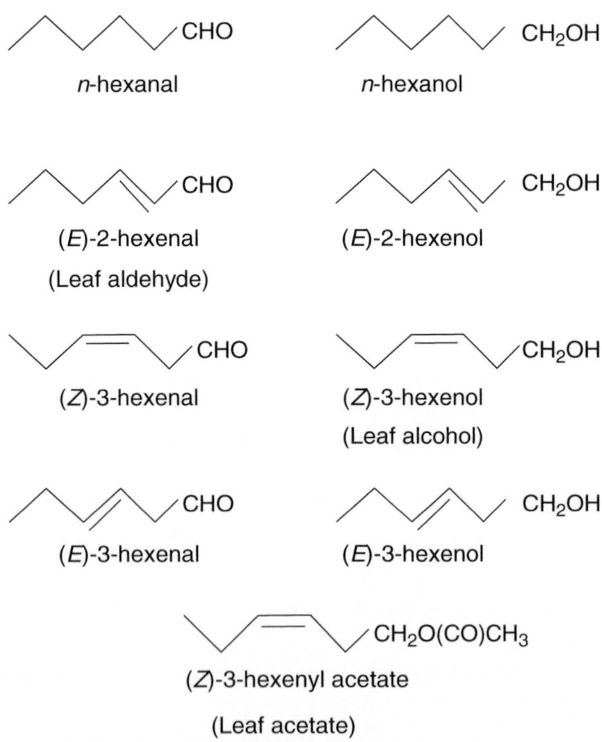

Fig. 17.1 Structures of green leaf volatiles (GLVs) commonly found in plants

also been noticed, especially because plants usually accumulate only small amounts of GLVs but quickly form them after mechanical wounding or herbivore attack (Scala et al. 2013).

In this chapter we discuss novel knowledge regarding the biosynthesis of GLVs and their evolution. Thereafter, we give up-to-date reviews of signaling and plant responses elicited by GLVs.

Biosynthesis of GLVs

GLVs are formed from fatty acids through a dioxygenation reaction catalyzed by lipoxygenases (LOXs) to yield fatty acid hydroperoxide, and a subsequent rearrangement reaction to cleave fatty acid hydroperoxides by hydroperoxide lyase (HPL) to form C6 aldehydes and 12 carbon oxo acids as counterparts (Fig. 17.2). The general outline of the GLV-forming pathway shares similarities with those for other oxylipin products, including jasmonates, and each branch in the oxylipin pathway could have diverged from one ancestral pathway. Based on the structure of allene oxide synthase (AOS) and HPL, and also on the results of the interconversion of AOS to HPL, Lee et al. (2008) proposed that the HPL pathway was established first and that the AOS pathway diverged thereafter. This view might be changed through the accumulation of knowledge regarding the distribution of HPL and AOS among algae, bryophytes, ferns, or gymnosperms as described in section "Evolution of genes involved in GLV formation".

The HPL and AOS pathways share the first part of a metabolic pathway, and they also use the same substrate. HPL and AOS should catalyze their reactions without competition to meet the demands for GLVs and JAs, respectively, under certain growth conditions because they have distinct physiological functions. Most HPL and AOS enzymes identified so far have chloroplast transit peptides. Their chloroplast localizations have been established through investigations using marker proteins, such as green fluorescence protein (GFP), and the fractionation of chloroplasts. In some plants, HPLs are localized to the lipid bodies (Mita et al. 2005), outer envelopes of chloroplasts (Froehlich et al. 2001), stroma (Bonaventure 2014), and, in some cases, no specific localization is observed (Phillips and Galliard 1978; Shibata et al. 1995; Noordermeer et al. 2000). Rice HPL3 (OsHPL3), which has the shortest extension on its N-terminal among the three rice HPLs, was transported to chloroplasts when a fusion protein of the transit peptide of OsHPL3 with GFP was expressed in Arabidopsis leaves (Savchenko et al. 2014). This is also the case with Arabidopsis, and Arabidopsis HPL fused with GFP was transported to the plastids (Mwenda et al. 2015). Although it is still possible for the two CYP74 enzymes to be segregated at the level of sub-chloroplast membrane, or even within the same membrane (Mita et al. 2005), the close localization of two enzymes sharing the same substrate would cause disordered competition, especially when the enzymes form their products in the disrupted tissues during rapid oxylipin bursts (Matsui 2006;

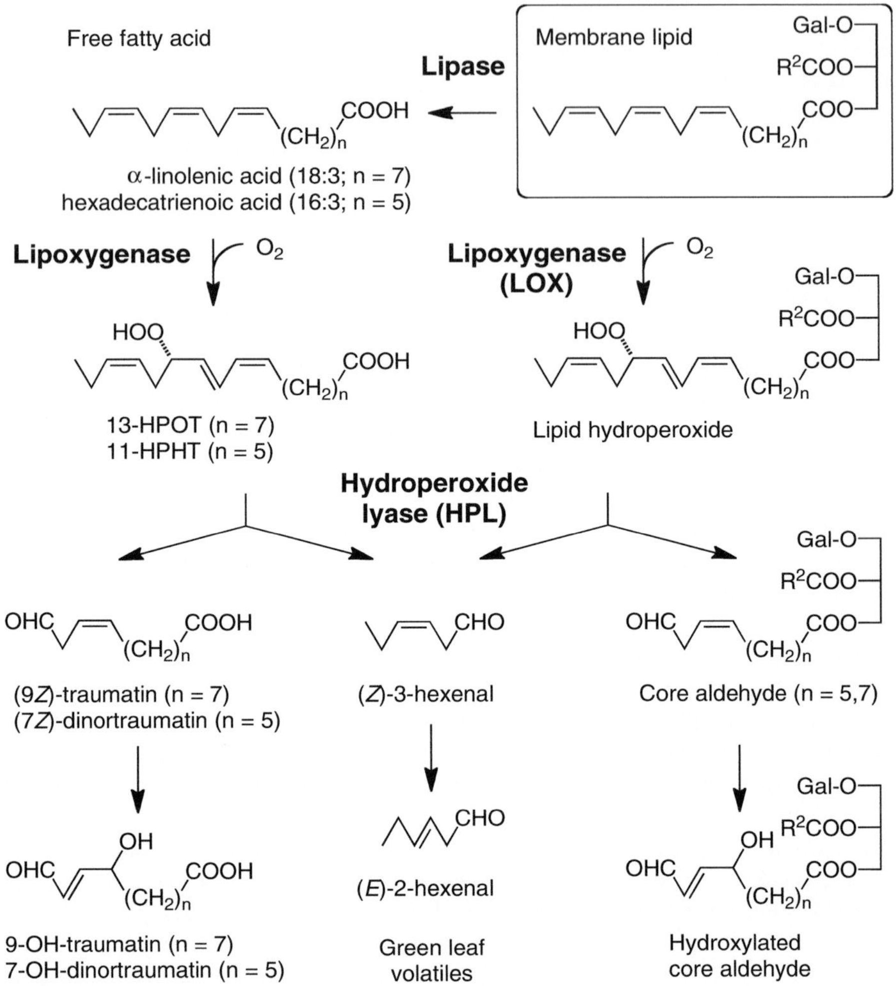

Fig. 17.2 Biosynthetic pathway to form green leaf volatiles (GLVs). It has been believed that a lipase is essential in forming free fatty acids from esterified lipids in the first committed step to form GLVs; however, recently we found that GLVs are formed even without the lipase reaction. In that case, lipoxygenase is the first enzyme that acts on the esterified lipids to form lipid hydroperoxides, which in turn is subsequently cleaved by hydroperoxide lyase to form the 6 carbon volatile aldehyde and 12 carbon oxo acid esterified to the glycerol backbone

Glauser et al. 2008, 2009). To avoid such competition, they should have different spatiotemporal expression patterns.

With a transgenic Arabidopsis harboring the GUS gene downstream of Arabidopsis HPL promoter, an intense expression of GUS activity was found in floral organs in a different manner than when the GUS activity was controlled by the Arabidopsis AOS promoter (Mwenda et al. 2015). The GUS activity under the control of the HPL promoter in intact cotyledons was low, but extensively enhanced after mechanical wounding, especially at their rims. However, high GUS activity under the control of the AOS promoter was detected in the vascular tissues of cotyledons (Kubigsteltig et al. 1999). Therefore, the promoter activities of HPL and AOS are distinctly spatially regulated after mechanical wounding, which may allow them to avoid competition. The distribution of the ability to form C6 volatiles roughly correlated with the profile of HPL promoter activity; however, the abilities showed little change after mechanical wounding. An inconsistency between the AOS promoter activity and JA levels was also evident. Accordingly, an additional factor, other than spatial segregation of HPL and AOS, also controls the GLV- and JA-forming abilities.

The most prominent feature of the HPL reaction is the rapid ‘burst’ of GLV formation after mechanical wounding. In intact leaf tissues, the amounts of GLVs are usually low, but after tissue disruption, the extensive formation of GLVs is induced. In Arabidopsis leaves, the amount of (*Z*)-3-hexenal went up to 1.5 μmol g FW^{-1} within 5 min after disruption (Matsui et al. 2012). This value corresponded to as much as ~30 % of the total amount of trienoic fatty acids in the leaf tissues. Because of the short period needed for the burst and the cells disruption, which made it impossible to activate the gene expression operating in intact cells, this burst should depend on enzymes and substrates that already exist in the cells. One possible explanation for the burst is a rapid mixing of enzymes and substrates; however, the situation is not that simple because some of the enzymes involved in GLV formation, such as lipoxygenases and HPLs, are usually localized in chloroplasts where their substrates, such as galactolipids, are abundant. A lipase that liberates free fatty acids from lipids might play a role in GLV formation by turning on the burst, but the fact that GLVs are formed even without free fatty acids (Nakashima et al. 2013) makes the possibility less likely. Another possibility is the segregation of enzymes and substrates at the cellular level, as found for glucosinolates and myrosinase in Brassicaceae plants (Kissen et al. 2009), but the distributions of LOX and HPL appear rather uniform in tissues when observed using their promoter::GUS constructs in reporter plants (Mwenda et al. 2015). Therefore, there must be a system to activate enzymes after tissue disruption. Such a regulatory mechanism is known for mammalian phospholipase A2 and lipoxygenases, and it has been shown that calcium ions play a significant role in the regulation (Murakami 2011). However, again, the disruption of leaf tissues in the presence of a calcium ion-chelating reagent, such as EGTA, showed no effect on the GLV formation rate. Apparently, the system to turn on the GLV burst is still unknown, and further studies are needed.

GLV formation is induced in occasions other than tissue disruption. When tomato plants were exposed to high temperatures (>46 °C), GLVs were extensively formed (Copolovici et al. 2012). Because several GLVs and their related carbonyl species, harboring α,β-unsaturated carbonyl groups, induce resistance to high temperature stress by inducing several genes involved in responses against abiotic stresses (Yamauchi et al. 2015), the formation of GLVs after heat stress may be a plant defense response. Under heat stress, plant cells generally encounter oxidative stress, which in turn, causes the deterioration of the membrane organization through oxidation of membrane components (Suzuki et al. 2012). This might induce GLV formation. GLV bursts following light-dark transitions have also been reported (Jardine et al. 2012). With this GLV burst, which was enhanced by darkening, there was a positive relationship between the amount of GLVs formed and the photosynthetic activity prior to darkening. This implies the involvement of a photosynthetic electron transport system in the GLV burst; however, details have not been elucidated. By investigating the mechanism behind the GLV burst that is elicited after darkening, insights into the regulatory mechanisms of GLV formation could be revealed.

Evolution of Genes Involved in GLV Formation

The CYP74 family is a class of enzymes, including HPL and AOS, that uses unsaturated fatty acid hydroperoxides derived from linoleic acid or α-linolenic acid as substrates. The CYP74B subfamily contains HPLs, which are widely distributed in higher plants (Grechkin 2002; Matsui 2006). Bell pepper HPL was the first CYP74B isolated in the 1990s, and since then an increasing number of HPL isoforms have been identified in the complete genome sequences of *Arabidopsis thaliana*, rice, and other plant species (Matsui et al. 1996; Bate et al. 1998). Recently, it has been reported that the moss *Physcomitrella patens* contains three CYP74 orthologs (Stumpe et al. 2006; Scholz et al. 2012). One of the three orthologs in moss has been identified as *bona fide* HPL and the others as AOSs. A database search revealed that liverwort *Marchantia* and Charophyta *Klebsormidium* encode two and one CYP74 genes, respectively (Table 17.1). However, they are AOSs, not HPLs (Koeduka et al. 2015). These observations raise the important questions of when and how an ancestral CYP74 may have functioned as an HPL. The investigation of the genetic context of CYP74 uncovered that the *Selaginella* genome includes ten CYP74 orthologs, whereas *Chlamydomonas* and *Nostoc* do not contain CYP74 genes.

Table 17.1 Survey of the CYP74 orthologs involved in green leaf volatiles

Plant species	Hydroperoxide lyase	CYP74 orthologs
Arabidopsis	+[a]	2
Rice	+	5
Amborella	–[b]	2
Selaginella	–	10
Physcomitrella	+	3
Marchantia	0	2
Klebsormidium	0	1
Chlamydomonas	0	0
Nostoc	0	0

[a]Hydroperoxide lyase genes were present in the public databases and biochemically characterized

[b]Hydroperoxide lyase orthologs were identified but have not been characterized

Thus, the acquisition of CYP74 genes may have occurred during the evolutionary process from *Chlamydomonas* to *Klebsormidium*, and then a CYP74 gene duplication event and the functional divergence of AOS to HPL, or vice versa, arose multiple times during plant evolution. This is a totally distinct view of the evolution of CYP74s from that proposed by Lee et al. (2008).

Physiological and Ecological Function of GLVs

Do Plants Sense GLVs?

Since the confirmation of GLVs as aroma constituents in green leaves, scientists have been considering their physiological functions. At the beginning, the effect of GLVs on insect behavior was investigated. For example, in 1967, Riddiford (1967) found that (*E*)-2-hexenal was an essential mating stimulant for polyphemus moths (*Antheraea polyphemus*). At present, the effects of just one GLV, (*E*)-2-hexenal, for example, acts as an attractant, allomone, kairomone, or pheromone on 186 different species of arthropods, which are listed in The Pherobase (www.pherobase.com). A review on the involvement of GLVs in plant–herbivore and plant–pathogen interactions was published recently (Scala et al. 2013). GLVs also exert several kinds of physiological effects on mammals. For example, *n*-hexanal increases the dopamine release from rat brains (Kako et al. 2011).

It has been repeatedly reported that the volatile organic chemicals formed by stress-treated plants induce several defense responses in neighboring plants (Dicke and Baldwin 2010). Because GLVs are common and abundant volatiles emitted from stressed plants, the effects of GLVs on plants were examined. Vapors from a series of alkenals and alkanals, including (*E*)-2-hexenal, induced the formation of phytoalexins, such as cadalene and scopoletin, in developing cotton bolls (Zeringue 1992). Among the aldehydes used in his study, (*E*)-2-alkenal induced a higher production of phytoalexins than saturated alkanals. Therefore, the structure of compounds might be important in eliciting the phytoalexin formation in cotton bolls. Aerial treatments of Arabidopsis seedlings with (*E*)-2-hexenal at 10 μM induced a subset of defense-related genes, including chalcone synthase, lipoxygenase, and AOS, all of which are somehow involved in the biosynthesis of secondary metabolites (Bate and Rothstein 1998). (*E*)-2-Hexenal and the other (*E*)-2-alkenals used to treat cotton bolls are reactive carbonyl species (RES) because they contain α,β-unsaturated carbonyl moieties that have the potential to inactivate biological molecules through a spontaneous reaction (Michael addition reaction) with nucleophilic substances containing amino or sulfhydryl groups. Therefore, it is assumed that (*E*)-2-alkenal induced defense responses because of the stress response elicited by the toxicity of (*E*)-2-alkenal. In the case of Arabidopsis, the response was observed in the *jar1-1* mutant that had a deficiency in JA signaling (Bate and Rothstein 1998), and thus, the response was thought to use a signaling system different from that used by JA. Even though α,β-unsaturated carbonyl moieties are important factors in eliciting defense responses in plants, (*Z*)-3-hexen-1-yl acetate, which is a much less harmful compound because of the absence of the α,β-unsaturated carbonyl moiety, was also effective to inducing the lipoxygenase gene in Arabidopsis seedlings. This implies that plants also sense some GLVs in a way other than as a response to toxic compounds containing α,β-unsaturated carbonyl groups.

We also found that treating with GLVs or an isoprenoid at 10 μM in the vapor phase elicited the induction of chalcone synthase, caffeic acid-*O*-methyltransferase, diacylglycerol kinase1, glutathione-*S*-transferase1, and lipoxygenase2 in Arabidopsis (Kishimoto et al. 2005). As a result of the induction of these defense-related genes, Arabidopsis acquired a higher resistance to the necrotrophic fungal pathogen, *Botrytis cinerea*. We also noticed that (*E*)-2-hexenal was a powerful elicitor, but at the same time, (*Z*)-3-hexenal, (*Z*)-3-hexen-1-ol, and *allo*-ocimene, also induced defense genes and resistance against *B. cinerea* to a level similar to or even higher than (*E*)-2-hexenal. The responses of Arabidopsis to these volatile compounds were partially suppressed in *jar1-1* and *etr1* mutants, thus, the involvement of JA signaling and ethylene signaling was hypothesized. The pre-treatment of Arabidopsis with okadaic acid also suppressed the response, which indicated the involvement of protein phosphatases in the system that sensed the volatiles. Treating with aerial GLVs also induced defense responses in lima bean (Arimura et al. 2000) and corn (Engelberth et al. 2004; Farag et al. 2005). In corn seedlings, GLVs primed the plants for the higher production of JAs and sesquiterpene volatiles after subsequent herbivore attacks (Engelberth et al. 2004).

Some claimed that the concentrations of volatiles used in these studies (10 μM in vapor phase, which corresponds to 224 ppmV) was unrealistically high and that the responses observed were not physiologically and ecologically relevant (Dicke et al. 2003). However, most green plants have the ability to form massive amounts of GLVs after mechanical wounding, and, in the case of Arabidopsis leaves, the local concentration in the disrupted leaf tissue could go up to 1 mM (in the aqueous phase in plant tissues) (Matsui et al. 2012). Also, volatiles usually diffuse into the atmosphere in a non-concentric manner as fragmented plumes directed by a turbulent flow (Baldwin et al. 2006). Therefore, it is possible for the plants to encounter relatively high concentrations of GLVs. Because it is difficult to simulate the diffusion or fate of volatiles once they are released into the atmosphere, the best way to know if the response is physiologically or ecologically relevant is to observe the effects of volatiles on plants in a natural environment or in an environment equivalent to nature.

When volatiles released by hybrid poplar (*Populus deltoids* × *P. nigra*) after herbivore damage were introduced into undamaged adjacent leaves, the exposed leaves had elevated defensive responses against feeding by gypsy moth larvae (*Lymantria dispar* L.) (Frost et al. 2008). Even though a detailed analysis of volatiles emitted by herbivore-damaged leaves was not performed in this study, it should contain substantial amounts of GLVs (Frost et al. 2008). Lima bean plants (*Phaseolus lunatus*) secrete extrafloral nectar to recruit carnivorous ants that feed on herbivores. Exposing intact lima bean plants to volatiles released from herbivore-damaged conspecifics induced the secretion of the extrafloral nectar (Kost and Heil 2006). For an interaction mediated by herbivore-induced volatiles, (*Z*)-3-hexen-1-yl acetate being the most abundant, in lima beans, their effects were reproduced using a low concentration of synthetic (*Z*)-3-hexen-1-yl acetate. Also, in an open chamber experiment conducted using transgenic *Nicotiana attenuata* that had a lowered HPL activity through an antisense technique, it was indicated that GLVs were responsible for the induction of some defense genes in plant–plant interactions (Paschold et al. 2006). We also observed that intact tomato plants acquired higher defense responses against herbivores after receiving volatiles released from damaged plants (Sugimoto et al. 2014). In the volatiles emitted from the herbivore-damaged tomato plants, (*Z*)-3-hexen-1-ol was the most abundant GLV. Because of these results, many scientists started to consider that plants perceive GLVs even under non-stressed conditions, and that, in some instances, the plants that sensed GLVs in their surrounding atmosphere changed their behavior to achieve a higher fitness in their environment.

How Do Plants Sense GLVs?

Perception as Toxic Xenobiotics

Even though the involvement of GLVs in plant–plant communication has been mostly established in section "Do plants sense GLVs?", we still do not know the mechanism behind how plants sense GLVs. To dissect the mechanism, it is better to separate members of the GLVs into two groups based on their chemical reactivity, RES and the others (Fig. 17.3). (*E*)-2-Hexenal is a representative RES-type GLV because it contains an α,β-unsaturated carbonyl group. (*E*)-2-Hexenal is formed by the isomerization of (*Z*)-3-hexenal, which is the first product of HPL. The ability to isomerize (*Z*)-3-hexenal to (*E*)-2-hexenal varies among plant species, and as a result, the ratio between the two hexenals is also different. The isomerization might have physiological and ecological relevance because the herbivore *Manduca sexta* has the isomerase in its oral secretion, and it decreases the (*Z*)/(*E*) ratio of GLVs formed by *Nicotiana attenuata* (Allmann and Baldwin 2010). The parasite of *M. sexta* is recruited by monitoring the (*Z*)/(*E*) ratio.

Because (*E*)-2-hexenal is a RES, its reactivity against biological substances is higher than the other GLVs, such as (*Z*)-3-hexenal, (*Z*)-3-hexen-1-ol, and (*Z*)-3-hexen-1-yl acetate. However, (*Z*)-3-hexenal is sensitive to enzymatic and spontaneous oxygenation to form 4-hydroperoxy-(*E*)-2-hexenal, which in turn, is converted into 4-hydroxy-(*E*)-2-hexenal or 4-oxo-(*E*)-2-hexenal (Matsui et al. 2012; Bonaventure et al. 2011) (Fig. 17.3). Because of this, (*Z*)-3-hexenal should also be considered as a 'potential' RES-type GLV.

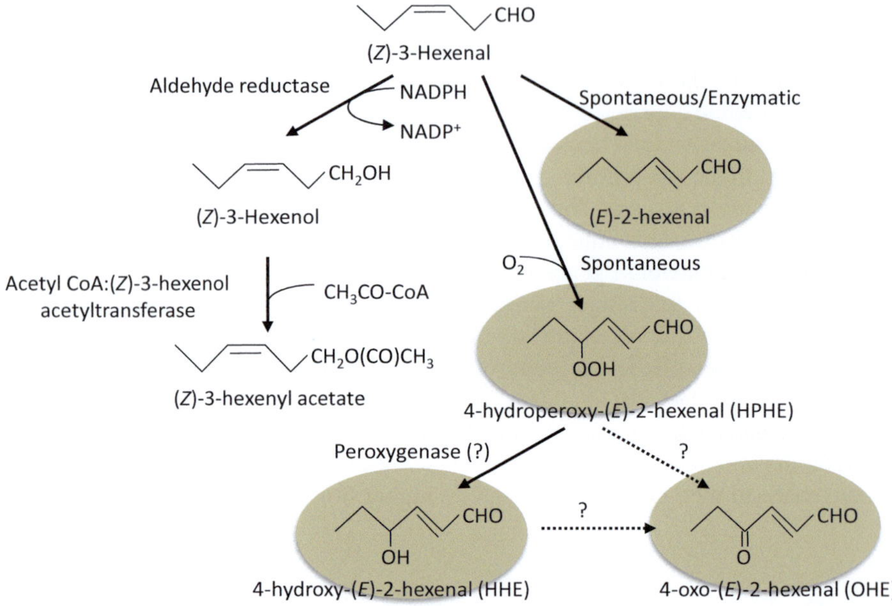

Fig. 17.3 (*Z*)-3-Hexenal is a 'potential' reactive electrophile species (RES)-type green leaf volatile (GLV). Because (*Z*)-3-hexenal does not contain a α,β-unsaturated carbonyl group, it is not a member of the RES-type GLVs. However, spontaneous and enzymatic isomerization, or oxygenation, results in the formation of RES-type structures as shown in the chemical structures with the *gray* background

12-Oxophytodienoic acid (OPDA) functions not only as a precursor of JA, but also as a signal molecule per se, inducing a distinct set of genes other than those induced by the JA (Taki et al. 2005). Phytoprostanes, which are formed non-enzymatically by the reaction of polyunsaturated fatty acid with reactive oxygen species (ROS), are structurally related to the OPDA group, especially because of the presence of the α,β-unsaturated carbonyl group found in the conjugated cyclopentenone group. Phytoprostanes also induce a subset of genes in plants, and it is assumed that this ability is largely because of their α,β-unsaturated carbonyl group. In Arabidopsis, TGA transcription factors are essential factors in mediating the response to OPDA and phytoprostanes (Mueller et al. 2008). Additionally, cyclophilin 20-3 in plastids was identified as a protein capable of binding OPDA, relaying the signal to induce gene expression (Park et al. 2013). Because the α,β-unsaturated carbonyl group is the essential structural requirement for the induction of this distinct signal transduction pathway, it is probable that GLVs having α,β-unsaturated carbonyl groups also activate the same signaling pathway.

Another important (bio)chemical aspect of RES is its high reactivity with nucleophiles in cells. Glutathione (GSH) reacts with (*E*)-2-hexenal either enzymatically using glutathione *S*-transferases or non-enzymatically at a physiological pH like 7.0. In tobacco leaves where hypersensitive responses were elicited by treating with cryptogein, several species of GSH adducts, such as OPDA-GSH, ketooctadecadienoic acid-GSH, 4-hydroxy-(*E*)-2-nonenal-GSH, and hexenal-GSH, are formed, probably as a detoxification reaction, to cope with potentially active and even toxic substances (Davoine et al. 2006). Among these adducts, the GSH adduct with (*E*)-2-hexenal accumulated to the highest level, as much as 100 nmol g^{-1} dry weight. When plants were exposed to volatile compounds having α,β-unsaturated carbonyl species, the volatiles would be partitioned into the plant tissues under the gas phase/water phase (inside of tissues) equilibrium determined by the Henry's law. Even though the partition would be much more complicated because volatiles could cross cell walls in apoplasts and plasma membranes, this indicated that it is inevitable for plants to accumulate such reactive xenobiotics in cells during normal gas-exchanges in photosynthesis through stomata. If the compounds were not appropriately detoxified, they would react with GSH to form adducts. Accordingly, endogenous GSH would be consumed, and the consumption of GSH would lead to an imbalance in the redox state (GSH/GSSG), which might cause the induction of several genes under redox regulation.

The involvement of GSH in responding to (*E*)-2-hexenal and (*Z*)-3-hexenal was confirmed using the *pad2-1* Arabidopsis mutant that had a lower level of GSH because of a defect in GSH biosynthesis. The expression of *PDF1.2* elicited by hexenals was not affected by the *pad2-1* mutation, but that of *VSP1* was totally eliminated in the mutant (Kishimoto et al. 2006). This, again, indicated that redox regulation, mediated by GSH, should play a crucial role in responding to GLVs containing α,β-unsaturated carbonyl groups.

Because the reactivity of RES is based on a simple chemical reaction, it is also possible for RES to react with the other biological components, such as proteins, nucleotides, and membrane components. Of course, there might be a distinct set of molecules that show a specificity, depending on the chemical nature of RES, and nucleophiles that might be involved in the reaction (Mano 2012; Mano et al. 2014). For GLVs containing α,β-unsaturated carbonyl groups, their moderate hydrophobicity allows them accumulate in a hydrophobic environment, such as a biological membrane. If this was the case, then a protein in the membrane would be a primary target for the GLVs. Also, the nucleophilicity of residues on the surface of biological molecules would be a determinant for the target. Signaling mediated by covalent binding is well studied in mammalian systems, and the Nrf/Keap1 system plays a central role (Higdon et al. 2012). There has been no report of Nrf/Keap1 homologs in plants and our preliminary BLAST search also failed to find the counterparts in plants, although it is still possible for plants to use a similar, covalent binding-mediated signaling pathway.

Interactions with Membranes

When volatiles released from herbivore-damaged tomato plants were blown across the surface of intact tomato leaves, their plasma membrane potential jumped to a depolarized value (from −114 to −76 mV) within 10 or 20 s (Zebelo et al. 2012).

GLVs were the most abundant volatiles released by the herbivore-infested tomato. As expected, a similar membrane potential depolarization was observed by blowing vapors of (*Z*)-3-hexenal, (*E*)-2-hexenal, or (*Z*)-3-hexen-1-yl acetate. (*Z*)-3-Hexen-1-yl acetate, even though it is not the RES-type, causes depolarization as much as, or even to higher degree than, (*E*)-2-hexenal, a RES, or (Z)-3-hexenal, a potential RES. The reactivity caused by the α,β-unsaturated carbonyl group should not be the prerequisite of depolarization.

It was expected that the volatile affected the nature of ion channels located in the plasma membrane. The α,β-unsaturated carbonyl group might be partially involved in the modulation of ion channels, but volatiles without the reactive moiety could induce almost the same depolarization, which indicates that the moiety does not play the critical role. These GLVs also induced a calcium influx into the cytosols of epidermal cells (Zebelo et al. 2012). The ability to promote calcium influx apparently varied depending on the molecular species used, and (*Z*)-3-hexen-1-yl acetate was the most potent. Using fluorescence microscopy, the integrity of chloroplasts in parenchymal cells could also be examined because of the autofluorescence of chlorophyll. Interestingly, exposure to a gas containing (*E*)-2-hexenal or (*Z*)-3-hexenal seemed to be toxic to the cells, and the integrity of the chloroplast was extensively damaged by the treatment. However, the (*Z*)-3-hexen-1-yl acetate treatment resulted in no apparent change in the chloroplasts' integrity. From this observation, again, it is proposed that the reactivity belonging to α,β-unsaturated carbonyl group does not always correlated to the ability to induce a calcium influx into the cytoplasm.

Heil et al. (2008) investigated the structure-activity relationship of a series of alka(e)nyl acetates of different chain lengths, and different degrees and positions of unsaturation by monitoring the amount of extrafloral nectar secreted by lima bean plants after exposure to the vapor. They found a relatively broad spectrum of activity for each compound, and apparently there was no essential structural requirement to exert the effect. Accordingly, they proposed that a physicochemical processes based on the amphiphilic nature of the compound was the important factor for the effect. Because plasma membranes should be the first site of penetration for exogenously supplied volatiles, the deposition of volatiles in the plasma membrane would cause the effect by distorting the membrane's organization. Alternatively, it is possible that some proteins, such as odorant binding proteins or lipid transfer proteins, are involved in the system.

Metabolism

Most plants also have the ability to detoxify reactive xenobiotics such as RES. Cucumber plants and Arabidopsis contain several reductases that detoxify them. An alkenal/one oxidoreductase (AOR) catalyzes the reduction of the α,β-unsaturated bond in RES (Yamauchi et al. 2011). Aldo-keto reductase (AKR) and aldehyde reductase catalyze the reduction of aldehyde to alcohol. In cucumber, acrolein was efficiently reduced to form propionaldehyde by two distinct AORs localized in the chloroplasts and cytosol, respectively. Yamauchi et al. (2011) also detected the reducing activity to form alcohol from aldehyde. For example, Arabidopsis has both AKR and aldehyde reductase in its chloroplasts. When intact Arabidopsis was exposed to a vapor containing (*Z*)-3-hexenal, the aldehyde was taken up by tissues, efficiently converted into (*Z*)-3-hexen-1-ol in a NADPH-dependent manner, and released in its alcohol form into the atmosphere (Matsui et al. 2012). This reduction is also an important detoxification system to Arabidopsis, where the excess amount of (*Z*)-3-hexenal above the reducing capacity of Arabidopsis resulted in the suppression of PSII activity estimated by PAM. This implies that exposing plant tissues to (*Z*)-3-hexenal at a level where the tissues employ a reductive detoxification system as much as they can would cause an imbalance in the redox state maintained by the NADPH/NADP$^+$ ratio. The imbalance caused by the volatile might result in the modulation of gene expression levels in tissues. This might be a scenario caused by GLVs during plant–plant interactions.

Even though the alcohol, (*Z*)-3-hexen-1-ol, is a rather inert compound, it is sometimes converted into (*Z*)-3-hexen-1-yl acetate, which is more volatile than (*Z*)-3-hexen-1-ol according to their Henry's law constants (25 and 1 M atm^{-1}, respectively). Because this conversion requires acetyl-CoA, which would otherwise be used in growth, this definitely is an active process. This suggests that (*Z*)-3-hexen-1-ol is also an active compound, and a surplus supply during the reduction of (*Z*)-3-hexenal or from the atmosphere would cause an imbalance in the homeostasis of plant cells (Farag et al. 2005). When tomato plants were exposed to volatile compounds released from herbivore-damaged conspecifics, the receiver plants adsorb (*Z*)-3-hexen-1-ol, and converted it to a glycoside, (*Z*)-3-hexenyl vicianocide (Fig. 17.4). Because the glycoside has a slight but distinct activity to suppress the growth of herbivore, the glycosidation and subsequent accumulation of the compound from

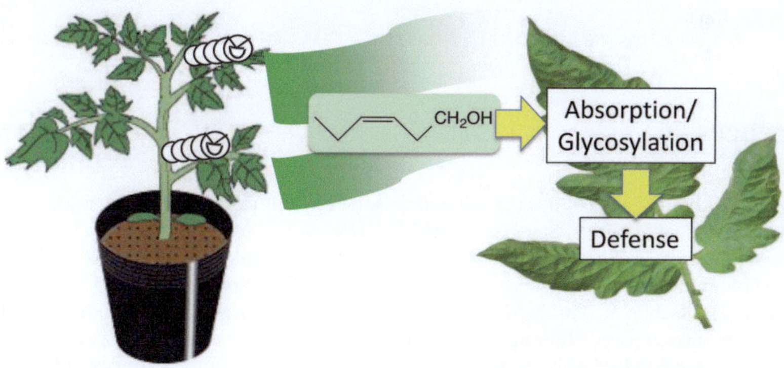

Fig. 17.4 Glycosidation is one way to 'sense' green leaf volatiles (GLVs). A herbivore-damaged tomato plant releases (*Z*)-3-hexen-1-ol into atmosphere. The neighboring plant takes up the GLV and converts it to (*Z*)-3-hexenyl vicianoside. Because the glycoside has an insecticidal activity, the 'receiver' plant acquires a higher resistance against the herbivore. Even though this is the metabolism of the plant responding to an exogenous compound, as a whole, this metabolism can be considered as an example of plant–plant interaction

the atmosphere benefitted the receiver plants. Therefore, glycosidation could be one way for tomato plants to carry out plant–plant communications. All flowering plants examined so far accumulated glycosides after exposure to vapor containing (*Z*)-3-hexen-1-ol. Therefore, the perception of volatiles through glycosidation is a general property of plants.

Concluding Remarks

Because of their abundance in nature, GLVs are familiar plant secondary metabolites. We enjoy their smells when we eat foods made from plants or when we walk through forests. Of course, plants form GLVs because it is beneficial. Since the discovery of the volatile sex pheromones of silkworms, we have been accumulating examples of the physiological and ecological functions of volatiles. GLVs are largely involved in the defense responses of plants against herbivores and pathogens, or even abiotic stresses. We are interested in when plants first acquired the ability to form GLVs because this knowledge would give us insights into how plants survive in environments where they interact with the other organisms. Also, the recent finding that plants respond to GLVs opens many areas to further study. Because they lack the nervous and olfactory systems of animals, plants may employ their own original systems to 'sense' GLVs. Chemical reactivity, physicochemical nature, and the nature of metabolization are the ways plants detect GLVs, but these fail to explain the phenomena found in nature completely. We are still beginning to decipher how plants 'sense' volatiles, and further extensive studies await.

Acknowledgement This study was partly supported by a Grant-in-aid for Scientific Research (K.M., 25282234 and 90199729) from the Ministry of Education, Culture, Sports, Science and Technology, Japan.

References

Allmann S, Baldwin IT (2010) Insects betray themselves in nature to predators by rapid isomerization of green leaf volatiles. Science 329:1075–1078

Arimura G, Ozawa R, Shimoda T, Nishioka T, Boland W, Takabayashi J (2000) Herbivory-induced volatiles elicit defense gene in lima bean leaves. Nature 406:512–515

Baldwin IT, Halitschke R, Paschold A, von Dahl CC, Preston CA (2006) Volatile signaling in plant-plant interactions: "Talking trees" in the genomics era. Science 311:812–815

Bate NJ, Rothstein SJ (1998) C6-volatiles derived from the lipoxygenase pathway induce a subset of defense-related genes. Plant J 16:561–569

Bate NJ, Sivasankar S, Moxon C, Riley JM, Thompson JE, Rothstein SJ (1998) Molecular characterization of an Arabidopsis gene encoding hydroperoxide lyase, a cytochrome P-450 that is wound inducible. Plant Physiol 117:1393–1400

Bonaventure G (2014) Lipases and the biosynthesis of free oxylipins in plants. Plant Signal Behav 3:9

Bonaventure G, Schuck S, Baldwin IT (2011) Revealing complexity and specificity in the activation of lipase-mediated oxylipin biosynthesis: a specific role of the *Nicotiana attenuata* GLA1 lipase in the activation of jasmonic acid biosynthesis in leaves and roots. Plant Cell Environ 34:1507–1520

Copolovici L, Kännaste A, Pazouki L, Niinemets Ü (2012) Emissions of green leaf volatiles and terpenoids from *Solanum lycopersicum* are quantitatively related to the severity of cold and heat shock treatments. J Plant Physiol 169:664–672

Curtius T, Franzen H (1912) Über die chemischen Bestandteile grüner Pflanzen. Über den Blätteraldehyd. Justus Liebigs Annalen der Chemie 390:89–121

Davoine C, Falletti O, Douki T, Iacazio G, Ennar N, Montillet JL, Triantaphylides C (2006) Adducts of oxylipin electrophiles to glutathione reflect a 13 specificity of the downstream lipoxygenase pathway in the tobacco hypersensitive response. Plant Physiol 140:1484–1493

Dicke M, Baldwin IT (2010) The evolutionary context for herbivore-induced plant volatiles: beyond the 'cry for help'. Trends Plant Sci 15:167–175

Dicke M, Agrawal AA, Bruin J (2003) Plants talk, but are they deaf? Trends Plant Sci 9:403–405

Engelberth J, Alborn HT, Schmelz EA, Tumlinson JH (2004) Airborne signals prime plants against insect herbivore attack. Proc Natl Acad Sci U S A 101:1781–1785

Farag MA, Fokar M, Abd H, Zhang H, Allen RD, Paré PW (2005) (*Z*)-3-Hexenol induces defense genes and downstream metabolites in maize. Planta 220:900–909

Froehlich JE, Itoh A, Howe GA (2001) Tomato allene oxide synthase and fatty acid hydroperoxide lyase, two cytochrome P450s involved in oxylipin metabolism, are targeted to different membranes of chloroplast envelope. Plant Physiol 125:306–317

Frost CJ, Mescher MC, Dervinis C, Davis JM, Carlson JE, De Moraes CM (2008) Priming defense genes and metabolites in hybrid poplar by the green leaf volatile *cis*-3-hexenyl acetate. New Phytol 180:722–734

Glauser G, Grata E, Dubugnon L, Rudaz S, Farmer EE, Wolfender J-L (2008) Spatial and temporal dynamics of jasmonate synthesis and accumulation in Arabidopsis in response to wounding. J Biol Chem 283:16400–16407

Glauser G, Dubugnon L, Mousavi SA, Rudaz S, Wolfender J-L, Farmer EE (2009) Velocity estimates for signal propagation leading to systemic jasmonic acid accumulation in wounded Arabidopsis. J Biol Chem 284:34506–34513

Grechkin AN (2002) Hydroperoxide lyase and divinyl ether synthase. Prostaglandins Other Lipid Mediat 68:457–470

Grechkin AN, Brühlmann F, Mukhtarova LS, Gogolev YV, Hamberg M (2006) Hydroperoxide lyases (CYP74C and CYP74B) catalyze the hemolytic isomerization of fatty acid hydroperoxides into hemiacetals. Biochim Biophys Acta 1761:1419–1428

Heil M, Lion U, Boland W (2008) Defense-inducing volatiles: in search of the active motif. J Chem Ecol 34:601–604

Higdon A, Diers AR, Oh JY, Landar A, Darley-Usmar VM (2012) Cell signaling by reactive lipid species: new concepts and molecular mechanisms. Biochem J 442:453–464

Jardine K, Barron-Gafford GA, Norman JP, Abrell L, Monson RK, Meyers KT, Pavao-Zuckerman M, Dontsova K, Kleist E, Werner C, Huxman TE (2012) Green leaf volatiles and oxygenate metabolite emission bursts from mesquite branches following light-dark transitions. Photosynth Res 113:321–333

Kako H, Kobayashi Y, Yokogoshi H (2011) Effects of n-hexanal on dopamine release in the striatum of living rats. Eur J Pharmacol 651:77–82

Kishimoto K, Matsui K, Ozawa R, Takabayashi J (2005) Volatile C6-aldehydes and allo-ocimene activate defense genes and induce resistance against *Botrytis cinerea* in *Arabidopsis thaliana*. Plant Cell Physiol 46:1093–1102

Kishimoto K, Matsui K, Ozawa R, Takabayashi J (2006) ETR1-, JAR1- ad PAD2-dependent signaling pathways are involved in C6-aldehyde-induced defense responses of Arabidopsis. Plant Sci 171:415–423

Kissen R, Rossiter JT, Bones AM (2009) The 'mustard oil bomb': not so easy to assemble?! Localization, expression and distribution of the components of the myrosinase enzyme system. Phytochem Rev 8:69–86

Koeduka T, Ishizaki K, Mwenda CM, Hori K, Sasaki-Sekimoto Y, Ohta H, Kohchi T, Matsui K (2015) Biochemical characterization of allene oxide synthases from the liverwort *Marchantia polymorpha* and green microalgae *Klebsormidium flaccidum* procides insight into the evolutionary divergence of the plant CYP74 family. Planta 242:1175–1186

Kost C, Heil M (2006) Herbivore-induced plant volatiles induce an indirect defence in neighbouring plants. J Ecol 94:619–628

Kubigsteltig I, Laudert D, Weiler E (1999) Structure and regulation of *Arabidopsis thaliana* allene oxide synthase gene. Planta 208:463–471

Lee DS, Nioche P, Hamberg M, Raman CS (2008) Structural insights into the evolutionary paths of oxylipin biosynthetic enzymes. Nature 455:363–368

Mano J (2012) Reactive carbonyl species: their production from lipid peroxides, action in environmental stress, and the detoxification mechanism. Plant Physiol Biochem 59:90–97

Mano J, Nagata M, Okamura S, Shiraya T, Mitsui T (2014) Identification of oxidatively modified proteins in salt-stressed Arabidopsis: a carbonyl-targeted proteomics approach. Plant Cell Physiol 55:1233–1244

Matsui K (2006) Green leaf volatiles. Hydroperoxide lyase pathway of oxylipin metabolism. Curr Opin Plant Biol 9:274–280

Matsui K, Shibutani M, Hase T, Kajiwara T (1996) Bell pepper fruit fatty acid hydroperoxide lyase is a cytochrome P450 (CYP74B). FEBS Lett 394:21–24

Matsui K, Sugimoto K, Mano J, Ozawa R, Takabayashi J (2012) Differential metabolisms of green leaf volatiles in injured and intact parts of a wounded leaf meet distinct ecophysiological requirements. PLoS ONE 7:e36433. doi:10.1371/journal.pone.0036433

Mita G, Quarta A, Fasano P, De Paolis A, Di Sansebastiano GP, Perrotta C, Iannacone R, Belfield E, Hughes R, Tsesmetzis N et al (2005) Molecular cloning and characterization of an almond 9-hydroperoxide lyase, a new CYP74 targeted to lipid bodies. J Exp Bot 56:2321–2333

Mueller S, Hilbert B, Dueckersjhoff K, Roitsch T, Krischke M, Mueller M, Berger S (2008) General detoxification and stress responses are mediated by oxidized lipids through TGA transcription factors in Arabidopsis. Plant Cell 20:768–785

Murakami M (2011) Lipid mediators in life science. Exp Anim 60:7–20

Mwenda CM, Matsuki A, Nishimura K, Koeduka T, Matsui K (2015) Spatial expression of the Arabidopsis hydroperoxide lyase is controlled differently from that of the allene oxide synthase gene. J Plant Interact 10:1–10

Nakashima A, von Reuss SH, Tasaka H, Nomura M, Mochizuki S, Iijima Y, Aoki K, Shibata D, Boland W, Takabayashi J, Matsui K (2013) Traumatin- and dinortraumatin-containing galactolipids in *Arabidopsis*: their formation in tissue-disrupted leaves as counterparts of green leaf volatiles. J Biol Chem 288:26078–26088

Noordermeer MA, Van Dijken AJ, Smeekens SC, Veldink GA, Vliegenthart JF (2000) Characterization of three cloned and expressed 13-hydroperoxide lyase isoenzymes from alfalfa with unusual N-terminal sequences and different enzyme kinetics. Eur J Biochem 267:2473–2482

Park SW, Li W, Viehhauser A, He B, Kim S, Nilsson AK, Andersson MX, Kittle JD, Ambavaram MMR, Luan S, Esker AR, Tholl D, Cimini D, Ellerström M, Coaker G, Mitchell TK, Pereira A, Dietz KJ, Lawrence CB (2013) Cyclophilin 20-3 relays a 12-oxo-phytodienoic acid signal during stress responsive regulation of cellular redox homeostasis. Proc Natl Acad Sci U S A 110:9559–9564

Paschold A, Halitschke R, Baldwin IT (2006) Using 'mute' plants to translate volatile signals. Plant J 45:275–291

Phillips DR, Galliard T (1978) Flavor biogenesis: partial purification and properties of a fatty acid hydroperoxide cleaving enzyme from fruits of cucumber. Phytochemistry 17:355–358

Riddiford LM (1967) Trans-2-hexenal: mating stimulant for polyphemus moths. Science 157:139–141

Savchenko T, Kolla VA, Wang CQ, Nasafi Z, Hicks DR, Phadungchob B, Chehab WE, Brandizzi F, Froehlich J, Dehesh K (2014) Functional convergence of oxylipin and abscisic acid pathways controls stomatal closure in response to drought. Plant Physiol 164:1151–1160

Scala A, Allman S, Mirabella R, Haring MA, Schuurink RC (2013) Green leaf volatiles: a plant's multifunctional weapon against herbivores and pathogens. Int J Mol Sci 14:17781–17911

Scholz J, Brodhun F, Hornung E, Herrfurth C, Stumpe M, Beike AK, Faltin B, Frank W, Reski R, Feussner I (2012) Biosynthesis of allene oxides in Physcomitrella patens. BMC Plant Biol 12:228

Shibata Y, Matsui K, Kajiwara T, Hatanaka A (1995) Purification and properties of fatty acid hydroperoxide lyase from green bell pepper fruits. Plant Cell Physiol 36:147–156

Stumpe M, Bode J, Göbel C, Wichard T, Schaaf A, Frank W, Frank M, Reski R, Pohnert G, Feussner I (2006) Biosynthesis of C9-aldehydes in the moss Physcomitrella patens. Biochim Biophys Acta 1761:301–312

Sugimoto K, Matsui K, Iijima Y, Akakabe Y, Muramoto S, Ozawa R, Uefune M, Sasaki R, Alamgir KM, Akitake S et al (2014) Intake and transformation to a glycoside of (*Z*)-3-hexenol from infested neighbors reveals a mode of plant odor reception and defense. Proc Natl Acad Sci U S A 111:7144–7149

Suzuki N, Koussevitzky S, Mittler R, Miller G (2012) ROS and redox signaling in the response of plants to abiotic stress. Plant Cell Environ 35:259–270

Taki N, Sasaki-Sekimoto Y, Obayashi T, Kikuta A, Kobayashi K, Ainai T, Yagi K, Sakurai N, Suzuki H, Masuda T, Takamiya K, Shibata D, Kobayashi Y, Ohta H (2005) 12-Oxo-phytodienoic acid triggers expression of a distinct set of genes and plays a role in wound-induced gene expression in Arabidopsis. Plant Physiol 139:1268–1283

Yamauchi Y, Hasegawa A, Taninaka A, Mizutani M, Sugimoto Y (2011) NADPH-dependent reductases involved in the detoxification of reactive carbonyls in plants. J Biol Chem 286:6999–7009

Yamauchi Y, Kunishima M, Mizutani M, Sugimoto Y (2015) Reactive short-chain leaf volatiles act as powerful inducers of abiotic stress-related gene expression. Sci Rep 5:8030. doi:10.1038/srep08030

Zebelo A, Matsui K, Ozawa R, Maffei ME (2012) Plasma membrane potential depolarization and cytosolic calcium flux are early events involved in tomato (*Solanum lycopersicon*) plant-to-plant communication. Plant Sci 196:93–100

Zeringue HJ Jr (1992) Effects of C6–C10 alkenals and alkanals on eliciting a defence response in the developing cotton boll. Phytochemistry 31:2305–2308

Part III
Lipids in Industrial Application

Chapter 18
Omics in *Chlamydomonas* for Biofuel Production

Hanna R. Aucoin*, Joseph Gardner*, and Nanette R. Boyle

Abstract In response to demands for sustainable domestic fuel sources, research into biofuels has become increasingly important. Many challenges face biofuels in their effort to replace petroleum fuels, but rational strain engineering of algae and photosynthetic organisms offers a great deal of promise. For decades, mutations and stress responses in photosynthetic microbiota were seen to result in production of exciting high-energy fuel molecules, giving hope but minor capability for design. However, '-omics' techniques for visualizing entire cell processing has clarified biosynthesis and regulatory networks. Investigation into the promising production behaviors of the model organism *C. reinhardtii* and its mutants with these powerful techniques has improved predictability and understanding of the diverse, complex interactions within photosynthetic organisms. This new equipment has created an exciting new frontier for high-throughput, predictable engineering of photosynthetically produced carbon-neutral biofuels.

Keywords Nutrient limitation • Triacylglycerols • Biohydrogen • Stress response • Gene discovery

Abbreviations

Enzymes

ACCase	Acetyl-CoA carboxylase, E.C. 6.4.1.2
ACK1	Non-specific protein-tyrosine kinase, E.C. 2.7.10.2
ACK2	(Acetyl-CoA carboxylase) kinase 2, E.C. 2.7.11.27
ADH1	Alcohol dehydrogenase 1/Formaldehyde dehydrogenase (FDH1), E.C. 1.1.1.1

---

*These authors contributed equally.

H.R. Aucoin • J. Gardner • N.R. Boyle (✉)
Department of Chemical and Biological Engineering, Colorado School of Mines, Golden, CO 80401, USA
e-mail: nboyle@mines.edu

Y. Nakamura, Y. Li-Beisson (eds.), *Lipids in Plant and Algae Development*, Subcellular Biochemistry 86, DOI 10.1007/978-3-319-25979-6_18

ADH2	Alcohol dehydrogenase (NADP+), E.C. 1.1.1.2
ADH3	Alcohol dehydrogenase Isoform, E.C. 1.1.1.1
ADP-Glc PPase	ADP-glucose pyrophosphorylase, E.C. 2.7.7.27
DGAT	Diacylglycerol acyltransferase, E.C. 2.3.1.20
FAS	Fatty acid synthase, E.C. 2.3.1.85
GPAT	Glycerol-3-phosphate acyltransferase, E.C. 2.3.1.15
HYD1	Ferredoxin hydrogenase, E.C. 1.12.7.2
HYD2	Ferredoxin hydrogenase Isoform, E.C. 1.12.7.2
ICL	Isocitrate lyase, E.C. 4.1.3.1
LDH	Lactate dehydrogenase, E.C. 1.1.1.27
LPAT	Lysophosphatidic acid acyltransferase, E.C. 2.3.1.51
MCT	Malonyl acyl carrier protein transferase, E.C. 2.3.1.39
MLDP	Major lipid droplet protein
PAT1	Phosphate acetyltransferase, E.C. 2.3.1.8
PAT2	Phosphate acetyltransferase Isoform, E.C. 2.3.1.8
PDC3	Pyruvate decarboxylase, E.C. 4.1.1.1
PDAT	Phospholipid diacylglycerol acyltransferase, E.C. 2.3.1.158
PDH	Pyruvate dehydrogenase, E.C. 1.2.1.51
PFL1	Formate C-acetyltransferase, E.C. 2.3.1.54

Transcripts

DGAT1	Diacylglycerol acyltransferase
DGTT1	Diacylglycerol acyltransferase
GPD2	Glycerol-3-phosphate dehydrogenase
GPD4	Glycerol-3-phosphate dehydrogenase
LHCBM9	Light Harvesting Complex Chlorophyll a-b binding protein
LHCSR2	Light Harvesting Complex stress-related 2
LPAT1	Lysophosphatidic acid acyltransferase
MLDP1	Major lipid droplet protein
PDAT1	Phospholipid diacylglycerol acyltransferase

Molecules

DAG	Diacylglycerol
FA	Fatty acid
FDox	Ferredoxin (oxidized)
FDred	Ferredoxin (reduced)
G3P	Glycerol-3-phosphate
PFR1ox	Pyruvate ferredoxin (oxidized)
PFR1red	Pyruvate ferredoxin (reduced)
SQDG	Sulfoquinovovosyl diacylglycerol
TAG	Triacylglycerol

Structures and Proteins

$Cytb_6$	Cytochrome b6 (small subunit of cytochrome b6f complex)
Cytf	Cytochrome f (large subunit of cytochrome b6f complex)
Fd	Ferredoxin
FNR	Ferredoxin – NADP(+) reductase
LHC	Light Harvesting Complex
NDH	Plastidial NAD(P)H dehydrogenase complex
P680	Chlorophyll a P680 (680 nm)
P700	Chlorophyll a P700 (700 nm)
Pc	Plastocyanin
$PQ(H)_2$	Plastoquinone (reduced)
PSI	Photosystem I
PSII	Photosystem II
PQ_0	Plastoquinone (oxidized)

Miscellaneous

2-DE	2-Dimensional Gel Electrophoresis
GC/MS	Gas Chromatography/Mass Spectroscopy
HS	Heat Shock
LC/MS	Liquid Chromatography/Mass Spectroscopy
MALDI	Matrix-Assisted Laser Desorption/Ionization
NMR	Nuclear Magnetic Resonance
MFA	Metabolite Flux Analysis
TOF	Time of Flight

Introduction

For long term sustainability, there is a pressing need to develop renewable sources of fuels which are efficient and compatible with current infrastructure. There are a number of alternative energy sources including wind, solar and hydroelectric which can be used to replace our dependence on fossil fuels for electricity; however, the ability to convert these energy sources to a form easily used for transportation is neither straightforward nor energy efficient. Biomass derived biofuels are particularly well-suited to displace some (or all) of our dependency on fossil fuels for the transportation sector because metabolites in the cell, such as starch or lipids, can easily be converted into fuels using current technologies. Advances in molecular biology and synthetic biology have allowed researchers to redesign metabolism and tailor-make molecules for fuel production (Oliver et al. 2013, 2014; Yim et al. 2011; Li and Liao 2013; Nakamura and Whited 2003; Kurian 2005). Most currently available biofuels are produced in *E. coli* or yeast using starch or cellulosic sugars as feedstocks. Another alternative, which has the potential to be carbon neutral, is the production of biofuels from photosynthetic microorganisms. Microalgae naturally accumulate lipids and starch or produce hydrogen when certain nutrients are depleted in the media (Miller et al. 2010; Li et al. 2010a; Moellering and Benning 2010; Blaby et al. 2013; Boyle et al. 2012; Timmins et al. 2009), making them particularly attractive as sources of biofuels. The exact mechanisms that control the accumulation or production of these important fuel precursors are still being investigated but great strides have been made to understand these mechanisms using '-omics' technologies.

With advances in analytical technologies, it is now possible to collect large amounts of data to characterize both model and non-model organisms. These data sets are coined '-omics'; meant to imply "quantification of a whole" (Jamers et al. 2009), '-omics' are divided into observing cellular behavior at different levels of regulation: predominantly at the genomic (DNA), transcriptomic (RNA), proteomic (protein/enzyme), and metabolomic (metabolism) levels. By comparing data sets from different growth conditions, environmental stimuli, or genetic backgrounds, a more complete picture of how the cell responds to different stimuli can be gained. Since little is known about the genes and regulatory mechanisms which control the pathways involved in fuel-relevant molecule production, '-omics' analyses are needed. In the discussion that follows, we will describe how '-omics' technologies have enabled advances in the understanding of biofuel production in algae, particularly *Chlamydomonas reinhardtii*.

Chlamydomonas reinhardtii *as a Model Organism*

Chlamydomonas reinhardtii, a soil-dwelling, unicellular green alga, has served as a model organism since its discovery and isolation 70 years ago. *C. reinhardtii* is particularly interesting because it can function photoautotrophically, mixotrophically, or heterotrophically on acetate while maintaining its photosynthetic apparatus (Kempa et al. 2009; Chen and Johns 1996). Its flexible metabolism is complemented by a complex structure reminiscent of higher order plants: it has multiple mitochondria, a chloroplast, flagella (Harris and Stern 2009), and a cell wall of extensin-like hydroxyproline-rich glycoproteins (Woessner and Goodenough 1994). There is also an emerging toolbox of molecular techniques for genetic manipulation of *C. reinhardtii* and libraries of knock-out mutants are available (Gonzalez-Ballester et al. 2011). The availability of knock out libraries enables further research into gene/function relationships. One of the most significant mutants to be isolated is the cell wall-less mutation as it demonstrates higher permeability to exogenous DNA (Harris 2001) and is therefore more genetically tractable. Likewise, knowing the genes participating in cell walls and vacuolization can help *C. reinhardtii* model organism acclimation to hyperosmotic environments. *C. reinhardtii* has also been studied to understand flagella dysfunction in complex eukaryotes, especially for heritable disease research in cilia and other filamentous structures (Cole 1999; Pazour et al. 1999a, b, 2000). *C. reinhardtii* can thus be used to model structural elements to determine mechanical and chemical stress responses in simpler, more controlled situations.

While understanding of the cellular machinery is crucial to strain design, deconvolution of metabolism has been possibly the largest goal in rational biosynthetic design. *C. reinhardtii* has been at the forefront of this research not only because it has metabolic flexibility, but because its metabolic flexibility is relatively well understood. As such, subjecting *C. reinhardtii* to nutrient stress has been a consistent tactic for investigation: nutrient deplete phenotypes and phenotype rescues are used as controls for conclusive, rigorous experimentation. Conclusions can be drawn from these experiments by using knowledge and connecting it to observed phenotypes; for example, removing copper – a cofactor in plastocyanin (Li et al. 1996) – resulted in a remodeling of the photosynthetic apparatus and identification of the *Crd1* gene product as a response to oxygen deficiency (Moseley et al. 2000). Carbon dioxide, sulfur deprivation, and other nutrient studies have demonstrated observable metabolic remodeling through starch accumulation, hydrogen accumulation, and various transcriptional changes, respectively (Buléon et al. 1997; Melis et al. 2000; Navarro et al. 2000; Quesada et al. 1998; Zhou et al. 2000; Quiñones et al. 1999). These stresses are now being studied at molecular levels to draw even deeper conclusions about stress response.

C. reinhardtii may not be the optimal organism for large-scale production of biofuels, but its research has been extremely valuable to the field as a whole (Merchant et al. 2007). In a lot of ways, the diversity of mutants and biological characterization afforded by the community has outpaced the sophistication of characterization techniques. As '-omics' analytical methods become more powerful, the links between phenotypes and genes can be further unveiled, resulting in concise, predictable genetic targets for strain and bioprocess design. Not only that, but the burgeoning molecular toolbox (Beer et al. 2009), including high-fidelity transgene expression (Neupert et al. 2009), gene regulation manipulation through riboswitches (Croft et al. 2007), inducible promoters (Neupert et al. 2009, Shao and Bock 2008), and chloroplast manipulation (Surzycki et al. 2007), can improve the experimental power in *C. reinhardtii*. Overall, the move towards carbon neutral fuels has been aided by *C. reinhardtii* and its degree of characterization, and further analysis into the precise molecular mechanisms is becoming available.

'-OMICS' Approaches

Genomics and Transcriptomics

Knowledge of the DNA and RNA sequences within a cell provides a picture of the cells capabilities and gene expression under different conditions. Genomics and transcriptomics can, then, be used to determine conserved behaviors across organisms, to infer transcript functions, or to determine gene function from phenotypes.

Since genomics is the analysis of the whole information reservoir for an organism, it establishes the boundaries of the cellular landscape. Transcriptomics augments this analysis by cooperatively restricting the landscape through finding which parts of the genome are expressed and by elucidating complex gene interactions. By varying conditions, both changes and constancy in transcript levels can provide insight into regulatory pathways and the mechanisms by which fitness is maintained. Understanding this passage of information aids in determining and assessing the viability of genetic targets while establishing the boundaries of the cellular landscape.

The similar structures for information storage in DNA and RNA correspond to similar methods of determining base pair order. Replacing classical Sanger sequencing (Sanger and Coulson 1975), the most common methods begin with shotgun which require DNA or RNA to be cut or sheared randomly into smaller fragments. From there, classic methods like the chain termination method or Maxam-Gilbert sequencing are used to determine the sequence of individual fragments for DNA (Sanger et al. 1977; Maxam and Gilbert 1977). Exciting advances in high-throughput, cost-controlled sequencing have resulted in next-generation sequencing methods like single-molecule real-time sequencing (Thompson and Steinmann 2010), ion semiconductor (Rusk 2011), pyrosequencing (Ronaghi et al. 1996), Applied Biosystems' SOLiD Sequencing (Valouev et al. 2008), 454 Life Science pyrosequencing (Rothberg and Leamon 2008), and sequencing by synthesis (Illumina) (Brenner et al. 2000). The main drawback of these methods is the high capital investment (Liu et al. 2012) but the cost to sequence DNA continues to fall rapidly (https://www.genome.gov/sequencingcosts). Sequencing of RNA (RNA-seq) is more challenging due to the nature of RNA: rapidly changing and easily degraded. Current methods allow for cheap, single, and simultaneous library generation reactions (Shishkin et al. 2015) and improved coverage of RNA functionality through separating different types of RNA (by size exclusion or similar routines) and consequent sequencing (Morin et al. 2008). Direct RNA Sequencing (without conversion to cDNA), is hypothetically less error-prone but is still in its infancy due to the instability of RNA (Ozsolak et al. 2009). Once the sequence of these small DNA or RNA sequences is known, computational methods need to be used to assemble sequences into longer pieces. Manual annotation of sequenced genomes is not possible given how quickly genomes can be sequenced today. Therefore, computational tools have been created which are capable of predicting gene structure and function, such as AUGUSTUS (now u10) (Specht et al. 2011). Genome browsers are available for a number of photosynthetic organisms through the Phytozome Genome Browser (Goodstein et al. 2012), which allows searching genomes for genes and supports further annotation efforts. A large amount of *Chlamydomonas reinhardtii* transcriptome data from various experiments is also available on the UCSC genome browser (https://genome.ucsc.edu/); this gives excellent snapshots of how transcript abundance varies in any given condition.

Base pairs sequences on their own offer little insight into cell behavior; additional computation is used to find patterns between genomic and transcriptomic data. While earlier studies were conducted slowly by referencing primary literature, software tools (Overbeek et al. 2014; Curwen et al. 2004; Zerbino 2010; Haas et al.

2013) in combination with genome and transcriptome databases (Kanehisa and Goto 2000; Caspi et al. 2008; Pruitt et al. 2012; Nordberg et al. 2014) have significantly streamlined the process. Most tools operate by automatically patching (if necessary) and correlating sequences of highly conserved genetic regions. This allows examination of the actual regions expressed in the genome and can be used to determine the structure of genes and possibly their function based on conserved sequences. These sequences can also be correlated across related strains to construct phylogenetic trees (Qi et al. 2004).

Proteomics

Proteomic analysis is conducted with the same shotgun approach as the previous two methods but with a different mean of quantifying the fragments. Similarly to genomics and transcriptomics, proteomics begins with deconstruction into peptides via proteolytic digesting. Once done, chromatography, electrophoresis, or similar separations tactics are used to separate proteins. Tandem mass spectrometry, Matrix-Assisted Laser Desorption/Ionization (MALDI) with time-of-flight (TOF) spectrometry, or gas chromatography/mass spectrometry are used to determine the peptide "fingerprints" (Karas et al. 1987). The peptide fragments can be sequenced using this method and analyzed by comparison to databases to identify concentrations and functions of the associated proteins. One drawback of this method is that degenerate peptides are indistinguishable with this technique (Alves et al. 2007). Computational and manual techniques can be used to link proteins to transcriptomics and genomics. This can be useful for several reasons: it identifies cell functions, protein conservation, and helps determine regulatory networks. In the first result, it builds on previous data from transcriptomics and genomics with either computational correlation (Förster et al. 2006) or more easily measured protein interactions. Furthermore, proteomics can be used in comparison to mutants or other organisms to determine conserved active sites and the effects of mutation (Morgan et al. 2014). Lastly, by determining which RNAs are translated into proteins, regulatory networks and RNAs can start to be decoded. Proteomics can build on other '-omics' analyses to determine further cellular behavior and parse the inner mechanisms contributing to it.

Metabolomics

Metabolomics, which measures the presence and/or concentration of metabolites within the cell, is the '-omics' method which is closest to measuring the phenotype of the cell. Metabolomics shows both conditional responses and the functional mechanisms in the cell which are not directly evident – simply implied – from protein concentrations and transcript levels. Methods in these studies are much different from those encountered in either of the other three '-omics': instead of defragmenting and sensing individual compounds, metabolites are quantified in the cell using gas chromatography/mass spectrometry (GC/MS), liquid chromatography/mass spectrometry (LC/MS) or nuclear magnetic resonance (NMR) (Griffiths and Wang 2009). Knowledge of the intracellular metabolite concentrations can be used for metabolic modeling efforts, such as metabolic flux analysis (MFA) (Edwards et al. 1999; Stephanopoulos et al. 1998; Wiechert 2001), as well. As with the other '-omics' technologies, metabolomics can be used as a comparative tool to gain a better understanding of how the cell functions at all levels of control (DNA, RNA, protein and metabolism).

Understanding Biofuel Production Using '-OMICS' Analyses

C. reinhardtii has played a significant role in cataloguing cellular machinery and stress responses in photosynthetic organisms, but molecular characterization of these processes has lagged behind observable phenomena. A symptom of this stress is the accumulation of energy storage and/or potential fuel molecules. '-Omics' approaches have revolutionized the field by offering insight into these complex cellular processes and giving researchers targets for further engineering efforts to maximize production.

Production of High Energy Carbon Storage Molecules

Several studies have illustrated that *C. reinhardtii* accumulates carbon storage molecules (starch and triacylglycerols (TAGs)) during periods of macronutrient (sulfur, nitrogen, and phosphorus) (Matthew et al. 2009; Hu et al. 2008; Miller et al. 2010; Moellering and Benning 2010; Goodson et al. 2011; Weers and Gulati 1997) or micronutrient (zinc, copper, and iron) (Kropat et al. 2011) limitations (See Fig. 18.1 for TAG synthesis pathway). It has also been reported that by removing the ability to store starch, cells accumulate much higher levels of TAGs (Wang et al. 2009; Li et al. 2010b; Work et al. 2010). Unfortunately, one undesirable side effect of nutrient depletion is reduced production of biomass (Sheehan et al. 1998). In addition to micro- and macronutrient conditions, exposing *C. reinhardtii* cells to heat shock also induces increased lipid accumulation (Hemme et al. 2014). These nutrient limitations and environmental stresses used to induce TAG accumulation result in global changes within the cell which can be measured by evaluating changes in gene expression, protein concentrations and metabolite abundance. Information gathered from '-omics' studies in *C. reinhardtii* when exposed to unfavorable environmental conditions can be used to understand the regulatory mechanisms of lipid production in *C. reinhardtii*. This in turn can be used to direct genome editing of other algal species more amenable to genetic manipulation or possessing traits more desirable for large scale production to improve the production of biofuels while maintaining normal growth rates and optimal cellular health.

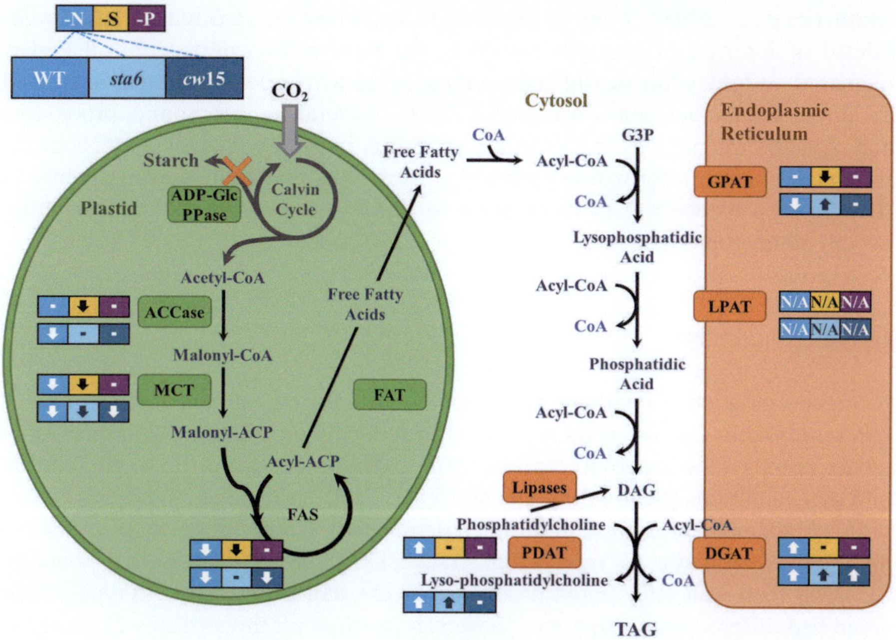

Fig. 18.1 Reported changes in transcript abundance in biosynthetic pathways of triacylglycerols in *Chlamydomonas reinhardtii*. Transcripts corresponding to enzymes involved in triacylglycerol (TAG) synthesis have varying transcript levels when starved of nitrogen (–N), sulfur (–S), or phosphorous (–P). In general, phosphorous starvation had little influence on TAG transcript levels, while sulfur starvation caused transcripts for MCT, ACCase, GPAT, and the subunits of FAS to decrease. Nitrogen starvation caused transcript levels in the plastid to decrease (ACCase, MCT and FAS), while PDAT and diacylglycerol acyltransferase DGAT transcripts increased. The *sta6* mutant does not produce starch due to inactivation of ADP-Glc PPase in the starch synthesis pathway indicated by ×. ↑: 1-log$_2$(fold change) transcript increase; ↓: 1-log$_2$(fold change) transcript decrease; – : no transcript change. –N data from Schmollinger et al. (Schmollinger et al. 2014); –P and –S data from González et al. (González-Ballester et al. 2010)

Nitrogen Starvation

'-Omics' data of *C. reinhardtii* cells grown under nitrogen deplete conditions provide a fundamental understanding of the regulatory mechanics that induce TAG accumulation during stress conditions. Cells starved of nitrogen can no longer synthesize new nucleic acids or proteins and therefore they are forced to slow (or cease) growth; however, to take advantage of ample carbon supply they divert their metabolism into storage of carbon as lipids and starches. Blocking the assimilation of starches for energy storage forces *C. reinhardtii* to accumulate more lipids. Strains of *C. reinhardtii* have been developed that carry an insertional deletion of either the *sta1* or *sta6* gene encoding the large and small subunit of ADP-glucose pyrophosphorylase, respectively (ADP-Glc PPase). Critical to starch synthesis, ADP-Glc PPase converts glucose-1-phosphate to ADP-glucose, the precursor to starch

(Ballicora et al. 2003). Wang et al. showed that when the *sta6* mutant of *C. reinhardtii* is deprived of nitrogen for 48 h, the number of cytoplasmic oil bodies increased 30-fold, while the oil body content in the wild type only increased 15-fold under the same conditions (Wang et al. 2009). Several transcriptomic, proteomic, and metabolomic studies exist for this condition in both wild type as well as the starchless mutant. Information gathered at each level may be used congruently to shed light on the regulatory mechanism for lipid accumulation without hindering overall biomass production.

Overall Metabolism

Nitrogen limitation or deprivation drastically alters the overall metabolism of all *C. reinhardtii* strains as the cells are forced to reduce their nitrogen to carbon ratio, reduce photosynthetic activity, and increase respiration to acclimate to the reduced nitrogen conditions (Schmollinger et al. 2014). To determine the underlying causes of this fundamental response to nitrogen deprivation, Schmollinger et al. measured the '-omics' of three *C. reinhardtii* strains: CC-4348 (starchless *sta6* mutant), CC-4349 (cell wall-less, starch producing *cw15*), and CC-4532 (wild type). Each strain had different transcriptomes; however, amongst all three strains 27 % of the total transcripts were shared. As expected, *cw15* and *sta*6 shared 35 % of the transcripts measured. Unsurprisingly, all three strains increase transcripts for nitrogen transportation and assimilation; the proteins translated by the increased transcripts were also found to have increased concentrations. Additionally, proteins with low nitrogen content also increased in abundance during nitrogen starvation, while those of high nitrogen content decreased (Schmollinger et al. 2014). This coincides with observations made by Miller et al.: up-regulation of genes involved in nitrogen metabolism, uptake, or consumption; down-regulation of protein biosynthesis genes due to decreased availability of amino acids; and down-regulation of photosynthetic genes to regulate the metabolic state of the cell (Miller et al. 2010). Consistent with transcriptomic predictions, Wase et al. found that the abundance of proteins involved in nitrogen and amino acid metabolism, oxidative phosphorylation, as well as glycolysis/gluconeogenesis were increased, while the abundance of proteins involved with protein synthesis and photosynthesis decreased in nitrogen starved cells. Of those glycolysis/gluconeogenesis enzymes with increased abundance, pyruvate carboxylase, citrate synthase, and ATP-citrate lyase may assist lipid synthesis by providing acetyl-CoA. Interestingly, of the 18 metabolites observed whose levels changed, amino acids phenylalanine, tryptophan, and aspartic acid as well as the amino acid degradation product putrescine were among those whose abundance decreased, possibly to serve as nitrogen sources for other metabolic needs (Wase et al. 2014). This mechanism of adaptation is responsible for the reduced biomass and cell arrest observed in cells starved of nitrogen. Thus, the regulatory mechanism by which these genes interact with those involved with lipid synthesis should be further investigated to identify specific genes that can be changed without causing cell death.

Photosynthetic activity is also reduced in nitrogen starved cells, which limits their growth and ability to fix carbon. Enzymes and pigments required for photosynthesis have relatively high nitrogen content, thus, under nitrogen deplete conditions, the *C. reinhardtii* cells are limited in how many of these proteins they can produce (Blaby et al. 2013). This limitation reduces the need to produce chloroplast ribosomes to synthesize these photosynthetic proteins and is apparent by the reduced chloroplast ribosome transcript levels (~75 % reduction) observed by Schmollinger et al. Transcripts associated with chlorophyll degradation however remained constant after N-deprivation, therefore cells preferentially maintained chlorophyll despite the high nitrogen content (Boyle et al. 2012). Additionally, transcripts for the Calvin-Benson cycle were reduced two to eightfold as well as a 20 % decrease in the corresponding proteins (Schmollinger et al. 2014). From this, we see that cells subjected to nitrogen starvation are severely hampered in their ability to grow normally and re-organize their metabolism to maintain growth as much as possible.

TAG Synthesis

Of great interest to biofuel production are the '-omics' data associated with TAG biosynthesis. Contrary to phenotypic results (TAG accumulation), most transcripts involved in TAG synthesis remain stable, and if anything, decrease; however, some of the reduced transcript levels returned to similar levels during nitrogen replete conditions. The exceptions to this trend were transcripts for the genes encoding two acyl-CoA dependent diacylglycerol acyltransferases *DGAT1* and *DGTT1* and two glycerol-3-phosphate dehydrogenase isozymes (*GPD2* and *GPD4*) whose transcript levels showed significant increase (8.2, 529.9, 146.0, and 20.1 fold respectively in *cw15*) (Schmollinger et al. 2014). Boyle et al. also identified the transcripts of *DGAT1* and *DGTT1* as being highly up-regulated (6.3 and 30.8 fold respectively) in wild type *C. reinhardtii* after exposure to nitrogen deplete conditions for 48 h. Transcripts for the gene *PDAT1* encoding another acyltransferase, phosphatidylcholine dependent acyltransferase (PDAT) were also reported to increase (2.9 fold) in nitrogen deplete conditions (Boyle et al. 2012). In the *sta6* mutant, the diacylglycerol acyltransferases, *DGTT2*, was reported by Blaby et al. to have ~4 times the number of transcripts than *cw15* during nitrogen deplete conditions. Of additional interest, transcript levels for the major lipid droplet protein gene (*MLDP1*) were also ~4 times higher in *st6* than in *cw15* (Blaby et al. 2013). The major lipid droplet protein (MLDP) is an important structural protein found in lipid bodies. Repression of *MLDP1* in *C. reinhardtii* using RNA interference has been shown to cause lipid bodies to increase in size, but not in TAG content (Moellering and Benning 2010). Although few major changes in TAG transcripts were observed in the aforementioned studies, TAG accumulation still occurs. Thus, proteomic studies are necessary to determine the amount of protein actually resulting from these transcripts.

Although transcript abundance for some TAG synthesis genes increased, their corresponding protein concentrations did not show corresponding changes in nitrogen starved *C. reinhardtii*. Indeed, protein abundance levels of most enzymes involved with lipid synthesis either remained the same or decreased indicating different levels of regulation involved with lipid synthesis. Of those enzymes involved with lipid synthesis whose associated transcript levels increased identified above (Boyle et al. 2012; Schmollinger et al. 2014; Blaby et al. 2013), Wase et al. were unable to identify any corresponding change in protein abundance. Indeed, only the following proteins associated with lipid synthesis were found to have increased abundance during nitrogen starvation: long-chain acyl-CoA synthetase, plastid lipid associated protein, and triglyceride lipases (Wase et al. 2014). Similar results were observed by Schmollinger et al. in addition to the reduced abundance of the acetyl-CoA carboxylase (ACCase) subunits (Schmollinger et al. 2014). Of great interest is the decreased abundance of the AMP-activated protein kinase under nitrogen deplete conditions also observed by Wase et al. This kinase, responsible for inhibiting ACCase, could cause the observed TAG accumulation despite the reduced abundance of ACCase subunits (Wase et al. 2014). Such discrepancies between the transcript and protein abundance levels indicate that lipid synthesis is possibly regulated by variations in enzyme kinetics at the protein level caused by substrate deviations at the metabolite level and/or post-translational modifications of those enzymes to promote lipid storage under nitrogen deplete conditions.

Other Macronutrient Conditions: Phosphorus and Sulfur

It has been shown that phosphorous and sulfur deplete conditions result in similar genetic regulatory results. According to Mosely, et al., *C. reinhardtii* cells respond similarly when they are exposed to phosphorous deplete or sulfur deplete conditions. Interestingly, when cells acclimated to phosphorous deplete conditions are exposed to sulfur deplete conditions, the cellular responses associated with sulfur deficiency become inhibited. Thus, the regulatory responses involved in sulfur deplete, phosphorous deplete, and possibly other nutrient deplete conditions are intimately dependent on one another (Moseley et al. 2009). Additional sulfur deplete studies indicate that the accumulation of lipids is a result of the conversion of phospholipids in the membranes into neutral lipids. Sugimoto et al. showed that when *C. reinhardtii* cells are starved of sulfur, they degrade almost 85 % of the sulfoquinovovosyl diacylglycerol (SQDG) present in their chloroplast membranes to reallocate the sulfur to increase the sulfur content in proteins (Sugimoto et al. 2007). Further studies from the same group indicate that the degradation of SQDG is coupled to the synthesis of phosphatidylglycerol, another acidic lipid present in chloroplasts; however, the increase in phosphatidylglycerol synthesis is most likely due to the SQDG loss, not the sulfur deplete conditions (Sugimoto et al. 2008). This insight indicates an opportunity to modify SQDG content in the membranes via gene deletion of those enzymes involved with SQDG synthesis to induce phosphatidylglycerol synthesis.

In addition to nitrogen response analyses, Schmollinger et al. also compared transcript levels of *C. reinhardtii* under nitrogen deplete conditions with those transcript levels measured by González-Ballester et al. during phosphorus and sulfur deprivation (Schmollinger et al. 2014; González-Ballester et al. 2010). Despite similar physiological responses between sulfur and phosphorous deplete conditions, the transcriptome of *C. reinhardtii* in phosphorous deplete conditions experienced fewer changes than the transcriptome of sulfur deprived cells. This observation is intriguing as phosphorous and sulfur deplete conditions exhibit a similar response as discussed above. FA synthesis transcripts varied amongst the conditions as illustrated in Fig. 18.2. The only consistent changes were found *DGAT1* and *DGTT1* which increased in all macronutrient deplete conditions. Again, this discrepancy between varied transcript data and the resulting TAG accumulation over all of the macronutrient deplete conditions illustrates the need for further investigation of the regulatory mechanisms involved in lipid production during nutrient deplete conditions by gathering more information on these conditions at the proteomic and metabolomic levels.

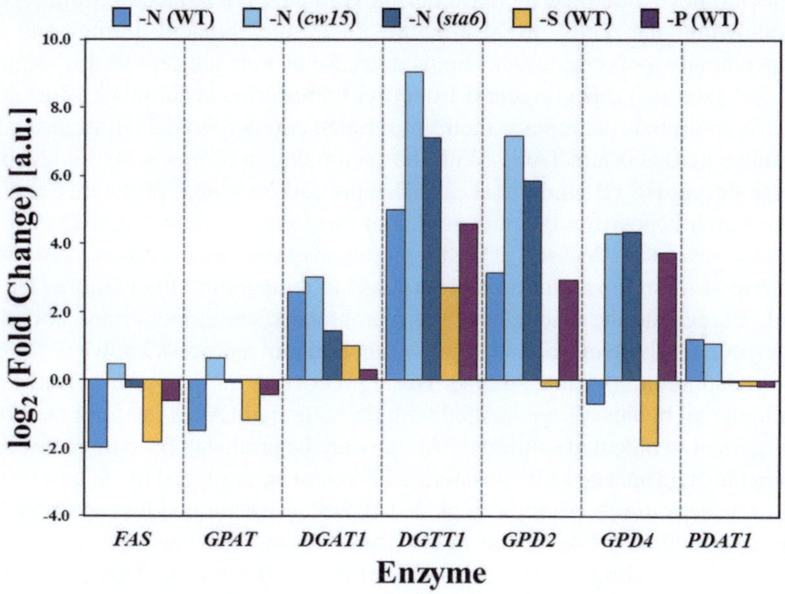

Fig. 18.2 Transcript fold changes of genes associated with enzymes involved in triacylglycerol synthesis of *Chlamydomonas reinhardtii* subjected to macronutrient deprivation. In general, most transcripts remain either unchanged or decrease when starved of nitrogen (-N), sulfur (-S), or phosphorus (-P). The response to -N causes relatively similar TAG synthesis transcript level changes in WT, *cw15*, and *sta6*. All three strains saw drastic transcript level increase for *DGAT1*, *DGTT1*, and *GPD2*. Even –S and –P conditions caused some transcript level changes in WT. Both –S and –P conditions caused *DGAT1* and *DGTT1* transcripts to increase. Only –P caused transcript levels for *GPD2* and *GPD4* to increase. –N data from Schmollinger et al. (2014); –P and –S data from González et al. (González-Ballester et al. 2010)

Heat Stress

Proteomic and metabolomic information from *C. reinhardtii* under heat stress (HS) conditions provides additional insight to lipid accumulation. Hemme et al. performed a large-scale proteomic and metabolomic analysis of cells before HS, during various times at 42 °C for 24 h, and during the 8 h recovery after HS conditions (Hemme et al. 2014). Although transcript levels were not analyzed, many predictions can be made about how cells accumulate lipids when heat stressed from the protein and metabolite variations observed. Like other stress conditions, *C. reinhardtii* experiences a restructuring of their central metabolism during and after HS that specifically induces cell arrest. Proteins critical to the Calvin-Benson cycle as well as those involved in gluconeogenesis and the TCA/glyoxylate cycle decreased, and their respective metabolites had a sharper decrease indicating reduced enzyme activity (Hemme et al. 2014). This decline in metabolic function illustrates that the cells undergo severe cell arrest, which may contribute to lipid accumulation. Similar to macronutrient deficient conditions, little variation in abundance was observed for *de novo* FA synthesis proteins during and after HS. Membrane lipid metabolism proteins did have a marked increase during HS and were reduced during recovery thus allowing the cells to restructure their membranes during and after HS. Specifically, polyunsaturated lipids decrease as they are replaced by saturated lipids and saturated diacylglycerol trimethyl homoserine accumulates during heat shock. As saturated FAs replace their unsaturated counterparts, the unsaturated FAs accumulate as DAGs and TAGs. With the accumulation of lipids, starch levels also decrease during HS (Hemme et al. 2014). This carbon source preference indicates that the starch is possibly being degraded to form glycerol and acetyl-CoA to synthesize the saturated FAs used in the membrane. By shifting the ratio of saturated to unsaturated FAs in the membrane in response to temperature fluctuations, the cells are able to regulate the viscosity of the membrane keeping membrane bound proteins, especially those of photosynthesis, functional in a process known as "homeoviscous adaptation" (Michael 1974). During recovery, the polyunsaturated FAs of TAGs in the oil bodies are exchanged with the saturated FAs in the lipid membrane. It is important to note that saturated FAs can only be produced from *de novo* biosynthesis in the chloroplast, while unsaturated FAs can be produced by *de novo* biosynthesis or by desaturases (Hemme et al. 2014). Because saturated FAs are most likely replacing unsaturated FAs in the lipids, these unsaturated FAs accumulate in the form of TAGs creating oil bodies as HS inhibits cell division. This FA exchange could be manipulated in conjunction with FA pathway modifications to optimize the yield of unsaturated TAGs.

Biohydrogen Production

Industrial hydrogen production occurs via steam reforming of small hydrocarbons and carbon monoxide, providing a potentially more efficient fuel source through use of low-energy molecules with a water. However, as indicated by the requirement for hydrocarbons and steam, this is not a carbon neutral process; in fact, it has been shown to have even more of a net negative impact than fossil fuels (Fishtik et al. 2000). Biological sources of hydrogen offer an advantage: the hydrogenase enzyme functions cooperatively with photosynthesis to convert protons into hydrogen gas as an electron sink instead of oxygen. '-Omics' has demonstrated that stress conditions affecting the photosynthetic apparatus cause activation of the hydrogenase enzyme. In particular, acclimation to sulfur deplete and anoxic environments result in hydrogen gas being formed to compensate for cellular overprotonation (acidification through proton accumulation) and electron evolution. Since these environments are suboptimal for *C. reinhardtii*, a corresponding loss of cellular fitness is observed. Analysis of these phenomena is crucial to eliminating the stress-induced cellular inefficiencies while maintaining the desired increase of biohydrogen.

Anoxic Conditions

Anoxic conditions in photosynthetic organisms result in the more hydrogenase enzyme to provide an electron sink and resist overprotonation of the thylakoid space. Transcriptome level studies support this, showing an increase in the concentration of *HYD1* and *HYD2* transcripts – coding for hydrogenase enzymes (Mus et al. 2007). Knockout studies indicate that absence of hydrogenase genes in anoxia display a diversion of resources to relieving oxidative stress via other pathways such as succinate production (Dubini et al. 2009) which can potentially provide targets for maintenance of cellular fitness during anoxia (Fig. 18.3).

In dark anaerobiosis, responses to the electron sink void are further exacerbated by the requirement to mitigate toxic byproduct formation while still providing cellular energy. These result from the inability to fully oxidize carbon substrates to CO_2, diverting glycolysis toward 2-carbon molecules (acetate and ethanol) and toward hydrogen from pyruvate. The transcriptomic and metabolomic studies display an increase in transcripts encoding proteins associated with fermentation as well as increases in ethanol and hydrogen production (Mus et al. 2007). This fermentative behavior is coupled with a slight remodeling of carbon metabolism to remove bottlenecks and conserve energy. Notably, the Calvin Cycle is subject to slowing due to accumulation of cycle intermediates and electron dense molecules in anoxia while upregulating pentose phosphate transcripts in order to adjust. Meanwhile, glyoxylate-related enzymes are synthesized to aid in conversion from acetyl-CoA – resulting from lipid catabolism – to organic acids (Mus et al. 2007). Isocitrate lyase (*ICL*) is one enzyme which bypasses α-ketoglutarate, a key precursor for chlorophyll biosynthesis, and has been implicated in cell progression towards senescence (Terashima et al. 2010). Reducing activity of this enzyme may aid in reducing senescent activity while retaining increased cellular growth rates. Glyoxylate related enzymes could also be downregulated to retain lipids, which represent much more valuable commodities than sugars and organic acids. Fermentation pathways upregulate important metabolites while providing insight into maintenance routines that can be exploited to maintain high growth rates (Fig. 18.4).

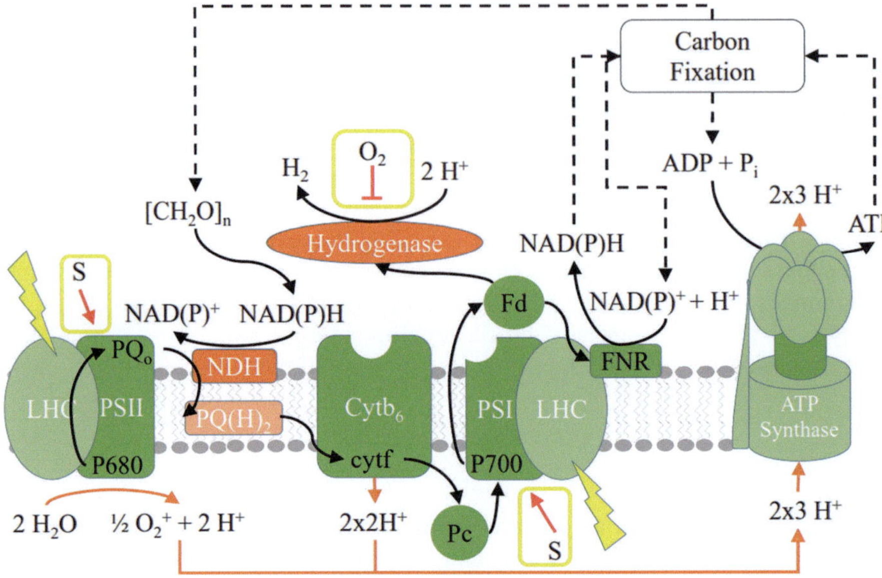

Fig. 18.3 Photosynthetic response to anoxia in light replete conditions to induce hydrogen formation and maintain photosynthesis. Biohydrogen in *Chlamydomonas reinhardtii* is generated via photosynthetic diversion of electrons into hydrogen protons. Sulfur deplete conditions reduce function of the photopigments as repair and protein synthesis are stunted. Oxygen deplete conditions upregulate hydrogenase related transcripts to ensure function of the hydrogenase enzyme and limit accumulation of protons. *Orange arrows* indicate molecular flow; *black arrows* indicate electron flow; *black* labeled structures are electron transporters; *white* labeled structures are photosynthetic macrostructures; *orange* structures are bound proteins. *Yellow* boxes are conditions leading to biohydrogen generation

Not only do energy metabolism cycles change, transcripts and proteins associated with amino acid synthesis show an accumulation of serine and isocitrate amino acids in conjunction with an upregulation of genes encoding enzymes associated with the glycine decarboxylase system. Proteomics imply that, by induction of glycolytic enzymes, intracellular nitrogen reorganization occurs via amino acid degradation as opposed to traditional methods of assimilation (Terashima et al. 2010). This determines that lower protein synthesis is present in these conditions and likewise a remission of cellular fitness. Adjustment of these pathways to more efficiently retain nitrogen may help in maintaining cellular fitness within biohydrogen production endeavors.

Sulfur Deplete Conditions

Sulfur depletion in *C. reinhardtii* and other photosynthetic organisms has been indicated to decrease photosystem II (PSII) activity and increase in hydrogenase activity in a similar way to light anoxia. It does this by halting the methionine repair

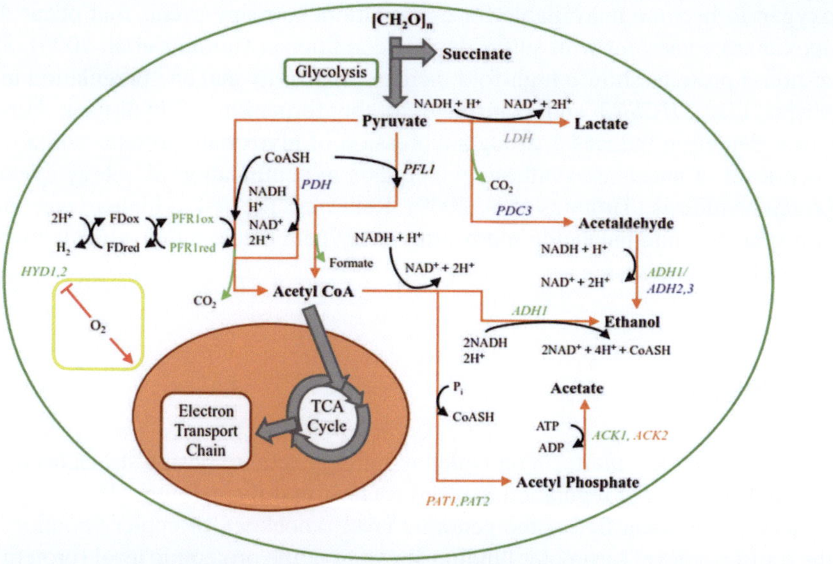

Fig. 18.4 Dark anaerobiosis to induce hydrogen and lipid accumulation in *C. reinhardtii* to maintain energy generation and reduce overprotonation. Biohydrogen accumulates via fermentation in anoxic/dark environments. The lack of oxygen cooperatively ceases function of the electron transport chain while halting inhibition of hydrogenase enzymes. Hydrogen is reduced to diatomic gas for lack of oxygen as an electron sink. Green enzymes are expressed in the chloroplasts, orange in the mitochondria, blue in the cytoplasm, black in both chloroplast and mitochondria, and gray is undetermined. *Orange arrows* indicate metabolite flow, *black arrows* indicate electron flow, *green arrows* indicate exported metabolites, and large *gray arrows* indicate summary reactions. The orange structure is the mitochondrion, the *green* circle is the cell, and the *yellow* box is the exterior conditions

system that responds to photo-damage in *C. reinhardtii*, causing the photosynthetic apparatus to remodel, protein synthesis to slow, and intracellular scavenging activities to gain more importance. These cellular diversions cause significant remodeling of PSII proteins and reapportioning of metabolites through different carbon pathways in nearly identical ways to those observed in anoxia (Zhang et al. 2004). The 2-DE coupled MALDI-TOF and MALDI-TOF/TOF-MS and 2-D gel maps displayed proteome and transcriptome changes including an overall decrease in PS related transcripts except *LCHBM9* and declined proclivity carbon-fixation. Meanwhile, metabolomic surveys determine that acetate is increased with hydrogen as an energy spilling mechanism without sulfur (Matthew et al. 2009). This correlates to the hypothesis that photosynthesis repair is no longer a feasible activity for energy harvesting and cell biosynthesis except for the proteins related to photoprotection and singlet oxygen reduction (Grewe et al. 2014). Protein repair operations, especially methionine dependent repair proteins, indicate a lower level of synthesis and the increase of chaperone proteins indicates more care taken to synthesize the required proteins for photosynthesis (Nguyen et al. 2008). These deficits cause oxygen to become unavailable for assimilation, so many events that occur during anoxia are consistent with sulfur deplete conditions (Timmins et al. 2009). Stress-response proteins show a high-fold increase in activity and an upregulation in transcripts like *LHCSR2* and contribute to the formation of hydrogen. However, sulfur-depletion induces a strong suppression of glyoxylate proteins and a heightened level of succinate synthesis which aids in continuation of energy cycles and protein synthesis (Timmins et al. 2009). Returning the cell to higher levels of protein synthesis and balancing redox stress may help create more valuable hydrogen producing processes.

Conclusion

'-Omics' provide a platform on which to build genetic re-design strategies by identifying key points of regulation in response to stressful conditions. These regulatory targets can be identified at the genomic level (knockout, promoter variation, etc.), the transcript level (promoter binding sites), or at the proteomic level (protein engineering). '-Omics' can then affirm these targets at the metabolomic level. Although directed genetic manipulation of *C. reinhardtii* is difficult to achieve, the insights provided by '-omics' can be used to predict the regulatory responses of other more genetically tractable algae. Combining these multi-level regulatory targets could pave the way towards creation of an optimized algal strain able to grow at operational capacities and maximize flux through biofuel pathways.

There are still shortcomings in knowledge and execution of these systems. For example, these '-omics' techniques are primarily diagnostic: they provide the means to determine enzymatic regulation and cellular tactics for maintaining homeostasis. Moreover, there have not been forward-genetic studies that seek to improve cellular performance in response to these mechanisms. That is to say, studies need to be done on how genetic modification for chlorophyll upregulation in concert with limiting oxygen might affect cellular dynamics and biohydrogen formation. In this regard, the knowledge that has been accumulated needs to be put to the test by inducing feasible cellular modifications. Computational methods are also helping spur this field forward but still have a lot of room for development. Current techniques hinge on the conservation of proteins between species, but rely on regression between similar DNA or peptide sequences. Predictive modeling of structural motifs could go a long way toward use of computational genomics as viable discovery platforms. DNA, RNA, and protein reconstitution in sequencing can be improved to augment throughput and accuracy of those studies. Other concerns involve studying these organisms in scaled-up environments. It has been demonstrated consistently that behaviors change in industrial size processes and this is a major concern; since these '-omics' studies are only performed in a lab-controlled environment, steps have to be taken to ensure application to industry. For this reason, nutrient fluctuations, combinations of stressful conditions, and larger population studies should be conducted to understand how microalgae respond to scale-up.

Fortunately, '-omics' approaches have shown a wide variety of applications and are appealing especially as the technological prowess of the field progresses. Even though there are shortcomings in the discoveries to this point, '-omics' for biofuel production in *C. reinhardtii* has given an incredible amount of targets for microbial development. Since the hardest part of creative discovery is asking the right questions, it can be considered that '-omics' in *C. reinhardtii* is a major victory for the industry as a whole.

References

Alves P, Arnold RY, Novotny MV, Radivojac P, Reilly JP, Tang H (2007) Advancement in protein inference from shotgun proteomics using peptide Detectability. Pac Symp Biocomput 12:409–420

Ballicora MA, Iglesias AA, Preiss J (2003) ADP-glucose pyrophosphorylase, a regulatory enzyme for bacterial glycogen synthesis. Microbiol Mol Biol Rev 67:213–225

Beer LL, Boyd ES, Peters JW, Posewitz MC (2009) Engineering algae for biohydrogen and biofuel production. Curr Opin Biotechnol 20:264–271

Blaby IK, Glaesener AG, Mettler T, Fitz-Gibbon ST, Gallaher SD, Liu B, Boyle NR, Kropat J, Stitt M, Johnson S, Benning C, Pellegrini M, Casero D, Merchant SS (2013) Systems-level analysis of nitrogen starvation-induced modifications of carbon metabolism in a Chlamydomonas reinhardtii starchless mutant. Plant Cell 25:4305–4323

Boyle NR, Page MD, Liu B, Blaby IK, Casero D, Kropat J, Cokus SJ, Hong-Hermesdorf A, Shaw J, Karpowicz SJ, Gallaher SD, Johnson S, Benning C, Pellegrini M, Grossman A, Merchant SS (2012) Three acyltransferases and nitrogen-responsive regulator are implicated in nitrogen starvation-induced triacylglycerol accumulation in chlamydomonas. J Biol Chem 287:15811–15825

Brenner S, Johnson M, Bridgham J, Golda G, Lloyd DH, Johnson D, Luo S, Mccurdy S, Foy M, Ewan M (2000) Gene expression analysis by massively parallel signature sequencing (MPSS) on microbead arrays. Nat Biotechnol 18:630–634

Buléon A, Gallant DJ, Bouchet B, Mouille G, D'Hulst C, Kossmann J, Ball S (1997) Starches from a to C (chlamydomonas reinhardtii as a model microbial system to investigate the biosynthesis of the plant amylopectin crystal). Plant Physiol 115:949–957

Caspi R, Foerster H, Fulcher CA, Kaipa P, Krummenacker M, Latendresse M, Paley S, Rhee SY, Shearer AG, Tissier C (2008) The MetaCyc database of metabolic pathways and enzymes and the BioCyc collection of pathway/genome databases. Nucleic Acids Res 36:D623–D631

Chen F, Johns MR (1996) Heterotrophic growth of chlamydomonas reinhardtii on acetate in chemostat culture. Process Biochem 31:601–604

cole DG (1999) Kinesin-II, coming and going. J Cell Biol 147:463–466

Croft MT, Moulin M, Webb ME, Smith AG (2007) Thiamine biosynthesis in algae is regulated by riboswitches. Proc Natl Acad Sci 104:20770–20775

Curwen V, Eyras E, Andrews TD, Clarke L, Mongin E, Searle SM, Clamp M (2004) The Ensembl automatic gene annotation system. Genome Res 14:942–950

Dubini A, Mus F, Seibert M, Grossman AR, Posewitz MC (2009) Flexibility in anaerobic metabolism as revealed in a mutant of chlamydomonas reinhardtii lacking Hydrogenase activity. J Biol Chem 284:7201–7213

Edwards J, Ramakrishna R, Schilling C, Palsson B (1999) Metabolic flux balance analysis. Metab Eng (eds. Lee, S.Y. & Papoutsakis, E.T.) 13–57 (Marcel Dekker)

Fishtik I, Alexander A, Datta R, Geana D (2000) A thermodynamic analysis of hydrogen production by steam reforming of ethanol via response reactions. Int J Hydrog Energy 25:31–45

Förster B, Mathesius U, Pogson BJ (2006) Comparative proteomics of high light stress in the model alga chlamydomonas reinhardtii. Proteomics 6:4309–4320
González-Ballester D, Casero D, Cokus S, Pellegrini M, Merchant SS, Grossman AR (2010) RNA-seq analysis of sulfur-deprived chlamydomonas cells reveals aspects of acclimation critical for cell survival. Plant Cell Online 22:2058–2084
Gonzalez-Ballester D, Pootakham W, Mus F, Yang W, Catalanotti C, Magneschi L, De Montaigu A, Higuera JJ, Prior M, Galván A (2011) Reverse genetics in chlamydomonas: a platform for isolating insertional mutants. Plant Methods 7:1–13
Goodson C, Roth R, Wang ZT, Goodenough U (2011) Structural correlates of cytoplasmic and chloroplast lipid body synthesis in chlamydomonas reinhardtii and stimulation of lipid body production with acetate boost. Eukaryot Cell 10:1592–1606
Goodstein DM, Shu S, Howson R, Neupane R, Hayes RD, Fazo J, Mitros T, Dirks W, Hellsten U, Putnam N (2012) Phytozome: a comparative platform for green plant genomics. Nucleic Acids Res 40:D1178–D1186
Grewe S, Ballottari M, Alcocer M, D'Andrea C, Blifernez-Klassen O, Hankamer B, Mussgnug JH, Bassi R, Kruse O (2014) Light-harvesting complex protein LHCBM9 is critical for photosystem II activity and hydrogen production in chlamydomonas reinhardtii. Plant Cell Online 26:1598–1611
Griffiths WJ, Wang W (2009) Mass spectrometry: from proteomics to metabolomics and lipidomics. Chem Soc Rev 38:1882–1896
Haas BJ, Papanicolaou A, Yassour M, Grabherr M, Blood PD, Bowden J, Couger MB, Eccles D, Li B, Lieber M (2013) De novo transcript sequence reconstruction from RNA-seq using the trinity platform for reference generation and analysis. Nat Protoc 8:1494–1512
Harris EH (2001) Chlamydomonas as a model organism. Annu Rev Plant Biol 52:363–406
Harris EH, Stern, D. (Ed.) (2009) The chlamydomonas sourcebook: introduction to chlamydomonas and its laboratory use. Academic Press, Oxford/Burlington, MA/San Diego
Hemme D, Veyel D, Mühlhaus T, Sommer F, Jüppner J, Unger A-K, Sandmann M, Fehrle I, Schönfelder S, Steup M, Geimer S, Kopka J, Giavalico P, Schroda M (2014) Systems-wide analysis of acclimation responses to long-term heat stress and recovery in the photosynthetic model organism Chlamydonomonas reinhardtii. Plant Cell 26:4270–4297
Hu Q, Sommerfeld M, Jarvis E, Ghirardi M, Posewitz M, Seibert M, Darzins A (2008) Microalgal triacylglycerols as feedstocks for biofuel production: perspectives and advances. Plant J 54:621–639
Jamers A, Blust R, De Coen W (2009) Omics in algae: paving the way for a systems biological understanding of algal stress phenomena? Aquat Toxicol 92:114–121
Kanehisa M, Goto S (2000) KEGG: kyoto encyclopedia of genes and genomes. Nucleic Acids Res 28:27–30
Karas M, Bachmann D, Bahr U, Hillenkamp F (1987) Matrix-assisted ultraviolet laser desorption of non-volatile compounds. Int J Mass Spectrom Ion Process 78:53–68
Kempa S, Hummel J, Schwemmer T, Pietzke M, Strehmel N, Wienkoop S, Kopka J, Weckwerth W (2009) An automated GCxGC-TOF-MS protocol for batch-wise extraction and alignment of mass isotopomer matrixes from differential 13C-labelling experiments: a case study for photoautotrophic-mixotrophic grown chlamydomonas reinhardtii cells. J Basic Microbiol 49:82–91
Kropat J, Hong-Hermesdorf A, Casero D, Ent P, Castruita M, Pellegrini M, Merchant SS, Malasarn D (2011) A revised mineral nutrient supplement increases biomass and growth rate in chlamydomonas reinhardtii. Plant J 66:770–780
Kurian J (2005) A new polymer platform for the future — Sorona® from corn derived 1,3-propanediol. J Polym Environ 13:159–167
Li H, Liao JC (2013) Engineering a cyanobacterium as the catalyst for the photosynthetic conversion of CO2 to 1,2-propanediol. Microbiol Cell Fact 12:4
Li HH, Quinn J, Culler D, Girard-Bascou J, Merchant S (1996) Molecular genetic analysis of plastocyanin biosynthesis in chlamydomonas reinhardtii. J Biol Chem 271:31283–31289
Li Y, Han D, Hu G, Dauvillee D, Sommerfeld M, Ball S, Hu Q (2010a) Chlamydomonas starchless mutant defective in ADP-glucose pyrophosphorylase hyper-accumulates triacylglycerol. Metab Eng 12:387–391

Li Y, Han D, Hu G, Sommerfeld M, HU Q (2010b) Inhibition of starch synthesis results in overproduction of lipids in chlamydomonas reinhardtii. Biotechnol Bioeng 107:258–268

Liu L, Li Y, Li S, Hu N, He Y, Pong R, Lin D, Lu L, Law M (2012) Comparison of next-generation sequencing systems. Biomed Res Int 2012:11

Matthew T, Zhou W, Rupprecht J, Lim L, Thomas-Hall SR, Doebbe A, Kruse O, Hankamer B, Marx UC, Smith SM (2009) The metabolome of chlamydomonas reinhardtii following induction of anaerobic H2 production by sulfur depletion. J Biol Chem 284:23415–23425

Maxam AM, Gilbert W (1977) A new method for sequencing DNA. Proc Natl Acad Sci 74:560–564

Melis A, Zhang L, Forestier M, Ghirardi ML, Seibert M (2000) Sustained photobiological hydrogen gas production upon reversible inactivation of oxygen evolution in the green AlgaChlamydomonas reinhardtii. Plant Physiol 122:127–136

Merchant SS, Prochnik SE, Vallon O, Harris EH, Karpowicz SJ, Witman GB, Terry A, Salamov A, Fritz-Laylin LK, Marechal-Drouard L, Marshall WF, Qu L-H, Nelson DR, Sanderfoot AA, Spalding MH, Kapitonov VV, Ren Q, Ferris P, Lindquist E, Shapiro H, Lucas SM, Grimwood J, Schmutz J, Team CA, Team JA, Grigoriev IV, Rokhsar DS, Grossman AR (2007) The Chlamydomonas genome reveals the evolution of Key animal and plant functions. Science 318:245–251

Michael S (1974) Homeoviscous adaptation – a homeostatic process that regulates the viscosity of membrane lipids in Escherichia coli. Proc Natl Acad Sci 71:522–525

Miller R, Wu G, Deshpande RR, Vieler A, Gärtner K, Li X, Moellering ER, Zäuner S, Cornish AJ, Liu B, Bullard B, Sears BB, Kuo M-H, Hegg EL, Shachar-Hill Y, Shiu S-H, Benning C (2010) Changes in transcript abundance in chlamydomonas reinhardtii following nitrogen deprivation predict diversion of metabolism. Plant Physiol 154:1737–1752

Moellering ER, Benning C (2010) RNA interference silencing of a major lipid droplet protein affects lipid droplet size in chlamydomonas reinhardtii. Eukaryot Cell 9:97–106

Morgan AD, Ness RW, Keightley PD, Colegrave N (2014) Spontaneous mutation accumulation in multiple strains of the green alga, chlamydomonas reinhardtii. Evolution 68:2589–2602

Morin RD, Bainbridge M, Fejes A, Hirst M, Krzywinski M, Pugh TJ, Mcdonald H, Varhol SJ, Marra MA (2008) Profiling the HeLa S3 transcriptome using randomly primed cDNA and massively parallel short-read sequencing. BioTechniques 45:81–94

Moseley J, Quinn J, Eriksson M, Merchant S (2000) The Crd1 gene encodes a putative di-iron enzyme required for photosystem I accumulation in copper deficiency and hypoxia in chlamydomonas reinhardtii. EMBO J 19:2139–2151

Moseley JL, Gonzalez-Ballester D, Pootakham W, Bailey S, Grossman AR (2009) Genetic interactions between regulators of chlamydomonas phosphorus and sulfur deprivation responses. Genetics 181:889–905

Mus F, Dubini A, Seibert M, Posewitz MC, Grossman AR (2007) Anaerobic acclimation in chlamydomonas reinhardtii: anoxic gene expression, Hydrogenase induction, and metabolic pathways. J Biol Chem 282:25475–25486

Nakamura CE, Whited GM (2003) Metabolic engineering for the microbial production of 1,3-propanediol. Curr Opin Biotechnol 14:454–459

Navarro MAT, Guerra E, Fernández E, Galván A (2000) Nitrite reductase mutants as an approach to understanding nitrate assimilation in chlamydomonas reinhardtii. Plant Physiol 122:283–290

Neupert J, Karcher D, Bock R (2009) Generation of chlamydomonas strains that efficiently express nuclear transgenes. Plant J 57:1140–1150

Nguyen AV, Thomas-Hall SR, Malnoe A, Timmins M, Mussgnug JH, Rupprecht J, Kruse O, Hankamer B, Schenk PM (2008) Transcriptome for photobiological hydrogen production induced by sulfur deprivation in the green alga Chlamydomonas reinhardtii. Eukaryot Cell 7:1965–1979

Nordberg H, Cantor M, Dusheyko S, Hua S, Poliakov A, Shabalov I, Smirnova T, Grigoriev IV, Dubchak I (2014) The genome portal of the department of energy joint genome institute: 2014 updates. Nucleic Acids Res 42:D26–D31

Oliver JWK, Machado IMP, Yoneda H, Atsumi S (2013) Cyanobacterial conversion of carbon dioxide to 2,3-butanediol. Proc Natl Acad Sci 110(4):1249–1254
Oliver JWK, Machado IMP, Yoneda H, Atsumi S (2014) Combinatorial optimization of cyanobacterial 2,3-butanediol production. Metab Eng 22:76–82
Overbeek R, Olson R, Pusch GD, Olsen GJ, Davis JJ, Disz T, Edwards RA, Gerdes S, Parrello B, Shukla M (2014) The SEED and the rapid annotation of microbial genomes using subsystems technology (RAST). Nucleic Acids Res 42:D206–D214
Ozsolak F, Platt AR, Jones DR, Reifenberger JG, Sass LE, Mcinerney P, Thompson JF, Bowers J, Jarosz M, Milos PM (2009) Direct RNA sequencing. Nature 461:814–818
Pazour GJ, Dickert BL, Witman GB (1999a) The DHC1b (DHC2) isoform of cytoplasmic dynein is required for flagellar assembly. J Cell Biol 144:473–481
Pazour GJ, Koutoulis A, Benashski SE, Dickert BL, Sheng H, Patel-King RS, King SM, Witman GB (1999b) LC2, the chlamydomonas homologue of the tComplex-encoded protein Tctex2, is essential for outer Dynein Arm assembly. Mol Biol Cell 10:3507–3520
Pazour GJ, Dickert BL, Vucica Y, Seeley ES, Rosenbaum JL, Witman GB, Cole DG (2000) Chlamydomonas IFT88 and its mouse homologue, polycystic kidney disease gene tg737, are required for assembly of cilia and flagella. J Cell Biol 151:709–718
Pruitt KD, Tatusova T, Brown GR, Maglott DR (2012) NCBI reference sequences (RefSeq): current status, new features and genome annotation policy. Nucleic Acids Res 40:D130–D135
Qi J, Luo H, Hao B (2004) CVTree: a phylogenetic tree reconstruction tool based on whole genomes. Nucleic Acids Res 32:W45–W47
Quesada A, Hidalgo J, Fernandez E (1998) Three Nrt2 genes are differentially regulated in chlamydomonas reinhardtii. Mol Gen Genet MGG 258:373–377
Quiñones MA, Galván A, Fernández E, Aparicio PJ (1999) Blue-light requirement for the biosynthesis of an NO2− transport system in the chlamydomonas reinhardtii nitrate transport mutant S10*. Plant Cell Environ 22:1169–1175
Ronaghi M, Karamohamed S, Pettersson B, Uhlén M, Nyrén P (1996) Real-time DNA sequencing using detection of pyrophosphate release. Anal Biochem 242:84–89
Rothberg JM, Leamon JH (2008) The development and impact of 454 sequencing. Nat Biotechnol 26:1117–1124
Rusk N (2011) Torrents of sequence. Nat Methods 8:44–44
Sanger F, Coulson A (1975) A rapid method for determining sequences in DNA by primed synthesis with DNA polymerase. J Mol Biol 94:447–448
Sanger F, Nicklen S, Coulson AR (1977) DNA sequencing with chain-terminating inhibitors. Proc Natl Acad Sci 74:5463–5467
Schmollinger S, Mühlhaus T, Boyle NR, Blaby IK, Casero D, Mettler T, Moseley JL, Kropat J, Sommer F, Strenkert D, Hemme D, Pellegrini M, Grossman AR, Stitt M, Schroda M, Moseley JL (2014) Nitrogen-sparing mechanisms in chlamydomans affect the transcriptome, the proteome, and photosynthetic metabolism. Plant Cell 26:1410–1435
Shao N, Bock R (2008) A codon-optimized luciferase from gaussia princeps facilitates the in vivo monitoring of gene expression in the model alga chlamydomonas reinhardtii. Curr Genet 53:381–388
Sheehan J, Dunahay T, Benemann J, Roessler P (1998) A look back at the U.S. Department of Energy's Aquatic Species Program – biodiesel from algae. National Renewable Energy Laboratory (NREL), Golden, Colorado
Shishkin AA, Giannoukos G, Kucukural A, Ciulla D, Busby M, Surka C, Chen J, Bhattacharyya RP, Rudy RF, Patel MM (2015) Simultaneous generation of many RNA-seq libraries in a single reaction. Nat Methods 12(4):323–325
Specht M, Stanke M, Terashima M, Naumann-Busch B, JANßEN I, Höhner R, Hom EF, Liang C, Hippler M (2011) Concerted action of the new genomic peptide finder and AUGUSTUS allows for automated proteogenomic annotation of the chlamydomonas reinhardtii genome. Proteomics 11:1814–1823
Stephanopoulos G, Aristidou A, Nielsen J (1998) Metabolic engineering principles and methodologies. Marcel Dekker, New York/Basel

Sugimoto K, Sato N, Tsuzuki M (2007) Utilization of a chloroplast membrane sulfolipid as a major internal sulfur source for protein synthesis in the early phase of sulfur starvation in chlamydomonas reinhardtii. FEBS Lett 581:4519–4522

Sugimoto K, Midorikawa T, Tsuzuki M, Sato N (2008) Upregulation of PG synthesis on sulfur-starvation for PS I in chlamydomonas. Biochem Biophys Res Commun 369:660–665

Surzycki R, Cournac L, Peltier G, Rochaix J-D (2007) Potential for hydrogen production with inducible chloroplast gene expression in Chlamydomonas. Proc Natl Acad Sci 104:17548–17553

Terashima M, Specht M, Naumann B, Hippler M (2010) Characterizing the anaerobic response of chlamydomonas reinhardtii by quantitative proteomics. Mol Cell Proteomics 9:1514–1532

Thompson JF, Steinmann KE (2010) Single molecule sequencing with a heliscope genetic analysis system. Curr Protocol Mol Biol 92:7.10:7.10–7.10.14

Timmins M, Zhou W, Rupprecht J, Lim L, Thomas-Hall SR, Doebbe A, Kruse O, Hankamer B, Marx UC, Smith SM, Schenk PM (2009) The Metabolome of *chlamydomonas reinhardtii* following induction of anaerobic H_2 production by sulfur depletion. J Biol Chem 284:23415–23425

Valouev A, Ichikawa J, Tonthat T, Stuart J, Ranade S, Peckham H, Zeng K, Malek JA, Costa G, Mckernan K (2008) A high-resolution, nucleosome position map of C. Elegans reveals a lack of universal sequence-dictated positioning. Genome Res 18:1051–1063

Wang ZT, Ullrich N, Joo S, Waffenschmidt S, Goodenough U (2009) Algal lipid bodies: stress induction, purification, and biochemical characterization in wild-type and starchless Chlamydomonas reinhardtii. Eukaryot Cell 8:1856–1868

Wase N, Black PN, Stanley BA, Dirusso CC (2014) Integrated quantitative analysis of nitrogen stress response in chlamydomonas reinhardtii using metabolite and protein profiling. J Proteome Res 13:1373–1396

Weers PM, Gulati RD (1997) Growth and reproduction of daphnia galeata in response to changes in fatty acids, phosphorus, and nitrogen in chlamydomonas reinhardtii. Limnol Oceanogr 42:1584–1589

Wiechert W (2001) 13C metabolic flux analysis. Metab Eng 3:195–206

Woessner J, Goodenough UW (1994) Volvocine cell walls and their constituent glycoproteins: an evolutionary perspective. Protoplasma 181:245–258

Work VH, Radakovits R, Jinkerson RE, Meuser JE, Elliott LG, Vinyard DJ, Laurens LM, Dismukes GC, Posewitz MC (2010) Increased lipid accumulation in the chlamydomonas reinhardtii sta7-10 starchless isoamylase mutant and increased carbohydrate synthesis in complemented strains. Eukaryot Cell 9:1251–1261

Yim H, Haselbeck R, Niu W, Pujol-Baxley C, Burgard A, Boldt J, Khandurina J, Trawick JD, Osterhout RE, Stephen R, Estadilla J, Teisan S, Schreyer HB, Andrae S, Yang TH, Lee SY, Burk MJ, van Dien S (2011) Metabolic engineering of Escherichia coli for direct production of 1,4-butanediol. Nat Chem Biol 7:445–452

Zerbino DR (2010) Using the velvet de novo assembler for short-read sequencing technologies. Curr Protocol Bioinform 12: Chapter: Unit-11.5

Zhang Z, Shrager J, Jain M, Chang C-W, Vallon O, Grossman AR (2004) Insights into the survival of chlamydomonas reinhardtii during sulfur starvation based on microarray analysis of gene expression. Eukaryot Cell 3:1331–1348

Zhou J-J, Fernández E, Galván A, Miller AJ (2000) A high affinity nitrate transport system from chlamydomonas requires two gene products. FEBS Lett 466:225–227

Chapter 19
Microalgae as a Source for VLC-PUFA Production

Inna Khozin-Goldberg, Stefan Leu, and Sammy Boussiba

Abstract Microalgae present a huge and still insufficiently tapped resource of very long-chain omega-3 and omega-6 polyunsaturated fatty acids (VLC-PUFA) for human nutrition and medicinal applications. This chapter describes the diversity of unicellular eukaryotic microalgae in respect to VLC-PUFA biosynthesis. Then, we outline the major biosynthetic pathways mediating the formation of VLC-PUFA by sequential desaturation and elongation of C18-PUFA acyl groups. We address the aspects of spatial localization of those pathways and elaborate on the role for VLC-PUFA in microalgal cells. Recent progress in microalgal genetic transformation and molecular engineering has opened the way to increased production efficiencies for VLC-PUFA. The perspectives of photobiotechnology and metabolic engineering of microalgae for altered or enhanced VLC-PUFA production are also discussed.

Keywords Biotechnology • Diversity • Lipid metabolism • Microalgae • Omega-3 LC-PUFA • Omega-6 LC-PUFA

Introduction

Eukaryotic microalgae, an evolutionary diverse group of unicellular and predominantly aquatic photosynthetic organisms, have been widely investigated for their potential to produce high-value health-beneficial compounds, such as carotenoids, polyunsaturated fatty acids (PUFA) and very-long-chain PUFA (VLC-PUFA) (Borowitzka 1988; Cohen and Khozin-Goldberg 2010; Leu and Boussiba 2014). Fatty acids are carboxylic acids with a terminal methyl group positioned at the so-called ω or *n* end. PUFA refer to unsaturated fatty acids containing two and more double bonds. C18 PUFA are 18-carbon fatty acids (C18 stands for the number of

I. Khozin-Goldberg (✉) • S. Leu • S. Boussiba
Microalgal Biotechnology Laboratory, French Associates Institute for Agriculture and Biotechnology of Drylands, J. Blaustein Institutes for Desert Research, Ben-Gurion University of the Negev, Sede Boqer Campus 84990, Israel
e-mail: khozin@bgu.ac.il

Y. Nakamura, Y. Li-Beisson (eds.), *Lipids in Plant and Algae Development*, Subcellular Biochemistry 86, DOI 10.1007/978-3-319-25979-6_19

carbons in the fatty-acyl chain), among them linoleic acid (LA, 18:2 *n*-6) and α-linolenic acid (ALA, 18:3 *n*-3) which are synthetized by plants and essential for animals. In many land plants and microalgae, both C18 PUFA and C16 PUFA are abundant constituents of the chloroplast-membrane glycerolipids. Most acyl groups associated with glycerolipids in plants and algae contain double bonds in a *cis* configuration. Microalgae are primary natural producers of VLC-PUFA, most commonly C20 and C22 highly unsaturated fatty acids containing four to six double bonds, including arachidonic acid (AA, $20{:}4^{\Delta 5,8,11,14}$, *n*-6), eicosapentaenoic acid (EPA, $20{:}5^{\Delta 5,8,11,14,17}$, *n*-3) and docosahexaenoic acid (DHA, $22{:}6^{\Delta 4,7,10,13,16,19}$, *n*-3), which are mostly known for their nutritional importance. The position of the C–C double bond is denoted by Δ nomenclature: the double-bond location is counted from the first carbon atom in the carbon chain located at the carboxylic group (designated α). PUFA are divided into two major families according to the position of the last double bond proximal to the methyl end (designated *n* or ω), known as omega-3 (ω-3) and omega-6 (ω-6) (alternatively *n*-3 and *n*-6), and their biosynthetic origin will be discussed in the following sections.

Photosynthetic planktonic microalgae are major contributors to the global carbon and nitrogen cycles, as well as primary producers of VLC-PUFA supporting productive marine food webs. Various eukaryotic marine microalgae possess the biosynthetic machinery for sequentially alternating desaturation and elongation of C18-PUFA acyl groups, mediating the formation of VLC-PUFA. The vascular land plants lack the enzymes for C18-PUFA elongation and do not produce VLC-PUFA. In contrast, in microalgae of numerous taxonomic groups, VLC-PUFA are abundant components of the membrane acyl lipids. Situated at the base of the marine food web as primary producers, phytoplanktonic microalgae supply VLC-PUFA to marine fish, which are impaired in the capacity to synthesize VLC-PUFA from the essential PUFA LA and ALA (Tocher 2003). Hence, in marine ecosystems, microalga-derived VLC-PUFA play a pivotal role in enriching aquatic organisms in these essential nutritional ingredients, which are incorporated into membrane lipids and accumulate in the oils and flesh of many fish species. Cultured marine fish require a dietary supply of VLC-PUFA for optimal growth and welfare (Bell and Sargent 2003).

The critical role of the ω-3 VLC-PUFA EPA and DHA in providing adequate dietary and health benefits has been widely investigated and publicly acknowledged (Simopoulos 2002, 2008; Calder 2006; Gil et al. 2012) with novel studies underpinning the complex mechanisms of action (Calder 2013). We do not intend, here, to provide a comprehensive overview of the health-beneficial effects of VLC-PUFA, which have been widely described in numerous dedicated works. Rather, we will emphasize the essential nature of VLC-PUFA as vital components of the human diet and draw the reader's attention to the significance of these molecules as the major subject of this chapter.

VLC-PUFA are essential nutrients in the development and functioning of the brain and visual systems (reviewed in Janssen and Kiliaan 2014). EPA and DHA play essential roles in regulating immune and inflammatory responses, neural development, and exerting beneficial effects in cardiovascular, neurological, and proliferative diseases (Marszalek and Lodish 2005; Le et al. 2009; Adkins and Kelley 2010; Crawford and Broadhurst 2012; Rangel-Huerta et al. 2012; Yang et al. 2012; Laviano et al. 2013; Betancor et al. 2015). Studies on the atheroprotective role of the ω-3 PUFA EPA and DHA in regulating macrophage foam-cell formation have implicated them as potential therapeutic agents for the treatment of clinical atherosclerosis (McLaren et al. 2011). The importance of adequate uptake of ω-3 PUFA has also been demonstrated in other metabolic disorders, for instance, in some studies enhanced ω-3 PUFA intake was associated with a reduced risk of amyotrophic lateral sclerosis and with lowering inflammation in adults with one or more features of metabolic syndrome (for details please refer in Robinson and Mazurak 2013). A role for dietary EPA in cardioprotection relates to its content in membrane lipids of the myocardium and to its role as a metabolic precursor in the synthesis of anti-inflammatory eicosanoid molecules, such as three-series prostaglandins, with proinflammatory properties. DHA and its precursor docosapentaenoic acid (DPA, 22:5 *n*-6) are precursors for eicosanoid-like lipid mediators, such as resolvins and protectins, playing a role in regulating inflammation (Serhan et al. 2002; Dangi et al. 2009). DHA potentially reduces the levels of neural inflammation through prevention of microglial activation by proinflammatory cytokines or through production of anti-inflammatory and antioxidative compounds (Bazan et al. 2011), and it is a vital determinant of neuronal migration, neurogenesis and the expression of several genes involved in brain development and function (Crawford and Broadhurst 2012). DHA originating from the marine food web is presumed to have played a critical role in human evolution, serving specific needs in the development of neural cell signaling and, globally, in maintaining human mental health.

The ω-6 VLC-PUFA AA is an important acyl component of brain-membrane glycerolipids and the major VLC-PUFA in breast milk; it is essential for infants' neuronal and cognitive development, particularly in preterm and non-breast-fed babies. This VLC-PUFA has been included in baby formulas in many countries worldwide to meet the requirements of infant nutrition and brain development. Intriguing new discoveries have revealed a role for AA and its precursor dihomo-γ-linolenic acid (DGLA; $20{:}3^{\Delta 5,8,11}$, *n*-6) in promoting resistance to fasting and extending life span under conditions of abundant food (O'Rourke et al. 2013). Involvement of the autophagy response has been implicated in this phenomenon, as supplementation with AA and its precursor DGLA, but not EPA, activated the autophagous surveillance responses that normally occur in calorically restricted animals. Of note, DGLA has attracted considerable interest as a pharmaceutical and nutraceutical compound with potent anti-inflammatory, anti- proliferative and anti-vascular activities related to production of prostaglandin E_1 (reviewed in Wang et al. 2012).

Indeed, with increasing public attention being drawn to their beneficial roles in human health and nutrition, the demand for the nutritionally important VLC-PUFA is on the rise. Moreover, an optimal supply of EPA and DHA, as well as adequate provision of AA, are also essential for farmed fish (Bell and Sargent 2003), especially in the larval stage and under stress (Koven et al. 2001, 2003). Natural marine sources of VLC-PUFA are continuously diminishing as fish catches reach a plateau and in some cases even exceed maximal sustainable levels (Shepherd and Jackson 2013). Thus, natural marine resources cannot satisfy an ever-growing demand for the salubrious VLC-PUFA and cannot fill the considerable gap in the actual versus recommended intake of DHA and EPA in most of the world's population (Winwood 2013). Furthermore, current natural sources of ω-3 VLC-PUFA are either limited or of unsatisfactory quality for human nutrition, with most of the VLC-PUFA being supplied from fish oil. The problem of fish-oil contamination with heavy metals and pesticides could potentially be solved by producing refined fish oil, but this would result in even higher prices due to greater input and operating costs. Hence, providing the recommended VLC-PUFA levels to a growing world population calls for additional sustainable supplies of appropriate quality for the human neutraceutical market. Today, most industrially produced DHA is resourced from the fermentation of a number of heterotrophic microalgae and marine unicellular protists (Barclay et al. 1994; Apt and Behrens 1999; De Swaaf et al. 2003; Winwood 2013). Similarly, AA-rich oil for the infant-formula industry is produced heterotrophically by fungal fermentation (Kyle 1997). A novel source of EPA is genetically modified oleaginous yeast *Yarrowia lipolytica*, which is cultivated at commercial scale (Xie et al. 2015). Recent years have witnessed increased interest in phototrophic VLC-PUFA production by microalgae, with several companies developing photobiotechnologies to produce EPA and DHA for human nutrition (Khozin-Goldberg et al. 2011; Leu and Boussiba 2014). Many marine planktonic species are cultivated in mariculture to supply marine fish with essential ω-3 VLC-PUFA.

Interest in microalgal lipids has further increased with the recent focus on renewable resources for biodiesel production, emerging from biochemical and genomic discoveries related to lipid-biosynthesis pathways in microalgae (Liu and Benning 2013). Photosynthetic microalgae have been identified as a potential feedstock for biofuels due to their capacity to produce high biomass and accumulate over 50 % of their dry weight as storage oil (triacylglycerol, TAG), generally under stressful conditions (Hu et al. 2008). Concurrent accumulation of oil and its high enrichment in PUFA and VLC-PUFA, however, is a rare phenomenon in microalgae (Cohen and Khozin-Goldberg 2010). Nevertheless, some species are able to deposit VLC-PUFA in their TAG under certain cultivation conditions (Bigogno et al. 2002b, c; Tonon et al. 2002; Guihéneuf and Stengel 2013), but only *Lobosphaera incisa* (formerly *Parietochloris*) has so far been reported to accumulate AA as a major proportion of its TAG (Bigogno et al. 2002b, c; Khozin-Goldberg et al. 2002a).

This chapter covers microalgal diversity in relation to PUFA and VLC-PUFA production, provides mechanistic insights into the biosynthesis of PUFA and VLC-PUFA in eukaryotic microalgal species, highlights biotechnological applications, and describes recent genetic-engineering approaches to manipulating VLC-PUFA production in microalgae.

Microalgal Diversity with Respect to PUFA and VLC-PUFA Biosynthesis

The term "microalgae" actually denotes a huge, ecologically and evolutionarily diverse group of eukaryotic, unicellular, primarily photosynthetic, aquatic organisms. However, in addition to aquatic habitats, microalgal classes of the core green algal lineage (such as Chlorophyceae and Trebouxiophyceae) are often encountered in aero-terrestrial and soil ecosystems (Holzinger and Karsten 2013). Diatoms, belonging to the Heterokonta (Stramenopiles), comprise the dominant and most abundant marine phytoplankton group in nature but may be present in freshwater ecosystems as well (Cermeño and Falkowski 2009; Mann and Vanormelingen 2013), and even terrestrial species are known. The enormous diversity of the eukaryotic microalgae is thought to stem from the evolutionary origin of the plastids that carry out photosynthesis in these organisms (Bhattacharya et al. 2004; Archibald 2005). The evolutionary history of photosynthetic eukaryotes consists of three endosymbiotic events (primary, secondary and tertiary) that are recognized by the origin of the endosymbiont engulfed by the eukaryotic host (Fig. 19.1). The presence and functional roles of distinct VLC-PUFA may possibly give further insight into the evolutionary events in the different microalgal groups.

Much information on VLC-PUFA biosynthesis in microalgae has been obtained by mining genome information in an increasing number of sequenced species from different phyla (Tonon et al. 2005; Sayanova et al. 2011a; Petrie et al. 2010, b; Ahman et al. 2011; Vieler et al. 2012b; Vaezi et al. 2013), as well as from case and global studies on these organisms' fatty-acid profiles (Alonso et al. 1998; Viso and Marty 1993; Dunstan et al. 1992;1994; Zhukova and Aizdaicher 1995; Lang et al. 2011). An exceptionally large-scale study was carried out on the systematic analysis of fatty-acid profiles of more than 2000 species of microalgae (predominantly freshwater) deposited in the SAG Culture Collection of Algae (Lang et al. 2011). Fatty-acid profiling revealed phylogenetic relationships at the phylum and class levels, but at lower taxonomic levels, fatty-acid distribution could not differentiate closely related species (Lang et al. 2011). Studies with radiolabeled precursors of VLC-PUFA biosynthesis provided some mechanistic insights into the pathways of PUFA and VLC-PUFA biosynthesis in a number of species from divergent microalgal groups (Henderson and Mackinlay 1992; Arao et al. 1994; Schneider and Roessler 1994; Khozin et al. 1997; Makewicz et al. 1997; Bigogno et al. 2002a; Khozin-Goldberg et al. 2002b).

A brief glimpse on a complex evolutionary history of microalgae is necessary to outline the diversity of these organisms in relation to their fatty acid composition and occurrence of VLC-PUFA. The primary plastid in the microalgal cells of the supergroup Plantae (Rhodophyta, Glaucophyta, and Viridiplantae, comprising green algae and land plants) is of cyanobacterial (prokaryotic) origin and ultimately, is the ancestor of all eukaryotic photosynthetic organisms. The primary endosymbiosis event traces back to a single event that probably occurred over one billion years ago (Yoon et al. 2004). As a result of secondary endosymbiosis, the

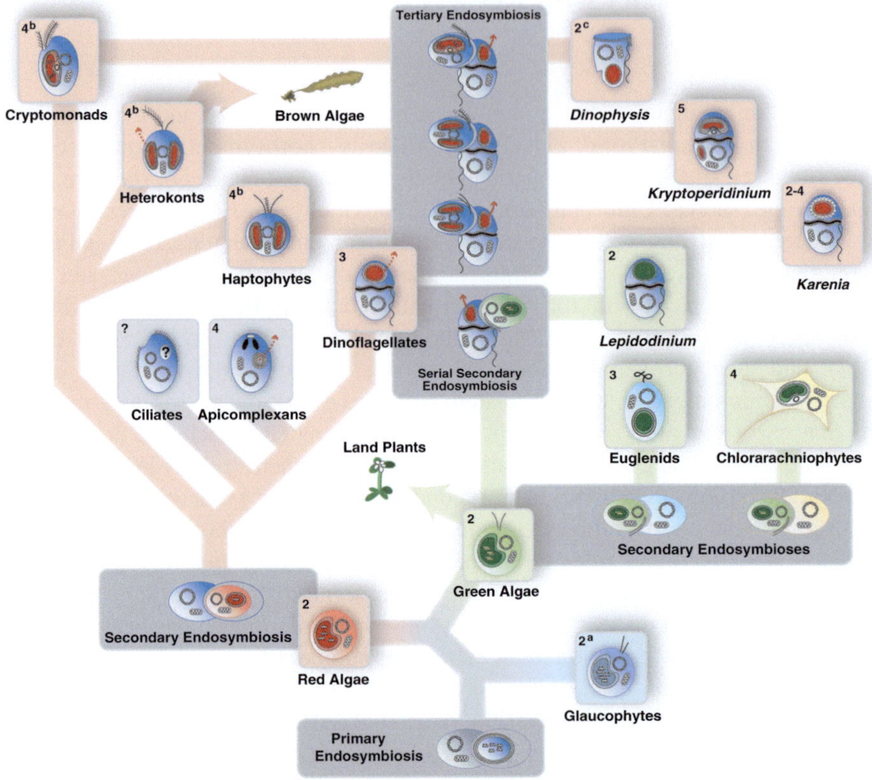

Fig. 19.1 A hypothesis for the origin and spread of plastids based on a consensus of molecular, morphological, and biochemical data (Courtesy of Prof. J. Archibald and John Wiley and Sons). For further details, please refer to Archibald (2005)

photosynthetic organelles of red and green algae dispersed to unrelated recipient eukaryotes, providing enormous diversity of microalgae (Archibald 2005; Nakayama and Archibald 2012) (Fig. 19.1). The complex plastid in the cells of "chromalveolates" [Heterokonta (Stramenopiles), Cryptophyta, Haptophyta)] is thought to have evolved via secondary or tertiary endosymbiosis of a red alga with a secondary eukaryotic host. In these unicellular organisms, the plastid is surrounded by three or four membranes, implying complexity of nuclear-organelle communication and protein trafficking. The Stramenopiles clade includes diverse photosynthetic microalgae with secondary plastids of red-lineage origin that contain chlorophyll *c* in addition to chlorophyll *a*, and feature complex chloroplast architecture (Bhattacharya et al. 2004). According to current classification, the Stramenopiles clade unites of a broad group of unicellular secondary plastid-bearing photosynthetic and heterotrophic VLC-PUFA-producing microalgae, such as diatoms, eustigmatophytes and dinoflagellates. Heterotrophic oomycetes, thraustochytrids, and labyrinthulids, producing VLC-PUFA, are also positioned in the

Stramenopiles, and these organisms are related to the recipient eukaryotic hosts (the heterotrophic ancestors) of the aforementioned eukaryotic microalgae. A secondary plastid of the extant euglenids and chlorarachniophytes, as one small group of dinoflagellates, is of the green algal origin (Keeling 2010).

In the green lineage of microalgae (Chlorophyta) with a primary plastid, the evolutionary history seems to be reflected in the fatty-acid composition of their glycerolipids. The marine picoplanktonic microalgae, belonging to the phylum Prasinophyceae, represent an early-diverging branch in green plant evolution (Leliaert et al. 2012; Lemieux et al. 2014), and they have the ability to synthesize the VLC-PUFA EPA and DHA (Domergue et al. 2005; Petri et al. 2010b, 2013; Hoffmann et al. 2008) and vary from core green algae in various aspects of lipid metabolism. In this respect, the picoplanktonic marine microalgae of the genera *Ostreococcus* and *Micromonas* have been the most studied, with a number of genes involved in the VLC-PUFA-biosynthesis pathway having been cloned and characterized (Meyer et al. 2004; Domergue et al. 2005; Petri et al. 2010a, b; Ahman et al. 2011; Vaezi et al. 2013).

The microalgae of the red algal lineage (Rhodophyta), e.g. *Porphyridium*, *Rhodosorus* and *Rhodella*, evolved in the course of the primary endosymbiosis of an ancestor cyanobacterium with a nonphotosynthetic eukaryotic host, presumably carrying the enzymes necessary for C18 PUFA elongation and desaturation, and able to produce C20 AA and EPA, but not C22 VLC-PUFA, such as DHA (Dunstan et al. 2005). This is exemplified by the fatty-acid composition of the well-studied *Porphyridium purpureum* (formerly *P. cruentum*) (Porphyridiophyceae) (Khozin et al. 1997; Cohen 1999), and several other unicellular rhodophytes (Dunstan et al. 2005). Similarly to Rhodophyta, only AA and EPA are encountered in Glaucophyta (Leblond et al. 2010a, b; Lang et al. 2011), evolved in the course of primary endosymbiosis and retaining the functional cyanobacterium as a photosynthetic organelle.

In contrast, the unicellular representatives of the core green lineage (such as Ulvophyceae, Chlorophyceae and Trebouxiophyceae) more closely related evolutionarily to the vascular plants have only been rarely characterized for the presence of VLC-PUFA (Lang et al. 2011). In general, core chlorophytes are most similar to vascular plants with respect to the fatty-acid composition of their acyl lipids; however, many representatives of this group synthesize Δ6 C18 PUFA, such as γ-linolenic acid (GLA, $18{:}3^{\Delta 6,9,12}$, *n*-6) and stearidonic acid (SDA, $18{:}4^{\Delta 6,9,12,15}$, *n*-3), fatty acids that are restricted to some specialized higher plants (e.g. of the family Boraginaceae) and certain species of cyanobacteria. Methylene-interrupted C18 PUFA with an unusual arrangement of double bonds are found in *Chlamydomonas reinhardtii*, namely pinolenic acid (PA; $18{:}3^{\Delta 5,9,12}$, *n*-6) and coniferonic acid (CA; $18{:}4^{\Delta 5,9,12,15}$), as well as enzyme implicated in their formation had been identified (Kajikawa et al. 2006). Furthermore, VLC-PUFA, such as AA and EPA, have been identified in the Chlorophyceae, Trebouxiophyceae and Ulvophyceae, albeit at low frequencies (Bigogno et al. 2002b; Lang et al. 2011). Notably, Lang et al. (2011) reported on a few cases of extraordinary high abundancies of AA, namely in the chlorophyte *Palmodictyon varium* (SAG 3.92), the chlorophyte *Trochisciopsis tetraspora* (SAG 19.95) and in the trebouxiophyte *Myrmecia bisecta* (SAG 2043) (synonym to *Lobosphaera incisa*). In *L. incisa*, relevant enzymes to AA biosynthesis have been identified and characterized (Iskandarov et al. 2009, 2010). Low concentrations of AA and EPA are detectable in the green microalga *Haematococcus pluvialis* (Chlorophyceae) (Peled et al. 2011). An evidence of a rare presence of EPA in green algae was supported by a large-scale analysis (Lang et al. 2011) but three strains demonstrated a high EPA proportion of about 20 % of total fatty acids (*Chlorella* sp. SAG 242.80; *Chlamydomonas allensworthii* SAG 28.98; *Cylindrocapsa involuta* SAG 314-1). The detailed examination of the above-mentioned strains with unusually high content of AA and EPA may shed more light on their exceptional ability to synthetize those fatty acids. Notably, in the Euglenoids – the secondary endosymbionts with the plastid of green alga origin – VLC-PUFA are found at high proportions of total fatty acids (Lang et al. 2011). It can be speculated that synthesis of VLC-PUFA is an ancient trait, maintained in the early diverging green algae Prasinophyceae and Rhodophyta, which seems to have been either completely lost or retained at low frequencies during the evolution of green lineage and higher plants. The degree to which the extensive replacement of VLC-PUFA by the shorter PUFA is related to the significant changes in thylakoid and chloroplast membrane architecture of the green lineage (chlorophylls *a*/*b* associated photosynthesis and LHCP complexes) (Wolfe et al. 1994; Keeling 2004; Rodriguez-Ezpeleta et al. 2005) and the functional consequences of this replacement remains to be investigated.

Multiple scenarios can be proposed in the non-monophyletic evolution of the secondary and tertiary endosymbionts and in spreading the genes associated with the VLC-PUFA biosynthesis in order to delineate the evolutionary origin and cellular topology of VLC-PUFA biosynthesis in the organisms with three- and four-plastid membranes, and much remains to be elucidated (Petroutsos et al. 2014). Certain groups of the heterokont microalgae (Stramenopiles) synthesize the VLC-PUFA EPA and DHA to different extents, as well as do haptophytes that form a sister group to the clade containing the heterokonts. However, DHA is not found in Eusigmatophytes and Xanthophytes, and EPA represents the major VLC-PUFA in these microalgal classes. Diatoms comprise an enormously abundant and highly dispersed group of phytoplanktonic microalgae, with the number of extant species as estimated by Mann and Vanormelingen (2013) to be at least 30,000 and even ca. 100,000, by extrapolation from an eclectic sample of genera and species complexes. A clear distinction between the different examined diatoms cannot be deduced based on fatty acid profile, as many diatoms have both EPA and DHA, however their proportions vary. Furthermore, the evolutionary history of the diatoms has been postulated to be even more complicated by likely multiple endosymbioses involving the capture of foreign cells, horizontal gene transfer into the host genome and recruitment of foreign genes (Chan et al. 2012). The so-called "green" genes recently discovered in diatom genomes seem to be evolutionarily related to the ancestor prasinophytes (Moustafa et al. 2009). This could have profound consequences on lipid metabolism in general and repertoire of VLC-PUFA biosynthesis genes, in particular.

Besides frequently mentioned in above sections, EPA and DHA, highly unsaturated C18 fatty acids with four or five double bonds are ubiquitous in some microalgal groups. The marine coccolithophore *Emiliania huxleyi* (Haptophyta) synthetizes high proportions of octadecapentaenoic acid (OPA, $18{:}5^{\Delta 3,6,9,12,15}$, *n*-3) (Sayanova et al. 2011a). Highly unsaturated C18 fatty acids with four or five double bonds are abundant in dinoflagellates (Dinophyceae), which have been shown to contain OPA as well (Leblond et al. 2003, 2010a) along with EPA and DHA. The precursor of DHA, DPA ($22{:}5^{\Delta 7,10,13,16,19}$, *n*-3), was identified in particularly high proportions in a polar lipid fraction of representatives of the class Chlorarachniophyceae (Leblond et al. 2005), a group of unicellular eukaryotic algae has apparently evolved as a result of secondary endosymbiosis of a green alga by a nonphotosynthetic amoeba or amoeboflagellate. The blue–green *Chroomonas placoidea* (Cryptophyta) is characterized by a low proportion of DHA but a significant proportion of DPA (Dunstan et al. 2005). Highly unsaturated VLC-PUFA with even longer carbon chains—octacosaoctaenoic acid (28:8 *n*-3) and octacosaheptaenoic acid (28:7 *n*-6)—have been identified in the toxic dinoflagellate *Karenia brevis* (Dinophyceae) (Leblond et al. 2003).

C16 PUFA, with number and location of double bonds differing from higher plants, may also occur in groups of microalgae. For instance, in contrast to hexadecatrienoic acid ($16{:}3^{\Delta 7,10,13}$), which is an abundant trienoic fatty acid in chloroplast glycerolipids in higher plants, the diatom *Phaeodactylum tricornutum* contains its isomer $16{:}3^{\Delta 6,9,12}$ with distinct location of its double bonds (Domergue et al. 2003b). The hexadecatetraenoic acid $16{:}4^{\Delta 4,7,10,13}$ is the dominant acyl component of the major chloroplast galactolipid in the green alga *Chlamydomonas reinhardtii* (Zäuner et al. 2012), and in many other chlorophytes.

Biosynthesis of VLC-PUFA in Microalgae

General Description of Desaturases Involved in PUFA and VLC-PUFA Biosynthesis

Double bonds are introduced into fatty-acyl chains by the activity of fatty-acid desaturases (FAD). An alternative pathway for the biosynthesis of VLC-PUFA that is catalyzed by polyketide synthase (PKS) and was discovered in heterotrophic heterokonts will be noted later in this section. Desaturases are non-heme monooxygenases that catalyze aerobic C–C double-bond formation by hydrogen abstraction with a parallel reduction of one molecule of oxygen to water. Desaturase activity requires molecular oxygen, reducing equivalents and electron donors. Desaturases feature three conserved histidine-box motifs [HX_3-4H, $HX_{2\text{-}3}HH$, and $(H/Q)X_{2\text{-}3}HH$], generally with eight essential histidines that are important for the formation and coordinate functioning of the di-iron center essential for FAD activity (Heinz 1993; Shanklin and Cahoon 1998). The resultant double bonds are in the *cis* configuration, but some exceptions are known, such as in the formation of $16{:}1^{\Delta 3}$*trans* in plastidic phosphatidylglycerol. Notably, the latter reaction is catalyzed by a distinct type of desaturase featuring only two histidine boxes and belonging to a different superfamily (Gao et al. 2009). The desaturation reactions in the plastid and in the endoplasmic reticulum (ER) are generally catalyzed by membrane-bound desaturases that use acyl groups bound to glycerolipids; they are therefore referred to as acyl-lipid desaturases. The exception is the desaturation of stearoyl-acyl carrier protein (ACP) which is mediated by a soluble desaturase in the stroma of chloroplast. Importantly, acyl-CoA-dependent desaturases engaged in VLC-PUFA biosynthesis have been identified in the Prasinophyceae microalgae (Domergue et al. 2005).

The soluble flavoprotein ferredoxin is used as the electron donor for the chloroplast desaturases, whereas cytochrome b5 is used as the primary electron donor in fatty-acid desaturation in the ER. Some desaturases in the ER, such as the Δ9, Δ12 and ω-3 desaturases, rely on free cytochrome b5, but cytochrome b5 may appear fused with the desaturase domain in a single protein at either the N or C terminus. The N-terminal cytochrome b5 domain is a typical feature for the front-end desaturases (Δ4, Δ5, Δ6 and Δ8 desaturases), introducing the double bond between the preexisting double bond and the carboxyl group in the VLC-PUFA-biosynthesis pathway (Napier et al. 2003). These desaturases are deemed to rely on their own electron donor (Sperling and Heinz 2001).

Biosynthetic Origin of the Chloroplast Galactolipids and Plastidic Desaturation

We will further cover the pathways involved in VLC-PUFA biosynthesis, but first, we provide a brief general description of the initial steps of fatty-acid synthesis and modifications, including the plastidic desaturases. In the course of de-novo biosynthesis, fatty acids are produced in the plastid by a type II fatty-acid synthase (FAS), which is a dissociable enzymatic complex that generally produces C16 and C18 fatty acids. This process is initiated by condensation of acetyl-CoA with malonoyl-ACP, and is terminated by the release of fatty acids in the form of ACP adducts. However, in many microalgae of secondary endosymbiosis origin (hereafter termed secondary endosymbionts), fatty-acyl chains with a smaller number of carbons (C14 and even C12) may be present as components of their glycerolipids. Recent genomic studies have indicated the likely presence of the eukaryotic type I FAS (a heterodomain multifunctional enzyme) operating in microalgal cells (Vieler et al. 2012b; Blanc et al. 2012) and, perhaps, in some other secondary and tertiary endosymbionts. The occurrence of the eukaryote-like FAS system can be evolutionarily related to the ancestral recipient organism, and might have evolved independently in different taxa. Two types of FAS have been revealed in the photosynthetic secondary endosymbiont *Euglena gracilis* (Euglenophyceae); the organism has a complex plastid bearing chlorophyll *a* and *b* and enveloped in three membranes due to the secondary endosymbiosis with a green alga.

The fatty acids remaining in the plastid—palmitic acid (16:0) and oleic acid (18:1)—are incorporated into the chloroplast glycerolipids and undergo sequential lipid-linked desaturation in the so-called prokaryotic pathway (indicating the evolutionary origin of the plastid). As was elucidated in mutant studies in higher plants (*Arabidopsis thaliana*), in the prokaryotic pathway of glycerolipid synthesis, which is entirely accomplished in the chloroplast, 16:0 and 18:1, attached to the *sn*-2 and *sn*-1 positions of the glycerol backbone (*sn* stands for stereochemical numbering of the glycerol backbone) of chloroplast-membrane lipids, respectively, are sequentially desaturated by the plastid-localized desaturases (Somerville and Browse 1996). In higher plants, as well as in the core green microalgae (Chlorophyceae, Trebouxiophyceae), this pathway yields chloroplast galactolipids with C16 fatty acids attached to the *sn*-2 position of the glycerol moiety. In plants, the major chloroplast membrane lipid monogalactosyldiacylglycerol (MGDG) is formed in this pathway with the *n*-3 trienoic acids hexadecatrienoic acid ($16{:}3^{\Delta7,10,13}$) and ALA ($18{:}3^{\Delta9,12,15}$) located at positions *sn*-2 and *sn*-1, respectively. It is thought that only the prokaryotic pathway of chloroplast galactolipid synthesis operates in the green microalga *Chlamydomonas reinhardtii*, which lacks phosphatidylcholine (PtdCho), the main phospholipid implicated in the eukaryotic pathway outside the chloroplast (see further on). In *C. reinhardtii*, MGDG is mainly comprised of the molecular species $18{:}3^{\Delta9,12,15}$ and $16{:}4^{\Delta4,7,10,13}$ (at *sn*-1 and *sn*-2, respectively), whereas digalactosyldiacylglycerol (DGDG) contains mostly $18{:}3^{\Delta9,12,15}$/16:0 (*sn*-1/*sn*-2) species (Giroud and Eichenberger 1988, 1989; Riekhof et al. 2005; Zäuner et al. 2012).

The eukaryotic pathway for the biogenesis of chloroplast membrane lipids involves the extraplastidial glycerolipids, such as the phospholipid PtdCho which plays a central role in providing the glycerol backbone and unsaturated acyl groups for chloroplast membrane lipids formed by this route. According to the current view, in higher plants, the de-novo-produced fatty acid 18:1 that is exported from the plastid are incorporated into the sn-2 position PtdCho. The exclusive presence of C18 fatty acids at the *sn*-2 position is the hallmark of the eukaryotic pathway in plants, while 16:0 is generally restricted to the *sn*-1 position of the glycerole backbone. A large fraction of the unsaturated acyl groups are exported back to the chloroplast for final ω-3 desaturation and formation of the eukaryotic type 18:3 *n*-3/18:3 *n*-3 MGDG (*sn*-1/*sn*-2). This scheme presumes tight coordination between compartments, involving many players, dedicated translocators, acyl-editing mechanisms and trafficking of intermediates, such as phosphatidic acid (for further information see Benning 2009; Bates et al. 2013). As can be envisaged, the accepted higher plant models can exhibit differences in microalgae due to the latter's distinct evolution, cellular architecture and versatility of lipid-biosynthesis routes (Harwood and Guschina 2009; Liu and Benning 2013; Petroutsos et al. 2014; Abida et al. 2015).

In several studies of VLC-PUFA-producing heterokonts and rhodophytes, the C20 PUFA AA and EPA have been identified in the chloroplast and extraplastidial membrane lipids in the same stereo-specific arrangement as C18 PUFA in higher plants, with C20 PUFA exclusively occupying the *sn*-1 position of the prokaryotic-like galactolipids, whereas these fatty acids are present in the C20 PUFA/C20 PUFA arrangement in the eukaryotic-like galactolipids (Khozin et al. 1997; Khozin-Goldberg et al. 2002b). C16 PUFA are almost excluded or present at very low levels in the galactolipids of rhodophytes (*Porphyridium cruentum*). In secondary endosymbionts with plastids originating from red algae, such as the heterokont microalgae *Nannochloropsis* and *Monodus* (Eustigmatophyceae), 16:0 and/or 16:1 represent the bulk of the acyl groups at the *sn*-2 position of MGDG, with virtual absence of C16 PUFA. On the other hand, C16 PUFA are abundant at the *sn*-2 position in galactolipids of the primary endosymbionts of the green lineage, such as prasynophytes, as exemplified by the MGDG of *Tetraselmis* sp. with the prokaryotic-like 18:5/16:3, 18:5/16:4 and 18:3/16:4 MGDG (Leblond and Lasiter 2009), and of the secondary endosymbionts, as exemplified by the diatom *P. tricornutum* (Abida et al. 2015). The dinoflagellate *Lepidodinium chlorophorum* (tertiary endosymbiont) shares, to a certain extent, the composition of MGDG with the free-living *Tetraselmis* sp., presumably acquiring this biochemical feature from the ancestral photosynthetic organism, related to the Prasinophyceae, in the course of the tertiary endosymbiosis (Leblond and Lasiter 2009).

As noted above, the plastidial desaturation pathway is initiated with a soluble plastidial Δ9-stearoyl desaturase (SAD). All other desaturation steps in the chloroplast occur when the acyl groups are attached to membrane glycerolipids. Microalgal plastidial ACP desaturases are capable of placing a double bond in other positions of the saturated fatty acid, such as the Δ7 position to form monounsaturated $16{:}1^{\Delta 7}$, as can be deduced from the acyl-group composition of chloroplast lipids and the preliminary genome annotation in the diatom *Phaeodactylum tricornutum* (Dolch and Marechal 2015). Distinctly, formation of monounsaturated fatty acids in the ER is catalyzed by the acyl-CoA-dependent desaturases. The ultra-small red alga *Cyanidioschyzon merolae*, with its very compact cell anatomy, lacks the entire plastidial desaturation pathway but possesses the ER-localized stearoyl-CoA desaturase featuring the C-terminal cytochrome b5 domain (Itoh et al. 1998; Sato and Moriyama 2007). Four isoforms of the Δ9 desaturase containing a C-terminal cytochrome b5 domain were also identified in the oleaginous diatom *Fistulifera* sp. along with two plastidial Δ9 acyl-ACP desaturases (Muto et al. 2013). An unusual desaturase, containing a cytochrome b5-like domain at its N terminus, from the marine diatom *Thalassiosira pseudonana* was shown to convert 16:1 to $16{:}1^{\Delta 11}$ (Tonon et al. 2004).

The chloroplastic lipids of the diatom *P. tricornutum* contain an unusual Δ6 C16 PUFA, hexadecatrienoic acid ($16{:}3^{\Delta 6,9,12}$, *n*-4) (Domergue et al. 2003b), whose formation is commenced by activity of the chloroplast-localized PtFAD6 which inserts the Δ6 double bond into $16{:}1\Delta^{9}$. This desaturase was functionally characterized in the photosynthetic cyanobacterium *Synechococcus* and was localized to the chloroplast. The N terminus of PtFAD6 bore a bipartite sequence typical of the nuclear-encoded and chloroplast-targeted diatom proteins required for transport into the complex diatom plastid.

The second and third double bonds are inserted by Δ12 and ω-3 lipid-linked desaturases, respectively. Whereas at least two isoforms of Δ12 desaturase (FAD6 in the plastid and FAD2 in the ER) are present in plants and algae, and two differentially localized isoforms of ω-3 desaturase are present in higher plants (FAD7/FAD8 in the plastid and FAD 3 in the ER)——only a single ω-3 desaturase (CrFAD7) is encoded by the *C. reinhardtii* genome (Nguyen et al. 2013). Experiments with green fluorescent protein-tagged CrFAD7 confirmed its localization to the chloroplast. Intriguingly, a comparative lipidomic analysis of the *crfad7* mutant and wild type revealed that the mutation impacts glycerolipids in both the chloroplast and extraplastidial membranes. There are two likely explanations for this phenomenon: either CrFAD7 accesses both plastidial and ER-located acyl-lipid substrates through ER–plastid contact sites, or the trafficking of precursors between compartments might be involved. Furthermore, the presence of a single ω-3 desaturase seems not to be restricted to *C. reinhardtii*; database searches have identified a single homolog in some other sequenced green algae (*Chlorella variabilis* NC64A, *Coccomyxa subellipsoidea* C-169 and *Volvox carteri*), as well as in the genomes of the VLC-PUFA-producing diatom *P. tricornutum*, the eustigmatophyte *Nannochloropsis*, and the chlorophyte *Lobosphaera incisa* (personal communication). This implies that ω-3 desaturase duplication occurred when the land plants diverged, and the microalgal desaturases act as an ω-3 desaturase counting from the ω-end (and not as Δ15 or Δ17) and may potentially be able to desaturase C18 and C20 substrates, depending on the organism's evolutionary origin. Regarding the cellular localization of the ω-3-desaturase-catalyzed reaction, the occurrence of the so-called chloroplast ER in heterokont microalgae (Kroth 2007) and recent studies in green microalgae support functional continuity between the chloroplast and ER (Peled et al. 2012; Davidi et al. 2014). This situation is more complicated in secondary and tertiary endosymbiont microalgae with their complex multienveloped plastid, particularly in diatoms with the so-called chloroplast–ER-presenting continuum of the outermost chloroplast membrane and the ER (Dolch and Maréchal 2015). Of course, these assumptions require experimental validation. The deduced ω-3 desaturase protein of the haptophyte *Emiliania huxleyi* featured a Δ6 desaturase-like domain, variant histidine boxes and a chloroplast-transit sequence, and formed a monophyletic group with its putative orthologs, exclusively from prasinophytes and the cryptophyte *Guillardia theta* [secondary endosymbiont with the descendant nucleus (nucleomorph) and cytoplasm of the donor red alga] (Kotajima et al. 2014). Horizontal gene transfer might explain this phenomenon. Expression of *EhDES15* increased in response to a decrease in temperature (from 25 to 15 °C) in line with the increase in C18 PUFA in *E. huxleyi*. Functional expression of *EhDES15* in the mutant strain of the cyanobacterium sp. *PCC* 6803, lacking endogenous Δ15 desaturase gene (*DesB*), led to an increase in the proportions of C18 PUFA 18:3 *n*-3 and 18:4 *n*-3, supporting the predicted function in ω-3 (Δ15) desaturation of 18:2 *n*-6 in the chloroplast.

A plastidic Δ4-MGDG-specific desaturase of *Chlamydomonas reinhardtii* (CrΔ4FAD) bears a chloroplast-targeting signal along with the N-terminal cytochrome b5 domain (Zäuner et al. 2012), as is typical for the ER-located Δ4 desaturases involved in VLC-PUFA biosynthesis (Meyer et al. 2003; Tonon et al. 2003). The functional role of the fused cytochrome b5 domain in the chloroplastic desaturase remains to be elucidated. Interestingly, manipulation of *CrΔ4FAD* expression (overexpression and downregulation) revealed that the amount of CrΔ4FAD protein affects the overall content of MGDG and the composition of its molecular species (Zäuner et al. 2012).

VLC-PUFA Biosynthesis at a Glance

At the fatty-acid level, the biosynthesis of VLC-PUFA in the ER is initiated by Δ12 desaturation of oleic acid ($18{:}1^{\Delta 9}$), generating LA ($18{:}2^{\Delta 9,12}$), which in turn might be further desaturated by an ω-end-specific desaturase to give ALA ($18{:}3^{\Delta 9,12,15}$) (Fig. 19.2). The respective pathways commencing with either LA or ALA (the so-called ω-6 and ω-3 pathways) comprise the *n*-6 or *n*-3 PUFA intermediates, respectively. LA and ALA are processed stepwise via the Δ6 desaturase/Δ6 elongase pathway, giving rise to AA (20:4 *n*-6) and EPA (20:5 *n*-3), respectively. These routes commence with the Δ6 desaturase and engage alternating desaturation and elongation

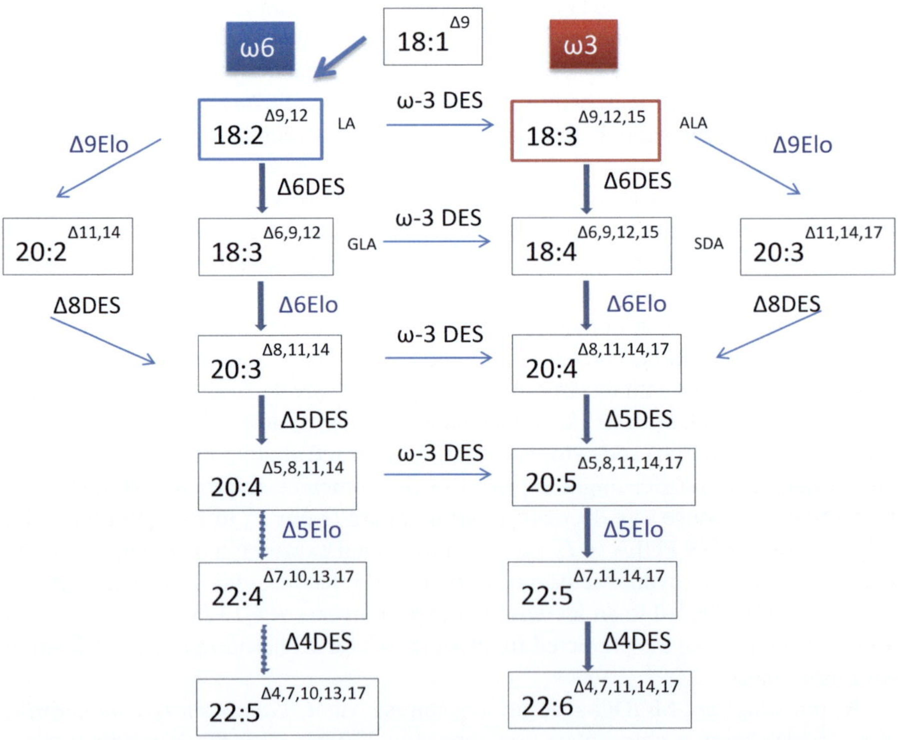

Fig. 19.2 Pathways for the biosynthesis of VLC-PUFA in microalgae

steps mediated by Δ6 PUFA elongase and Δ5 desaturase (Khozin-Goldberg et al. 2011). The Δ9 pathway, bypassing Δ6 desaturation, initiates with Δ9-specific elongation of LA or ALA to eicosadienoic acid (EDA, $20{:}2^{\Delta 11,14}$) or eicosatrienoic acid ($20{:}3^{\Delta 11,14,17}$), respectively, followed by sequential Δ8 and Δ5 desaturations. This alternative pathway appeared to occur in a number of secondary endosymbiont microalgae, such as the haptophytes *Isochrysis galbana* (Qi et al. 2002, 2003) and *Pavlova salina* (Zhou et al. 2007; Petri et al. 2010a), the coccolithophore *Emiliania huxleyi* (Coccolithophyceae) (Sayanova et al. 2011a), and the euglenophyte *Euglena gracilis* (Wallis and Browse 1999).

An ω-3 desaturation can link two routes (*n*-6 and *n*-3) by conversion of *n*-6 to *n*-3 fatty acids. Based on radiolabeling studies, AA, which is produced via the *n*-6 pathway, is assumed to be converted to EPA by ω-3 desaturation in the eustigmatophytes *Nannochloropsis* sp. (Schneider and Roessler 1994; Vieler et al. 2012b) and *Monodus subterraneus* (Khozin-Goldberg et al. 2002b), and the red microalga *Porphyridium cruentum* (Khozin et al. 1997).

The generation of DHA (22:6 *n*-3) from EPA requires two additional steps, mediated by the Δ5-specific C20 PUFA elongase, which catalyzes the 2-C chain elongation of EPA to docosapentaenoic acid (DPA, 22:5 *n*-3), and Δ4 desaturase. The activity of Δ4 desaturase accomplishes DHA biosynthesis in microalgae. In mammals and fish, biosynthesis of DHA from EPA involves the formation of polyunsaturated C24 intermediates, followed by chain shortening via β-oxidation of 24:6 *n*-3 to 22:6 *n*-3 in the peroxisomes (Sprecher 2000).

We should emphasize that in the heterotrophic marine eukaryotes of the family Thraustochytriaceae, the alternative PKS pathway catalyzes the anaerobic formation of VLC-PUFA (Metz et al. 2001). The PKS pathway does not involve aerobic desaturation, while the double bonds are introduced during the process of fatty-acid synthesis on a single multidomain polypeptide. The PKS pathway is predominant in *Schizochytrium*, whereas a desaturation/elongation pathway acts in *Thraustochytrium* from the same family (Qiu 2003). The homology between the prokaryotic *Shewanella* and eukaryotic *Schizochytrium PKS* genes suggests that the PUFA PKS pathway has undergone lateral gene transfer (Metz et al. 2001). The PKS-like-encoding genes are annotated in the genomes of photosynthetic microalgae. Future research may perhaps reveal their functions.

Desaturases of VLC-PUFA Biosynthesis

The "front-end" desaturases involved in the above described VLC-PUFA-biosynthesis pathways (Δ4, Δ5, Δ6, and Δ8) are cytochrome b5-fusion proteins, which introduce a new double bond between the preexisting double bond and the carboxyl end of the fatty acid in a methylene-interrupted manner, and differ from the methyl-end desaturases, such as Δ12 and ω-3 (Δ15) desaturases, inserting the next double bond toward the methyl end (Napier et al. 1999; Sperling and Heinz 2001; Sperling et al. 2003; Napier et al 2003; Meesapyodsuk and Qiu 2012). The front-end desaturases are quite similar in their primary amino-acid sequence and are likely to have evolved from a common ancestor – an ancient fusion desaturase with Δ6-regioselectivity (Sperling et al. 2003; Meesapyodsuk and Qiu 2012); however, there are characteristic amino acid substitutions in their histidine boxes (López Alonso et al. 2003). A common structural feature of the front-end desaturases is replacement of the first histidine in the third histidine box by glutamine, which is essential for the desaturase activity (Sayanova et al. 2001). Phylogenetic analysis performed on the front-end desaturases from different organisms demonstrated a remarkable and highly significant separation between the Δ6- and the Δ5-desaturases in fungi, mosses and algae (Sperling et al. 2003). It is assumed that multiple independent gene duplication and diversification events might have occurred in the course of evolution. The independent evolutionary history of cytochrome b5-fused desaturases with Δ5, Δ8 and Δ4 regioselectivity in eukaryotic organisms, including VLC-PUFA-producing microalgae, might have traced back to the progenitor Δ6 desaturase (Meesapyodsuk and Qiu 2012). However, given that sphingolipid desaturases are closely related to the front-end acyl lipid desaturases, the evolutionary origin in the former type of enzymes is not excluded.

Functional expression and characterization of the putative desaturase is thus essential to assigning its function (Sayanova et al. 2011a). This is commonly performed by expressing the algal protein of interest in a heterologous eukaryotic system, such as the yeast *Saccharomyces cerevisiae* which has a single Δ9 desaturase that is involved in production of monounsaturated fatty acids. *Saccharomyces cerevisiae* can be fed various fatty-acid substrates, and the respective products analyzed to monitor the substrates' conversion into respective products. To date, a great number of front-end desaturases with different regiospecificities have been cloned from microalgae, and their substrate specificity has been investigated in this manner (Wallis and Browse 1999; Domergue et al. 2002, 2003, 2005; Pereira et al. 2004; Tonon et al. 2005; Iskandarov et al. 2010; Sayanova et al. 2011b; Ahman et al. 2011, etc.). A few examples of the functionally characterized front-end desaturases are given here in relation to their substrate preference in a heterologous system. Coexpression of the Δ5 and Δ6 acyl-lipid desaturases from the EPA-producing diatom *P. tricornutum* in *S. cerevisiae* in the presence of either *n*-3 or *n*-6 C18 substrates yielded AA and EPA, demonstrating that they can function with similar efficiency in both the *n*-3 and *n*-6 pathways (Domergue et al. 2002). The Δ6 desaturase of the diatom *Thalassiosira pseudonana* exhibited a slight preference for the exogenously fed 18:3 *n*-3 over 18:2 *n*-6 (Tonon et al. 2005). Both Δ6 desaturase and Δ5 desaturase of the green microalga *L. incisa*, extraordinarily rich in AA produced in the ω-6 pathway but poor in EPA, did not discriminate *n*-3 and *n*-6 fatty acids when expressed in yeast, and acted on both types of PUFA with similar efficiency (Iskandarov et al. 2010). Δ6 desaturase of *L. incisa* acted on 18:2 *n*-6 and 18:3 *n*-3, but did not desaturate $20{:}3^{\Delta 11,14,17}$, implying the importance of the preceding Δ9 double bond; Δ5 desaturase of *L. incisa* inserted a double bond at the Δ5 position of its principal substrate $20{:}3^{\Delta 8,11,14}$ *n*-6 and of $20{:}4^{\Delta 8,11,14,17}$ *n*-3, and with less activity, $20{:}3^{\Delta 11,14,17}$.

A number of genes encoding Δ4 desaturases have been cloned from DHA-producing microalgal species (Meyer et al. 2003; Pereira et al. 2004; Zhou et al. 2007). The PsD4Des of *Pavlova salina* desaturated both $22{:}4^{\Delta7,10,13,16}$ *n*-6 and $22{:}5^{\Delta7,10,13,16,19}$ *n*-3 at the Δ4 position when expressed in *S. cerevisiae* (Pereira et al. 2004). The Δ4 desaturase of *Euglena gracilis* showed strict Δ4 regioselectivity and required the presence of a Δ7-double bond in the substrate acyl group; however, it featured broad substrate specificity, acting on C22 PUFA and C16 PUFA, such as $16{:}3^{\Delta7,10,13}$ *n*-3 (Meyer et al. 2003). Similarly, the Δ4 desaturase cloned from *Ostreococcus lucimarinus* was found to act on both *n*-3 and *n*-6 substrates, such as 22:4 *n*-6, 22:5 *n*-3, and 16:3 *n*-3, and to preferentially desaturate acyl chains bound to PtdCho (Ahmann et al. 2011).

Domergue et al. (2005) identified the first instance of acyl-CoA-dependent desaturase involved in VLC-PUFA biosynthesis in a photosynthetic organism, the picogreen prasynophyte microalga *O. tauri*. Later, a few more microalgal Δ6 and Δ5 desaturases (all from the Prasinophyceae: *Ostreococcus lucimarinus*, *Mantoniella squamata*, *Micromonas pusilla*) were shown to act on CoA-activated acyl groups, similar to mammalian front-end desaturases (Hoffmann et al. 2008; Petrie et al. 2010b). This biochemical feature offers great promise in plant biotechnology for the metabolic engineering of oilseed plants to produce VLC-PUFA because it allows efficient substrate exchange between desaturases and elongases in the cytoplasmic acyl-CoA pool (for further information see Sayanova et al. 2011b; Ruiz-Lopez et al. 2013). Furthermore, some of the characterized microalgal acyl-CoA-dependent desaturases have been shown to be highly specific for the ω-3 substrates and thus provide a useful gene resource for reconstruction of ω-3 VLC-PUFA biosynthesis in higher plants (Sayanova et al. 2011b; Petrie et al. 2012).

To date, the still enigmatic ω-3 desaturases acting on ω-6 C20 VLC-PUFA in the pathway of ω-3 VLC-PUFA biosynthesis have not been cloned or characterized in microalgae; in contrast, enzymes with Δ17 activity and a preference for AA have been characterized in phyto- and fish-pathogenic VLC-PUFA-producing oomycetes (Xue et al. 2013a), which are suggested to be related to the ancient eukaryotic host in evolution of heterokont microalgae. The cloned ω-3 desaturases of oomycetes showed a broad spectrum of ω-6 fatty-acid substrates, including both ω-6 C18 and C20 PUFA, but with an apparent preference for the C20 substrates, showing strong Δ17 desaturase activity.

PUFA Elongation

Two-carbon elongation of polyunsaturated acyl groups is an essential step in VLC-PUFA biosynthesis; it is comprised of four sequential enzymatic reactions: condensation of acyl-CoA with malonyl-CoA, ketoreduction, dehydration and enoyl reduction. Microalgal PUFA elongases are structurally similar to the ELO family of eukaryotic integral membrane proteins that catalyze the condensation step of fatty-acid elongation in animals and fungi (Meyer et al. 2004). These enzymes differ in their primary sequence and function from the higher plant condensing enzymes that participate in microsomal elongation of saturated and monounsaturated fatty acids in the synthesis of long-chain fatty acids. Three major types of PUFA elongases, which differ in substrate specificity, have been characterized from VLC-PUFA-producing microalgae, depending on the nature of their principal substrate: Δ6 C18-PUFA-specific elongases are involved in the elongation of C18 PUFA (18:3 *n*-6 and 18:4 *n*-3) and production of the C20 VLC-PUFA AA and EPA; Δ5 C20-PUFA-specific elongases are engaged in the elongation of 20:5 *n*-3 (EPA) in the pathway of DHA biosynthesis. in an alternative route, Δ9-specific elongation of 18:2 *n*-6 and 18:3 *n*-3 to the respective C20 intermediates precedes sequential Δ8 and Δ5 desaturations to form AA and EPA, respectively (Qi et al. 2002, 2003; Fraser et al. 2004). For functional characterization, the putative elongase is generally expressed in yeast where the recombinant proteins are suggested to achieve a condensation step in conjunction with the host enzymatic activities required for elongation. ELO-type enzymes have been intensively characterized in a number of microalgal species. Two types of PUFA elongases have been cloned from the prasynophyte *O. tauri* and the diatom *T. pseudonana*: a Δ6 C18-PUFA-specific elongase involved in the elongation of 18:3 *n*-6 and 18:4 *n*-3, and a Δ5 C20-PUFA-specific elongase involved in the elongation of 20:5 *n*-3 (Meyer et al. 2004). When heterologously expressed, the Δ6 C18 PUFA-specific elongase from the green alga *L. incisa*, involved in AA synthesis, was active on the ω-6 and ω-3 C18 PUFA 18:3 *n*-6 and 18:4 *n*-3, and discriminated against C20 PUFA (Iskandarov et al. 2009). C20 PUFA elongases cloned from the marine haptophytes of the *Pavlova* genus catalyzed the conversion of EPA to DPA (22:5 *n*-3), showing specificity toward ω-6 and ω-3 C20 PUFA substrates, with virtually no activity toward C18 or C22 PUFA (Pereira et al. 2004; Robert et al. 2009). A recently characterized, unusual multifunctional Δ6 C18 PUFA/Δ5 C20 PUFA elongase from *Emiliania huxleyi* generated DPA (22:5 *n*-3) from EPA as well as DGLA (20:3 *n*-6) from GLA (18:3 *n*-6) (Sayanova et al. 2011a). On the other hand, the alternative pathway of VLC-PUFA biosynthesis in this alga seems to start with the action of Δ9 elongase, which showed a typical preference for LA and ALA, converting them to 20:2 *n*-6 and 20:3 *n*-3, respectively. Given the presence of the Δ6-desaturated SDA ($18{:}4\Delta^{6,9,12,15}$, *n*-3) in *E. huxleyi*, and two candidate elongase systems, the authors inferred possible metabolic compartmentalization of SDA and DHA biosynthesis in the cell of this secondary endosymbiont, with its complex internal structure.

Cellular Localization and Compartmentalization of VLC-PUFA Biosynthesis in Microalgae

Based on a genomic survey of the secondary endosymbionts and existing literature data, Petroutsos et al. (2014) affiliated the VLC-PUFA biosynthesis in chromalveolates to the extraplastidial compartment, whereas their major cellular deposition is within major chloroplast galactolipids. Lipid-linked desaturation seems to be the dominant route in the ER of VLC-PUFA-producing microalgae, analogous to the eukaryotic C18 PUFA-biosynthesis pathway in the ER of higher plant cells, where acyl groups are sequentially desaturated while bound to PtdCho; however in microalgae, two phospholipids, PtdCho and phosphatidylethanolamine (PtdEtn), can be engaged in different desaturation steps. Moreover, positional distribution of the acyl groups in microalgal extraplastidial lipids seems to vary from the conventional models in higher plants (Davidi et al. 2014; Abida et al. 2015). The VLC-PUFA biosynthesis routes in microalgae are likely entail intermittent acyl-group exchange, presumably involving an acyl-editing mechanism, and elongation of CoA-activated acyl moieties in the cytoplasm (Schneider and Roessler 1994; Khozin et al. 1997; Khozin-Goldberg et al. 2002b; Shrestha 2005). Nevertheless, additional experimental data are needed for different taxonomic groups, in particular for prasinophytes with their peculiar and incompletely disclosed features of lipid metabolism, such as the acyl-CoA dependent desaturation. VLC-PUFA biosynthesis in microalgae is assumed to involve cooperation between cellular compartments—the chloroplast and the ER—such that the VLC-PUFA produced in the ER (AA and EPA) can be exported to the plastid (Khozin et al. 1997; Sukenik 1999; Bigogno et al. 2002a; Khozin-Goldberg et al. 2002b); the underlying mechanisms remain unknown, but are probably analogous to the trafficking of C18 PUFA in the ER of higher plant cells or the export of VLC-PUFA-moieties for incorporation into chloroplast lipids.

In the rhodophyte *Porphyridium cruentum*, immediately after a 30-min pulse with [^{14}C]18:2 *n*-6, the label was preferentially incorporated into the *sn*-2 position of PtdCho, as well as appearing in the TAG. With progression of 18:2 *n*-6 conversion to AA via the ω-6 pathway, the *sn*-1 position of PtdCho gradually became labeled, implying the activity of a dedicated acyltransferase with a preference for 20:4 *n*-6 and operation of lipid-remodeling mechanisms. Indeed, [^{14}C]20:4 *n*-6 was readily incorporated into both *sn*-1 and *sn*-2 positions of PtdCho. These observations are in line with PtdCho's central role as the primary acceptor of the de-novo-synthesized C18 fatty acids exported from the chloroplast, and as a major player in acyl-lipid remodeling as has recently become evident in higher plants (Bates et al. 2013). The labeling pattern suggested a flux of fatty acids from cytoplasmic to chloroplastic lipids, mainly from PtdCho and TAG to galactolipids, a characteristic of eukaryotic pathways. Overall interpretation of the labeling kinetics of individual lipids and acyl groups, as well as molecular species analysis, raised the suggestion that the final ω-3 desaturation of 20:4 *n*-6 to 20:5 *n*-3 is predominantly chloroplastic, involving eukaryotic-like MGDG and prokaryotic-like MGDG and DGDG molecular species, and a putative chloroplast-localized ω-3 desaturase with Δ17 activity (Khozin et al. 1997). The existence of this ω-3 desaturase was reinforced by the recently released genome data of *Porphyridium purpureum* (the taxonomic synonym of *P. cruentum*), where a sequence putatively encoding for the plastidial ω-3 desaturase was identified (Bhattacharya et al. 2013).

Remarkable labeling of TAG in numerous experiments with [^{14}]C-labeled fatty-acid precursors to VLC-PUFA illuminates the important role played by TAG as the dynamic depot of acyl groups in microalgal cells (Cohen et al. 2000; Khozin-Goldberg et al. 2000, 2005), as corroborated by the label turnover from TAG to membrane glycerolipids in the course of the experiments. We emphasize this point given the recent immense interest in the mechanisms regulating TAG assembly and remobilization, and experimental evidence for the role played by lipases in algal lipid metabolism (Boyle et al. 2012; Liu and Benning 2013; Trentacoste et al. 2013).

In addition to phospholipids, the non-phosphorous betaine membrane lipids, such as 1,2-diacylglyceryl-*O*-4-(*N*,*N*,*N*-trimethyl) homoserine (DGTS) and 1,2-diacylglyceryl-3-O-2′-(hydroxymethyl)-(N,N,N-trimethyl)-β-alanine (DGTA), are abundant extraplastidial lipids in many classes of microalgae and, in certain instances, completely replace the phospholipids (Araki et al. 1991; Eichenberger 1993; Kato et al. 1996; Dembitsky 1996; Harwood 1998; Eichenberger and Gribi 1997; Vogel and Eichenberger 1992). Betaine lipids are likely participating in acyl-lipid-linked desaturation outside the chloroplast. This assumption is primarily based on the results of radiolabeling studies with fatty-acid precursors; direct molecular evidence is however missing. Radiolabeling studies in *L. incisa*, *Nannochloropsis* sp. and *Monodus subterraneous* revealed that three extraplastidial lipids are involved in various steps of AA (20:4 *n*-6) biosynthesis: PtdCho and DGTS were involved in the Δ12 and subsequent Δ6 desaturations, whereas PtdEtn was a principal substrate for the Δ5 desaturation as, to some extent, PtdCho (Bigogno et al. 2002a; Khozin-Goldberg et al. 2002b).

In the freshwater trebouxiophyte *L. incisa*, the initial label from [^{14}C]18:1 *n*-9 was incorporated into PtdCho, DGTS and TAG. Labeling kinetics indicated the involvement of PtdCho and DGTS in Δ12 and Δ6 desaturations, and of PtdEtn in Δ5 desaturation; little label from synthetized [^{14}C]20:4 *n*-6 was turned over to the galactolipids, in agreement with the low proportion of AA in the chloroplast lipids of this alga (Bigogno et al. 2002a). The principal role of PtdEtn in AA biosynthesis as a substrate for Δ5 desaturation in *L. incisa* was also reflected in its molecular species composition, displaying a major 20:4/20:4 species followed by 16:0/20:4 and 18:1 *n*-7/20:4 species (Shrestha 2005). The conversion of AA to EPA by a putative ω-3 desaturase is scarce, and may be governed by a single enzyme acting on both chloroplast MGDG and extraplastidial lipids.

The final step in EPA biosynthesis by ω-3 desaturation may involve different lipid substrates and be localized to different cellular compartments. Intermediates of both the ω-3 and ω-6 pathways may contribute to the biosynthesis of EPA in the diatom *P. tricornutum* (Arao et al. 1994) and likely scenarios have been suggested (Domergue et al. 2002). However, the synthesis of EPA in the ER has not been unambiguously demonstrated in *P. tricornutum* (Abida et al. 2015), nor has the cellular localization of the cryptic single ω-3 desaturase that is identified in the genome, possibly acting in proximity to chloroplast and extraplastidial lipids, as can be postulated based on the membrane continuum between the ER and the outermost membrane of the plastid in this diatom (Kroth et al. 2008). Given that PtdCho, PtdEtn, and the betaine lipid featured C16 PUFA at the *sn*-2 position, and ultimately all species of MGDG were of the 20:5/C16 (*sn*-1/*sn*-2) structure, it was postulated that EPA synthesized in the endomembrane system has to be released from a phospholipid into the cytosolic acyl-CoA pool, and then transported to the chloroplasts to be attached to the glycerol backbone of chloroplast lipids (Abida et al. 2015). This is in agreement with a high proportion of EPA-CoA in the acyl-CoA pool of *P. tricornutum* as determined by Hamilton et al. (2014). In an alternative scenario, trafficking of the entire esterified glycerol moiety through the complex membrane system can be postulated. Two candidate ω3-desaturases, both predicted to be localized in the chloroplast, were identified by bioinformatics analysis of the genome of the oleaginous diatom *Fistulifera* sp. strain JPCC DA0580 (Liang et al. 2013). It was suggested that the ω-6 pathway is utilized by this organism for the synthesis of AA in the ER, but the final step in the synthesis of EPA is likely carried out in the chloroplast, similar to the rhodophyte *P. cruentum*. Distinctly from *P. cruentum* however, in two eustigmatophytes, PtdEtn was implicated in the ω-3 desaturation, followed by export of EPA to the chloroplast—possibly as a diacylglycerol-bearing scaffold—for the eukaryotic-like MGDG and as 20:5 for the *sn*-1 position of the prokaryotic-like galactolipids (Schneider and Roessler 1994; Khozin-Goldberg et al. 2002b). How this trafficking occurs in microalgae and which acyl carriers are involved remains to be elucidated; future studies should provide novel insights into the intricate machinery of VLC-PUFA biosynthesis and its cellular compartmentalization in evolutionary diverse microalgae.

On the Role of VLC-PUFA in Microalgae

The unsaturation level of membrane lipids alters the biological membrane's conformation and affects its physical properties, such as fluidity (Los and Murata 2004; Los et al. 2013). Alterations in membrane fluidity, in turn, impact the activities of various integral membrane proteins, such as translocators and ion channels, among others (reviewed in Los et al. 2013). Furthermore, unsaturation of the major chloroplast glycerolipids is critical for the functioning of light-harvesting complexes and photosystems (Dörmann 2005), prevention of photoinhibition at low temperatures (Gombos et al. 1997; Mizusawa and Wada 2012), and regulation of chilling sensitivity of plants and cyanobacteria (Nishida and Murata 1996). The importance of fatty-acid unsaturation level in maintaining membrane fluidity in response to temperature was studied in depth in two transformable cyanobacteria by employing site-directed mutagenesis of the genes encoding acyl-lipid desaturases (*Synechocystis*) (e.g. Inaba et al. 2003) or by engineering enhanced unsaturation level through the expression of additional desaturase activity (*Synechococcus* sp. PCC 7942) (Mironov et al. 2012; Los et al. 2013). Influential mutant studies in *A. thaliana* not only helped deduce the biosynthetic pathway of chloroplast glycerolipids but also illuminated the importance of unsaturation level of chloroplast lipids for plant development and the temperature response (Wallis and Browse 2002). Genetic modifications of unsaturation level by recombinant expression of ω-3 desaturase resulted in enhanced tolerance to chilling temperatures (Khodakovskaya et al. 2006; Yu et al. 2009; Dominguez et al. 2010). Applying heat stress to the wild type and an ω-3 desaturase mutant of *Chlamydomonas reinhardtii* revealed the negative effect of a decreased level of the *n*-3 PUFA on maintenance of photosynthetic activity at high temperatures (Nguyen et al 2013). Similar studies are still lacking in VLC-PUFA-producing microalgae.

To the best of our knowledge, the only reported case of a low-temperature-sensitive mutant in the VLC-PUFA-biosynthesis pathway was obtained by chemical mutagenesis in the AA-producing microalga *L. incisa* (Cohen and Khozin-Goldberg 2010; Iskandarov et al. 2011). Growth of the Δ5 desaturase mutant, lacking AA, was impaired at 10–15 °C as compared to the wild type, in agreement with substitution of AA with the less unsaturated trienoic C20 PUFA DGLA in all acyl-lipid classes. *Lobosphaera incisa* is a psychrotolerant microalga, isolated from alpine environments; it contains AA not only in the membrane lipids but predominantly accumulates in the storage TAG (Bigogno et al. 2002b, c; Khozin-Goldberg et al. 2005). Upon temperature shift from 25 to 4 °C, a fraction of the AA is transferred from TAG to polar lipids, implying the significance of AA stored in TAG in the provision of buffering capacity for the rapid increase in membrane lipids' unsaturation level (Bigogno et al. 2002c). Furthermore, in response to nitrogen replenishment, AA accumulated in TAG was swiftly relocated to MGDG and incorporated into the eukaryotic-type 20:4 *n*-6/C18 PUFA molecular species, likely to boost the unsaturation level in chloroplast lipids and to support photosynthetic efficiencies, as required for active growth recovery (Khozin-Goldberg et al. 2005). It is worth noting that a similar phenomenon, albeit at normal temperature, was described in the mutant of the red microalga *P. cruentum* impaired in the utilization of VLC-PUFA from TAG for chloroplast eukaryotic-like molecular species of AA- and EPA-rich galactolipids (Cohen et al. 2000). In general, VLC-PUFA accumulation in TAG in photosynthetic organisms and, in particular, under nitrogen starvation, is not at all common; it can be enhanced with progression of culture aging in some species or by modifying the growth conditions (Tonon et al. 2002; Guihéneuf and Stengel 2013).

Given the high abundance of VLC-PUFA in chloroplast-membrane glycerolipids in various microalgae, one can assume their importance in maintaining membrane fluidity at optimal growth temperatures and preventing membrane rigidification at suboptimal low temperatures. Indeed, the general trend of increasing VLC-PUFA content with decreasing ambient temperature has been shown in a number of studies with microalgae inhabiting temperate waters. For example, lowering the temperature increased EPA and DHA contents in the diatom *P. tricornutum* (Jiang and Gao 2004; Hamilton et al. 2014). An opposite trend of decreased unsaturation of MGDG and DGDG was determined in two diatom species, the model pennate diatom *P. tricornutum* and the marennine-producing *Haslea ostrearia*; C20 PUFA were replaced by C18 PUFA with a reduced number of double bonds at the *sn*-1 position of these galactolipids upon an increase in temperature from 20 to 30 °C (Dodson et al. 2014). A reduction in the growth temperature led to an increase in the proportion of EPA with corresponding rearrangement of molecular species of the major galactolipid MGDG in the red microalga *Porphyridium cruentum* (Adlerstein et al. 1997). The eukaryotic-like molecular species of MGDG, in particular, the highly unsaturated 20:5 *n*-3/20:5 *n*-3, increased in proportion over the prokaryotic-like 20:5 *n*-3/16:0 MGDG in response to temperature decline. These results suggested that highly unsaturated eukaryotic-like molecular species of MGDG in rhodophytes may play a role in the adaptation of cells to low temperatures. DGDG in this alga is predominantly of the prokaryotic-like type, with 16:0 enriched at the *sn*-2 position, so the main modification at low temperature was an increase in the AA-to-EPA conversion leading to predominance of 20:5 *n*-3/16:0 DGDG. In dinoflagellates from the genus *Pyrocystis*, the acyl composition of DGDG was more responsive to changes in temperature than that of MGDG, displaying more unsaturated forms of the *sn*-2-bound fatty acids at lower temperatures (Leblond et al. 2010).

Maintenance of a high unsaturation level of chloroplast glycerolipids is imperative for the survival of extremophile microalgae, such as polar diatoms and chlorophytes, whose natural habitats are characterized by extreme conditions, such as freezing temperatures, along with fluctuating exposure to solar, osmotic, oxidative and nutrient stresses (reviewed in Lyon and Mock 2014). VLC-PUFA and PUFA were shown to increase in abundance in response to low temperature in polar diatoms and chlorophytes, respectively (Teoh et al. 2012). In polar diatoms adapted to very low light intensities and temperatures, VLC-PUFA play a prominent role in protecting the membranes from rigidification or even solidification, and support effective electron-transport chains in chloroplast at low temperatures and a very low intensity of incident light (2 μE m^{-2} s^{-1}) (Mock and Kroon 2002). Of note, desaturases involved in C18 PUFA biosynthesis in recently isolated Antarctic chlorophytes demonstrated increased mRNA expression at low temperatures (Lu et al. 2009; Zhang et al. 2011; An et al. 2013).

Another important biochemical trait of major ecological significance, linking PUFA and VLC-PUFA metabolism, is the diatoms' ability to produce a plethora of oxygenated derivatives of C16 PUFA and C20 PUFA with biological activities and defense potential against their predators and grazers under natural conditions (Wichard et al. 2005; Fontana et al. 2007; Andreou et al. 2009).

Microalgal Biotechnology for VLC-PUFA Production

With the continuous decline and deterioration of fishery and seafood resources, microalgae are seen as potent renewable sources of VLC-PUFA. These photosynthetic eukaryotic organisms use sunlight to produce biomass and oxygen from CO_2, require marginal water, land resources and mineral nutrients. Mass cultivation of photosynthetic microalgae offers the unique opportunity to shift significant agricultural production volumes to unproductive land using marginal or saline water (Leu and Bouusiba 2014). Because microalgae can be cultivated in marginal or saline water on nonproductive drylands, they do not compete with agriculture for arable land or fresh water. While many photoautrophic species are routinely cultivated in mariculture, the production of VLC-PUFA-rich oil from photosynthetic algae has begun relatively recently. This section provides information on the most important

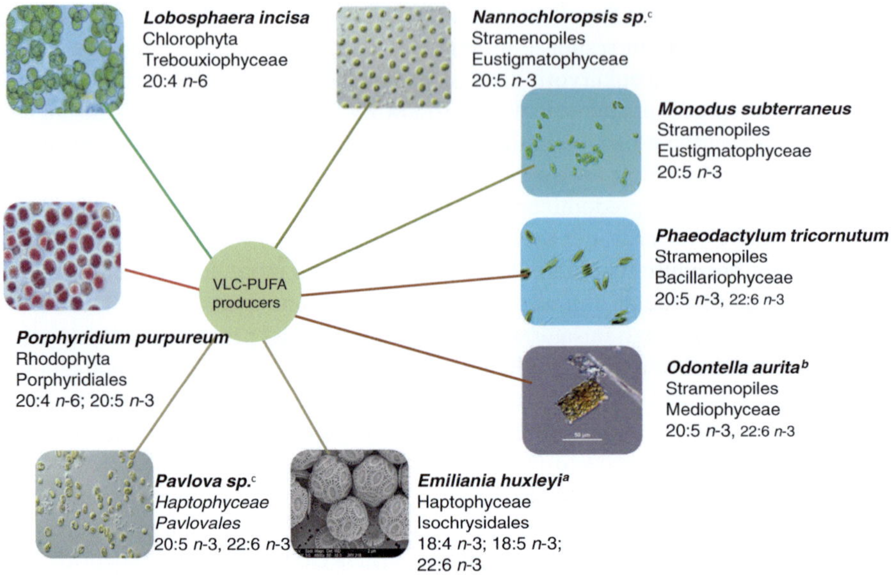

Fig. 19.3 Examples of VLC-PUFA producing algae of different lineages. Micrographs were taken in the authors' laboratory under light microscope. Additional sources are gratefully acknowledged ([a]http://genome.jgi-psf.org/Emihu1/Emihu1.home.html;[b]http://www.eos.ubc.ca/research/phytoplankton/diatoms/centric/odontella/o_aurita.html; [c]http://www.scienceimage.csiro.au/image/10697/microalgae/)

microalgal species in biotechnology for VLC-PUFA production. Micrographs of some microalgae frequently quoted in this chapter are shown on Fig. 19.3. Photosynthetic marine and freshwater species of the genus *Nannochloropsis* (*Nannochloropsis* sp., *N. gaditana*, *N. oculata*, *N. limnetica*) (Eustigmatophyceae) are rich in EPA, mainly due to its high content in the chloroplast glycerolipids (Sukenik 1999; Simionato et al. 2013; Pal et al. 2013). Another recently characterized eustigmatophyte, *Trachydiscus minutus* (Bourr.) H. Ettl, produces EPA, deposits it in large amounts in storage lipids and presents a novel potent EPA resource for photobiotechnology (Řezanka et al. 2010, 2011). Interestingly, molecular phylogenetic analysis has indicated that the Eustigmatophyceae are a more diverse class than previously thought, with many new species, genera, and families awaiting taxonomic treatment and biochemical characterization (Fawley et al. 2014). Thus, it can be anticipated that more species, likely enriched in EPA, will be characterized in the future. Photoautotrophic diatoms and haptophyte microalgae have also been widely explored in marine aquaculture owing to their good nutritional properties and contents of the ω-3 VLC-PUFA EPA and DHA (Orcutt and Patterson 1975; Lebeau and Robert 2003). The pennate diatom microalga *P. tricornutum* (Bacillariophyceae) and the centric diatom *T. pseudonana* (Coscinodiscophyceae) serve as model diatom species in basic research and show promise in biotechnology.

Haptophytes, e.g. *Isochrysis galbana* (Coccolithophyceae) and *Pavlova lutheri* (Pavlovophyceae), are widely used in aquaculture.

The green microalga *L. incisa* is the best-studied photosynthetic producer of AA with the quite rare ability to deposit large amounts of VLC-PUFA in storage lipids, TAG, under nitrogen starvation (Bigogno et al. 2002b, c; Khozin-Goldberg et al. 2002a). Furthermore, the ability to sequester and accumulate VLC-PUFA in TAG is considered a biotechnologically valuable trait, providing the traditional dietary form of VLC-PUFA and allowing oil extraction by conventional methods. *L. incisa* increases AA biosynthesis under nitrogen starvation, shuffling the flux of de-novo-produced AA into the machinery of TAG biosynthesis and assembly. A Δ5 desaturase mutant strain derived from the wild-type strain by chemical mutagenesis produces another nutritionally valuable ω-6 VLC-PUFA, DGLA. Similar to the wild type, the mutant strain deposits DGLA in TAG, making it a potent and rare photosynthetic source of DGLA (Iskandarov et al. 2011; Abu-Ghosh et al. 2015).

As already noted, VLC-PUFA are generally abundant acyl components of major chloroplast glycerolipids in diverse microalgae with both primary and secondary plastids. Accordingly, stressful conditions that are often encountered in mass microalgal cultures (variations in incident and external light intensities and temperatures, increased salinities, nutrient depletion) generally impair their VLC-PUFA content, while promoting the accumulation of saturated and monounsaturated fatty acids (Sukenik et al. 1989; Adlerstein et al. 1997; Solovchenko et al. 2008, 2010; Pal et al. 2011; Simionato et al. 2013, among others). In this respect, finding physiological conditions that enable adequate VLC-PUFA production in terms of biomass and culture contents without impairing VLC-PUFA proportions is of major significance.

Commercial companies have developed the technologies for obtaining "photosynthetic" EPA from *Nannochloropsis* algae. Aurora Algae, for instance, has commercialized a high-EPA content (>65 %) oil under the product name A2 EPA Pure, targeted for use in the supplements and pharmaceuticals sectors (Winwood 2013). A marine photosynthetic diatom rich in EPA, *Odontella aurita* (Guihéneuf et al. 2010), is currently approved for use as a dietary supplement for human nutrition with proven beneficial effects on dyslipidemia, platelet function and oxidative stress in high-fat-fed rats (Haimeur et al. 2012). Heterotrophic production of EPA and DHA by fermentation of nonphotosynthetic eukaryotes is being commercialized on a large scale: the dinoflagellate *Crypthecodinium cohnii* (Dynophyta) is used by DSM Nutritional Products to produce oil with 40–45 % DHA, and virtually no EPA, for the infant-formula market. Fermentation of another heterotrophic organism, *Schizochytrium* (Stramenopiles, Thraustochytriaceae), yields oil with 35–40 % DHA, and <2 % EPA (Winwood 2013). A recent study has indicated the superior oxidative stability of VLC-PUFA-rich lipid extracts from photosynthetic microalgae relative to those of fish and heterotrophic organisms (Ryckebosch et al. 2013). Another study has reported no genetic or acute oral toxicity effects in Sprague-Dawley rats of ethyl ester derivatives containing 23.9 % EPA and 22.3 % palmitoleic acid (Algal-EE,

Aurora Algae, Inc.) from *Nannochloropsis* sp. (Collins et al. 2014). Earlier works examining the nutritional value of *Nannochloropsis* biomass in poultry and aquaculture should also be noted (e.g. Sukenik et al. 1994).

Modification of VLC-PUFA Content and Composition in Microalgae by Genetic Engineering: A Recent Status

The genomes and transcriptomes of numerous VLC-PUFA-producing microalgal species have been sequenced, and many important molecular pathways, including those of VLC-PUFA biosynthesis, have been annotated (Kroth et al. 2008; Fabris et al. 2012; Vieler et al. 2012b; Jinkerson et al. 2013). The recent development of genetic-transformation systems in *Nannochloropsis* species (Radakovits et al. 2012; Vieler et al. 2012b), *O. tauri* (Djouani-Tahri et al. 2011), several biotechnologically relevant diatoms including *P. tricornutum*, *T. pseudonana*, *T. weissflogii*, *Navicula saprophilia*, and *Cylindrotheca fusiformis* (for further information see Bozarth et al. 2009), and *L. incisa* (Zorin et al. 2014) have enabled metabolic engineering for the enhancement of VLC-PUFA contents. The prospects of these approaches are supported by successful studies on reconstruction of the VLC-PUFA-production pathway in oilseed plants (Sayanova et al. 2011b; Petrie and Singh 2011; Ruiz-Lopez et al. 2013) and in oleaginous yeast *Yarrowia lipolytica* (Xue et al. 2013b; Xie et al. 2015) by multistep iterative approaches using, among others, microalgal desaturases and elongases engaged in VLC-PUFA biosynthesis. Remarkably, the entire pathway of DHA-rich TAG biosynthesis was reconstructed in *Nicotiana benthamiana* leaf tissue by transient expression of the plasmid constructs bearing exclusively microalgal genes coding for consequent steps of DHA biosynthesis, namely *Micromonas pusilla* Δ6-desaturase, *Pyramimonas cordata* (Prasinophyceae) Δ6-elongase, *P. salina* Δ5-desaturase, *P. cordata* Δ5-elongase and *P. salina* Δ4-desaturase, along with diacylglycerol acyltransferase (Petrie and Singh 2011). An apparent shift toward DHA production through enhancing of EPA elongation was observed with the highly efficient in this expression system *P. cordata* Δ 5-elongase, whereas expression of the constructs with *P. salina* Δ5-elongase lead to significant accumulation of EPA. Hence, a careful selection and testing of transgene is critical for efficient production of VLC-PUFA by metabolic engineering approaches.

The results of the first engineering approaches in microalgae are highly promising in this regard, with single-gene modifications able to significantly increase VLC-PUFA contents or composition (Hamilton et al. 2014; Kaye et al. 2015). However, recent research efforts have revealed that engineering approaches for VLC-PUFA biosynthesis in microalgae may differ significantly from those developed for higher plants. Similarly, in oilseed plants engineered for the reconstruction of VLC-PUFA production, the step mediated by the phospholipid-dependent Δ6 desaturase presents a bottleneck in the efficient flow of fatty-acid intermediates toward further elongation and desaturation. This phenomenon, termed substrate dichotomy, is associated with the different substrates used in the two sequential steps mediated by the lipid-linked Δ6 desaturase (acting on acyl groups of PtdCho) and the acyl-CoA-dependent elongation of GLA to DGLA (Fig. 19.2). This dichotomy can be circumvented by expressing an acyl-CoA-dependent Δ6 desaturase, such as of *O. tauri*, acting in the same cytosolic acyl-CoA pool as the Δ6 PUFA elongase (Sayanova et al. 2011b; Ruiz-Lopez et al. 2013). Whereas this approach has been attempted in the diatom *P. tricornutum* to enhance the content of VLC-PUFA, no significant alterations were observed in the fatty-acid profile (Hamilton et al. 2014). This may reflect efficient trafficking of precursors in the VLC-PUFA-biosynthesis route in microalgae that intrinsically evolved to efficiently process unsaturated fatty-acid intermediates between different polar lipid classes and the acyl-CoA pool, as was earlier shown in radiolabeling studies with fatty-acid precursors (Schneider and Roessler 1994; Khozin et al. 1997; Bigogno et al. 2002a; Khozin-Goldberg et al. 2002b). Indeed, the acyl-CoA pool of *P. tricornutum* is highly enriched in EPA-CoA, the fatty acid comprising the major VLC-PUFA acyl constituent of membrane glycerolipids (Hamilton et al. 2014). Based on this metabolic trait, another approach was taken, employing the expression of a C20 Δ5 elongase from *O. tauri* (OtElo5) that converts C20 EPA to C22 DPA. In this way, a significant increase in DHA content was observed, along with intermediate accumulation of DPA accompanied by a concurrent decrease in EPA. These results imply that C20 Δ5-elongating activity in conjunction with the endogenous Δ4-desaturase activity are sufficient to shift the pathway toward the synthesis of DHA, which is not accumulated at high levels in this diatom. Furthermore, lipidomics analysis of the transgenic lines revealed a number of novel DHA-containing molecular species of TAG that were produced upon expression of C20 Δ5 elongase in the stationary-phase cells. Coexpression of OtDes6 and OtElo5 using a vector construct engineered to express two genes simultaneously demonstrated an additional increase in DHA and the feasibility of iterative engineering of VLC-PUFA biosynthesis in the diatom (Hamilton et al. 2014).

This represents the first solid experimental evidence for the feasibility of modifying VLC-PUFA biosynthesis via the expression of foreign activities from distantly related green algal species. Enhanced deposition of DHA in TAG opens the door to further studies of targeted incorporation of VLC-PUFA into TAG by recombinant expression of, for example, diacylglycerol acyltransferases, mediating the final step of TAG synthesis.

Enhanced VLC-PUFA partitioning into TAG and subsequently to lipid droplets (LDs), together with additional modifications to enhance biomass and TAG productivity, could potentiate this success for the creation of microalgal strains that can compete on an economic level with current PUFA-production technologies such as heterotrophic fermentation (Leu and Boussiba 2014). Such studies may target suitable modulation of endogenous activities to maintain VLC-PUFA biosynthesis in microalgae during nutrient depletion or starvation, thus producing strains with significant VLC-PUFA content in their TAG fraction commensurate with overall biomass enrichment with VLC-PUFA. This biochemical trait is characteristic of *L. incisa*, an algal species that upregulates the AA-biosynthesis machinery under conditions triggering VLC-PUFA overproduction and accumulation in TAG (Khozin-Goldberg et al. 2002a; Iskandarov et al. 2009, 2010). The recently developed and described transformation system for this strain, along with the first use of metabolic engineering to restore an AA-deficient phenotype in the mutant strain by complementation with the functional Δ5 desaturase, promises further progress in genetic modifications of this alga (Zorin et al. 2014).

Given its industrial potential, significant effort has been invested in developing *Nannochloropsis* into a model oleaginous microalga. This is due to its ability to accumulate storage TAG to over half of its dry weight under conditions of nitrogen starvation (Rodolfi et al. 2009; Pal et al. 2011). While about 35 % or more EPA out of total fatty acids has been reported in *Nannochloropsis* microalgae under nutrient-sufficient growth (Sukenik 1999; Pal et al. 2011, 2013), under nitrogen depletion or starvation, VLC-PUFA biosynthesis is arrested and the shorter-chain, more saturated fatty acids (16:0, 16:1 and 18:1) synthesized *de novo* in the chloroplast are incorporated into the TAG fraction which can represent over half of the cell's dry weight. Accordingly, EPA contents decrease under nitrogen starvation in line with a concomitant reduction in thylakoid membrane lipids, which are the primary EPA reservoir in *Nannochloropsis* cells (Sukenik 1999; Pal et al. 2013; Simionato et al. 2013). Only low proportions of EPA were detected in *Nannochloropsis* TAG, made up of a few of EPA-containing molecular species (Pal et al. 2011; Li et al. 2014). Aiming to modify VLC-PUFA content of TAG under nitrogen-starvation conditions, the gene encoding endogenous Δ12 desaturase (the putative ER-localized isoform), *NoD12*, was overexpressed in *N. oceanica* (Kaye et al. 2015). The rationale for this work was to uncouple the expression of the gene from its downregulation under nitrogen starvation in a manner that occurs intrinsically in *L. incisa*, as well as to use only homologous genes and promoters. The expression of *NoD12* was driven by the stress-induced promoter of the lipid droplet surface protein-encoding gene (Vieler et al. 2012a) to increase the flow of available 18:1 toward VLC-PUFA synthesis under conditions of oil accumulation. The promoter appeared to be an excellent tool to express homologous genes under nitrogen-starvation conditions. Indeed, this study provided the first evidence of successful stress-induced overexpression of an endogenous desaturase in *N. oceanica*, as demonstrated by significant increases in its product 18:2, predominantly in PtdCho, accompanied by increases in AA and EPA concentrations in TAG (Kaye et al. 2015). This study paves the way for overexpression or dowregulation of additional homologous genes engaged in TAG assembly and fatty-acid desaturation as a means of enhancing the VLC-PUFA content in TAG of *N. oceanica*, coupled with fine-tuning of physiological conditions.

Conclusions

Microalgae present a huge and still insufficiently tapped resource of VLC-PUFA for human nutritional and health-related applications. In photobiotechnology, VLC-PUFA biosynthesis can be modulated by using informed approaches based on a thorough investigation of physiological responses to the key cultivation parameters and environmental cues. The ongoing research focuses on understanding the molecular mechanisms by which microalgae of the different evolutionary groups synthetize, relocate and incorporate VLC-PUFA into complex lipids in the context of their multifaceted cellular organization. This research is streamlined by the availability of "omics" techniques and genome information in a growing number of species. Recent progress in microalgal genetic transformation and molecular engineering has opened the way to increased production efficiencies for VLC-PUFA. Such studies will lay the foundation for increasing VLC-PUFA content of microalgae and introducing these traits into those species with plant-like fatty-acid composition, ultimately producing nutritional and medicinal products with high levels of essential VLC-PUFA.

Acknowledgements The authors acknowledge Professor Zvi Cohen for his great contribution to the field, constant guidance and support, Dr. Umidjon Iskandarov for reading the manuscript and his valuable suggestions. Financial support from the European Commission's Seventh Framework Program for Research and Technology Development (FP7), project GIAVAP, Grant No. 266401 is acknowledged.

References

Abida H, Dolch LJ, Meï C, Villanova V, Conte M et al (2015) Membrane glycerolipid remodeling triggered by nitrogen and phosphorus starvation in *Phaeodactylum tricornutum*. Plant Physiol 167:118–136. doi:10.1104/pp.114.252395

Abu-Ghosh S, Didi-Cohen S, Markovitch D et al (2015) A novel source of dihomo-γ-linolenic acid: possibilities and limitations of DGLA production in the high-density cultures of the Δ5 desaturase-mutant microalga *Lobosphaera incisa*. Eur J Lipid Sci Technol.117: 760–766 doi:10.1002/ejlt.201400430

Adkins Y, Kelley DS (2010) Mechanisms underlying the cardioprotective effects of omega-3 polyunsaturated fatty acids. J Nutr Biochem 21:781–792

Adlerstein D, Bigogno C, Khozin I, Cohen Z (1997) The effect of growth temperature and culture density on the molecular species composition of the galactolipids in the red microalga *Porphyridium cruentum* (Rhodophyta). J Phycol 33:975–979. doi:10.1111/j.0022-3646.1997.00975.x

Ahman K, Heilmann M, Feussner I (2011) Identification of a Δ4-desaturase from microalga *Ostreococcus lucimarinus*. Eur J Lipid Sci Technol 113:832–840

Alonso DL, Belarbi EH, Rodriguez-Ruiz J et al (1998) Acyl lipids of three microalgae. Phytochemistry 47:1473–1483. doi:10.1016/S0031-9422(97)01080-7

An M, Mou S, Zhang X et al (2013) Temperature regulates fatty acid desaturases at a transcriptional level and modulates the fatty acid profile in the Antarctic microalga *Chlamydomonas* sp. ICE-L. Bioresour Technol 134:151–157

Andreou A, Brodhun F, Feussner I (2009) Biosynthesis of oxylipins in non-mammals. Prog Lipid Res 48:148–170

Apt KE, Behrens PW (1999) Commercial developments in microalgal biotechnology. J Phycol 35:215–226

Araki S, Eichenberger W, Sakurai T, Sato N (1991) Distribution of diacylglycerylhydroxymethyltrimethyl-β-alanine (DGTA) and phosphatidylcholine in brown algae. Plant Cell Physiol 32:623–628

Arao T, Sakaki T, Yamada M (1994) Biosynthesis of polyunsaturated lipids in the diatom, *Phaeodactylum tricornutum*. Phytochemistry 36:629–635

Archibald J (2005) Jumping genes and shrinking genomes-probing the evolution of eukaryotic photosynthesis using genomics. IUBMB Life 57:539–547

Barclay WR, Meager KM, Abril JR (1994) Heterotrophic production of long chain omega-3 fatty acids utilizing algae and algae-like microorganisms. J Appl Phycol 6:123–129

Bates PD, Stymne S, Ohlrogge J (2013) Biochemical pathways in seed oil synthesis. Curr Opin Plant Biol 16:358–364

Bazan NG, Molina MF, Gordon WC (2011) Docosahexaenoic acid signalolipidomics in nutrition: significance in aging, neuroinflammation, macular degeneration, Alzheimer's, and other neurodegenerative diseases. Annu Rev Nutr 31:321–351

Bell JG, Sargent JR (2003) Arachidonic acid in aquaculture feeds: current status and future. Aquaculture 218:491–499

Benning C (2009) Mechanisms of lipid transport involved in organelle biogenesis in plant cells. Annu Rev Cell Dev Biol 25:71–91

Betancor MB, Sprague M, Usher S et al (2015) A nutritionally-enhanced oil from transgenic *Camelina sativa* effectively replaces fish oil as a source of eicosapentaenoic acid for fish. Sci Rep 5:8104. doi:10.1038/srep08104

Bhattacharya D, Yoon HS, Hackett JD (2004) Photosynthetic eukaryotes unite: endosymbiosis connects the dots. Bioessays 26:50–60

Bhattacharya D, Price DC, Chan CX et al (2013) Genome of the red alga *Porphyridium purpureum*. Nat Commun 4:1941. doi:10.1038/ncomms2931

Bigogno C, Khozin-Goldberg I, Adlerstein D, Cohen Z (2002a) Biosynthesis of arachidonic acid in the oleaginous microalga *Parietochloris incisa* (Chloropyceae): radiolabeling studies. Lipids 37:209–216

Bigogno C, Khozin-Goldberg I, Boussiba S et al (2002b) Lipid and fatty acid composition of the green alga *Parietochloris incisa*. Phytochemistry 60:497–503

Bigogno C, Khozin-Goldberg I, Cohen Z (2002c) Accumulation of arachidonic acid-rich triacylglycerols in the microalga *Parietochloris incisa* (Trebuxiophyceae, Chlorophyta). Phytochemistry 60:135–143

Blanc G, Agarkova I, Grimwood J et al (2012) The genome of the polar eukaryotic microalga Coccomyxa subellipsoidea reveals traits of cold adaptation. Genome Biol 13:R39. doi:10.1186/gb-2012-13-5-r39

Borowitzka MA (1988) Microalgae as sources of pharmaceuticals and other biologically active compounds. J Appl Phycol 4:267–279

Boyle NR, Page MD, Liu B et al (2012) Three acyltransferases and nitrogen-responsive regulator are implicated in nitrogen starvation-induced triacylglycerol accumulation in Chlamydomonas. J Biol Chem 287:15811–15825. doi:10.1074/jbc.M111.334052

Bozarth A, Maier UG, Zauner S (2009) Diatoms in biotechnology: modern tools and applications. Appl Microbiol Biotechnol 82:195–201. doi:10.1007/s00253-008-1804-1808

Calder PC (2006) n-3 polyunsaturated fatty acids, inflammation, and inflammatory diseases. Am J Clin Nutr 83(6 suppl):1505S–1519S

Calder PC (2013) n-3 fatty acids, inflammation and immunity: new mechanisms to explain old actions. Proc Nutr Soc 72:326–336. doi:10.1007/s11745-003-1068-y

Cermeño P, Falkowski PG (2009) Controls on diatom biogeography in the ocean. Science 325:1539–1541. doi:10.1126/science.1174159

Chan CX, Bhattacharya D, Reyes-Prieto A (2012) Endosymbiotic and horizontal gene transfer in microbial eukaryotes: impacts on cell evolution and the tree of life. Mob Genet Elem 2:101–105

Cohen Z (1999) Production of polyunsaturated fatty acids by the microalga *Porphyridium cruentum*. In: Cohen Z (ed) Production of chemicals by microalgae. Taylor and Francis, London, pp 1–24

Cohen Z, Khozin-Goldberg I (2010) Searching for PUFA-rich microalgae. In: Cohen Z, Ratledge C (eds) Single cell oils, 2nd edn. American Oil Chemists' Society, Champaign, pp 201–224

Cohen Z, Khozin-Goldberg I, Adlerstein D et al (2000) The role of triacylglycerol as a reservoir of polyunsaturated fatty acids for the rapid production of chloroplastic lipids in certain microalgae. Biochem Soc Trans 28:740–743

Collins ML, Lynch B, Barfield W, Bull A, Ryan AS, Astwood JD (2014) Genetic and acute toxicological evaluation of an algal oil containing eicosapentaenoic acid (EPA) and palmitoleic acid. Food Chem Toxicol 72:162–168. doi:10.1016/j.fct.2014.07.021

Crawford MA, Broadhurst CL (2012) The role of docosahexaenoic and the marine food web as determinants of evolution and hominid brain development: the challenge for human sustainability. Nutr Health 21:17–39. doi:10.1177/0260106012437550

Dangi B, Obeng M, Nauroth JM et al (2009) Biogenic synthesis, purification, and chemical characterization of anti-inflammatory resolvins derived from docosapentaenoic acid (DPA *n*-6). J Biol Chem 284:14744–14759.

Davidi L, Shimoni E, Khozin-Goldberg I, Zamir A, Pick U (2014) Origin of β -carotene-rich plastoglobuli in Dunaliella bardawil. Plant Physiol 164:2139–56

De Swaaf ME, Sijtsma L, Pronk JT (2003) High-cell-density fed-batch cultivation of the docosahexaenoic acid producing marine alga *Crypthecodinium cohnii*. Biotechnol Bioeng 81:666–672

Dembitsky VM (1996) Betaine ether-linked glycerolipids: chemistry and biology. Prog Lipid Res 35:1–51

Djouani-Tahri EB, Sanchez F, Lozano J-C, Bouget F-Y (2011) A phosphate-regulated promoter for fine-tuned and reversible overexpression in *Ostreococcus*: application to circadian clock functional analysis. PLoS One 6(12):e28471. doi:10.1371/journal.pone.0028471

Dodson VJ, Mouget JL, Dahmen JL, Leblond JD (2014) The long and short of it: temperature-dependent modifications of fatty acid chain length and unsaturation in the galactolipid profiles of the diatoms *Haslea ostrearia* and *Phaeodactylum tricornutum*. Hydrobiologia 727:95–107. doi:10.1007/s10750-013-1790-4

Dolch L-J, Maréchal E (2015) Inventory of fatty acid desaturases in the pennate diatom *Phaeodactylum tricornutum*. Mar Drugs. 13(3):1317–39. doi:10.3390/md13031317

Domergue F, Lerchl J, Zahringer U, Heinz E (2002) Cloning and functional characterization of *Phaeodactylum tricornutum* front-end desaturases involved in eicosapentaenoic acid biosynthesis. Eur J Biochem 269:4105–4113

Domergue F, Abbadi A, Ott C et al (2003a) Acyl carriers used as substrates by the desaturases and elongases involved in very long-chain polyunsaturated fatty acids biosynthesis reconstituted in yeast. J Biol Chem 278:35115–35126

Domergue F, Spiekermann P, Lerchl J et al (2003b) New insight into *Phaeodactylum tricornutum* fatty acid metabolism. Cloning and functional characterization of plastidial and microsomal Delta 12-fatty acid desaturases. Plant Physiol 131:1648–1660

Domergue F, Abbadi A, Zähringer U et al (2005) In vivo characterization of the first acyl-CoA Δ6-desaturase from a member of the plant kingdom, the microalga *Ostreococcus tauri*. Biochem J 389:483–490

Dominguez T, Hernandez ML, Pennycooke JC et al (2010) Increasing omega-3 desaturase expression in tomato results in altered aroma profile and enhanced resistance to cold stress. Plant Physiol 153:655–665

Dörmann P (2005) Membrane lipids. In: Murphy D (ed) Plant lipids. Biology, utilization and manipulation. Blackwell Publishing, CRC Press, Oxford, pp 123–160

Dunstan GA, Volkman JK, Jeffrey SW, Barrett SM (1992) Biochemical composition of microalgae from the green algal classes Chlorophyceae and Prasinophyceae 2. Lipid classes and fatty acids. J Exp Mar Biol Ecol 161:115–134

Dunstan GA, Volkman JK, Barrett SM et al (1994) Essential polyunsaturated fatty-acids from 14 species of diatom (Bacillariophyceae). Phytochemistry 35:155–161

Dunstan GA, Brown MR, Volkman JK (2005) Cryptophyceae and rhodophyceae; chemotaxonomy, phylogeny, and application. Phytochemistry 66:2557–2570

Eichenberger W (1993) Betaine lipids in lower plants: distribution of DGTS, DGTA and phospholipids, and the intracellular localization and site of biosynthesis of DGTS. Plant Physiol Biochem 31:213–221

Eichenberger W, Gribi C (1997) Lipids of *Pavlova lutheri* (Haptophyceae): cellular site and metabolic role of DGCC. Phytochemistry 45:1561–1567

Fabris M, Matthijs M, Rombauts S et al (2012) The metabolic blueprint of *Phaeodactylum tricornutum* reveals a eukaryotic Entner-Doudoroff glycolytic pathway. Plant J 70:1004–1014

Fawley KP, Eliáš M, Fawley MW (2014) The diversity and phylogeny of the commercially important algal class Eustigmatophyceae, including the new clade Goniochloridales. J Appl Phycol 26:1773–1782. doi:10.1007/s10811-013-0216-z

Fontana A, d'Ippolito G, Cutignano A et al (2007) LOX-induced lipid peroxidation mechanism responsible for the detrimental effect of marine diatoms on zooplankton grazers. Chembiochem 8:1810–1818

Fraser TC, Qi B, Elhussein S, Chatrattanakunchai S, Stobart AK, Lazarus CM (2004) Expression of the *Isochrysis* C18-delta9 polyunsaturated fatty acid specific elongase component alters Arabidopsis glycerolipid profiles. Plant Physiol 135:859–866

Gao J, Ajjawi I, Manoli A et al (2009) FATTY ACID DESATURASE4 of Arabidopsis encodes a protein distinct from characterized fatty acid desaturases. Plant J 60:832–839

Gil A, Serra-Majem L, Calder PC et al (2012) Systematic reviews of the role of omega-3 fatty acids in the prevention and treatment of disease. Br J Nutr 107:S1–S2

Giroud C, Eichenberger W (1988) Fattyacids of *Chlamydomonas reinhardtii*: structure, positional distribution and biosynthesis. Biol Chem 369:18–19

Giroud C, Eichenberger W (1989) Lipids of *Chlamydomonas-reinhardtii* – incorporation of ^{14}C-acetate, ^{14}C-palmitate and ^{14}C-oleate into different lipids and evidence for lipid-linked desaturation of fattyacids. Plant Cell Physiol 30:121–128

Gombos Z, Kanervo E, Tsvetkova N et al (1997) Genetic enhancement of the ability to tolerate photoinhibition by introduction of unsaturated bonds into membrane glycerolipids. Plant Physiol 115:551–559

Guihéneuf F, Stengel DB (2013) LC-PUFA-enriched oil production by microalgae: accumulation of lipid and triacylglycerols containing n-3 LC-PUFA is triggered by nitrogen limitation and inorganic carbon availability in the marine haptophyte *Pavlova lutheri*. Mar Drugs 11:4246–4266. doi:10.3390/md11114246

Guihéneuf F, Fouqueray M, Mimouni V et al (2010) Effect of UV stress on the fatty acid and lipid class composition in two marine microalgae *Pavlova lutheri* (Pavlovophyceae) and *Odontella aurita* (Bacillariophyceae). J Appl Phycol 22:629–638

Haimeur A, Ulmann L, Mimouni V et al (2012) The role of *Odontella aurita*, a marine diatom rich in EPA, as a dietary supplement in dyslipidemia, platelet function and oxidative stress in high-fat fed rats. Lipids Health Dis 11:147. doi:10.1186/1476-511X-11-147

Hamilton ML, Haslam RP, Napier JA, Sayanova O (2014) Metabolic engineering of *Phaeodactylum tricornutum* for the enhanced accumulation of omega-3 long chain polyunsaturated fatty acids. Metab Eng 22:3–9. http://dx.doi.org/10.1016/j.ymben.2013.12.003

Harwood JL (1998) Membrane lipids in algae. In: Siegenthaler PA, Murata N (eds) Lipids in photosynthesis: structure, function and genetics. Kluwer Academic Publishers, Dordrecht, pp 53–64

Harwood JL, Guschina IA (2009) The versatility of algae and their lipid metabolism. Biochimie 91:679–684

Heinz E (1993) Biosynthesis of polyunsaturated fatty acids. In: Moore TS (ed) Lipid metabolism in plants. CRC Press, Boca Raton, pp 34–89
Henderson RJ, Mackinlay EE (1992) Radiolabeling studies of lipids in the marine cryptomonad *Chroomonas salina* in relation to fatty acid desaturation. Plant Cell Physiol 33:395–406
Hoffmann M, Wagner M, Abbadi A et al (2008) Metabolic engineering of omega3-very long chain polyunsaturated fatty acid production by an exclusively acyl-CoA-dependent pathway. J Biol Chem 283:22352–22362. doi:10.1074/jbc.M802377200
Holzinger A, Karsten U (2013) Desiccation stress and tolerance in green algae: consequences for ultrastructure, physiological and molecular mechanisms. Front Plant Sci 4:327. doi:10.3389/fpls.2013.00327
Hu Q, Sommerfeld M, Jarvis E et al (2008) Microalgal triacylglycerols as feedstocks for biofuel production: perspectives and advances. Plant J 54:621–639
Inaba M, Suzuki I, Szalontai B et al (2003) Gene-engineered rigidification of membrane lipids enhances the cold inducibility of gene expression in Synechocystis. J Biol Chem 278:12191–12198. doi:10.1074/jbc.M212204200
Iskandarov U, Khozin-Goldberg I, Ofir R, Cohen Z (2009) Cloning and characterization of the Δ6 polyunsaturated fatty acid elongase from the green microalga *Parietochloris incisa*. Lipids 44:545–554. doi:10.1007/s11745-009-3301-y
Iskandarov U, Khozin-Goldberg I, Cohen Z (2010) Identification and characterization of Δ12, Δ6, and Δ5 desaturases from the green microalga *Parietochloris incisa*. Lipids 45:519–530. doi:10.1007/s11745-010-3421-4
Iskandarov U, Khozin-Goldberg I, Cohen Z (2011) Selection of a DGLA-producing mutant of the microalga *Parietochloris incisa*: I. Identification of mutation site and expression of VLC-PUFA biosynthesis genes. Appl Microbiol Biotechnol 90:249–256. doi:10.1007/s00253-010-3005-5
Itoh R, Toda K, Takahashi H, Takano H, Kuroiwa T (1998) Delta-9 fatty acid desaturase gene containing a carboxyl-terminal cytochrome b5 domain from the red alga *Cyanidioschyzon merolae*. Curr Genet 33:165–170
Janssen CI, Kiliaan AJ (2014) Long-chain polyunsaturated fatty acids (LCPUFA) from genesis to senescence: the influence of LCPUFA on neural development, aging, and neurodegeneration. Prog Lipid Res 53:1–17. doi:10.1016/j.plipres.2013.10.002
Jiang H, Gao K (2004) Effects of lowering temperature during culture on the production of polyunsaturated fatty acids in the marine diatom *Phaeodactylum tricornutum*. J Phycol 40:651–654
Jinkerson RE, Radakovits R, Posewitz MC (2013) Genomic insights from the oleaginous model alga *Nannochloropsis gaditana*. Bioengineered 4:37–43
Kajikawa M, Yamato KT, Kohzu Y et al (2006) A front-end desaturase from *Chlamydomonas reinhardtii* produces pinolenic and coniferonic acids by omega13 desaturation in methylotrophic yeast and tobacco. Plant Cell Physiol 47:64–73
Kato M, Sakai S, Adachi K et al (1996) Distribution of betaine lipids in marine algae. Phytochemistry 42:1341–1345
Kaye Y, Grundman O, Zorin B et al (2015) Metabolic engineering toward enhanced LC-PUFA biosynthesis in Nannochloropsis oceanica: I. Overexpression of endogenous Δ12 desaturase driven by stress-inducible promoter leads to enhanced deposition of polyunsaturated fatty acids in TAG. Algal Res 11:387–398
Keeling PJ (2004) Diversity and evolutionary history of plastids and their hosts. Am J Bot 91:1481–1493
Keeling PJ (2010) The endosymbiotic origin, diversification and fate of plastids. Philos Trans R Soc Lond B Biol Sci 365:729–748. doi:10.1098/rstb.2009.0103
Khodakovskaya M, McAvoy R, Peters J et al (2006) Enhanced cold tolerance in transgenic tobacco expressing a chloroplast omega-3 fatty acid desaturase gene under the control of a cold-inducible promoter. Planta 223:1090–1100

Khozin I, Adlerstein D, Bigongo C et al (1997) Elucidation of the biosynthesis of eicosapentaenoic acid in the microalga *Porphyridium cruentum*. II. Studies with radiolabeled precursors. Plant Physiol 114:223–230

Khozin-Goldberg I, Yu HZ, Adlerstein D, Didi-Cohen S, Heimer YM, Cohen Z (2000) Triacylglycerols of the red microalga *Porphyridium cruentum* can contribute to the biosynthesis of eukaryotic galactolipids. Lipids 35:881–889

Khozin-Goldberg I, Bigogno C, Shrestha P, Cohen Z (2002a) Nitrogen starvation induced accumulation of arachidonic acid in the freshwater green alga *Parietochloris incisa*. J Phycol 38:991–994

Khozin-Goldberg I, Didi-Cohen S, Shayakhmetova I, Cohen Z (2002b) Biosynthesis of eicosapentaenoic acid (EPA) in the freshwater eustigmatophyte *Monodus subterraneus* (Eustigmatophyceae). J Phycol 38:745–756

Khozin-Goldberg I, Shrestha P, Cohen Z (2005) Mobilization of arachidonyl moieties from triacylglycerols into chloroplastic lipids following recovery from nitrogen starvation of the microalga *Parietochloris incisa*. Biochim Biophys Acta 1738:63–71

Khozin-Goldberg I, Iskandarov U, Cohen Z (2011) LC-PUFA from photosynthetic microalgae: occurrence, biosynthesis, and prospects in biotechnology. Appl Microbiol Biotechnol 91:905–915

Kotajima T, Shiraiwa Y, Suzuki I (2014) Functional screening of a novel Δ15 fatty acid desaturase from the coccolithophorid *Emiliania huxleyi*. Biochim Biophys Acta 1841:1451–1458

Koven W, Barr Y, Lutzky S et al (2001) The effect of dietary arachidonic acid (20:4*n*-6) on growth, survival and resistance to handling stress in gilthead seabream (*Sparus aurata*) larvae. Aquaculture 193:107–122

Koven W, van Anholt R, Lutzky S et al (2003) The effect of dietary arachidonic acid on growth, survival, and cortisol levels in different-age gilthead seabream larvae (*Sparus auratus*) exposed to handling or daily salinity change. Aquaculture 228:307–320

Kroth P (2007) Molecular biology and the biotechnological potential of diatoms. Springer, Berlin

Kroth PG, Chiovitti A, Gruber A et al (2008) A model for carbohydrate metabolism in the diatom *Phaeodactylum tricornutum* deduced from comparative whole genome analysis. PLoS One 3:e1426

Kyle DJ (1997) Production and use of a single cell oil highly enriched in arachidonic acid. Lipid Technol 9:116–121

Lang I, Hodac L, Friedl T, Feussner I (2011) Fatty acid profiles and their distribution patterns in microalgae: a comprehensive analysis of more than 2000 strains from the SAG culture collection. BMC Plant Biol 11:124. doi:10.1186/1471-2229-11-124

Laviano A, Rianda S, Molfino A et al (2013) Omega-3 fatty acids in cancer. Curr Opin Clin Nutr Metab Care 16:156–161. doi:10.1097/MCO.0b013e32835d2d99

Le HD, Meisel JA, de Meijer VE et al (2009) The essentiality of arachidonic acid and docosahexaenoic acid. Prostaglandins Leukot Essent Fatty Acids 81:165–170

Lebeau T, Robert JM (2003) Diatom cultivation and biotechnologically relevant products. Part I: cultivation at various scales. Appl Microbiol Biotechnol 60:612–623

Leblond JD, Evens TJ, Chapman PJ (2003) The biochemistry of dinoflagellate lipids, with particular reference to the fatty acid and sterol composition of a Karenia brevis bloom. Phycologia 42:324–331

Leblond JD, Dahmen JL, Seipelt RL, Elrod-Erickson MJ, Kincaid R, Howard JC, Evens TJ, Chapman PJ (2005) Lipid composition of chlorarachniophytes (Chlorarachniophyceae) from the genera Bigelowiella, Gymnochlora and Lotharella. J Phycol 41:311–321

Leblond JD, Lasiter AD (2009) Mono- and digalactosyldiacylglycerol composition of dinoflagellates. II. *Lepidodinium chlorophorum*, *Karenia brevis*, and *Kryptoperidinium foliaceum*, three dinoflagellates with aberrant plastids. Eur J Phycol 44:199–205

Leblond JD, Dahmen JL, Evens TJ (2010a) Mono- and digalactosyldiacylglycerol composition of dinoflagellates. IV. Temperature-induced modulation of fatty acid regiochemistry as observed by electrospray ionization/mass spectrometry. Eur J Phycol 45:13–18

Leblond JD, Timofte HI, Roche SA, Porter NM (2010b) Mono and digalactosyldiacylglycerol composition of glaucocystophytes (Glaucophyta): a modern interpretation using positive ion electrospray ionization/mass spectrometry/mass spectrometry. Phycol Res 58:222–229

Leliaert F, Smith DR, Moreau H et al (2012) Phylogeny and molecular evolution of the green algae. Crit Rev Plant Sci 31:1–46

Lemieux C, Otis C, Turmel M (2014) Six newly sequenced chloroplast genomes from prasinophyte green algae provide insights into the relationships among prasinophyte lineages and the diversity of streamlined genome architecture in picoplanktonic species. BMC Genomics 15:857. doi:10.1186/1471-2164-15-857

Leu S, Boussiba S (2014) Advances in the Production of High-Value Products by Microalgae. Ind Biotechnol (New Rochelle N Y) 10(3):169–183. doi:10.1089/ind.2013.0039

Li J, Han D, Wang D et al (2014) Choreography of transcriptomes and lipidomes of Nannochloropsis reveals the mechanisms of oil synthesis in microalgae. Plant Cell 26:1645–1665

Liang Y, Maeda Y, Sunaga Y et al (2013) Biosynthesis of polyunsaturated fatty acids in the oleaginous marine diatom *Fistulifera sp.* strain JPCC DA0580. Mar Drugs 11(12):5008–5023. doi:10.3390/md11125008

Liu B, Benning C (2013) Lipid metabolism in microalgae distinguishes itself. Curr Opin Biotechnol 24:300–309

López Alonso D, Garcıá-Maroto F, Rodriguez-Ruiz J et al (2003) Evolution of the membrane-bound fatty acid desaturases. Biochem Syst Ecol 31:1111–1124

Los DA, Murata N (2004) Membrane fluidity and its roles in the perception of environmental signals. Biochim Biophys Acta 1666:142–157

Los DA, Mironov KS, Allakhverdiev SI (2013) Regulatory role of membrane fluidity in gene expression and physiological functions. Photosynth Res 116:489–509

Lu Y, Chi X, Yang Q, Li Z et al (2009) Molecular cloning and stress-dependent expression of a gene encoding Δ12-fatty acid desaturase in the antarctic microalga *Chlorella vulgaris* NJ-7. Extremophiles 13:875–884

Lyon BR, Mock T (2014) Polar microalgae: new approaches towards understanding adaptations to an extreme and changing environment. Biology (Basel) 3:56–80. doi:10.3390/biology3010056

Makewicz A, Gribi C, Eichenberger W (1997) Lipids of *Ectocarpus fasciculatus* (Phaeophyceae). Incorporation of [l-^{14}C]oleate and the role of TAG and MGDG in lipid metabolism. Plant Cell Physiol 38:952–960

Mann DG, Vanormelingen P (2013) An inordinate fondness? The number, distributions, and origins of diatom species. J Eukaryot Microbiol 60:414–420. doi:10.1111/jeu.12047

Marszalek JR, Lodish HF (2005) Docosahexaenoic acid, fatty acid-interacting proteins, and neuronal function: breastmilk and fish are good for you. Annu Rev Cell Dev Biol 21:633–657

McLaren JE, Michael DR, Guschina IA et al (2011) Eicosapentaenoic acid and docosahexaenoic acid regulate modified LDL uptake and macropinocytosis in human macrophages. Lipids 46:1053–1061

Meesapyodsuk D, Qiu X (2012) The front-end desaturase: structure, function, evolution and biotechnological use. Lipids 47:227–237

Metz JG, Roessler P, Facciotti D et al (2001) Production of polyunsaturated fatty acids by polyketide synthases in both prokaryotes and eukaryotes. Science 293:290–293

Meyer A, Cirpus P, Ott C et al (2003) Biosynthesis of docosahexaenoic acid in *Euglena gracilis*: biochemical and molecular evidence for the involvement of a Δ4-fatty acyl group desaturase. Biochemistry 42:9779–9788

Meyer A, Kirsch H, Domergue F et al (2004) Novel fatty acid elongases and their use for the reconstitution of docosahexaenoic acid biosynthesis. J Lipid Res 45:1899–1909

Mironov KS, Maksimov EG, Maksimov GV, Los DA (2012) Feedback between fluidity of membranes and transcription of the desB gene for the ω3-desaturase in the cyanobacterium *Synechocystis*. Mol Biol 46:134–141. doi:10.1134/S002689331201013X

Mizusawa N, Wada H (2012) The role of lipids in photosystem II. Biochim Biophys Acta 1817:194–208
Mock T, Kroon BMA (2002) Photosynthetic energy conversion under extreme conditions—II: the significance of lipids under light limited growth in Antarctic sea ice diatoms. Phytochemistry 61:53–60
Moustafa A, Beszteri B, Maier UG et al (2009) Genomic footprints of a cryptic plastid endosymbiosis in diatoms. Science 324:1724–1726. http://dx.doi.org/10.1126/science.1172983
Muto M, Kubota C, Tanaka M et al (2013) Identification and functional analysis of delta-9 desaturase, a key enzyme in PUFA synthesis, isolated from the oleaginous diatom *Fistulifera*. PLoS One 8(9):e73507. doi:10.1371/journal.pone.0073507
Nakayama T, Archibald JM (2012) Evolving a photosynthetic organelle. BMC Biol 10:35. doi:10.1186/1741-7007-10-35
Napier JA, Sayanova O, Sperling P, Heinz E (1999) A growing family of cytochrome b5 fusion desaturases. Trends Plant Sci 4:2–5
Napier JA, Michaelson LV, Sayanova O (2003) The role of cytochrome b5 fusion desaturases in the synthesis of polyunsaturated fatty acids. Prostaglandins Leukot Essent Fatty Acids 68:135–143
Nguyen HM, Cuiné S, Beyly-Adriano A et al (2013) The green microalga *Chlamydomonas reinhardtii* has a single ω-3 fatty acid desaturase which localizes to the chloroplast and impacts both plastidic and extraplastidic membrane lipids. Plant Physiol 163:914–928. doi:10.1104/pp.113.223941
Nishida I, Murata N (1996) Chilling sensitivity in plants and cyanobacteria: the crucial contribution of membrane lipids. Annu Rev Plant Physiol Plant Mol Biol 47:541–568
O'Rourke EJ, Kuballa P, Xavier R, Ruvkun G (2013) ω-6 polyunsaturated fatty acids extend life span through the activation. Genes Dev 27:429–440. doi:10.1101/gad.205294.112
Orcutt DM, Patterson GW (1975) Sterol, fatty acid and elemental composition of diatoms grown in chemically defined media. Comp Biochem Physiol B 50:579–583
Pal D, Khozin-Goldberg I, Cohen Z et al (2011) The effect of light, salinity and nitrogen availability on lipid production by *Nannochloropsis* sp. Appl Microbiol Biotechnol 90:1429–1441
Pal D, Khozin-Goldberg I, Didi-Cohen S et al (2013) Growth, lipid production and metabolic adjustments in the euryhaline eustigmatophyte *Nannochloropsis oceanica* CCALA 804 in response to osmotic downshift. Appl Microbiol Biotechnol 97:8291–8306
Peled E, Leu S, Zarka A et al (2011) Isolation of a novel oil globule protein from the green alga *Haematococcus pluvialis* (Chlorophyceae). Lipids 46:851–861. doi:10.1007/s11745-011-3579-4
Peled E, Pick U, Zarka S, Shimoni E, Leu S, Boussiba S (2012) Light-induced oil globule migration in Haematococcus pluvialis (Chlorophyceae). J Phycol 48:1209–1219
Pereira SL, Huang Y-S, Bobik EG et al (2004) A novel omega3-fatty acid desaturase involved in the biosynthesis of eicosapentaenoic acid. Biochem J 378:665–671
Petrie JR, Singh SP (2011) Expanding the docosahexaenoic acid food web for sustainable production: engineering lower plant pathways into higher plants. AoB Plants 2011:plr011. doi:10.1093/aobpla/plr011
Petrie J, Mackenzie AM, Shrestha P et al (2010a) Isolation of three novel long-chain polyunsaturated fatty acid detla9-elongases and the transgenic assembly of the entire *Pavlova* salina docosahexaenoic acid pathway in *Nicotiana benthamiana*. J Phycol 46:917–925. doi:10.1111/j.1529-8817.2010.00870.x
Petrie JR, Liu Q, Mackenzie AM et al (2010b) Isolation and characterisation of a high-efficiency desaturase and elongases from microalgae for transgenic VLC-PUFA production. Mar Biotechnol 12:430–438
Petrie JR, Shrestha P, Mansour MP et al (2010c) Metabolic engineering of omega-3 long-chain polyunsaturated fatty acids in plants using an acyl-CoA Δ6-desaturase with omega3-preference from the marine microalga *Micromonas pusilla*. Metab Eng 12:233–240

Petrie JR, Shrestha P, Mansour MP, Nichols PD, Liu Q, Singh SP (2012) Metabolic engineering plant seeds with fish oil-like levels of DHA. PLoS One 7(11):e49165. doi:10.1371/journal.pone.0049165

Petroutsos D, Amiar S, Abida H et al (2014) Evolution of galactoglycerolipid biosynthetic pathways – from cyanobacteria to primary plastids and from primary to secondary plastids. Prog Lipid Res 54:68–85. doi:10.1016/j.plipres.2014.02.001

Qi B, Beaudoin F, Fraser T, Stobart AK, Napier JA, Lazarus CM (2002) Identification of a cDNA encoding a novel C18-Δ9 polyunsaturated fatty acid-specific elongating activity from the docosahexaenoic acid (DHA)-producing microalga, *Isochrysis galbana*. FEBS Lett 510:159–165

Qi B, Fraser TC, Bleakley CL, Shaw EM, Stobart AK, Lazarus CM (2003) The variant 'his-box' of the C18-Delta9-PUFA-specific elongase IgASE1 from *Isochrysis galbana* is essential for optimum enzyme activity. FEBS Lett 547:137–139.

Qiu X (2003) Biosynthesis of docosahexaenoic acid (DHA, 22:6-4, 7,10,13,16,19): two distinct pathways. Prostaglandins Leukot Essent Fatty Acids 68:181–186

Radakovits R, Jinkerson RE, Fuerstenberg SI et al (2012) Draft genome sequence and genetic transformation of the oleaginous alga *Nannochloropsis gaditana*. Nat Commun 3:686. doi:10.1038/ncomms1688

Rangel-Huerta OD, Aguilera CM, Mesa MD, Gil A (2012) Omega-3 long-chain polyunsaturated fatty acids supplementation on inflammatory biomakers: a systematic review of randomised clinical trials. Br J Nutr 107(Suppl 2):S159–S170. doi:10.1017/S0007114512001559

Řezanka T, Petránková M, Cepák V, PŘibyl P, Sigler K, Cajthaml T (2010) *Trachydiscus minutus*, a new biotechnological source of eicosapentaenoic acid. Folia Microbiol 55:265–269

Rezanka T, Lukavsky J, Nedbalova L et al (2011) Effect of nitrogen and phosphorus starvation on the polyunsaturated triacylglycerol composition, including positional isomer distribution, in the alga *Trachydiscus minutus*. Phytochemistry 72:2342–2351

Riekhof WR, Sears BB, Benning C (2005) Annotation of genes involved in glycerolipid biosynthesis in *Chlamydomonas reinhardtii*: discovery of the betaine lipid synthase BTA1(Cr). Eukaryot Cell 4:242–252

Robert S, Petrie J, Zhou X-R, Mansour M, Blackburn S, Green A et al (2009) Isolation and characterisation of a Δ5-fatty acid elongase from the marine microalga *Pavlova salina*. Mar Biotechnol 11:410–418

Robinson LE, Mazurak VC (2013) *n*-3 polyunsaturated fatty acids: relationship to inflammation in healthy adults and adults exhibiting features of metabolic syndrome. Lipids 48:319–332. doi:10.1007/s11745-013-3774-6

Rodolfi L, Chini Zittelli G, Bassi N et al (2009) Microalgae for oil: strain selection, induction of lipid synthesis and outdoor mass cultivation in a low-cost photobioreactor. Biotechnol Bioeng 102:100–112

Rodriguez-Ezpeleta N, Brinkmann H, Burey SC et al (2005) Monophyly of primary photosynthetic eukaryotes: green plants, red algae, and glaucophytes. Curr Biol 15:1325–1330

Ruiz-Lopez NR, Haslam RP, Usher SL, Napier JA, Sayanova O (2013) Reconstitution of EPA and DHA biosynthesis in *Arabidopsis*: iterative metabolic engineering for the synthesis of n-3 VLC-PUFAs in transgenic plants. Metab Eng 17:30–41. doi:10.1016/j.ymben.2013.03.001

Ryckebosch E, Bruneel C, Termote-Verhalle R et al (2013) Stability of omega-3 LC-PUFA-rich photoautotrophic microalgal oils compared to commercially available omega-3 LC-PUFA oils. J Agric Food Chem 61:10145–10155. doi:10.1021/jf402296s

Sato N, Moriyama T (2007) Genomic and biochemical analysis of lipid biosynthesis in the unicellular rhodophyte *Cyanidioschyzon merolae*: lack of a plastidic desaturation pathway results in the coupled pathway of galactolipid synthesis. Eukaryot Cell 6:1006–1017

Sayanova O, Beaudoin F, Libisch B et al (2001) Mutagenesis and heterologous expression in yeast of a plant Δ6-fatty acid desaturase. J Exp Bot 52:1581–1585

Sayanova O, Haslam RP, Calerón MV, López NR, Worthy C, Rooks P et al (2011a) Identification and functional characterisation of genes encoding the omega-3 polyunsaturated fatty acid biosynthetic pathway from the coccolithophore *Emiliania huxleyi*. Phytochemistry 72(7):594–600. doi:10.1016/j.phytochem.2011.01.022

Sayanova O, Ruiz-Lopez N, Haslam RP et al (2011b) The role of Δ6-desaturase acyl-carrier specificity in the efficient synthesis of long-chain polyunsaturated fatty acids in transgenic plants. Plant Biotechnol J 10:195–206

Schneider JC, Roessler P (1994) Radiolabeling studies of lipids and fatty acids in *Nannochloropsis* (Eustigmatophyceae), an oleaginous marine alga. J Phycol 30:594–598

Serhan CN, Hong S, Gronert K et al (2002) Resolvins: a family of bioactive products of omega-3 fatty acid transformation circuits initiated by aspirin treatment that counter proinflammation signals. J Exp Med 196:1025–1037

Shanklin J, Cahoon EB (1998) Desaturation and related modifications of fatty acids. Annu Rev Plant Physiol Plant Mol Biol 49:611–641

Shepherd CJ, Jackson AJ (2013) Global fishmeal and fish-oil supply: inputs, outputs and markets. J Fish Biol 83:1046–1066

Shrestha P (2005) Biosynthesis and mobilization of arachidonic-acid-rich triacylglycerols in the green alga *Parietochloris incisa*. PhD dissertation, Ben-Gurion University of the Negev

Simionato D, Block MA, La Rocca N et al (2013) The response of *Nannochloropsis gaditana* to nitrogen starvation includes *de novo* biosynthesis of triacylglycerols, a decrease of chloroplast galactolipids, and reorganization of the photosynthetic apparatus. Eukaryot Cell 12:665–676

Simopoulos AP (2002) Omega-3 fatty acids in inflammation and autoimmune diseases. J Am Coll Nutr 21:495–505

Simopoulos AP (2008) The importance of the ω-6/ω-3 fatty acid ratio in cardiovascular disease and other chronic diseases. Exp Biol Med 233:674–688

Solovchenko A, Khozin-Goldberg I, Didi-Cohen S et al (2008) Effects of light intensity and nitrogen starvation on growth, total fatty acids and arachidonic acid in the green microalga *Parietochloris incisa*. J Appl Phycol 20:245–251

Solovchenko A, Merzlyak M, Khozin-Goldberg I et al (2010) Coordinated carotenoid and lipid syntheses induced in *Parietochloris incisa* (Chlorophyta, Trebouxiophyceae) mutant deficient in Δ5 desaturase by nitrogen starvation and high light. J Phycol 46:763–772

Somerville CR, Browse JA (1996) Dissecting desaturation; plants prove advantageous. Trends Cell Biol 6:148–153

Sperling P, Heinz E (2001) Desaturases fused to their electron donor. Eur J Lipid Sci Technol 103:158–180

Sperling P, Ternes P, Zank TK, Heinz E (2003) The evolution of desaturases. Prostaglandins Leukot Essent Fatty Acids 68:73–95

Sprecher H (2000) Metabolism of highly unsaturated *n*-3 and *n*-6 fatty acids. Biochim Biophys Acta 1486:219–231

Sukenik A (1999) Production of eicosapentaenoic acid by the marine eustigmatophyte *Nannochloropsis* sp. In: Cohen Z (ed) Chemicals from microalgae. Taylor and Francis, London, pp 41–56

Sukenik A, Carmeli Y, Berner T (1989) Regulation of fatty-acid composition by irradiance level in the eustigmatophyte *Nannochloropsis* sp. J Phycol 25:686–692

Sukenik A, Takahashi H, Mokady S (1994) Dietary lipids from marine unicellular algae enhance the amount of liver and blood omega-3 fatty acids in rats. Ann Nutr Metab 38:85–96

Teoh M-L, Phang S-M, Chu W-L (2012) Response of Antarctic, temperate, and tropical microalgae to temperature stress. J Appl Phycol 1:1–13

Tocher DR (2003) Metabolism and functions of lipids and fatty acids in teleost fish. Rev Fish Sci 11:107–184

Tonon T, Harvey D, Larson TR, Graham IA (2002) Long chain polyunsaturated fatty acid production and partitioning to triacylglycerols in four microalgae. Phytochemistry 61:15–24

Tonon T, Harvey D, Larson TR, Graham IA (2003) Identification of a very long chain polyunsaturated fatty acid Δ4-desaturase from the microalga *Pavlova lutheri*. FEBS Lett 553:440–444. doi:10.1016/S0014-5793(03)01078-0

Tonon T, Harvey D, Qing R et al (2004) Identification of a fatty acid Delta11-desaturase from the microalga *Thalassiosira pseudonana*. FEBS Lett 563:28–34

Tonon T, Sayanova O, Michaelson LV et al (2005) Fatty acid desaturases from the microalga *Thalassiosira pseudonana*. FEBS J 272:3401–3412

Trentacoste EM, Shrestha RP, Smith SR et al (2013) Metabolic engineering of lipid catabolism increases microalgal lipid accumulation without compromising growth. Proc Natl Acad Sci U S A 110:19748–19753. doi:10.1073/pnas.1309299110

Vaezi R, Napier JA, Sayanova O (2013) Identification and functional characterization of genes encoding omega-3 polyunsaturated fatty acid biosynthetic activities from unicellular microalgae. Mar Drugs 11:5116–5129. doi:10.3390/md11125116

Vieler A, Brubaker SB, Vick B, Benning CA (2012a) Lipid droplet protein of Nannochloropsis with functions partially analogous to plant oleosins. Plant Physiol 158:1562–1569. http://dx.doi.org/10.1104/pp.111.193029

Vieler A, Wu G, Tsai CH et al (2012b) Genome, functional gene annotation, and nuclear transformation of the heterokont oleaginous alga *Nannochloropsis oceanica* CCMP1779. PLoS Genet 8(11):e1003064. doi:10.1371/journal.pgen.1003064

Viso A-C, Marty J-C (1993) Fatty acids from 28 marine microalgae. Phytochemistry 34:1521–1533

Vogel G, Eichenberger W (1992) Betaine lipids in lower plants. Biosynthesis of DGTS and DGTA in *Ochromonas danica* (Chrysophyceae) and the possible role of DGTS in lipid metabolism. Plant Cell Physiol 33:427–436

Wallis JG, Browse J (1999) The Δ8-desaturase of *Euglena gracilis*: an alternate pathway for synthesis of 20-Carbon polyunsaturated fatty acids. Arch Biochem Biophys 365(2):307–316

Wallis JG, Browse J (2002) Mutants of Arabidopsis reveal many roles for membrane lipids. Prog Lipid Res 41:254–278. doi:10.1016/S0163-7827(01)00027-3

Wang X, Lin H, Gu Y (2012) Multiple roles of dihomo-γ-linolenic acid against proliferation diseases. Lipids Health Dis 11:25. doi:10.1186/1476-511X-11-25

Wichard T, Poulet SA, Halsband-Lenk C et al (2005) Survey of the chemical defence potential of diatoms: screening of fifty one species for alpha, beta, gamma, delta-unsaturated aldehydes. J Chem Ecol 31:949–958

Winwood RJ (2013) Recent developments in the commercial production of DHA and EPA rich oils from micro-algae. OCL 20(6):D604. http://dx.doi.org/10.1051/ocl/2013030

Wolfe GR, Cunningham FX, Durnfordt D et al (1994) Evidence for a common origin of chloroplasts with light-harvesting complexes of different pigmentation. Nature 367:566–568

Xie D, Jackson EN, Zhu Q (2015) Sustainable source of omega-3 eicosapentaenoic acid from metabolically engineered *Yarrowia lipolytica*: from fundamental research to commercial production. Appl Microbiol Biotechnol 99:1599–1610. doi:10.1007/s00253-014-6318-y

Xue Z, He H, Hollerbach D et al (2013a) Identification and characterization of new Δ-17 fatty acid desaturases. Appl Microbiol Biotechnol 97:1973–1985. doi:10.1007/s00253-012-4068-2

Xue Z, Sharpe PL, Hong SP et al (2013b) Production of omega-3 eicosapentaenoic acid by metabolic engineering of *Yarrowia lipolytica*. Nat Biotechnol 31:734–740. doi:10.1038/nbt.2622

Yang T, Fang S, Zhang HX et al (2012) N-3 PUFAs have antiproliferative and apoptotic effects on human colorectal cancer stem-like cells *in vitro*. J Nutr Biochem 24:744–753. doi:10.1016/j.jnutbio.2012.03.023

Yoon HS, Hackett JD, Ciniglia C, Pinto G, Bhattacharya D (2004) A molecular timeline for the origin of photosynthetic eukaryotes. Mol Biol Evol 21:809–818

Yu C, Wang HS, Yang S et al (2009) Overexpression of endoplasmic reticulum omega-3 fatty acid desaturase gene improves chilling tolerance in tomato. Plant Physiol Biochem 47:1102–1112

Zäuner S, Jochum W, Bigorowski T, Benning C (2012) A cytochrome b5-containing plastid-located fatty acid desaturase from Chlamydomonas reinhardtii. Eukaryot Cell 11:856–863

Zhang P, Liu S, Cong B et al (2011) A novel omega-3 fatty acid desaturase involved in acclimation processes of polar condition from Antarctic ice algae *Chlamydomonas* sp. ICE-L. Mar Biotechnol 13:393–401

Zhou XR, Robert SS, Petrie JR et al (2007) Isolation and characterization of genes from the marine microalga *Pavlova salina* encoding three front-end desaturases involved in docosahexaenoic acid biosynthesis. Phytochemistry 68:785–796

Zhukova NV, AizdaicheR NA (1995) Fatty-acid composition of 15 species of marine microalgae. Phytochemistry 39:351–356. doi:10.1016/0031-9422(94)00913-E

Zorin B, Grundman O, Khozin-Goldberg I et al (2014) Development of a nuclear transformation system for oleaginous green alga *Lobosphaera* (*Parietochloris*) *incisa* and genetic complementation of a mutant strain, deficient in arachidonic acid biosynthesis. PLoS One 9(8):e105223

Chapter 20
Understanding Sugar Catabolism in Unicellular Cyanobacteria Toward the Application in Biofuel and Biomaterial Production

Takashi Osanai, Hiroko Iijima, and Masami Yokota Hirai

Abstract *Synechocystis* sp. PCC 6803 is a model species of the cyanobacteria that undergo oxygenic photosynthesis, and has garnered much attention for its potential biotechnological applications. The regulatory mechanism of sugar metabolism in this cyanobacterium has been intensively studied and recent omics approaches have revealed the changes in transcripts, proteins, and metabolites of sugar catabolism under different light and nutrient conditions. Several transcriptional regulators that control the gene expression of enzymes related to sugar catabolism have been identified in the past 10 years, including a sigma factor, transcription factors, and histidine kinases. The modification of these genes can lead to alterations in the primary metabolism as well as the levels of high-value products such as bioplastics and hydrogen. This review summarizes recent studies on sugar catabolism in *Synechocystis* sp. PCC 6803, emphasizing the importance of elucidating the molecular mechanisms of cyanobacterial metabolism for biotechnological applications.

Keywords Cyanobacteria • Metabolic engineering • Sigma factor • Sugar metabolism • Transcriptional regulator

Introduction

Biotechnology using photosynthetic organisms has attracted much public interests owing to lower emissions of CO_2 generated by such applications. The utilization of light energy and CO_2 by photosynthesis is expected to replace the usage of fossil

T. Osanai (✉) • H. Iijima
School of Agriculture, Meiji University, 1-1-1, Higashimita, Tama-ku,
Kawasaki, Kanagawa 214-8571, Japan
e-mail: tosanai@meiji.ac.jp

M.Y. Hirai
RIKEN Center for Sustainable Resource Science, 1-7-22 Suehiro-cho, Tsurumi-ku,
Yokohama, Kanagawa 230-0045, Japan

Y. Nakamura, Y. Li-Beisson (eds.), *Lipids in Plant and Algae Development*,
Subcellular Biochemistry 86, DOI 10.1007/978-3-319-25979-6_20

fuels to produce fuels and materials in the future, and therefore, the study of photosynthetic organisms has boomed in popularity all around the world.

Cyanobacteria are a group of bacteria that undergo oxygenic photosynthesis. They include morphologically varied species, that are spherical, rod-like, filamentous, or spiral in shape. Cyanobacteria possess both Photosystem I and II, similar to eukaryotic algae and plants, but these bacteria possess a different light-harvesting complex called phycobilisomes (Allen and Smith 1969). External CO_2 is assimilated via ribulose-1,5-bisphosphate carboxylase/oxygenase (RubisCO), an enzyme that is surrounded by structural protein under low CO_2 conditions in a supercomplex is called the carboxysome (Badger and Price 2003). Cyanobacteria retain circadian rhythms to alter their cellular metabolism and proliferation (Ishiura et al. 1998). These bacteria are divided into nitrogen-fixing and non-nitrogen-fixing species. In this review, we discuss only the non-nitrogen species, particularly the most widely used strain *Synechocystis* sp. PCC 6803 (hereafter designated *Synechocystis* 6803), to simplify the explanation of cyanobacterial metabolisms.

Sugar Catabolism during Light/Dark and Day/Night Conditions

Sugar metabolism occurs via anabolic and catabolic pathways. Sugar anabolism includes CO_2 fixation via the Calvin cycle, gluconeogenesis, and glycogen synthesis. The mechanism of sugar anabolism has been intensively studied because carbon fixation represents the characteristics of photosynthetic organisms. In contrast, there are fewer studies on sugar catabolism in photosynthetic organisms (including cyanobacteria) than there are on sugar anabolism. Genes encoding the enzymes of sugar catabolism are annotated, because the genomic information can be retrieved from several databases, namely CyanoBase (http://genome.microbedb.jp/cyanobase), KEGG (http://www.genome.jp/kegg/), and CYORF (http://cyano.genome.ad.jp/).

The metabolic map of sugar catabolism in *Synechocystis* 6803 is described in Fig. 20.1 (reproduced from Nakaya et al. 2015). *Synechocystis* 6803 stores glycogen as a carbon sources and the glycogen is converted to glucose-1-phosphate (G1P) by glycogen phosphorylase (encoded by *glgP*) and isoamylase (encoded by *glgX*) during glycogen catabolism (Fig. 20.1). The genome of *Synechocystis* 6803 contains two *glgP*s (sll1356 and slr1367) and two *glgXs* (slr0237 and slr1857) homologs (Osanai et al. 2007). G1P is converted into a glucose-6-phosphate (G6P) by phosphoglucomutase (encoded by *pgm*), and G6P is subsequently catabolized by glycolysis or the oxidative pentose phosphate (OPP) pathway (Fig. 20.1). The OPP pathway is a major route for glucose catabolism in *Synechocystis* 6803 under heterotrophic conditions (Yang et al. 2002). Both the flow of glycolysis and the OPP pathway are activated during mixotrophic conditions, compared with under photoautotrophic conditions (Yoshikawa et al. 2013). Integrated omics analysis combining transcriptome and metabolome data has suggested that post-transcriptional changes are important for altering the metabolic flux during the transition among different

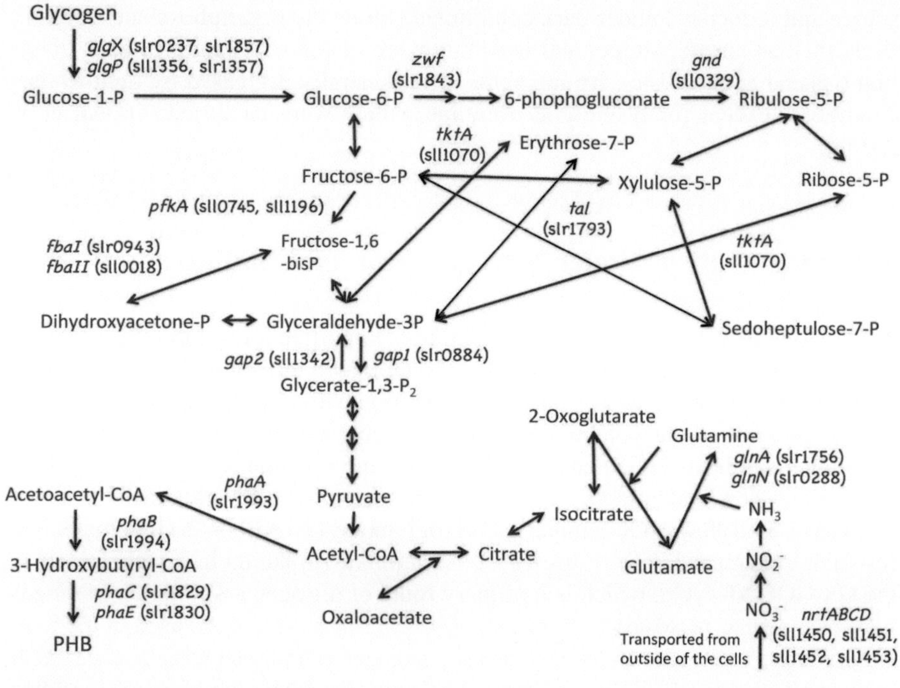

Fig. 20.1 Metabolic map of sugar catabolism in *Synechocystis* sp. PCC 6803 (Reproduced from Nakaya et al. 2015)

trophic conditions (Nakajima et al. 2014). Another group comparing mixotrophic and photoautotrophic conditions revealed that metabolite levels of sugar phosphates such as G6P, 6-phosphogluconate, and fructose-1,6-bisphosphate were higher under the mixotrophic conditions, whereas the levels of 3-phosphoglycerate, phosphoenolpyruvate, and pyruvate were lower (Takahashi et al. 2008).

Experiments measuring rhythmic changes in the transcript levels in *Synechocystis* 6803 indicated metabolic alterations during day/night cycles (Kucho et al. 2005). Genes encoding the enzymes of catabolism and respiration peaked during the dusk or night (Kucho et al. 2005). These results clearly suggest that sugar catabolism and respiration are activated at nighttime, while sugar anabolism and photosynthesis are activated at daytime. The expression of the OPP pathway genes and glycolytic genes was transiently up-regulated by light-to-dark transition, although these inductions were dependent on the experimental conditions (Osanai et al. 2005; Singh and Sherman 2005).

Recently, the metabolic changes in wild-type *Synechocystis* 6803 by light-to-dark transition (after 1 day) were shown by our group (Osanai et al. 2014c). The levels of sugar phosphates, such as G1P, G6P, fructose-6-phosphate, ribose-5-phosphate, ribulose-5-phosphate, and ADP-glucose, were reduced to <20 % of those under light conditions during 1 day of dark cultivation (Osanai et al. 2014c). These results indicate the activation of glycogen catabolism to provide a carbon source and reductants under dark conditions. Unlike the sugar phosphates, organic acids such as lactate, citrate, and isocitrate were increased by light-to-dark transition (Osanai et al. 2014c). Amino acids were generally decreased by light-to-dark conditions, except for lysine and ornithine, which were increased (Osanai et al. 2014c).

Sugar Catabolism Under Nitrogen-Starved Conditions

Acclimation to changes in nutrient conditions is important for cyanobacteria as well as for other organisms. Nitrogen is an essential macronutrient and is often lacking in the natural environment because of competition with other organisms for its use. Gene expression of the metabolic enzymes is widely altered by nitrogen conditions. Glutamine synthetase (GS) catalyzes the following reaction: Glutamate + $NH_3 \rightarrow$ Glutamine. The glutamine generated in then converted by glutamate synthase (GOGAT) as follows: Glutamine + 2-Oxoglutarate (2-OG) $\rightarrow$ 2 Glutamate. The resultant glutamate can be re-used for ammonium assimilation in a cycle known as the GS-GOGAT cycle, which is a primary route of nitrogen assimilation in unicellular cyanobacteria (Muro-Pastor et al. 2005). GS activity is decreased in the presence of ammonium ions and is increased by nitrogen starvation (Merida et al. 1991). Two small proteins (IF7 and IF17) interact with *Synechocystis* 6803 GS proteins, resulting in the inactivation of GS activity during nitrogen-replete conditions (Garcia-Dominguez et al. 1999).

Muro-Pastor et al. (2001b) found that 2-OG levels were increased within 15 min during nitrogen starvation, indicating that 2-OG is important for perceiving the nitrogen status in *Synechocystis* 6803 cells. 2-OG binds physically with two regulatory proteins NtcA, which is a CRP-type transcription factor, and PII, which is a deduced carbon/nitrogen balance sensor (Forchhammer 2004). 2-OG is thus a signaling metabolite, reflecting the nitrogen status of unicellular cyanobacteria. 2-OG is also a metabolite both the tricarboxylic acid cycle and GS-GOGAT cycle, indicating its importance in the interaction between carbon and nitrogen metabolism.

Technical advances in transcriptome analysis have enabled us to measure the expressions at the genome-wide level. A microarray experiment using nitrogen-starved *Synechocystis* 6803 cells was performed in 2006 (Osanai et al. 2006). The expression of nitrogen assimilatory genes and nitrogen transporter genes was up-regulated by nitrogen starvation for 4 h, whereas the expression of genes related to photosynthesis and carbon assimilation was repressed, consistent with previous results. Microarray analysis newly revealed that genes related to sugar catabolism and respiration were considerably activated by nitrogen starvation (Osanai et al. 2006). Time-course microarray experiments confirmed the up-regulation of genes related to sugar catabolism and respiration under nitrogen starvation condition (Aquirre von Wobeser et al. 2011).

The protein levels of two key enzymes of the OPP pathway, glucose-6-phosphate dehydrogenase (G6PD encoded by *zwf*) and 6-phosphogluconate dehydrogenase

(6PGD encoded by *gnd*), were increased more than doubled after 3 days of nitrogen depletion (Osanai et al. 2013a). The protein levels of two glycogen catabolic enzymes, GlgX (slr1857) and GlgP (sll1356), were increased by nitrogen depletion, although the total amount of glycogen was increased during nitrogen starvation (Osanai et al. 2013a). The up-regulation of the mRNAs and protein levels of enzymes related to glycogen catabolism and the OPP pathway thus seems to be contradictory. However, metabolome analysis revealed that organic acids such as malate, succinate, and fumarate, were markedly increased by nitrogen depletion (Osanai et al. 2014b). In addition, amino acids synthesized from pyruvate and acetyl-CoA, which include alanine, glycine, valine, leucine, isoleucine, phenylalanine, and tyrosine, were increased during nitrogen starvation (Osanai et al. 2014b). These results indicate that metabolites downstream of sugar catabolism accumulate during nitrogen starvation, which is consistent with the changes in the transcript and protein levels of sugar catabolic enzymes (Fig. 20.2). Thus, *Synechocystis* 6803 sugar catabolism is drastically altered during nitrogen starvation, leading to bioproduction, as discussed in section "Engineering of sugar catabolism in *Synechocystis* 6803 for bioplastics and hydrogen production".

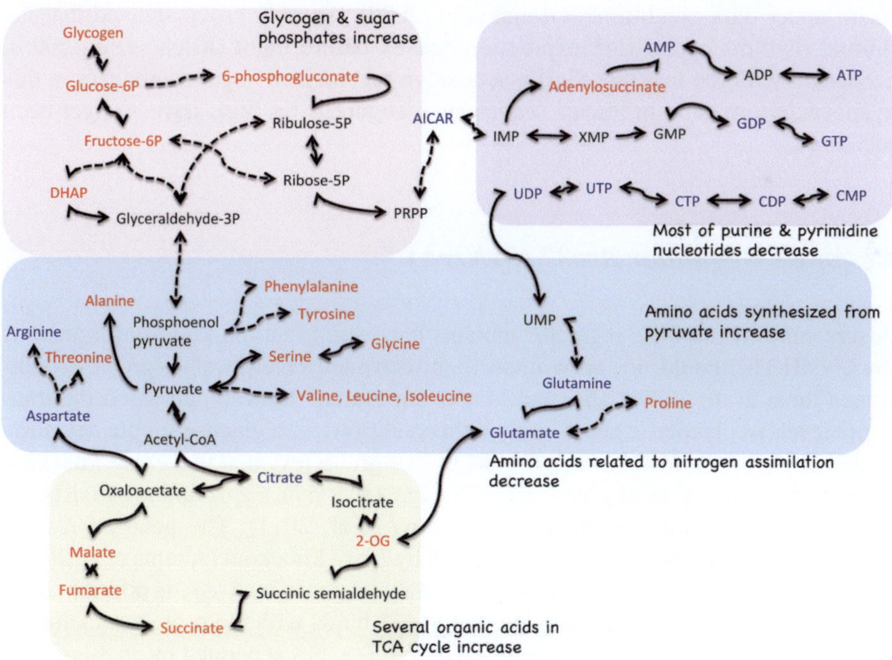

Fig. 20.2 Metabolic changes after nitrogen starvation for 4 h. Metabolites increased or decreased by nitrogen depletion are marked in *red* or *blue*, respectively (Reproduced from Osanai et al. 2014b)

Transcriptional Regulators Controlling Sugar Catabolism in *Synechocystis* 6803

RNA Polymerase Sigma Factor SigE

RNA polymerase is a basic transcriptional machinery, performing RNA synthesis from template DNAs. The enzyme consists of multiple subunits including a sigma factor that is involved in recognizing the promoter DNAs and initiating transcription. Generally, bacteria contain multiple sigma factors, of which *Synechocystis* 6803 possesses nine (SigA~SigI), to acclimate to different environmental conditions (Goto-Seki 1999). The expression of *sigE* was induced by nitrogen depletion, and NtcA (a global nitrogen transcription factor that is conserved among cyanobacteria) was found to bind to the promoter region of *sigE* (Muro-Pastor et al. 2001a). Microarray analysis revealed that the *sigE* knockout decreases the transcript levels of genes encoding enzymes of glycolysis, the OPP pathway, and glycogen catabolism, indicating that SigE is a global regulator of sugar catabolism in *Synechocystis* 6803 (Osanai et al. 2005). SigE transcriptional activity is repressed under light conditions by binding to ChlH, a subunit of Mg-chelatase (one of the chlorophyll biosynthetic enzymes), and SigE proteins are released from ChlH by light-to-dark transition, resulting in activation of the transcription of genes related to sugar catabolism under dark conditions (Osanai et al. 2009). Another group showed that the diurnal rhythms of the *sigE* expression peaked before night (Kucho et al. 2005), consistent with the fact that SigE is a positive regulator of sugar catabolism in this cyanobacterium. The promoter sequences recognized by SigE have not yet been identified.

Response Regulator Rre37 (SyNrrA)

A screening of response regulator mutants uncovered that the knockout mutant of *rre37* (sll1330) could not grow under light-activated heterotrophic growth conditions (Tabei et al. 2007). The *rre37* knockout mutant showed decreased the transcript levels of glycolytic genes under light conditions with glucose supplementation (Tabei et al. 2007). Microarray analysis by our group revealed that *rre37* knockout reduces the expression of *glgP* (slr1367), *glgX* (slr1857), *gap1*, and *pfkA* (sll1196) particularly after nitrogen starvation (Azuma et al. 2011). The gene expression related to the OPP pathway was not affected by *rre37* knockout (Azuma et al. 2011), which implies that the regulation of sugar catabolic genes by Rre37 is different from SigE. Biochemical analysis revealed that Rre37 binds with the promoter including the motif consisting of two direct repeats of TG(T/A)CA separated by an 8-bp A/T--rich spacer (Liu and Yang 2014). Involvement of the transcription of the enzymes related to cyanophycin, arginine, pyruvate and polyhydroxybutyrate (PHB) metabolism by Rre37 was also revealed from transcript analyses (Azuma et al. 2011;

Iijima et al. 2014; Liu and Yang 2014), which is discussed in section "Engineering of sugar catabolism in *Synechocystis* 6803 for bioplastics and hydrogen production".

Two Transcription Factors Belonging to the AbrB-Family

The *Synechocystis* 6803 genome contains two AbrB-type transcription factors, sll0359 and sll0822, that belong to a family of proteins known as transition state regulators in other heterotrophic bacteria (Klein and Marahiel 2002). AbrB (sll0359) proteins bind with the promoter region of the *hoxEFUYH* operon, which encodes hydrogenase subunits, and the knockout of *abrB* (sll0359) was shown to decrease *hoxE* expression (Oliveira and Lindblad 2008). Knockout of *abrB* (sll0822) decreased the nitrogen-regulated genes, including nitrate and ammonium transporters and *sigE* (Ishii and Hihara 2008). Biochemical analyses suggested that AbrB (sll0822) binds with the promoter regions of *sbtA* (whose protein is involved in inorganic carbon uptake) and the *hox* operon (Lieman-Hurwitz et al. 2009; Dutheil et al. 2012). The *abrB* (sll0822) knockout mutant showed pleiotropic phenotypes, which include slower growth, decreased chlorophylls and phycocyanins, increased glycogen, and increased cell diameter (Yamauchi et al. 2011). These results indicate that AbrB proteins play roles in carbon/nitrogen metabolism and cell proliferation, although the mechanistic implication remains to be solved.

Histidine Kinases (Hik8 and Hik31)

As mentioned in section "Sugar catabolism during light/dark and day/night conditions", sugar metabolism is closely related to the day/night cycle in cyanobacteria. A cyanobacterial circadian clock consists of three proteins (KaiA, KaiB, and KaiC), and the phosphorylation cycle and transcriptional control regulated by these three proteins are essential for circadian oscillation (Ishiura et al. 1998; Tomita et al. 2005). SasA, a histidine kinase interacted with KaiC in *Synechococcus* sp. PCC 7942, and *sasA* knockout abolished the rhythms of transcription during the day/night cycles (Iwasaki et al. 2000). Hik8 is an ortholog of SasA in *Synechocystis* 6803, and the *hik8* knockout mutant exhibited decreased mRNA levels of phosphofructokinase, G6PD, 6PGD, glyceraldehyde-3-phosphate dehydrogenase, and glycogen phosphorylase, and glucose-1-phosphate adenylyltransferase (encoded by *glgC*) (Singh and Sherman 2005). G6PD and 6PGD enzymatic activities were decreased by *hik8* knockout under photoautotrophic conditions (Singh and Sherman 2005). The expression of genes related to both sugar catabolism and sugar anabolism was also affected by *hik8* knockout (Singh and Sherman 2005). Recently, we revealed that the *hik8* overexpression diminished the glycogen levels under photoautotrophic conditions in *Synechocystis* 6803 (Osanai et al. 2014c). The *hik8* overexpression widely altered primary metabolism; particularly a decrease in sugar phosphates and an increase in several amino acids (Osanai et al. 2014c). In *Synechococcus*, RpaA is a cognate response regulator of SasA (Takai et al. 2006). The cognate response regulator of Hik8 in *Synechocystis* 6803 remains to be determined.

The *Synechocystis* 6803 genome contains a gene encoding the histidine kinase Hik31 (sll0790) in the chromosome, and the homolog of Hik31 (slr6041), which is 95.7 % identical to Hik31 (sll0790) at the amino acid level, is encoded by the plasmid in *Synechocystis* 6803 (Kahlon et al. 2006). The *hik31* knockout mutant could not grow under mixotrophic conditions (in the presence of 5 mM glucose), and the G6PD activity was increased to more than twice that of the wild-type strain (Kahlon et al. 2006). Importantly, the authors claimed the instability of phenotypes of the *hik31* mutant (Kahlon et al. 2006), a fact consistent with our transcriptome analysis; among all histidine kinases in *Synechocystis* 6803, only *hik31* exhibited a large value of standard deviations (Osanai et al. 2006). The deletion mutant of the plasmid operon of the *hik31* ortholog (slr6039-slr6041) lost its viability under dark conditions, and the expression levels of *glgP* and *zwf* were decreased, the expression of *glgC*, *fbaI*, *fbaII* (encoding fructose-1,6-bisphosphate aldolases), *gap2* (encoding glyceraldehyde-3-phosphate dehydrogenase), and *pdh* (encoding pyruvate dehydrogenase) were increased (Nagarajan et al. 2014). Cell division and copper resistance were abolished by the deletion of the *hik31* operon (Giner-Lamia et al. 2012; Nagarajan et al. 2012), suggesting that Hik31 plays multiple roles in primary and secondary metabolism and cell proliferation. Hik31 is localized to both the plasma and thylakoid membranes (Giner-Lamia et al. 2012) and Rre34/CopR, whose gene is included in the *hik31* operon, is a probable cognate response regulator of Hik31 in *Synechocystis* 6803.

Engineering of Sugar Catabolism in *Synechocystis* 6803 for Bioplastics and Hydrogen Production

Metabolic engineering involves the alteration of cellular metabolisms by genetic modification to increase products of interest, and has a long history of studies. Overexpressing and/or increasing the activities of metabolic enzymes are straightforward methods of metabolic engineering; however, researchers often encounter only marginal increases in productivity. Since multiple genes should be modified simultaneously to alter metabolism and increase the products of interest, Alper and Stephanopoulos (2007) put forward global transcription machinery engineering (gTME), which alters transcription factors and sigma factors rather than metabolic enzymes.

As mentioned in section "Transcriptional regulators controlling sugar catabolism in *Synechocystis* 6803", several transcriptional regulators that control the expression of genes of sugar catabolism have been identified in *Synechocystis* 6803, and therefore, our group tried to use these factors for metabolic engineering. To generate the *sigE*-overexpressing strain, a *sigE* open reading frame was fused with the *psbAII* promoter and integrated into a neutral site of the *Synechocystis* 6803 genome. The *sigE*-overexpressing strain exhibited an increased expression of enzymes of glycogen catabolism and the OPP pathway, and decreased glycogen levels under photoautotrophic conditions (Osanai et al. 2011). Metabolome analysis revealed that acetyl-CoA, citrate, and isocitrate levels were increased by *sigE*-overexpression, indicating enhanced sugar catabolism by *sigE* overexpression (Osanai et al. 2011).

Synechocystis 6803 accumulates PHB, one of the biodegradable polyesters grouped in polyhydroxyalkanoates (PHAs), during nitrogen starvation (Hein et al. 1998). PHB is synthesized from acetyl-CoA by three reactions: acetoacetyl-CoA synthesized from acetyl-CoA by ketothiolase (encoded by *phaA*), 3-hydroxybutylyl-CoA synthesized from acetoacetyl-CoA by acetoacetyl-CoA reductase (encoded by *phaB*), and PHB synthesized from 3-hydroxybutylyl-CoA by PHA synthase (encoded by *phaC* and *phaE*) (Hein et al. 1998). Microarray analysis suggested that the expression of *phaC* and *phaE* was up-regulated by *sigE* overexpression (Osanai et al. 2011). The levels of PHB in the *sigE*-overexpressing strain after nitrogen deprivation were 2.5 times that in the wild-type strain (Osanai et al. 2013b). Another microarray analysis indicated that the transcripts of *phaA* and *phaB* were increased by *rre37* overexpression, and the PHB levels were doubled in the *rre37*-overexpresing strain compared with the wild-type strain (Osanai et al. 2014a). PHB was additively increased by the double overexpression of *sigE* and *rre37*, suggesting that SigE and Rre37 activate the PHB biosynthetic pathway in parallel (Osanai et al. 2014a) (Fig. 20.3). Thus, overexpression of the transcriptional regulators increased the PHB levels in *Synechocystis* 6803, demonstrating that gTME is also an effective method for metabolic engineering in cyanobacteria.

Microarray analysis also suggested that the expression of *hoxEFUYH* was up-regulated by *sigE* overexpression, which was confirmed by quantitative real-time PCR (Osanai et al. 2011, 2013a). Hydrogen is accumulated during anaerobic

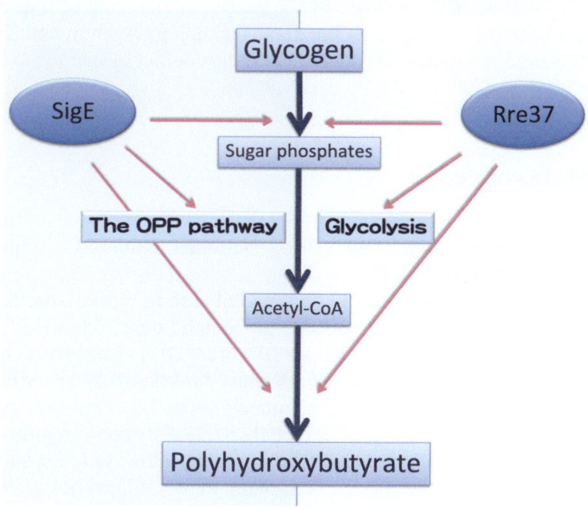

Fig. 20.3 Schematic model of activation of sugar catabolism and polyhydroxybutyrate biosynthesis by SigE and Rre37 (*red arrows* indicate the transcriptional activation by SigE and Rre37)

conditions, and it was found that the levels of accumulated hydrogen were higher in the *sigE*-overexpressing strain than in the wild-type strain under both light and dark conditions (Osanai et al. 2013a). Thus, SigE regulates the genes related to sugar catabolism, PHB biosynthesis, and hydrogen production. As mentioned in section "Two transcription factors belonging to the AbrB-family", AbrB transcription factors regulates the expression of enzymes related to primary metabolism and hydrogen production, suggesting the close relationship among carbon, nitrogen, and hydrogen metabolism in *Synechocystis* 6803, although its physiological meaning is unclear at present.

Conclusion

In this review, we have summarized the recent progress of studies on the primary metabolism of unicellular cyanobacteria by focusing on *Synechocystis* 6803 sugar catabolism and transcriptional regulators. Several transcriptional regulators of sugar metabolism in *Synechocystis* 6803 have been identified and molecular mechanisms of the signal transduction of light/dark, day/night, and nutrient status have been elucidated. In addition to basic science, there are examples of which mutants of transcriptional regulators of sugar catabolism increased high-value products, opening up the novel strategy of the metabolic engineering of *Synechocystis* 6803. The advances in the applied sciences of cyanobacteria will contribute to generations of renewable energy and alternative materials, and the basic knowledge about cyanobacterial transcription and metabolism will be indispensable for the sustainable development of human society.

Acknowledgments This work was supported by the Ministry of Education, Culture, Sports, Science, and Technology, Japan; by a grant to T.O. by funds from ALCA (Project name "Production of cyanobacterial succinate by the genetic engineering of transcriptional regulators and circadian clocks") from the Japan Science and Technology Agency, and CREST from the Japan Science and Technology Agency. The authors have no conflict of interest to declare.

References

Allen MMM, Smith AJ (1969) Nitrogen chlorosis in blue-green algae. Arch Microbiol 69:114–120

Alper H, Stephanopoulos G (2007) Global transcription machinery engineering: a new approach for improving cellular phenotype. Metab Eng 9:258–267

Aquirre von Wobeser E, Ibelings BW et al (2011) Concerted changes in gene expression and cell physiology of the cyanobacterium *Synechocystis* sp. strain PCC 6803 during transitions between nitrogen and light-limited growth. Plant Physiol 155:1445–1457

Azuma M, Osanai T, Hirai MY et al (2011) A response regulator Rre37 and an RNA polymerase sigma factor SigE represent two parallel pathways to activate sugar catabolism in a cyanobacterium *Synechocystis* sp. PCC 6803. Plant Cell Physiol 52:404–412

Badger MR, Price GD (2003) CO_2 concentrating mechanisms in cyanobacteria: molecular components, their diversity and evolution. J Exp Bot 54:609–622

Dutheil J, Saenkham P, Sakr S et al (2012) The AbrB2 autorepressor, expressed from an atypical promoter, represses the hydrogenase operon to regulate hydrogen production in *Synechocystis* strain PCC6803. J Bacteriol 194:5423–5433

Forchhammer K (2004) Global carbon/nitrogen control by PII signal transduction in cyanobacteria: from signals to targets. FEMS Microbiol Rev 28:319–333

Garcia-Dominguez M, Reyes JC, Florencio FJ (1999) Glutamine synthetase inactivation by protein-protein interaction. Proc Natl Acad Sci U S A 96:7161–7166

Giner-Lamia J, Lopez-Maury L, Reyes JC, Florencio FJ (2012) The CopRS two-component system is responsible for resistance to copper in the cyanobacterium *Synechocystis* sp. PCC 6803. Plant Physiol 159:1806–1818

Goto-Seki A, Shirokane M, Masuda S et al (1999) Specificity crosstalk among group 1 and group 2 sigma factors in the cyanobacterium *Synechococcus* sp. PCC7942: in vitro specificity and a phylogenetic analysis. Mol Microbiol 34:473–484

Hein S, Tran H, Steinbuchel A (1998) *Synechocystis* sp. PCC6803 possesses a two-component polyhydroxyalkanoic acid synthase similar to that of anoxygenic purple sulfur bacteria. Arch Microbiol 170:162–170

Iijima H, Watanabe A, Takanobu J et al (2014) *rre37* overexpression alters gene expression related to the tricarboxylic acid cycle and pyruvate metabolism in *Synechocystis* sp. PCC 6803. Sci World J 2014, 921976

Ishii A, Hihara Y (2008) An AbrB-like transcriptional regulator, Sll0822, is essential for the activation of nitrogen-regulated genes in *Synechocystis* sp. PCC 6803. Plant Physiol 148:660–670

Ishiura M, Kutsuna S, Aoki S et al (1998) Expression of a gene cluster *kaiABC* as a circadian feedback process in cyanobacteria. Science 281:1519–1523

Iwasaki H, Williams SB, Kitayama Y et al (2000) A *kaiC*-interacting sensory histidine kinase, SasA, necessary to sustain robust circadian oscillation in cyanobacteria. Cell 101:223–233

Kahlon S, Beeri K, Ohkawa H et al (2006) A putative sensor kinase, Hik31, is involved in the response of *Synechocystis* sp. strain PCC 6803 to the presence of glucose. Microbiology 152:647–655

Klein W, Marahiel MA (2002) Structure-function relationship and regulation of two *Bacillus subtilis* DNA-binding proteins, HBsu and AbrB. J Mol Microbiol Biotechnol 4:323–329

Kucho K, Okamoto K, Tsuchiya Y et al (2005) Global analysis of circadian expression in the cyanobacterium *Synechocystis* sp. strain PCC 6803. J Bacteriol 187:2190–2199

Lieman-Hurwitz J, Haimovich M, Shalev-Malul G et al (2009) A cyanobacterial AbrB-like protein affects the apparent photosynthetic affinity for CO_2 by modulating low-CO_2-induced gene expression. Environ Microbiol 11:927–936

Liu D, Yang C (2014) The nitrogen-regulated response regulator NrrA controls cyanophycin synthesis and glycogen catabolism in the cyanobacterium *Synechocystis* sp. PCC 6803. J Biol Chem 289:2055–2071

Merida A, Candau P, Florencio FJ (1991) Regulation of glutamine synthetase activity in the unicellular cyanobacterium *Synechocystis* sp. strain PCC 6803 by the nitrogen source: effect of ammonium. J Bacteriol 173:4095–4100

Muro-Pastor AM, Herrero A, Flores E (2001a) Nitrogen-regulated group 2 sigma factor from *Synechocystis* sp. strain PCC 6803 involved in survival under nitrogen stress. J Bacteriol 183:1090–1095

Muro-Pastor MI, Reyes JC, Florencio FJ (2001b) Cyanobacteria perceive nitrogen status by sensing intracellular 2-oxoglutarate levels. J Biol Chem 276:38320–38328

Muro-Pastor MI, Reyes JC, Florencio FJ (2005) Ammonium assimilation in cyanobacteria. Photosynth Res 83:135–150

Nagarajan S, Sherman DM, Shaw I, Sherman LA (2012) Functions of the duplicated *hik31* operons in central metabolism and responses to light, dark, and carbon sources in *Synechocystis* sp. strain PCC 6803. J Bacteriol 194:448–459

Nagarajan S, Srivastava S, Sherman LA (2014) Essential role of the plasmid *hik31* operon in regulating central metabolism in the dark in *Synechocystis* sp. PCC 6803. Mol Microbiol 91:79–97

Nakajima T, Kajihata S, Yoshikawa K et al (2014) Integrated metabolic flux and omics analysis of *Synechocystis* sp. PCC 6803 under mixotrophic and photoheterotrophic conditions. Plant Cell Physiol 55:1605–1612

Nakaya Y, Iijima H, Takanobu J et al (2015) One day of nitrogen starvation reveals the effect of sigE and rre37 overexpression on the expression of genes related to carbon and nitrogen metabolism in *Synechocystis* sp. PCC 6803. J Biosci Bioeng 120(2):128–1234

Oliveira P, Lindblad P (2008) An AbrB-like protein regulates the expression of the bidirectional hydrogenase in *Synechocystis* sp. strain PCC 6803. J Bacteriol 190:1011–1019

Osanai T, Kanesaki Y, Nakano T et al (2005) Positive regulation of sugar catabolic pathways in the cyanobacterium *Synechocystis* sp. PCC 6803 by the group 2 sigma factor SigE. J Biol Chem 280:30653–30659

Osanai T, Imamura S, Asayama M et al (2006) Nitrogen induction of sugar catabolic gene expression in *Synechocystis* sp. PCC 6803. DNA Res 13:185–195

Osanai T, Azuma M, Tanaka K (2007) Sugar catabolism regulated by light- and nitrogen-status in the cyanobacterium *Synechocystis* sp. PCC 6803. Photochem Photobiol Sci 6:508–514

Osanai T, Imashimizu M, Seki A et al (2009) ChlH, the H subunit of the Mg-chelatase, is an anti-sigma factor for SigE in *Synechocystis* sp. PCC 6803. Proc Natl Acad Sci U S A 106:6860–6865

Osanai T, Oikawa A, Azuma M et al (2011) Genetic engineering of group 2 sigma factor SigE widely activates expressions of sugar catabolic genes in *Synechocystis* species PCC 6803. J Biol Chem 286:30962–30971

Osanai T, Kuwahara A, Iijima H et al (2013a) Pleiotropic effect of *sigE* over-expression on cell morphology, photosynthesis and hydrogen production in *Synechocystis* sp. PCC 6803. Plant J 76:456–465

Osanai T, Numata K, Oikawa A et al (2013b) Increased bioplastics production with an RNA polymerase sigma factor SigE during nitrogen starvation in *Synechocystis* species PCC 6803. DNA Res 20:525–535

Osanai T, Oikawa A, Numata K et al (2014a) Pathway-level acceleration of glycogen catabolism by a response regulator in the cyanobacterium *Synechocystis* species PCC 6803. Plant Physiol 164:1831–1841

Osanai T, Oikawa A, Shirai T et al (2014b) Capillary electrophoresis-mass spectrometry reveals the distribution of carbon metabolites during nitrogen starvation in *Synechocystis* sp. PCC 6803. Environ Microbiol 16:512–524

Osanai T, Shirai T, Iijima H et al (2014c) Alteration of cyanobacterial sugar and amino acid metabolism by overexpression *hik8*, encoding a KaiC-associated histidine kinase. Environ Microbiol. doi:10.1111/1462-2920.12715

Singh AK, Sherman LA (2005) Pleiotropic effect of a histidine kinase on carbohydrate metabolism in *Synechocystis* sp. strain PCC 6803 and its requirement for heterotrophic growth. J Bacteriol 187:2368–2376

Tabei Y, Okada K, Tsuzuki M (2007) Sll1330 controls the expression of glycolytic genes in *Synechocystis* sp. PCC 6803. Biochem Biophys Res Commun 355:1045–1050

Takahashi H, Uchimiya H, Hihara Y (2008) Difference in metabolite levels between photoautotrophic and photomixotrophic cultures of *Synechocystis* sp. PCC 6803 examined by capillary electrophoresis electrospray ionization mass spectrometry. J Exp Bot 59:3009–3018

Takai N, Nakajima M, Oyama T et al (2006) A KaiC-associating SasA-RpaA two-component regulatory system as a major circadian timing mediator in cyanobacteria. Proc Natl Acad Sci U S A 103:12109–12114

Tomita J, Nakajima M, Kondo T, Iwasaki H (2005) No transcription-translation feedback in circadian rhythm of KaiC phosphorylation. Science 307:251–254

Yamauchi Y, Kaniya Y, Kaneko Y, Hihara Y (2011) Physiological roles of the cyAbrB transcriptional regulator pair Sll0822 and Sll0359 in *Synechocystis* sp. strain PCC 6803. J Bacteriol 193:3702–3709

Yang C, Hua Q, Shimizu K (2002) Metabolic flux analysis in Synechocystis using isotope distribution from ^{13}C-labeled glucose. Metab Eng 4:202–216

Yoshikawa K, Hirasawa T, Ogawa T et al (2013) Integrated transcriptomic and metabolomic analysis of the central metabolism of *Synechocystis* sp. PCC 6803 under different trophic conditions. Biotechnol J 8:571–580

Index

A
Aborted microspore (AMS), 328
AbrB-family, 517
Abscisic acid (ABA), 342
Abscisic acid insensitive4 (ABI4), 182
Acetyl-CoA carboxylase (ACCase) subunits, 8, 9, 458
Acetyl-CoA synthetase (ACS), 8
Acyl-CoA-binding proteins (ACBPs)
  AtACBPs (*see* AtACBP)
  in eukaryotes (*see* Eukaryotes)
  OsACBP6, peroxisomal fatty acid β-oxidation, 384–385
  in seed oil biosynthesis, 377–384
  in systemic transport, 392
Allene oxide synthase (AOS), 408
Annexin, 347
Applied Biosystems SOLiD Sequencing, 452
Arabidopsides, 412
Arabidopsis mgd1-1 mutant, 28
AtACBP
  in cuticle formation, 388–389
  in embryo development, 387–388
  in leaf senescence, 391
  in pollen development, 389–390
  seed germination and development, 385–386
AtMYB103, 329
AUGUSTUS, 452
Auxins, 342
Azelaic acid (AzA), 343
Azelaic acid induced 1 (AZI1), 344

B
Basic helix-loop-helix (bHLH) transcription factors, 321
Biomolecular fluorescence complementation (BiFC), 371
Blue native polyacrylamide gel electrophoresis (BN-PAGE) analysis, 31
Botrytis cinerea, 412
Bryonia, 410

C
Calvin-Benson cycle, 457
Calvin cycle, 461
Chain termination method, 452
Chemical genetics
  advantages, 160
  direct
    phosphate, 170
    PTN, 171
    screening strategies, 169–170
  disadvantages, 161
  reverse
    acyl profile, 166, 167
    fatty acid biosynthesis, 162–164
    galvestine-1, 166–168
    gavestine-1 (Ki), 166
    lyso-PA and PA, 164–166
    phosphatidylcholine (PC), 168
    piperidinyl-benzimidazolidinone scaffold, 168
    screening strategies, 161–162

Y. Nakamura, Y. Li-Beisson (eds.), *Lipids in Plant and Algae Development*,
Subcellular Biochemistry 86, DOI 10.1007/978-3-319-25979-6

Chlamydomonas
  classic hypothesis, 190
  DCMU, 191
  endogenous genes, 193
  factors, 192
  global transcript analysis, 192
  insertional mutagenesis, 189
  lipid droplet formation, 192
  mechanism, 189
  nitrogen deprivation, 189, 190
  PDAT, 190
  peripheral distribution, 191
  PGD1, 190, 191
  photosynthesis/flagella biogenesis, 189
  reverse genetic screening, 189
  SQD2 promoter, 193
  SQUAMOSA promoter-binding-protein-domain, 193
  TBARS, 191
Chlamydomonas reinhardtii
  biohydrogen production, 460
    anoxic conditions, 461, 462
    sulfur deplete conditions, 462–464
  high energy carbon storage molecules
    C. reinhardtii strains, 456
    glycolysis/gluconeogenesis enzymes, 456
    heat stress conditions, 460
    lipid synthesis, 456
    nitrogen metabolism, 456
    nitrogen starvation, 455, 457
    phosphorus and sulfur, 458, 459
    photosynthetic activity, 457
    protein synthesis, 456
    TAG synthesis, 454, 457, 458
  model organism
    knock out libraries, 450
    nutrient stress, 451
Chl fluorescence analysis, 29
Chloroplast biogenesis
  galactolipid biosynthesis, 116–117
    regulation of, 111–113
  photosynthesis, 107–108
  thylakoid lipid biogenesis, 106–107
  thylakoid lipid biosynthesis, 117–118
  thylakoid lipid synthesis
    alternative galactolipid pathway, 114
    DLA2, 116
    MGDG biosynthesis, 114
    P deficiency, 116
    pgp1-2 mutant, 114
    photoprotection machinery, 113
    photosynthesis-associated genes, 116
    plastid nucleoids, 114
  thylakoid membrane lipids, 108–110
Chloroplast galactolipids
  CrΔ4FAD, 483, 484
  EhDES15 expression, 483
  eukaryotic pathway, 481
  extraplastidial membranes, 483
  fatty-acid synthesis and modifications, 480
  glycerolipid synthesis, 481
  heterokonts and rhodophytes, 481
  MGDG, 481
  ω-3 desaturase, 483
  Phaeodactylum tricornutum, 482
  plastidial desaturation pathway, 482
  prokaryotic pathway, 481
  tertiary endosymbiosis, 482
Circium arvense, 412
Cis-jasmone (CJ), 412, 413
Cis-oxo-phytodienoic acid (OPDA), 410–412
Cold stress signaling, 276–277
Crystallography analysis, 33
CURVATURE THYLAKOID 1 (CURT1), 89, 134
Cuticular waxes, 289
Cutin, 289
Cyanobacterial circadian clock, 517
CYP94 enzymes, 417, 418
Cyt b_6/f, 26, 27
Cytochrome P450 family, 324

D

Defective in cuticular ridges (DCR), 293
Defective in induced resistance 1 (DIR1), 344, 347
Defective pollen wall (DPW), 323
Deficient in Cutin Ferulate (DCF), 293
Dehydroabietinal (DA), 343
Delayed tapetum degeneration (DTD), 328
Desorption electrospray ionization (DESI), 14
DGD1, 59
Diacylglycerol (DAG)
  acetyl-CoA carboxylase, 208
  biosynthesis, 208
  Chlamydomonas chloroplast lipid biosynthesis, 209
  eukaryotic pathway, 209
  fatty acid synthase, 208
  origin of, 211–212
  PA dephosphorylation reactions, 209
  phospholipid, 209
  prokaryotic pathway, 208
  sequential G-3-P acylation, 209
  synthesize extraplastidic membrane lipids, 209
Diacylglyceryl-trimethylhomoserine (DGTS), 52

3-(3,4-dichlorophenyl)-1,1-dimethylurea (DCMU), 191
Digalactosyldiacylglycerol (DGDG)
 cyanobacteria
  BN-PAGE analysis, 33
  crystallography analysis, 33
  in vitro PSII activity, 32
  oxygen-evolving complex, 33
  photodamage, 32
  repair process, 32, 33
  Synechocystis dgdA mutant, 33
 in plants
  BN-PAGE analysis, 31
  complementation analysis, 31
  deficiency, 32
  DGD1 and *DGD2 genes*, 30
  dgd1 mutation, 31
  fluorometric analysis, 31
  in vitro protein analysis, 31
  LHCII levels, 31
  photoinhibition, 31
  PSI, structure and function of, 31
  thermostability, 32
  transgenic expression, 30
Direct RNA sequencing, 452
DNA-binding with one finger (DOF) box, 374

E

Embryo Cuticle Functionality (ECF) pathway, 304
Epidermal cells
 clonal analyses, 303
 cuticle mutants
  fatty elongation complex, 295–298
  knock-out mutations, 298
  pleiotropic defects, 298–300
 cuticular barrier
  ABC transporters and LTPs, 294
  alcohol forming pathway, 291
  alkane-forming pathway, 292
  composition, 288
  depolymerization reactions, 292
  fatty acid elongase complex, 289–291
  fluorescent fusion proteins, 293
  lcr and *att1* mutants, 293
 dermatogen stage, 303
 double *gso1 gso2* mutant, 304
 ECF pathway, 304, 306
 epidermal specification pathway, 306
 juxtaposed cuticles, 303
 LACS2, *FDH*, and *BDG1* pathway, 303
 mathematical modeling, 306
 molecular and cellular mechanisms, 304
 molecular and genetic analysis, 306
 non-cell autonomous component, 304
 pavement cells, 288
 in restricting plant growth, 288
 syncytial endosperm, 303
 transcriptional regulation, 300–302
Ethylene response factor (ERF), 371
Eukaryotes
 Arabidopsis
  amino acid determinants, 371
  AtACBP2-AtEBP interaction, 371
  detoxification, 372
  differential subcellular localization and ligand specificities, 371
  DOF box and GT-1 motif, 374
  ERF, 371
  GFP tagged proteins, 370
  HAB1, regulation of, 373
  hypoxia, heavy metal and oxidative stresses, 373
  hypoxia-responsive gene expression, 372
  in vitro binding, 372
  Northern blot analysis, 372
  RNA gel blot analysis, 374
  sequence alignment, 370
  six paralogs, 370
  stress-responsiveness, 374
  subcellular fractionation experiments, 370
  transgenic expression, 374
 combination techniques, 366
 prototype form, 365
 rice, 374–376

F

Fatty acid methyl esters (FAMEs), 12
Fatty-acid desaturases (FAD), 479–480
Fatty acid synthase (FAS), 9
Fish oil components, 7
Flexible surface model (FSM), 146
FT signals, 347
Fucoxanthin chlorophyll protein (FCP) complexes, 136

G

Galactolipid
 biosynthesis, 92–94
 localization, 95
 substrate specificity, 94–95

GAMYB, 329
Gas chromatography (GC), 12–13, 453
Gas chromatography-flame ionization detector (GC-FID) analysis, 230
Glucosylgalactosyldiacylglycerol (GlcGalDG), 30
Glutamate synthase (GOGAT), 514
Glycerol-3-phosphate-derivative, 343
Glycolipids
algae, lipid composition, 52
Δ9-acyl-ACP desaturase, 53
in eukaryotic algae, 70
chlorophyta, 65–68
complex plastid, 64
glaucophyta, 65
Paulinella, 74
phosphate, 75
plastidial glycolipid metabolism, 65
primary endosymbiosis, 64
primary plastids, 64
rhodophyta, 69–70
secondary endosymbiosis, 65 (*see also* Secondary endosymbiosis)
in plants
6-acyl-MGDG, 63
angiosperms, 54
DGDG synthases, 53, 54, 59
chloroplast development and morphology, 61
"18:3"-plants, 53
lipid composition, 52
lipid trafficking, 63–64
mgd1 mutant deficient, 61
MGDG synthase, 53, 54
MGD2/MGD3/DGD2 pathway, 62
oxylipins, 62
pgp1 mutant, 62
phosphate, 62
Picea abies, 54
Pinus sylvestris, 54
Pleurozium schreberi and *Ceratodon purpureus* (Bryophyta), 54
Polypodium vulgare/Pteridium aquilinum, 54
16\:3-plants, 53
Sphagnum fimbriatum, 54
sqd2 mutant, 61, 62
SQDG synthesis, 53, 54, 59
MGlcDG, 53
plastidial lipids (prokaryotic pathway), 53
prokaryotic/eukaryotic pathways, 53
Glycosylglycerides, 3
Green fluorescence protein (GFP), 429
Green leaf volatiles (GLVs)
biosynthesis
'burst' of GLV formation, 431
fatty acids, 429, 430
GUS activity, 430
heat stress, 431
HPL and AOS pathways, 429
gene evolution, 432, 433
sensing by plants
Arabidopsis, 433, 434
on insect behavior, 433
interactions with membranes, 437, 438
metabolism, 438–440
perception as toxic xenobiotics, 435–437
physiological effects on mammals, 433
stressed plants, 433
volatiles, 434, 435
structure, in plants, 428
GS-GOGAT cycle, 514

H

Henry's law constants, 439
High-throughput genetics strategies
chlamydomonas, 223–224
growth conditions, 226
mutation and mutant phenotype
genetic complementation, 241
genetic linkage analysis, 240–241
locus, 240
multiple alleles, 240
primary screen
flow cytometry, 229–230
fluorescent dye, 228
GC-FID analysis, 230
lipid bio-markers, 230
permeabilizing cells, 228–229
plate reader, 229–230
SCRS, 231
quantitative analysis and microscopy, 233
electron microscopy, 232
GC-FID, 232
insertional mutagenesis (*see* Insertional mutagenesis)
mapped insertional mutant library, 238–240
MS, 232
point mutations, 233
TAG metabolism, 225–226
Histidine kinases (Hik8 and Hik31), 517, 518
Hydrophobic/amphipathic small molecules, 2
Hydroxylated fatty acids, 324
Hypersensitive response (HR), 276
Hypersensitive to ABA1 (*HAB1*), 373

I

Insertional mutagenesis
  Agrobacterium tumefaciens, 235
  antibiotic resistance markers, 234
  auxotrophy markers, 234
  biolistic particle delivery, 235
  drug-selectable markers, 234
  electroporation, 235
  glass beads, 235
  insertion site, 237
  number of insertions per mutant, 235
  plasmid rescue/PCR-based methods, 235–237
  promoter, 234
Isocitrate lyase (ICL), 461
In vitro protein analysis, 31

J

JA-amino acid conjugates, 414
Jasmonate metabolic grid, 419, 420
Jasmonic acid (JA)
  anti-cancer properties of, 415
  arabidopsides, 412
  biosynthesis, 408
  biosynthetic pathways/biological activities, 409, 410
  cis-jasmone, 412, 413
  diversity, structures and biological activities, 407
  glucosylated/sulfated derivatives, 414
  herbivores attacks, 406
  jasmonoyl-isoleucine catabolism, 414, 415
    amido-hydrolases, 418
    CYP94 enzymes, 417, 418
    Jasmonate metabolic grid, 419, 420
  JA-triggered plant immunity, 406
  and methyl jasmonate (MJ), 413
  OPDA, 410, 411
  in reproduction, 406
  signaling pathway, 409
  tuberonic acid, 414
  unifying signaling model, 406
Jasmonoyl-isoleucine (JA-Ile), 414, 415

K

3-ketosphinganine reductase (KSR), 260–261

L

Leaf closing factor (LCF), 414
LEAFY COTYLEDON1 (*LEC1*), 182
Light-harvesting complex of photosystem II (LHCII), 27
Lipid biosynthesis, transcriptional regulation, 300–302
Lipids
  algae and higher plants
    compositions, 6–7
    structure of, 3–5
  analyses
    GC, 12, 13
    lipidomics, 13
    MSI, 14
    TLC, 11
  biosynthesis
    carbon source and two-carbon, 8–9
    cellular functions, 9–10
    glycerolipid assembly, 9
  biotechnological applications, 14, 15
  classes of, 2
  definition, 2
  membrane structures, 2
  signaling molecules, 2
  storage lipids, 2
  surface coverings, 2
Lipid signaling
  conductive system, 340
  lipophilic compounds
    JA, 344–345
    oxylipins, 344–345
    phosphatidyl inositol and phosphates, 349–351
    (phospho-)glycerolipids, 345–348
    phospholipases, 349
    plant hormones, 342
    PtdOH, 351–353
    SAR, 342–344
    small lipophilic metabolites, 342–344
  in long-distance signaling, 340
  pressure flow hypothesis, 340
  signaling compounds, 340
Lipid transfer protein (LTP), 347
Long-chain bases (LCB)
  biosynthesis
    KSR, 260
    serine palmitoyltransferase complex, 252–259
  ceramides phosphorylation/dephosphorylation, 269
  modifications
    C-4 hydroxylation, 261
    Δ4 desaturation, 262
    Δ8 unsaturation, 261
    hydroxylation and desaturation, 262–263

Long-distance developmental signals, 346
Lyso-phosphatidic (lyso-PA), 164–166

M

Major lipid droplet protein (MLDP), 230, 457
MALE STERILITY 1 (MS1), 327
MALE STERILITY2 (MS2), 323
Mass spectrometry imaging (MSI), 14
Mass spectroscopy (MS), 232
Matrix-assisted laser desorption/ionization (MALDI), 14, 453
Maxam-Gilbert sequencing, 452
Metabolomics, 453
Methyl jasmonate (MJ), 413
Microalgae
  active growth recovery, 492
  ALA, 484
  C16 PUFA and C20 PUFA, 493
  cellular localization and compartmentalization
    betaine lipids, 490
    chloroplast lipids, 489
    cytoplasm, 489
    EPA-CoA, 491
    genomic survey, 488
    L. incisa, 490
    ω-3 desaturation, 490
    phospholipids, 490
    Porphyridium cruentum, 489
    PtdCho, 489
    TAG, 489, 490
  Chlamydomonas reinhardtii, 492
  chloroplast-membrane glycerolipids, 492
  Δ4 desaturases, 487
  Δ6 and Δ5 desaturases, 486, 487
  DGDG, 493
  DHA, 473, 485
  diversity
    C16 PUFA, 479
    Chlorarachniophyceae, 479
    chromalveolates, 476
    core green lineage, 477
    diatoms, 478
    freshwater ecosystems, 475
    glimpse, 475
    green algal lineage, 475
    green lineage, 477, 478
    Lobosphaera incisa, 478
    non-monophyletic evolution, 478
    octadecapentaenoic acid, 479
    primary endosymbiosis, 475
    primary plastid, 475
    red algal lineage, 477
    secondary endosymbiosis, 475
    secondary plastid, 477
    Stramenopiles clade, 476
    VLC-PUFA biosynthesis, 475
  double-bond location, 472
  EPA, 473
  eukaryotic microalgae, 471
  FAD, 479–480
  fish-oil contamination, 474
  "front-end" desaturases, 485
  genetic engineering
    acyl-CoA pool, 497
    genomes and transcriptomes, 496
    lipid droplets, 497
    lipidomics analysis, 497
    Nannochloropsis, 496, 498
    NoD12 expression, 498
    oilseed plants, 496
    polar lipid classes, 497
    TAG, 497
  lipid-biosynthesis pathways, 474
  low-temperature-sensitive mutant, 492
  ω-3 desaturases, 487
  ω-6 VLC-PUFA AA, 473
  photosynthetic organism, 487
  photosynthetic planktonic microalgae, 472
  plastidic desaturation (*see* Chloroplast galactolipids)
  polar diatoms and chlorophytes, 493
  PUFA elongation, 487–488
  putative desaturase, 486
  Thraustochytriaceae, 485
  unsaturation level, 491
  VLC-PUFA production, 472
    heterotrophic production, 495
    L. incisa, 495
    marine photosynthetic diatom, 495
    mass cultivation, 493
    micrographs, 494
    photoautotrophic diatoms, 494
    primary and secondary plastids, 495
  Yarrowia lipolytica, 474
Microspore mother cells (MMCs), 320
Monogalactosyldiacylglycerol (MGDG), 53
  CURT1, 134
  cyanobacteria, 29
  in plants
    amiR-MGD1, 29
    Arabidopsis mgd1-1 mutant, 28
    Chl fluorescence analysis, 29
    isoform, 28
    photoprotective mechanism, 28
    synthase genes, 28
    thylakoid membrane biogenesis, 28

EPR measurements, 134
extended ordered protein arrays, 134
membrane fusions, 135
protein complexes
cytochrome *b6f*, 139
in LHCII structures, 135–136
light-harvesting complexes, 136–137
PSI, structure and function, 139
PSII, structure and function, 137–139
xanthophyll cycle
localization and operation, 143–144
non-bilayer phases, 142, 143
pigments, 141–142
solubilisation capacity, 142, 143
MYB transcription factors, 321

N

NINJA, co-repressors, 409
Non-bilayer lipids, 134
flexible surface model, 146
fluid mosaic membrane model, 146
hypotheses, 148
lateral pressure bilayer membrane model, 146
MGDG (*see* Monogalactosyldiacylglycerol (MGDG))
phases, 147
refined model, 147
VDE and lipocalin proteins, 149
Nuclear magnetic resonance (NMR), 454

O

-Omics approaches, 458
Chlamydomonas reinhardtii (*see Chlamydomonas reinhardtii*)
genomics and transcriptomics, 451–453
metabolomics, 453, 454
proteomics, 453
Oxylipins, 62, 344–345

P

Pavement cells, 288
Peptide “fingerprints”, 453
Phaseolus, 410
PHD-finger proteins, 321
Phloem-localized lipid-associated protein (PLAFP), 348
Phosphatidic acid (PA), 164–166, 208, 351–353
Phosphatidylglycerol (PG), 3
cyanobacteria, 35–39
in plants, 34–35
Phosphoinositides, 350
Photosynthesis
core photosynthetic protein–cofactor complexes, 22
DGDG
cyanobacteria, 32–33
in plants, 30–32
MGDG
cyanobacteria, 29
in plants, 27–29
PG
cyanobacteria, 35–39
in plants, 34–35
photosynthetic protein–cofactor complexes
Cyt b_6/*f*, 26
Cyt *f*, 27
LHCII, 27
PSI, 25
PSII, 25, 26
SQDG
cyanobacteria, 40
in plants, 39–40
thylakoid glycerolipids, 22
Photosystem (PS) I, 25
Photosystem (PS) II, 25, 26
Physcomitrella patens, 411
Phytooxylipins, 406
Phytozome Genome Browser, 452
Plant pollen exine development
angiosperms, life cycle, 315
biological processes, 317
chemical staining experiments, 318
colonization, 329
development stages, 319
exine biosynthesis
de novo FAS, 323–324
fatty acid synthesis, 323
phenolic compounds, 324
functions and enzymatic activities, 330
inner intine, 316
morphological analysis, 317
morphological features, 320
outer exine, 316, 317
phylogenetic studies, 330
pollen coat, 316
regulation network
AMS, 328
AtMYB103, 329
GAMYB, 329
MS1, 327
TAZ1, 328
TDR/bHLH5, 328
reverse genetic analysis, 330

Plant pollen exine development (*cont.*)
SAPs, 318
sporopollenin precursors, 317
ABC transporters, 326
lipid transfer proteins, 326–327
tapetal cells, 319
tapetum, 320–321
Plastidial fatty acid biosynthesis, 8
Plastidial lipids (prokaryotic pathway), 53
Processing-Associated Tetratricopeptide Repeat protein (PratA), 90
Programmed cell death (PCD), 274–276
Proteomics, 453
Pyrosequencing, 452
Pyruvate dehydrogenase (PDH), 8

Q
Quantitative real-time PCR (qRT-PCR), 372

R
Reactive electrophile species (RES), 411, 436
Reactive oxygen species (ROS), 436
Relief of repression model, 409
Response regulator Rre37 (SyNrrA), 516

S
Salicylic acid (SA), 342, 343
Sanger sequencing, 452
Secondary endosymbiosis
with chlorophyta, 70
with rhodophyta
in alveolata, 73–74
haptophyte, 71
in heterokonts (Strameopiles), 71–73
Secondary ion MS (SIMS), 14
Sequencing by synthesis (Illumina), 452
Serine palmitoyltransferase (SPT) activity, 275
sigE-overexpression, 518–520
Sigma factor (SigE), 516
Single-cell Raman spectroscopy (SCRS), 231
Sphingolipids
acid, 271
alkaline ceramidases, 271
ceramide synthesis, 265–266
endomembrane trafficking, 273
fatty acid synthesis and structural modifications, 263–264
glucosylceramide synthesis, 267
inositolphosphoceramide synthesis, 267–268
LCB (*see* Long-chain bases (LCB))
membrane function, 271–273
net content and composition, 270
neutral, 271
physiological mediators
ABA-dependent guard cell closure, 274
cold stress signaling, 276–277
pathogen resistance, 276
PCD, 274–276
structure, 250–252
Sphingosine-analog mycotoxins (SAMs), 275
Sporopollenin acceptor particles (SAPs), 318
Sugar catabolism
bioplastics, 519
hydrogen production, 519
light/dark and day/night conditions, 512, 513
metabolic engineering, 518
metabolic map, 513
nitrogen-starved conditions, 515
glutamine synthetase, 514
microarray analysis, 514
2-OG levels, 514
protein levels, OPP pathway, 514
PHB biosynthesis, 519
sigE-overexpression, 519, 520
transcriptional regulators
AbrB-family, 517
histidine kinases (Hik8 and Hik31), 517, 518
response regulator, Rre37 (SyNrrA), 516
RNA polymerase SigE, 516
Sulfoquinovose transferase (SQD1), 59
Sulfoquinovosyldiacylglycerol (SQDG), 458
cyanobacteria, 40
in plants, 39–40
Sulphoquinovosyldiacylglycerol 2 (SQD2), 193
Syncytial endosperm, 303
Synechocystis 6803, 512–514, (*see also* Sugar catabolism)

T
Tapetum degeneration retardation (TDR), 328
TAPETUM DEVELOPMENT ZINC FINGER PROTEIN1 (*TAZ1*), 328
Thin layer chromatography (TLC), 11–12
Thiobarbituric acid reactive substances (TBARS), 191
Thylakoid membranes
biogenesis, 88–90

of cyanobacteria, 90–92
de novo biogenesis, 96–97
diatom, 131
galactolipid MGDG, 130
lipid composition, 129
negatively charged PG, 130
non-bilayer propensity, 130
phase behaviour, 131–134
photosystem I, mutant of, 96
proteoliposome experiments, 131
SQDG, 130
structure of, 86–88
ultrastructure, 86–88
Time-of-flight (TOF) spectrometry, 453
TOPLESS, co-repressors, 409
Triacylglycerols (TAGs)
environmental stresses, 183–185
microalgae, 189–193
acyl chains, 186
acyl-CoA-dependent and independent pathways, 210–211
acyl remodeling, 213–215
bioethanol production, 186
biofuel products vs. crops, 186
Brefeldin A, 188
chemical/non-chemical, 188
chlamydomonas (*see* Chlamydomonas)
cost effective, 187
culture aging/senescence, 189
culture and aging, 188
DAG, 208–212
glycerolipid synthesis, 208–210
lipid droplets, 215–217
nitrogen deficiency, 188
non-chemical based stresses, 188
phosphorus starvation, 188
Nannochloropsis, 193–196
non-seed plant tissues
genetic engineering, 181–183
plant oils, 181
Trigalactosyldiacylglycerol (TGDG), 91
Tuberonic acid (TA), 413

U
UDP-glucose pyrophsophorylase (UGP3), 59
UDP-sulfoquinovose synthase (SQD1), 59

V
Very long chain polyunsaturated fatty acids (VLCPUFAs), 53
Very long-chain omega-3 and omega-6 polyunsaturated fatty acids (VLC-PUFA). *See* Microalgae
Vesicle-inducing protein in plastids 1 (VIPP1), 63, 88

W
Western blot analysis, 370
Wintermans enzyme or galactolipid:galactolipid galactosyltransferase, 92
WRINKLED1 (*WRI1*), 182

X
Xanthophyll cycle (XC), 134
higher plants and algae, 140–141
localization and operation of, 143–144
non-bilayer phase, 142
non-bilayer phases, 143
pigments, 141–142
solubilisation, 143
solubilisation capacity, 142

Y
Yeast 2-hybrid (Y2H) assays, 371

MIX
Papier aus verantwortungsvollen Quellen
Paper from responsible sources
FSC® C105338

If you have any concerns about our products,
you can contact us on
ProductSafety@springernature.com

In case Publisher is established outside the EU,
the EU authorized representative is:
Springer Nature Customer Service Center GmbH
Europaplatz 3, 69115 Heidelberg, Germany

Printed by Libri Plureos GmbH
in Hamburg, Germany